FIRST LATIN AMERICAN SYMPOSIUM ON HIGH ENERGY PHYSICS

AND

VII MEXICAN SCHOOL OF PARTICLES AND FIELDS

FIRST LATIN AMERICAN SYMPOSIUM ON HIGH ENERGY PHYSICS

AND

VII MEXICAN SCHOOL OF PARTICLES AND FIELDS

Mérida, México October 1996

EDITORS

Juan Carlos D'Olivo
ICN-UNAM México

Martin Klein-Kreisler
IF-UNAM México

Héctor Méndez
CINVESTAV México

AIP CONFERENCE PROCEEDINGS 400

American Institute of Physics **Woodbury, New York**

L.C. Catalog Card No. 97-73971
ISBN 1-56396-686-7
ISSN 0094-243X
DOE CONF- 9610310

Printed in the United States of America

CONTENTS

Seminars

PART II.

COSMOLOGY AND QUANTUM FIELD THEORY

Tutorial Courses

Lectures

Seminars

PREFACE

The Seventh Mexican School of Particles and Fields (VII-EMPC) was the continuation of an enduring tradition in Mexico which has helped to promote the development of High Energy Physics within the scientific community. The School has always been addressed to the younger generation. This time it had an additional significance, as it was organized together with the First Latin American Symposium of High Energy Physics (I-SILAFAE). The symposium was dedicated to the memory of Juan José Giambiagi, a lifelong fighter for the development of Science in Latin America and co-founder of the Centro Latinoamericano de Física (CLAF).

The Division of Particles and Fields of the Mexican Physical Society was the organizer of both scientific events. Its full commitment to the task led to the successful completion of the meetings. Various institutions should be acknowledged for their special support: Centro de Investigación y de Estudios Avanzados del IPN (CINVESTAV), CLAF, CLAF-México, Consejo Nacional de Ciencia y Tecnología (CONACYT), UNESCO, and Universidad Nacional Autónoma de México (UNAM).

The meetings took place in the city of Mérida, in the Mexican state of Yucatán, between October 30 and November 6, 1996. Yucatán is world famous for having three wonderful things: one of the most exciting archaeological sites of the Maya culture; an original and delicious cuisine; and beautiful natural places and friendly people. We had the opportunity to experience all these aspects when the participants were taken for a day trip to the spectacular archaeological site of Uxmal, placed in the middle of a tropical forest, where we enjoyed a typical Yucatán meal.

Both the VII-EMPC and the I-SILAFAE brought together worldwide recognized experts which offered courses and plenary lectures of wide interest to students and researchers. Participants also took an active part by presenting research seminars on their work. The response of the Latin American physics community was most encouraging, as more than 150 particle physicists participated.

In preparing these proceedings, we have divided the content of the School and Symposium in two parts: Experimental and Phenomenological High Energy Physics, on the one hand, and Cosmology and Quantum Field Theory, on the other. In each part, we have placed the courses first, followed by the plenary lectures and the research seminars. We hope this will facilitate the reading of the volume.

Special thanks are due to the Local Organizing Committee (Axel de la Macorra, Lorenzo Díaz Cruz, Rodrigo Huerta, Piotr Kielanowsky, Miguel A. Pérez, and Gabriel Sánchez Colón) for these enjoyable and instructive scientific events. People who played a fundamental role in making the meetings possible are: Matías Moreno (IF-UNAM), Luis F. Urrutia (ICN-UNAM), and María Eugenia Rodríguez (Interclub). The help given by Sofia Alonso Lope (UM-Cinvestav), Javier Hernández (Cinvestav), and Jesús Martínez-Coronado (Cinvestav) is also acknowledged.

Juan Carlos D'Olivo
ICN-UNAM

Martin Klein-Kreisler
IF-UNAM

Héctor Méndez
CINVESTAV

ACKNOWLEDGMENTS

The I-SILAFAE and the VII-EMPC have been sponsored by the following institutions:

- CENTRO DE INVESTIGACIÓN Y DE ESTUDIOS AVANZADOS
  - Dirección General
  - Departamento de Física (Zacatenco)
  - Departamento de Física Aplicada (Unidad Mérida)
- CENTRO LATINOAMERICANO DE FÍSICA (CLAF)
- CLAF, MÉXICO
- CONSEJO NACIONAL DE CIENCIA Y TECNOLOGÍA, MÉXICO
- SOCIEDAD MEXICANA DE FÍSICA
- UNESCO
- UNIVERSIDAD NACIONAL AUTÓNOMA DE MÉXICO
  - Coordinación de la Investigación Científica
  - Coordinación General de Estudios de Posgrado
  - Dirección General de Asuntos del Personal Académico
  - Dirección General de Intercambio Académico
  - Instituto de Ciencias Nucleares
  - Instituto de Física
  - Facultad de Ciencias

List of Committees

SILAFAE International Advisory Committee

Aragao de Carvalho, Carlos
García Canal, Carlos
García, Augusto
Glashow, Sheldon
Grinstein, Benjamín
Hojvat, Carlos
Kane, Gordon
Lederman, León M.
Masperi, Luis
Nieves, José F.
Peccei, Roberto
Ponce, William
Richter, Burton
Santoro, Alberto
t'Hooft, Gerard
Schmidt, Iván
Urrutia, Luis F.

SILAFAE Scientific Committee

Alfaro, Jorge
Bediaga, Ignacio
D'Olivo, Juan Carlos
dos Anjos, Joao
Dova, María Teresa
Ferreira, Luiz A.
López, Angel
Lucio, José Luis
Martínez, Roberto
Restuccia, Alvaro
Méndez, Héctor
Méndez-Galain, Ramón
Pérez Rojas, Hugo
Quevedo, Fernando
Sampayo, Oscar Alfredo
Swain, John

Local Organizing Committee

de la Macorra, Axel
Díaz Cruz, Lorenzo
D'Olivo, Juan Carlos
Huerta, Rodrigo
Kielanowski, Piotr
Klein, Martin
Pérez, Miguel Angel
Sánchez Colón, Gabriel

IN MEMORIAM

Juan José Giambiagi, "Bocha" to those who knew him

In recent years, Physics itself has paid homage to Juan Jose Giambiagi. It was, in fact, when a quark, the *c-quark* was named *charm*. Bocha himself was a charmer in every aspect. He was certainly a charming person: as a friend, as a teacher, as a physicist, as an authority, as a man. Moreover, Giambiagi's physics contributions, so very important and transcendental, have plenty of charm. Nobody would disagree with the adjective of fascinating to describe the dimensional regularization method that he developed together with his comrade Carlos Bollini in the 70's.

It is after the ideas presented in the dimensional regularization method, that a Quantum Field Theory can be treated in a consistent way, maintaining all relevant symmetries and in particular the fundamental gauge symmetry. This method allowed the complete consistency of the Standard Model of the Fundamental Interactions to be shown.

The scientific activity of Giambiagi went far beyond the academic area. He undertook important responsibilities not only in Argentina, but at an international level. He was Director of the Physics Department of the University of Buenos Aires, were he demonstrated that apparent disorder based on freedom, as molecular chaos, can give rise to a first level scientific institution, in other words macroscopic quantities can be defined.... He was also Director of the National Research Council of Argentina, CONICET, from which he resigned as soon as the military secret services attempted to pass judgement on scientific merit. Giambiagi was a member of the Scientific Board of the International Centre for Theoretical Physics in Trieste an also the unforgettable Director of the Latin American Centre for Physics, CLAF, where he was able to rebuild an almost inexistent institution to produce a true renaissance. We also remember that he was the founder of the Latin American School of Physics, ELAF, together with Jose Leite Lopez from Brazil and Marcos Moshinsky from Mexico. Everybody knows the transcendental role played by ELAF in our present scientific development.

His recognition and prizes, listed in a previous talk, speak for themselves of the continuous presence of Bocha in our region. He was certainly very Argentinean in his manners, but he always felt himself a true Latin American.

We can only applaud the organizer's decision to give the name *Juan Jose Giambiagi* to the Mexican School of Particles and Fields 1996 and to the First Latin American Symposium on High Energy Physics that starts today. We all owe Bocha his Latin American sensibility, his continuous efforts to guarantee a real development of Physics in our region and his decision to impulse our science. From one side, towards a real presence in the world, and from the other one, to the present and future needs of our people.

I am tempted to end by reproducing here part of the Foreword of the Proceedings of the Latin American School of Physics of 1997 written by Giambiagi. There he says: "ELAF has survived 30 stormy years in the region. Its lectures and seminars have travelled throughout Latin America carrying interest and motivation to almost the whole physicists community. This strange stability is undoubtedly due to the fact that the School has fairly satisfied a genuine demand of this community from the scientific and technical point of view. That is a necessary but not sufficient reason for its stability. There are other reasons. One of them being, no doubt, the relevant support the School has received from Mexico. The other one is the fact that the School is a ghost walking in a virtual state and becoming real each year to jump to a new place....". It seems that there is no better *scientific blessing* to this starting Symposium.

Bocha, thanks a lot.

Mérida, November 1996

Professor Carlos García Canal
Laboratorio de Física Teórica
Departamento de Física
Universidad Nacional de La Plata
C.C. 67 - 1900 La Plata, Argentina

(Read by Carlos Hojvat, Fermilab)

Juan José Giambiagi, A Latin American Physicist

It was a very happy and legitimate idea to dedicate this First Latin American Symposium on High Energy Physics and the VII Mexican School of Particles and Fields to the memory of Juan José Giambiagi, or Bocha, as his friends used to call him. Many of us here have witnessed the influence he had in the development of Physics in Latin America. At the same time we have seen how the social and political events in Latin America have shaped his ideas, career and life.

Giambiagi's interest for science may have sparkled in high school at the Colegio Nacional de Buenos Aires. The good humanistic and scientific education given there have influenced him throughout his life. Some members of his family, including his father who worked for the railway company Ferrocarriles Pacífico, wanted him to become an engineer. However, in 1943 he entered the Universidad de Buenos Aires and there he did his graduate and postgraduate courses in Physics. His PhD thesis of 1950, entitled *Aplicación del método de Hadamard a algunos problemas de fisicomatemática*, was done under the supervision of the argentinean mathematician Alberto Gonzáles Domínguez. Giambiagi was influenced by Gonzáles' ideas and enthusiasm for research, and that may be felt in the future scientific contributions of Giambiagi, marked by the use of mathematical tools in the study of quantum field theories.

Giambiagi has spent some time at the Department of Electronic Engineering at the Instituto Radiotécnico, and in 1952 he went to the University of Manchester with a British Council grant. In 1953, interested in quantum field theory and invited by Prof. José Leite Lopes, Giambiagi goes to Centro Brasileiro de Pesquisas Físicas (CBPF) in Rio de Janeiro, where the research conditions were very favorable. In fact, the interest of Giambiagi in quantum field theory started with Gonzáles Domínguez and a contact he had with Leite Lopes in a meeting in Tucumán some years earlier.

In 1956, Giambiagi goes back to Argentina to work at the Comisión Nacional de Energia Atómica (CNEA), where he first meets Carlos Guido Bollini. That is the beginning of a very fertile collaboration that will last for the rest of his life. Bollini and Giambiagi have quite different personalities and it is difficult to imagine how they could get along so well. Being their student and then post-doc in the eighties at CBPF, I profited from these differences. Bollini's knowledge and discipline has guided me safely and well through the subjects, while Bocha's imagination was always at work proposing new, interesting and *dangerous* ways of proceeding. Those differences however, I believe, were the source of the important contributions they gave to theoretical physics.

In 1957, Giambiagi and Gonzáles Domínguez, went to the Universidad de Buenos Aires. There Giambiagi became full professor of Theoretical Physics and head of the Physics Department. Bollini joins them a bit later. After an interval visiting Caltech in California to work with Gell-Mann, comes one of the most important periods of Giambiagi's life. As the head of the Physics Department of the Universidad de Buenos Aires, from 1959 to 1966, he created perhaps the best theoretical physics group in Latin America at that time. He had the financial support of the university itself, the Ford Foundation and the Consejo Nacional de Investigaciones Científicas y Técnicas (CONICET) of Argentina, of which he was a council member. That support allowed the purchase of new equipment, and payment of salaries which allowed the scientists to work full time at the department. In fact, that was a new thing in Argentina, since before the fifties there were no full time positions in argentinean universities. In 1962, in a risky move, Giambiagi increases the number of faculty members by a factor three, attracting several senior and young physicists with experience in advanced research centers abroad. The quality and originality of the research work in the department, led by Bollini and Giambiagi, was very high, and it was concentrated in quantum field, Regge trajectories, quantum electrodynamics, etc. The postgraduate students were stimulated to develop a critical attitude towards the research activity, by discussing their work, the last publications, etc. At the same time it was given a great importance for the teaching to the undergraduate students. In fact, one can say that for the first time in Argentina it had been created a school of theoretical physics. Today one can find several prominent physicists around the world who were students at that group created by Giambiagi and collaborators.

However, in 1966 that important experience came to a tragic end. The government of the president Juan Carlos Onganía ordered the invasion of the Universidad de Buenos Aires. The faculty members were removed from the buildings in a quite violent way, in an episode that became known as *la noche de los bastones largos*. Giambiagi luckily escaped *los bastones*, since he had gone out for a coffee with a friend minutes before the invasion. In a meeting, the faculty members decided for a collective resignation. Giambiagi voted against it, but soon realized that it was inevitable. Something like 200 physicists left Argentina by that time, a good proportion of the scientific human capital of the country. The great majority, 80% went to USA, and the remaining to countries in Latin America.

Bollini and Giambiagi decided to stay, even though they have received invitations from California and the University of Paris. Then, they made a quite peculiar move. With some collaborators they turned a one bedroom flat which belonged to Giambiagi, in what they ironically called the *Instituto Juan Carlos Onganía*, and continued their research work. They were financially supported by Bariloche Foundation. The funds lasted a year and a half, and then Bollini got a position at the Universidad de La Plata, and Giambiagi in 1968, went to the Universidade de São Paulo for a year, invited by Prof. Jayme Tiomno. That year he also resigned as a member of CONICET because he was against the new rule imposed by the military that the assignment of any position to a scientist should be submitted to the argentinean secret service. Returning from São Paulo, Giambiagi joined Bollini again in La Plata where he stayed six years. Even though they were working in La Plata, the *Instituto Juan Carlos Onganía* was still active in Buenos Aires. It was on the same building where Giambiagi lived with his family. In fact, the family still owes that flat, and in October this year they have donated to the Universidad de Buenos Aires and Universidad de La Plata, his desk and private library that were still there.

That was perhaps one of the most fruitful period for them. Among other works they wrote, in October 1971, the first celebrated paper introducing the method of dimensional regularization. They used the dimension of space time as a continuous variable to develop an analytic regularization of the divergencies appearing in calculations in quantum field theories. The paper appeared in August 1972 in Physics Letters 40B (1972), 566-568, under the title *Lowest order "divergent" graphs in ν-dimensional space.* The second paper on the same subject appeared in Nuovo Cimento 12B (1972), 20-26, entitled *Dimensional renormalization: the number of dimensions as regularization parameter.* The first paper, due to the referee's point of view, took almost a year to be published, and that delay allowed the appearance of a paper by 't Hooft and Veltman on the same subject which got published a bit earlier in Nuclear Physics, even though it was written later. The difficulty in publishing these results is perhaps an indication of the originality of the ideas involved for that time. The method of dimensional regularization is nowadays widely used, and is explained in any textbook on quantum field theory. One of the beauties of the method is that it respects some symmetries of the theories including gauge invariance. Perhaps it is right to say that the skills of Giambiagi and Bollini for using mathematical tools in physics, goes back to the collaboration with Gonzáles Domínguez. Indeed, in 1964 the trio wrote a paper entitled *Analytic Regularization and the divergencies of quantum field theories* (Nuovo Cimento 31 (1964), 550), where in a calculation they used the power of the D'Alembertian as a continuous variable to make an analytic regularization. I must say that the ideas involving the dimension of space time have occupied Giambiagi's imagination until the end of his life.

Then came troubled periods for Giambiagi. In 1973, he was arrested by a paramilitary group and interrogated at the Federal Police department in Buenos Aires. The charges were very curious indeed. He was accused of being an israeli agent, perhaps because of his contact with many jewish physicists, and also for putting forward a *sovietization* of the university. After being released, he gave an ironic talk at La Plata about the arguments used by the police. How could Ford Foundation have given such financial support to a communist?

In 1976 there was a new arrest and, due to the political situation, he decided it was time to leave Argentina. He went to Rio de Janeiro where he first stayed at the Pontifícia Universidade Católica (PUC), and soon after went to CBPF. Bollini left Argentina some time later, for political reasons too, and went to the Instituto de Física Teórica (IFT) in São Paulo for about a year. He then joined Giambiagi at CBPF. Giambiagi was the head of the Departamento de Campos e Partículas at CBPF from 1978 to 1985, and again in 1994. He was a member of the Physics and Astronomy Scientific Council of CNPq (Conselho Nacional de Desenvolvimento Científico e Tecnológico) from 1980 to 1982. Bollini and Giambiagi had a very influential role in the brazilian physics. At CBPF they developed very important works in several areas of theoretical physics like Wilson loops, wave equations in arbitrary dimensions, Huygen's principle for non local operators, higher derivative field equations, etc. Under their scientific leadership a very good theoretical physics group has been created at CBPF. They have supervised several students and their influence is still felt nowadays.

Giambiagi always had a participation in the Centro Latinoamericano de Física (CLAF), an institution with offices in Rio de Janeiro. It is funded by some Latin American countries with the objective of promoting the research in physics in Latin America, supporting interchange programs, meetings, scientific visits, etc. He was a member of CLAF's council from 1968 to 1972. In 1986 he became the president of CLAF and stayed until 1994, when he was substituted by his close friend and collaborator Prof. Carlos Aragão de Carvalho. During his term, Giambiagi gave a new impetus to CLAF's activities. He started supporting research in areas with a direct role in the economic and social development of Latin American countries. He promoted the study of the atmospheric phenomenon called El Niño, a change in temperature in some regions of the southern Pacific ocean, which affects the climate of all three Americas. He also supported research about soil physics, geophysics, the Antarctic continent, recovering of oil fields, and also created a scholarship program in collaboration with CNPq for Latin American post graduate students. Giambiagi's interest in the role that physics must play for the society did not start with his term as president of CLAF. His experience in Argentina and Brazil made him realize that one can not think of stable scientific institutions if the scientific activity is not coupled to the economic reality. For that, it is necessary that the society itself demands the creation of know-how. That is perhaps one of the great puzzles for science in Latin American countries, where the technology demanded by the society is supplied mostly by multinational companies which just import it from their branches in developed countries. When Giambiagi took office at CLAF he tried to find issues that were demanded by society and which could also satisfy the academic aspiration of the university researchers.

Another important fact about Gimabiagi's role in the physics of developing countries, is his activities at the International Centre for Theoretical Physics (ICTP) in Trieste. Giambiagi was a close friend of Prof. Abdus Salam, the founder of ICTP and Nobel Laureate in Physics (1979). He participated from the beginning, helping Salam in many issues, among them to overcome the resistance of some diplomats against the creation of the Centre in Trieste. Giambiagi was, for several years, an Associate Member of ICTP and also a member of its Scientific Council. In 1995, under the presidency of Prof. Miguel Virasoro at ICTP, his former student at the Universidad de Buenos Aires, he was reelected to the Scientific Council, but died before he could take office.

Giambiagi has been awarded very important prizes and distinctions. He received the Prize Ricardo Ganz from Universidad de La Plata in 1985, the National Consagration Prize from the Argentinean Government in 1989, the Mexico Prize for Science and Technology in 1991 and the Order of Scientific Merit from the Brazilian Government in 1994. He was a member of the Argentinean, Brazilian, Latin American and Third World Academies of Science. He was the founder, together with José Leite Lopes and Marcos Moshinsky, of the Latin American School of Physics.

In October 1995, Bocha gave a colloquium at our institute in São Paulo where he presented his views of the developments in Physics during this century, and where he shared his expectations for the next century. It was a charming and touching talk. Although suffering with the illness, he gave an enthusiastic and lucid account of his ideas. In his own words, a man's typical length is of order of the meter and his heart beats about every second, and the important physics is that at the scale of us humans. It was his farewell to the Institute he so much respected. He has given us a lot and we are grateful to him. I am sure Bocha would like to be remembered as a Latin American. He loved very much this land and was very proud, despite of all difficulties, of never having left it.

Giambiagi died in Rio de Janeiro on 8th January 1996 at the age of 71.

Mérida, November 1996

Luiz Agostinho Ferreira
Instituto de Física Teórica - IFT/UNESP
Rua Pamplona 145
01405-900, São Paulo - SP, Brazil

PART I.

EXPERIMENTAL AND PHENOMENOLOGICAL HIGH ENERGY PHYSICS

Spectroscopy and Decays of Charm and Bottom

Joel N. Butler*

*Fermi National Accelerator Laboratory[1]
Batavia Illinois 60510

Abstract. After a brief review of the quark model, we discuss our present knowledge of the spectroscopy of charm and bottom mesons and baryons. We go on to review the lifetimes, semileptonic, and purely leptonic decays of these particles. We conclude with a brief discussion B and D mixing and rare decays.

INTRODUCTION

In the Standard Model(SM) of Elementary Particle Physics, matter is composed of two kinds of particles – quarks and leptons – whose interactions are mediated by vector gauge bosons and who receive their mass via the Higgs mechanism [1].

The quarks undergo all the known kinds of elementary particle interactions: strong, weak, and electromagnetic. While the u and d quarks make up nearly all of ordinary matter, the heavy quarks, charm and bottom (or beauty), are good laboratories for the study of the weak interaction and its interplay with the strong interaction. [2] They open up a whole field of study of spectroscopy and weak decays that challenges our detailed understanding of particle physics. Hidden within their decays may be examples of important phenomena that fall within the SM, such as CP violation, and possibly new phenomena, such as certain rare decays, which may lie outside the SM and provide us with clues for extending it.

The goal of this paper is to review several topics involving the spectroscopy and weak decay of two flavors of heavy quarks – charm and bottom. For each topic, we show how the Standard Model of elementary particle physics

---

[1]) Fermi National Accelerator Laboratory is operated by Universities Research Association under contract to the U.S. Department of Energy.

[2]) The top quark is so massive that it actually undergoes weak decay before it ever has a chance to form a hadron. This gives it very different behavior from that of charm and bottom. The properties of the top quark are the subject of another lecture at this school.

CP400, *First Latin American Symposium on High Energy Physics/ VII Mexican School of Particles and Fields*, edited by D'Olivo/Klein-Kreisler/Mendez

adequately accounts for the similarities and differences between these two systems.

We begin with a brief review of the quark model and then discuss the spectroscopy of charm and bottom mesons and baryons, emphasizing excited meson states. We then enter a discussion of weak decays of charm and bottom. We discuss the lifetimes of the various hadrons - what is expected from theory and what has been measured in experiments. Then, we review our knowledge of semileptonic decays and touch briefly on purely leptonic decays. Finally, we touch on flavor mixing and rare decays within the context of the Standard Model.

I A BRIEF REVIEW OF THE QUARK MODEL

In our current understanding of nature, matter is composed of quarks and leptons. The three 'generations' of quarks and leptons are:

$$\begin{pmatrix} u \\ d \end{pmatrix} \begin{pmatrix} \nu_e \\ e \end{pmatrix} ; \begin{pmatrix} c \\ s \end{pmatrix} \begin{pmatrix} \nu_\mu \\ \mu \end{pmatrix} ; \begin{pmatrix} t \\ b \end{pmatrix} \begin{pmatrix} \nu_\tau \\ \tau \end{pmatrix} . \tag{1}$$

The leptons appear in nature as isolated particles or, in the case of electrons (and muons) can be bound into atoms. The quarks, because of the nature of the color forces which mediate the strong interaction, never appear in isolation but are bound into color singlets. The allowed configurations are

- a quark and an anti-quark – these are the mesons; and
- three quarks (three anti-quarks) – these form the baryons (antibaryons).

Heavy quarks, such as charm and bottom, do not appear in ordinary matter but are produced in high energy collisions. The mesons and baryons which are produced decay via the weak interaction. The decay rate may be estimated by analogy with another weak decay – the decay of the muon – depicted in figure 1.

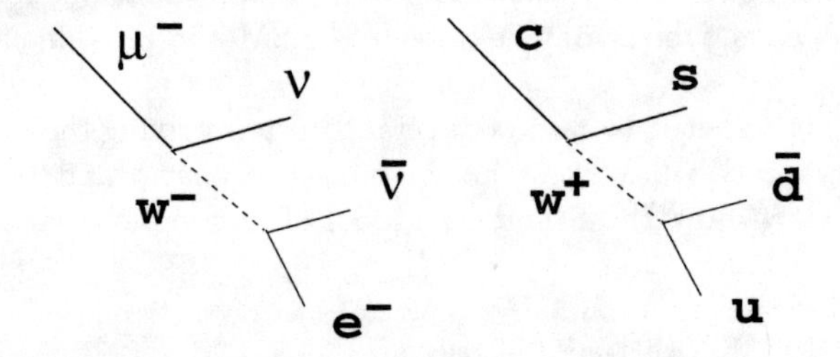

FIGURE 1. Weak Decay of a) muon and b) a charm quark

The rate for muon decay is given by

$$\Gamma_\mu = \frac{G_F^2}{192\pi^3} m_\mu^5 \qquad (2)$$

The dependence on the mass to the fifth power is a consequence of phase space and the spin structure of the decay and carries over to the analogous quark decay. The coupling constant, G_F, must have the unit of $(\text{Mass})^{-2}$.

One expects heavy quarks to decay much more rapidly than muons. The muon's mean lifetime is 2.2×10^{-6}s. The lightest particles containing the charm quark, with a mass of approximately 1.5 GeV/c^2, all have mean lifetimes between 1×10^{-12}s and 1×10^{-13}s. Applying the formula above directly, the ratio of the charm, bottom, and top quark [3] decay rates should be 1:500:2×10^{10}.

However, the story is a bit more complicated than this. The coupling 'strength' of the W bosons to quarks and leptons is not required to be the same. It turns out that if we take the strength to be ONE for muons, then it is the same for electrons and τ's. However, it is not always the same for the various quarks. The W connects pairs of quarks whose electric charge differs by one unit. For each possible decay there is a 'weak charge' which describes the strength of the W's coupling and we denote these charges by the symbols $V_{q_{2/3}q_{1/3}}$, where q_i represents a quark of electric charge i. If the W's connected only the quarks within each generation, then the lightest quark in each generation would be stable. However, this is not what happens. The charge 1/3 quarks may be viewed as 'mixed'. That is, the W couples to linear combinations of the 'physical' quark states – that is the quark states produced in the strong interaction. This gives rise to the so-called CKM [2] matrix which effectively gives the nine coupling strengths of the W bosons to the various combinations of quark decays that it mediates. The CKM matrix is written as:

$$V = \begin{pmatrix} V_{ud} & V_{us} & V_{ub} \\ V_{cd} & V_{cs} & V_{cb} \\ V_{td} & V_{ts} & V_{tb} \end{pmatrix} \qquad (3)$$

The semileptonic decay width of a heavy quark Q to a light quark q would be, by way of analogy with muon decay:

$$\Gamma_{Q\to q\mu\nu} = \frac{G_F^2}{192\pi^3} m_Q^5 \times |V_{Qq}|^2 \times F(\epsilon) \qquad (4)$$

where $F(\epsilon)$ is a phase space factor with $\epsilon = \frac{M_q}{M_Q}$.

---

3) The number for top is not really correct. Because the top is massive enough to decay into a real W boson, there is only one weak interaction decay vertex and the decay width, by dimensional analysis, depends on the top quark mass to the third power. However, it depends on G_F rather than G_F^2 so its decay goes even faster than indicated above.

The nine CKM matrix elements must form a unitary matrix so the elements are not all independent. Within the SM, the CKM elements are fundamental constants by which we mean there is no prescription for calculating what they are. The determination of the elements of the CKM matrix is a fundamental problem for particle physics and must be done by making measurements of the various properties of weak decays. The methods for determining the CKM elements that involve the heavy quarks are a central theme of this paper. However, we cannot make observations directly upon the quarks themselves since they are always bound into hadrons by the strong interaction. The challenge is to understand the strong interaction well enough to disentangle it from the weak interaction effects so that the CKM elements can be extracted. This situation is depicted in figure 2 which shows a B-meson decaying semileptonically to a D-meson, electron, and neutrino. It is through study of weak decays and lifetimes that one concludes that V_{cb} is small compared to V_{cs} which gives the B-mesons longer lifetimes than charmed mesons even though they are much heavier. In the next section, we will review the status of charm and bottom spectroscopy and we will introduce some ideas which can be used to help deal with the strong interaction effects.

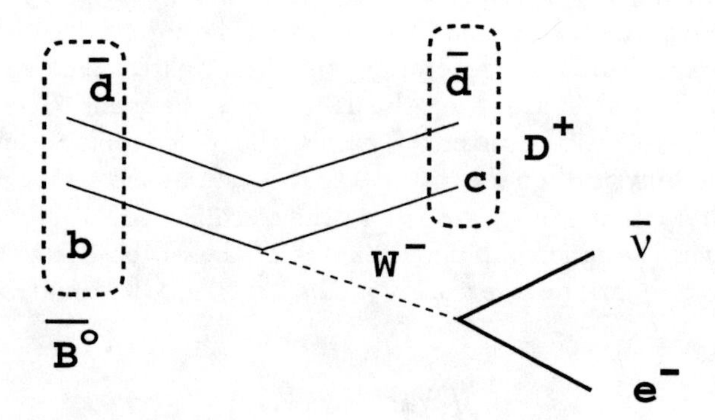

FIGURE 2. Weak semileptonic decay: $\bar{B}^o \rightarrow D^+ e^- \bar{\nu}$

II THE SPECTROSCOPY OF CHARM AND BOTTOM HADRONS

A Ground-State Charmed Particles

The charmed mesons, the D^o, D^+, and D_s, are composed of a charmed quark in association with a $\bar{u}$,$\bar{d}$, and $\bar{s}$ antiquark respectively. These lowest mass mesons are pseudoscalars. The quarks are in a total spin state of 0 and a relative orbital angular momentum state $L = 0$. The vector mesons, D^{*o},

D^{*+}, and D_s^{*+}, have a total spin of 1 but the charmed quark and light anti-quark still have relative angular momentum 0. The charmed mesons and the light quark mesons form SU(4) 16-plets shown below in figure 3.

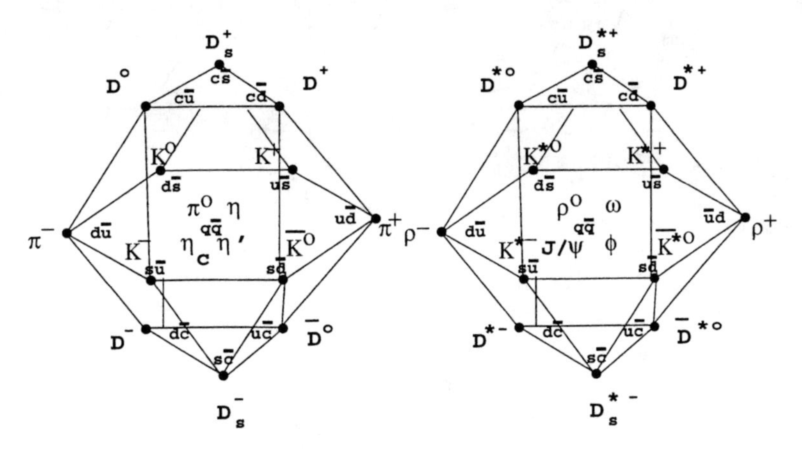

FIGURE 3. SU(4) 16-plets for lowest mass charmed mesons. left side: pseudoscalar mesons; right side: vector mesons. This figure is adapted from reference [3]

The charmed baryons consist of at least one charmed quark and two other quarks. The baryons made of only light quarks obey SU(3) and fall into an octet and a decuplet. When the charmed quark is brought into the picture, SU(4) symmetry results in two 20-plets shown below in figure 4.

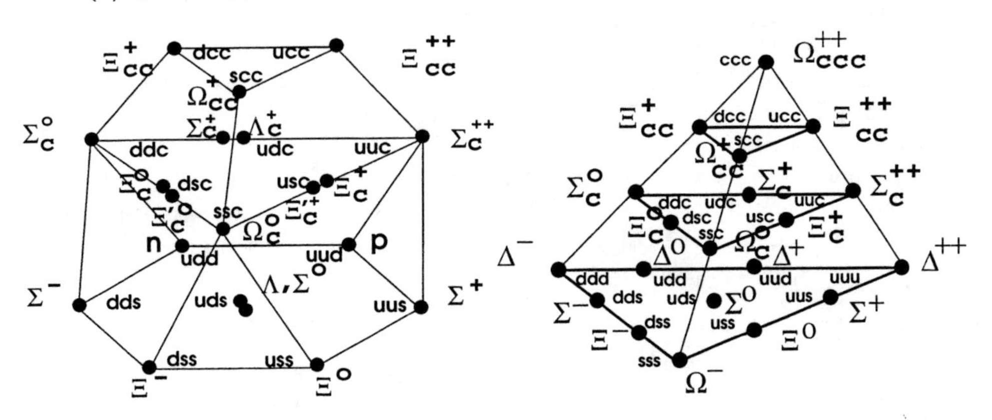

FIGURE 4. SU(4) 20-plets for lowest mass charmed baryons. left side: spin 1/2 baryons; right side: spin 3/2 baryons. This figure is adapted from reference [4].

The nine states on the middle level the spin 1/2 20-plet consist of one charm quark and two light quarks. These can be decomposed into 6 states which are symmetric under interchange of the light quarks and three which are antisymmetric. This decompostion is shown in figure 5.

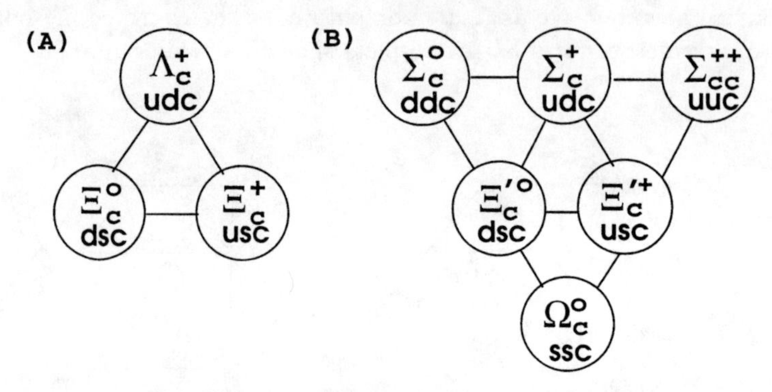

FIGURE 5. Exchange symmetry of the nine ground state spin 1/2 singly charmed baryons: The three states on the left have wavefunctions that are antisymmetric with respect to the interchange of the light quarks; the six on the right are symmetric under interchange. This figure is adapted from reference [5].

The comparable asymmetric states do not exist in the spin 3/2 20-plet because the light quarks must be in a symmetric state to satisfy the requirement that the overall state be antisymmetric. Similarly, the ccc state exists only in the spin 3/2 multiplet.

Not all of the charm baryon states have yet been discovered. In fact, only states with one charmed quark have been convincingly observed and even a few of them have not yet been observed. Table 1 shows the current experimental situation for these states.

Figure 6 shows the evidence [7] for the Ω_c, a baryon whose quark composition is css,decaying into $\Sigma^+ K^- K^- \pi^+$. The mass of this state is 2700 MeV/c^2.

The Ξ_c^{*+} decays to $\Xi_c^o \pi^+$. The evidence [8] for this excited state of Ξ_c is shown in figure 7.

Figure 8 shows the mass spectra obtained by the CLEO Collaboration [9] for $\Lambda_c^+\pi^+$ and $\Lambda_c^+\pi^-$. In each plot, there are two prominent peaks: the one at the lower mass is the Σ_c; and the one at the higher mass is identified as the Σ_c^*. The characteristics of the Σ_c states are

$$\begin{aligned} \Delta M(\Sigma_c^{*++} - \Lambda_c^+) &= 234.5 \pm 1.1 \pm 0.8\,\text{MeV} \\ \Gamma(\Sigma_c^{*++}) &= 17.9^{+3.8}_{-3.2} \pm 4.0\,\text{MeV} \\ \Delta M(\Sigma_c^{*o} - \Lambda_c^+) &= 232.6 \pm 1.0 \pm 0.8\,\text{MeV} \\ \Gamma(\Sigma_c^{*o}) &= 13.0^{+3.7}_{-3.0} \pm 4.0\,\text{MeV} \\ \Delta M(\Sigma_c^{*++} - \Sigma_c^{*o}) &= 1.9 \pm 1.4 \pm 1.0\text{MeV} \end{aligned} \tag{5}$$

By analogy with atomic physics, we also expect excited states of the charmed mesons and baryons which have angular momentum between the quarks or feature higher radial excitations. These will be discussed in detail

TABLE 1. Charmed $1/2^+$ baryon states. $[a, b]$ and $\{a, b\}$ denote antisymmetric and symmetric flavor index combinations. This table is adapted from reference [6]. Values in parentheses on states labelled 'not seen' are taken from this reference and are included to gently guide searches for them.

Notation	Quark content	SU(3)	(I, I_3)	S	C	Mass (MeV)
Λ_c^+	$c[ud]$	3*	(0,0)	0	1	2285.1±0.6
Ξ_c^+	$c[su]$	3*	(1/2,1/2)	-1	1	2465.1±1.6
Ξ_c^o	$c[sd]$	3*	(1/2,-1/2)	-1	1	2470.3±1.8
Σ_c^{++}	cuu	6	(1,1)	0	1	2453.1±0.6
Σ_c^+	$c\{ud\}$	6	(1,0)	0	1	2453.8±0.9
Σ_c^o	cdd	6	(1,-1)	0	1	2452.4±0.7
$\Xi_c^{+'}$	$c\{su\}$	6	(1/2,1/2)	-1	1	not seen(2561)
$\Xi_c^{o'}$	$c\{sd\}$	6	(1/2,-1/2)	-1	1	not seen(2561)
Ω_c^o	css	6	(0,0)	-2	1	2705.0±1.0
Ξ_{cc}^{++}	ccu	3	(1/2,1/2)	0	2	not seen(3616)
Ξ_{cc}^+	ccd	3	(1/2,-1/2)	0	2	not seen(3616)
Ω_{cc}^+	ccs	3	(0,0)	-1	2	not seen(3706)

later. However, to round out our tour of the charmed baryons, we anticipate this discussion by showing candidates for such excited baryon states, the $\Lambda_c^{*+}(2593)$ and $\Lambda_c^{*+}(2630)$ [10]. These appear as resonances in the invariant mass spectra $\Lambda_c^+\pi^+\pi^-$shown in figure 9. Two peaks are seen. These states are interpreted as consisting of an $L = 0$, $S = 0$ light diquark orbiting around a c-quark with relative orbital angular momentum L equal to 1. This state is shown schematically in figure 10. The lighter state is identified with total angular momentum $J = 1/2$ state and the heavier with $J = 3/2$. The light one decays through $\Sigma_c\pi$ but the heavier one cannot decay to its favored mode $\Sigma_c^*\pi$ which is too massive so it goes through a decay to a virtual $\Sigma_c^*\pi$ state with the virtual Σ_c^* decaying to $\Lambda_c\pi$.

B Ground State Bottom Particles

The ground state B-meson family is very similar to the D-meson family and consists of three $S = 0$ and three $S = 1$ states:

$$\begin{array}{c} B^o, B^+, B_s^o \\ B^{*o}, B^{*+}, B_s^{*o} \end{array} \tag{6}$$

All these states have been observed. In addition, there is the as yet unobserved B_c, a state consisting of a $\bar{b}$ quark and a charm quark. This state is truly fascinating because it is a flavor-bearing state consisting of two heavy quarks. Nonrelativistic potential models should provide an excellent description of its

TABLE 2. Charmed $3/2^+$ baryon states. This table is adapted from reference [6]. Values in parentheses on states labelled 'not seen' are taken from this reference and are included to gently guide searches for them. For observed states, mass values are from PDG or, if not listed, my estimates (Ξ_c^*'s and Σ_c^*'s).

Notation	Quark content	SU(3)	(I, I_3)	S	C	Mass (MeV)
Σ_c^{*++}	*cuu*	6	(1,1)	0	1	2519.6 ± 2
Σ_c^{*+}	*cud*	6	(1,0)	0	1	not seen(2545)
Σ_c^{*o}	*cdd*	6	(1,-1)	0	1	2517.7 ± 2
Ξ_c^{*+}	*cus*	6	(1/2,1/2)	-1	1	2645 ± 2
Ξ_c^{*o}	*cds*	6	(1/2,-1/2)	-1	1	2644 ± 2
Ω_c^{*o}	*css*	6	(0,0)	-2	1	not seen(2778)
Ξ_{cc}^{*++}	*ccu*	3	(1/2,1/2)	0	2	not seen(3744)
Ξ_{cc}^{*+}	*ccd*	3	(1/2,-1/2)	0	2	not seen(3744)
Ω_{cc}^{*+}	*ccs*	3	(0,0)	-1	2	not seen(3838)
Ω_{ccc}^{*++}	*ccc*	1	(0,0)	0	3	not seen(4797)

wave function. A very rich collection of excited states exist and they decay by radiative and hadronic transitions. A study of these would provide insight into the interquark potential when the quarks are very close to being at rest. The spectroscopy of the B_c state is discussed below.

The mass separation between the B^*'s and the B's is smaller than between the D^*'s and D's: about 45 MeV/c^2 as compared to 145 MeV/c^2. The reason, which will be discussed more later on, is due to the fact that the hyperfine splitting, which is responsible for the difference between the $S = 0$ and $S = 1$ states, goes like $\frac{1}{M_Q}$ where M_Q is the mass of the heavy quark. Because of this, all B^*'s are below threshold for decay through $B\pi$ and decay radiatively:

$$B^* \to B\gamma \qquad (7)$$

The b-baryon family is quite large. Very little is known about b-baryons except for the Λ_b which has been observed through its decay to $\psi\Lambda$ and its semileptonic decay to Λ_c. Figure 11 shows the Λ_b signal obtained by the CDF experiment at Fermilab [11]. The mass values obtained by various experiments [12] are given in table 3.

TABLE 3. Mass values for Λ_b.

Exp.	Mass (MeV/c^2)
CDF	5621 ± 4(stat) ± 3(syst)
ALEPH	5614 ± 21(stat) ± 4(syst)
DELPHI	5668 ± 16(stat) ± 8(syst)
UA1	5640 ± 50(stat) ± 30(syst)

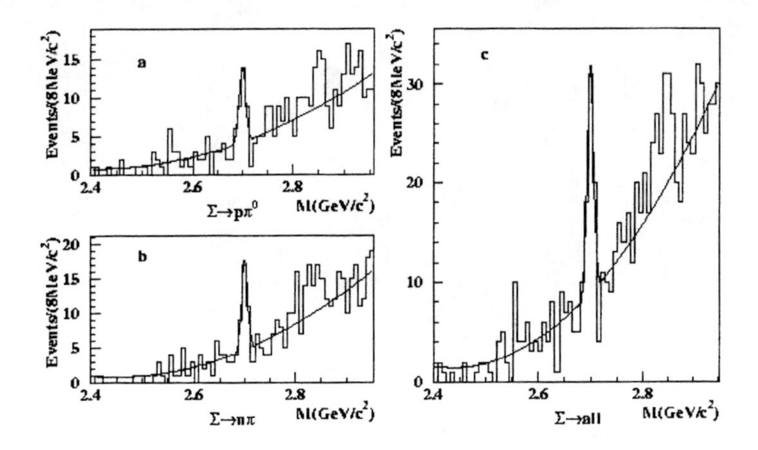

FIGURE 6. Invariant mass spectrum for the final state $\Sigma^+K^-K^-\pi^+$, showing a narrow peak around 2700 MeV/c from Fermilab experiment 687. a) shows the spectrum when the Σ decays through $p\pi^o$; b) shows the spectrum when the Σ decays through $n\pi^+$; and c) is the combined spectrum.

C The Physics of Excited States of Charm and Bottom

In addition to the states we have discussed so far, there are also states with orbital angular momentum between the quarks. These states are often referred to as generically D^{**}'s and B^{**}'s. The existence of these states is established for both charm and bottom. Taken together, the spectra of these states permits some important tests of key ideas about heavy quark states and how the light degrees of freedom behave in these states. These ideas carry over to the discussion of how the hadronic degrees of freedom influence the extraction of weak interaction parameters in the study of the semileptonic decays of hadrons.

Excited D mesons have higher values of the radial or orbital quantum number and have higher mass. The first such excited state was seen by the ARGUS collaboration in 1986 [13]. Since then, 5 additional excited charmed meson states have been seen [14]. The main contributors to the experimental observations have been the ARGUS experiment, CLEO, and Fermilab experiments E691 and E687. LEP experiments and Fermilab Experiment 791 are also beginning to contribute to our knowledge of these states.

The first set of excited states are expected to have L = 1 , shown schematically in figure 12. **These states all have positive parity.** Since the spin of the light quark, $\vec{s}_{q=u,d,s}$, and that of the heavy quark, $\vec{s}_{Q=c,b}$, can add to a

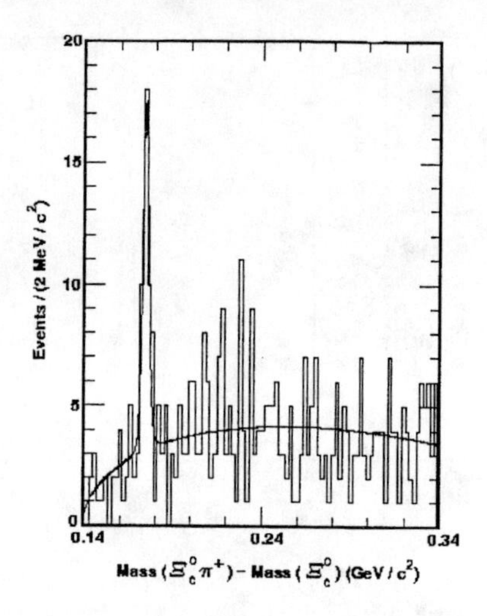

FIGURE 7. Invariant mass spectrum for the final state $\Xi_c^o\pi^+$, showing a narrow peak about 174 MeV/c^2 above the Ξ_c from CLEO.

total spin $\vec{S}$ of 0 or 1, the total angular momentum, J, can be 0, 1, or 2.

Assuming that these states are heavy enough to decay **strongly** into a ground state charmed meson and a pseudoscalar meson (π or K), then conservation rules such as angular momentum, parity, and isospin lead to twelve states with the pattern of quantum numbers and allowed decays shown here [4]:

TABLE 4. Quantum numbers and allowed decays of excited charmed mesons, using total quark spin, orbital angular momentum, and total angular momentum to describe the states.

$^{2S+1}L_J$	$c\bar{u}$	cd	$c\bar{s}$
$^3P_2$	$D\pi$,$D^*\pi$	$D\pi$,$D^*\pi$	D^*K, DK
$^3P_1$	$D^*\pi$	$D^*\pi$	D^*K
$^3P_0$	$D\pi$	$D\pi$	DK
$^1P_1$	$D^*\pi$	$D^*\pi$	D^*K

While these quantum numbers may be an 'appropriate choice' for charmonium where the masses of the two quarks are equal, they may not be 'appropriate' to a 'heavy-light' system. By 'appropriate choice' we mean that the quantum numbers best express the symmetry of the system (leaving the rest

[4] One complication is that the two 1^+ states can mix. We shall ignore this in the following discussion.

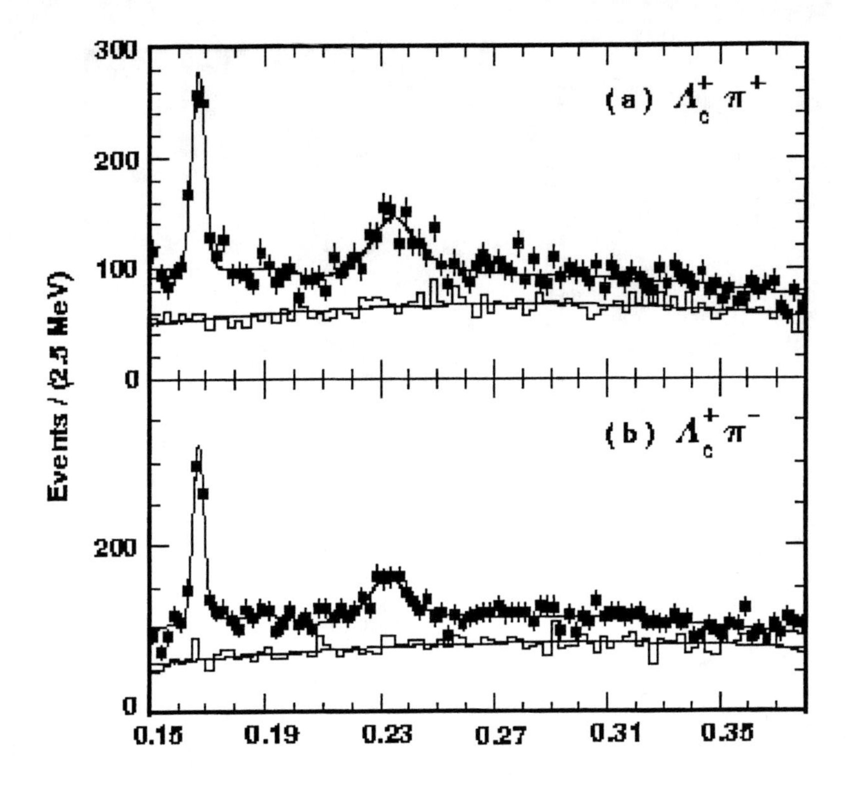

FIGURE 8. Invariant mass spectrum for the final states $\Lambda_c^+\pi^+$ (upper), and $\Lambda_c^+\pi^-$(lower) from CLEO.

as a small perturbation). For example, in the hydrogen atom, we do not worry about the nuclear spin – it largely decouples from the spectroscopy and enters only as a 'hyperfine' effect.

One very relevant symmetry is 'Heavy Quark Symmetry' or HQS which is supported by a 'Heavy Quark Effective Theory' or HQET [15]. According to this, in the limit of infinitely heavy quark mass, the heavy and light degrees of freedom decouple and the light degrees of freedom determine the quantum states, level spacing, and decay rates (and hence widths) of the heavy-light mesons. This same approach leads to relations between the transition matrix elements which appear, for example, in semileptonic decays of these systems.

The model is expected to be a good approximation when

$$M_Q >> \Lambda_{QCD}$$

and, therefore should apply to the b-quark and hopefully the c-quark. Corrections would be expected to be of order $\frac{\Lambda_{QCD}}{M_Q}$ for these finite mass quarks.

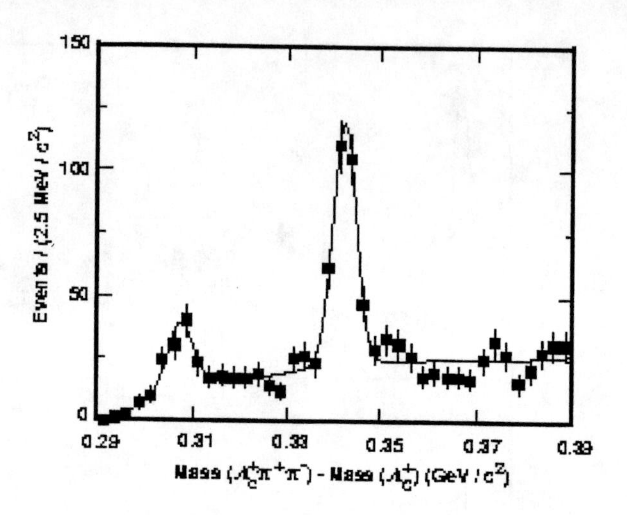

FIGURE 9. Invariant mass spectrum for the final states $\Lambda_c^+\pi^+\pi^-$ from CLEO.

In this picture, the best choice for the quantum numbers would be the spin of the heavy quark,

$$\vec{S}_Q$$

and the total angular momentum of the light degrees of freedom:

$$\vec{j} = \vec{L}_{\bar{q}} + \vec{s}_{\bar{q}}.$$

According to Heavy Quark Symmetry, the quantum number j and the quantum number S_Q are separately conserved and the static properties and decay rates of the particles will depend only on the light degrees of freedom. For any value of L there are two values of the angular momentum of the light degrees of freedom: $j = L \pm \frac{1}{2}$. For L=1, this gives two sets of levels: $j = \frac{3}{2}$ and $j = \frac{1}{2}$. In the heavy quark limit, each level consists of two degenerate states corresponding to the different orientations of the heavy quark spin. For finite (c,b) quark masses, the degeneracy will be broken to the order of $\frac{\Lambda_{QCD}}{M_{b,c}}$. The $j = \frac{3}{2}$, 2^+ state decays through a D-wave so it is expected to be relatively narrow. Because the quantum numbers of heavy and light degrees of freedom are independently conserved, $j = \frac{3}{2}$, 1^+ state also decays by a D-wave. This means that HQET predicts that the $j = \frac{3}{2}$ states are relatively narrow. Similarly, the $j = \frac{1}{2}$, 0^+ state must decay via an S-wave and therefore is expected to be very broad – with widths of 100 MeV/c^2 or more. Because the quantum numbers of heavy and light degrees of freedom are independently conserved, HQET predicts that the $j = \frac{1}{2}$, 1^+ state decays purely by S-wave also. Taken together, this means that the $j = \frac{3}{2}$ states are relatively narrow while the

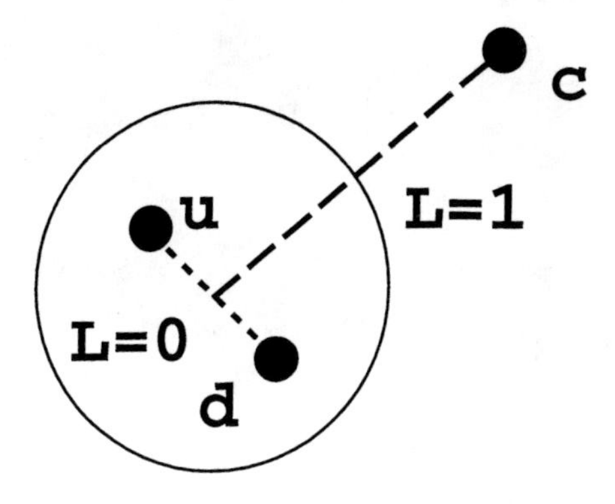

FIGURE 10. Schematic representation of quarks forming Λ_c^* states.

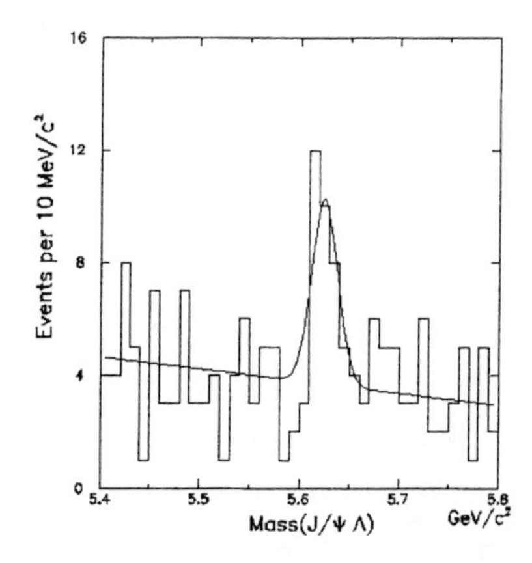

FIGURE 11. Invariant mass spectrum of $\psi\Lambda^0$ obtained by CDF experiment

$j = \frac{1}{2}$ states are quite broad. Experimental backgrounds make the identification of broad states very difficult and none has so far been convincingly observed. The rest of this paper will confine its attention to the $j = \frac{3}{2}$ states. To differentiate the two $j = \frac{3}{2}$ states, we refer to them as the 1^+ and the 2^+ from now on. The predicted spectrum is shown in figure 13.

HQS predicts specific relationships between the level spacing of the D^{**}'s and B^{**}'s. Symmetry breaking effects in the Hamiltonian can be used to predict the splitting of the 3/2 and 1/2 states. Attempts have even been made to extend the applicability of the symmetry to strange mesons.

Quantum number restrictions similar to the ones shown above also exist. The non-strange D^{**} 2^+ state can decay into $D^*\pi$ or $D\pi$. The 1^+ only decays through $D^*\pi$. Similarly, the D_s^{**} 2^+ state can decay into D^*K or DK while the 1^+ state can only decay through D^*K. In addition, the model makes specific predictions on the decay rates into charm particles and light mesons.

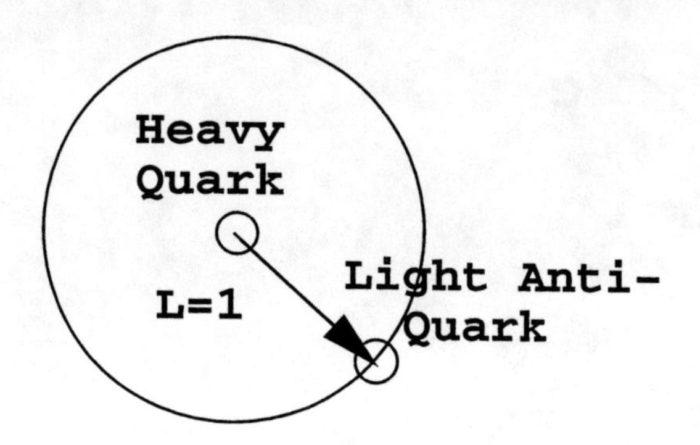

FIGURE 12. Schematic representation of an excited D meson

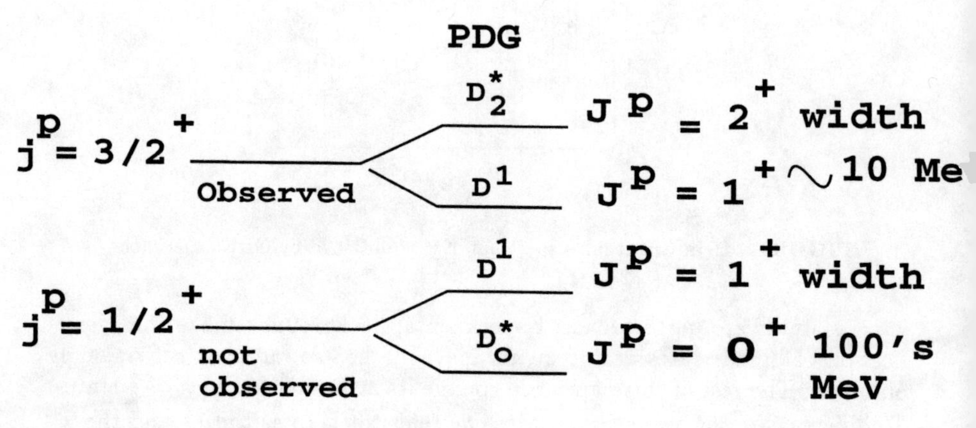

FIGURE 13. Level diagram for $L = 1$ excited mesons according to Heavy Quark Symmetry

HQS also predicts that the decay rates from each excited state into each final state, such as $D\pi$, $D^*\pi$, $D\rho$, and $D\eta$, are independent of the heavy quark mass.

Using HQS and some arguments about the hadronic form factors, it is possible to estimate the widths of the various excited mesons into various final states and to add them up to get estimates of the total widths. Pionic transitions between any two heavy-light states should be identical independent of the heavy quark mass so charm transitions can be used to predict B transitions. One prediction is that $\frac{\Gamma(D_2^* \to D\pi)}{\Gamma(D_2^* \to D^*\pi)}$ is about 1.8.

If one accepts the validity of the model, then relative rates into different final states, total widths, and certain angular correlations can help associate particular mass bumps with particular states predicted by the HQS model.

D Experimental Results on Excited Charm States

Figure 14 shows the invariant mass difference distributions $M_{D^+\pi^-} - M_{D^+}$ and $M_{D^o\pi^+} - M_{D^o}$ from E687 [16]. (It is conventional in searches for D^{**}'s to show the mass difference distribution since certain systematic errors are cancelled on an event-by-event basis.) There are clear peaks near 600 MeV/c^2. These distributions are obtained by taking D-meson decays into simple final states and combining them with pions of the appropriate sign that emerge from the primary interaction vertex. There is a lot more background in such plots than typically occurs in the spectra of weakly decaying mesons. This is because of the unavoidable background that comes from combining the real D signal (in experiments where the D moves in the lab and a vertex detector is available, like E687, the states forming the D signal can be required to be well-separated from the interaction vertex so that there is little background to the D-meson itself) with pions coming from the underlying hadronic interaction and having nothing to do with the decay of a strong resonance. The generation of this 'combinatoric background' is shown schematically in figure 15. The high levels of background make it difficult to isolate the 'broad' states predicted by the theory.

The identification of various peaks with particular quantum states is based on model calculations of the masses, widths, and decay rates. However, in some cases, it is actually possible to extract the quantum numbers of the state. This is true when the state can decay into a D^{*+} and a π, since the angular distribution between the daughter pion and the pion from the decay of the D^* in the D^* rest frame conveys information on the overall angular momentum. For more details on these very beautiful analyses, see reference [18].

Tables 5 and 6 [17] summarize the present knowledge of the masses and widths of the six $j = \frac{3}{2}$ excited mesons which have been observed. While the overall picture is satisfactory, there are differences in the mass values that lie outside of the quoted statistical and systematic uncertainties. I attribute

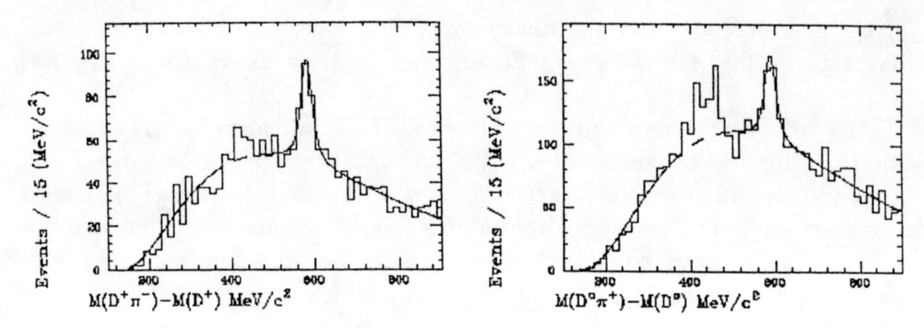

FIGURE 14. Examples of signals for D^{**} mesons from E687. Left side: Invariant mass difference distribution, $M(D^+\pi^-) - M(D^+)$ showing a peak around 600 MeV/c^2, or a total mass of about 2460 MeV/c^2; and right side: Invariant mass difference distribution, $M(D^o\pi^+) - M(D^o)$ showing a peak at almost the same mass. The peak near 400 MeV/c^2 is believed to be a feed down from the $D^{*o}\pi^+$ decay mode of the same state, where the $D^{o*} \rightarrow D^o\pi^o$ since the π^o is not reconstructed in this analysis. These two plots are identified as the two isospin partners of the 2^+ excited D-mesons.

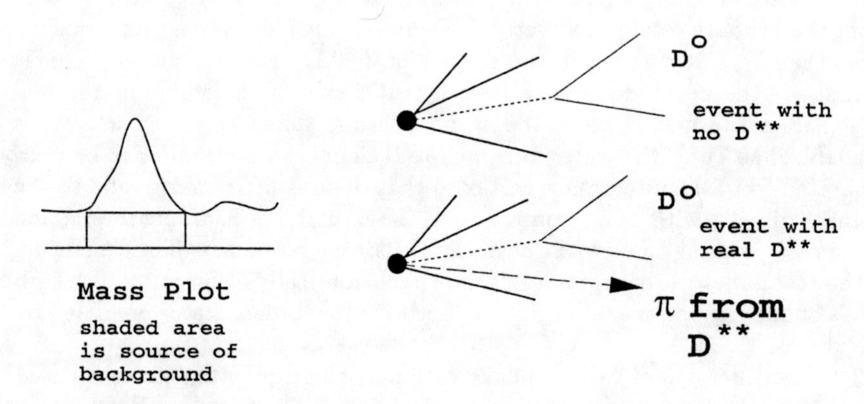

FIGURE 15. Mechanisms for background generation in searches for excited mesons. left side: depicts background faking the D, which can combine with π's from the primary to generate background to the D^{**}; right side upper: an event with a real D but no D^{**} can produce background combinations with π's coming from the primary vertex; and lower right side: a real D^{**}.

TABLE 5. Properties of the 2^+ States

experiment	Mass (MeV/c^2)	Width (MeV/c^2)
$D^{**o} \to D^+\pi^-$:		
E687	$2453 \pm 3 \pm 2$	$25 \pm 10 \pm 5$
E691	$2459 \pm 3 \pm 2$	$20 \pm 10 \pm 5$
ARGUS	$2455 \pm 3 \pm 5$	$15\ ^{+13+5}_{-10-10}$
CLEO 1.5	$2461 \pm 3 \pm 1$	$20\ ^{+9+9}_{-12-10}$
CLEO II	$2465 \pm 3 \pm 3$	$28\ ^{+8}_{-7} \pm 6$
$D^{**+} \to D^o\pi^+$:		
E687	$2453 \pm 3 \pm 2$	$23 \pm 9 \pm 5$
ARGUS	$2469 \pm 4 \pm 6$	27 ± 12
CLEO II	$2463 \pm 3 \pm 3$	$27\ ^{+11}_{-8} \pm 5$
$D_s^{**+} \to D^oK^+$:		
CLEO II	$2573.3\ ^{+1.7}_{-1.6} \pm 0.9$	$16\ ^{+5}_{-4} \pm 3$

TABLE 6. Properties of the 1^+ States

experiment	Mass (MeV/c^2)	Width (MeV/c^2)
$D^{**o} \to D^{*+}\pi^-$:		
ARGUS	$2414 \pm 2 \pm 5$	$13\ ^{+6+10}_{-6-5}$
CLEO 1.5	$2428 \pm 3 \pm 2$	$23\ ^{+8+10}_{-6-4}$
CLEO II	$2421\ ^{+1}_{-2} \pm 2$	$20\ ^{+6+3}_{-5-3}$
E687	$2422 \pm 2 \pm 2$	$15 \pm 8 \pm 4$
$D^{**+} \to D^{*o}\pi^+$:		
CLEO II	$2425 \pm 2 \pm 2$	$26\ ^{+8}_{-7} \pm 4$
$D_s^{**+} \to D^{*o}K^+$:		
ARGUS	$2535.5 \pm 0.4 \pm 1.3$	<3.9 (90%CL)
CLEO II	$2535.1 \pm 0.2 \pm 0.5$	<2.3 (90%CL)
E687	$2535.0 \pm 0.6 \pm 1.0$	<3.2 (90%CL)

this to the large and highly structured backgrounds. The structures limit the region over which fits can be performed. In some cases, subtle variations of the fit near the signal region can cause shifts in the mass values. In general, I feel that the quoted systematic errors tend to be optimistic.

E Prediction of Excited B-meson Spectra from Charm Meson Spectra Using an HQS-motivated Approach

Eichten, Quigg, and Hill [19] write the equations for the mass difference between the ground state and $j = 3/2$ excited state strange and charmed mesons as:

TABLE 7. Masses (in MeV) predicted for the $2P(\frac{3}{2})$ levels of the B, D_s, and B_s systems. Underlined entries are spin-averaged quantities (derived from Particle Data Group masses) used as inputs.

Meson Family	K	D	B	D_s	B_s
M(1S)	794.3	1973.2	5313.1	2074.9	5403.0
M($2^+(\frac{3}{2})$)	1429 ± 6	2459.4 ± 2.2	5771	2561	5861
M($1^+(\frac{3}{2})$)	1270 ± 10	2424 ± 6	5759	2526	5849
M($2^+(\frac{3}{2})$)-M($1^+(\frac{3}{2})$)	159	35	12	35	12

$$\begin{aligned} M(2P_2)_K - M(1S)_K &= E(2P)_K + \frac{C(2P_2)}{m_s} \\ M(2P_1)_K - M(1S)_K &= E(2P)_K + \frac{C(2P_1)}{m_s} \\ M(2P_2)_D - M(1S)_D &= E(2P)_D + \frac{C(2P_2)}{m_c} \\ M(2P_1)_D - M(1S)_D &= E(2P)_D + \frac{C(2P_1)}{m_c} \end{aligned} \tag{8}$$

For the masses and levels, the notation is nL_J, where n is the principal quantum number. Here, $M(1S) = [3M(1S_1) + M(1S_o)]/4$ is the spin-averaged mass of the ground state. The difference $E(2P)_K - E(2P)_D$ is determined by potential models to be 32 MeV. This leaves 5 unknowns: $E(2P)_D$, $C(2P_2)$, $C(2P_1)$, and m_s and m_c. The charm mass was fixed at various values and particular states were used to determine the mass splittings on the left side of the equations. This leaves four equations in four unknowns.

The parameters so determined are used to predict the $j = \frac{3}{2}$ excited B meson states. The results are given in table 7 [19].

F Experimental Results on Excited B-Mesons

The predicted excited states of the B (non-strange) mesons are listed in table 8. In this table, we use notation motivated by HQS and also use the names which the PDG has adopted for these states.

The B^{**} states were first observed by LEP experiments. Figure 16 below shows the signal for B^{**}'s from the ALEPH experiment [20]. The analysis is done by identifying B-jets produced in Z decays using the ALEPH vertex detector and then combining the jet momentum vector with that of pions not associated with the other B-jet. The analysis cannot distinguish between a B and a B^* so the notation $B^{(*)}$ is used to denote a state which could contain either. The masses of these states obtained from ALEPH, DELPHI [21], and OPAL [22] are given in table 9. It can be seen that the predictions for the level spacings are well born out. DELPHI and OPAL have observed the B_s^{**}

TABLE 8. Quantum numbers of excited (nonstrange) B mesons, using orbital angular momentum, the total angular momentum of the light degrees of freedom, and total angular momentum and parity to describe the states.

L	j	J^P	State	Dominant Decay Mode
0	$\frac{1}{2}$	0^-	B	
0	$\frac{1}{2}$	1^-	B^*	$B\gamma$
1	$\frac{1}{2}$	0^+	B_o^*	$B\pi$ (S-Wave)
1	$\frac{1}{2}$	1^+	B_1^*	$B^*\pi$ (S-Wave)
1	$\frac{3}{2}$	1^+	B_1	$B^*\pi$ (D-Wave)
1	$\frac{3}{2}$	2^+	B_2^*	$B^*\pi$, $B\pi$ (D-Wave)

TABLE 9. Masses of the B^{**} states

experiment	Mass MeV/c^2
$B_{u,d}^{**}$:	
ALEPH(91-94)	$5734 \pm 4 \pm 10$
DELPHI(91-94)	$5734 \pm 5 \pm 17$
OPAL(91-94)	5712 ± 11 (stat)
LEP Average	5727 ± 6
B_s^{**}:	
DELPHI(B_{s1})	$5888 \pm 4 \pm 8$
DELPHI(B_{s2}^*)	$5914 \pm 4 \pm 8$
OPAL	5884 ± 15

decaying into $B^{(*)}K$. The DELPHI result is shown in figure 17. The two states observed by DELPHI are identified as the B_{s1} and the B_{s2}^*.

DELPHI [23] has also observed *two* resonances in the state $B^{(*)}\pi^+\pi^-$. The data are shown in figure 18. The bump at $Q = 301\,\mathrm{MeV/c^2}$ is interpreted as evidence for a radial excitation of the B at a mass value of about 5860 MeV/c^2. The predicted spectrum is shown in figure 19 and supports this identification.

G The B_c Meson

Finally, we discuss a particle that has not yet been observed, the B_c^+ meson. This particle is especially interesting because it can be treated by means of nonrelativistic potential models. Such a treatement [24] gives a prediction for the mass value of the lowest mass states of this family:

$$M_{B_c} = 6258 \pm 20\,MeV/c^2 \qquad M_{B_c^*} - M_{B_c} = 73\,MeV/c^2 \tag{9}$$

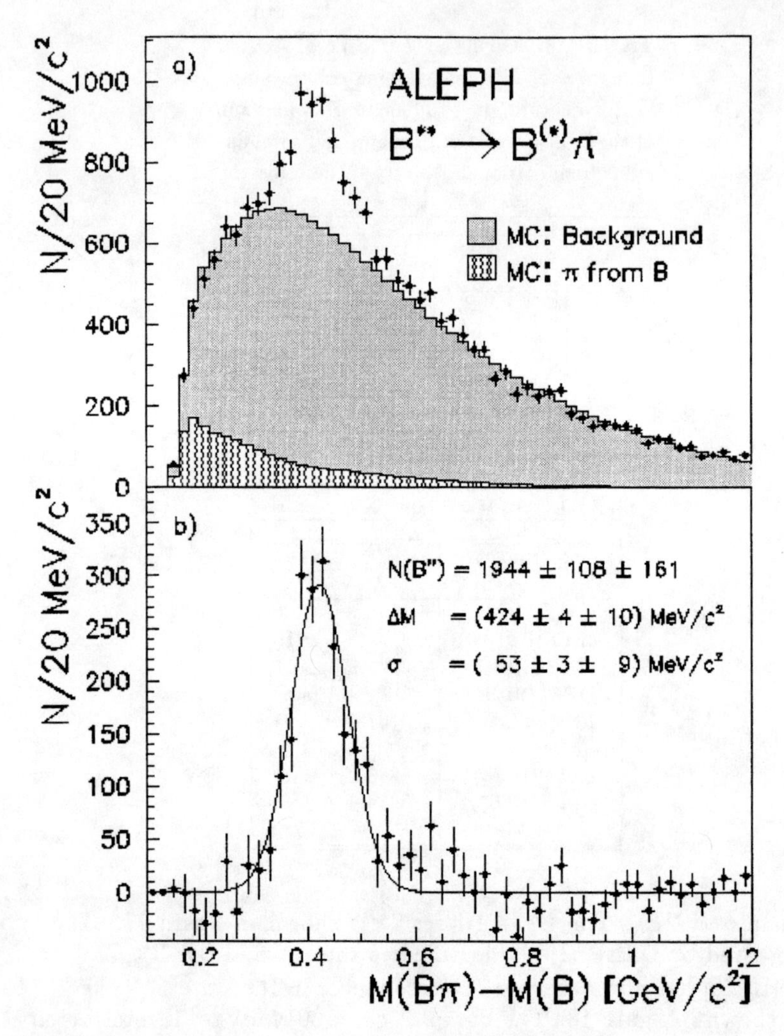

FIGURE 16. a) The $(B\pi) - B$ mass difference distribution from 1991 through 1994 data from ALEPH. The background estimated from the Monte Carlo simulation is shown by the hatched area. b) The background subtracted signal for the decay $B^{**} \to B^{(*)}\pi^{\pm}$ fitted with a Gaussian(curve).

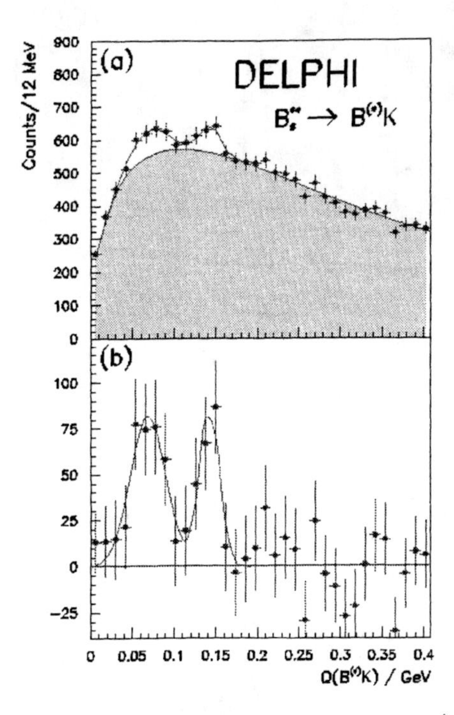

FIGURE 17. Invariant mass difference distribution for $B_s^{**} \to B^{(*)}K$ from DELPHI. a) Inclusive distribution with estimated background from Monte Carlo showed as shaded area; and b) Background subtracted distribution.

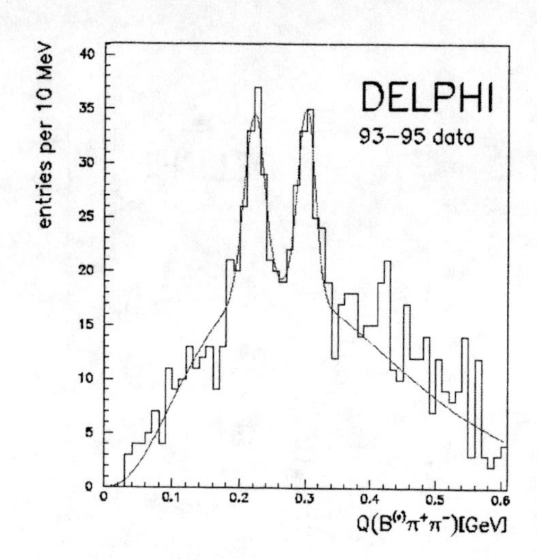

FIGURE 18. Observation of resonances in the $B^{(*)}\pi^+\pi^-$ mass distribution from DELPHI. The quantity plotted on the lower axis, $Q(B^{(*)}\pi^+\pi^-)$ is $M(B^{(*)}\pi^+\pi^-) - M(B^{(*)}) - 2M(\pi)$.

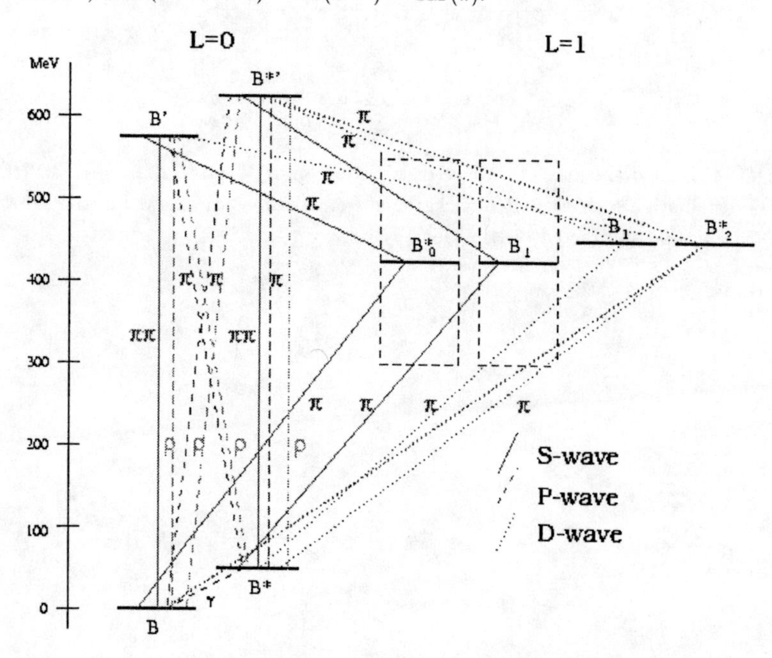

FIGURE 19. Spectrum of excited B mesons showing the radial excitations B' and $B^{*\prime}$ decaying to ground state B and B^* mesons through the sequential emission of pions.

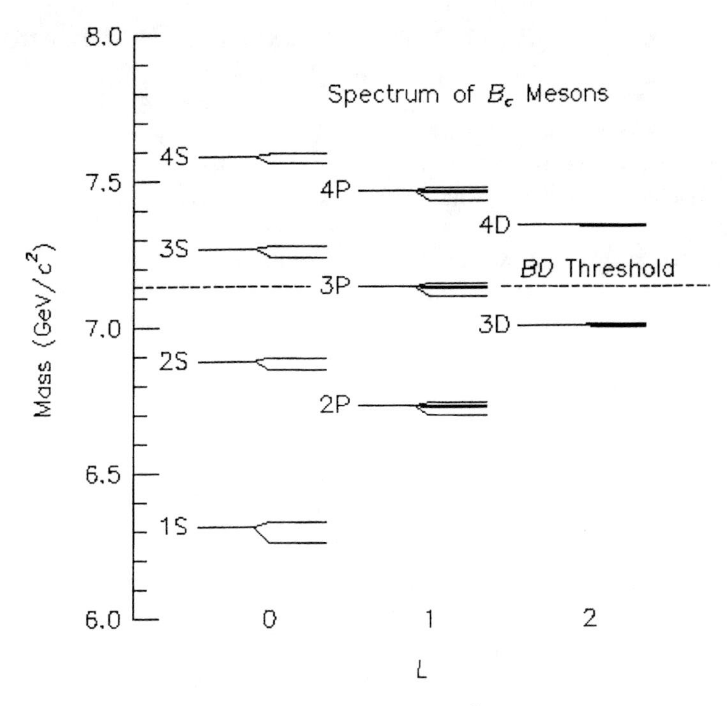

FIGURE 20. Spectrum of $c\bar{b}$ states.

The models predict a spectrum of excited states many of which are below BD threshold and thus decay mainly through radiative transitions. The spectrum of these states is shown in figure 20.

These states are expected to decay into states containing ψ mesons and this may provide a convenient set of decay modes in which to search for these particles. Predictions of branching fractions for promising decay modes in which to observe these states are given in table 10.

TABLE 10. Promising Decay Modes for the Observation of B_c States

Mode	Br
$B_c \rightarrow \psi\pi^+$	4.0×10^{-3}
$B_c \rightarrow D_s\psi$	5.0×10^{-2}
$B_c \rightarrow \psi l\nu$	10%

H Concluding Remarks on Excited Charm and Bottom Mesons

The spectroscopy of heavy quarks, while it may be considered to be a mature field which is reasonably well described by theoretical models, nevertheless still presents subtle and exciting challenges both to experimentalists and to theorists.

The excited states of D and B mesons provide important insights into the properties of hadrons containing heavy quarks through the following relations, which are expected from HQS:

$$M_{D^{**}} - M_D \sim M_{B^{**}} - M_B \tag{10}$$
$$M_{D^*} - M_D \sim \frac{M_b}{M_c} \times (M_{B^*} - M_B)$$

There are many more measurements one can make including detailed studies of other decay modes and a comparison of their rates relative to the ones already observed and to the predictions of HQS.

Below, it will be seen that an understanding of the hadronic structure of charmed and bottom mesons is necessary to interpret semileptonic (and non-leptonic) decays and to extract Standard Model Parameters, especially CKM elements, from them. HQS will be seen to greatly simplify the problem of disentagling the weak interaction properties of the heavy quarks from the influence of the strong interactions with the accompanying light quarks.

III LIFETIMES OF CHARMED AND BOTTOM PARTICLES

The lifetimes of the charm and bottom particles provide a good 'overview' of several important features of heavy quark decays. This may seem surprising since they involve a summation over all of the many decay channels which are allowed. Nevertheless, many important insights have been gained by a study of lifetimes.

When charmed mesons were first observed, it was generally expected that their decay rate would be roughly that of a free charm quark. This gave rise to the 'spectator model' – the quark decayed via a W emission and the W had no direct interaction with the original light quark which was named the spectator quark because it did not change during the decay process. A typical spectator decay is shown in figure 21.

In this picture, all charmed particles would have similar lifetimes. As soon as lifetime measurements became available, it was clear that the spectator model was not correct. The D^+ lifetime was more than twice the D^o lifetime. Moreover, the Λ_c^+ lifetime was much shorter than the D^o lifetime. Table 11 shows the world-average measurements of the charmed meson lifetimes [25].

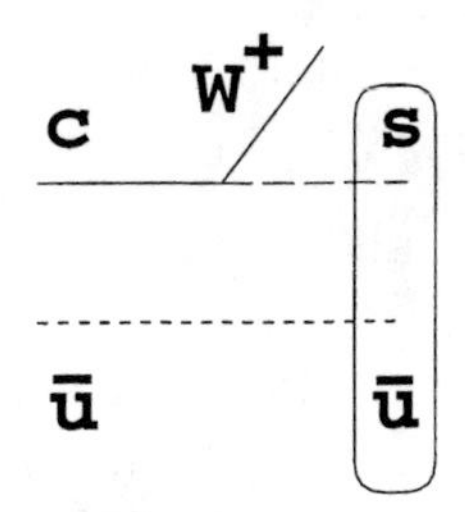

FIGURE 21. Spectator decay of the D^o meson.

TABLE 11. Lifetimes of charmed mesons

Meson	lifetime ps
D^+	1.057 ± 0.015
D^o	0.415 ± 0.004
D_s^+	0.467 ± 0.017

Many attempts were made to explain the pattern of lifetimes of the charmed mesons and baryons. The solution to the problem required taking into account additional quark level processes which can contribute to the weak decays and the quantum effects associated with them. There are six such processes whose diagrams are shown in figure 22.

These diagrams have different dependences on the heavy quark mass. The spectator diagram goes like the mass to the fifth power, M_Q^5, which is required by dimensional analysis and which is analogous to muon decay. The other processes all require the two valence quarks to be within the range of the virtual W and therefore depend on the wave function squared of the initial state heavy meson or baryon at the origin. Since the wave function squared has the dimension

$$|\psi(0)|^2 \propto \frac{1}{length^3} \propto M_Q^3 \qquad (11)$$

the dependence of all these diagrams on the heavy quark mass goes like M_Q^2.[5] Therefore, as the quark mass gets heavy, the relative importance of the non-spectator diagrams to the spectator diagram should decrease.

In fact, there are two diagrams which are classified as 'spectator diagrams' in figure 22:

- One is just the simple diagram shown above with the W decaying into quarks which form hadrons without any communication with other fi-

[5] Various QCD effects like gluon emission which can relieve the helicity suppression or color suppression can change this dependence to M_Q^2.

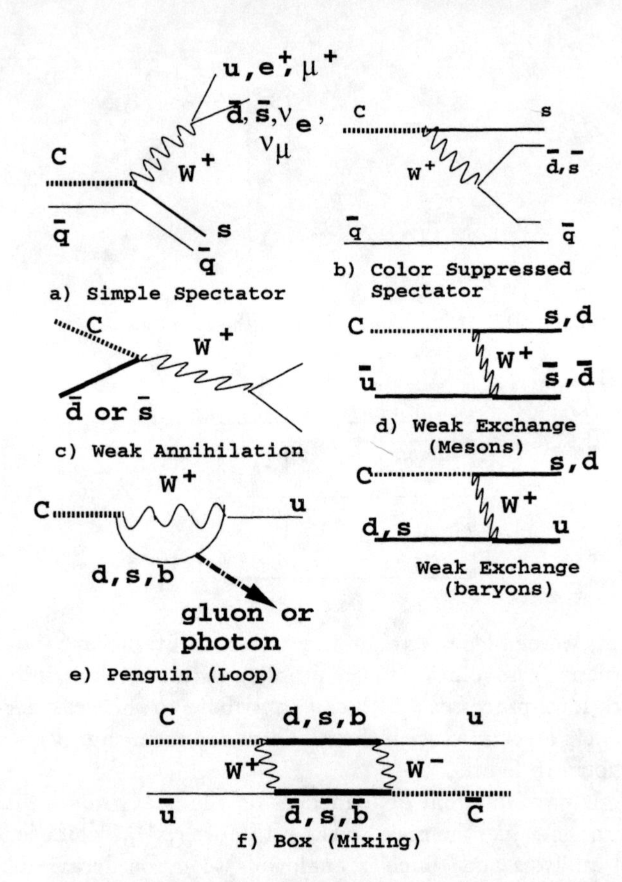

FIGURE 22. Quark level diagrams for weak decays

nal state quarks. This is called the 'external W emission' or 'external spectator' diagram.

- In the second diagram, the quarks from the W decay combine with the spectator quark and with the final state quark that coupled directly to the W. This is the so-called 'internal W emission' or 'internal spectator' diagram.

For final states where both diagrams are possible, if one of the light antiquarks in the W decay is the same as the spectator antiquark, then interference effects are possible. Another important consideration is that the quarks from the 'internal W emission' must match color of the other two quarks in the final state. This restricts the number of possibilities and reduces the overall contribution to decay amplitude by a factor of 3. Because of this, the 'internal emission' diagram is often referred to as the 'color-suppressed' diagram. However, gluon radiation in the initial or final state can disrupt the color suppression so the degree of suppression is not well-predicted and the two amplitudes could turn out to be 'comparable'.

A Charmed Meson Lifetimes

What is responsible for the lifetime differences among the charmed mesons? In particular, why is the D^+ lifetime so much longer than the D^o lifetime? It has to be the result of some kind of destructive interference effect between the amplitudes described above. In meson decay, the W-exchange and annihilation amplitudes are suppressed by helicity and color considerations and are believed to be too small to be the cause of the observed differences. It is natural to look for the answer in the two spectator decay diagrams.

A key difference between D^+ and D^o decays is that the Cabibbo-favored D^+ decays can experience light quark interference but the (CF) D^o decays cannot [26]. The situation is shown in figure 23 for two body decays. The two diagrams for the D^+ can interfere destructively reducing the decay rate and increasing the lifetime. In the figure, the decay final states are divided into three types:

- TYPE I: Decays which can only proceed through external W emission with an amplitude a_1.;

- TYPE II: Decays which can only proceed through internal W emission with an amplitude a_2; and

- TYPE III: Decays which can proceed through both internal and external W emission and will depend on the magnitude a_1 and a_2 and their relative sign which determines whether the interference is constructive or destructive.

Only the D^+'s have Cabibbo-favored TYPE III decays. This results in a suppression of some Cabibbo-favored two body decay modes. Since a large fraction of all charm decays are two body or quasi-two body, this could result in an overall increase in the D^+ lifetime relative to the D^o. However, it is not very satisfying to extrapolate from an argument based on two body decays to the full decay rate.

It is also possible to make a simlar 'inclusive' argument based on quark level dynamics alone. For the Cabibbo-favored decays, we have $D^o \to su\bar{d}\bar{u}$ and $D^+ \to su\bar{d}\bar{d}$. Simlarly, we have $B^o \to cd\bar{u}\bar{d}$ and $B^+ \to cd\bar{u}\bar{u}$. In each case, the internal and external diagrams can interfere. Let c_1 be the amplitude for external W emission and c_2 be the amplitude for internal emission. Then for the D^o (and B^o), the (nonleptonic) decay width,Γ_{NL}, is given by

$$\Gamma_{NL}(D^o \,\text{or}\, B^o) \;\propto 3(c_1^2 + c_2^2 + 2\xi' c_1 c_2) G_F^2 M_{c\,\text{or}\,b}^5 / 192\pi^3 \tag{12}$$

where ξ' is the degree of color coherence between the diagrams. However, for the $D^+(B^+)$, the $\bar{d}(\bar{u})$ spectator antiquark and antiquark from the W decay can also interfere adding extra terms to the decay rate. The expression for the decay rate is:

$$\begin{aligned}\Gamma_{NL}(D^+ \,\text{or}\, B^+) \propto 3G_F^2 M_{c\,\text{or}\,b}^2 [(c_1^2 + c_2^2 + 2\xi' c_1 c_2) M_{c\,\text{or}\,b}^3 / 192\pi^3 \\ + (2c_1 c_2 + \xi'(c_1^2 + c_2^2)) |\psi(0)|^2 / \pi]\end{aligned} \tag{13}$$

The extra term is proportional to M_Q^2 and the value of the wave function at the origin and which is destructive, at least in the case of charm. This contribution, called 'Pauli interference', can explain the lifetime differences.

Since some two body decays proceed only via external emission, some by only internal emission, and some through both, it is possible to determine the parameters a_1 and a_2. The study of two body charm decays gives the result [27] that:

$$\frac{a_2}{a_1} \;=\; -(0.4 \pm 0.1) \tag{14}$$

These are related to c_1 and c_2 as follows:

$$\begin{aligned} a_1 &= c_1 \,+\, \xi c_2 \\ a_2 &= c_2 \,+\, \xi c_1 \end{aligned} \tag{15}$$

The quantity ξ is not necessarily the same as ξ' in the inclusive treatment. The amplitudes do indeed interfere **destructively** and are 'comparable'. Models based on light quark interference in D^+ decays can explain the lower D^+ decay rate (longer lifetime) and this is considered to be the correct explanation for this puzzle.

For B-decays, a recent analysis [28] of the measured two body decays gives

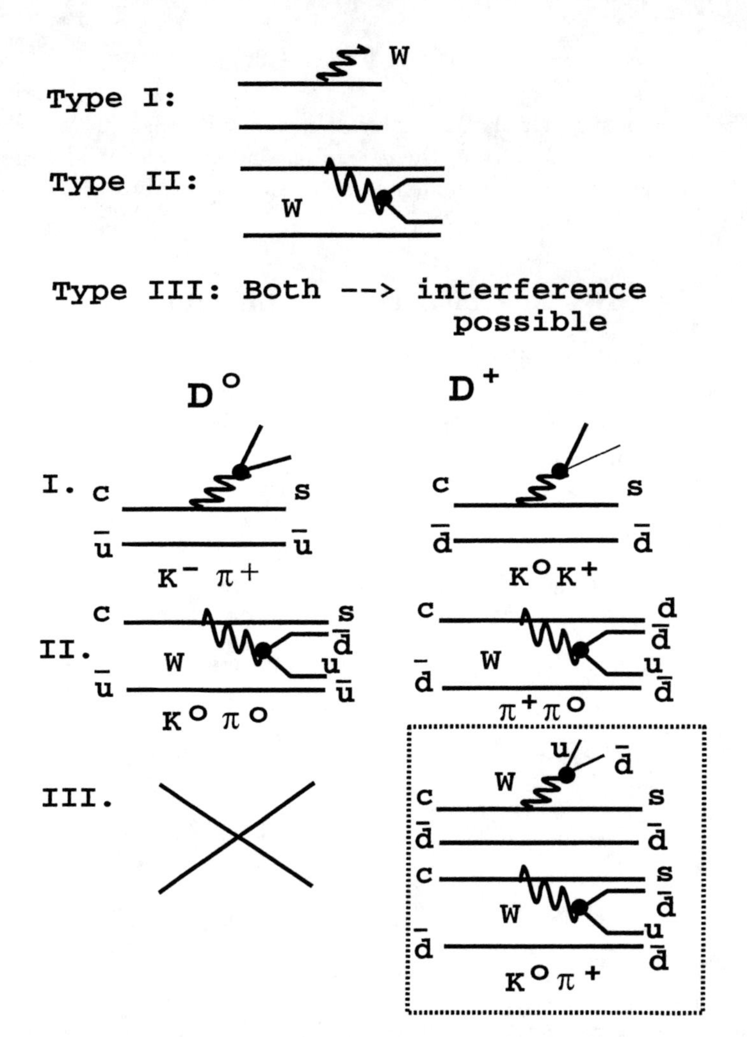

FIGURE 23. Comparison of the Cabibbo-favored spectator diagrams contributing to the decay of the D^o and D^+ mesons.

$$\frac{a_2}{a_1} = +(0.26 \pm 0.08 \pm 0.06) \tag{16}$$

The positive sign suggests that the interference may be constructive which is very surprising. If that is true, the B^+ might have a shorter lifetime than the B^o. More accurate measurements are needed to verify this unexpected prediction.

B Charmed Baryon Lifetimes

There are significant differences between meson decay and baryon decay. The meson decays are constrained by color considerations and by helicity suppression. All three colors are present in baryons as are three spins so neither color suppression nor helicity suppression play a role whereas they are important factors in meson decays. Because the nonspectator diagrams are unsuppressed, their effect is expected to show up more dramatically in charmed baryon decays than in charmed meson decays.

The nonleptonic decays of the four weakly decaying singly charmed baryons may be written, in the spirit of the inclusive analysis described above, as:

$$\Gamma_{NL}(\Lambda_c^+) = \Gamma^{dec}(\Lambda_c^+) + \Gamma^{exch}(\Lambda_c^+)_d + \Gamma_-^{int}(\Lambda_c^+)_u \tag{17}$$

$$\Gamma_{NL}(\Xi_c^+) = \Gamma^{dec}(\Xi_c^+) + \Gamma_-^{int}(\Xi_c^+)_u + \Gamma_+^{int}(\Xi_c^+)_s \tag{18}$$

$$\Gamma_{NL}(\Xi_c^o) = \Gamma^{dec}(\Xi_c^o) + \Gamma^{exch}(\Xi_c^o)_d + \Gamma_+^{int}(\Xi_c^o)_s \tag{19}$$

$$\Gamma_{NL}(\Omega_c^o) = \Gamma^{dec}(\Omega_c^o) + \Gamma_+^{int}(\Omega_c^o)_s \tag{20}$$

Here, Γ^{dec} is just the spectator decay. The 'exchange' contribution and the 'constructive' light quark interference term labelled Γ_+^{int} add to the width and decrease the lifetime and the 'destructive' light quark interference term Γ_-^{int} increases the lifetime. With these expressions, one can deduce the 'lifetime hierarchy' of the charmed baryons. For example, Guberina, Ruckl, and Trampetic [29] estimate

$$\Gamma^{exch} : \Gamma_+^{int} : \Gamma_-^{int} \simeq 1 : 0.5 : -0.3 \tag{21}$$

and, in a more detailed calculation, predict

$$\tau(\Xi_c^o) : \tau(\Lambda_c^+) : \tau(\Xi_c^+) \propto 0.6, 1.0, 1.6 \tag{22}$$

E687 at Fermilab has measured all four of these lifetimes. The world-average results and E687 results are presented in table 12. It should be noted that the Ω_c appears to be the shortest lived of the charmed baryons. Also,

$$\frac{\tau(\Xi_c^+)}{\tau(\Lambda_c^+)} = 2.15 \pm 0.59 \text{ and } \frac{\tau(\Lambda_c^+)}{\tau(\Xi_c^o)} = 1.89^{+0.35}_{-0.48} \tag{23}$$

Obviously, more statistics will help sharpen this comparison but the theory seems to reproduce the trend of the data. In particular, significant contributions from exchange diagrams and interference effects are required to explain the charmed baryon lifetime hierarchy.

TABLE 12. World average and E687 values of the lifetimes of charmed baryons [30].

Baryon	lifetime (PDG) ps	lifetime (E687) ps
Λ_c^+	0.206 ± 0.012	$0.215 \pm 0.01 \pm 0.008$
Ξ_c^+	$0.35^{+0.07}_{-0.04}$	$0.41^{+0.11}_{-0.08} \pm 0.02$
Ξ_c^o	$0.098^{+0.023}_{-0.015}$	$0.101^{+0.025}_{-0.017} \pm 0.005$
Ω_c^o	0.064 ± 0.020	$0.086^{+0.027}_{-0.020} \pm 0.028$

C B Meson and Baryon Lifetimes

The B meson lifetimes have been a topic of intense investigation at e^+e^- machines and at CDF at the Fermilab Tevatron. From the picture presented above, one expects a further suppression of nonspectator effects of about a factor of 10-20 because of the large mass of the b-quark. The lifetimes of the B-mesons should all be very similar, with the B^+ being slightly longer lived than the B^o due to interference. However, the recent analysis of two body B-meson decays from CLEO, described above, suggests that the Pauli interference is constructive in B^+ decays. If this surprising result is true, then the B^+ may turn out to be shorter lived than the B^o! The baryons are expected to be about 10% shorter lived than the mesons. The general picture is certainly borne out by the data given in table 13, at least to the extent that the fractional lifetime differences among the various species of B-hadrons are certainly much smaller than the differences among the corresponding species of charmed hadrons. More accurate measurements will tell us whether the detailed predictions of the small differences among the B-hadron species are borne out.

IV SEMILEPTONIC AND LEPTONIC DECAYS

About 20% of the time, a $D^{o,+}$ or a $B^{o,+}$ decays semileptonically. Weak semileptonic decays are among the simplest decays to describe theoretically because:

TABLE 13. Results on lifetimes of B mesons and baryons [31].

	LEP Avg	CDF	SLD	World Avg
b-quark	1.54 ± 0.02	1.51 ± 0.03	1.56 ± 0.05	1.53 ± 0.02
B^-	1.63 ± 0.06	1.68 ± 0.07		1.65 ± 0.05
B^o	1.52 ± 0.06	1.58 ± 0.09		1.55 ± 0.05
B_s	1.60 ± 0.10	1.36 ± 0.12		1.50 ± 0.08
Λ_b	1.21 ± 0.07	1.32 ± 0.17		1.23 ± 0.06
Ξ_b	$1.39^{+0.34}_{-0.28}$			

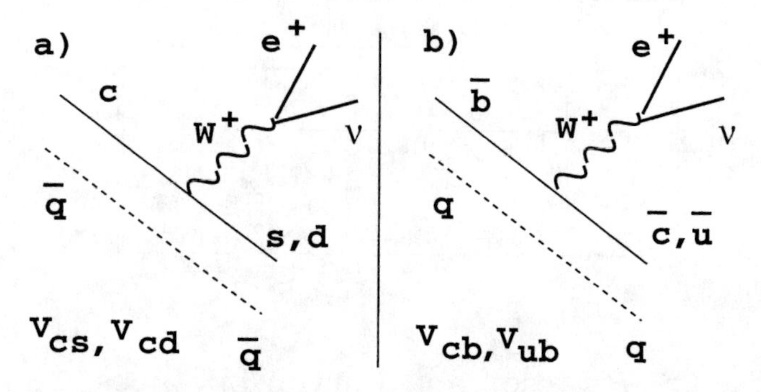

FIGURE 24. Schematic representation on the decay of a) a charmed or b) a bottom meson.

- Only the spectator diagram contributes because the leptons coming from the virtual W do not interact with the remaining quarks at all;
- The leptonic vertex is completely determined by the theory of the electroweak interaction; and
- The strong interaction effects, which come from the requirement that the initial and final quarks be bound into hadrons, can be described by one or more form factors which depend only on q^2, the mass of the virtual W which is responsible for the decay.

These decays are shown schematically in figure 24. The number of form factors which describe the hadronic vertex in the decay of a pseudoscalar meson depends on the spin of the outgoing hadron.

The matrix element for a semileptonic decay is given by

$$A(M \rightarrow me\nu) = \frac{G_F}{\sqrt{2}} V_{Qq} L^\mu H_\mu, \text{ where} \tag{24}$$

$$L^\mu = \bar{u}_e \gamma^\mu (1 - \gamma_5) v_\nu, \text{ and} \tag{25}$$

$$H^\mu = < m | J^\mu_{had}(0) | M > . \tag{26}$$

The semileptonic decay rate is proportional to the square of a CKM matrix element. It is possible to extract CKM elements from measurements of the semileptonic width and this is, in fact, the preferred method. However, the value of a CKM matrix element always appears in a product with the hadronic form factor. To obtain the value of the CKM parameter, we must have a model which predicts the form factor and allows us to divide out its effect. This represents a challenge for nonperturbative QCD. Today, extraction of CKM parameters from data is often limited by the uncertainties in the theoretical estimate of the form factors rather than by the statistical or systematic errors of the measurement of the the semileptonic width.

In the charm system, the CKM matrix elements are given by the Cabibbo angle which can be determined from K meson decay. This permits us to use charm semileptonic decays to extract the hadronic matrix elements without relying on models. We then have the following game plan:

- Use charm semileptonic decays where we 'know' V_{cs} and V_{cd} to isolate the properties of the hadronic vertex;
- Compare these results to the predictions of various quark models and lattice gauge calculations;
- Establish confidence in at least some of these approaches; and
- Armed with the experience, use theoretical calculations to extract V_{cb} and V_{ub} from measurements of B semileptonic decay.

Since the quantity q^2 plays an important role in the study of semileptonic decays, it is worth understanding the range which can be assumed by this variable and the physical configuration corresponding to its various values. Figure 25 shows the minimum q^2 configuration and the maximum q^2 configuration. In the minimum q^2 configuration ($q^2 \sim 0$ if the lepton is an electron), the recoiling final state meson is moving fast and is relativistic which causes problems for theory. In the maximum q^2 configuration, the final state hadron is at rest in the rest frame of the parent meson and is nonrelativistic so theory is happy. Unfortunately, it is often difficult for experiments to measure out to high q^2. The maximum value of q^2 is given by

$$q^2_{max} = (M - m)^2 \tag{27}$$

where M is the mass of the parent hadron and m is the mass of the daughter hadron. Table 14 gives the value of this important quantity for typical decays.

A Pseudoscalar to pseudoscalar decay

The formula for the semileptonic decay rate of pseudoscalar meson to a lighter pseudoscalar meson is [32]:

$$\frac{d^2\Gamma}{dq^2} = \frac{G_F^2|V_{Qq}|^2|h|^3}{24\pi^3}[|f_+(q^2)|^2 + m_l^2|f_-(q^2)|^2 ...] \tag{28}$$

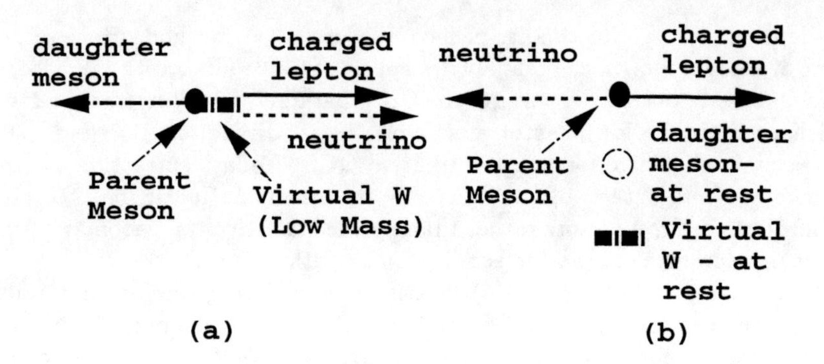

FIGURE 25. Final state particle configurations for a) minimum q^2 and b) maximum q^2.

TABLE 14. Maximum value of q^2 for various semileptonic decays

Decay	q^2_{max} (GeV/c^2)
$D^o \to K^- l^+ \nu$	1.88
$D^o \to \pi^- l^+ \nu$	3.00
$D^+ \to K^{*o} l^+ \nu$	0.95
$B^+ \to D^o l^+ \nu$	11.65
$B^o \to D^{*+} l^+ \nu$	10.60

where

$$h = \frac{M}{2}[(1 - \frac{m^2}{M^2} - y)^2 - 4\frac{m^2}{M^2}y]^{1/2} \tag{29}$$

and $y = q^2/M^2$, M is the mass of the parent meson, and m is the mass of the daughter meson. h is the magnitude of the momentum of the daughter meson m in the rest frame of the parent meson M. $|h|^3$ is large when q^2 is small and small when q^2 is large. The f_- form factor contribution vanishes for zero mass leptons and is negligible for electrons but may begin to be felt for muons.

The form factor is usually approximated by a function like

$$f_+(q^2) = \frac{f_+(0)}{(1 - q^2/M_p^2)} \tag{30}$$

or something like

$$f_+(q^2) = f_+(0) \times \exp^{-\alpha q^2} \tag{31}$$

The pole mass, M_p, is expected, by a duality argument, to be the lowest mass charmed vector or axial vector meson whose quantum numbers correspond to those of the hadron vertex. For example, figure 26 shows why the D_s^*, whose mass is 2112 MeV/c^2, is expected to be the dominant pole contributing to the

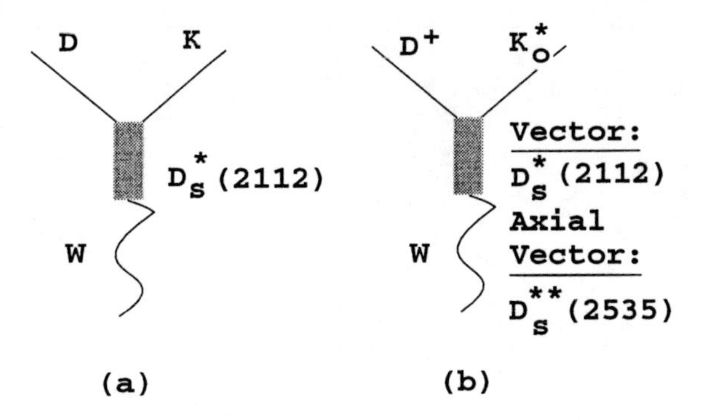

FIGURE 26. Illustration of pole contribution to hadronic vertex

TABLE 15. $Kl\nu$ Form factor measurements [35]

Exp.	mode	M_{pole}	$f_+(0)$
E691	$K^-e^+\nu_e$	$2.1^{+0.4}_{-0.2} \pm 0.2$	$0.79 \pm 0.05 \pm 0.06$
CLEO(91)	$K^-e^+\nu_e$	$2.1^{+0.4+0.3}_{-0.2-0.2}$	$0.81 \pm 0.03 \pm 0.06$
CLEO(93)	$K^-l^+\nu_l$	$2.0 \pm 0.12 \pm 0.18$	$0.77 \pm 0.01 \pm 0.04$
MKIII	$K^-e^+\nu_e$	$1.8^{+0.5+0.3}_{-0.2-0.2}$	$\|V_{cs}\|(0.72 \pm 0.05 \pm 0.04)$
E687	$K^-\mu^+\nu_\mu$(tag)	$1.97^{+0.43+0.07}_{-0.22-0.06}$	$0.71 \pm 0.05 \pm 0.03$
E687	$K^-\mu^+\nu_\mu$(inc)	$1.87^{+0.11+0.07}_{-0.08-0.06}$	$0.71 \pm 0.03 \pm 0.02$

decay $D^o \to K^-l^+\nu$. Note that the pole term increases as q^2 increases but the $|h|^3$ term falls off rapidly and wins out causing the q^2 distribution to be peaked at low q^2.

The semileptonic decay width is proportional to $|V_{Qq}|^2 \times |f_+(0)|^2$. To derive the CKM matrix element from the measured width one must have some model for the form factor. The form factor reflects the probability for the quark from the weak decay and the spectator quark to 'fit into' the wave function of the final state hadron.

1 $D^o \to K^-l^+\nu$

Figure 27 shows the q^2 distribution obtained by CLEO [33] and by E687 [34]. Table 15 lists recent results for the form factor values at $q^2 = 0$ and the pole masses obtained from fitting the q^2 distributions. The most recent measurements tend to favor values of the pole mass which are lower than D_s^*.

The measured value of $f_+(0)$ agrees well with a variety of quark model and lattice gauge calculations whose predictions range from 0.6 to 0.8.

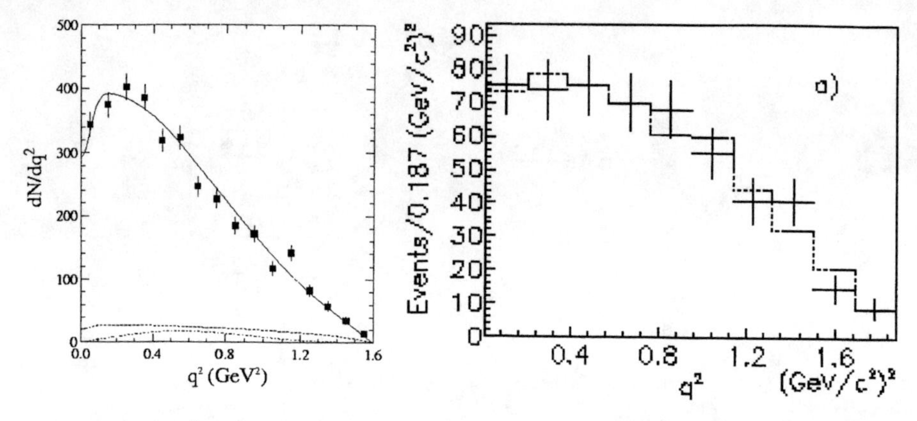

FIGURE 27. q^2 Dependence for the decay $D^o \rightarrow K^- l^+ \nu$ from left side: CLEO and right side: E687.

2 $D^o \rightarrow \pi^- \mu^+ \nu$

Figure 28 shows how different models of the pole behavior affect the predicted q^2 dependence. In the Cabibbo favored semileptonic decays, the q^2 range is limited to less than 2.0 and the predicted q^2 dependence differs only slightly for pole masses within the expected range. The Cabibbo suppressed decay has a larger allowed range of q^2 and, at large q^2, there is a big difference in the value of the form factor for various choices of the pole mass.

Measurements of $D^o \rightarrow \pi^- \mu^+ \nu$ are finally becoming available. Figure 29 shows the signals for these decays obtained from E687 [36]. The signal is

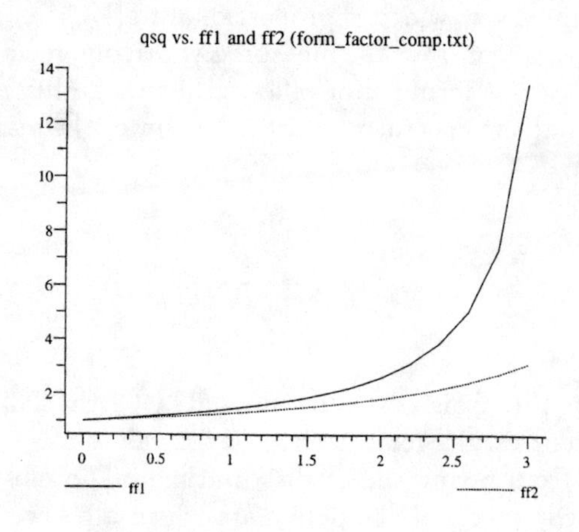

FIGURE 28. Behavior of form factors for two values of the pole mass: 1.8 GeV/c² (solid line) and 2.1 GeV/c² (dotted line).

TABLE 16. V_{cb} Measurements from $B^o \to D^+e^-\bar{\nu}$

Model	$f_+(0)$	$V_{cb}f_+(0) \times 10^3$	$V_{cb}f_+(0) \times 10^3$
WB	0.69	$25.7 \pm 1.4 \pm 1.7$	$37.3 \pm 2.0 \pm 2.5$
KS	0.70	$25.7 \pm 1.4 \pm 1.7$	$36.7 \pm 2.0 \pm 2.5$
Demchuk	0.68	$24.8 \pm 1.4 \pm 1.6$	$36.4 \pm 1.6 \pm 2.4$
Average			$36.9 \pm 3.7 \pm 0.5$

isolated by finding $\pi^-\mu^+$ or π^-e^+ vertices downstream of the primary vertex, then using the direction of the parent D^o obtained from the line connecting the primary and secondary vertex to balance p_t about the D^o flight direction. This determines the kinematics of the missing neutrino (up to a two-fold ambiguity). The signal is then confirmed by finding a pion from the primary vertex which forms a good D^{*+} with the D^o semileptonic candidate. The signal appears as a peak in the mass difference distribution between the mass of the candidate semileptonic decay and the mass of the candidate combined with a correctly signed pion from the primary vertex which forms the D^* parent.

A comparison of the branching fractions and form factors of $\pi l\nu$ to $Kl\nu$ is given in figure 30.

3 $\bar{B}^o \to D^+e^-\bar{\nu}$

Figure 31 shows the mass distribution obtained by the CLEO collaboration [38] for the $D^+e^-\bar{\nu}$ final state. In this analysis, the beam constraint and missing energy are used to obtain the neutrino momentum vector. We defer presentation of the q^2 dependence of the form factor until after a discussion below of Heavy Quark Effective Theory. The branching fraction for the decay is

$$1.78 \pm 0.20 \pm 0.24\% \tag{32}$$

Table 16 gives values obtained for V_{cb} from this measurement [39]. To obtain the partial width, one gets the total width from the measured value of the B^o lifetime, here taken to be 1.55±0.05 ps.

B Pseudoscalar to vector decay

The semileptonic decay of a heavy pseudoscalar meson, M_p, to a lighter vector meson, M_v, which subsequently decays into a pseudoscalar, m, and a pion via the strong interaction depends on four variables shown in figure 32:

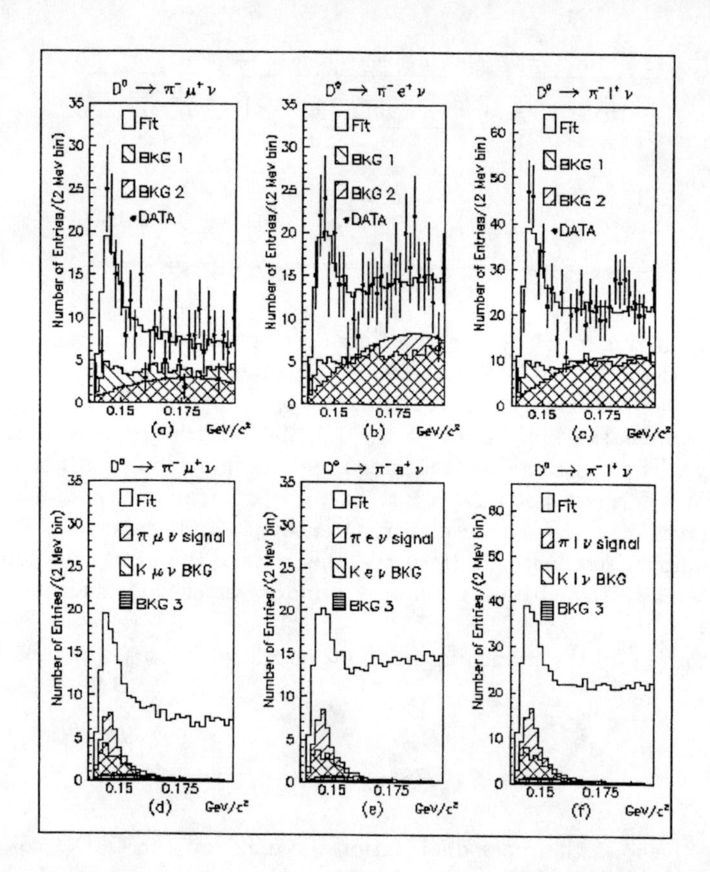

FIGURE 29. Signals for the decay $D^o \to \pi^- l^+ \nu$ from E687. a) shows the D^* signal for candidate events with a $\pi\mu$ secondary vertex. Also shown is the background from hadrons misidentified as leptons (BKG 1) and from random combinatoric background (BKG 2); b) shows the same distributions for πe candidates; c) shows the combined $\pi\mu$ and πe distributions; d) is the same as a) except that the background from misidentified $K\mu\nu$ and from $D^o \to K^{*-} l^+ \nu$ (BKG 3) is shown; e) shows these two backgrounds for πe; and f) shows these two backgrounds for the combined distribution.

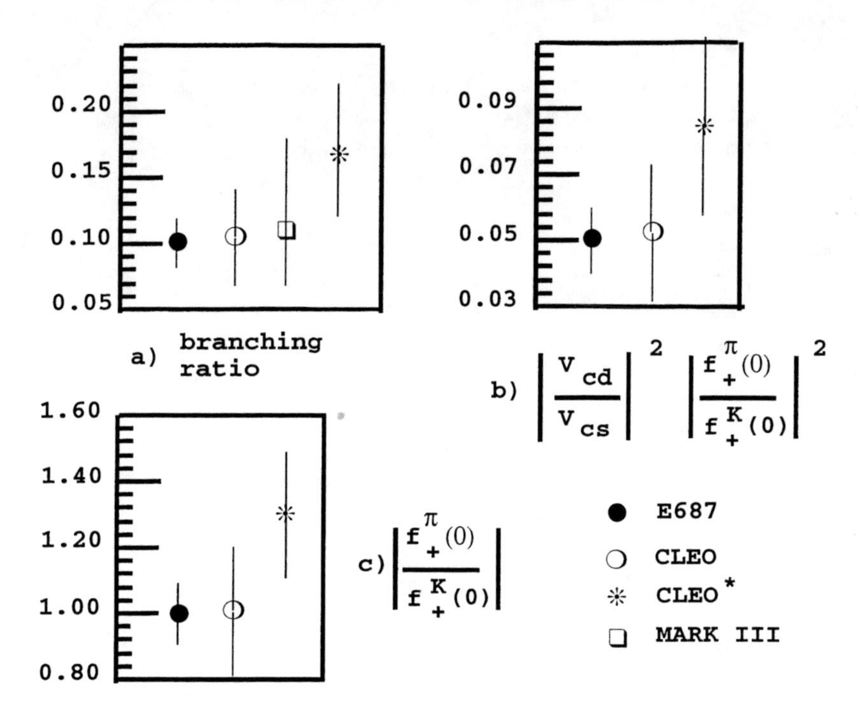

FIGURE 30. Comparison of $D^o \to K^- l^+ \nu$ and $D^o \to \pi^- l^+ \nu$: a) Branching ratios; b) CKM elements times form factors at $q^2 = 0$; and c) form factors at $q^2 = 0$ assuming Cabibbo angle values for CKM elements.In this plot, CLEO* refers instead to the ratio $\frac{D^+ \to \pi^o l^+ \nu}{D^+ \to K^o l^+ \nu}$. Data are from reference [37].

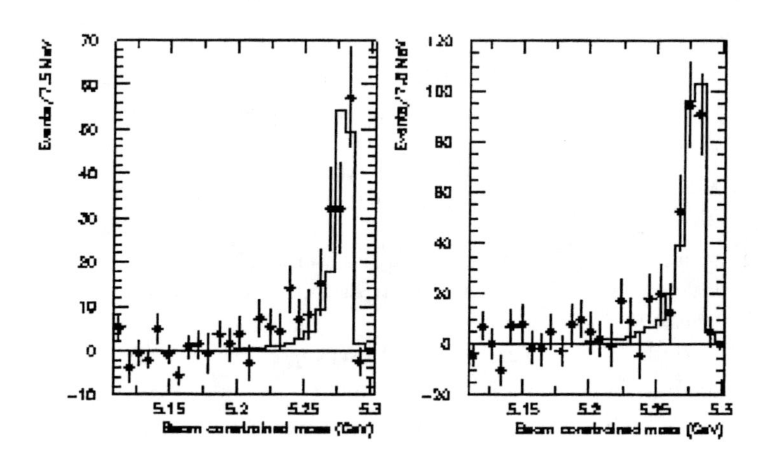

FIGURE 31. Signals for the decay $\bar{B}^o \to D^+ e^- \bar{\nu}$ from CLEO

θ_v: polar angle between M_p and m in M_v rest frame
θ_l: polar angle between l and M_p in $l\nu$ (W) rest frame
t (q^2): mass of virtual W ($M^2(l\nu)$)
χ: angle between the two decay planes height

The differential cross section is given by

$$\frac{d\Gamma}{dM_{m\pi}^2\, dt\, d\cos\,\theta_v\, d\cos\,\theta_l} \propto PS \times BW \times |M|^2. \tag{33}$$

The phase space term, PS, is given by

$$PS \;\propto\; K(1-\frac{M_l^2}{t}); \tag{34}$$

where K is the $m\pi$ momentum in the parent meson rest frame. The Breit-Wigner term, BW, for the vector resonance, V, is

$$BW \;=\; \frac{\Gamma M_V}{(M_{m\pi}^2\;-\;M_V^2)^2+M_V^2\Gamma^2}; \tag{35}$$

The matrix element squared $|M|^2$ is given by

$$\begin{aligned}|M|^2 = t(1-M_l^2/t) \times \\ \sin^2\theta_v[(1+\cos\theta_l)^2|H_+(t)|^2-(1-\cos\theta_l)^2|H_-(t)|^2 \\ +4\cos^2\theta_v\sin^2\theta_l|H_o(t)|^2 \\ +\frac{M_l^2}{t}\times\text{[lepton mass terms]}\end{aligned} \tag{36}$$

The quantities H_+, H_-, and H_o are the positive and negative helicity form factors and the longitudinal form factor respectively.

In the limit of $M_l \to 0$, these form factors are related as follows:

$$H_\pm = \alpha A_1(t) \;\mp\; \beta V(t) \tag{37}$$
$$h_o = \delta A_1(t) \;-\; \epsilon A_2(t) \tag{38}$$

where the functions α, β, δ, and ϵ are functions of t, $M_{m\pi}$, and K.

Given the level of statistics of current measurements, it is conventional to parameterize the form factors by the pole approximation described above and to express the result as two form factor ratios at $q^2 = 0$:

$$R_v \;=\; \frac{V(0)}{A_1(0)}\;\;,\;R_2\;=\;\frac{A_2(0)}{A_1(0)} \tag{39}$$

Applying the same argument we used above, the relevant poles are expected to be the D_s^* and the D_s^{**}'s so it is conventional to take $M_V = 2.1$, $M_{A_1} = 2.5$, and $M_{A_2} = 2.5$ GeV/c^2.

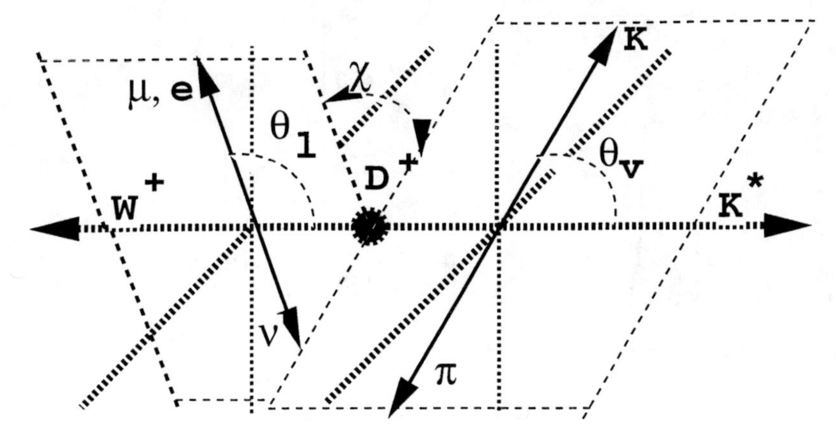

FIGURE 32. Angles used in the description of the semileptonic decay of a pseudoscaler meson to a vector meson.

1 The Decay $D^+ \to \bar{K}^{*o}(890)\mu^+\nu$

Figure 33 shows the K^* signal obtained from events with a $K\pi\mu$ detached vertex in photoproduction experiment E687 [40]. The signal appears as a strong $K^*(890)$ peak in the invariant mass distribution of the $K^-\pi^+$ from a downstream three body $K^-\pi^+\mu$ vertex. The world data [41] on the quantities R_v, R_2, Γ_L/Γ_T[6], and the branching fraction of $K^{*o}\mu\nu$ relative to $K\mu\nu$ are shown in figure 34.

2 The Decay $\bar{B}^o \to D^{*+}l^-\bar{\nu}$

This decay is similar to $D^+ \to K^{*o}l^+\nu$. V_{cb} can be extracted from it. So far, this has been done by using the branching fractions. The data [42] on the branching fraction to this mode is shown in table 17. To extract the value of V_{cb} one must use various models of the form factors. The results for V_{cb} using various models are shown in table 18 [43].

C Heavy Quark Effective Theory (HQET) and Form Factors

Above, we discussed Heavy Quark Symmetry in connection with the level spacing of excited charm and bottom mesons. This symmetry can also be applied to hadronic form factors and it results in relationships among the

---

[6] This is the ratio of the longitudinal to transverse polarization. See reference [32] for its relation to the form factors.

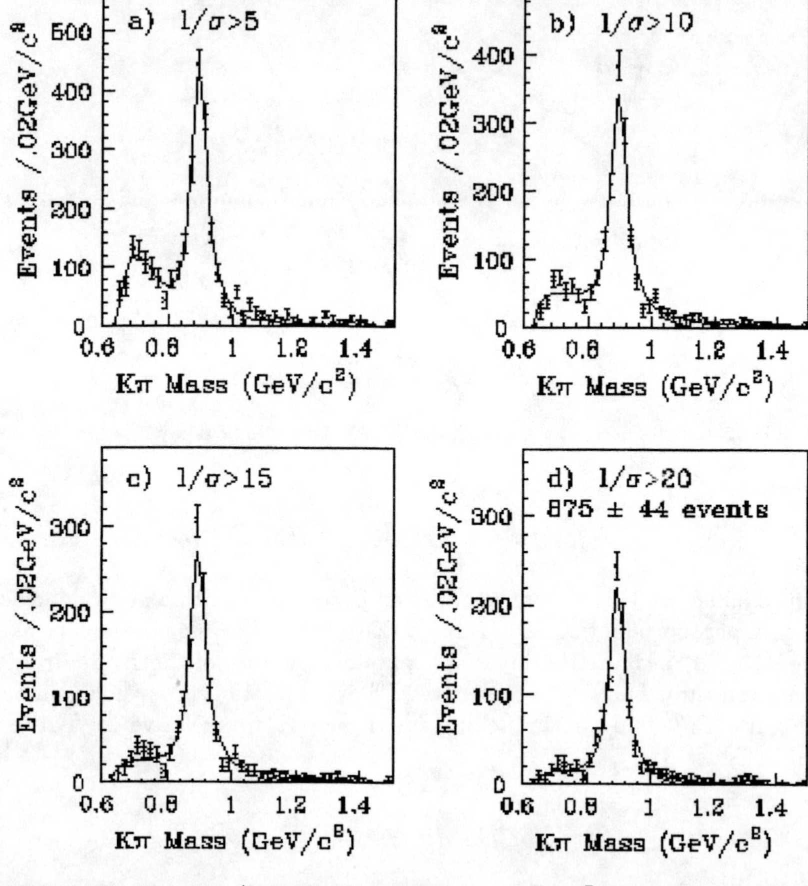

FIGURE 33. The $K^-\pi^+$ distribution showing a strong $\bar{K}^{*o}$ signal from the decay $D^+ \to \bar{K}^{*o}\mu^+\nu$ from E687. a), b), c), and d) show the distribution for a 'significance of detachment', $\frac{l}{\sigma_l}$, of 5, 10, 15, and 20, respectively. The last distribution, d), shows very little background.

TABLE 17. Branching fractions for $B \to D^{*+}l^-\nu(\%)$

Experiment	B(%)
CLEO	$4.1 \pm 0.5 \pm 0.7$
ARGUS	$4.7 \pm 0.6 \pm 0.6$
CLEO II	$4.50 \pm 0.44 \pm 0.44$
ALEPH	$5.18 \pm 0.30 \pm 0.62$
DELPHI	$5.47 \pm 0.16 \pm 0.67$
Average	4.90 ± 0.35

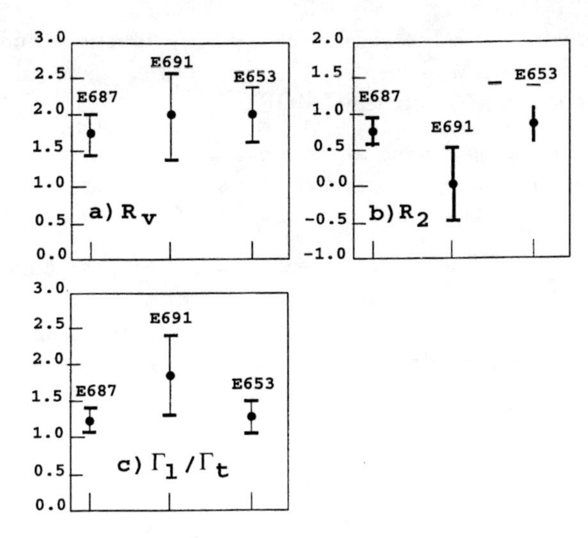

FIGURE 34. Results for the decay $D^+ \to \bar{K}^{*o}\mu\nu$: a) R_v; b) R_2; and c) $\frac{\Gamma_l}{\Gamma_t}$.

TABLE 18. Predictions for V_{cb} using various form factors

Model	$\Gamma(B \to D^* l\nu)(\text{ps}^{-1})$	V_{cb}
ISGW	$25.2\lvert V_{cb}\rvert^2$	0.0352 ± 0.0014
ISGW II	$24.8\lvert V_{cb}\rvert^2$	0.0355 ± 0.0014
KS	$25.7\lvert V_{cb}\rvert^2$	0.0348 ± 0.0014
WBS	$21.9\lvert V_{cb}\rvert^2$	0.0378 ± 0.0015
Jaus1	$21.7\lvert V_{cb}\rvert^2$	0.0379 ± 0.0015
Jaus2	$21.7\lvert V_{cb}\rvert^2$	0.0379 ± 0.0015

form factors which reduce the number of unknown functions and simplifies the extraction of the CKM elements.

The following is a brief summary of HQET:

- HQET should be applicable to the b-quark. Application to the c-quark is a bit less justified and corrections for symmetry breaking effects are necessary for accurate comparisons between the two flavors.
- Since all spin and flavor related forces go like $\frac{1}{M_Q}$, the only force that governs the behavior of the light degrees of freedom in a meson with a heavy quark in it is the (strong) color force. The light quarks are viewed as moving in a static color field.
- In the limit where the mass of the heavy quark goes to infinity, hadronic systems which differ only in the flavor or spin of the heavy quark have the same configuration of their light degrees of freedom.
- One can show that all form factors in 'heavy quark to heavy quark' decays reduce to one 'universal' function which depends only on the invariant four velocity transfer from the heavy quark to its lighter (but still heavy) daughter quark in the decay. This velocity transfer is related to the meson masses (here taken to be for the decay of a B-meson to a D^*, a case which has actually been studied experimentally) and q^2 as:

$$y = \frac{M_B^2 + M_{D^*}^2 - q^2}{2M_B M_{D^*}} \tag{40}$$

  This quantity is large when q^2 is small and goes to 1 at q^2_{max} which occurs when the B decays to a D^* which is at rest in the B rest frame.
- The universal function $\xi(y)$ is not known a priori but may be derived from models.
- In the lowest order QCD, $\xi(1) = 1$, and higher order terms lower this by around 10%.

To see this, consider the decay of a meson containing heavy quark at q^2_{max} shown in figure 35. The parent meson is initially travelling along with some velocity and its light quark is orbiting around it. We have seen that the heavy quark just acts as a static color source and otherwise does not couple to the light degrees of freedom. Suddenly, the heavy quark turns into a different heavy quark, which is an (almost) equivalent color source. Thus, in the heavy quark limit, the light degrees of freedom are already in the right configuration to form the daughter meson. This picture corresponds to a value of $y = 1$, where the value of the universal form factor $\xi(1) = 1$ in lowest order.

More generally,the light constituents of a heavy-light system are not affected by the replacement of the heavy quark, $Q(\vec{v}_Q, \vec{S})$ with another heavy quark

$Q'(\vec{v}_{Q'}, \vec{S}')$ where the velocities of the two quarks are the same: $\vec{v}_Q = \vec{v}_{Q'}$. The spins do not matter.

This leads to the idea of a single universal form factor, ξ, that depends only on y. In the case of pseudoscalar to vector semileptonic decays where in the classical formalism presented above there are several form factors, all of them will be related to this one universal function.

While the symmetry predicts the existence of this universal function, it does not tell us what the function is. The function must be obtained from QCD-motivated models or calculations. There are several versions on the market. Some of the functions proposed are

$$\begin{aligned} \xi(y) &= 1 - \rho^2(y-1) \\ &= \frac{2}{y+1} \times \exp^{-(2\rho^2-1)\frac{y-1}{y+1}} \\ &= (\frac{2}{y+1})^{2\rho^2} \\ &= \exp^{-\rho^2(y-1)} \end{aligned} \tag{41}$$

The value of ξ at 1 must be corrected for gluon emission and finite mass effects. If these corrections can be brought under control, then the systematic error from theory in the extraction of the CKM element, V_{cb}, will be reduced.

Figure 36 shows $\xi(y)$ measured in by the CLEO collaboration [44]. Stone [45] has recently fit the CLEO data to a variety of forms for ξ and has derived the result

$$|V_{cb}|\xi(1) \times 10^{-3} = 36.3 \pm 1.6 \pm 1.0 \tag{42}$$

giving a value of

$$|V_{cb}| = 0.0397 \pm 0.002 \pm 0.0017 \tag{43}$$

This result is slightly higher than previous results.

D Charmless Semileptonic B Decays

Semileptonic B decays which produce non-charmed mesons are the result of the b-quark transforming to a u-quark and a W^-. The amplitude is proportional to the CKM parameter V_{ub}. CLEO [46] has studied the decays

$$\begin{aligned} &B^- \rightarrow (\pi^o, \rho^o, \omega^o) - l\nu; \text{and} \\ &\bar{B}^o \rightarrow (\pi^+, \rho^+) - l\nu \end{aligned} \tag{44}$$

To get a clean signal, they reconstruct the energy and momentum of the ν by using their knowledge of the total energy and by exploiting their excellent

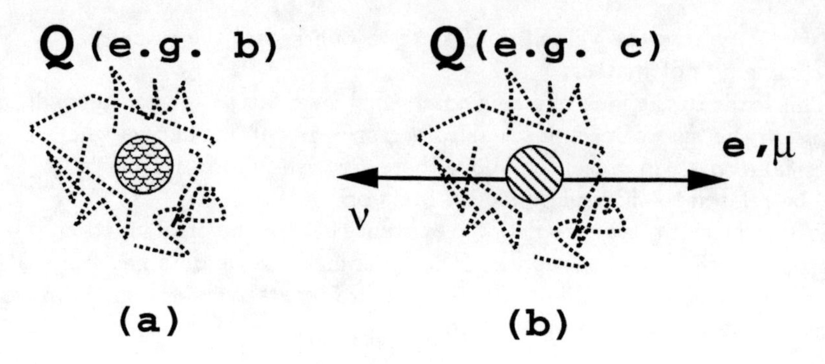

FIGURE 35. Decay of a meson containing a heavy quark

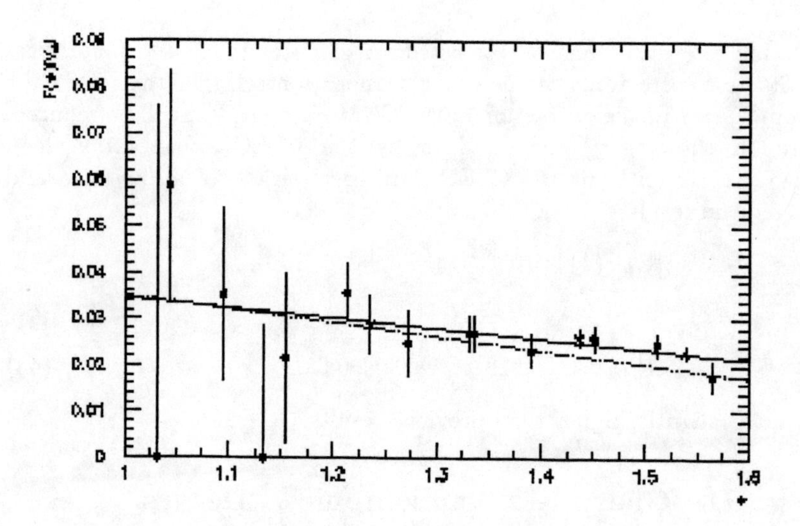

FIGURE 36. Form Factor for $\bar{B}^o \rightarrow D^+ e^- \bar{\nu}$ plotted vs the Isgur-Wise variable.

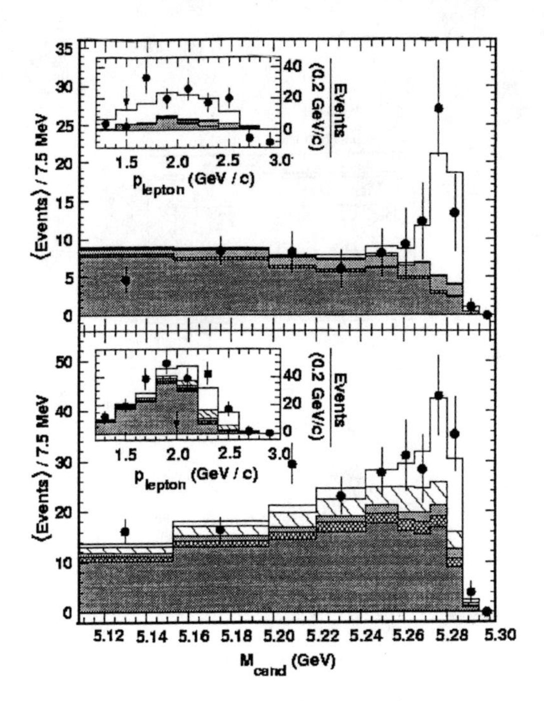

FIGURE 37. Evidence for $B \to \pi e\nu$ and $B \to \rho(\omega)e\nu$. The B candidate mass distributions, M_{cand}, for the sum of the scalar $\pi^+ l\nu$ and $\pi^o l\nu$ (top) and the vector modes (ρ and ω) (bottom). The points are the data after continuum and fake background subtractions. The unshaded histogram is the signal while the dark shaded area shows the $b \to cX$ background estimate, the cross hatched area the estimated $b \to ul\nu$ feedown. For the π(vector) modes, the light-shaded and hatched histograms are $\pi \to \pi$ (vector $\to$ vector) and vector $\to \pi$ ($\pi \to$ vector) crossfeed, respectively. The insets show the lepton momentum spectra for the events in the B mass peak (the arrows indicate the momentum cuts).

angular coverage to get the magnitude and direction of the missing energy in the event.

The experimental evidence for the observation by CLEO [46] of exclusive charmless semileptonic decay is given in figure 37. A rather complicated background subtraction is required to prove that the $\pi\pi$ system really is dominated by the ρ. Results for V_{ub} using various models for the form factor are given in table 19 taken from reference [39].

E Leptonic Decays

Figure 38 shows an example of the purely leptonic decay of a pseudoscalar meson. The decay rate is:

TABLE 19. Values of V_{ub}/V_{cb} from exclusive $\pi l\nu$ and $\rho l\nu$ analyses combined and taking $V_{cb} = 0.0381$

Model	V_{ub}/V_{cb}
ISGW II	$0.089^{+0.010}_{-0.011}$
WBS	0.076 ± 0.010
KS	$0.058^{+0.006}_{-0.008}$
Melnikov	$0.105^{+0.018}_{-0.019}$

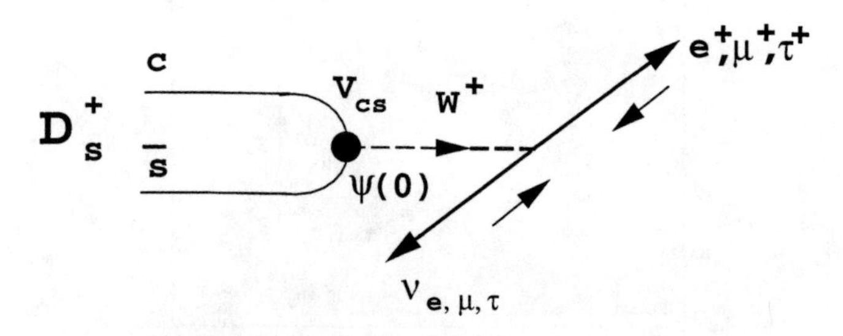

FIGURE 38. Purely leptonic decay of D_s meson

$$\Gamma = \frac{G_F^2 |V_{Qq}|^2 f_X^2}{8\pi} \times m_X m_l^2 (1 - \frac{m_l^2}{m_X^2})^2 \tag{45}$$

The 'decay constant', f_X, gives the probability that the two valence quarks in the parent meson, 'X', are close enough to experience the short range weak interaction and is proportional to the wave function squared at the origin.

There are several points to note. First is the dependence on the final state lepton mass squared. This is a result of helicity suppression. The neutrino forces the positron to be mainly in a helicity configuration that does not couple to the W. The positron then can couple to the W only through its 'small' helicity component, leading to the suppression. The heavier μ and τ have larger 'small' components and so have larger leptonic widths. Thus, decays to heaviest lepton, the τ, are heavily favored and decays to positrons are very rare. This is true even though there is some suppression for decay to heavy quarks coming from the last term. Second, since the purely leptonic decay is proportional to the mass of the parent state and the total decay rate goes roughly like the fifth power, the branching fraction for purely leptonic decays will be very tiny for mesons containing heavy quarks.

The reason for interest in this decay is that it provides a relatively direct and clean measure of f_X. The quantity f_B is needed to extract the CKM element V_{td} from a measurement of B^o mixing, which will be discussed briefly below. The leptonic B_u decay is nearly impossible to observe because it depends on V_{ub} and because leptonic decays in general are suppressed relative to other decays for heavy quarks as explained above. For now, we must depend on a theoretical calculation. There is a large variation in the results of the various calculations. Any information that would give us confidence in some of the theoretical results would help reduce our uncertainty in this important quantity and in our ability to use theory to extract V_{td}. For that, we must look to charm. There has recently been an observation [47] of the decay $D_s \to \mu\nu$. The resulting value for f_{D_s} is:

$$f_{D_s} = 344 \pm 37 \pm 52 \pm 42 \text{ MeV} \tag{46}$$

where the first two errors are the statistical and systematic errors from the measurement and the third uncertainty is from the error on the absolute branching fraction of $D_s \to \phi\pi$ which is used to convert the signal to a width. This is to be compared with the value of Aoki [48] et al.:

$$f_{D_s} = 232 \pm 45 \pm 20 \pm 48 \text{ MeV} \tag{47}$$

Theoretical predictions have been made using lattice gauge calculations, potential models, sum rules, and other methods which involve using both experimental and theoretical inputs. Typical results lie in the range from 200 to 300 MeV.

V RARE DECAYS

While this topic is far too broad and complex to receive a complete treatment in these lectures, I do want to illustrate briefly why 'rare processes' – by which I mean those involving loops or box diagrams at the quark level – play a much larger role in B decays than in charm decays.

The CKM matrix has been written by Wolfenstein [49] in the following form:

$$V = \left(\begin{array}{ccc|c} u & c & t & \\ \hline 1 - \lambda^2/2 & \lambda & A\lambda^3[\rho - i\eta] & d \\ -\lambda & 1 - \lambda^2/2 & A\lambda^2 & s \\ A\lambda^3(1 - \rho - i\eta) & -A\lambda^2 & 1 & b \end{array} \right)$$

The quantity λ is the *sine* of the normal Cabibbo angle. Since its value is approximately 0.22, and since A, ρ, and η are all known to be less than 1.0 [50], it is clear that mixing between quarks of different generations is small. This fact, combined with the observed value of the quark masses, leads to the conclusion that rare decays and mixing, as defined above, are much smaller for charmed particles than for bottom particles within the Standard Model.

Figure 39 shows the box diagrams responsible for flavor mixing and Penguin (loop) diagrams which are responsible for a flavor changing neutral currents for the case of bottom and charm. The expression for B-mixing is

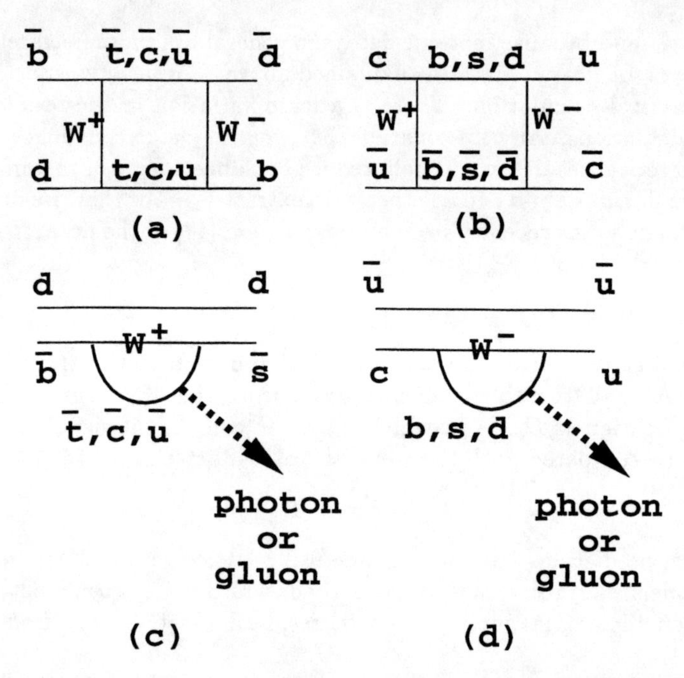

FIGURE 39. Rare processes: a) B^o mixing; b) D^o mixing; c) A penguin diagram for $B^o \to K^{*o}\gamma$; and d) A penguin diagram for $D^o \to \rho^o\gamma$.

$$x_B = \frac{\Delta M_B}{\Gamma_B} = \frac{G_F^2}{6\pi^2}\tau_B B_B f_B^2 m_B m_t^2 |V_{tb}^* V_{td}|^2 F(x)\eta_{QCD} \tag{48}$$

where $x = \frac{m_t^2}{M_W^2}$. B-mixing is a large effect and has been extensively studied.

It can be seen that the corresponding expression for D^o mixing will depend on the masses of the objects in the box diagram and the CKM elements as:

$$x_D = \frac{\Delta M_D}{\Gamma_D} \propto |V_{cb}^* V_{ub}|^2 m_b^2 \tag{49}$$

and the ratio is

$$\frac{x_D}{x_B} \sim \lambda^4 \frac{M_b^2}{M_t^2} \tag{50}$$

which is much smaller than B-mixing because of the appearance of small CKM elements and because the b-quark in the box is so much lighter than the t-quark which dominates the B-mixing. In fact, the CKM suppression is so strong that the s-quark may be the dominant short range contributor to D-mixing. So-called long range effects, that is hadrons appearing in the diagram in place of the box of quarks, may increase the mixing by up to an order of magnitude but it is still far below the level of B-mixing and has never been observed [51].

For similar reasons, penguin (loop) diagrams appear at the level of $10^{-4}\%$ of B decays and have been observed [52], but loop diagrams have not been observed in D decays and SM predictions are that they are very small and unlikely to be observed [53]. For example, the SM prediction for the branching fraction for $D^o \to \rho^o \gamma$ is of order of 10^{-17}! However, long distance contributions may increase this by a large factor.

CP violating asymmetries in weak decays are expected to arise from the interference between direct decays and flavor mixing followed by decays into the same states [54]. Since D-mixing is so small that it has yet to be observed and is predicted to lie well below existing limits, it is clear that the B system is a much more promising one for studies of CP violation. Another way to look at this is that the CKM elements that carry the weak phase and therefore are responsible for CP violation are the ones that involve third generation quarks. Since the b-quark is itself a third generation quark, it is strongly connected to CP violation. The c-quark is not a third generation quark and is therefore only weakly connected.

B-decays offer a good opportunity to study rare decays which occur within the context of the Standard Model. The outlook for studying these processes for charm is bleak if the only mechanisms by which they can occur are Standard Model processes. Let me end this paper by reminding everyone that, while the B system is very exciting because it allows one to address many fundamental issues, the charm system should not be neglected. If mixing, penguin processes, or CP violation show up at a level significantly higher than predicted by the Standard Model, it might be an indication of new physics!

VI CONCLUSION

In these lectures, we have shown that in some cases particles with charm and bottom quarks are very similar – in their spectroscopy and semileptonic decays – and in other cases are very different – in their lifetime patterns and likelihood of undergoing rare decays. In all cases, the Standard Model offers a sound explanation for the observed behavior.

Acknowledgements:
I would like to thank the conference organizers for their hospitality and for giving me the opportunity to participate as both a student and a lecturer in the VII Mexican School of Particles and Fields. I want to thank Sheldon Stone and James E. Wiss from whose lectures at the NATO Advanced Institute and the SLAC Summer School I borrowed copiously. I also want to thank all my colleagues on the E687 and FOCUS/E831 experiments and the whole B Study Group at Fermilab who have taught me much of the physics presented here. Finally, I would like to thank my colleague Hector Mendez

for his help and 'gentle' encouragement in getting me to write up these lectures. This article was written at and the work supported by the Fermi National Accelerator Laboratory, which is operated by Universities Research Association, Inc., under contract DE-AC02-76CH03000 with the U.S. Department of Energy.

REFERENCES

1. See, for example, Chris Quigg, *Gauge Theories of the Strong, Weak, and Electromagnetic Interactions*, Frontiers in Physics, Benjamin-Cummings Press, Menlo Park (1983).
2. M. Kobayashi and K. Maskawa, *Prog. Theor. Phys. 49*, 652(1973).
3. Review of Particle Properties, *Phys. Rev. D 50*, p1173-1826(1994). See page 1319.
4. See reference [3], page 1321.
5. See reference [3], page 1782.
6. J.G. Korner, "Charm Baryons, Theory and Experiment", *Ann. Rev. Nucl. Part. Sci. 41*, 511-45(1991).
7. Frabetti et al., *Phys. Lett. B* 323, 459(1994).
8. L. Gibbons et al., "Observation of an Excited Charmed Baryon Decaying into $\Xi_c^0\pi^+$", *Phys.Rev. Lett. 77*, 810(1996) and P. Avery et al., "Observation of a Narrow State Decaying into $\Xi_c^+\pi^-$", *Phys. Rev. Lett. 75*, 4364(1995).
9. G. Brandenburg et al., "Observation of Two Excited Charmed Baryons Decaying into $\Lambda_c^+\pi^\pm$", clns 96/1427, cleo 96-113(1996).
10. K.W. Edwards et al., "Observation of Excited Charmed Baryon States Decaying to $\Lambda_c^+\pi^+\pi^-$", *Phys. Rev. Lett. 74*, 3331(1995).
11. F. Abe et al., "Observation of $\Lambda_b^o \to J/\psi\Lambda$ at the Fermilab Proton-Antiproton Collider", *Phys. Rev. D* 55, 1142 (1997)
12. CDF result is reference [11]; C. Albajar et al., *Phys. Lett. B 273*, 540(1991); P. Abreu et al., *Phys. Lett. B 374*, 351(1996); and D. Buskulic et al., *Phys. Lett. B 380*, 442(1996).
13. Albrecht H. et al., ARGUS Collaboration, *Phys. Rev. Lett.* 56, 549 (1986).
14. See John Bartelt and Shekhar Shukla, "Charmed Meson Spectroscopy", *Annu. Rev. Nucl. Part. Sci.* 45, 133-161(1995).
15. N. Isgur and M. Wise, "Heavy Quark Symmetry", in **B Decays** *revised second edition*, p231-282, edited by Sheldon Stone, World Scientific (1994).
16. Frabetti et al., "Measurement Of The Masses And Widths Of L = 1 Charm Mesons", *Phys. Rev. Lett. 72*, 324(1994).
17. See reference [14] references 18, 21, 20, 22 16, 17, 19 for $D_2^{*o,+}$; references 5, 20, 16, 22, and 19 for $D_1^{o,+}$; references 5, 20, 25, 19, and 28 for D_{s1}^+; and reference 23 for D_{s2}^{*+}.
18. See John Bartelt and Shekhar Shukla, "Charmed Meson Spectroscopy", *Annu. Rev. Nucl. Part. Sci. 45*, 133-161(1995).

19. E. Eichten, C.T. Hill, and C. Quigg, "Properties of Orbitally Excited Heavy-Light Mesons", *Phys. Rev. Lett. 71*, 4116(1993).
20. D. Buskulic et al., *Z. Phys. C 69*, 393(1996).
21. P. Abreu et al., *Phys. Lett. B 345*, 598(1995).
22. R. Akers et al., *Z. Phys. C 66*, 19(1995).
23. M. Feindt and O. Podobrin, DELPHI 96-93 CONF-22, 25 June 1996 (paper submitted to ICHEP96).
24. C. Quigg and E. Eichten, "Mesons with Beauty Charm" *Phys. Rev. D 49*, 5845(1994).
25. R.M. Barnett et al., *Phys. Rev. D54*, 1 (1996).
26. M. Wirbel. B. Stech, and M. Bauer, *Z. Phys. C 29*, 637(1985), M. Bauer and M. Wirbel, *Z. Phys. C 42*, 671(1989).
27. T.E. Browder, K. Honscheid, and D. Pedrini, "Nonleptonic Decays and Lifetimes of Charm and Beauty Particles" in *Annual Review of Nuclear and Particle Science Vol. 46*, p458-461(1996).
28. K. Honscheid, "Hadronic Decays of B Mesons", in the proceedings of The Advanced Study Conference on Heavy Flavors, p 61-76, Pavia (Italy), 1993.
29. B. Guberina, R Ruckl, and J. Trampetic, "Charmed Baryon Lifetime Differences", *Z. Phys. C 33*, 297-305(1986).
30. a) Frabetti et al., "Measurement of the Mass and Lifetime of the Ξ_c^+", *Phys. Rev. Lett. 70*, 1381(1993); b) Frabetti et al., "A Measurement of the Λ_c^+ Lifetime", *Phys. Rev. Lett. 70*, 1755(1993); c) Frabetti et al., "Measurement of the Lifetime of the Ξ_c^o", *Phys. Rev. Lett. 70*, 2058(1993); and d)Frabetti et al, "Measurement of the Ω_c^o Lifetime", *Phys. Lett. B 357*, 67(1995).
31. F. Abe et al., *Phys. Rev Lett. 77*, 1439(1996); D. Buskulic et al., *Phys. Lett. B 357*, 685(1995); P. Abreu et al., *Z. Phys. C 71*, 199 (1996); and R. Akers et al., *Z. Phys. C. 69*, 195(1966).
32. An excellent reference is Gilman, F.J. and Singleton, R.L., "Analysis of semileptonic decays of mesons containing heavy quarks", *Phys. Rev. D 41*, 142(1990).
33. A. Bean et al., "Measurements of Exclusive Semileptonic Decays of D Mesons", *Phys. Lett. B 317*, 647(1993).
34. Frabetti et al, "Analysis of the Decay Mode $D^o \to K^-\mu^+\nu$", *Phys. Lett. B 364*, 127(1995).
35. A. Bean et al., "Measurements of Exclusive Semileptonic Decays of D Mesons", *Phys. Lett. B 317*, 647(1993).
36. Frabetti et al., "Analysis of the Cabibbo Supressed Decay $D^o \to \pi^- l\nu$", *Phys. Lett. B 382*, 312(1996).
37. F. Butler et al., "Measurement of the Ratio of Branching Fractions $(D^0 \to \pi^- e^+\nu_e)/(D^0 \to K^- e^+\nu_e)$", *Phys. Rev. D 52*, 2656 (1995) and M.S. Alam et al., "Measurement of the Ratio $\mathcal{B}(D^+ \to \pi^0 l^+\nu)/\mathcal{B}(D^+ \to \bar{K}^0 l^+\nu)$", *Phys. Rev. Lett. 71*, 1311 (1993).
38. T. Bergfeld et al., "Measurement of $B(\bar{B}^o \to D^+ l^- \bar{\nu})$ and Extraction of V_{cb}", CLEO-CONF 96-3, ICHEP-96 PA05-78(1996).
39. S. Stone, "Prospects for B Physics in the Next Decade", Presented at NATO Advanced Study Institute on Techniques and Concepts of High Energy Physics,

Virgin Islands, July 1996. Syracuse Preprint HEPSY 96-01, October 1996.
40. Frabetti et al., "Analysis of the Decay Mode $D^+ \to K^{*o}\bar{\mu}^+\nu$", *Phys. Lett. B 307*, 262(1993).
41. E687 data are from reference [40]; E691 Collaboration, J.C. Anjos et al., *Phys. Rev. Lett. 65*, 2630(1990); and E653 Collaboration, K. Kodama et al., *Phys. Lett. B 274*, 246(1992).
42. D. Bortoletto et al., *Phys. Rev. Lett. 16*, 1667(1989) and J.E. Duboscq et. al., "Measurement of the Form Factors for $\bar{B}^0 \to D^{*+}l^-\bar{\nu}$", *Phys. Rev. Lett. 76*, 3898 (1996).
43. B. Grinstein, N. Isgur and M.B. Wise, *Phys. Rev. Lett. 56*, 258(1986); N. Isgur, D. Scora, B. Grinstein, and M.B. Wise, *Phys. Rev. D 39*, 799(1989); W. Jaus, *Phys. Rev. D 41*, 3394(1990); M. Wirbel, B. Stech, and M. Bauer, *Z. Phys. C 29*, 637(1985); and M. Bauer and M. Wirbel, *Z. Phys. C 42*, 671(1989); and J.G. Korner and G.A. Schuler, *Z. Phys. C 38*, 511(1988). ibid,(erratum)C 41, 690(1989).
44. see reference [39], p 16 -20.
45. reference [39], p 19.
46. J.P. Alexander et al., "First Measurement of the $B \to \pi l\nu$ and $B \to \rho(\omega)l\nu$ Branching Fractions", *Phys. Rev. Lett. 77*, 5000(1996).
47. D. Acosta et al., "First Measurement of $\Gamma(D_s^+ \to \mu^+\nu)/\Gamma(D_s^+ \to \phi\pi^+)$", *Phys. Rev. D 49*, 5690(1994).
48. S. Aoki et al., *Progress of Theoretical Physics 89*, 131(1993).
49. L. Wolfenstein, *Phys. Rev. Lett. 51*, 1945 (1983).
50. B Decays, J. L. Rosner, "The Cabibbo-Kobayashi-Masakawa Matrix", in **B Decays**, *revised second edition*, edited by Sheldon Stone, World Scientific, Singapore(1994).
51. Gustavo Burdman, "Charm Mixing and CP Violation in the Standard Model", Proceedings of the CHARM2000 Workshop, Fermilab, June 7-9, 1994, Editors: Daniel M. Kaplan and Simon Kwan and *ibid.* Jim Wiss, "Semileptonic Decays, Absolute Branching Ratios and Charm Mixing".
52. R. Ammar et al., "Radiative Penguin Decays of the B Meson", CLEO-CONF 96-6 (1996) and M.S. Alam et al., *Phys. Rev. Lett. 74*, 288(1995).
53. Sandip Pakvasa, "Charm as a Probe of New Physics", Proceedings of the CHARM2000 Workshop, Fermilab, June 7-9, 1994, Editors: Daniel M. Kaplan and Simon Kwan.
54. see for example, I.I.Bigi, V.A. Khoze, N.G. Uraltsev, and A.I. Sanda, "The Question of CP Noninvariance – As Seen Through The Eyes of Neutral Beauty", in *CP VIOLATION*, editor C. Jarlskog, World Scientific, Singapore, 1989.

Experimental Top Quark Physics

Jacobo Konigsberg

University of Florida,
Gainesville, Florida 32611

Abstract. These proceedings are a summary of the three lectures on "Experimental Top Quark Physics" presented at the VII Mexican School of Particles and Fields in Merida, Mexico. More elaborate details on top quark physics can be found in the references provided in the text, in references therein and in reviews such as [1–5].

INTRODUCTION

A The Top Quark in the Standard Model

The top quark discovery at the Fermilab Tevatron was announced about two years ago (March 1995) by the CDF and the D0 collaborations [6,7]. With the exception of the not yet directly observed tau neutrino, the top quark completes the three generations of discovered fermions that are the fundamental building blocks of matter. These particles, together with the bosons which transmit the fundamental forces of nature comprise what we call the "Standard Model" (SM) of elementary particles and interactions.

The bottom quark was discovered in 1975, also at Fermilab [8]. The existence of this fifth quark made a strong case for a sixth quark to exist. The 1/3 absolute value of the electrical charge of the b-quark was measured in 1982 [9] in e^+e^- collisions by measuring the relative rate of $e^+e^- \to hadrons$ to that of $e^+e^- \to \mu^+\mu^-$. Later on, in the mid-eighties [10] it was established that the weak isospin of the b-quark was $I_3 = -1/2$, deduced from the forward-backward asymmetry in $e^+e^- \to b\bar{b}$ implying the existence of a $I_3 = +1/2$ isospin partner. The case for the top quark was further strengthened by searches for $B \to \ell^+\ell^- X$. If only five quarks exist then it can be calculated theoretically that $BR(B \to \ell^+\ell^- X) \geq 0.12$ [11]. However experiments have set upper limits for this decay well below this value, therefore ruling out this flavor-changing neutral-current exchange. Furthermore, a third generation of a quark doublet, together with the three lepton doublet generations are needed

CP400, *First Latin American Symposium on High Energy Physics/ VII Mexican School of Particles and Fields,*
edited by D'Olivo/Klein-Kreisler/Mendez

in order to cancel out diagrams that contribute to triangle anomalies which would prevent the electroweak theory from being renormalizable.

The Standard Model therefore predicted (and needed) the existence of a quark with electrical charge of 2/3 and with $I_3 = 1/2$. The mass of this quark is a free parameter. The top quark mass, M_{top}, can be constrained from measurements in the electroweak sector. Radiative corrections to the W and Z boson masses, and other electroweak precision measurements are sensitive to M_{top} through loop corrections to the tree-level processes. These corrections go as $\propto \frac{M_{top}^2}{M_Z^2}$. Consistency between all these measurements is what yields constraints on the value of M_{top}. These corrections are also sensitive to the Higgs mass value, but only logarithmically ($\propto ln(\frac{M_{Higgs}}{M_Z})$). Figure 1 shows the quality of this consistency (a chi-squared of the fit) as a function of M_{top} for different assumptions of M_{Higgs}. Also shown is the measured value of the top quark mass together with its uncertainty. The fit was done to data from LEP, SLD, $p\bar{p}$ colliders and νN scattering experiments. The smaller the uncertainties in M_{top} the more accurate a prediction for M_{Higgs} can be made.

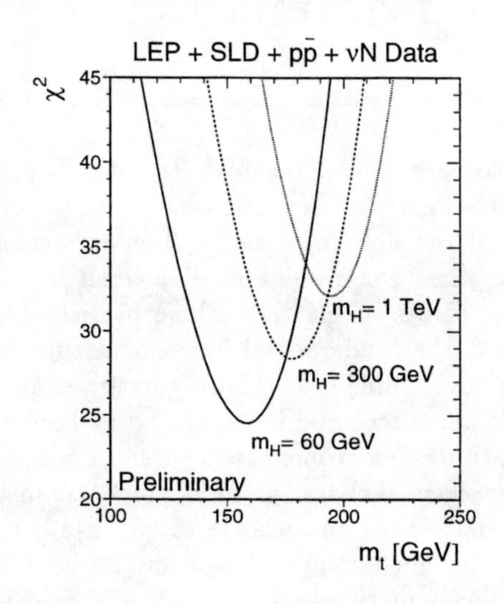

FIGURE 1. The χ^2 curves for the SM fit to the electro-weak precision measurements from LEP, SLD, CDF, D0 and neutrino scattering experiments (The LEP Collaboration, 1995).

The top quark also plays an important role in $B^o - \bar{B}^o$ mixing, $K^o - \bar{K}^o$ and CP-violation (through "box" diagrams), and in "direct" CP-violation in the Kaon system and in the B_S lifetime through "penguin" diagrams.

B Searches for the Top Quark

Particle colliders have been the main accelerators at which the top quark was sought. For a beam energy E_{beam} the available energy to create new particles in the center of mass of the collision is $E_{CM} \propto E_{beam}$. In fixed target experiments the available energy is proportional to $\sqrt{E_{beam}}$. Even though searches for heavier particles are more readily done at collider experiments, the price one pays is in collision rates. Particle beams with very high intensities are needed for these searches to be viable. The following is a list of searches at various colliders and center of mass energy ranges ($\sqrt{s} = E_{CM}$) prior to the top discovery. Included are the top quark mass lower limits resulting from those searches:

- e^+e^- Colliders:
  - 1979-1984 at PETRA (Germany), $12 \leq \sqrt{s} \leq 47\ GeV$ $\Rightarrow M_{top} < 23.3\ GeV$
  - 1988-1990 at TRISTAN (Japan), $\sqrt{s} = 61\ GeV \Rightarrow M_{top} < 30.2\ GeV$
  - 1989-1990 at SLD (U.S.) and at LEP (CERN) $\sqrt{s} \sim M_Z$ $\Rightarrow M_{top} < 45.8\ GeV$
- Hadron Colliders:
  - 1988-1990 at UA1 (CERN), $\sqrt{s} = 630\ GeV \Rightarrow M_{top} < 60\ GeV$
  - 1988-1990 at UA2 (CERN), $\sqrt{s} = 630\ GeV \Rightarrow M_{top} < 69\ GeV$
  - 1987-1990 at CDF (Fermilab), $\sqrt{s} = 1800\ GeV \Rightarrow M_{top} < 91\ GeV$
  - 1993 at D0 (Fermilab), $\sqrt{s} = 1800\ GeV \Rightarrow M_{top} < 131\ GeV$
  - 1994 at CDF (Fermilab), $\sqrt{s} = 1800\ GeV$, first evidence for top with $M_{top} = 174(\pm 16)\ GeV$ and $\sigma(p\bar{p} \rightarrow t\bar{t}) = 14(\pm 6)pb$

C Discovery of the Top Quark

On March 2nd, 1995, the discovery of the top quark was finally announced at Fermilab by the CDF and D0 Collaborations [6,7]. The total amount of data analyzed corresponded to a total integrated luminosity of 67 and 50 pb^{-1} respectively. By the Fall of 1996 the Collider run ended and both experiments gathered a dataset corresponding to about 110 pb^{-1}. These datasets which, as we shall see, do not contain that many top quark events, mark the beginning of the era of top quark physics. In these lectures we'll describe the measurements made with these data.

I TOP QUARK PRODUCTION

Figure 2 shows the various accelerators at Fermilab that lead to the $p\bar{p}$ collisions at the Tevatron [12].

- Cockroft Walton - Adds e^- to gaseous Hydrogen atoms and accelerates the H^- ions to 750 KeV.
- Linac - Accelerates H^- ions to 400 MeV and strips off the electrons when passing through Carbon foils, leaving just protons.
- Booster - Accelerates the protons to 8 GeV.
- Main Ring - Accelerates protons to 150 GeV. Will be replaced by the Main Injector in 1999.
- Anti-proton Storage Ring - 120 GeV protons from the Main Injector are collided into a target to produce anti-protons which are then accumulated in this ring for later injection into the Tevatron
- Tevatron - Accelerates protons and anti-protons to 900 GeV and collides them at two interaction regions located at the center of the CDF and D0 detectors.

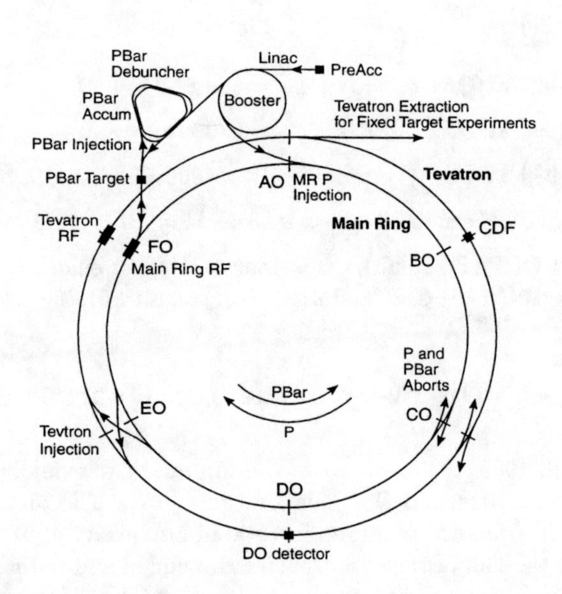

FIGURE 2. The Fermilab accelerator complex.

A Production Cross-Section

In $p\bar{p}$ collisions the production of heavy quarks can be calculated within the formalism of the parton model. In this model the cross-section for a given sub-process, $\hat{\sigma_{ij}}$, is calculated (to a given order in perturbation theory) for two incoming constituents of the proton (partons), i and j, each carrying a given fraction of the proton and anti-proton momenta, x_i and x_j. This process is calculated at the center-of-mass energy $E_{CM}^2 = \hat{s}$, where $\hat{s} = x_i x_j \cdot s$, s being the total available center-of-mass energy in the proton anti-proton collision. Then the probabilities for these momentum fractions to occur for a given parton, $F_i(x_i), F_j(x_j)$ are folded in with the sub-process cross-section to give the total production cross-section [13]:

$$\sigma(p\bar{p} \rightarrow t\bar{t}) = \sum_{i,j} \int dx_i F_i(x_i) \int dx_j F_j(x_j) \cdot \hat{\sigma_{ij}}(\hat{s}) \tag{1}$$

At the Tevatron, for large M_{top} the dominant top quark production process is of $t\bar{t}$ pairs via $q\bar{q}$ annihilation (about 80% of the cross-section). Gluon-gluon diagrams contribute the remaining of the $t\bar{t}$ cross-section. Figure 3 shows the Feynman diagrams for these processes.

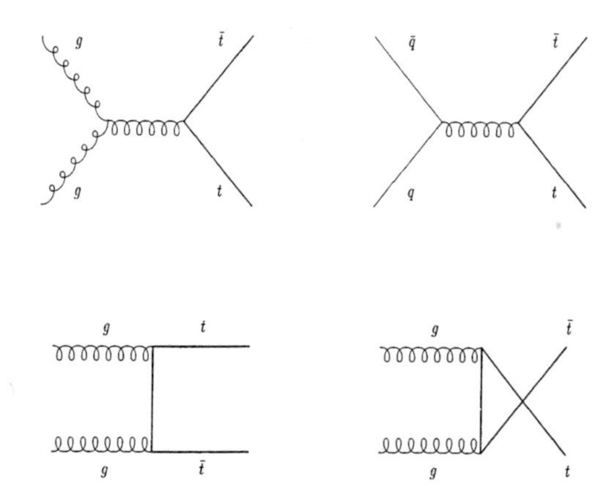

FIGURE 3. Lowest order Feynman diagrams for production of $t\bar{t}$ pairs in $p\bar{p}$ collisions.

The $t\bar{t}$ production cross-section has been calculated to order α_s^3 (next-to-leading order, NLO) [14–17]. Fig.4 shows the Feynman diagrams for the corrections. The cross-sections need also to be corrected for the effects of soft gluon emission ("soft gluon resummation").

At large M_{top} the production of a single top quark is possible through W-gluon fusion and through a virtual W-boson as depicted in the diagrams in

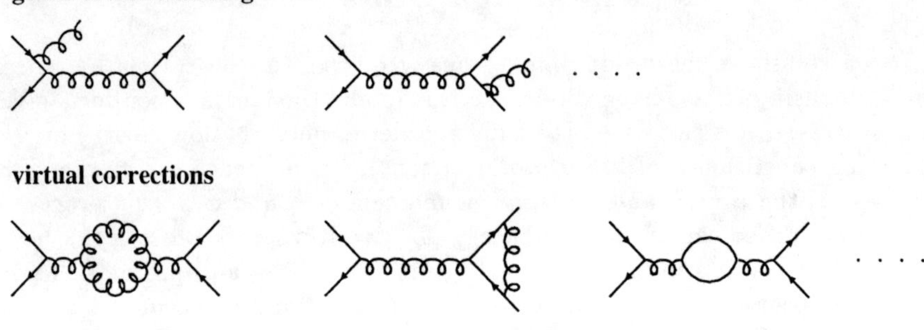

FIGURE 4. NLO Feynman diagrams for production of $t\bar{t}$ pairs in $p\bar{p}$ collisions.

Figure 5. Single top production is about 20% of the total top production cross-section at the Tevatron. The rest of these lectures deal only with $t\bar{t}$ production.

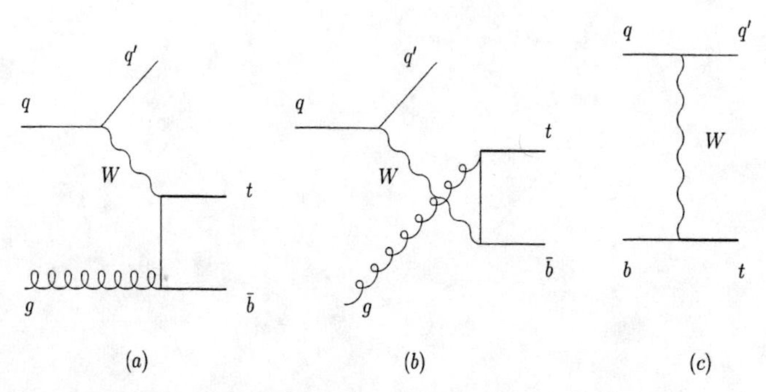

FIGURE 5. Lowest order Feynman diagrams for production of single top via W-gluon fusion in $p\bar{p}$ collisions.

Table 1 compares the production cross-section for various physics processes at the Tevatron with the cross-section for $t\bar{t}$ production. About 10^{10} $p\bar{p}$ collisions are needed in order to produce a single $t\bar{t}$ pair.

In order to calculate the number of $t\bar{t}$ events expected it is necessary to understand the accelerator parameters that determine this number. All these

TABLE 1. $t\bar{t}$ production rates at the Tevatron vs. other processes.

Physics process cross-section	Production rates at $\mathcal{L} = 10^{31}$ cm^{-1}s^{-1}	Num. of events produced at $\int \mathcal{L} \cdot dt = 100\text{pb}^{-1}$
$\sigma(p\bar{p} \to X) \sim 60$ mb	600 KHz	$6 \cdot 10^{12}$
$\sigma(p\bar{p} \to b\bar{b}) \sim 20$ μb	200 Hz	$2 \cdot 10^{9}$
$\sigma(p\bar{p} \to W + X) \sim 20$ nb	0.2 Hz	$2 \cdot 10^{6}$
$\sigma(p\bar{p} \to Z + X) \sim 6$ nb	0.06 Hz	$6 \cdot 10^{5}$
$\sigma(p\bar{p} \to t\bar{t}) \sim 5$ nb ($M_{top} = 175$ GeV)	4/day	500

parameters are combined into one, the instantaneous luminosity $\mathcal{L}$. The number of expected events is proportional to the instantaneous luminosity and to ϵ which is the product of three factors; the specific branching fraction of interest for a given physics process, the efficiency for the detector to record these events and the efficiency of the selection criteria used in the data analysis.

$$N_{t\bar{t}\ \text{evts}} = \sigma(p\bar{p} \to t\bar{t}) \cdot \mathcal{L} \cdot \epsilon \tag{2}$$

$$\mathcal{L} = \frac{N_p \cdot N_{\bar{p}} \cdot B \cdot f_o}{4\pi\sigma^2} \tag{3}$$

Where:

- $N_{t\bar{t}\ \text{evts}}$ is the number of detected $t\bar{t}$ events.
- B is the number of bunches circulating in the accelerator (=6).
- N_p is the number of protons per bunch ($\sim 2 \times 10^{11}$).
- $N_{\bar{p}}$ is the number of anti-protons per bunch ($\sim 6 \times 10^{10}$).
- f_o is the frequency of bunch crossings (50 KHz, 3.5 μsec/revolution).
- σ^2 is the physical beam cross-section (3×10^{-5} cm^2, for ~ 100 μm beam diameter).

During the 1992-1995 collider run the average instantaneous luminosity was $\sim 1 \times 10^{31}$ cm^{-2}s^{-1}. Table 1 shows the production rates for various process at the Tevatron at this luminosity together with the total number of events produced when the total integrated luminosity is 100 pb^{-1}.

To collect a dataset of 100 pb^{-1} would take of the order of 4 months of running at 100% efficiency. In reality it took about 2 years to gather the data. The inefficiencies come from various sources related to either the accelerator or the detectors. The accelerator can be down due to quenches in the superconducting magnets, due to scheduled studies, maintenance, construction, etc.. Sometimes nature contributes significantly to the down-time; excessive heat can cause equipment malfunctions, floods and small animals can also interfere with the accelerator operations. The detectors also need time without collisions for calibrations and repairs. Overall, thousands of components have to work together, synchronously, for the experiments to be able to log-in good quality data.

II TOP LIFETIME AND WIDTH

In the Standard Model, assuming V-A coupling with a CKM mixing parameter $|V_{tb}| = 1$, the width for the $t \rightarrow Wb$ decay is given in lowest order [18] by:

$$\Gamma(t \rightarrow Wb) \sim 175 \text{ MeV} \times (\frac{M_{top}}{M_W})^3 \qquad (\text{for } M_{top}, M_W >> M_b) \tag{4}$$

Figure 6 shows the width and lifetime of the top quark as a function of M_{top} [18]. For $M_{top} = 175$ GeV this results in a width of 1.5 GeV and a lifetime of 4×10^{-25} seconds (0.4 yoctoseconds). Non-perturbative QCD hadronization is estimated to take place in a time of order $\Lambda_{QCD}^{-1} \sim O(100 \text{ MeV}^{-1}) \sim O(10^{-23}\text{sec})$ [18,19]. Therefore **the top quark decays as a free quark.** This fact has significant implications in top quark physics; no hadronic states containing top and no toponium spectroscopy can be expected.

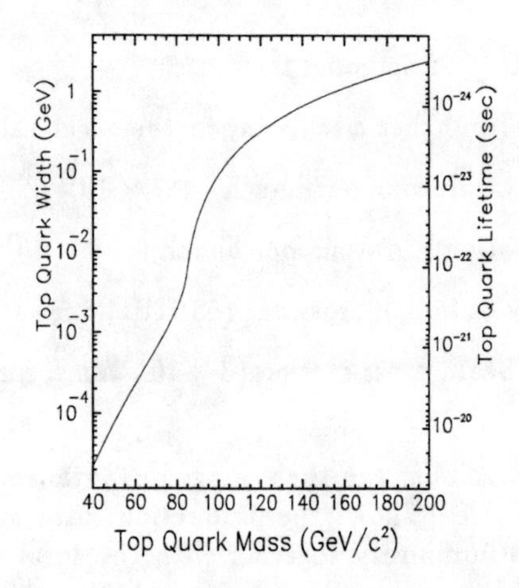

FIGURE 6. The Standard Model width of the top quark as a function of its mass.

Other top decays are allowed in the standard model, $t \rightarrow Ws$ and $t \rightarrow Wd$ but these are suppressed by factors of $|V_{ts}|^2/|V_{tb}|^2 \sim 10^{-3}$ and $|V_{td}|^2/|V_{tb}|^2 \sim 5 \times 10^{-5}$ respectively.

III DETECTION OF THE TOP QUARK

A Decay signatures

Figure 7 shows the production and decay diagram for $p\bar{p} \to t\bar{t}$. The different decay signatures are determined by the decay of each of the two W-bosons in the event. Table 2 shows the probabilities for the possible W decays. The $t\bar{t}$ decay channels are labeled according to the number of leptons that are detected in the final state. Table 3 has the branching fractions for each of these decay channels. In the following sections, unless stated differently, when we refer to leptons, we are exclusively referring to electrons and muons. Taus are more difficult to detect and a special section is devoted to channels with taus in them.

TABLE 2. Branching fractions for W decays

W Decay mode:	$W \to e\nu$	$W \to \mu\nu$	$W \to \tau\nu$	$W \to jets(ud, cs)$
Branching fraction:	1/81	1/81	1/81	6/81

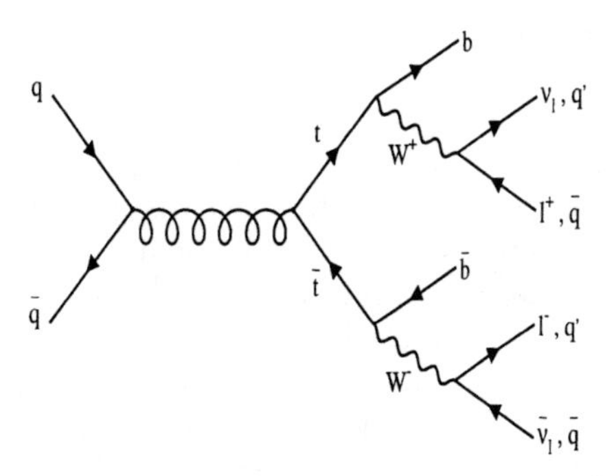

FIGURE 7. $t\bar{t}$ decay chain.

TABLE 3. $t\bar{t}$ signatures ($\ell = \mu, e$).

Channel	Signature	Branching Fraction (BR)
Dilepton	$t\bar{t} \rightarrow \ell\bar{\nu}b\ell\nu\bar{b}$	$4/81 = 5\%$
	$t\bar{t} \rightarrow \ell\bar{\nu}b\tau\nu\bar{b}$	$4/81 = 5\%$
	$t\bar{t} \rightarrow \tau\bar{\nu}b\tau\nu\bar{b}$	$1/81 = 1\%$
Lepton + jets	$t\bar{t} \rightarrow q\bar{q}b\ell\nu\bar{b}$	$24/81 = 30\%$
	$t\bar{t} \rightarrow q\bar{q}b\tau\nu\bar{b}$	$12/81 = 15\%$
All jets	$t\bar{t} \rightarrow q\bar{q}bq'\bar{q}'\bar{b}$	$36/81 = 44\%$

B Event detection

Two large international collaborations have built detectors that are optimized for the $t\bar{t}$ signatures; the CDF and D0 experiments. Figures 8 and 9 show a schematic of these detectors. A detailed description of these detectors can be found in [6,7].

The system of coordinates used is one in which the proton beam direction defines the z axis (see Fig. 8). The azimuth angle ϕ is the angle in the plane transverse to the beam and θ is the angle with respect to the z axis. The approximately Lorentz invariant quantity η is used together with ϕ and the radial coordinate r to locate particles in the detector ($\eta = -ln(tg(\frac{\theta}{2}))$. The decay products of the $t\bar{t}$ events, for heavy top quarks, will typically have large momenta and will lie in the "central" region of the detectors, $|\eta| < 1.5$.

Charged particles are identified by tracks in the tracking chambers in the central region of the detectors. The CDF detector has its tracking volume embedded in a large magnetic field (1.4 tesla) and can therefore measure accurately the particle's charge and momentum. Electron identification is provided by the calorimeters surrounding the tracking devices. These calorimeters are made of lead and scintillating plastic in CDF and of uranium and liquid argon in D0. Finely segmented chambers inside the calorimeters help identify the relatively contained electromagnetic shower characteristic of an electron. Extensions of these calorimeters, iron and scintillator in CDF and more liquid argon and uranium in D0, help contain the nuclear showers created by hadrons in the event. Quarks in $t\bar{t}$ decays hadronize into "jets" or collimated bundles of charged and neutral particles. Jets are detected as clumps of energy deposited in the calorimeters using cone-clustering algorithms. Jet energies, and therefore their parent parton energy, are measured with relatively poor resolution due to lost energy in the cone from gluon splitting and radiation in the fragmentation process. In addition, extra energy not belonging to the parton, from the "underlying event", can be picked up in the cone. The underlying event are the remnants of the $p\bar{p}$ collision that also hadronize. It is a poorly understood non-perturbative QCD process which ultimately is a perturbation

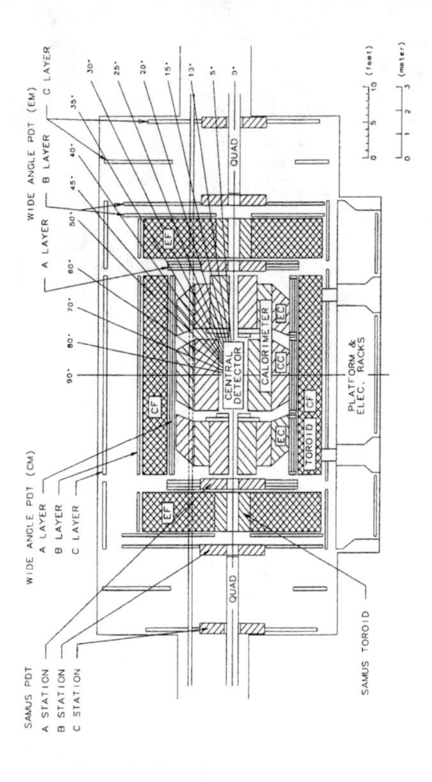

FIGURE 8. The D0 detector.

to all measurements in the detector. However, this effect is not big and can be corrected for.

Outside the calorimeters are wire chambers which are used to detect the passage of muons which typically traverse the calorimeters without interacting. The presence of neutrinos is detected indirectly. For a given event, the total energy in the parton-parton collision that produced the $t\bar{t}$ pair is unknown. However the total momentum transverse to the collision line is zero, therefore in the final state, the total momentum transverse to the collision line should be zero. If the energy/momentum is found to be non-zero we call it "missing-E_T" or $\not{E}_t$. The presence of significant $\not{E}_t$ is identified as the presence of neutrinos in the event.

In all the $t\bar{t}$ decay modes there are two b-quarks present and therefore the ability to efficiently identify these b-quarks is a very important feature of the detectors. Both CDF and D0 look for b's by identifying the softer leptons in their semi-leptonic decays: $b \to \ell\bar{\nu}X$ (BR$\sim$ 20%) or $b \to c \to \ell\bar{\nu}X$

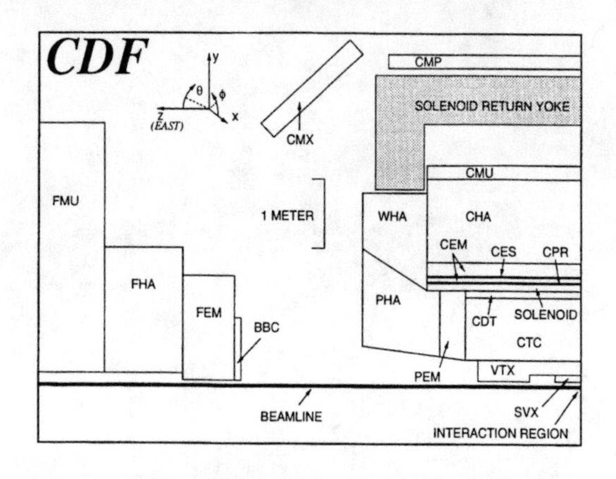

FIGURE 9. The CDF detector.

(BR$\sim$ 20%). The "soft lepton tagging" (SLT) method can identify b's in about 20% of the $t\bar{t}$ events and can mis-tag (identify wrongly as a b) about 0.5% of generic (non heavy-flavor) jets. CDF has an additional powerful tool for identifying b-quarks; a silicon-vertex detector. This detector sits very closely to the collision point and is very finely segmented so it can reconstruct tracks which originate from a vertex away from the primary vertex of the parton-parton collision. These secondary vertices are associated with long-lived particles such as mesons with heavy flavor content, i.e. b or c quarks. The b lifetime is measured to be about 1.5 ps, which in top events translates to displacements of secondary vertices of the order of a few millimeters. The probability of identifying a b quark by this method in a $t\bar{t}$ event is about 42%. The probability of mis-tagging a generic QCD jet is about 0.5%.

C Monte-Carlo

The various physics processes of interest in the analysis are simulated in the computer by combining several tools. The combinations of these tools are generically referred to as the "Monte-Carlo". These Monte-Carlo programs generate events using the theoretical calculation of a given matrix element to some order, usually tree-level (although some go as far as α_s^3). Then partons are made to hadronize, and hadrons made to fragment into showers of observable particles. All these processes are repeated over and over by sampling randomly from the physical distributions of all relevant variables (kinematical, spin etc.). These "generators" therefore produce events with well defined observable particles.

Another Monte-Carlo is then used to simulate the detector response to the passage of these particles. One can have very fast detector simulations in which the detector response is parametrized in various forms from well known interaction processes (electromagnetic showers, multiple-scattering etc.) or slower ones which track down every single interaction and create fake data in every single detector element.

The fake data created by these Monte-carlo programs is then analyzed with the same programs used to analyze the real data from the detector. A lot of care is taken to understand all the detector effects and incorporate them into the simulation programs.

IV TOP QUARK EVENTS IN THE DATA

In this section we discuss the data taken in the Tevatron during a period of about 2 years in 1994-1995. During that period about 5×10^{12} $p\bar{p}$ collisions occurred in each of the CDF and D0 experiments. By selecting the most interesting events using sophisticated fast electronics (called "triggers") each experiment recorded about 50 million events on magnetic tape. Each event contains about 180 Kbytes of information.

In what follows we discuss the details of the data analysis done in order to obtain a sample of data enriched in top quarks. The details are not identical in both experiments, however they are similar enough. Therefore the rest of this discussion will attempt to give an understanding of the criteria used in analyzing the data without attempting to differentiate, except when necessary, between the CDF and D0 experiments.

Kinematics dictate that the decay products of a massive top quark will be of high transverse momentum ($P_T > 15$ GeV) and will be "central" in the detector ($|\eta| < 2.0$). A preliminary selection of events with central, high-P_T, electrons or muons yields about 200,000 events per experiment. This sample is reduced through further selection criteria ("cuts") which are aimed at maximizing the acceptance of top events and minimizing the acceptance of "backgrounds" (events from other physics processes which mimic the signature of a $t\bar{t}$ event).

When CDF published evidence [20] for the top quark and later CDF and D0 published the observation of the top quark [6,7], the papers demonstrated a significant excess of lepton+jets and dilepton events over the expected number of backgrounds. Furthermore the presence of b-quarks in these events was demonstrated by soft-lepton tags in jets by CDF and D0 and by finding displaced secondary vertices, consistent with the b-lifetime, by CDF. In addition, the presence of $W \to \ell\nu$ was evident from the reconstruction of the transverse invariant mass of the $\not{E}_t$ and the lepton. The excess of events resulted in a $t\bar{t}$ production cross-section consistent with that expected from theory. The top quark mass was reconstructed from all the decay products and a mass peak was observed. In the remainder of this section we'll discuss the cuts used to create the datasets for these measurements and the measurements themselves as they stand today, a year after the completion of the Tevatron run, when most of the measurements are near final. The total integrated luminosity is about 110 pb^{-1} per experiment.

A The lepton+jets channel

This channel ($t\bar{t} \to W^+b, W^-\bar{b} \to \ell^+\nu b, q\bar{q}'\bar{b}$) is $\sim$ 35% of the $t\bar{t}$ branching ratio. The two b's and two quarks from the second W decay produce 4-jets. The main background for this channel comes from the production of "W+jets" (see Fig 10):

$$p\bar{p} \to W + \text{n jets} \to \ell\nu + \text{n jets} \tag{5}$$

The ratio of the W+jets cross-section to the $t\bar{t} \to$ lepton + jets cross-section is $\sim$ 100 : 1, 20 : 1, 5 : 1, and 1 : 1 for ≥ 1, ≥ 2, ≥ 3 and ≥ 4 jets respectively.

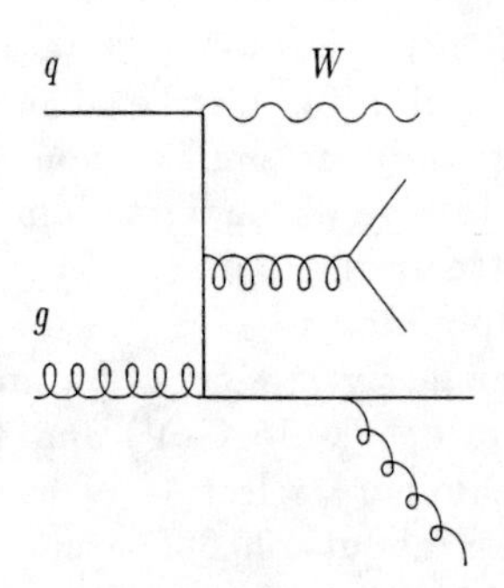

FIGURE 10. Example of Feynman diagram for $W+$ 4 jet production in $p\bar{p}$ collisions. Other diagrams also contribute.

A lepton+jets sample is selected by requiring:

- A high E_T lepton ($>$ 20 GeV) which is central and isolated from other activity in the detector.
- $\not\!\!E_t > 20$ GeV which signals the possible presence of a neutrino.
- ≥ 3 central jets with high-E_T ($>$ 15 GeV).

After these cuts about 300 events remain in the data. Monte-Carlo studies indicate that about 60 events are expected from $t\bar{t}$ production in this channel. The signal to background is therefore $\sim 1/5$ and further cuts are necessary to enrich the sample with top quarks. For a heavy top quark ($M_{top} > 130$ GeV) "event-shape" cuts are useful in this task. These can include the event aplanarity, various likelihood functions formed with kinematical variables such as the E_T of the jets in the event, and others. Most of the backgrounds are poor in b-quark content and the requirement of "b-tagged" jets is a powerful tool in reducing the number of backgrounds. Table 4 shows the number of events in the data for the lepton+jets channel for both CDF and D0, together with the estimated number of events expected from background.

TABLE 4. Summary of Lepton+jets events

Sample	D0	CDF
Event shape		
Num. observed	21	22
Num. background	9.2 ± 2.4	7.2 ± 2.1
Num. top ($M_{top} = 175$) GeV	12.9 ± 2.1	-
SLT		
Num. observed	11	40
Num. background	2.5 ± 0.4	24.3 ± 3.5
Num. top ($M_{top} = 175$) GeV	5.2 ± 1.0	9.6 ± 1.7
SVX		
Num. observed	-	34
Num. background	-	8.0 ± 1.4
Num. top ($M_{top} = 175$) GeV	-	19.8 ± 4.0

The background estimates are the most difficult part of the analysis because a precise understanding of the probabilities that generic QCD jets from gluons and light quarks in W+jets events would fake a b-quark tag is needed. These backgrounds are referred to as "mis-tags". Another important part of the background, which is irreducible, is when a gluon in W+jets events splits to a $b\bar{b}$ pair. It is important to understand well the probability for this to happen. The mis-tags are obtained from the data itself using jet-enriched event samples. The $W + b\bar{b}$ fraction of events is estimated from theoretical calculations and Monte-Carlo programs. Other background sources such as $W + c\bar{c}$, $W + c$, QCD multi-jets, WW, WZ etc. yield smaller contributions to the background and are estimated from Monte-Carlo.

Figure 11 shows the number of lepton+jets events as a function of the number of jets in the event. It also shows the number of SVX b-tags in each of the jet multiplicity bins. Superimposed is the expected number of tags from all background sources. A clear excess is observed in the 3-jet and $\geq$ 4-jet bins. In Figure 12 the expected number of tags from a $t\bar{t}$ Monte-Carlo with $M_{top} = 175$ GeV is superimposed on the expected background. The data fits very well to the top plus background hypothesis. Figure 13 shows the corresponding jet-multiplicity distribution for the sample tagged by soft leptons. Figure 14 shows an electron plus 4-jet event from CDF. This event is particularly striking because two of the jets are b-tagged by the SVX. The displaced vertices are clearly seen in the display of SVX tracks.

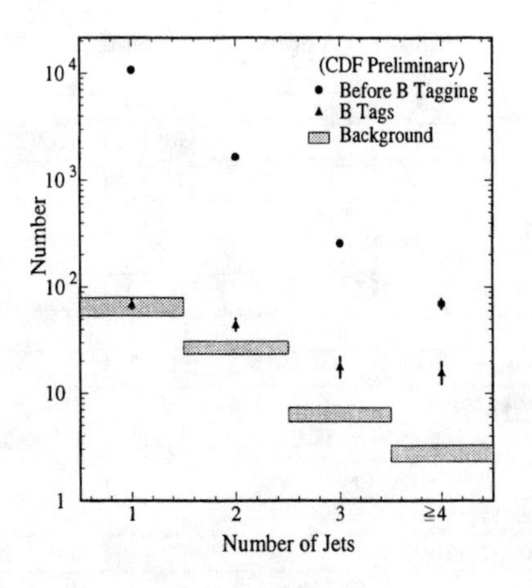

FIGURE 11. Number of events and SVX b-tags as a function of jet multiplicity in $W+$jets events (CDF).

The following figures, also from the CDF data, demonstrate the top-like features of the lepton+jets dataset. Figure 15 shows the presence of W-bosons decaying to $\ell\nu$. A jacobian peak near M_W is obtained when reconstructing the transverse invariant mass of the $\not\!\!E_t$ and the lepton. Figure 16 shows, for those events that have another b-tagged jet in the event (with looser requirements), the invariant mass of the two non-tagged jets. A peak at the W mass is evident. Figure 17 shows the reconstructed $c\tau$ distribution for the b-tags in the lepton+jets events. The distribution is consistent with the measured lifetime of the b.

B The dilepton channel

This channel ($t\bar{t} \to W^+b, W^-\bar{b} \to \ell^+\nu b, \ell^-\bar{\nu}\bar{b}$) is $\sim 5\%$ of the $t\bar{t}$ branching ratio. The two b quarks produce 2-jets and the two neutrinos generate $\not\!\!E_t$.

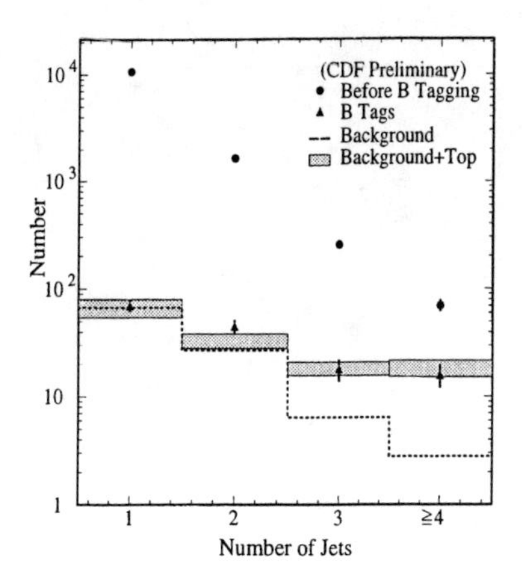

FIGURE 12. Number of events and SVX b-tags as a function of jet multiplicity in $W+$jets events. The top contribution is included (CDF).

Several physics processes can yield two leptons in the final state. Table 5 lists the cross-sections for these processes and compares them to the $t\bar{t} \to \ell^+\ell^-$ cross-section. The appearance of two jets in the event comes through higher order QCD processes and therefore reduces the backgrounds significantly. Requiring large $\not{E}_t$ also reduces the backgrounds.

TABLE 5. Summary of Dilepton Events

Process	$\frac{\sigma}{\sigma(t\bar{t}\to\ell^+\ell^-)}$	cuts reduction factor
$p\bar{p} \to b\bar{b} \to c\ell^-\bar{\nu}, c\ell^+\nu$	$\sim 5\times10^4$	10^6
$p\bar{p} \to Z^o \to \ell^+\ell^-$	$\sim 2,000$	$\sim 10^5$
$p\bar{p} \to \gamma Z^{o*} \to \ell^+\ell^-$	~ 200	$\sim 10^3$
$p\bar{p} \to W + jets \to \ell\nu + fake(\ell)$	~ 1	$\sim 10^4$
$p\bar{p} \to \gamma Z^{o\to} \tau\tau \to \ell^+\nu\bar{\nu}, \ell^-\bar{\nu}\nu$	~ 1	~ 20
$p\bar{p} \to W^+W^- \to \ell^+\nu\ , \ell^-\bar{\nu}$	~ 1	~ 20

A dilepton sample is selected by requiring:

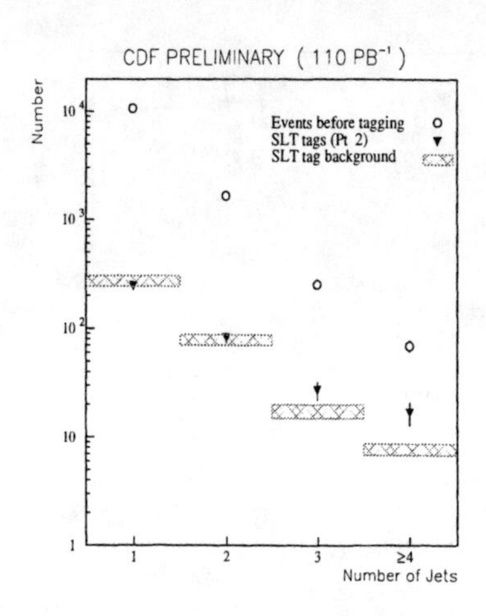

FIGURE 13. Number of events and SLT b-tags as a function of jet multiplicity in $W+$jets events (CDF).

- Two high E_T opposite-charge leptons (> 20 GeV), central and isolated in the detector.
- $\not{E}_t > 25$ GeV which signals the possible presence of the two neutrinos.
- ≥ 2 jets with high-E_T (> 10 GeV) and central.
- Dileptons with invariant mass close to the Z^o mass are rejected ($75 < M_{\ell\ell} < 175$ GeV)
- Azimuthal separation between the $\not{E}_t$ and the leptons and jets is required.
- A cut requiring large H_T (the sum of the E_T of the various objects in the event) can also be imposed ($> 100-200$ GeV, depending on the channel).

All these cuts reduce the background drastically by several orders of magnitude (see Table 5) to a point where the signal over background is about 2:1. The final background contribution is roughly the same for WW, Drell-Yan, $Z \to \tau\tau$ and fakes. Table 6 lists the number of observed dilepton events in CDF and D0 together with the expectations from backgrounds and from $t\bar{t}$. The efficiency of these cuts for $t\bar{t} \to W^+b, W^-\bar{b} \to \ell^+\nu b, \ell^-\bar{\nu}\bar{b}$ events is about 20% which when combined with the 5% branching fraction results in only about 1% of all produced $t\bar{t}$ events being detected in the dilepton channel. Using a value for the theoretical production cross-section of 5 pb (for

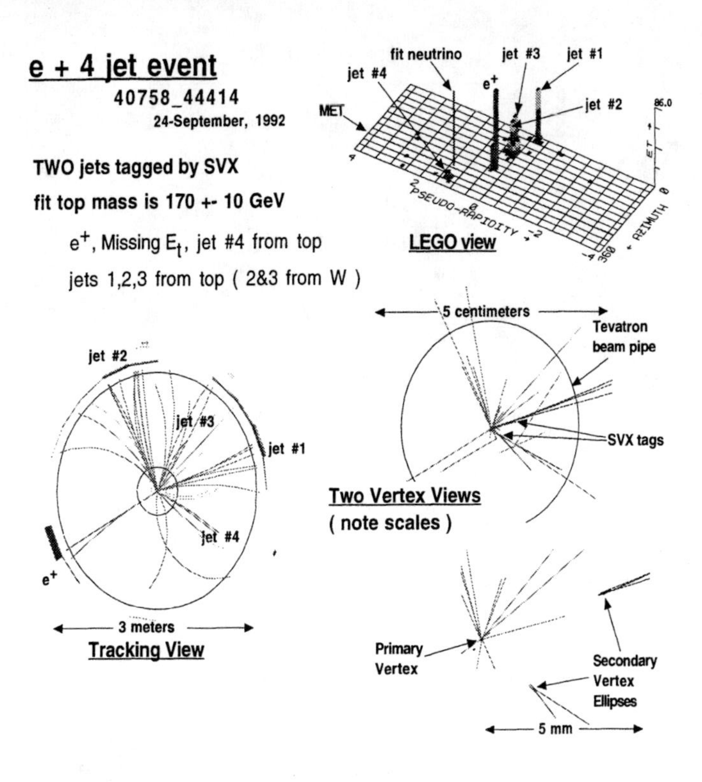

FIGURE 14. Golden lepton+4-jets event from CDF.

$M_{top} \sim 170$ GeV) and a luminosity of 100 pb^{-1} one expects to detect only about 5 events. It is worth noticing that of the nine CDF dilepton candidates, four have jets that are b-tagged. Figure 18 shows the CDF dilepton data for events with two or more jets as a scatter plot of the azimuthal angle between the $\not{E}_t$ and the closest lepton or jet versus the $\not{E}_t$. The events that pass the cuts are to the right of the line.

C The all-jets channel

This channel ($t\bar{t} \to W^+b, W^-\bar{b} \to q\bar{q}'b, q\bar{q}'\bar{b}$) is $\sim 44\%$ of the $t\bar{t}$ branching ratio. The two b quarks produce 2-jets and the two W bosons 4 more. Jet merging due to the limitations of reconstruction algorithms, jets not traversing the active detector regions and the effects of initial and final state QCD

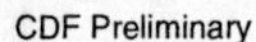

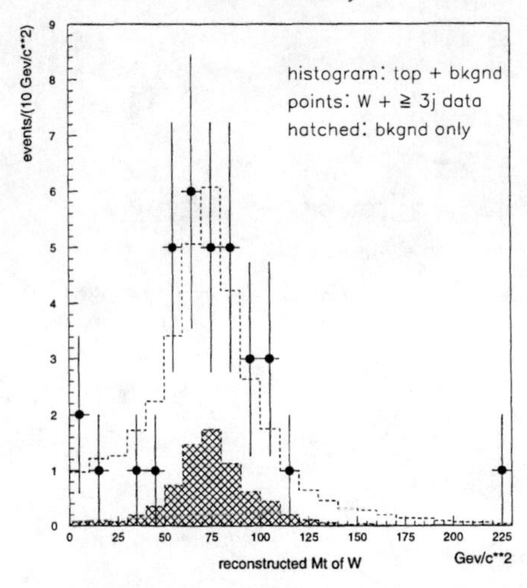

FIGURE 15. Invariant mass for $\not{E}_t$ and lepton in W+jets events (CDF).

TABLE 6. Summary of Dilepton Events

	D0	CDF
Num. observed	5	9
Num. background	1.6 ± 0.5	2.1 ± 0.4
Num. top ($M_{top} = 175$) GeV	3.1 ± 0.3	4.0 ± 0.3

radiation can result in some fraction of the events having not exactly six jets.

This channel suffers from a copious QCD multi-jet production background. In the data there are more than 200,000 events with four or more jets, more than 60,000 with five or more jets and more than 12,000 with 6 or more jets. At this level any top production is overwhelmed by QCD.

The cuts applied to enhance the signal to background ratio are listed below and take advantage of the difference in topology between QCD jet production and heavy top production. A b-tag is essential to enhance the significance of the signal.

- ≥ 5 high-E_T (> 15 GeV) central jets.
- $\sum E_T$(jets) > 300 GeV (works for heavy top)

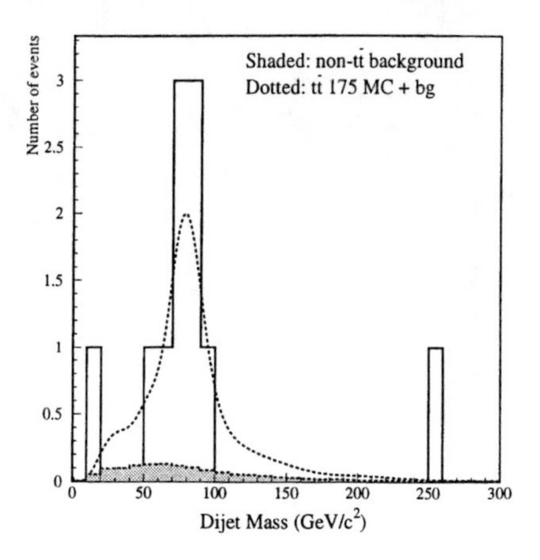

FIGURE 16. Invariant mass for non-tagged jets in W+jets double tagged events (CDF).

- Aplanarity of the event
- Centrality of energy flow
- At least one (SVX) b-tag (CDF)

Table 7 lists the number of events in the data together with the expectation from top and background. With the b-tag CDF shows an excess of events over background of about 3 standard deviations [21].

TABLE 7. Summary of All-Jets Channel

	D0	CDF
Num. observed	15	192
Num. background	11 ± 2	137 ± 11
Num. top ($M_{top} = 175$) GeV	4.5 ± 0.5	27 ± 9

D Channels with taus

In the dilepton and lepton+jets channels discussed above, the leptons where either electrons or mouns. Taus that decayed leptonically already contributed to these signatures (about 10% of the total dilepton acceptance). In order to increase the acceptance for channels with taus one has to look for the hadronic

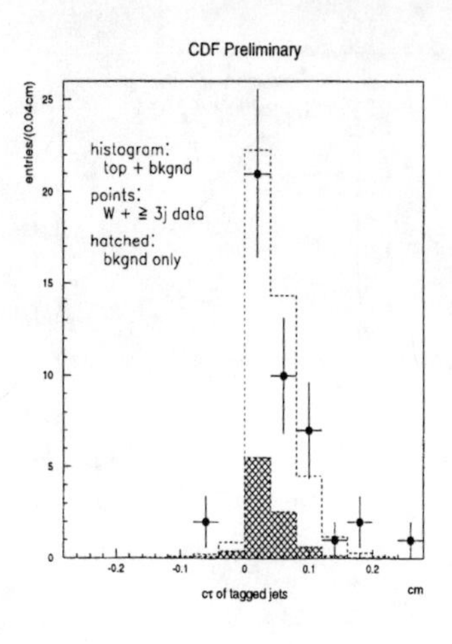

FIGURE 17. Reconstructed $c\tau$ distribution for CDF SVX b-tagged jets.

tau decays. Both CDF and D0 have recently looked at dileptons where one of the leptons is an electron or a muon and the second one is a tau that decayed hadronically. The branching fraction for hadronic tau decays is 64% (50% in 1-prong and 14% three-prong). Because of the extra neutrinos in tau decays the detectable P_T of the observables is smaller and therefore the acceptance is smaller. In addition this channel suffers from a large background from events with real high-P_T leptons and jets that can fake the hadronic tau decays (W +jets and Z+jets).

The event selection for this channel is complicated [22] and involves taking full advantage of the tracking and calorimetry in the detectors. Charged daughters (mainly pions) from the hadronic tau decays are more isolated than tracks in QCD jets and therefore various tracking isolation requirements are made. A cut in track multiplicity, to have only a one or three-prong is very useful as well. Taus form narrower clusters in the calorimeters and it is also useful to to make this requirement. In the shower max detectors embedded in the calorimeters one can look for small clusters of electromagnetic energy associated with accompanying π^o's. The invariant mass of the tracks in the cluster of tracks associated to a calorimeter cluster is also a useful variable for separating QCD jets from taus. Typically a selection for this channel will include:

- A high E_T lepton (e or μ), central and isolated.
- A hadronically decaying tau as described above.
- Two or more central jets ($E_T > 10$ GeV)
- Large $\not{E}_t$ (~ 50 GeV).

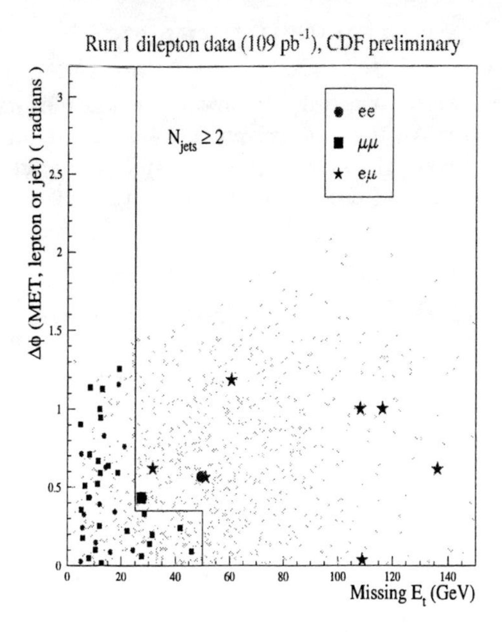

FIGURE 18. CDF dilepton candidates.

- A cut on the sum of the E_T of the leptons, jets and $\not{E}_t$.
- Removal of Z^o's.

Table 8 lists the observed number of events together with the expectations from background and from $t\bar{t}$. The significance of this channel is still marginal with this level of statistics. However a couple of the CDF events have in addition a b-tag, further evidence for top decays.

TABLE 8. Summary of dilepton channel with taus

	D0	CDF
Num. observed	2	4
Num. background	1.4 ± 0.5	2.0 ± 0.4
Num. top ($M_{top} = 175$) GeV	1.4 ± 0.1	0.7 ± 0.1

V MEASUREMENTS

In this Section we describe the various measurements performed with the datasets described above. All these measurements are statistically limited and will benefit enormously from the increased luminosity expected in the coming collider run in 1999 (Run 2). About 20 times more data ($\sim 2\ fb^{-1}$) is expected to be logged-in.

A $t\bar{t}$ Production Cross-Section

One can measure the $p\bar{p}$ production cross-section from each of the channels using:

$$\sigma(p\bar{p} \rightarrow t\bar{t} \rightarrow channel_i) = \frac{N^i_{obs} - N^i_{bck}}{\epsilon^i \cdot \int Ldt} \tag{6}$$

Where:

- ϵ^i is the detection efficiency for channel "i".
- $\int Ldt$ is the total integrated luminosity.
- N^i_{obs} is the observed number of events.
- N^i_{bck} is the expected number of background events.

The cross-section can also be calculated for any given combination of channels by maximizing a likelihood function which takes into account all uncertainties and possible correlations between channels. Table 9 lists both CDF and D0's results. Because ϵ depends on the top mass the cross-section needs to be stated at a given M_{top}. CDF uses 175 GeV and D0 uses 170 GeV. Each experiment measures the cross-section with about 25% uncertainty. A naive average of the two experiments yield a cross-section of $6.4^{+1.3}_{-1.2}$ pb (about 20% uncertainty). Figure 19 shows the comparison between the experimental cross-sections and the theoretical calculations. Good agreement is observed within the uncertainties.

B Measurement of V_{tb}

Here we describe briefly a measurement of the CKM matrix element V_{tb} under the assumption that the Standard Model holds and that the top quark decays only to b, c, s. If we define R as the fraction of top quarks that decay to b quarks then:

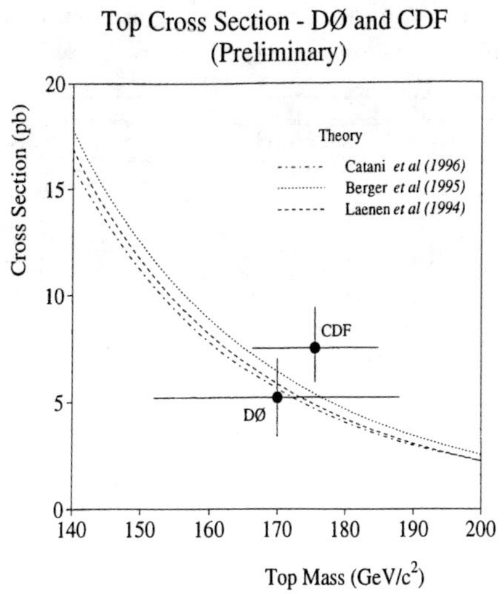

Top Cross Section - DØ and CDF
(Preliminary)

FIGURE 19. CDF and D0 mass and cross-section vs theory.

$$R \equiv \frac{BR(t \to Wb)}{BR(t \to Wb, Ws, Wd)} = \frac{|V_{tb}|^2}{|V_{tb}|^2 + |V_{ts}|^2 + |V_{td}|^2} \tag{7}$$

From this definition R can be interpreted as the taggable fraction of $t\bar{t}$ decays. If f_n is the measured fraction of candidate events with n tags and ϵ is the tagging efficiency then:

$$f_2 = (R \cdot \epsilon)^2 \;, \quad f_1 = 2R \cdot \epsilon(1 - R \cdot \epsilon) \;, \quad f_0 = (1 - R \cdot \epsilon)^2 \tag{8}$$

And therefore:

$$R = \frac{2}{\epsilon(f_1/f_2 + 2)} = \frac{2}{\epsilon(N_1/N_2 + 2)}, \tag{9}$$

where N_i is the number of events in the data with i tags. Therefore $|V_{tb}|$ can be calculated from the measured ratios of the number of tags in the data. The measurement, made by CDF, takes into account the correlations between the SVX and SLT taggers, correlations in acceptances and systematics and combines the lepton+jets and the dilepton channels. All this is done by the maximum likelihood method in which the question: "which R gives the highest probability to the measurement of N_i/N_j ?" is answered. Figure 20 shows the likelihood as a function of R. We obtain:

TABLE 9. $\sigma(p\bar{p} \to t\bar{t})$ production cross-section measurements. All numbers are preliminary.

Experiment	Channel	cross-section (pb)
CDF	SVX	$6.8^{+2.3}_{-1.8}$
CDF	SLT	$8.0^{+4.4}_{-3.6}$
CDF	DIL	$9.3^{+4.4}_{-3.4}$
CDF	HAD	$10.7^{+7.6}_{-4.0}$
CDF	TAU	$15.6^{+18.6}_{-13.2}$
CDF	SVX+SLT+DIL	$7.5^{+1.9}_{-1.6}$
D0	DIL	4.7 ± 3.2
D0	HAD	4.6 ± 5.1
D0	L+JETS (shape)	4.2 ± 2.0
D0	L+JETS (SLT)	7.0 ± 3.3
World Avg.		$6.4^{+1.3}_{-1.2}$

$$R = 0.94 \pm 0.27(\text{stat}) \pm 0.13(\text{syst}) \Rightarrow R > 0.34 \ (95\% C.L.) \tag{10}$$

Assuming unitarity, $|V_{tb}|^2 + |V_{ts}|^2 + |V_{td}|^2 = 1$, a value for V_{tb} is obtained directly:

$$|V_{tb}| = 0.97 \pm 0.15 \pm 0.07 \Rightarrow |V_{tb}| > 0.58 \ (95\% C.L.) \tag{11}$$

If one relaxes unitarity and allows V_{ts} and V_{td} to vary within their allowed ranges from low energy measurements [23] then one gets:

$$|V_{tb}| > 0.022 \ (95\% C.L.) \tag{12}$$

C The top quark mass

Given the size and purity of the samples of various channels described above, the lepton+jets channel is the process in which the top quark mass is best measured. It is however very important to determine that all channels yield consistent values for M_{top}.

1 M_{top} from the lepton+jets channel

The CDF measurement is done on a sample with 4 or more jets and at least one b-tag. The fourth jet is allowed to be softer (> 8 GeV) to increase the acceptance. The sample consists of 34 events. For each event a value for M_{top} is obtained by performing a kinematic fit of the measured objects (lepton, $\not{E}_t$, jet$_1$ to jet$_4$) to the hypothesis that it comes from $t\bar{t}$ production

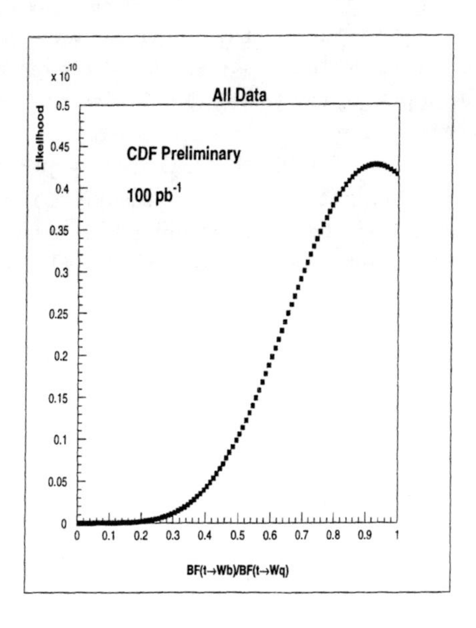

FIGURE 20. CDF V_{tb} measurement.

through $t\bar{t} \rightarrow (W^+b)(W^-b) \rightarrow (\ell\bar{\nu}b)(q\bar{q}'b)$. There are constraints to this fit that come from 4-momentum conservation at each vertex of the decay process; $M(\ell,\nu) = M_W$, $M(j_1,j_2) = M_W$ and $M_t = M_{\bar{t}}$. Because of all the possible assignments of the jets to the partons in the process, there are several combinatoric solutions. The solution with the best χ^2 is chosen. Tags are associated to b-quarks and both solutions for the neutrino P_z are considered. This method has some complications; the best combination is not always the correct one. Monte-Carlo studies show that due to poor resolution in the jet energy measurements and due to jets not coming from top decays (initial state radiation for example) only about 50% of the cases have the observed objects associated to the correct partons. If one takes only these cases, the resulting distribution is Gaussian-like, peaked at the correct Monte-Carlo value for the top mass and its width is due to the resolutions in the measurements. The wrong combinations will broaden this distribution and create tails, although the peak is still close to the generated top mass. Once the distribution for the 34 events is obtained, a likelihood fit is performed to a superposition of the mass distribution from the Monte-Carlo for a given top mass and the distribution resulting from the kinematic fits to background events. Many templates are tried for a range of top masses and the top mass that results in the maximum likelihood is chosen as the measured top mass in the sample.

The likelihood method yields a statistical uncertainty in this measurement. Figure 21 shows the reconstructed top mass for the events from the CDF data, the various subsamples of the lepton + $\leq$4-jets are highlighted and the inset has the fit to the negative log of the likelihood. The resulting value for the top mass is: $M_{top} = 176.8 \pm 4.4(\text{stat})$ GeV. The systematic uncertainties in this measurement are dominated by the knowledge of the energy scale for jets in the calorimeter and by the understanding of extra gluon radiation in top events. These uncertainties all add up to about 5 GeV, and the final result is:

$$M_{top} = 176.8 \pm 4.4(\text{stat}) \pm 4.8(\text{syst}) \qquad \text{(CDF preliminary)} \qquad (13)$$

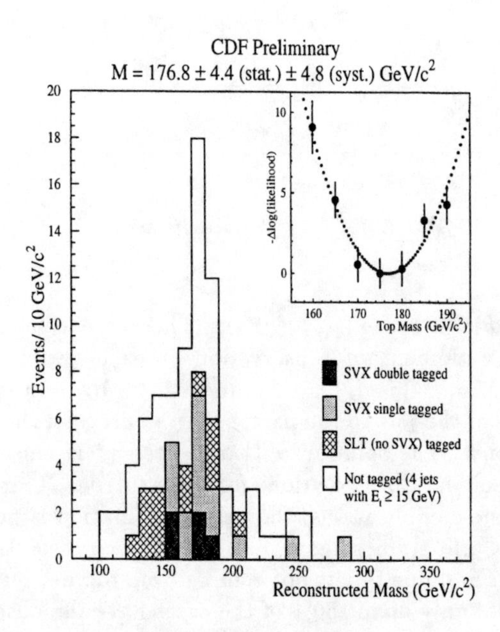

FIGURE 21. Top quark mass measurement from CDF.

The D0 method starts by making a top enriched sample for the mass measurement by also requiring 4-jets in the event. In addition a cut is made in a likelihood variable which combines several kinematic variables; the $\not{E}_t$, the aplanarity of the W+jets, the fraction of the E_T of the W+jets that is carried by the W and the E_T-weighted r.m.s η of the W and jets. Once this cut is applied, 34 events survive and 30 of them can be kinematically fit to a top mass using the same kinematic fit described above for CDF. The rest of the procedure is then similar to CDF's; templates are constructed from $t\bar{t}$ Monte-Carlo samples having the same cuts as the data and from background and a likelihood fit is made. The systematic uncertainties are in essence coming from similar sources as CDF's even though the details are somewhat different. The D0 top mass measurement is (Fig. 22):

$$M_{top} = 170 \pm 8(\text{stat}) \pm 8(\text{syst}) \qquad \text{(D0 preliminary)}. \tag{14}$$

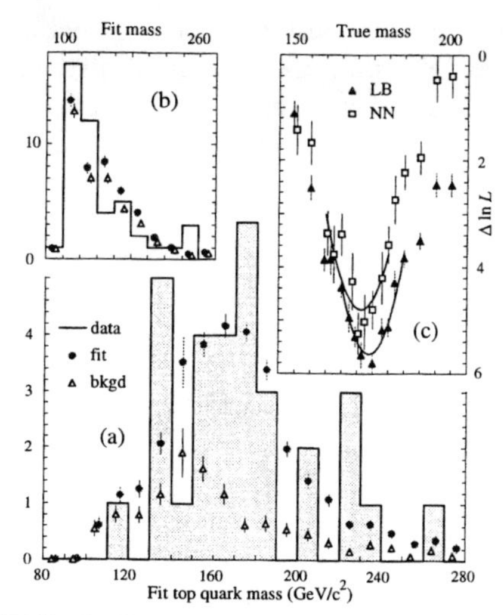

FIGURE 22. Top quark mass measurement from D0.

2 *M_{top} from dilepton and all-jets channels*

For top dilepton decays, the problem of reconstructing the top mass from the measured final state is that the kinematics are under-constrained due to the presence of the two neutrinos. This can be overcome by taking the kinematical distributions from the decay products that are sensitive to the top mass and fitting them to various M_{top} hypotheses (the background distributions are included as well). CDF has made a measurement using the energy of the two jets in the dilepton candidates and also using the invariant mass of the b-jet and the lepton ($M_{\ell b}$). The mass obtained is $M_{top} = 160 \pm 28$ GeV (for the jet energies) and $M_{top} = 162 \pm 22$ GeV (for the ($M_{\ell b}$ case). The uncertainties are dominated by the small statistics of the sample. D0, on the other hand, has used a technique which weights each possible solution of the unconstrained kinematical fit with the expected distribution of "dynamical" variables in $t\bar{t}$ production. The measured mass in this channel is $M_{top} = 151 \pm 23$ GeV.

CDF has also performed a top mass measurement from the all-jets sample using essentially the same method used in the lepton+jets channel. Combinatorics, jet energies are again a large source of the uncertainties together with the more copious background. THe mass obtained is: $M_{top} = 187 \pm 8(\text{stat}) \pm 13(\text{syst})$.

D Kinematic properties

From the constrained fits used in determining the mass of the top quark in the lepton+jets channel one can obtain the 4-vectors for all the $t\bar{t}$ decay products and therefore determine the kinematic distributions of several variables. Figures23 and 24 show for example the distribution of the P_T of the top quark and the distribution of the invariant mass of the t and $\bar{t}$. The detailed measurement of the various kinematic distributions is important to test our understanding of $t\bar{t}$ production.

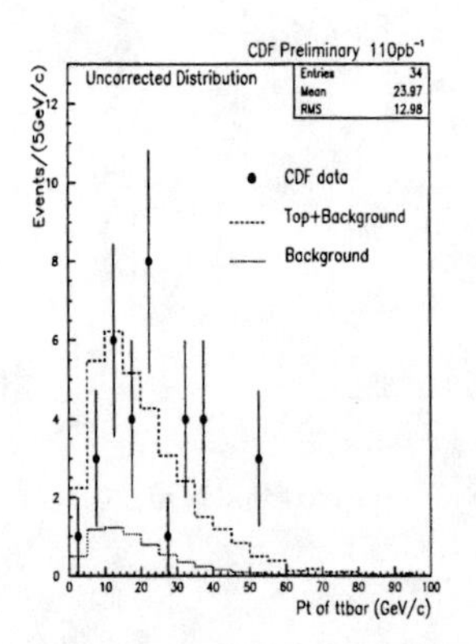

FIGURE 23. Top quark P_T measurement from CDF.

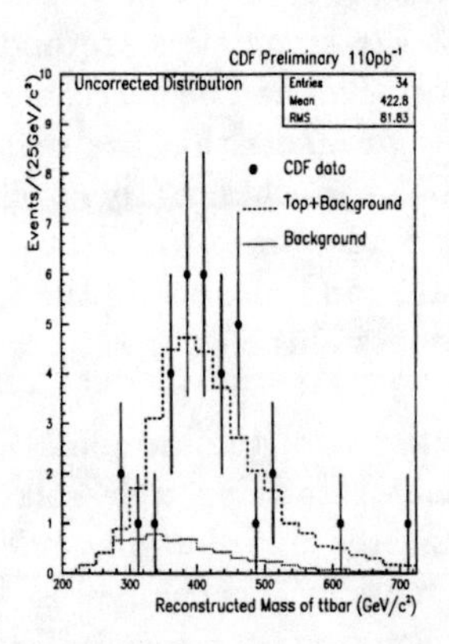

FIGURE 24. $t\bar{t}$ invariant mass measurement from CDF.

VI SEARCHES

A Rare Decays

CDF has searched for the flavor-changing neutral current decays $t \to qZ$ and $t \to q\gamma$. The search includes the cases in which one of the top quarks decays in the standard fashion ($t \to Wb$) and the other in a rare mode. Some theoretical models [24] predict a branching ratio as high as 0.1% for the qZ decay. For this mode, the signature sought out is $Z \to e^+e^-$ and four jets. Backgrounds come from WZ and ZZ production with extra jets. These are estimated to be $0.60 \pm 0.14 \pm 0.12$ events. The background from standard $t\bar{t}$ decays yields an additional 0.5 events. One event is observed in the data, resulting in a limit:

$$BR(t \to qZ) < 0.41 \ (90\% C.L.) \tag{15}$$

The branching ratio for $t \to q\gamma$ is predicted to be about 10^{-10} [25], therefore an observation of this decay would indicate the presence of new, unexpected, physics. The search is for the case in which one of the tops decays to Wb and the other to $q\gamma$. In the leptonic channel $W \to \ell\nu$ and the signature is $\ell\nu\gamma$ + 2 or more jets. The expected background is only about 0.06 events and one candidate event is observed. In the hadronic channel $W \to q\bar{q}'$ and the signature is a photon + 4 or more jets. The estimated background is 0.5 events and no events are seen. The limit obtained is:

$$BR(t \to q\gamma) < 0.029 \ (95\% C.L.) \tag{16}$$

B Charged Higgs

CDF has searched for the non-Standard Model top decays $t \to H^+b$ [26]. The charged Higgs exists in SM extensions with expanded Higgs sectors. For example SUSY can have two Higgs doublets, ϕ_1, ϕ_2. If the charged Higgs decay is kinematically available to the top quark, the probability for $t \to H^+b$ compared to the standard $t \to W^+b$ decay is governed by a parameter, $tan\beta$ which is the ratio of the vacuum expectation values of the two charged Higgs doublets. This same parameter governs the decay of the charged Higgs, which can go mainly into either $\tau\nu$ or $c\bar{s}$. Figure 25 shows these branching fractions as a function of $tan\beta$.

The signature of the $t \to H^+b$ decay depends on M_{top}, M_{H^+} and $tan\beta$. The search is made for the cases when at least one of the top quarks in $t\bar{t}$ decays to H^+b and optimized for the $H^+ \to \tau\nu$ channel where the tau decays hadronically. The backgrounds come from various sources such as $Z \to \tau\tau$ +jets, $W \to \tau\nu$ +jets and events where a generic QCD jet fakes a

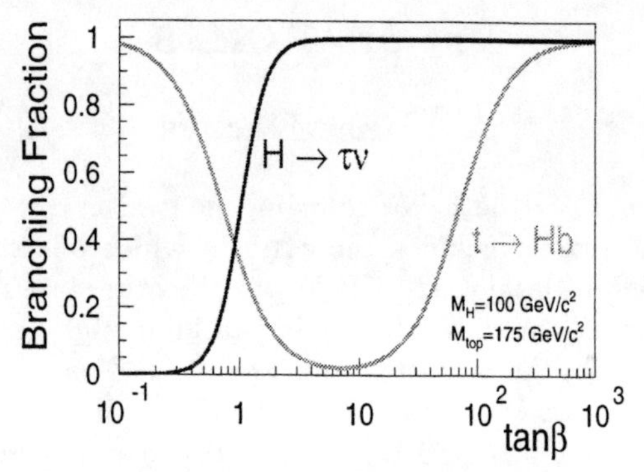

FIGURE 25. Charged Higgs branching ratios.

tau. The backgrounds with real taus are estimated from Monte-Carlo and are about 25% of the total background. The fake background is estimated directly from the data. The total expected number of background events is 8.1 ± 2.2. there are six events observed in the data and any model where more than 8.5 signal events are expected can be therefore excluded at 95% C.L. Figure 26 shows the exclusion regions in $\{M_{H^+}, tan\beta\}$ space depending on the assumed $t\bar{t}$ production cross-section. Notice that $tan\beta > 70$ is almost completely excluded for $M_{H^+} < 150$ GeV.

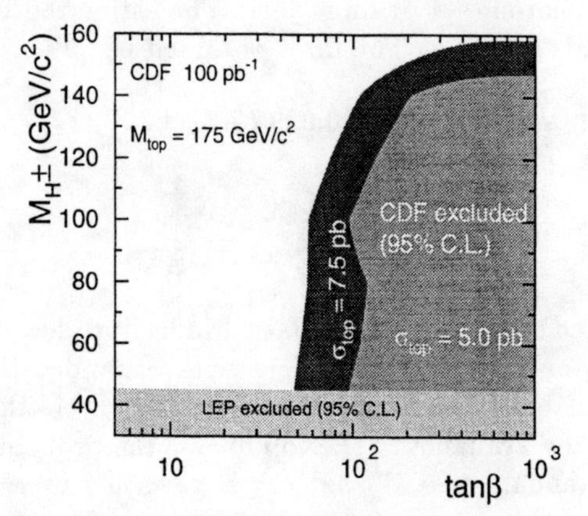

FIGURE 26. CDF limits for charged higgs search.

VII CONCLUSIONS

The top quark has been observed in $p\bar{p}$ collisions through its various decay channels and the $t\bar{t}$ production cross-section and M_{top} have been measured with the small samples available. The precision of these measurements will increase significantly with the data expected from the Tevatron Run 2 in 1999. So far, nothing observed in the production and decays of the top quark appears to be inconsistent with the Standard Model. In the next 10 or so years the Tevatron at Fermilab will be the only place to study this peculiarly heavy elementary particle. The expected yields, both from improvements in the accelerator and in the CDF and D0 detectors, are about 20 times those collected in Run 1. We are looking forward to an exciting period of top quark physics.

VIII ACKNOWLEDGMENTS

I would like to thank the members of the CDF and D0 collaborations for discussions and help, especially with gathering the figures for this document. In particular I'd like to thank C. Campagnari and M. Franklin for making the figures that they gathered in Ref. [1] available over the Web, and also M. Kruse for reading a draft of this document and making useful suggestions. Finally I'd like to thank the members of the Organizing Committee of the School, especially J.C. D'Olivo, H. Mendez and M.A. Perez Angon for their invitation and for making my stay in Merida an extremely pleasant one.

REFERENCES

1. C. Campagnari and M. Franklin, "The Discovery of the Top Quark", Rev. Mod. Phys. 69:137-212, 1997.
2. S.J. Wimpenny and B.L. Winer "The Top Quark", Ann. Rev. Nucl. Part. Sci. 46:149-195, 1996.
3. C. Quigg, "Topo-logy", FERMILAB-PUB-97-091-T, Apr 1997. Submitted to Phys.Today, e-Print Archive: hep-ph/9704332.
4. Sinervo P.K., "Top Quark Studies at Hadron Colliders", To be published in the proceedings of 23rd Annual SLAC Summer Institute on Particle Physics: The Top Quark and the Electroweak Interaction (SSI 95), Stanford, CA, 10-21 Jul 1995. e-Print Archive: hep-ex/9608005.
5. D. Gerdes, "Top Quark Physics: Results from CDF and D0". FERMILAB-CONF-96-342-E, Jun 1996. 11pp. Presented at 1996 DPF/DPB Summer Study on New Directions for High-energy Physics (Snowmass 96), Snowmass, CO, 25-Jun to 12-Jul 1996. e-Print Archive: hep-ex/9609013.
6. F. Abe *et al.* (CDF Collaboration), Phys. Rev. Lett. **74**, 2626 (1995).
7. S. Abachi *et al.* (D0 Collaboration), Phys. Rev. Lett. **74**, 2632 (1995).
8. Herb, S.W. *et al.*, 1977, Phys. Rev Lett. **39**, 252.
9. Berger, Ch. *et al.*, 1978, Phys. Lett. **B76**, 243. Bienlein, J.K. *et al.*, 1978, Phys. Lett. **78B**, 360. Darden, C.W. *et al.*, 1978, Phys. Lett. **76B**, 246. Rice, E. *et al.*, 1982, Phys.Rev. Lett. **48**, 906.

10. Bartel, W. *et al.*, 1984, Phys. Lett. **146B**, 437.
11. Kane, G.L. and M.E. Peskin, 1982, Nucl. Phys. **B195**, 29.
12. Thompson W.J., 1994, Ph.D. thesis, SUNY-Stony Brook (unpublished).
13. Collins, J.C., D.E. Soper and G. Sterman, 1986 Nucl. Phys. **B263**, 37.
14. Ellis R.K., 1991, Phys. Lett. **B259**, 492.
15. Laenen, E.J. *et al.*, 1994, Phys. Lett. **B321**, 254.
16. Berger, E.L. and H. Contopanagos, 1995, Phys. Lett. **B361**, 115.
17. Catani, S. *et al.*, 1996, CERN-TH/96-21.
18. I. Bigi *et al.* Phys. Lett. **B181**:157, 1986.
19. Orr, L.H., 1991, Phys. Rev. **D44**, 88.
20. F. Abe *et al.* (CDF Collaboration), Phys. Rev. D**50**, 2966 (1994); Phys. Rev. Lett. **73**, 225 (1994); Phys. Rev. D**51**, 949 (1995).
21. F. Abe *et al.* FERMILAB-PUB-97-075-E, Mar 1997. 15pp. Submitted to Phys.Rev.Lett.
22. F. Abe *et al.* FERMILAB-PUB-97-096-E, Apr 1997. 5pp. Submitted to Phys.Rev.Lett.
23. R.M. Barnett *et al.* Physical Review D54 1, (1996).
24. H. Fritzsch, Phys. Lett. **B224** 423 (1989).
25. S. Parke, in Proceedings of DPF'94, University of New Mexico, Albuquerque, NM (August, 1994).
26. F. Abe *et al.* FERMILAB-PUB-97-058-E, Apr 1997. 6pp. Submitted to Phys.Rev.Lett. e-Print Archive: hep-ex/9704003.

Photoproduction of Charm Particles at Fermilab

John P. Cumalat

University of Colorado
Department of Physics
Boulder, Colorado 80309

Abstract. A brief description of the Fermilab Photoproduction Experiment E831 or FOCUS is presented. The experiment concentrates on the reconstruction of charm particles. The FOCUS collaboration has participants from several Central American and Latin American institutions; CINVESTAV and Universidad Autonoma de Puebla from Mexico, University of Puerto Rico from the United States, and Centro Brasileiro de Pesquisas Fisicas in Rio de Janeiro from Brasil.

INTRODUCTION

The charm quark is likely to be the only quark for which one can observe Cabibbo favored, Cabibbo suppressed, and doubly Cabibbo suppressed hadronic decays. Doubly Cabibbo suppressed decays are likely to only be studied in the charm sector. These hadronic decays test our detailed understanding of the weak current and the contributions from non-perturbative and perturbative QCD. These decays should be tested for both mesons and baryons. In the charm sector it should also be possible to observe second order weak processes via D^0-$\bar{D}^0$ mixing and perhaps even CP violating transitions. It is important to continue the studies of rare and anomalous weak interactions as couplings to new particles may be flavor dependent. To investigate these decays and processes one simply needs larger charm samples. This talk will concentrate on the successor of Fermilab Photoproduction Experiment E687 called FOCUS [1]. The new experiment FOCUS has a goal to produce 10^6 reconstructed charm particles.

Progress in particle physics requires parallel experiments at the high energy frontier and at the high precision frontier. At the high energy frontier one searches for new particles and new interactions. At the high precision frontier one studies CP violation and makes precision studies of the present quarks and

CP400, *First Latin American Symposium on High Energy Physics/ VII Mexican School of Particles and Fields,*
edited by D'Olivo/Klein-Kreisler/Mendez

leptons. One strengthens the underpinning of the Standard Model by measuring quantities that are not defined such as the Cabibbo-Kobayashi-Maskawa matrix elements and one searches for rare processes and subtle deviations from the Standard Model. Such measurements may lead to both consolidation and extension of the Standard Model.

Fermilab fixed target experiments using incident photons and hadrons have played a major role in the understanding of charm decays. Over the past 10 years the charm yield in these experiments has increased by 3 orders of magnitude [2]. Some of the highlights of the Fermilab program have been the best measurements of the lifetimes for all weakly decaying charm particles [3], the observation of several new charm baryon channels including the Ω_c^0, large samples of semileptonic decays which have led to the best value for the CKM matrix element $|V_{cs}|$ and charm form factors [4], and the best limits for $D^0 - \overline{D^0}$ mixing [5].

As Fermilab looks toward the future, approved experiments (E781 [6] and FOCUS) believe that they can increase the existing charm statistics by at least a factor of 10 with improved signal to noise. The FOCUS experiment plans for an even larger increase ($\times 30$) for semileptonic decays due to an improved detector. The goal is to have at least 1 million fully reconstructed charm particles. The increased statistics will allow these experiments to improve many programmatic measurements and push limits or observe states on topics that would indicate new physics. The principal competition will likely be from CLEO and possibly from BEPC on the D_s meson.

FOCUS PHYSICS GOALS

The physics goals of the FOCUS experiment can be placed in two categories, programmatic measurements and searches for new physics.

The programmatic measurements involve high precision studies of D semileptonic decays, measurements of leptonic decays to measure the pseudoscalar decay constant, better determinations of absolute branching fractions, high statistics Dalitz plot analyses, and studies of charm production dynamics. In addition improvements will be made in production cross sections for the J/ψ and ψ' and hopefully, the first measurement of the Υ-nucleon cross section will be completed. FOCUS will also create over 10,000 double charm events, 20,000 $D_s^+ \rightarrow K^+K^-\pi^+$, 20,000 $\Lambda_c^+ \rightarrow pK^-\pi^+$, and large, clean samples of p wave D states. The FOCUS experiment is unique in its ability to identify neutrons and K_L^0's. Searches will be made for doubly-charmed baryons and for excited charmed baryon states decaying to Λ_c^+.

The new physics studies involve searches for CP violation, D^0-$\bar{D}^0$ mixing, rare decays of $D^0 \rightarrow \mu^+\mu^-$ and $D^0 \rightarrow e^+e^-$, and forbidden decays such as $D^0 \rightarrow \mu^+e^-$.

In order to accomplish the goals of the experiment a modification to the

E687 beamline [7] and spectrometer [8] are required. In fact the plan was to have a 5× increase in photon flux and a factor of 2 improvement in the data acquisition livetime and the detector efficiency. Each of these areas will be discussed. A photon beam is selected over a hadron beam because there are fewer primary tracks created in a photon interaction than in a hadron interaction. It is also better to have a neutral particle passing through the detector rather than a charged particle. Another advantage is the relatively larger percentage of charm production compared to the total hadronic cross section. The main drawback of a photon beam is the intense background of e^+e^- pairs.

FOCUS BEAMLINE

The FOCUS beamline is presented in figure 1. It is a multi-step beamline in order to produce a clean photon beam. The beamline begins by impacting 800 GeV/c protons onto a liquid deuterium productin target. The charged

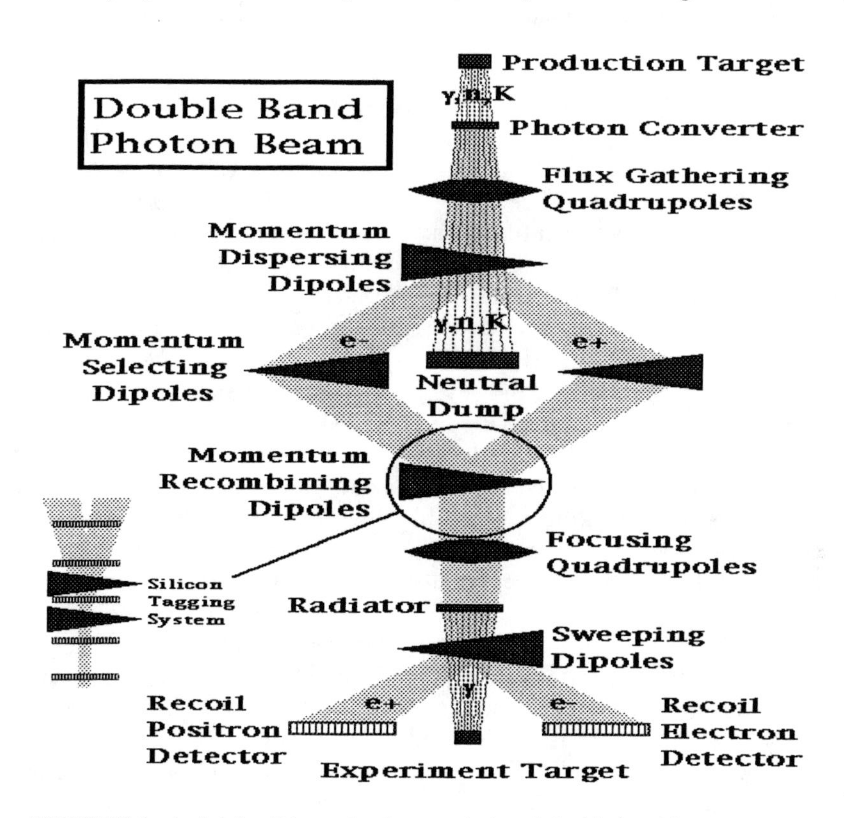

FIGURE 1. A sketch of the main characteristics of double band beam of FOCUS.

particles created in this interaction are swept away leaving the neutral particles of γ, n, K_L^0, and a small background from Λ's. These neutral particles strike a lead converter where the photons are converted to e^+ and e^-. The uninteracted neutral particles are stopped in the 'neutral' dump. The converted electrons and positrons are gathered and symmetrically transported around the neutral dump. The electrons and positrons are momentum selected, and then placed back together after the neutral dump. At this point the electrons' and positrons' momenta are measured with silicon strip detectors located in the beam. The electrons and positrons are focussed to the experimental target, but when the beams strike the radiator bremmstrahlung photons are created. These bremmstrahlung photons form the high energy photon beam. By measuring on each event the incoming electron beam momentum and the outgoing recoil electron momentum, the energy of the photon beam is "tagged."

By running the secondary beam at a 250 GeV setting, we achieve an electron/positron flux of approximately 10^{-4} per incident 800 GeV proton. The factor of 5 increase in photon flux over the E687 experiment (which only used an electron beam) is achieved by transporting both positrons and electrons, by lowering the secondary energy, and by running at higher intensities.

FOCUS SPECTROMETER

The E687 spectrometer had 300 threshold Cerenkov cells arranged in 3 counters; a 10-meter neutral-Vee decay volume; gamma and pizero identification; electron and muon identification; a large charged particle acceptance covering the entire forward hemisphere; and most importantly, 12 microstrip planes with a total of 8256 pulse-height-analyzed strips. More than 40 papers from the E687 collaboration have been published on charm results from the 1990-1991 Fixed Target run at Fermilab.

The E687 spectrometer required several upgrades to handle the increased rate. To reduce the accidental background caused by the coincidence of very high rate of e^+e^- pairs with beam halo muons, the inner muon system was changed to fast scintillator detectors and the outer muon system was changed to fast resistive plate chambers. An added benefit is that each system has been more highly segmented than in E687. The gas Iarocci tubes used for the E687 hadron calorimeter were replaced with scintillator. The proportional wire chambers (PWC's) were rebuilt to have lower gain and less noise, and the gate width for the flash-ADC system has been reduced from 150ns to 120ns in the silicon-microstrip system. Timing information was added to each Cerenkov cell to achieve single bucket resolution. Yet, the greatest change has occurred in the data acquisition electronics. The slow E687 time to digital converters (TDC's) for the PWC's were replaced by fast latches. This change facilitated a faster readout. The readout time is now completed in under 70μs and we have plans to reduce it further. A schematic drawing of the FOCUS

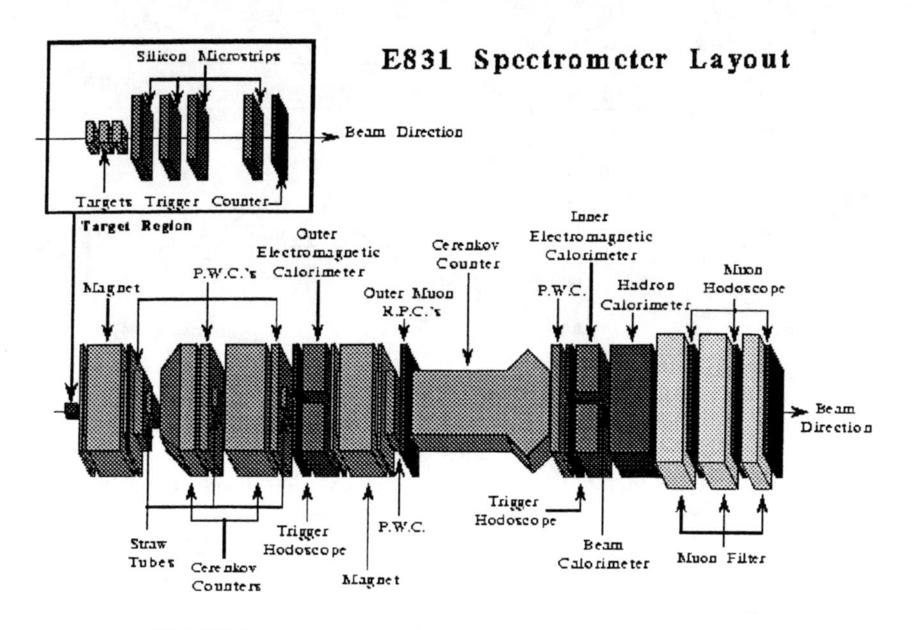

FIGURE 2. A schematic layout of the FOCUS spectrometer.

spectrometer is presented in Fig. 2.

Other significant upgrades to improve efficiency involved replacing the inner lead-scintillator sandwich electromagnetic calorimeter with an 802 element lead glass array arranged in a tower geometry, segmenting the experimental target, and inserting silicon microstrip planes into the target region.

The target segmentation is a major improvement. We discovered that decays outside the target are almost entirely due to charmed particles. Inside the target we have additional secondary interactions and background from strange-particle production. Outside the target there is a factor of four or more improvement in signal to noise for most charm final states. In E687 we used a 4 cm long solid-beryllium target. In this configuration only 14% of the reconstructed D^0's decay in a material-free region. Table 1 shows the percentage of decays for 3 different charmed particles that decay outside of the target material for five different target configurations. The columns labelled element and gap refer to the material thickness and space(gap) between material. Due to the higher density of diamond, it is possible to increase the number of D^0's decaying outside of the target by more than a factor of three. We can also increase the number of D^+'s decaying outside of the target material by a factor of two and the number of Λ_c^+'s decaying outside of the target material by a factor of four. With a diamond target we could significantly improve the signal-to-noise ratios for our signals, but BeO is much cheaper and nearly as good in maximizing the number of decays in air. Hence, we have now installed

TABLE 1. Percentage of Decays Outside of the Target

Material	D^0 Decays	D^+ Decays	Λ_c^+	Element	Gap
Be	14.3%	32.9%	5.0%	4.0	0
Be(2 seg)	23.5%	44.8%	9.73%	2.0	2.0
Be(4 seg)	37.7%	54.7%	18.8%	1.0	1.0
Dia(4 seg)	52.4%	70.2%	28.6%	0.625	1.5
BeO(4 seg)	50.1%	68.3%	26.7%	0.675	1.5

BeO targets.

The target microstrips are also a major improvement in FOCUS. The construction of the target microstrip detectors was particularly challenging as there was very little available space in which to extract the electronic signals. Two x-y doublets, one after each pair of target BeO segments have been installed. Each doublet consists of two nearly identical planes. The silicon chips are wire bonded onto a quartz substrate and flexible FR4 ribbons carrying the signals are wire bonded to the quartz. Each silicon chip (plane) is 5 cm long and contains 1024 25-μm strips. The additional information from these doublets reduces the experiment's transverse spatial resolution at the target by more than a factor of two and the mass resolution by 10% or more. These additional silicon planes enable us to better isolate the charm decays and to extend the geometrical acceptance. The planes further allow us to track a portion of the long-lived charged D decays. Table 2 shows the efficiency for fully reconstructing $D^0 \rightarrow K^-\pi^+\pi^+\pi^-$ as a function of l/σ (significance of separation of the secondary decay vertex from the primary vertex) for the new microstrip system and for the E687 system. These numbers have been determined from a full simulation using the well-tested E687 Monte Carlo program. With the improved spatial resolution we expect gains as large as a factor of three for severe lifetime cuts. An added benefit is that the background is lower and the mass resolution is improved.

TABLE 2. The FOCUS system compared to the E687 system for $D^0 \rightarrow K^-\pi^+\pi^+\pi^-$

	E687			FOCUS		
l/σ cut	Yield	Effic. (%)	Mass Res. (MeV)	Yield	Effic. (%)	Mass Res. (MeV)
5	771.3±31.7	14.7	12.3	788.6±30.2	15.0	10.7
10	469.5±23.2	9.0	13.0	647.5±26.6	12.3	11.5
20	189.5±14.3	3.6	14.4	425.3±21.2	8.1	11.8
30	89.8±9.7	1.7	14.2	264.1±16.8	5.0	12.8

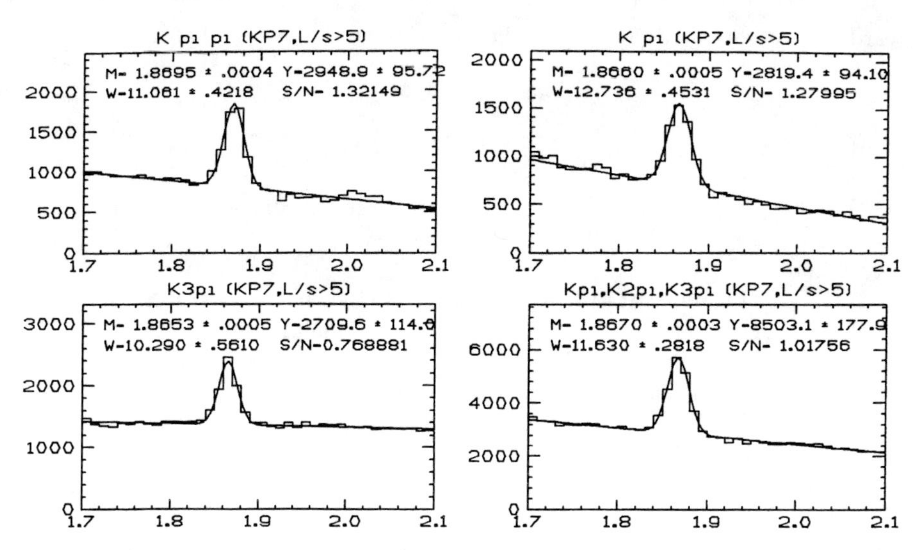

FIGURE 3. Invariant mass distributions for three charmed decay channels plus the summed plot.

PRELIMINARY RESULTS

FOCUS began accumulating data in September, 1996 and at the time of the conference the experiment had accumulated about $3\times$ the data of E687. The experiment plans to run continuously through September, 1997 except for a scheduled Fermilab shutdown during Christmas, 1996 and in late March, 1997. Already the experiment is observing large charm signals and we are well on our way to achieving our goal of $\times 10$ increase over the reconstructed charm of E687. It seems highly likely that we will be able to accomplish our physics goals mentioned earlier. In figure 3 we present charm signals based on on-line express results. The signals are observed using the same l/σ cuts for each channel and are based on the processing of 108 Million events. We anticipate that the signals will be come clearer and narrower as our analysis techniques are refined.

CONCLUSIONS

At the time this talk was written the FOCUS experiment had already accumulated more than $3\times$ the charm of E687. It is very likely the FOCUS experiment will be able to fully reconstruct 1 million charmed particle decays and to address the physics issues discussed earlier. The experiment looks forward to presenting the results of the experiment at the next Latin American Symposium.

REFERENCES

1. Fermilab Proposal P831, "A High Statistics Study of States Containing Heavy Quarks Using the Wideband Photon Beam and the E687 Multiparticle Spectrometer."
2. J. Appel, "Experimental Issues in High Sensitivity Charm Experiments", in the Proceedings of the CHARM2000 Workshop at Fermilab June 7-9, 1994.
3. J. Cumalat, Proceedings of the LAFEX International School in High Energy Physics in Rio de Janeiro in 1995 summarizing charm lifetime data.
4. J.C. Anjos *et al*, Phys. Rev. Lett. **62**(1989) 1587; P.L. Frabetti *et al* , Phys. Lett. **B315**,203 (1993); P.L. Frabetti *et al*, Phys. Lett. **B364**,127(1995).
5. E791 Collaboration, E.M. Aitala, Fermilab PUB-96/214-E; Fermilab PUB 96-109-E.
6. J. Russ *et al*, Fermilab Proposal P781,"Study of Charm Baryon Physics"; This experiment uses a hyperon beam to enhance production of charm-strange particles.
7. P.L. Frabetti *et al*, Nucl. Inst. Meth. **A329**, 62(1993)
8. P.L. Frabetti *et al*, Nucl. Inst. Meth. **A320**, 519-547(1992)

Possible Evidence for Sparticles from Collider Data, and Some Implications

G.L. Kane[1]

Randall Lab of Physics, University of Michigan, Ann Arbor, MI 48109-1120

Abstract. We examine the CDF $ee\gamma\gamma \not{E}_T$ event as a candidate for sparticle production. Possible connections to other observables such as R_b, α_s, $b \to s\gamma$, cold dark matter are briefly considered, and also implications for present and future data at FNAL and LEP, including the possibility that stops, gluinos and squarks are already being detected at FNAL. An analysis of the resulting Higgsino-like LSP as a cold dark matter candidate gives very encouraging results.

INTRODUCTION

The theoretical motivation for nature being supersymmetric on the weak scale is very strong. The possibility of unifying the SM forces, of relating them to gravity, and of explaining the Higgs mechanism are exciting. Perhaps the most important aspect of SUSY is connecting theory at the unification and Planck scales with experimentation at colliders. Close behind is that SUSY provides a non-baryonic cold dark matter candidate, which the SM does not.

There are several reasons why it is important to pursue all hints of direct sparticle production. (a) SUSY manifests itself in rather subtle ways, and often the confirmation of one signal can be a different one related by the theory. (b) If light sparticles exist there are major implications for utilization of FNAL and LEP, and for planning for future facilities. (c) Once there is data on sparticle masses and couplings we can test whether the LSP can indeed be a cold dark matter candidate. (d) Most important, perhaps, is that with data we can begin to construct the effective Lagrangian of the theory at the electroweak scale. Then we can relate it to the effective Lagrangian at the unification scale and begin to make detailed experimental contact with fundamental theories at the Planck scale.

Recently a candidate for sparticle production has been reported [1] by the CDF group. This has been interpreted in several ways [2–5] and later with

---

[1] Based on work in collaboration with S. Ambrosanio, C. Kolda, G. Kribs, S. Martin, S. Mrenna, and J. Wells.

CP400, *First Latin American Symposium on High Energy Physics/ VII Mexican School of Particles and Fields,* edited by D'Olivo/Klein-Kreisler/Mendez

additional variations [6–8]. The main two paths are whether the LSP is the lightest neutralino [2], [9], or a nearly massless gravitino [2-7] or axino [8]. In the gravitino or axino case the LSP is not a candidate for cold dark matter, SUSY can have no effect on R_b or α_s^Z or $BR(b \to s\gamma)$, and stops and gluinos are not being observed at FNAL. In the case where the lightest neutralino is the LSP (NLSP) the opposite holds for all of these observables, so it is very interesting to pursue this case in detail and that is what I will do for the remainder of this talk.

Recently the CDF group has suggested that the "positron" in the above event may be a τ. That does not have much effect on the interpretation, because the main conclusions come from the presence of the photons and missing energy. The photons tell us that the second lightest neutralino is the one produced, and that it has a large branching ratio to the lightest neutralino and a photon. The missing energy constrains the masses, and is a marker for SUSY.

The SUSY Lagrangian depends on a number of parameters. Since all of them have the dimension of mass that should not be viewed as a weakness because at present we have no theory of the origin of mass parameters. Probably getting such a theory will depend on understanding how supersymmetry is broken. When there is no data on sparticle masses and couplings it is appropriate to make theoretical simplifying assumptions to reduce the number of parameters, but once there may be data it is important to measure the parameters and to see what patterns emerge. We will proceed by making no assumptions about soft-breaking parameters. In practice even though the full theory has over a hundred such parameters that is seldom a problem since any given observable depends on at most a few.

After describing the data and its implications we can examine to the properties of the "observed" LSP and whether it can be the cold dark matter of the universe. The answer is yes.

NLSP INTERPRETATION OF THE $ee\gamma\gamma$ EVENT

The CDF event [1] has a 36 GeV e^- , a 59 GeV e^+ (or τ), photons of 38 and 30 GeV, and $\not{E}_T = 53$ GeV (these are transverse energies). The first question of course is whether there can be a SM interpretation — the largest contribution thought of so far is that the event is $W(\to e\nu)W(\to e\nu)\gamma\gamma$. We [2] estimate the rate is a little over 10^{-4} for such an event in 100 pb^{-1}. Further, no other events have been reported with $\gamma\gamma$ accompanied by possible WW decay products. On the other hand, $\sigma \times BR$ for selectron pair production is of order one event for 100 pb^{-1} (more precisely, $\sigma \times BR$ is over $\frac{1}{2}$ for a significant part of parameter space) and the detection efficiency for such an event is 5-25%

The SUSY interpretation is then $q\bar{q} \to \gamma, Z \to \tilde{e}^+\tilde{e}^-$, followed by each

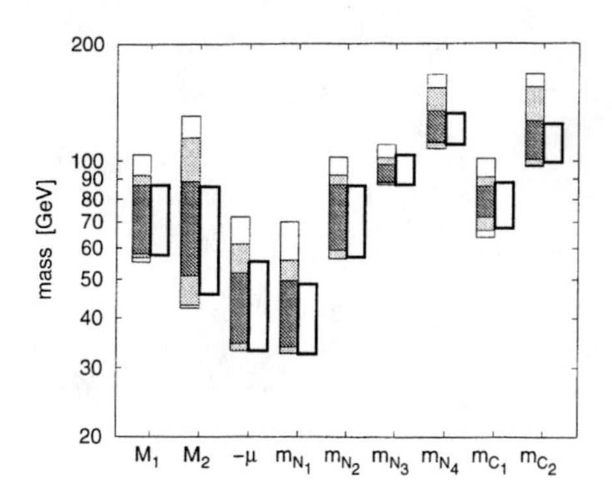

FIGURE 1. The allowed mass spectrum is shown for all models (shaded bands on the left) and for $\tilde{e}_R$ models only (thick solid outline on the right). The increasingly darker shades in the left-hand column correspond to increasing stricter cuts on $\mathcal{A} = 5$, 7.5, 10 fb. As for $\tan\beta$, the allowed range in all models is $1.0 < \tan\beta < (2.8, 2.6, 1.8)$ for $\mathcal{A} = 5$, 7.5, 10 fb respectively. The allowed range of $\tan\beta$ in $\tilde{e}_R$ models only is $1.0 < \tan\beta < 2.0$. $\mathcal{A} \equiv \sigma BR^2$. N_i are the four neutralino mass eigenstates and C_i the two chargino mass eigenstates.

$\tilde{e}^{\pm} \to e^{\pm} N_2$, followed by $N_2 \to \gamma N_1$. The second lightest neutralino, N_2, must be photino-like since it couples strongly to $\tilde{e}e$. Then the LSP=N_1 must be higgsino-like [10–12] to have a large $BR(N_2 \to N_1\gamma)$. Ref. 9 summarizes the resulting masses and couplings for charginos and neutralinos in detail, under various assumptions. Fig. 1 shows allowed ranges (keep in mind that values are correlated).

If the event is $e\tau\gamma\gamma$, there are several possibilities. It could be $q\bar{q} \to C^+C^-$, with $C^+ \to W^+(\to \tau^+\nu)N_2(\to \gamma N_1)$, $C^- \to W^-(\to e^-\nu)N_2(\to \gamma N_1)$, or $C^+ \to \tilde{\tau}(\to \tau^+ N_2(\to \gamma N_1))\nu_\tau$, $C^- \to \tilde{e}(\to e N_2(\to \gamma N_1))\nu_e$. The W's can be virtual.

PREDICTIONS FOR FNAL, LEP

If light superpartners indeed exist, as in Fig. 1, FNAL and LEP will produce thousands of them, and measure their properties very well. It will be important to ensure these facilities are fully utilized.

The first thing to check at FNAL is whether the produced selectron is $\tilde{e}_L$ or $\tilde{e}_R$. If $\tilde{e}_L$, then the charged current channel $u\bar{d} \to W^+ \to \tilde{e}_L\tilde{\nu}$ has 5-10 times the rate of $\tilde{e}_L^+\tilde{e}_L^-$, so some events might be in the present sample. We expect $\tilde{e}_L \to eN_2(\to \gamma N_1)$. If $\tilde{\nu}$ is heavier than C_1, [9] $\tilde{\nu} \to eC_1$. If $m_{\tilde{t}} < m_{C_1}$, then $C_1 \to \tilde{t}(\to cN_1)b$ so $\tilde{\nu} \to ebcN_1$; if $m_{\tilde{t}} > m_{C_1}$ then $C_1 \to W^*(\to jj)N_1$ so $\tilde{\nu} \to ejjN_1$, where $j = u, d, s, c$. Either way, dominantly $\tilde{e}_L\tilde{\nu} \to ee\gamma JJ\not{E}_T$ where J may be light or heavy quarks. If no such signal appears probably the produced selection was $\tilde{e}_R$. Also, $\sigma(\tilde{\nu}\tilde{\nu}) \cong \sigma(\tilde{e}_L\tilde{e}_L)$. Alternatively, if $\tilde{\nu}$ is lighter than C_1, $\tilde{\nu} \to N_1\nu$ and is invisible (or $\tilde{\nu} \to N_2\nu$ if N_1 has no zino component).

The most interesting channel at FNAL is $u\bar{d} \to W^+ \to C_i^+ N_2$. This gives a signature $\gamma J J \not{E}_T$, for which there is no parton-level SM background. If $\tilde{t} < C_i$ one of J is a b. If $\tilde{\nu} < C_i$ then $C_i^\pm \to \tilde{\nu}\ell^\pm$ also. If $t \to \tilde{t}N_2$ (expected about 10% of the time) and if $\tilde{q}$ are produced at FNAL there are additional sources of such events (see below). They could be the best way to confirm SUSY in existing data. Fig. 2 shows the cross sections for charginos and neutralinos at FNAL.

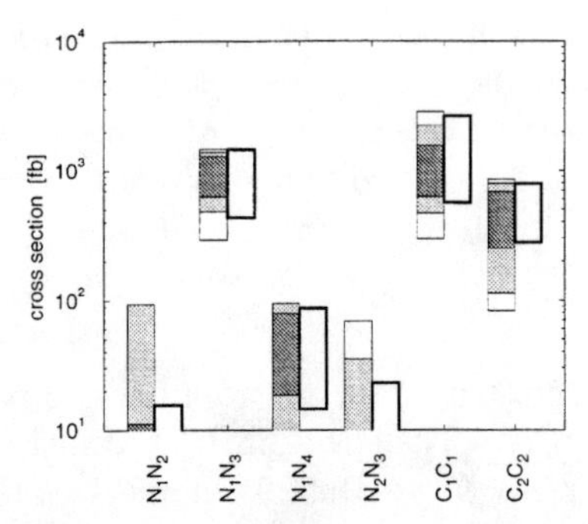

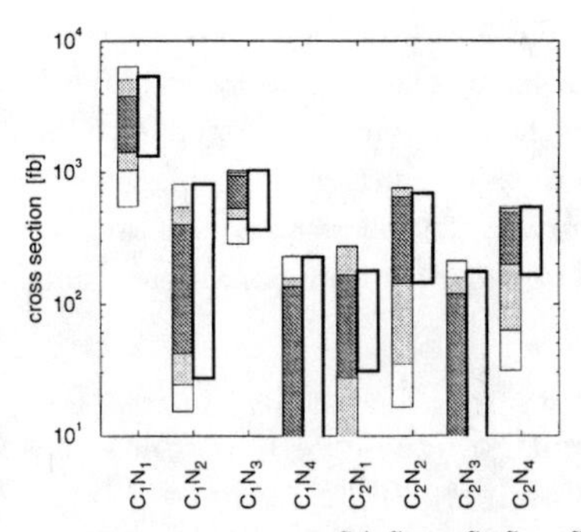

FIGURE 2. As in Fig. 1, for Tevatron $\sqrt{s}$ = 1.8 TeV with all $\tilde{C}_i^\pm\tilde{C}_j^\mp$, $\tilde{N}_i\tilde{N}_j$, $\tilde{C}_i^\pm\tilde{N}_j$ processes shown that can have cross sections larger than about 50 fb.

Several processes could occur at LEP161, and many must occur at LEP190 [9]. The most likely one is $e^+e^- \to N_1N_3(\to Z^*(\to q\bar{q})N_1)$, giving $q\bar{q}\not{E}$ events. The $m(q\bar{q})$ is at most $m_{N_3} - m_{N_1} \lesssim$ 65 GeV. However, if N_3 is heavier than $\tilde{\nu}$, $N_3 \to \tilde{\nu}\nu$ dominates. Then if C_1 is heavier than $\tilde{\nu}$, $\tilde{\nu}$ may decay invisibly to $N_1\nu$, or to $N_2\nu$ followed by $N_2 \to N_1\gamma$. If the C_1 mass is in the allowed region $e^+e^- \to C_1^+C_1^-$ will occur, often giving $JJJJ\not{E}$ events ($b\bar{b}c\bar{c}$ if $\tilde{t} < C_1$, otherwise $jjjj$ with $j = u, d, s, c$). However, if C_1 is heavier than $\tilde{\nu}$ other channels are open.

The process $e^+e^- \to N_2N_2$ gives $\gamma\gamma\not{E}$ with missing mass $\gtrsim m_Z$. Fig. 3 shows the LEP cross sections. This has a very good signature. It is very interesting, because the neutralino-LSP world gives excess events in this channel, all with invariant missing mass above the Z, while the gravitino-LSP world gives events with invariant missing mass of any value. There should be no background below the Z, so any confirmed events there both imply new physics and the very light LSP world. There is some background above the Z, but a substantial number of events there would imply new physics and the neutralino-LSP world.

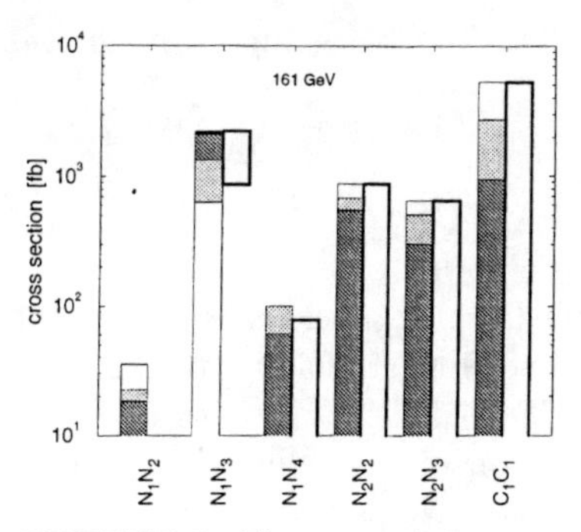

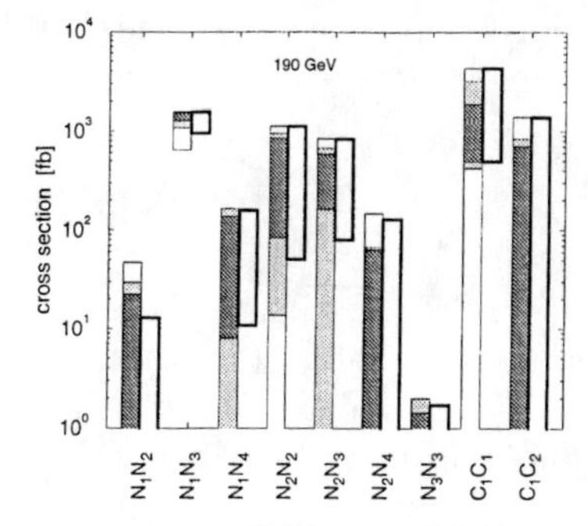

FIGURE 3. The range of the non-negligible cross sections at LEP161 and LEP190.

GLUINOS, SQUARKS AT FNAL?

If charginos and neutralinos and selectrons are light then gluinos and squarks should not be too heavy. If stops are light, then $BR(t \to \tilde{t}N_i) \lesssim 1/2$ [13], in which case extra tops must be produced because $\sigma \times BR(t \to Wb)$ is near or above its SM value with $BR(t \to Wb) = 1$. With these motivations, we [14] have suggested that one assume $m_{\tilde{g}} \geq m_t + m_{\tilde{t}}$ and $m_{\tilde{q}} \geq m_{\tilde{g}}$. Then there are several pb of top production via channels $\tilde{q}\tilde{g}, \tilde{g}\tilde{g}, \tilde{q}\bar{\tilde{q}}$ with $\tilde{q} \to q\tilde{g}$, and $\tilde{g} \to t\tilde{t}$ since $t\tilde{t}$ is the gluino's only two-body BR. Analysis shows that this scenario is at least as consistent as the SM with published data, and perhaps more so in that the distribution for $P_T(t\bar{t})$ should peak at small P_T for the SM but at larger P_T for the SUSY case since the system is recoiling against extra jets in the SUSY case. To be quantitative, the fraction of events expected for $P_T(t\bar{t}) > 15$ GeV is 35% for the SM vs. 63% for SUSY; the reported CDF data has 71% beyond $P_T(t\bar{t}) = 15$ GeV. The SUSY case suggests that if m_t or $\sigma_{t\bar{t}}$ are measured in different channels one will obtain different values, which is also consistent with reported data. This analysis also shows that the present data is consistent with $BR(t \to \tilde{t}N_i) = 1/2$. That this analysis does not lead to contradictions with data adds additional support to the whole picture.

$R_b, \alpha_s, BR(b \to s\gamma)$

At present R_b and $BR(b \to s\gamma)$ differ from their SM predictions by 1.5-2σ, and (in my opinion) α_s measured by the Z width differs by about 1.5-2σ from its value measured other ways. If these effects are real they can be explained by $C_i - \tilde{t}$ loops [16]. What is particularly relevant here, and exciting, is that they can be explained by precisely the SUSY parameters deduced from the $ee\gamma\gamma$ event (+ a light, mainly right handed, stop). Although $\tan\beta, \mu$, and M_2 a priori could be anything, they come out the same from the analysis of these loops and from $ee\gamma\gamma$ ($\tan\beta \leq 1.5, \mu \sim -m_Z/2, M_2 \sim 60-80$ GeV). While we cannot fully interpret this until these effects are better determined, this agreement is very encouraging.

COLD DARK MATTER

The LSP=N_1 apparently escapes the CDF detector in the $ee\gamma\gamma$ event, suggesting it is stable (though only proving it lives longer than $\sim 10^{-8}$ sec). If so it is a candidate for cold dark matter (CDM). The properties of N_1 are deduced from the analysis [9] so the calculation of the relic density [17] is a "no-parameter" one. The analysis shows that the s-channel annihilation through the Z dominates, so the needed parameters are $\tan\beta$, m_{N_1} and the higgsino fraction for N_1, which is large. The results are very encouraging, giving $0.1 \leq \Omega h^2 \leq 1$, as shown in Fig. 4.

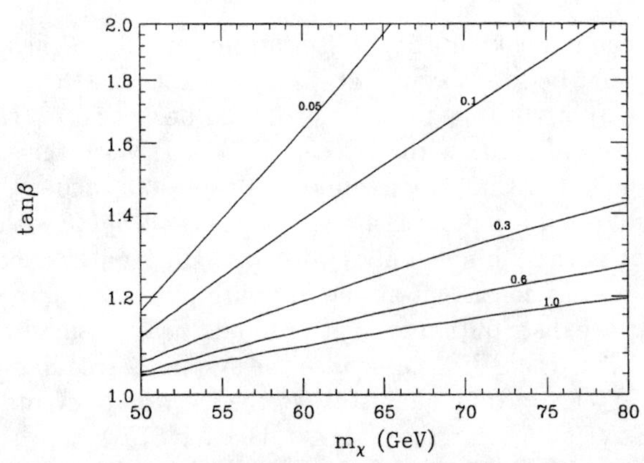

FIGURE 4. Contours of constant Ωh^2 for the Higgsino-like LSP described in the text (solid).

The central value is $\Omega h^2 \simeq 1/4$. Thus the CDM of the universe may have been observed at FNAL — the particle that makes up the CDM is approximately the superpartner of the Higgs boson. Such calculations can give Ωh^2 very large compared to 1 or very small; that this gives about the right relic density for a flat universe is further encouragement that the entire picture may be correct.

At FNAL, of course, it can be demonstrated that N_1 escapes the detector but not that it is stable. We have checked [17] that the prospects for detecting N_1 in "direct" experiments are not bad. For example, in ^{73}Ge events would be seen in the worst case at a few hundred kg-days, and for some values of m_{N_1} and $\tan\beta$ in the preferred region the event rate could be 10 times larger. Flourine and other nuclei give larger rates. While such rates are not large, they are probably eventually detectable in direct experiments, and the direct detection of cold dark matter is so important that is is worth a lot of effort.

Freese and Kamionkowski [18] have looked at detection via N_1 annihilation to energetic neutrinos in the Sun, and conclude the rate for underground detectors could be within an order of magnitude of today's limits. See also the recent paper of ref. [19] for study of labroatory detections of the LSP; however, beware of increased rates from rescaling cases with small Ωh^2.

SUMMARY

The $ee\gamma\gamma$ (or $e\tau\gamma\gamma$) event may be the first direct observation of superpartners. It is encouraging that the higgsino-like LSP deduced from the analysis of the event gives a relic density $\Omega h^2 \sim 1/4$, that one can explain the 1.5-2σ effects in R_b, α_s, and $BR(b \to s\gamma)$, and that there may be additional evidence for $\tilde{t}, \tilde{g}$, and $\tilde{q}$ production at FNAL. The consistency of this picture need not have occurred. We have given a number of predictions for existing and future FNAL and LEP data to test the correctness of this picture.

ACKNOWLEDGMENTS

This research was supported in part by DoE. I am very grateful to my collaborators Sandro Ambrosanio, Graham Kribs, Steve Martin, Steve Mrenna and Jim Wells and to A. Blondel, A. Buras, M. Carena, S. Pokorski and C. Wagner for helpful conversations. I wold like to thank L. Diaz Cruz, M. Perez, A. Garcia, and R. Huerta very much for their warm hospitality.

REFERENCES

1. S. Park, "Search for New Phenomena in CDF", 10th Topical Workshop on Proton–Anti-proton Collider Physics, edited by Rajendran Raja and John Yoh, AIP Conf. Proc. No. 357 (AIP, New York, 1996); L. Nodulman, "New Particle Searches at CDF", Proceedings of the International Europhysics Conference on High Energy Physics, edited by J. Lemonne et al. Brussels, 1995.
2. S. Ambrosanio, G. L. Kane, G. D. Kribs, S. P. Martin, and S. Mrenna, Phys. Rev. Lett. **76** (1996) 3498.
3. S. Dimopoulos, M. Dine, S. Raby, and S. Thomas, Phys. Rev. Lett. **76** (1996) 3494.
4. S. Dimopoulos, S. Thomas, and J. D. Wells, Phys. Rev. **D54** (1996) 3283.
5. S. Ambrosanio, G. L. Kane, G. D. Kribs, S. P. Martin, and S. Mrenna, Phys. Rev. D **54** (1996) 5395.
6. K. Babu, C. Kolda, and F. Wilczek, Phys. Rev. Lett. **77** (1996) 3070.
7. J. L. Lopez and D. V. Nanopoulos, Mod. Phys. Lett. **A10** (1996) 2473.
8. J. Hisano, K. Tobe, and T. Yanagida, Phys. Rev. **D55** (1997) 411.
9. S. Ambrosanio, G.L. Kane, G. Kribs, S. Martin, S. Mrenna, hep-ph/9607414.

10. H: Komatsu and J. Kubo, Phys. Lett. **157B** (1985) 90; Nucl. Phys. **B263** (1986) 265; H. E. Haber, G. L. Kane, and M. Quirós, Phys. Lett. **160B** (1985) 297; Nucl. Phys. **B273** (1986) 333.
11. H. E. Haber and D. Wyler, Nucl. Phys. **B323** (1989) 267.
12. S. Ambrosanio and B. Mele, Phys. Rev. **D52** (1995) 3900; Phys. Rev. **D53** (1996) 2541
    S. Ambrosanio and B. Mele, ROME1-1148/96.
13. J.D. Wells, C. Kolda, G. L. Kane, Phys. Lett. B **338** (1994) 219.
14. G. L. Kane and S. Mrenna, Phys. Rev. Lett. **77** (1996) 3502.
15. See the Rapporteur talks of A. Blondel and A. Buras at the XXVIII International Conference on HEP, Warsaw, July 1996.
16. A. Djouadi et al. Nucl. Phys. **B349** (1991) 48; M. Boulware, D. Finnell, Phys. Rev. **D44** (1991) 2054; J. D. Wells, C. Kolda, G. L. Kane, Phys. Lett. B **338** (1994) 219; D. Garcia, J. Sola, Phys. Lett. B **354** (1995) 335; G. L. Kane, R. G. Stuart, J. D. Wells, Phys. Lett. B **354** (1995) 350; A. Dabelstein, W. Hollik, W. Mösle, hep-ph/9506251; M. Carena, H. E. Haber, and C. E. M. Wagner, hep-ph/9512446. P. H. Chankowski, S. Pokorski, Phys. Lett. B **366** (1996) 188; J. Ellis, J. Lopez, D. Nanopoulos, Phys. Lett. B **372** (1996) 95; J. D. Wells, G. L. Kane, Phys. Rev. Lett. **76** (1996) 869; E. H. Simmons, Y. Su, hep-ph/9602267; P. H. Chankowski, S. Pokorski, hep-ph/9603310.
17. G. L. Kane and J. D. Wells, Phys. Rev. Lett. **76** (1996) 4458.
18. K. Freese and M. Kamionkowski, hep-ph/9609370; K. Freese, private communication.
19. A. Bottino, N. Forneugo, G. MIgrida, M. Olechowski, and S. Scopel, astro-ph/9611030.

Francium Spectroscopy: Towards a Low Energy Test of the Standard Model

L. A. Orozco, J. E. Simsarian, G. D. Sprouse and W.Z. Zhao

Department of Physics, State University of New York, Stony Brook, NY 11794-3800, USA.

Abstract.
An atomic parity non-conservation measurement can test the predictions of the standard model for the electron-quark coupling constants. The measurements, performed at very low energies compared to the Z^0 pole, can be sensitive to physics beyond the standard model. Francium, the heaviest alkali, is a viable candidate for atomic parity violation measurements. The extraction of weak interaction parameters requires a detailed knowledge of the electronic wavefunctions of the atom. Measurements of atomic properties of francium provide data for careful comparisons with *ab initio* calculations of its atomic structure. The spectroscopy, including energy level location and atomic lifetimes, is carried out using the recently developed techniques of laser cooling and trapping of atoms.

INTRODUCTION

The standard model of the fundamental interactions has been extremely successful in describing experimental observations. Not only can it explain qualitatively a wealth of phenomena, but its quantitative predictions are impressive. The precision tests of the standard model performed at CERN, SLAC and Fermi Lab by looking at the $Z^0, W^\pm$ resonances have increased our knowledge of the electro-weak interactions to an unprecedented level (see for example the recent book edited by P. Langacker [1]). The search for small quantitative deviations that hint to new physics is the center of the continuing effort of many experimental groups in the high-energy community.

Despite the wealth of data that can be explained by the standard model, it can not be considered the final theory. The model has more than fifteen free parameters. New physics is likely to exist.

A phenomenon that provides an interesting test of the standard model is atomic parity non-conservation, a manifestation of the Z^0 heavy neutral gauge

CP400, *First Latin American Symposium on High Energy Physics/ VII Mexican School of Particles and Fields*, edited by D'Olivo/Klein-Kreisler/Mendez

boson. In recent years, the field has become a valuable source of information about electroweak unification and it has become an important chapter of non-accelerator particle physics [2]. Ongoing experiments by the group of Prof. Carl Wieman [3] of the University of Colorado and JILA at Boulder have achieved experimental precisions of a fraction of a percent. The results obtained in atomic parity non-conservation are important tests of the radiative electroweak corrections and are very sensitive to extensions of the standard model [4,5]. Despite being a low-energy experiment, atomic parity non-conservation is sensitive to additional gauge bosons. Such experiments can provide evidence for Z' bosons via deviations from standard model predictions. Rosner [5] has stressed the importance of atomic parity non-conservation and shows how a measurement with different isotopes can directly provide a competitive number for $\sin\theta_W$.

We present here our advances towards the implementation of an atomic parity non-conservation experiment in francium. We have focused our efforts on the atomic spectroscopy of francium. This atom is the heaviest alkali and the least studied of them. Its nucleus is the most unstable of the first 103 elements of the periodic table. The experiments utilize a unique mixture of atomic physics, optics, nuclear and particle physics. While other-low energy approaches are classified as non-accelerator, this experiment uses radioactive atoms produced in an accelerator facility, eroding the demarcation between low- and high-energy physics goals and techniques.

THE ELECTROWEAK THEORY AND PARITY NON-CONSERVATION

The parity violating interaction between the nucleus and an electron leads to a mixing of electronic states with opposite parity, in particular between S and P states. For example an S state will have a very small amount of a P state mixed with it (a typical number for a heavy atom is 1×10^{-11}). This results in an electric dipole transition amplitude between two states of the same parity, which would otherwise be forbidden.

We restrict the discussion to a vector-axial vector type weak force (V-A). Parity violation has its origin in the fact that the Z^0 gauge boson can couple to the vector or axial-vector components of the electronic and nucleonic currents. The coupling of the electronic axial vector current with the hadronic vector current dominates since all quarks contribute coherently. Only the valence nucleons contribute to the axial hadron current resulting in the nuclear-spin dependent part of this force. Strong interactions inside the nucleons and the nucleus renormalize the hadronic axial current, while the conserved vector current hypothesis guarantees that the hadronic vector current remains unmodified. For the remainder of this section we will consider the couplings $C_{1u,d}$ that describe the interaction of the electronic axial-vector current with the hadronic vector current.

Following Bouchiat [2], the weak charge Q_W of the atomic nucleus is proportional to the sum of the weak charges of all the nuclear constituents, the $(2Z+N)$ up quarks and $(Z+2N)$ down quarks:

$$Q_W = -2Z(2C_{1u} + C_{1d}) - 2N(C_{1u} + 2C_{1d}). \tag{1}$$

The tree-level Standard Model predictions for the coupling constants are:

$$C_{1u}^{tree} = -\frac{1}{2}\left(1 - \frac{8}{3}\bar{x}\right) \approx -0.19, \tag{2}$$

$$C_{1d}^{tree} = \frac{1}{2}\left(1 - \frac{4}{3}\bar{x}\right) \approx 0.34, \tag{3}$$

where $\bar{x} = \sin^2\theta_W(M_Z) \approx 0.23$. which yields

$$Q_W \approx -Z(0.08) - N(0.98). \tag{4}$$

The term proportional to the number of neutrons is the dominant term in the weak charge. Any experiment aiming to measure parity non-conservation in atoms tries to maximize the number of neutrons available.

The amount of P state mixing is proportional to the weak charge times an atomic matrix element, $\langle\gamma^5\rangle$. The interaction scales a little bit faster than Z^3 [2,6], favoring heavy atoms. The enhancement comes for various reasons: A heavy nucleus has more nucleons for the electron to exchange a Z^0, the electron spends more time in a larger nucleus with a larger charge, and has a higher momentum. The interaction Hamiltonian V is then:

$$V = \frac{G_F}{2\sqrt{2}} Q_W \gamma^5 \rho_N(\mathbf{r}) \tag{5}$$

which in the non-relativistic limit becomes:

$$V = \frac{G_F}{2\sqrt{2}} Q_W \frac{\mathbf{s_e}\cdot\mathbf{p_e}}{m_e c} \rho_N(\mathbf{r}) + h.c. \tag{6}$$

G_F is the Fermi constant, $\mathbf{s_e}, \mathbf{p_e}, m_e$ refer to the electron spin, momentum and mass, $\rho_N(\mathbf{r})$ is the nucleon density normalized to unity. All the quarks and nucleons inside the nucleus add their contributions coherently. The parity violating amplitude is proportional to Q_W via a coefficient that must come from an atomic calculation.

Although experiments in heavy non-alkali atoms have achieved very small uncertainties, the extraction of information relevant to the standard model is difficult due to the complexity of the atomic structure. Nevertheless, impressive progress has been made by the Berkeley, Oxford and Seattle groups

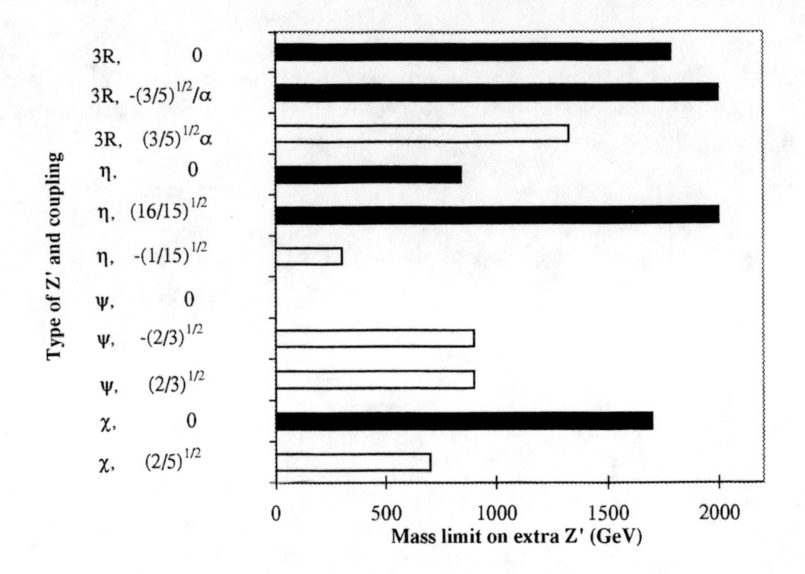

FIGURE 1. Range of sensitivity to the mass of extra Z' bosons for a future atomic parity non-conservation experiment. The Z' boson type and coupling appears next to each limit. The values come from Langacker *et al.* [12]. The solid bars indicate particular cases where atomic parity-non conservation can set the best limit.

working with non-alkali atoms [7–9]. However, the theory of atomic physics necessary to calculate the matrix element for the interaction is complicated. Parity non-conservation in alkali elements has the advantage of being quantitatively accountable. Theorists can compute better the structure of the alkali atoms with a core and a single electron in its outer shell, and have shown remarkable success in recent years [10,11].

We have chosen francium since it is the heaviest of the alkalis. Although it is not naturally available in large quantities, it is now possible to capture it on-line from an accelerator in a laser trap. It has a variety of isotopes where the effect can be measured. The calculated [11] enhancement factor with respect to cesium is eighteen due to its larger weak charge and to extra relativistic corrections.

The extensive analysis presented by Langacker *et al.* in 1992 [12] on high-precision electroweak experiments, shows the high sensitivity of parity non-conservation experiments to extra intermediate vector bosons Z'. The influence of these particles enters at the tree level. The sensitiviy of atomic parity non-conservation measurements can be higher than any other experimental method. The amount depends on the model studied and the coupling between the extra Z' boson to the Z^0. Fig. 1 compiles the limits imposed on the mass of extra Z' bosons by a future atomic parity non-conservation experiment using different isotopes of the same atom. The filled bars indicate that atomic parity non-conservation provides the best limit available for that particular boson and coupling. The values plotted come from the sensitivity of the parameter $C_{1+} = 0.666C_{1u} + 0.747C_{1d}$ assuming a future precision of 0.2%. The possible limits are tabulated by type of Z' and coupling as stated in Ref. [12].

For higher order corrections, Marciano and Rosner have analyzed the effects of physics beyond the standard model on electroweak observables. They use the Peskin-Takeuchi isospin-conserving S, and isospin-breaking T, parametrization of 'new' quantum loop corrections of the oblique kind [4]. They show atomic parity-violating experiments to be particularly sensitive to S. Lynn and Sandars [13] continues the analysis using a slightly different approach and obtains numbers similar to those of Marciano and Rosen. Atomic parity non-conservation depends very little on either the top or the Higgs mass.

An important generalization by C. P. Burgess and coworkers [14] is not restricted to the special class of effective interactions known as 'oblique' corrections. They have carried out a study of model-independent global constraints on new physics. Their analysis of atomic parity violation includes expressions for the weak charge, S and T corrections, as well as contributions due to flavor changing neutral currents. Taking the work of Burgess [14] as guide, expressions for the weak charges, with radiative and other corrections included, in francium and cesium are:

$$Q_W({}^{133}_{55}Cs) = -73.2 - 0.796S - 0.011T \pm 0.2 \tag{7}$$

$$Q_W({}^{210}_{87}Fr) = -115.4 - 1.258S - 0.015T \pm 0.3 \tag{8}$$

RADIOACTIVE ATOMS IN A LASER TRAP

Stony Brook has lead the field [15–17] of injecting nuclear reaction products into a magneto-optic trap (MOT). These recent developments have provided the possibility to study in detail properties of radioactive atoms such as francium. A cold, dense cloud of neutral atoms in a MOT, has many orders of magnitude compression in phase space. It is an ideal sample on which to perform detailed spectroscopic measurements, such as those needed for an atomic parity non-conservation measurement.

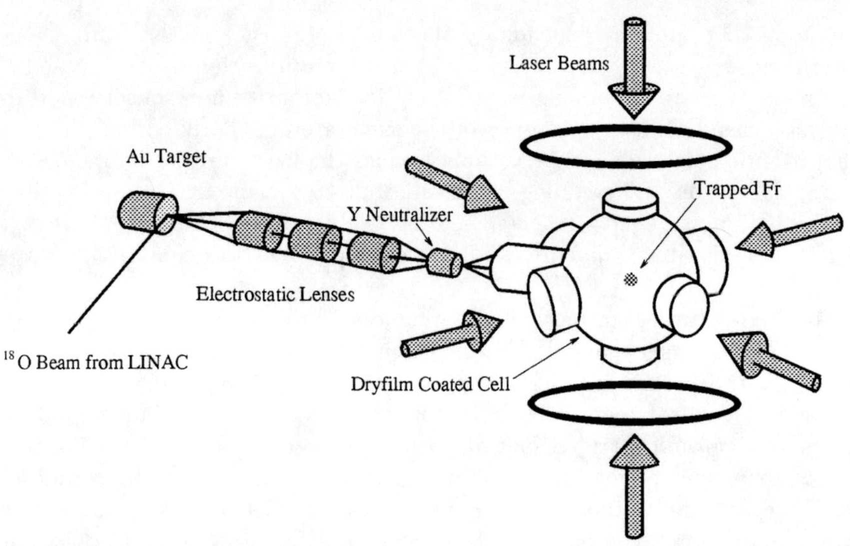

FIGURE 2. Schematic view of target, ion transport system, and magneto-optic trap.

Magneto-Optic Trap

The operation of a MOT utilizes three properties of the interaction between atoms and photons. Photons can transfer momentum to the atoms. Selective absorption of photons with appropriate angular momentum can be tuned by a spatially varying magnetic field to provide a position dependent force. Excitation of the atoms with photons slightly detuned to the red side of the resonance can lower the kinetic energy of the atom. The laser detuning sets the order of magnitude of the well depth of the trap, typically about 0.5 mK. Collisions with background gas are more important in a MOT than in the deeper ion traps. It is possible to capture from the Maxwell-Boltzmann distribution of the vapor pressure of an element in a glass cell [18]. As the slow atoms fall into the trap, the walls thermalize the depleted distribution allowing more atoms to fall into the well. This technique is used in our trap.

Francium Production

Heavy-ion fusion reactions can, by proper choice of projectile, target and beam energy, provide selective production of the neutron deficient francium isotopes. Gold is an ideal target because it is chemically inert, has clean surfaces, and a low vapor pressure. The ^{197}Au(^{18}O,xn) reaction at 100 MeV produces predominantly ^{210}Fr, which has a 3.2 min half-life.

The apparatus shown in Fig. 2 is based on the same principles of our earlier work [15].

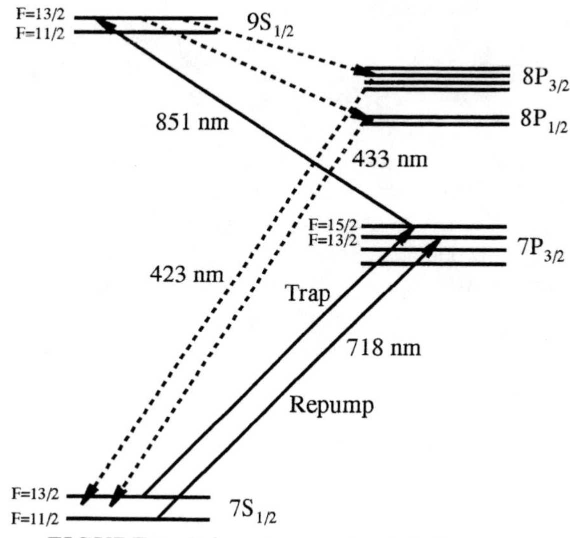

FIGURE 3. Relevant energy levels in francium.

A beam of 6.3×10^{11} ^{18}O ions/s on the Au produces $\approx 1 \times 10^6$/s ^{210}Fr in the target, with less than 10% of other isotopes. The target is heated to $\approx$ 1200 K by the beam power and by an auxiliary resistance heater. The elevated temperature is necessary for the alkali elements to rapidly diffuse to the surface and be surface ionized.

Separation of the production and the trapping regions is critical in order to operate the trap in a UHV environment. Extracted at 500 V, the francium ions travel about one meter where they are deposited on the inner surface of a cylinder coated with yttrium which is heated to 1000 K and located 0.3 cm away from the entrance of the cell. Neutral Fr atoms evaporate from the Y surface and form an atomic beam directed towards an aperture into the vapor cell MOT.

The Francium Trap

The physical trap consists of a 10 cm diameter Pyrex bulb with six 5 cm diameter windows and two viewing windows 3 cm in diameter. The MOT is formed by six intersecting laser beams each with $1/e^2$ (power) diameter of 4 cm and power of 50 mW, with a magnetic field gradient of 9 G/cm. The glass cell is coated with a non-stick Dry-film coating [19,20] to allow the atoms multiple passes through the trapping region. The trap and repump lasers operate in the D_2 line of francium, (see Fig. 3). We have successfully trapped francium in a MOT [21]. We estimate the number of atoms in the trap by comparing the measured fluorescence signal from the trap with the expected fluorescence/atom.

Spectroscopy

A wealth of information is now known about francium [22]. However, the energy of two of the three states ($8S$, $9S$ and $6D$) of primary interest for parity non-conservation have not been measured. At Stony Brook we have undertaken a program to locate them. The unique availability of large amounts of cold, trapped francium atoms permits us to probe the atomic structure.

Recently we located the 9S state [23]. We used a two photon resonant excitation process to excite the atom from the $7S_{1/2} \rightarrow 7P_{3/2} \rightarrow 9S_{1/2}$. The first photon is provided by the trapping laser and the second by a narrow diode laser operating at 851 nm. We detected the resonance by observing blue photons at 433 nm and 423 nm as the atom decayed from the $9S$ state via the $8P_{3/2}$ and $8P_{1/2}$ states, (see Fig. 3). The photons were detected with a photon counting system with very good discrimination from background. Our result for the energy difference between the $9S_{1/2}$ and the $7S_{1/2}$ ground state is 25671.021 ± 0.006 cm^{-1}. The uncertainty is comparable with the accuracy of the $7P_{3/2}$ energy [22]. The recent calculations by Dzuba *et al.* [11] using *ab initio* wave-functions for the location of the 9S were within 7 cm^{-1}, with a quoted uncertainty of ± 10 cm^{-1} accurate to four digits. This is a remarkable achievement and builds great confidence that this theoretical group has a very good understanding of the francium atomic structure. They have performed the calculations necessary to extract the weak charge from a parity non-conservation measurement in francium. Their conservative estimate of the accuracy of the calculated atomic parameters needed for the parity non-conservation measurement in francium is 1%.

Lifetime of the Electronic States

One of the most stringent tests for the atomic theory calculations is the prediction of the lifetime of an electronic state. For the case of francium and the other alkalis the transition from the first nP excited state to the nS ground state is a useful measurement since the lifetime is directly proportional to the square of the dipole matrix element between the two states. The calculation relies on having very good wavefunctions for both the S and the P states.

Great effort has been devoted to the cesium D_2 line lifetime both experimentally [24,25] and theoretically [26,27]. The results bring credibility to the atomic calculations and strengthen the experimental knowledge of the atom. The efforts are converging at a little better than 0.3 % uncertainty.

The Stony Brook group has made preliminary measurements of the atomic lifetimes of the $7P_{3/2}$ and $7P_{1/2}$ excite state decay into the $7S_{1/2}$ ground state in francium. We have used a time-correlated single-photon counting technique applicable to a cold sample of Fr atoms in a MOT. We expect the overall uncertainties in each of the measurements to be around 1%. This is smaller than the uncertainty in the calculations of these lifetimes [11,26].

CONCLUSIONS

The calculations of the theoretical groups using different techniques [11,26], agree with our measurements. The calculation of the needed matrix element to extract the weak charge from a parity non-conservation measurement uses the same wavefunctions as for the energy levels and lifetimes.

We plan to measure the location and lifetime of the $8S$ state in the near future. Those results will increase our knowledge of the francium atom and continue to test the calculations of the atomic structure.

Our advances in the spectroscopy of francium set the groundwork for the parity non-conservation measurement. The long term goal is becoming a reality as different techniques from atomic, nuclear and high energy physics combine to make francium a laboratory for precision studies of the weak force.

ACKNOWLEDGEMENTS

The present situation of the experiment would not have been possible without the past collaboration of J. A. Behr, G. Gwinner, W. Shi and P. A. Voytas. We have greatly profited from interactions with S. L. Rolston, C. E. Weiman, H. Metcalf, T. Bergeman, H. Ravn and W. Phillips. This work was supported in part by a Precision Measurement Grant from NIST and by the National Science Foundation. One of us, L. A. O., would like to thank CINVESTAV-IPN and the organizers of the First Latin American Symposium on High Energy Physics for support to participate in it.

REFERENCES

1. Langacker P. editor *Precision Tests of the Standard Electroweak Model* World Scientific, Singapore, 1995.
2. See for example Bouchiat M. A. in *Atomic Physics 12* Ed. by Lewis R. R. and Zorn J. C., AIP, New York, p. 399 (1991).
3. Noecker M. C., Masterson B. P., and Wieman C. E., Phys. Rev. Lett. **61**, 310 (1988). Also Wieman C. E. private communication (1996).
4. Marciano W. J. and Rosner J. L., *Phys. Rev. Lett.* **65**, 2963 (1990).
5. Rosner J. L., *Phys. Rev. D* **53**, 2724 (1996).
6. Khriplovich I. B. *Parity Nonconservation in Atomic Phenomena.* Gordon and Breach, Philadelphia (1991).
7. Vetter P. A., Meekhof D. M., Majumder P. K., Lamoreaux S. K., and Fortson E. N., *Phys. Rev. Lett.* **74**, 2658 (1995).

8. Budker D., DeMille D., Commins E. D., Zoltorev M. S., *Phys. Rev. Lett.* **70**, 3019 (1994).
9. Warrington R. B., Thompson C. D., Stacey D. N., *Europhys. Lett.* **24**, 641 (1993).
10. Blundell S. A., Johnson W. R., and Sapirstein J., *Phys. Rev. Lett.* **65**, 1411 (1990); Blundell S. A., Sapirstein J., and Johnson W. R., *Phys. Rev. D* **45**, 1602 (1992).
11. Dzuba V. A., Flambaum V. V., and Sushkov O. P., *Phys. Rev. A* **51**, 3454 (1995).
12. Langacker P., Luo M., and Mann A. K., *Rev. Mod. Phys.* **64**, 87 (1992).
13. Lynn B. W., Sandars P. G. H., *J. Phys. B* **27**, 1469 (1994).
14. Burgess C. P., Godfrey S., Köning H., London D., Maksymyk I., *Phys. Rev. D* **49**, 6115 (1994).
15. Gwinner G., Behr J. A., Cahn S. B., Ghosh A., Orozco L. A., Sprouse G. D., and Xu F., *Phys. Rev. Lett.* **72**, 3795 (1994).
16. Lu Z-T., Bowers C. T., Freedman S. J., Fujikawa B. K., Mortara J. L., Shang S-Q., Coulter K. P., and Young L., *Phys. Rev. Lett.* **72**, 3791 (1994).
17. Swanson T. B., *et al. Bull. Am. Phys. Soc.* **41** 1088 (1996).
18. Monroe C., Swann W., Robinson H., and Wieman C., *Phys. Rev. Lett.* **65**, 1571 (1990).
19. Swenson D. R. and Anderson L. W., *Nucl. Instr. and Methods B* **29**, 627 (1988).
20. Stephens M., Rhodes R., and Wieman C., *J. Appl. Phys.* **76**, 3479 (1994). Stephens M., and Wieman C., *Phys. Rev. Lett.* **72**, 3787 (1994).
21. Simsarian J. E., Ghosh A., Gwinner G., Orozco L. A. Sprouse G. D., and Voytas P. A., *Phys. Rev. Lett.* **76**, 3522 (1996).
22. Arnold E., Borchers W., Duong H. T., Juncar P., Lermé J., Lievens P., Neu W., Neugart R., Pellarin M., Pinard J., Vialle J. L., Wendt K., and the ISOLDE Collaboration, *J. Phys. B* **23**, 3511 (1990).
23. Simsarian J. E., Shi W., Orozco L. A., Sprouse G. D. and Zhao W. Z. *Opt. Lett.* **21**, 1939 (1996).
24. Tanner C. in *Atomic Physics 14* Ed. by Wineland D. J., Wieman C. E. and Smith S. J., AIP Conference Proceedings 323, p. 130, New York 1995.
25. Young L., Hill III W. T., Sibener S. J., Price S. D., Tanner C. E., Wieman C. E., and Leone S. R., *Phys. Rev. A.* **50**, 2174 (1994).
26. Johnson W. R., Liu Z. W., and Sapirstein J., *At. Data Nucl. Data Tables* **64**, 279 (1996).
27. Dzuba V. A., Flambaum V.V., Kraftmakher A. Ya., Sushkov O. P., *Phys. Lett A* **142** 373 (1989).

Antihydrogen

Iván Schmidt

Universidad Federico Santa María
Casilla 110-V
Valparaíso, Chile

Abstract. CERN announced in January 1996 the detection of the first eleven atoms of antimatter ever produced. The experiment was based on a method proposed earlier by S. Brodsky, C. Munger and I. Schmidt, and which furthermore predicted exactly the number of atoms that were detected for the particular conditions of the experiment. The study of antihydrogen affords science the opportunity to continue research on the symmetry between matter and antimatter. In this talk the importance of antihydrogen as a basic physical system is discussed. Different production methods that have been tried in the past are briefly presented, and the method that was used in the CERN experiment is analyzed in detail. It consists in producing antihydrogen by circulating a beam of an antiproton ring through an internal gas target. In the Coulomb field of a nucleus, an electron-positron pair is created, and antihydrogen will form when the positron is created in a bound rather that a continuum state about the antiproton. The theoretical calculation of the production cross section is presented in detail. A discussion of the detection systems used both in the CERN experiment and in another similar experiment that is right now underway at Fermilab are also given. Finally I present and discuss possible future experiments using antihydrogen, including the measurement of the antihydrogen Lamb shift.

Antihydrogen, the simplest atomic bound state of antimatter, $\bar{H} \equiv (e^+\bar{p})$, had never been observed experimentally until the CERN announcement of January 1996: a team of physicists had succeeded for the first time in synthesizing atoms of antimatter from their constituents antiparticles [1]. The particular method that was used had been suggested first in 1990 [2], and the detailed calculation presented later [3,4]. The news of this discovery appeared in all major newspapers and scientific journals in the world.

I WHY ANTIHYDROGEN ?

The hydrogen atom has been one of the most important physical systems for a wide variety of fundamental measurements related to the behavior

CP400, *First Latin American Symposium on High Energy Physics/ VII Mexican School of Particles and Fields*,
edited by D'Olivo/Klein-Kreisler/Mendez

of ordinary matter. The production of antihydrogen opens the door for a systematic exploration of the properties of antimatter, and give us unique possibilities of testing fundamental physical principles [5].

CPT Invariance: The CPT theorem is a basic result of quantum field theory, and it is the essential ingredient for knowing the properties of antimatter. CPT invariance implies that a particle and its antiparticle have: equal and opposite additive quantum numbers, such as electric charge, and equal masses, total lifetimes and gyromagnetic ratios. Nevertheless, CPT violation could occur when non-finite-dimensional representations of the Lorentz group are permitted, in curved spacetimes, or through non-localities.

In the case of antihydrogen, CPT invariance requires that the its spectrum has to be exactly equal to that of the hydrogen atom. Since the hyperfine splitting in hydrogen is known to a precision of roughly $6\ 10^{-12}$, comparison with antihydrogen will provide a superlative direct CPT test for baryons.

Equivalence principle for antiparticles: The question is whether antimatter might behave differently than ordinary matter in a gravitational field. Since transition frequencies are subject to gravitational red shifts when the system is moved through a gravitational potential difference, this could be the basis of experimental tests of the equivalence principle for antimatter.

II PRODUCTION

An important obstacle for the production of antihydrogen is the high energy of the antiproton beam. By using special devices such as Penning traps, researchers have been able to reduce substantially their energy. The next problem is to combine the antiprotons with positrons. Several sources are currently under development. For example [6]:

Radiative recombination , $\overline{p} + e^+ \to (\overline{p}e^+) + \gamma$. The problem is that the collision time ($\sim 10^{-15}s$) is much less than the typical radiative lifetime ($\sim 10^{-9}s$). Thus it takes a long time to radiate a photon compared to the collision time.

Laser assisted Recombination , $\overline{p} + e^+ + n\gamma \to (\overline{p}e^+) + (n+1)\gamma$. Unfortunately the verified gain is high only for states with $n > 8$.

Plasma Recombination , $\overline{p} + e^+ + e^+ \to (\overline{p}e^+) + e^+$. Three way collision, in which the extra particle would carry off the excess energy.

Capture from Positronium , $\overline{p} + (e^+e^-) \to (\overline{p}e^+) + e^-$.

From antiprotonic helium states , $(\overline{p}He) + (e^+e^-) \to (\overline{p}e^+) + (e^-He)$. The problem here is the annihilation of the antihydrogen within the liquid helium.

No antihydrogen has yet been made by any of these processes, yet curiously a working source had been producing antihydrogen for some time, but with the antihydrogens going out undetected. This is the process that was used in the CERN experiment, colliding relativistic antiprotons with ordinary atoms. An antiproton passing through the Coulomb field of a nucleus of charge Z will create electron-positron pairs; occasionally a positron will appear in a bound instead of a continuum state about the moving antiproton to form antihydrogen. The cross section for this capture process we calculate to be $\sigma(\bar{p}Z \rightarrow \bar{H}e^-Z) \sim 4\ Z^2 pb$ for antiproton momenta above ~ 6 GeV/c.

III CROSS SECTION CALCULATION

The production process for antihydrogen studied here is the exclusive two- to three-particle reaction $\bar{p}p \rightarrow \bar{H}e^-p'$. The equivalent photon approximation [7], applied in the antiproton rest frame, relates the cross section for pair creation with capture to $1s$ state, $\bar{p}Z \rightarrow \bar{H}(1s)e^-Z$, to the cross section for photon-induced capture using virtual photons, $\gamma^*\bar{p} \rightarrow e^-\bar{H}(1s)$. Consistent with this approximation we assume the $\bar{p}$ and the $\bar{H}$ in these processes to remain at rest in a common frame of reference, and neglect the electron and photon energies compared with the antiproton mass. We find

$$\sigma_{\bar{p}Z\rightarrow\bar{H}(1s)e^-Z} = \frac{Z^2\alpha}{\pi} \times$$
$$\int_{2m/E}^{1} \frac{dx}{x} \int_0^{q^2_{\perp max}} dq^2_\perp \frac{q^2_\perp}{(q^2_\perp + x^2M^2)^2} \left(\frac{1+(1-x)^2}{2}\right) \sigma_{\gamma^*\bar{p}\rightarrow\bar{H}(1s)e^-}(\omega, q^2). \quad (1)$$

Here $x = \omega/E = \omega/(M\gamma)$ is the photon energy fraction evaluated in the antiproton rest frame, $q_\perp$ is the photon's transverse momentum, and M is the antiproton mass. At large photon virtuality $Q^2 = -q^2$ the photoabsorption cross section falls off as $(Q^2 + 4m_e^2)^{-1}$, so the transverse momentum $q_\perp$ in the integrand is typically of order $2m_e$. The upper limit of integration $q_{\perp max}$ has the same order of magnitude as the photon energy ω because the photon tends to have $q^2 \approx 0$.

The contributions where q^2 is small dominate the integral and so we can set $q^2 \approx 0$ and substitute the cross section for real instead of virtual photons. Performing the integral over $q^2_\perp$ we find that the total cross section factors,

$$\sigma_{\bar{p}Z\rightarrow\bar{H}(1s)e^-Z}(\gamma) = Z^2F(\gamma)\,\frac{\alpha}{\pi}\bar{\sigma}, \quad (2)$$

where

$$F(\gamma) = \ln(\gamma^2+1) - \frac{\gamma^2}{\gamma^2+1}, \quad (3)$$

contains the dependence on the relative velocity of the antiproton and the nucleus, where γ is the Lorentz factor $\gamma = (1-\beta^2)^{-1/2}$, and where

$$\bar{\sigma} = \int_{2m_e}^{E} \frac{d\omega}{\omega} \sigma_{\gamma\bar{p} \to \bar{H}(1s)\, e^-}(\omega) \tag{4}$$

is a constant.

Crossing symmetry relates the matrix elements for photon-induced capture and for photoionization: the sum over initial and final spins of the squares of the matrix elements for the reactions $\gamma\bar{p} \to e^-\bar{H}(1s)$ and $\gamma\bar{H}(1s) \to e^-\bar{p}$ become equal if they are written in terms of the Mandelstam variables s, t, and u, and if the variables s and u in one element are exchanged. The squares of the matrix elements when the final spins are summed but the initial spins are averaged, $|M|^2$, are likewise related except that a factor of two is introduced:

$$|M|^2_{capture}(s, t, u) = 2|M|^2_{photo}(u, t, s). \tag{5}$$

The matrix element for photoionization is well known [7] so we easily obtain the matrix element for capture. In the $\bar{p}$ rest frame the Mandelstam variables for photoionization are equal to

$$\begin{aligned} s &= \quad M^2 + 2M\omega, \\ t &= m^2 - 2\omega\epsilon + 2\vec{k} \cdot \vec{p}, \\ u &= \quad M^2 - 2\epsilon M + m^2, \end{aligned} \tag{6}$$

where $\vec{p}$ and $\vec{k}$ are respectively the momentum vectors of the outgoing electron and incoming photon, and $\epsilon = \sqrt{|\vec{p}|^2 + m^2}$ is the electron energy. We define also $p = |\vec{p}|$ and $k = |\vec{k}| = \omega$.

After rewriting the matrix element for photoionization in terms of the Mandelstam variables, and exchanging s and u, we find in the $\bar{p}$ rest frame that

$$\begin{aligned} |M|^2_{capture} = \frac{64\pi^2\alpha^5 m^4}{\epsilon(\vec{k} - \vec{p})^4} \Bigg[& \tfrac{a^2 p^2 m}{\epsilon + m} + a(\vec{k} \cdot \vec{p} - p^2)\left(\tfrac{1}{k^2 - p^2} + \tfrac{1}{(\vec{k} - \vec{p})^2}\right) \\ & + \tfrac{\epsilon + m}{4m}(\vec{k} - \vec{p})^2 \left(\tfrac{1}{k^2 - p^2} + \tfrac{1}{(\vec{k} - \vec{p})^2}\right)^2 \\ & - \tfrac{\epsilon + m}{2m} \tfrac{k^2 p^2 - (\vec{k} \cdot \vec{p})^2}{(k^2 - p^2)(\vec{k} - \vec{p})^2 k^2} + a \tfrac{k^2 p^2 - (\vec{k} \cdot \vec{p})^2}{(\vec{k} - \vec{p})^2 k^2} \Bigg] \end{aligned} \tag{7}$$

where the quantity a is given by

$$a = \frac{1}{(\vec{k} - \vec{p})^2} + \frac{\epsilon}{m(k^2 - p^2)}. \tag{8}$$

The total cross section for photon-induced capture is

$$\sigma_{\gamma\bar{p} \to e^- \bar{H}(1s)}(\omega) = \frac{\alpha\epsilon p}{2\pi M^2} \int |M|^2_{capture} d\cos\theta. \tag{9}$$

Integrating the previous equations numerically yields $(\alpha/\pi)\bar{\sigma} = 1.42$ pb. We then get the cross section for $\bar{p}Z \to \bar{H}(1s)e^-Z$ as a function of the antiproton momentum; it is approximately $4Z^2$ pb for momenta above ~ 6 GeV/c. Capture into states of higher principal quantum number will increase the total cross section for capture by $\sim 10\% - 20\%$.

In the case of the CERN experiment, the momentum of the antiproton is 1.94 GeV/c, which gives a cross section of approximately $2\ pb \times Z^2 \approx 6 \times 10^{-33} cm^2$. With an integrated luminosity of $5 \times 10^{33} cm^{-2} \pm 50\%$, and an acceptance of $\epsilon = 0.3$ for the detection system, the expected number of events is about 10. The actual number was 11 ! Right now a similar experiment is underway at Fermilab. As of this writing they have detected 18 antihydrogen atoms [8]

Calculations of similar cross sections have been reviewed by Eichler [9], and give $\sigma(\bar{p}p \to \bar{H}pe^-) = 2.7 \ln(\gamma)$ pb, which is very close to our asymptotic result $\sigma = 2.8 \ln(\gamma)$ pb. Becker [10] computes the cross section for two different momenta; the ratio of the cross sections agrees remarkably well with our prediction, though his cross sections are lower than ours by a factor of 2.8.

IV DETECTION

The momentum transferred to the antiproton in the process $\bar{p}Z \to \bar{H}e^-Z$ is small, the order of $mc \sim 5 \cdot 10^{-4} GeV/c$ The momentum and position vectors of an antihydrogen atom are therefore the same as those of the antiproton from which it forms; a monoenergetic, small-divergence bunch of antiprotons exits a cloud of gas overlapped by an equivalent bunch of antihydrogen atoms.

It can easily be shown that the antihydrogen atoms escape the gas intact, and that it easily survives its escape through the dipole fields of the Accumulator ring [4]. Fast antihydrogen separates into a pair of free particles with a probability greater than 0.99 in a mere membrane of polyethylene 400 μgm cm^2 thick. An antihydrogen atom therefore generates in coincidence, from some point in a known, few-square-centimeter area of a membrane possibly tens of meters from the gas target, a positron and an antiproton with a common and tightly constrained velocity equal to the known velocity of the antiprotons circulating in the storage ring. So spectacular is this signature that the chief difficulty in designing an apparatus to detect antihydrogen is to choose which of many sufficient schemes is the simplest.

The CERN experimental detection system is roughly as follows. As a neutral object the antihydrogen, produced ten meters upstream in the center of the straight section of LEAR, will exit the accelerator ring tangentially and will be stripped in the first silicon counter. Then the resulting antiproton and positron hit simultaneously this first silicon counter which, together with a second one, measures the dE/dx of the antiproton plus the kinetic energy of the positron being stopped in these two detectors. The third silicon counter should give a signal proportional to the dE/dx of the $\overline{p}$ only. The two 511 KeV photons from the $e^+ - e^-$ annihilation are detected back to back in a NaI counter. The silicon counter telescope is located at the center of the NaI crystal arrangement. The $\overline{p}$'s resulting from the stripped $\overline{H}$ continue with a velocity of $\beta = 0.900$. They penetrate through a set of three start scintillators, a scintillator hodoscope and a group of stop scintillators. The deflection of the registered charged particles was measured with a spectrometer, and it should be within the predicted values for a $\overline{p}$ from $\overline{H}$. Several cuts were performed to the data in order to reach the conclusion that 11 $\overline{H}$ had been observed.

In the case of the Fermilab experiment the detection system is slightly different. An antihydrogen atom exits the Accumulator ring and strikes a 400 μgm cm^{-2} membrane, where it separates into a positron and an antiproton. Because the particles' magnetic rigidities differ by a factor of nearly 2000 the positron can be bent away and focussed while hardly affecting the antiproton, so the particles can be directed into separate detectors. We describe how these detectors function for an antihydrogen momentum of 3 GeV/c; the apparatus works equally well up to the maximum Accumulator momentum of 8.8 GeV/c merely by scaling various magnetic fields.

A positron spectrometer a few meters in length, consisting in succession of a solenoid lens, a sector magnet, and a second solenoid lens, separates the positrons and focuses them onto a few square-centimeter spot. A momentum resolution of a few percent matches the Fermi smear in the momentum of the positrons. The positrons stop in a scintillator ~ 1 centimeter thick whose total volume is only a few cubic centimeters. Light from the scintillator is guided into a single phototube; the rise and height of the output pulse give respectively a measure of the arrival time of the positron within 1 ns, and a measure of its kinetic energy to 20%. A 4π NaI detector surrounds the scintillator and intercepts the two 511-keV photons from the positron's annihilation; the detector also vetoes the passage of stray charged particles, which deposit far more energy than 511 keV. The collection efficiency of the spectrometer is 99%. Some 95% of the positrons come to rest in the scintillator and deposit their full energy; the 5% that backscatter nonetheless deposit enough energy to make a signal, and still come to rest and annihilate in the volume surrounded by NaI.

Tagging the antiproton is straightforward. Over a flight distance of 40 meters it passes through a pair of scintillator paddles that measure its velocity by time of flight; a sequence of multiwire proportional chambers (MWPC's) and bend magnets that measure its momentum to 0.2%, and a terminal Čerenkov threshold detector that provides a redundant velocity separation of antiprotons and other negative particles. The tiny beamspot, angular divergence, and unique and known velocity of the antiproton favor a differential-velocity Čerenkov detector , but for momenta as low as 3 GeV/c it is difficult to generate enough light for such a device without using a Čerenkov medium so thick that antiprotons do not survive their passage.

A candidate antiproton is defined by hits in the MWPC's consistent with a 3 GeV/c antiproton that originates in the right few square-centimeter area of the membrane, by the absence of a hit in the Čerenkov threshold detector, and by hits in the time-of-flight scintillators consistent with the passage of the particle of the right velocity. A candidate positron is defined by a hit in the positron scintillator with the right deposit of kinetic energy, coincident within a few nanoseconds with a hit in the surrounding NaI consistent with the absorption or Compton scatter of two 511 keV photons. An antihydrogen candidate is defined as a sub-nanosecond coincidence between antiproton and positron candidates.

Backgrounds: Despite the $\sim 10^{10}$ times higher cross section for $\bar{p}p$ annihilation than for antihydrogen production, the backgrounds from particles originating in the gas target will be zero. Here we consider only those processes which have such peculiar kinematics and large branching ratio as to mimic at least part of an antihydrogen signal, most importantly the antiproton, without requiring a failure of some part of the detector. We will consider just the Fermilab experiment, but a very similar analysis applies to the CERN experiment.

The only particle which can satisfy the momentum and velocity measurements made by the particle tracking and by the time-of-flight and Čerenkov detectors is an antiproton. Antiprotons lost from the ring are unimportant; few thread properly through the target and the wire chambers, and the passage of an antiproton through even an extra 2 cm of aluminium will slow the antiproton below the 0.2% momentum resolution of the particle tracking. The only plausible source of antiprotons is from antineutrons, made in the target by the charge-exchange reaction $\bar{p}p \to \bar{n}n$, that convert by a second charge-exchange reaction into an antiproton. The simplest possibility is for an antineutron generated in the gas target to pass undeflected through the Accumulator's dipole magnet and convert in the first MWPC; the new antiproton fools the rest of the detectors in our antiproton beam line. Of all processes that make antiprotons this one has the most dangerous combination of unfavorable kinematics and big relevant cross sections. The number of $\bar{n}$'s made in the gas target is large; both the neutral $\bar{n}$'s and the converted $\bar{p}$'s are thrown forward, the more easily to pass through the apparatus and the antiproton beamline; and the forward $\bar{p}$'s from this two-step process can have precisely the same velocity as the antiprotons circulating in the ring.

Nonetheless the beamline will count fewer antiprotons from this process than it will count antiprotons from antihydrogen. The count rate is small principally because of the small solid angle subtended by the known small spots (of order 3 cm^2) that legitimate antiprotons make on both the first and last MWPC's, compared to the typical solid angle over which particles from charge exchange are distributed. At a model momentum of $p = 3$ GeV/c the cross section for $\bar{p}p \rightarrow n\bar{n}$ is 2.0 mbarn; an integrated luminosity of 200 pb^{-1} will produce $4.0 \cdot 10^{11}$ antineutrons. In charge exchange the typical momentum transfer is the order of the pion mass, $\delta p \sim 135$ Mev/c, and so the outgoing antineutrons will be distributed over a solid angle that is the order of $\Delta\Omega \sim \pi(\delta p/p)^2 \sim 6.4 \cdot 10^{-3}$ sterradian. The thickest material in front of all the MWPC's is 0.3 cm of scintillator in the first time-of-flight detector. Reconstruction of the antiproton track limits the active area on this scintillator to roughly 3 cm^2; the solid angle of this spot as seen from the gas target, 20 meters away, is only $7.5 \cdot 10^{-7}$ sterradian, and the probability any antineutron from charge exchange hits the spot is only $1.2 \cdot 10^{-4}$. The probability of a hadronic interaction in the scintillator is only $3.8 \cdot 10^{-3}$; if an antineutron interacts, it will generate an antiproton with a probability equal to the ratio of the cross section for charge exchange to the total cross section, which for a proton target is equal to 2.0 mbarn/79.9 mbarn $= 2.5 \cdot 10^{-2}$. Any antiproton made must now pass though the antiproton beam line and strike a 3 cm^2 spot on the last MWPC, at least 20 meters away; because of the momentum transfer in charge exchange the probability it will do so is again only $1.2 \cdot 10^{-4}$. Collecting all the factors, the number of antiprotons from this process indistinguishable from the 760 antiprotons we expect from the separation of antihydrogen is at most

$$4.0 \cdot 10^{11} \times 1.2 \cdot 10^{-4} \times 3.8 \cdot 10^{-3} \times 2.5 \cdot 10^{-2} \times 1.2 \cdot 10^{-4} = 0.5$$

This order-of-magnitude argument is quite crude but sufficient to show that the rate of such false antiprotons, occurring in coincidence with a positron candidate, is certainly negligible.

The kinematics of an antineutron' ordinary β-decay, $\bar{n} \rightarrow \bar{p}e^+\nu$, are also unfavorable because the decay produces not only a 3 GeV/c antiproton but also a positron whose range of laboratory kinetic energy overlaps 1.200 MeV; fortunately its slow decay rate $\sim 10^{-3}\,\mathrm{s}^{-1}$ prevents a significant fraction from decaying within our apparatus.

V OTHER EXPERIMENTS WITH RELATIVISTIC ANTIHYDROGEN

To test CPT invariance it is best to study hydrogen and antihydrogen in the same apparatus. The polarity of the magnets in the Fermilab Accumulator can be reversed and protons circulated; as the protons pass through the target gas they pick up atomic electrons and make a neutral hydrogen beam that has the same optics as the antiproton beam. For protons above 3 GeV/c the dominant process is one in which an essentially free electron in the target falls into the $1s$ state and a photon carries off the binding energy. The cross section per target electron is ≥ 1.7 nanobarn, so for equal circulating currents through a hydrogen gas target the hydrogen beam will have ~ 430 times the intensity of the antihydrogen beam.

Two experiments seem practical with meager samples respectively of order 10^3 and $3 \cdot 10^4$ antihydrogen atoms. The first is a measurement of the rate of field ionization of the $n = 2$ states in an electric field provided by the Lorentz transform of a laboratory magnetic field. Roughly 10% of a $3GeV/c$ beam of antihydrogen in the $1s$ state can be excited into states with $n = 2$ by passing it through a thin membrane. If the membrane sits in a 20 kgauss transverse magnetic field, equation shows that the states with $n > 2$ will ionize instantly, that the states with $n = 2$ will ionize with $1/e$ decay lengths of order 10 cm, and that the $1s$ state will not ionize at all. The distance a state with $n = 2$ flies before ionizing is marked by the a deflection of the freed antiproton by the magnetic field by an amount between the zero deflection of the surviving $1s$ component of the beam and the large deflection of the antihydrogen that ionizes instantly or separates in the membrane. Ten centimeters of flight before ionization changes the deflection of the antiproton seen 3 meters away by 6.7 cm, many times the antiproton spot size of $\leq 1cm$; and changes the antiprotons angle by 22 mrad, many times both its original angular divergence of 0.2 mrad and the resolution of 0.1 mrad that can be provided by a pair of MWPC's with 1 mm resolution and 10 meters apart. The distance a state flies is also marked by the freed positron, whose orbit radius is only 2 mm in the transverse field, and which can be directed along the field lines into some sort of position-sensitive detector. The positron and antiproton have of course their usual known common velocity. A flux of a few thousand $\bar{H}$'s may be sufficient to measure the three distinct field ionization rates of the $n = 2$ states to $\sim 10\%$. Because ionization is a tunneling process its rate is surprisingly sensitive to details of the antihydrogen wavefunction; a 10% shift would require for example a change in $\langle r \rangle$ for the $n = 2$ states of only 0.24%.

Evidently ionization in a magnetic field can be used to count efficiently states with $n = 2$ without counting states of different principal quantum number. No other method is available; the Accumulator runs with long antiproton bunches, and no laser for example has sufficient continuous power to photo-ionize efficiently the relativistic antihydrogen beam. By driving the 1000 Mhz $2s - 2p$ transition and monitoring the surviving $2s$ population as a function of frequency the frequency of the antihydrogen Lamb shift can be measured. The $1/e$ decay length of the $2p$ states is 1.4 meters at 3 GeV/c, so a few meters from an excitation membrane only the $2s$ population will survive. The Doppler-shifted transition can be driven by chasing the beam with 6.1 GHz radiation aimed down a waveguide roughly that is 10 meters long and roughly 10 cm^2 in cross section; this guide may also serve as a beam pipe. The $2s - 2p$ resonance has a quality factor of only 10 because of the width of the $2p$ state, and so the Doppler broadening of the transition, or a misalignment of the axes of the beam and the guide, will not put the transition out of resonance. Because the $2s - 2p$ transition is electric dipole in character and has a large matrix element, modest laboratory powers of roughly 10 Watt/cm^2 suffice to mix the $2s$ state completely with the $2p$ and make the $2s$ state decay with a $1/e$ distance of 2.8 m. To prevent Stark mixing of the $2s$ and $2p$ states, transverse magnetic fields must be less than 0.1 gauss from the excitation membrane down the guide's length until the sharp rise of the transverse magnetic field that is used to ionize and count the $n = 2$ states. If the rise occurs over less than the fully mixed $2s$ decay length of 2.8 meters little of the $2s$ state will decay to the $1s$ instead of ionize. A sample of a 300 antihydrogen atoms in the $2s$ state would suffice to see a dip in the transmitted $2s$ population as a function of drive frequency, find its center to within 10% of its width, and so measure the antihydrogen Lamb shift to $\sim 1\%$. A total of $3 \cdot 10^4$ antihydrogen atoms is enough to provide such a sample if a membrane yields as expected 0.01 $2s$ states per incident $1s$. The experiment would be sensitive to a differential shift of the $2s$ and $2p$ states of hydrogen and antihydrogen equal to a fraction $\sim 2 \cdot 10^{-8}$ of the states' binding energy, and would test the CPT symmetry of the $e^+\bar{p}$ interaction at momentum scales characteristic of atomic binding, 10 eV/c.

REFERENCES

1. Baur G., et al., *Phys. Lett.* **B368**, 251 (1996).
2. Brodsky S. J., and Schmidt I., *Phys. Rev.***D43**, 179 (1990).
3. Munger C. T., Brodsky S. J., and Schmidt I., *Hyperfine Interactions* **76**, 175 (1993).
4. Munger C. T., Brodsky S. J., and Schmidt I., *Phys. Rev.* **D49**, 3228 (1994).
5. For a review see: Charlton M., et al., *Phys. Rep.* **241**, 65 (1994).
6. Proceedings of the Antihydrogen Workshop, Munich, July 1992.
7. Berestetskii V. B., Lifshitz E. M. and Pitaevskii L. P.,*Relativistic Quantum Theory, Part 1,* (second edition, Pergamon Press), page 212 (A missing factor of 2 has here been restored to their formulae; see for example *The Quantum Theory of Radiation, Third Edition*, Heitler W., Oxford, 1954, p. 209); and page 438.
8. Source: Fermilab experiment E862 Internet Home Page.
9. Eichler J., *Phys. Rep.* **193**, 165 (1990).
10. Becker U., *J. Phys.* **B20**, 6563 (1987).

Connections Between Inclusive and Exclusive Semileptonic B Decay

Mark B. Wise

California Institute of Technology, Pasadena, CA 91125, USA[1]

Abstract. B decay sum rules relate exclusive B semileptonic decay matrix elements to forward B-meson matrix elements of operators in HQET. At leading order the operators that occur are the b-quark kinetic energy λ_1 and chromomagnetic energy λ_2. The latter is determined by the measured $B^* - B$ mass splitting. The derivation of these sum rules is reviewed and perturbative QCD corrections are discussed. A determination of λ_1 and the energy of the light degrees of freedom in a B-meson, $\bar{\Lambda}$, from semileptonic B decay data is presented. Future prospects for improving these sum rules are discussed.

I INTRODUCTION

In this lecture I review some connections between inclusive and exclusive semileptonic B-meson decays. These arise from sum rules that relate the form factors for exclusive semileptonic decays to nonperturbative QCD matrix elements that occur in the inclusive semileptonic decay rate. Sum rules that relate inclusive B transitions to a sum over exclusive states were first derived by Bjorken [1,2] and Voloshin [3]. Then a general framework for B-decay sum rules was presented by Bigi, et al. in [4]. Since this very important work, there has been a considerable amount of theoretical activity in the area of B decay sum rules [5–12].

For inclusive decays it is possible using the operator product expansion and a transition to the heavy quark effective theory (HQET) [13] to show that at leading order in $\Lambda_{\rm QCD}/m_b$ the B semileptonic decay rate is equal to the b-quark decay rate [14,15]. There are no nonperturbative corrections to this at order $\Lambda_{\rm QCD}/m_b$ [14]. The first corrections arise at order $\Lambda_{\rm QCD}^2/m_b^2$, and are characterized by matrix elements [16] that are related to the b-quark kinetic energy

[1] Work supported in part by the U.S. Dept. of Energy under Grant No. DE-FG03-92-ER40701.

CP400, *First Latin American Symposium on High Energy Physics/ VII Mexican School of Particles and Fields,* edited by D'Olivo/Klein-Kreisler/Mendez

$$\lambda_1 = \frac{1}{2m_B}\langle B(v)|\bar{h}_v^{(b)}(iD)^2 h_v^{(b)}|B(v)\rangle, \tag{1}$$

and the color magnetic energy

$$\lambda_2 = \frac{1}{6m_B}\langle B(v)|\bar{h}_v^{(b)}\frac{g}{2}\sigma_{\mu\nu}G^{\mu\nu}h_v^{(b)}|B(v)\rangle. \tag{2}$$

The parameters λ_1 and λ_2 are independent of the heavy quark mass and occur in the formulas for the B, B^*, D, and D^* meson masses:

$$\begin{aligned} m_B &= m_b + \bar{\Lambda} - (\lambda_1 + 3\lambda_2)/2m_b, \\ m_{B^*} &= m_b + \bar{\Lambda} - (\lambda_1 - \lambda_2)/2m_b, \\ m_D &= m_c + \bar{\Lambda} - (\lambda_1 + 3\lambda_2)/2m_c, \\ m_{D^*} &= m_c + \bar{\Lambda} - (\lambda_1 - \lambda_2)/2m_c. \end{aligned} \tag{3}$$

The measured $B^* - B$ mass splitting $(46 \pm 0.6)\,\mathrm{MeV}$ implies that $\lambda_2 = 0.12\,\mathrm{GeV}^2$. The quantity $\bar{\Lambda}$ represents the energy of the light degrees of freedom for the ground state $s_\ell^{\pi_\ell} = \frac{1}{2}^-$ multiplet in the $m_{b,c} \to \infty$ limit. Note that in the average masses $\bar{m}_B = (m_B + 3m_{B^*})/4$ and $\bar{m}_D = (m_D + 3m_{D^*})/4$ the parameter λ_2 cancels out.

The leading order prediction of the operator product expansion for the B semileptonic decay rate involves quark masses, which are not known experimentally. What is measured are the hadron masses. It is possible using eq. (3) to express the quark masses, m_b and m_c, in terms of the hadron masses, $\bar{m}_B$ and $\bar{m}_D$, and the parameters λ_1 and $\bar{\Lambda}$. When this is done the semileptonic B-meson decay rate depends on the unknown parameters λ_1 and $\bar{\Lambda}$ that are of order Λ_{QCD}^2 and Λ_{QCD} respectively. In this way of looking at the predicted decay rate there are contributions of order $\Lambda_{\mathrm{QCD}}/m_{c,b}$, but they are given in terms of the single parameter $\bar{\Lambda}$.

The form factors for semileptonic $B \to D^{(*)}e\bar{\nu}_e$ decay are defined by

$$\begin{aligned} \frac{\langle D(v')|V^\mu|B(v)\rangle}{\sqrt{m_B m_D}} &= h_+(w)(v+v')^\mu + h_-(w)(v-v')^\mu, \\ \frac{\langle D^*(v',\epsilon)|V^\mu|B(v)\rangle}{\sqrt{m_B m_{D^*}}} &= i h_V(w)\varepsilon^{\mu\nu\alpha\beta}\epsilon_\nu^* v'_\alpha v_\beta, \\ \frac{\langle D^*(v',\epsilon)|A^\mu|B(v)\rangle}{\sqrt{m_B m_{D^*}}} &= h_{A_1}(w)(w+1)\epsilon^{*\mu} - h_{A_2}(w)(\epsilon^*\cdot v)v^\mu \\ &\quad - h_{A_3}(w)(\epsilon^*\cdot v)v'^\mu. \end{aligned} \tag{4}$$

Here $V^\mu = \bar{c}\gamma^\mu b$ and $A^\mu = \bar{c}\gamma^\mu\gamma_5 b$ are the vector and axial vector currents. The four-velocities of the initial and final states are denoted by v and v' respectively. The dot product of these four-velocities is $w = v \cdot v'$ and at the zero recoil point, where $v = v'$, $w = 1$. Up to corrections suppressed by powers of $\alpha_s(m_{c,b})$ and $\Lambda_{\rm QCD}/m_{c,b}$, $h_-(w) = h_{A_2}(w) = 0$ and $h_+(w) = h_V(w) = h_{A_1}(w) = h_{A_3}(w) = \xi(w)$, where the Isgur–Wise function [17] ξ is evaluated at a subtraction point around $m_{c,b}$. The differential decay rates are

$$\frac{d\Gamma(B \to D^* \ell \bar{\nu}_e)}{dw} = \frac{G_F^2 m_B^5}{48\pi^3} r^{*3}(1-r^*)^2 (w^2-1)^{1/2}(w+1)^2 \times \left[1 + \frac{4w}{w+1}\frac{1-2wr^*+r^{*2}}{(1-r^*)^2}\right] |V_{cb}|^2 |\mathcal{F}_{B\to D^*}(w)|^2,$$

$$\frac{d\Gamma(B \to D \ell \bar{\nu}_e)}{dw} = \frac{G_F^2 m_B^5}{48\pi^3} r^3 (1+r)^2 (w^2-1)^{3/2} |V_{cb}|^2 |\mathcal{F}_{B\to D}(w)|^2, \quad (5)$$

where $r^{(*)} = m_{D^{(*)}}/m_B$. The functions $\mathcal{F}_{B\to D^*}$ and $\mathcal{F}_{B\to D}$ are given in terms of the form factors of the vector and axial vector currents defined in eq. (4) as

$$|\mathcal{F}_{B\to D^*}(w)|^2 = \left[1 + \frac{4w}{w+1}\frac{1-2wr^*+r^{*2}}{(1-r^*)^2}\right]^{-1} \times \left\{\frac{1-2wr^*+r^{*2}}{(1-r^*)^2} 2\left[h_{A_1}^2(w) + \frac{w-1}{w+1} h_V^2(w)\right] + \left[h_{A_1}(w) + \frac{w-1}{1-r^*}\left(h_{A_1}(w) - h_{A_3}(w) - r^* h_{A_2}(w)\right)\right]^2\right\},$$

$$\mathcal{F}_{B\to D}(w) = h_+(w) - \frac{1-r}{1+r} h_-(w). \quad (6)$$

Note that $\mathcal{F}_{B\to D^*}(1) = h_{A_1}(1)$ and due to Luke's theorem [18]

$$h_{A_1}(1) = \eta_A + \mathcal{O}(\Lambda_{\rm QCD}^2/m_{cb}^2). \quad (7)$$

For the $B \to D$ case $\mathcal{F}_{B\to D}(1) = \eta_V + \mathcal{O}(\Lambda_{\rm QCD}/m_{c,b})$. The quantities η_A and η_V relate the axial and vector currents in the full theory of QCD to those in HQET at zero recoil. Including corrections of order $\alpha_s^2\beta_0$ [19,20]

$$\eta_A = 1 - \frac{\alpha_s(\sqrt{m_b m_c})}{\pi}\left(\frac{1+z}{1-z}\ln z + \frac{8}{3}\right) - \frac{\alpha_s^2(\sqrt{m_b m_c})}{\pi^2}\beta_0 \frac{5}{24}\left(\frac{1+z}{1-z}\ln z + \frac{44}{15}\right), \quad (8)$$

where $z = m_c/m_b$ and $\beta_0 = 11 - 2N_f/3$ is the 1-loop beta function. In eq. (8) and hereafter dimensional regularization with $\overline{\rm MS}$ subtraction is used. The full order α_s^2 expression for η_A is known [21] and the $\alpha_s^2\beta_0$ part presented in eq. (8) dominates it.

II SUM RULES

To derive the sum rules, we consider the time-ordered product

$$T_{\mu\nu} = \frac{i}{2m_B} \int \mathrm{d}^4x \, e^{-iq\cdot x} \, \langle B \,|\, T\{J_\mu^\dagger(x), J_\nu(0)\} | B \,\rangle \,, \tag{9}$$

where J_μ is a $b \to c$ axial or vector current, the B states are at rest, $\vec{q}$ is fixed, and $q_0 = m_B - E_{D^{(*)}} - \epsilon$. Here $E_{D^{(*)}} = \sqrt{m_{D^{(*)}}^2 + |\vec{q}\,|^2}$ is the minimal possible energy of the hadronic final states that can be created by the current J_μ at fixed $|\vec{q}\,|$. (We deal with cases where the lowest energy state is either a D or a D^*.) With this definition of ϵ, $T_{\mu\nu}$ has a cut in in the complex ϵ plane that lies along $0 < \epsilon < \infty$, corresponding to physical intermediate states with a charm quark. At the same value of $|\vec{q}|$ the cut at the parton level lies in the smaller region $\epsilon > \bar{\Lambda}(w-1)/w + \mathcal{O}(\Lambda_{\rm QCD}^2/m_{c,b}^2)$, were $\vec{q}^{\,2} = m_c^2(w^2-1)$. $T_{\mu\nu}$ has another cut corresponding to physical states with two b-quarks and a $\bar{c}$ quark that lies along $-\infty < \epsilon < -2E_{D^{(*)}}$. To separate out specific hadronic form factors, one contracts the currents in eq. (9) with a suitably chosen four-vector a. Inserting a complete set of states between the currents yields

$$a^{*\mu} \, T_{\mu\nu}(\epsilon) \, a^\nu = \frac{1}{2m_B} \sum_X (2\pi)^3 \, \delta^3(\vec{q} + \vec{p}_X) \, \frac{\langle B|J^\dagger \cdot a^*|X\rangle\langle X|J\cdot a|B\rangle}{E_X - E_{D^{(*)}} - \epsilon} + \ldots \,, \tag{10}$$

where the ellipses denote the contribution from the cut corresponding to two b-quarks and a $\bar{c}$ quark. The sum over X includes the usual phase space factors, $\mathrm{d}^3p/(2\pi)^3 2E_X$, for each particle in the state X.

While $T_{\mu\nu}(\epsilon)$ cannot be computed for arbitrary values of ϵ, its integrals with appropriate weight functions are calculable using the operator product expansion and perturbative QCD. Consider integrating the product of a weight function $W_\Delta(\epsilon)$ with $T_{\mu\nu}(\epsilon)$ along the contour C surrounding the physical cut shown in Fig. 1. Assuming W is analytic in the shaded region enclosed by this contour, we get

$$\begin{aligned} &\frac{1}{2\pi i} \int_C \mathrm{d}\epsilon \, W_\Delta(\epsilon) \, [a^{*\mu} \, T_{\mu\nu}(\epsilon) \, a^\nu] \\ &\qquad = \sum_X W_\Delta(E_X - E_{D^{(*)}}) \, (2\pi)^3 \, \delta^3(\vec{q} + \vec{p}_X) \, \frac{|\langle X|J\cdot a|B\rangle|^2}{2m_B} \,. \end{aligned} \tag{11}$$

The weight function is assumed to be positive along the cut and to satisfy the normalization condition $W_\Delta(0) = 1$. Then $W_\Delta \cdot |\langle X|J\cdot a|B\rangle|^2$ is positive for all states X, and eq. (11) implies an upper bound on the magnitude of form factors for semileptonic B decays to the ground states $D^{(*)}$.

$$\frac{|\langle D^{(*)}|J\cdot a|B\rangle|^2}{4m_B E_{D^{(*)}}} < \frac{1}{2\pi i} \int_C \mathrm{d}\epsilon \, W_\Delta(\epsilon) [a^{*\mu} T_{\mu\nu} a^\nu]. \tag{12}$$

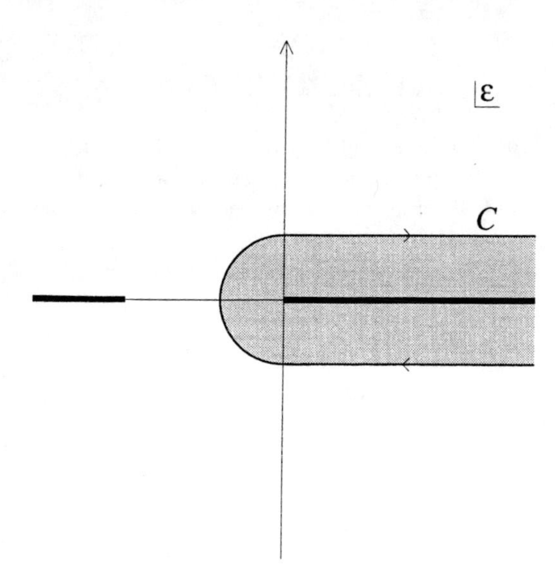

FIGURE 1. The integration contour C in the complex ϵ plane. The cuts extend to $\mathrm{Re}\,\epsilon \to \pm\infty$.

In eq. (12) a sum over D^* polarizations is understood. It is also possible to derive a lower bound if some model dependent assumptions concerning the spectrum of final states X are made.

A possible set of weight functions is [6],

$$W_\Delta^{(n)}(\epsilon) = \frac{\Delta^{2n}}{\epsilon^{2n} + \Delta^{2n}}, \qquad (n = 2, 3, \ldots). \tag{13}$$

They satisfy the following properties: (*i*) W_Δ is positive along the cut so that every term in the sum over X on the hadron side of the sum rule is positive; (*ii*) $W_\Delta(0) = 1$; (*iii*) W_Δ is flat near $\epsilon = 0$; and (*iv*) W_Δ falls off rapidly for $\epsilon > \Delta$. For values of n of order unity all the poles of $W_\Delta^{(n)}$ lie at a distance of order Δ away from the physical cut. As $n \to \infty$, $W_\Delta^{(n)}$ approaches $\theta(\Delta - \epsilon)$ for $\epsilon > 0$, which corresponds to summing over all final hadronic resonances up to excitation energy Δ with equal weight. In this limit the poles of $W_\Delta^{(n)}$ approach the cut, and the contour C is forced to pinch the cut at $\epsilon = \Delta$. Then the evaluation of the contour integrals using perturbative QCD relies on local duality at the scale Δ. In practice, for $n > 3$ the results obtained are very close to those for $n = \infty$ and for the remainder of this lecture I will only quote results obtained from the weight $W_\Delta^{(\infty)}(\epsilon) = \theta(\Delta - \epsilon)$.

III APPLICATION OF SUM RULES AT ZERO RECOIL

The sum rule bound in eq. (12) is made explicit by using the operator product expansion and perturbative QCD to evaluate the right-hand side. The most important kinematic point is the zero recoil point where $\vec{q} = 0$. Choosing a to be a spatial vector $a = (0, \hat{n})$ and averaging over directions of the unit vector $\hat{n}$, we obtain for the axial vector current

$$|\mathcal{F}_{B\to D^*}(1)|^2 \le \eta_A^2 - \frac{\lambda_2}{m_c^2} + \left(\frac{\lambda_1 + 3\lambda_2}{4}\right)\left(\frac{1}{m_c^2} + \frac{1}{m_b^2} + \frac{2}{3m_c m_b}\right) + \frac{\alpha_s(\Delta)}{\pi} X_{AA}(\Delta) + \frac{\alpha_s^2(\Delta)}{\pi^2}\beta_0 Y_{AA}(\Delta), \quad (14)$$

and for the vector current

$$0 \le \frac{\lambda_2}{m_c^2} - \left(\frac{\lambda_1 + 3\lambda_2}{4}\right)\left(\frac{1}{m_c^2} + \frac{1}{m_b^2} - \frac{2}{3m_c m_b}\right) + \frac{\alpha_s(\Delta)}{\pi} X_{VV}(\Delta) + \frac{\alpha_s^2(\Delta)}{\pi^2}\beta_0 Y_{VV}(\Delta). \quad (15)$$

Eqs. (14) and (15) include terms of order $\Lambda_{\rm QCD}^2/m_c^2$ coming from dimension five operators in the operator product expansion for the time ordered product of currents. The coefficients of these operators are evaluated at tree level. Also included is the contribution from the dimension 3 operator $\bar{h}_v^{(b)}\Gamma h_v^{(b)}$ evaluated to order $\alpha_s^2\beta_0$. There are two distinct sources of perturbative QCD corrections. Those in η_A correspond to a final state X_c that at the parton level is a single charm quark. These terms are independent of Δ and come from matching of the axial vector current onto its HQET counterpart, i.e., $A^\nu = \eta_A \bar{h}_v^{(b)}\gamma^\nu\gamma_5 h_v^{(c)}$. The part of the QCD correction involving X_{AA}, Y_{AA}, X_{VV} and Y_{VV}, comes at the parton level from states with a charm quark and a gluon or even more partons. These corrections depend on Δ. Since Δ is the cut off on the invariant mass of the final hadronic states it seems most natural to write these terms as a power series in $\alpha_s(\Delta)$. If one used $\alpha_s(\mu)$ with μ much different from Δ the coefficients Y_{AA} and Y_{VV} would contain large logarithms of Δ/μ. Analytic expressions for the order α_s corrections are known [6]

$$X_{AA}(\Delta) = \frac{\Delta(\Delta + 2m_c)[2(\Delta + m_c)^2 - 2m_b^2 - (m_b + m_c)^2]}{18m_b^2(\Delta + m_c)^2} + \frac{3m_b^2 + 2m_b m_c - m_c^2}{9m_b^2}\ln\left(\frac{\Delta + m_c}{m_c}\right), \quad (16)$$

$$X_{VV}(\Delta) = \frac{\Delta(\Delta + 2m_c)[2(\Delta + m_c)^2 - 2m_b^2 - (m_b - m_c)^2]}{18m_b^2(\Delta + m_c)^2} + \frac{3m_b^2 - 2m_b m_c - m_c^2}{9m_b^2}\ln\left(\frac{\Delta + m_c}{m_c}\right). \quad (17)$$

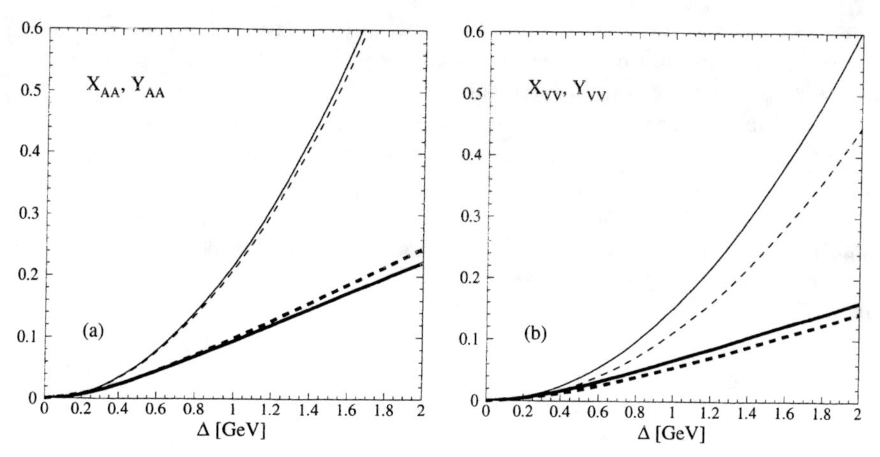

FIGURE 2. $X(\Delta)$ and $Y(\Delta)$ for the a) axial, and b) vector coefficients. Thick solid lines are X while thick dashed lines are Y. The thin solid and dashed lines are X and Y to order $\Delta^2/m_{c,b}^2$.

For small Δ, X_{AA} and X_{VV} are of order $\Delta^2/m_{c,b}^2$; however, even when $\Delta = 1\,\text{GeV}$, terms higher order in $\Delta/m_{c,b}$ are important (the small Δ approximation to X_{AA} and X_{VV} was calculated in Ref. [4]). The values of Y_{AA} and Y_{VV} have been determined numerically [6]. In Fig. 2, X_{AA}, Y_{AA}, X_{VV}, and Y_{VV} are plotted as functions of Δ in the region $\Delta < 2\,\text{GeV}$. The values of Y are quite close to X in this region.

The vector current sum rule bound in eq. (15) implies a bound on λ_1. This bound is strongest for $m_c \gg m_b \gg \Delta$. In that limit it becomes [4,6]

$$\lambda_1 \leq -3\lambda_2 + \frac{\alpha_s(\Delta)}{\pi}\Delta^2\left(\frac{4}{3}\right) + \frac{\alpha_s^2(\Delta)}{\pi^2}\beta_0\Delta^2\left(\frac{13}{9} - \frac{2\ln 2}{3}\right). \tag{18}$$

The parameter Δ must be chosen large enough that perturbative QCD is meaningful. However the bounds on λ_1 and $|\mathcal{F}_{B\to D^*}|^2$ become stronger the smaller the value of Δ. The smallest value of Δ for which one can imagine using perturbative QCD is 1 GeV. Using $\Delta = 1\,\text{GeV}$, $\alpha_s(1\text{GeV}) = 0.45$, $\lambda_2 = 0.12\,\text{GeV}^2$, eq. (18) implies

$$\lambda_1 \leq (-0.36 + 0.19 + 0.20)\,\text{GeV}^2. \tag{19}$$

The three terms on the right-hand side of eq. (19) correspond respectively to the contribution of λ_2, the perturbative part of order $\alpha_s(\Delta)\Delta^2/\pi$, and the perturbative part of order $[\alpha_s(\Delta)/\pi]^2\Delta^2$. Notice that with $\Delta = 1\,\text{GeV}$ the α_s^2 term is as large as the order α_s term. It may be a mistake to conclude from this that $\Delta = 1\,\text{GeV}$ is too low for QCD perturbation theory to be meaningful. It has been conjectured that λ_1 has a renormalon ambiguity of order Λ_{QCD}^2

(one does not see this from the usual sum of bubble graphs) [22]. Even though the renormalon ambiguity arises from large orders of perturbation theory, it is possible that the bad behavior of the first few terms in the perturbative series presented in eq. (18) is a reflection of this uncertainty.

In this lecture the matrix element λ_1 is defined using dimensional regularization and $\overline{\rm MS}$ subtraction. If λ_1 has a renormalon ambiguity (of order $\Lambda_{\rm QCD}^2$), the perturbative QCD series that relates it to a physical quantity, for example computed in lattice QCD, is not Borel summable. However, there is no evidence that this is a serious problem. Whenever λ_1 occurs in an expression for some measurable quantity, e.g., the bound on $|\mathcal{F}_{B\to D^*}(1)|^2$, there is another perturbative series that when combined with the series in λ_1 (e.g., from matching onto lattice QCD) probably has no renormalon ambiguity (of order $\Lambda_{\rm QCD}^2$) [23].

Next consider the bound on $|\mathcal{F}_{B\to D^*}(1)|^2$ in eq. (14). We can eliminate λ_1 from it by combining (14) and (15). This gives

$$|\mathcal{F}_{B\to D^*}(1)|^2 \le \eta_A^2 - \frac{\lambda_2}{m_c^2} + \frac{\alpha_s(\Delta)}{\pi}\left[X_{AA}(\Delta) + \frac{1}{3}\left(\frac{\Delta^2}{m_c^2} + \frac{\Delta^2}{m_b^2} + \frac{2\Delta^2}{3m_c m_b}\right)\right] \qquad (20)$$
$$+ \frac{\alpha_s^2(\Delta)}{\pi^2}\beta_0\left[Y_{AA}(\Delta) + \left(\frac{13}{36} - \frac{\ln 2}{6}\right)\left(\frac{\Delta^2}{m_c^2} + \frac{\Delta^2}{m_b^2} + \frac{2\Delta^2}{3m_c m_b}\right)\right].$$

Neglecting the nonperturbative correction factor of $-\lambda_2/m_c^2$ and again using $\Delta = 1\,\text{GeV}$, the above bound is

$$\begin{aligned}|\mathcal{F}_{B\to D^*}(1)|^2 &\le 1 - 0.074 - 0.020\\ &\quad + 0.044 + 0.046\\ &= 1 - 0.030 + 0.026. \end{aligned} \qquad (21)$$

Here we used $m_c = 1.4\,\text{GeV}$, $m_b = 4.8\,\text{GeV}$, $\alpha_s(\sqrt{m_c m_b}) = 0.28$, $\alpha_s(1\,\text{GeV}) = 0.45$, and $\beta_0 = 9$. The first row is the perturbative expansion of η_A^2, the second row is the order $\alpha_s(\Delta)$ and order $\alpha_s(\Delta)^2$ terms, and the third row sums the columns. There is a renormalon ambiguity of order $\Lambda_{\rm QCD}^2/m_{c,b}^2$ that cancels between the perturbative series for η_A^2 and the series in $\alpha_s(\Delta)$. This bound on the physical quantity $|\mathcal{F}_{B\to D^*}(1)|^2$ is not very strong (even when the factor of $-\lambda_2/m_c^2$ is included), and furthermore the third row of eq. (21) seems to indicate that with $\Delta = 1\,\text{GeV}$ QCD perturbation theory is not very well behaved. However, this does not mean that the sum rule for $|\mathcal{F}_{B\to D^*}(1)|^2$ in eq. (14) is not useful. Consider the perturbative part of eq. (14), neglecting for now both the terms of order $\lambda_1/m_{c,b}^2$ and $\lambda_2/m_{c,b}^2$. Numerically, with $\Delta = 1\,\text{GeV}$, this gives

$$\begin{aligned}|\mathcal{F}_{B\to D^*}(1)|^2 &\le 1 - 0.074 - 0.020\\ &\quad + 0.013 + 0.017\\ &= 1 - 0.061 - 0.003. \end{aligned} \qquad (22)$$

Again, the first row is the perturbative expansion of η_A^2 and the second row are the terms of order $\alpha_s(\Delta)$ and $\alpha_s^2(\Delta)$. The third row of eq. (22) does not indicate that there is any breakdown of QCD perturbation theory. If λ_1 can be determined experimentally from, for example, the electron spectrum in inclusive semileptonic B-decay then the sum rule in eq. (14) may lead to an important constraint. For example, with $\Delta = 1\,\mathrm{GeV}$ and $\lambda_1 = -0.2\,\mathrm{GeV}^2$, eq. (14) implies the bound

$$|F_{B\to D^*}(1)|^2 \leq 0.90, \tag{23}$$

which is smaller than $\eta_A^2 = 0.91$.

IV THE LEPTON ENERGY SPECTRUM AND THE PARAMETERS λ_1, $\bar{\Lambda}$

The CLEO collaboration has measured the lepton energy spectrum for inclusive $B \to X\ell\bar{\nu}_e$ decay, both demanding only one charged lepton (single tagged data) and two charged leptons (double tagged sample) [24,25]. In the double tagged sample the charge of the high momentum lepton determines whether the other lepton comes from a semileptonic B decay (primary lepton) or the semileptonic decay of a D-meson (secondary lepton). The single tagged sample is presented in 50 MeV bins while the double tagged data is presented in 100 MeV bins. The single tagged sample has much higher statistics, but is significantly contaminated by secondaries below $E_\ell = 1.5\,\mathrm{GeV}$.

The operator product expansion for semileptonic B decay does not reproduce the physical lepton spectrum point by point near the maximal electron energy. Near the endpoint, comparison of theory with data can only be made after smearing or integrating over a large enough region. The minimal size of this region has been estimated to be about 500 MeV. As was mentioned in the introduction, the theoretical prediction for the lepton energy spectrum depends on λ_1 and $\bar{\Lambda}$, so we can try to use the data to determine these quantities. We want to consider observables sensitive to $\bar{\Lambda}$ and λ_1, but we also want deviations from the b-quark decay rate to be small enough so that contributions from even higher dimension operators in the operator product expansion are small. Ref. [26] uses R_1 and R_2, where

$$R_1 = \frac{\int_{1.5\,\mathrm{GeV}} (\mathrm{d}\Gamma/\mathrm{d}E_\ell) E_\ell \,\mathrm{d}E_\ell}{\int_{1.5\,\mathrm{GeV}} (\mathrm{d}\Gamma/\mathrm{d}E_\ell)\,\mathrm{d}E_\ell}, \tag{24}$$

and

$$R_2 = \frac{\int_{1.7\,\mathrm{GeV}} (\mathrm{d}\Gamma/\mathrm{d}E_\ell)\,\mathrm{d}E_\ell}{\int_{1.5\,\mathrm{GeV}} (\mathrm{d}\Gamma/\mathrm{d}E_\ell)\,\mathrm{d}E_\ell}. \tag{25}$$

Here E_ℓ denotes the lepton energy. The variable R_1 has dimensions of mass and values for it will be given in GeV. Ratios are considered so that $|V_{cb}|$ cancels out. Before comparing the experimental data with theoretical predictions derived from the operator product expansion and QCD perturbation theory, it is necessary to include electromagnetic corrections and effects of the boost to the laboratory frame. This gives

$$\begin{aligned} R_1 = 1.8059 - 0.309\left(\frac{\bar\Lambda}{\bar m_B}\right) - 0.35\left(\frac{\bar\Lambda}{\bar m_B}\right)^2 - 2.32\left(\frac{\lambda_1}{\bar m_B^2}\right) - 3.96\left(\frac{\lambda_2}{\bar m_B^2}\right) \\ - \frac{\alpha_s}{\pi}\left(0.035 + 0.07\frac{\bar\Lambda}{\bar m_B}\right) + \left|\frac{V_{ub}}{V_{cb}}\right|^2\left(1.33 - 10.3\frac{\bar\Lambda}{\bar m_B}\right) \\ - \left(0.0041 - 0.004\frac{\bar\Lambda}{\bar m_B}\right) + \left(0.0062 + 0.002\frac{\bar\Lambda}{\bar m_B}\right), \end{aligned} \quad (26)$$

and

$$\begin{aligned} R_2 = 0.6581 - 0.315\left(\frac{\bar\Lambda}{\bar m_B}\right) - 0.68\left(\frac{\bar\Lambda}{\bar m_B}\right)^2 - 1.65\left(\frac{\lambda_1}{\bar m_B^2}\right) - 4.94\left(\frac{\lambda_2}{\bar m_B^2}\right) \\ - \frac{\alpha_s}{\pi}\left(0.039 + 0.18\frac{\bar\Lambda}{\bar m_B}\right) + \left|\frac{V_{ub}}{V_{cb}}\right|^2\left(0.87 - 3.8\frac{\bar\Lambda}{\bar m_B}\right) \\ - \left(0.0073 + 0.005\frac{\bar\Lambda}{\bar m_B}\right) + \left(0.0021 + 0.003\frac{\bar\Lambda}{\bar m_B}\right). \end{aligned} \quad (27)$$

In eqs. (26) and (27) the charm and bottom quark masses have been expressed in terms of $\bar m_B$, $\bar m_D$, $\bar\Lambda$, λ_1, and λ_2 using eq. (3), which is why $\bar\Lambda$ occurs in these formulas. The last two terms in eqs. (26) and (27) are from electromagnetic radiative corrections and from the boost to the laboratory frame respectively. The experimental values for R_1 and R_2 are $R_1 = 1.7831\,\mathrm{GeV}$ and $R_2 = 0.6159$. These were obtained from the single tagged data with a correction for the secondary leptons obtained from the double tagged sample. For R_1 this correction is 0.0001 GeV and for R_2 it is 0.0051. Comparing experiment with eqs. (26) and (27) gives the central values $\bar\Lambda = 0.39 \pm 0.11\,\mathrm{GeV}$ and $\lambda_1 = -0.19 \pm 0.10\,\mathrm{GeV}^2$. Fig. 3 shows the one sigma bands on the allowed values of $\bar\Lambda$ and λ_1 from R_1 and R_2. The narrower band corresponds to the R_1 constraint. The shaded ellipse is the one sigma allowed region for $\bar\Lambda$ and λ_1 including correlations between R_1 and R_2. The errors included in this analysis are just the statistical ones. An analysis of systematic errors has not been performed. However, they are only weakly energy dependent for $E_\ell \geq 1.5\,\mathrm{GeV}$ and it is hoped that for $R_{1,2}$ systematic errors are smaller than the statistical ones. Note that the bands from R_1 and R_2 are almost parallel, so even small corrections can significantly change the central values for $\bar\Lambda$ and λ_1 obtained from Fig. 3.

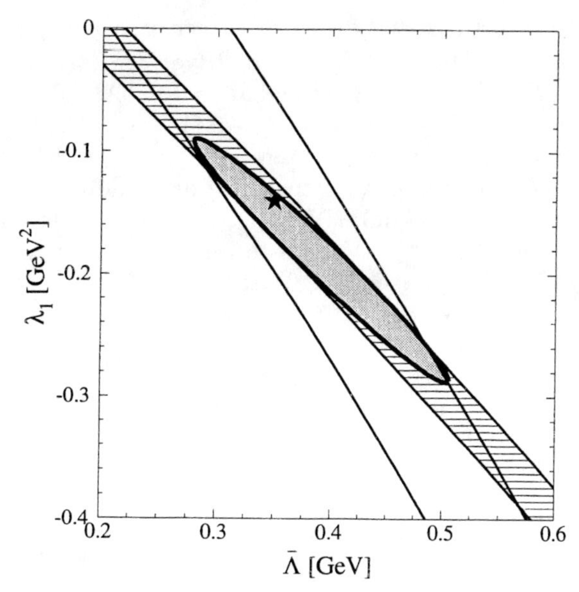

FIGURE 3. Allowed regions in the $\bar{\Lambda} - \lambda_1$ plane for R_1 and R_2. The bands represent the 1σ statistical errors, while the ellipse is the allowed region taking correlations into account. The star denotes where the order $\Lambda^3_{\rm QCD}/\bar{m}^3_B$ corrections discussed in the text would shift the center of the ellipse.

One such set of corrections comes from higher dimension operator in the operator product expansion. At order $\Lambda^3_{\rm QCD}/\bar{m}^3_B$ new terms occur characterized by two matrix elements ρ_1 and ρ_2 and two time ordered products. ρ_1 can be estimated by factorization, $\rho_1 = (2\pi\alpha_s/9)m_B f_B^2 \approx (300\,{\rm MeV})^3$, and ρ_2 is expected to be small [4]. Neglecting ρ_2 and the time ordered products gives the following order $\Lambda^3_{\rm QCD}/\bar{m}^3_B$ corrections to R_1 and R_2,

$$\begin{aligned}\delta R_1 &= -(0.4\bar{\Lambda}^3 + 5.7\bar{\Lambda}\lambda_1 + 6.8\bar{\Lambda}\lambda_2 + 7.7\rho_1)/\bar{m}^3_B,\\ \delta R_2 &= -(1.5\bar{\Lambda}^3 + 7.1\bar{\Lambda}\lambda_1 + 17.5\bar{\Lambda}\lambda_2 + 1.8\rho_1)/\bar{m}^3_B.\end{aligned} \tag{28}$$

Including these corrections shifts the central values for $\bar{\Lambda}$ and λ_1 to the location $\bar{\Lambda} = 0.35\,{\rm GeV}$ and $\lambda_1 = -0.15\,{\rm GeV}^2$ marked by the star in Fig. 3. For a more complete discussion of the order $\Lambda^3_{\rm QCD}/\bar{m}^3_B$ corrections, see Ref. [27].

The bands in Fig. 3 were determined using $|V_{ub}/V_{cb}| = 0.08$. This value is model dependent. If $|V_{ub}/V_{cb}| = 0.10$ is used then the central values shift to $\bar{\Lambda} = 0.42\,{\rm GeV}$ and $\lambda_1 = -0.19\,{\rm GeV}^2$. In Fig. 3, $\alpha_s = 0.22$ was used corresponding to a subtraction point near m_b. With $\alpha_s = 0.35$ the central values shift to $\bar{\Lambda} = 0.36\,{\rm GeV}$ and $\lambda_1 = -0.18\,{\rm GeV}^2$.

Theoretical uncertainty in this determination of $\bar{\Lambda}$ and λ_1 originate from the reliability of quark hadron duality at the limits of integration defining $R_{1,2}$, order $\Lambda_{\rm QCD}^3/\bar{m}_B^3$ corrections, and higher order perturbative QCD corrections. Recently the order $\alpha_s^2\beta_0$ terms in R_1 and R_2 have been computed [28]. They give corrections $\delta R_1 = -0.082\alpha_s^2\beta_0/\pi^2$ and $\delta R_2 = -0.098\alpha_s^2\beta_0/\pi^2$, moving the central values to $\bar{\Lambda} = 0.33\,\text{GeV}$ and $\lambda_1 = -0.17\,\text{GeV}^2$. Concerning duality, note that the lower limits on the lepton energy $E_\ell \geq 1.5\,\text{GeV}$ and $E_\ell \geq 1.7\,\text{GeV}$ used in $R_{1,2}$ correspond to summing over hadronic states X with masses less than 3.6 GeV and 3.3 GeV respectively. Changing the lower limit in the numerator of R_2 to 1.8 GeV leads to central values $\bar{\Lambda} = 0.47\,\text{GeV}$ and $\lambda_1 = -0.26\,\text{GeV}^2$. The plot in Fig. 3 uses electron data only. Using the muon sample instead gives compatible central values of $\bar{\Lambda} = 0.43\,\text{GeV}$ and $\lambda_1 = -0.21\,\text{GeV}^2$.

It would be nice to have another constraint on λ_1 and $\bar{\Lambda}$ that would be less parallel than R_1 and R_2 are. This can be provided by the photon spectrum in inclusive $B \to X_s\gamma$ decay [29] which, when the data improves, will give a band almost parallel to the λ_1 axis of Fig. 3. Lattice QCD can also be used to determine λ_1 and $\bar{\Lambda}$, although for λ_1 there are serious difficulties coming from mixing with the lower dimension operator $\bar{h}_v^{(b)} h_v^{(b)}$ [30]. This mixing does not occur in the continuum if dimensional regularization with $\overline{\rm MS}$ subtraction is used.

V CONCLUDING REMARKS

In this lecture I have reviewed the derivation of B-decay sum rules, discussed the perturbative QCD corrections, and reviewed the status of the determination of the nonperturbative QCD matrix element λ_1 that occurs in the sum rules.

If the contribution of the lowest lying excited states X on the right-hand side of eq. (11) were known then this would imply a better bound on the ground state matrix elements. The lowest lying excited states are nonresonant $D^{(*)}\pi$. Their contribution, for low $D^{(*)}\pi$ invariant mass, is calculable [4] in terms of the one coupling constant, g, of heavy hadron chiral perturbation theory [31]. This coupling also determines the D^* width, $\Gamma(D^{*+} \to D^0\pi^+) = (g^2/6\pi f_\pi^2)|\vec{p}_\pi|^3$ (for the neutral pion mode there is an additional factor of 1/2). Unfortunately at the present time there is only a limit on the D^* width and hence an upper bound on g. A measurement of the D^* width would give a direct determination of this coupling. Then we would know the contribution of these nonresonant states to the sum rules. (Determining g from various D^* and D_s^* branching ratios is discussed in Ref. [32]). Higher in mass is the $s_\ell^{\pi_\ell} = \frac{3}{2}^+$ doublet of excited charmed mesons $D_1(2420)$ and $D_2^*(2460)$. These states are narrow with widths around 20 MeV. A doublet with $s_\ell^{\pi_\ell} = \frac{1}{2}^+$ quantum numbers is expected to also exist, but these states are thought to be quite broad (i.e., widths greater than 100 MeV). Very recently the contribution of these excited charmed mesons to the sum rules has been discussed [33].

Perturbative QCD corrections to the sum rules have been calculated to order $\alpha_s^2\beta_0$. It is important to improve this to a full order α_s^2 calculation. Finally, it is interesting to note that B-decay sum rules may also be important for the $b \to u$ transition, giving valuable information on exclusive $B \to \pi\ell\bar{\nu}_\ell$ or $B \to \rho\ell\bar{\nu}_\ell$ form factors [8]. At this time perturbative corrections to these $b \to u$ transition sum rules have not been computed.

REFERENCES

1. Bjorken, J.D., Invited talk at Les Rencontre de la Valle d'Aoste (La Thuile, Italy), SLAC-PUB-5278 (1990);
   Bjorken, J.D., *et al.*, *Nucl. Phys.* **B371**, 111 (1992).
2. Isgur, N. and Wise, M.B., *Phys. Rev.* **D43**, 819 (1991).
3. Voloshin, M.B., *Phys. Rev.* **D46**, 3062 (1992).
4. Bigi, I.I., *et al.*, *Phys. Rev.* **D52**, 196 (1995).
5. Bigi, I.I., *et al.*, *Phys. Lett.* **B339**, 160 (1994).
6. Kapustin, A., *et al.*, *Phys. Lett.* **B375**, 327 (1996).
7. Grozin, A.G. and Korchemsky, G.P., *Phys. Rev.* **D53**, 1378 (1996).
8. Boyd, G. and Rothstein, I.Z., UCSP-TH-96-20 [hep-ph/9607418] (1996).
9. Boyd, C.G., *et al.*, *Phys. Rev.* **D55**, 3027 (1997).
10. Uraltsev, N., UND-HEP-96-BIG04 [hep-ph/9610425] (1996).
11. Koyrakh, L.A., Ph.D. Thesis, University of Minnesota [hep-ph/9607443] (1996).
12. Davoudiasl, H. and Leibovich, A.K., CALT-68-2098 [hep-ph/9702341] (1997).
13. Eichten, E. and Hill, B., *Phys. Lett.* **B234**, 511 (1990);
    Georgi, H., *Phys. Lett.* **B240**, 447 (1990).
14. Chay, J., *et al.*, *Phys. Lett.* **B247**, 399 (1990).
15. Voloshin, M. and Shifman, M., *Sov. J. Nucl. Phys.* **41**, 120 (1985).
16. Bigi, I.I., *et al.*, *Phys. Rev. Lett.* **71**, 496 (1993);
    Manohar, A.V. and Wise, M.B., *Phys. Rev.* **D49**, 1310 (1994);
    Blok, B., *et al.*, *Phys. Rev.* **D49**, 3356 (1994); (E) **50**, 35 (1994);
    Mannel, T., *Nucl. Phys.* **B413**, 396 (1994).
17. Isgur, N. and Wise, M.B., *Phys. Lett.* **B232**, 113 (1989); *Phys. Lett.* **B237**, 527 (1990).
18. Luke, M.B., *Phys. Lett.* **B252**, 447 (1990).
19. Voloshin, M. and Shifman, M., *Sov. J. Nucl. Phys.* **47**, 511 (1988).
20. Neubert, M., *Phys. Lett.* **B341**, 367 (1996).
21. Czarnecki, A., *Phys. Rev. Lett.* **76**, 4124 (1996).
22. Neubert, M., CERN-TH-96-282 [hep-ph/9610471] (1996);
    Martinelli, G., *et al.*, *Nucl. Phys.* **B461**, 238 (1996).
23. Beneke, M., *et al.*, *Phys. Rev. Lett* **73**, 3058 (1994);
    Luke, M., *et al.*, *Phys. Rev.* **D51**, 4924 (1995);
    Neubert, M. and Sachrajda, C.T., *Nucl. Phys.* **B438**, 235 (1995).
24. Bartlett, J., *et al.*, CLEO Collaboration, CLEO/CONF 93-19 (1993);
    Barish, B., *et al.*, CLEO Collaboration, *Phys. Rev. Lett.* **76**, 1570 (1996).

25. Wang, R., Ph.D. Thesis, University of Minnesota (1994).
26. Gremm, M., *et al.*, *Phys. Rev. Lett.* **77**, 20 (1996).
27. Gremm, M. and Kapustin, A., CALT-68-2042 [hep-ph/9603448] (1996).
28. Gremm, M. and Stewart, I., *Phys. Rev.* **D55**, 1226 (1997).
29. Kapustin, A. and Ligeti, Z., *Phys. Lett.* **B355**, 318 (1995).
30. Gimenez, V., *et al.*, *Nucl. Phys. Proc. Suppl.* **53**, 365 (1997).
31. Wise, M., *Phys. Rev.* **D45**, R2188 (1992);
Burdman, G. and Donoghue, J., *Phys. Lett.* **B280**, 287 (1992);
Cheng, H.Y., *et al.*, *Phys. Rev.* **D47**, 1030 (1992).
32. Amundsen, A., *et al.*, *Phys. Lett.* **B296**, 415 (1992);
Cho, P. and Wise, M.B., *Phys. Rev.* **D49**, 6228 (1994);
Cho, P. and Georgi, H., *Phys. Lett.* **B296**, 408 (1992).
33. Leibovich, A., *et al.*, CALT-68-2102 [hep-ph/9703213] (1997).

From CP-violation to hypernuclei: physics programme at the DAΦNE Φ-factory

Stefano Bianco

Laboratori Nazionali di Frascati dell'INFN, Frascati, I-00044

Abstract. Physics at DAΦNE , the new Frascati e^+e^- machine, is reviewed, as well as the experiments: DEAR - search for KN exotic atoms, FINUDA - spectroscopy and decays of hypernuclei, and KLOE - a multipurpose detector designed for detecting direct CP violation.

INTRODUCTION

The DAΦNE e^+e^- collider facility [1], [2], [3], proposed in 1990 [4] and rapidly approved, is now being built at the Laboratori Nazionali di Frascati of INFN. The DAΦNE collider [5] [6] will operate at the center of mass energy of the ϕ meson with an initial luminosity $\mathcal{L} = 1.3 \times 10^{32}\mathrm{cm}^{-2}\mathrm{s}^{-1}$ and a target luminosity $\mathcal{L} = 5.3 \times 10^{32}\mathrm{cm}^{-2}\mathrm{s}^{-1}$.

The $\phi(1020)$ meson [$s\bar{s}$ quark assignment, mass $m = 1019.413 \pm 0.008\,\mathrm{MeV}$, total width $\Gamma = 4.43 \pm 0.06\,\mathrm{MeV}$, electronic width $\Gamma_{ee} = 1.37 \pm 0.05\,\mathrm{keV}$, and $I^G(J^{PC}) = 0^-(1^{--})$ quantum numbers] is produced in e^+e^- collisions with a

$$\sigma(e^+e^- \to \phi) = 4.4\,\mu\mathrm{b}$$

peak cross section, which translates into a production rate of

$$R_\phi = (5.7 \times 10^2 \to 2.3 \times 10^3)\mathrm{s}^{-1}$$

from initial to target luminosity. The production cross section $\sigma(e^+e^- \to \phi)$ should be compared with the hadronic production, which, in this energy range and for $\beta \to 1$, is given by

$$\sigma(e^+e^- \to f\bar{f}) = \frac{4\pi\alpha^2}{3s}Q_f^2 = \frac{86.8\,Q_f^2\,\mathrm{nb}}{s[\mathrm{GeV}^2]} = 56\,\mathrm{nb},$$

where Q_f is the charge of the fermion in units of the proton charge, i.e., with an integral S/N ratio of about 40:1.

CP400, *First Latin American Symposium on High Energy Physics/ VII Mexican School of Particles and Fields,*
edited by D'Olivo/Klein-Kreisler/Mendez

TABLE 1. Quantum number assignments for kaons

	$I_3 = +1/2$	$I_3 = +1/2$
$S = +1$	$K^+(\bar{s}u)$	$K^0(\bar{s}d)$
$S = -1$	$\bar{K^0}(s\bar{d})$	$K^-(s\bar{u})$

The ϕ then decays at rest into K^+K^- (49%), $K_S^0 K_L^0$ (34%), $\rho\pi$ (13%), $\pi^+\pi^-\pi^0$ (2.5%) and $\eta\gamma$ (1.3%), which translate in 0.8 kHz $\phi \to K^0\bar{K^0}$, 1.1 kHz $\phi \to K^+K^-$, 300 Hz $\phi \to \rho\pi$, 60 Hz $\phi \to \pi^+\pi^-\pi^0$, 30 Hz $\phi \to \eta\gamma$. With a canonical efficiency of 30%, this corresponds to 2.3×10^{10} ϕ decays in one calendar year. Therefore, a ϕ-factory is a unique source of monochromatic (110 and 127 MeV/c, respectively), slow [1], collinear, quantum-defined (pure $J^{PC} = 1^{--}$ quantum states), and tagged [2] neutral and charged kaons.

With $\mathcal{O}(10^{10})$ kaons per year we can form kaonic atoms and study K-Nucleon interactions at low energy (DEAR), produce nuclei with strangeness (*hypernuclei*, FINUDA), study particle physics fundamental symmetries (KLOE).

A BRIEF HISTORY OF KAONS

The kaonic enigma [7], [8] began in 1947 with their very discovery in p-N interactions such as $\pi^- + p \to \Lambda + K^0$. The long lifetime $\sim 10^{-10}$s, and the fact they were always produced in pairs, was explained by postulating a new quantum number *strangeness* conserved by strong interactions (production) but not weak interactions (decay) (Tab.1). In 1956, Lee and Yang postulated first, and Wu et al. demonstrated next (1957), that the $\tau - \theta$ puzzle [the same K^+ particle decaying to opposite parity $(2\pi, 3\pi)$ states] was actually caused by P and C being violated in weak decays, with CP conserved. In 1964 Christenson, Cronin, Fitch and Turlay disproved this hypothesis by studying K^0 decays. K^0 and $\bar{K^0}$ have same quantum numbers except S, which is not conserved in weak interactions; therefore, they have common virtual decay channels and they can mix when decaying [3]:

$$K^0 \leftrightarrow 2\pi\,(CP = +1) \leftrightarrow \bar{K^0} \qquad K^0 \leftrightarrow 3\pi\,(CP = -1) \leftrightarrow \bar{K^0}$$

with the CP eigenstates being

$$|K^0_{1,2}\rangle = (|K^0\rangle \pm |\bar{K^0}\rangle)\sqrt{2}$$

1) Kaons are produced with $\beta_K \simeq 0.2$, which means they travel a short path before decaying, and can be stopped after traversing very little matter. Experimentally, this means small detectors and thin targets

2) Detection of one K out of the two produced in the ϕ decay determines the existence and direction of the other K.

3) The 3π state has $CP = -1$ when in the L=0 state, the L=1 state being depressed by centrifugal potential

and time evolution

$$P(K^0,t) = 1/4[e^{-\Gamma_1 t} + e^{-\Gamma_2 t} - 2e^{-(\Gamma_1+\Gamma_2)t/2}\cos\Delta m t].$$

Neutral kaons decay preferably (Cabibbo-unfavored) to pions: since the phase space available for 2π is much greater than for 3π, lifetimes will be largely different $\Gamma_1 >> \Gamma_2$. Quite fortuitously, the oscillation frequency is $\Delta m \equiv |m_1 - m_2|$, large and observable, since the mass difference is comparable to the width of the short-lived state $\frac{\Delta m}{\Gamma_1} \sim 0.5$. Christenson et al. indeed showed that a 10^{-3} fraction of K_2's decay to 2π. Physical states are therefore

$$|K_L\rangle \propto |K_2\rangle + \epsilon|K_1\rangle \quad \text{and} \quad |K_S\rangle \propto \epsilon|K_2\rangle + |K_1\rangle.$$

The ϵ parameter describes *indirect* CP violation (CPV in short), which originates from the mass matrix mixing ($\Delta S = 2$ transitions), not from the interaction responsible of the decay.

The central role played by the $K^0 - \bar{K}^0$ mixing was revived in 1970. The suppressed nature of neutral-current decays such as $K_L^0 \to \mu\mu$ was explained through GIM-like box-diagrams and predicted a fourth flavour - *charm* - with mass $m_C^2 \propto \Delta m$.

Together with CPV from mixing, there can be a *direct* ($\Delta S = 1$) transition from $|K_L(CP = -1)\rangle$ to $|2\pi(CP = +1)\rangle$, without $K^0 \leftrightarrow \bar{K}^0$ intermediate transition. Direct CPV is parametrized by

$$\epsilon' \propto \frac{\langle 2\pi\, I = 2|H|K^0\rangle}{\langle 2\pi\, I = 0|H|K^0\rangle}.$$

The issue of CPV is a most fundamental one. As first emphasized by Sakharov in 1967, the predominant asymmetry of baryons versus antibaryons in the universe could be explained by three requirements: baryon number non-conservation, a CPV baryon-creating process, with baryons out of the thermal equilibrium. See M. Gleiser's talk at this Symposium [9].

Models proposed so far can be schematically distinguished in two classes, milliweak and superweak. Milliweak models advocate that a tiny (10^{-3}) fraction of the weak interaction is indeed CPV at leading order ($\Delta S = 1$). They imply T violation to guarantee the CPT theorem, and they can explain both direct [one ($\Delta S = 1$) transition] and indirect [two ($\Delta S = 1$) transitions]. Superweak models (Wolfenstein, 1964), on the other hand, postulate a new ($\Delta S = 2$), CPV process which transforms $K_L \to K_S$. The strength of this interaction relative to weak coupling is only 10^{-10}, since $M_{K_S} \simeq M_{K_L}$ and, therefore, no direct CPV is predicted nor allowed, no CPV is expected in other systems. The standard model (SM), a milliweak-type theory, *naturally* allows for CPV: the angle δ in the CKM matrix allows for CPV transitions

$$V_{CKM} = \begin{pmatrix} 1-\lambda^2/2 & \lambda & A\lambda^3\sigma e^{-i\delta} \\ \lambda & 1-\lambda^2/2 & A\lambda^2 \\ A\lambda^3(1-\sigma e^{i\delta}) & -A\lambda^2 & 1 \end{pmatrix} + \mathcal{O}(\lambda^3) \qquad (1)$$

where $\lambda \equiv \sin\theta_C$, $\rho - i\eta \equiv \sigma e^{-i\delta}$.

The indirect CPV parameter ϵ is proportional in the SM picture to $\Delta S = 2$ box diagrams (Fig.1a) $\epsilon \propto \lambda^4 \sin\delta \simeq 10^{-3} \sin\delta$. Therefore, ϵ is small per se because of the suppression of interfamily transitions, while the CPV angle δ is not necessarily small. Experimentally, $|\epsilon| = 2.26(2) \times 10^{-3}$.

The direct CPV parameter ϵ', on the other hand, is proportional in the SM picture to $\Delta S = 1$ *penguin* diagrams (Fig.1b) which are difficult to compute, and suppressed by both the $\Delta I = 1/2$ and the Zweig rule [4]. Calculations [11] have shown how in the SM picture direct CPV is a decreasing function of the quark masses in the penguin loop, primarily the top quark mass. Prediction for $m_{top} = 175\,\mathrm{GeV}$ is $0 \leq \epsilon'/\epsilon \leq 0.001$.

One can try and use the precise measurement of $|\epsilon|$ to set bounds on δ, although along with the use of the experimental values from the $B_d^0 - \bar{B}_d^0$ mixing and the ratio of CKM elements $|V_{ub}|/|V_{cb}|$, obtaining [10] the qualitative picture in Fig.2.

Finally, the most precise measurements [5] of the relative strength of direct versus indirect CPV yield result in mild agreement: $\Re(\epsilon'/\epsilon) = (7.4 \pm 5.2 \pm 2.9) \times 10^{-4}$ (E731 at Fermilab) and $\Re(\epsilon'/\epsilon) = (23.0 \pm 3.6 \pm 5.4) \times 10^{-4}$ (NA31 at CERN), from the double-ratio

$$\left|\frac{\eta_{+-}}{\eta_{00}}\right|^2 = \frac{N(K_L \to \pi^+\pi^-)/N(K_S \to \pi^+\pi^-)}{N(K_L \to \pi^0\pi^0)/N(K_S \to \pi^0\pi^0)} \simeq 1 + 6\Re(\epsilon'/\epsilon) \qquad (2)$$

with other useful relations holding

$$\eta_{+-} \equiv \epsilon + \epsilon' \equiv \frac{\langle\pi^+\pi^-|H|K_L\rangle}{\langle\pi^+\pi^-|H|K_S\rangle} \quad \eta_{00} \equiv \epsilon - 2\epsilon' \equiv \frac{\langle\pi^0\pi^0|H|K_L\rangle}{\langle\pi^0\pi^0|H|K_S\rangle}.$$

From the body of information available, the CPV angle δ seems to be large ($\rho \simeq 0 \Rightarrow \delta \simeq \pi/2$); a large top mass (175 GeV) predicts $\Re(\epsilon'/\epsilon)$ to be in the 10^{-4} region [11]; a measurement of $\Re(\epsilon'/\epsilon) \neq 0$ will indicate the general validity of the CKM picture, since it requires the presence of a $\Delta S = 1$ phase.

KLOE

The KLOE experiment was proposed in April 1992; approved and funded at the beginning of 1993, construction of the detector began in 1994. The principal aim of KLOE is the detection of direct CP violation in K^0 decays

---

4) We do not discuss here the third piece of experimental information that can be used to set limits to the CPV angle δ, i.e. the experimental bounds on the electric dipole moments of the neutron and the electron. See [10] and references therein for a thorough review.

5) Experimentally, $|\eta_{+-}| \simeq |\eta_{00}| \simeq 2 \times 10^{-3}$ and $\phi_{+-} \simeq \phi_{00} \simeq 44^o$. Therefore, $\epsilon'/\epsilon \simeq \Re(\epsilon'/\epsilon)$.

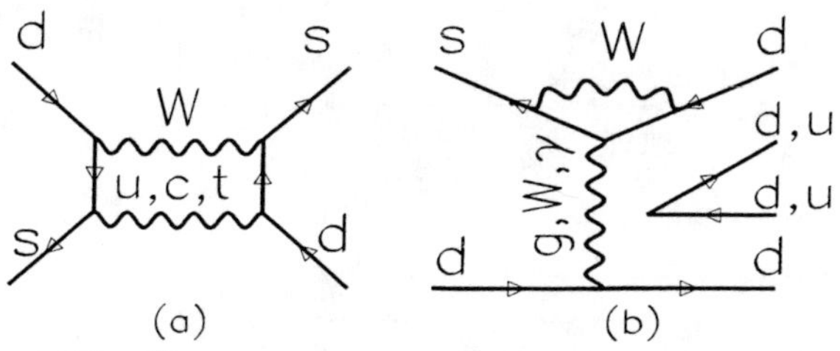

FIGURE 1. a) $K^0 - \bar{K^0}$ mixing b) Direct CP violation through *penguin* diagrams.

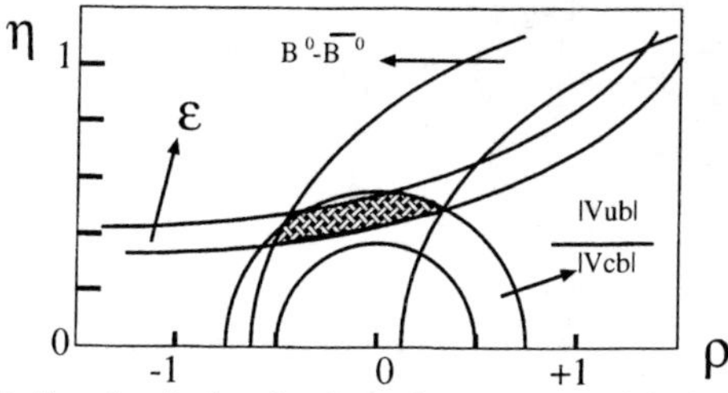

FIGURE 2. Qualitative sketch showing bounds on the CP violating angle of CKM matrix.

with a 10^{-4} sensitivity on $\Re(\epsilon'/\epsilon)$. For a comprehensive review see [12] and references therein, while an up-to-date status report can be found in [13]. The measurement in KLOE of $\Re(\epsilon'/\epsilon)$ with the required precision takes advantage of unique features of a ϕ-factory: numerous kinematical constraints exist, K_S and K_L are detected at the same time, clean environment typical of e^+e^- machines with respect to experiments at extracted hadron beams, built-in calibration processes. On the down side, the K_S and K_L are detected with different topologies in both the charged and the neutral decay modes.

Quite generally, KLOE should be regarded as a precision kaon interferometer. As already stressed, collinear kaon pairs are produced from the ϕ decay in a pure, coherent quantum state. If one of the two neutral kaons from the ϕ decays into a state f_1 at time t_1 and the other to a state f_2 at time t_2, the decay intensity to f_1 and f_2 as a function of $\Delta t \equiv t_1 - t_2$ is

$$I(f_1, f_2; \Delta t; \forall \Delta t > 0) = \frac{1}{2\Gamma}|\langle f_1|K_S\rangle\langle f_2|K_S\rangle|^2[|\eta_1|^2 e^{-\Gamma_L \Delta t} + |\eta_2|^2 e^{-\Gamma_S \Delta t} - 2||\eta_1||\eta_2|e^{-\Gamma\Delta/2}\cos(\Delta m \Delta t + \phi_1 - \phi_2)] \tag{3}$$

where $\eta_i \equiv \langle f_i|K_L\rangle / \langle f_i|K_S\rangle$. The interference term in the decay intensities above gives measurements of all sixteen parameters describing CP and CPT (if any) violations in the neutral kaon system [10]. As instance, when $f_1 = f_2$, one measures Γ_S, Γ_L and Δm; when $f_1 = \pi^+\pi^-$, $f_2 = \pi^0\pi^0$, one measures $\Re(\epsilon'/\epsilon)$ at large Δt, and $\Im(\epsilon'/\epsilon)$ near $\Delta t = 0$.

Other physics topics include rare K_S^0 decays (10^{10} kaons per year will improve the sensitivity to branching ratios down to the 10^{-8} range), tests of Chiral Perturbation Theories, radiative ϕ decays, and $\gamma\gamma$ physics [14]. Out of such a physics spectrum other than CPV, it should be underlined the im-

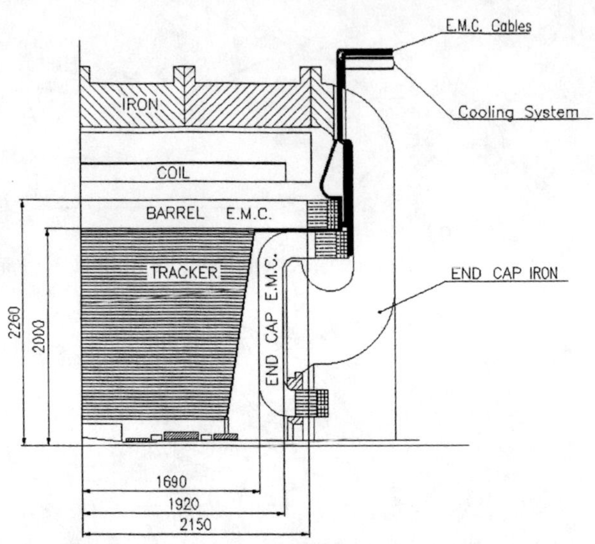

FIGURE 3. Cross-sectional schematics of the KLOE detector (one quadrant, side view). Dimensions are millimeters.

portance of a study of the $J^{PC} = 0^{++}$ $f_0(975)$ meson state, produced via the ϕ radiative decay with a branching ratio at least 1×10^{-6}. The nature of the lightest scalar is still unclear, whether a KK molecule, an exotic $qq\bar{q}\bar{q}$ or a glue-ball state. The decay $f_0 \to \pi^0\pi^0$ should be easily measured even at low luminosity during the machine commissioning [15], while the charged channel $\pi^+\pi^-$ suffers by backgrounds from continuum.

To reach the design sensitivity on the measurement of $\Re(\epsilon'/\epsilon)$ via the classical double-ratio method (Eq.2), a one-year statistics at full DAΦNE luminosity is necessary. It is also necessary to control the detector efficiency for the decays of interest and to reject backgrounds from the copious K_L^0 decays to states other than two-pion.

The KLOE detector [16] (fig.3) is a hermetic, 4π geometry apparatus: a cylindrical structure surrounding the beam pipe and consisting of a vertex chamber, a large drift chamber, an electromagnetic calorimeter with state-of-the-art energy and time resolution [17], [18], [19], and a 6 kG superconducting magnet. Counting of the four K_S, K_L decays used in the double-ratio formula (eq.2) is performed by defining a K_S fiducial volume (10 cm radius around the interaction point, about 17 lifetimes), and a K_L fiducial volume (145 cm, about one half lifetime). It is very important to avoid systematical errors in the determination of the boundaries of the fiducial volumes for neutral and charged decay channels. This is accomplished by using K_L decays where both the neutral and the charged vertex are measured, as in $K_L \to \pi^+\pi^-\pi^0$.

The KLOE calorimeter will reconstruct the $\pi^0\pi^0$ mode, determine the decay vertex space location, reject the $3\pi^0$ decay, and provide π/μ rejection. The technique employed is a Pb-scintillator sampling with 1-mm scintillating fibers embedded in very thin Pb grooved foils, which has been demonstrated to provide good energy ($\sigma_E/E = 4\%/\sqrt{E[\text{GeV}]}$) and space ($\sigma_x \simeq 0.5\,\text{cm}$, $\sigma_z \simeq 2\,\text{cm}$) resolution, acceptable efficiency for photon energies down to 20 MeV, and spectacular time resolution ($\sigma_t = 55\,\text{ps}/\sqrt{E[\text{GeV}]}$). The measurement of the time of arrival t_γ and the impact point (x, y, z) of one photon out of four at the calorimeter, and the knowledge of the flight direction $\cos\theta$ from the K_S tagging, gives the K_L decay length L_K to an accuracy of 0.6 cm, by using the formulae $t_\gamma = L_K/(\beta_K c) + L_\gamma/c$ and $L_\gamma^2 = D^2 + L_K^2 - 2DL_K\cos\theta$, where $D^2 \equiv x^2 + y^2 + z^2$.

A He-based, very transparent, drift chamber will reconstruct the $\pi^+\pi^-$ final state, reject the $K_{\ell 3}$ background, and determine the K_S^0 flight direction and the charged decay vertices. The KLOE drift chamber has space resolution $\sigma_{\rho,\phi} = 200\,\mu\text{m}$ and $\sigma_z = 3\,\text{mm}$, 0.5% relative resolution on the transverse momentum, and a $\geq$ 98% track reconstruction efficiency averaged over the $K_L^0 \to \pi^+\pi^-$ decay.

To achieve the required statistical sensitivity, the entire ϕ event rate (5 kHz) has to be written on tape. The calorimeter will provide triggering for most of the decay channels of interest. The rejection of three neutral-pion decays relies on the calorimeter's hermiticity and efficiency for photons down to 20 MeV energy. The rejection of the $K_{\mu 3}$ decay over the $\pi^+\pi^-$ decay of interest is based on several kinematical variables and constraints. The high-resolution features of the tracking limits the residual contamination to 4.5×10^{-4}, at a 6 kG solenoidal field, with a 99.8% $\pi^+\pi^-$ efficiency.

KLOE expects to be operational at beginning of 1998.

FINUDA

Nuclear physics topics will be investigated by the FINUDA (standing for *FIsica NUcleare a DAΦne*) experiment. A nuclear physics experiment carried out at an e^+e^- collider sounds contradictory in itself, but this is where the uniqueness of the idea lies [20]: charged kaons from ϕ decays are used as a monochromatic, slow (127 MeV/c), tagged, background-free, high-counting rate beam on a thin target surrounding the beam pipe. The possibility of stopping low-momentum monochromatic K^- with a thin target (typically $0.5\,\text{g}\,\text{cm}^{-2}$ of ^{12}C) is unique to DAΦNE : K^-'s can be stopped with minimal straggling very near the target surface, so that outgoing prompt pions do not cross any significant amount of the target and do not undergo any momentum degradation. This feature provides unprecedented momentum resolution as long as very transparent detectors are employed before and after the target.

The FINUDA detector is optimized to perform high-resolution studies of hypernucleus production and non-mesonic decays [21], by means of a spectrometer with the large acceptance typical of collider experiments.

Negative kaons stopping inside the target produce a Y-hypernucleus ($Y = \Lambda, \Sigma, ...$) via the process

$$K^-_{stop} + {}^{A}Z \rightarrow {}^{A}_{Y}Z + \pi^-, \tag{4}$$

where the momentum of the outgoing π^- is directly related to the level of the hypernucleus formed (two-body reaction). In the case of Λ hypernucleus formation, the following weak-interaction 'decays' are strongly favored in medium-heavy nuclei

$$\Lambda + n \rightarrow n + n \qquad\qquad \Lambda + p \rightarrow n + p,$$

with the nucleus undergoing the reactions

$${}^{A}_{\Lambda}Z \rightarrow {}^{(A-2)}Z + n + n \qquad\qquad {}^{A}_{\Lambda}Z \rightarrow {}^{(A-2)}(Z-1) + n + p,$$

which are interesting for studying the validity of the $\Delta I = 1/2$ rule [21].

Hypernuclei are a unique playground for both nuclear and particle physics [22]. A Λ embedded in a nucleus can occupy, because of the strangeness content, levels forbidden to a p and a n by the Pauli exclusion principle. Furthermore, the $I = 0$ assignment makes the ΛN interaction much weaker than the ordinary NN interaction. In the shell model of nuclei the quarks are confined in bags inside baryons. Baryons inside nuclei maintain their identity and interact by the exchange of mesons. At short distance the bags can fuse, quarks can get deconfined and begin interacting by exchanging gluons. For a ${}^{5}_{\Lambda}He$ hypernucleus the two models give different predictions: while in the baryon model the Λ occupies the $S_{\frac{1}{2}}$ orbital due to its strangeness quantum number, in the quark model the u and d quarks inside the Λ cannot stay in the $S_{\frac{1}{2}}$ but are obliged to move to the $P_{\frac{2}{3}}$ orbital by Pauli blocking. As a result, different binding energies are predicted by the models.

Besides studies on hypernucleus spectroscopy, interest exists for K-N interactions at low energy [23]. Recently, the possibility of improving the precision on the K_{e2} branching ratio by a factor of 3 over the present world average was also discussed [24].

FINUDA [25] is a magnetic spectrometer with cylindrical geometry, optimized in order to have large solid-angle acceptance, optimal momentum resolution (of the order of 0.3 % FWHM on prompt pions) and good triggering capabilities. The geometrical acceptance is 100% in the ϕ angle, and approximately $135^0 \leq \vartheta \leq 45^0$, thus naturally rejecting e^+e^- background from Bhabha scattering. FINUDA consists of (fig.4) an interaction/target region, external tracker, outer scintillator array, and superconducting solenoid.

The (K^+, K^-) pairs from ϕ decay emerge from the interaction region (fig.4), and are identified for triggering by exploiting the back-to-back event topology

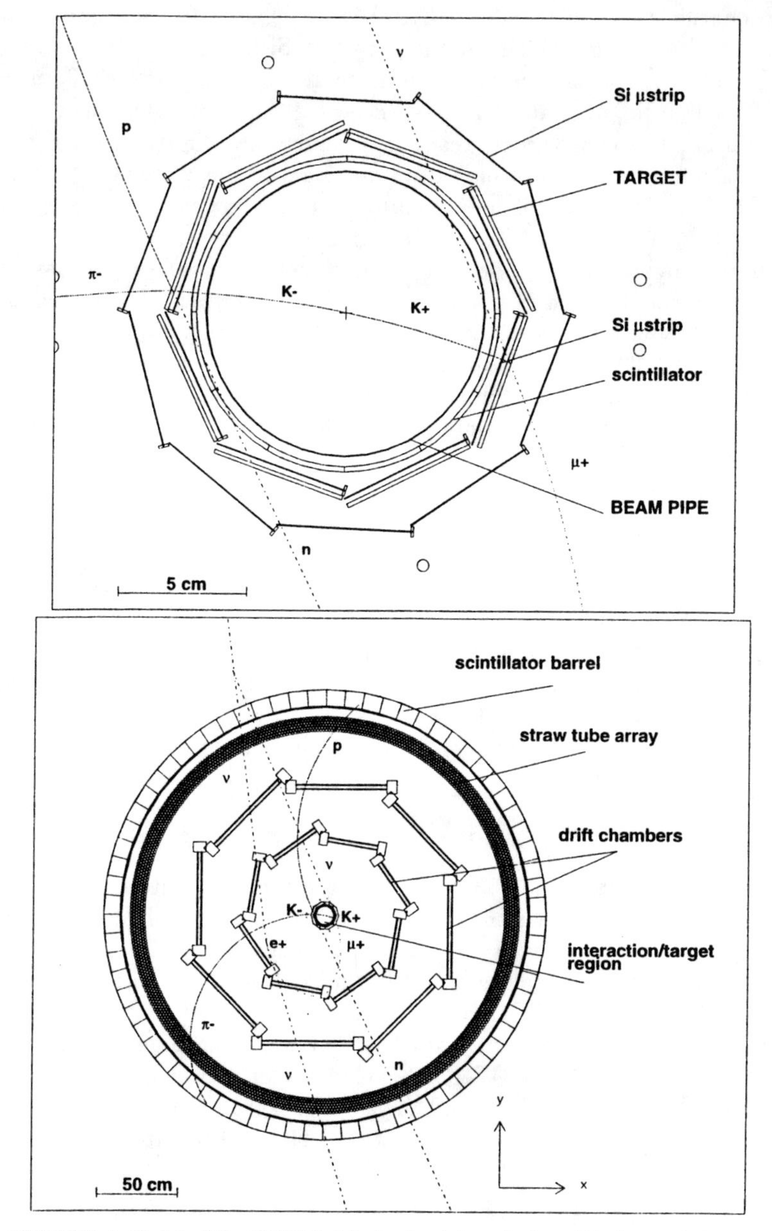

FIGURE 4. Sketch of the FINUDA detector front (beam) view. The interaction-target region (top); the outer tracker and scintillator barrel (bottom). An hypernuclear event is superimposed. A ϕ decays to K^+K^-. The K^+ decays to $\mu^+\nu_\mu$, followed by $\mu^+ \rightarrow e^+\nu_e\bar{\nu}_\mu$. The K^- crosses the inner trigger scintillator, is tracked by the inner Si-μstrip detector, stops in the target and forms an hypernucleus (eq.4). The momentum of the emitted prompt π^-, proportional to the energy of the hypernuclear level formed, is measured by the spectrometer. The hypernuclear Λ undergoes the process $\Lambda p \rightarrow n\, p$.

and by identification through $-(dE/dx)$ in the TOFINO triggering scintillator array. Before impinging on the target, a Si-μstrip array measure the (K^+, K^-) coordinates before the nuclear target with $\sigma \sim 30 - 50\mu$m and provides particle identification by $-(dE/dx)$. The monochromatic K^- is stopped inside the thin target, reaction products are emitted isotropically: the momentum of the prompt pion is proportional to the energy of the hypernuclear level formed, and it is measured by Si-μstrip, low-mass drift chambers [27], and straw tube arrays [28]. The signal in the outer scintillator array (TOFONE) provides a fast trigger logic coincidence. Baryons or mesons from Λ decays are also tracked in the spectrometer (protons, $\pi^\pm$) or detected in the TOFONE (neutrons). The expected FWHM resolution on the 270 MeV/c prompt pion is 0.25% for forward pions (i.e., those emitted towards the outer region of the spectrometer), which corresponds to a 0.7 MeV resolution on the hypernuclear level, to be compared to the best result achieved so far at fixed-target of about 2 MeV [29].

The rate of reconstructed hypernuclear events with resolution better than 1 MeV at $\mathcal{L} = 10^{32}\mathrm{cm}^{-2}\mathrm{s}^{-1}$ is given by the expression:

$$R({}_\Lambda Z) = R_\phi \times B_{K^+K^-} \times \frac{N_{{}_\Lambda Z}}{K_{stop}} \times \epsilon_{\pi^-} \times \epsilon_{transp} \simeq 2.5 \times 10^{-2}\mathrm{s}^{-1} \simeq 100\,\mathrm{evnts/hr}$$

where $R_\phi = 5 \times 10^2\mathrm{s}^{-1}$, $B(\phi \to K^+K^-) = 0.49$, $N_{{}_\Lambda Z}/K_{stop} = 10^{-3}$ is the capture rate, $\epsilon_{\pi^-} = 13\%$ is the total efficiency for forward (high-resolution) prompt pions, and $\epsilon_{transp} = 80\%$ is the chamber transparency.

The FINUDA experiment expects to be operational by end 1997.

DEAR

The DEAR experiment [31] [30] will measure the KN scattering length by studying the shift and the width of the energy levels of kaonic hydrogen atoms, i.e. atoms in which a K^- from the ϕ decay will replace the orbital electron in a hydrogen target. The negative kaon will then cascade down initially through Auger transitions, and finally through emission of X-rays. The energy of the X-rays will be measured by an array of CCD (Fig.5), detailed simulations based on a upgraded version of the GEANT3 code to treat photon energies down to 1 keV show that both machine and physical background can be kept under control. The DEAR Collaboration (a total of 11 Institutes from 6 countries) will take data as soon as the first beam circulates in the DAΦNE rings.

CONCLUSIONS

A wealth of physics ranging from CP violation to hypernuclear studies is expected from DAΦNE . The beginning of machine commissioning is planned for mid 1997; construction of the detectors is well underway.

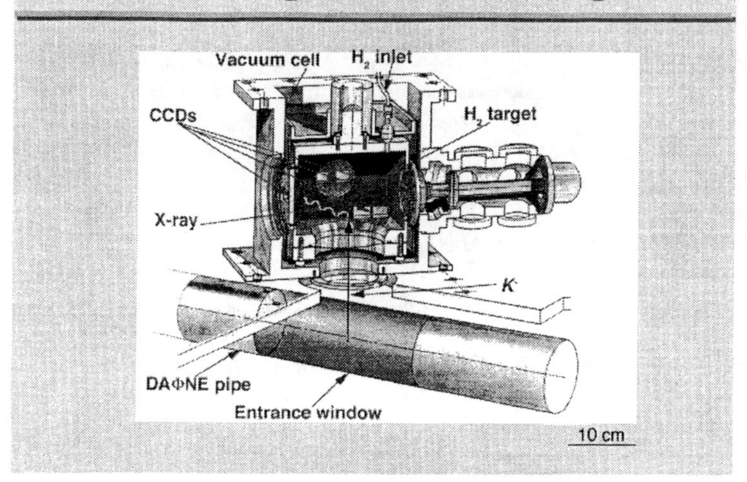

FIGURE 5. Cut-out sketch of the DEAR setup.

I gratefully acknowledge the help and information given by my colleagues on the DAΦNE project team, DEAR collaboration, FINUDA collaboration, and KLOE collaboration, particularly G. Vignola, C. Guaraldo, C. Petrascu, F.L. Fabbri, J. Lee-Franzini, P. Franzini and S. Miscetti. Finally, I wish to thank the Organizing Committee for a perfectly enjoyable Symposium.

REFERENCES

1. Maiani L., et al. (Eds.), *The Second* DAΦNE *Physics Handbook*, Frascati: SIS Laboratori Nazionali di Frascati (1995).
2. Baldini R., et al. (Eds.), *Proceedings II Work. on Phys. and Detector for* DAΦNE , Frascati Physics Series **4**, Frascati: SIS Laboratori Nazionali di Frascati (1996).
3. The Frascati home page is `http://www.lnf.infn.it/`.
4. The ϕ-factory group, *Proposal for a ϕ-factory*, Frascati Report 90/031(R) (1990).
5. The DAΦNE project team, DAΦNE *Machine Project*, Proc. EPAC 94, London, 1994, also Frascati report LNF-94/055(P) (1994).
6. Status Report on DAΦNE , G.Vignola, ref. [2].
7. Any textbook will discuss the basics of kaon physics. For a clear, elementary level discussion see: Perkins D.H., *Introduction to High Energy Physics*, Addison Wesley (1987).
8. For a more advanced and up-to-date review see: Leader E. and Predazzi P., *An introduction to gauge theories and modern particle physics*, Cambridge Monographs on Particle Physics **4**, Cambridge: Cambridge University Press (1996).

9. Gleiser M., these Proceedings.
10. Peccei R., *Proc. 23rd INS Int. Symp. on Nucl. and Part. Phys.*, Tokyo: Univ. Academy Press, Inc., 3 (1995).
11. Buchalla G., et al., *Nucl. Phys.* **B349**, 1 (1991).
12. Bertolucci S., et al., *Nucl. Phys. B (Proc. Suppl.)* **37A**, 43 (1994).
13. Lee-Franzini J., *Proceedings Heavy Quarks at Fixed Target, St.Goar (Germany), Oct. 3rd, 1996*, Frascati Physics Series **7**, 363, Frascati: SIS Laboratori Nazionali di Frascati (1997).
14. F. Anulli *et al.*, LNF Report 95-007(P) (1995).
15. Lee-Franzini J., Kim W., Franzini P. J., *Phys. Lett.* **B287**, 259 (1992).
16. The KLOE Collaboration, LNF Reports LNF-92/019 (1992), LNF-93/002 (IR) (1993), LNF-94/028 (1994), LNF-95/014 (1995), LNF-96/043 (1996).
17. Babusci D., et al., *Nucl.Instr. and Meth.* **A332**, 444 (1993).
18. Antonelli A., et al., *Nucl. Instr. and Meth.* **A354**, 352 (1995).
19. M. Antonelli et al., *Proc. ICHEP 96* , in press, Warsaw (1996).
20. Bressani T., *Proc. Work. on Phys. and Detector for* DAΦNE , Frascati, April 9-12 1991, 475 (G. Pancheri Ed.) (1991).
21. Bressani T., *Non-mesonic Decay of Hypernuclei and the* $\Delta I = 1/2$ *Rule*, ref. [1].
22. Fabbri F. L., Frascati Report 96/062(R) (1996).
23. Olin A., *Low-energy K-N Scattering with FINUDA*, ref. [2].
24. The FINUDA Collaboration (M. Agnello et al.), *The* K_{e2} *decay with FINUDA at DAFNE* , presented by T.Bressani at the Workshop on K Physics, Orsay, France, May 30-June 4 (1996).
25. The FINUDA Collaboration (M. Agnello *et al.*), LNF Reports LNF-93/021(IR) (1993), LNF-95/024(IR) (1995).
26. Bianco S., *Proc. 23rd INS Int. Symp. on Nucl. and Part. Physics*, Tokyo: Univ. Academy Press, Inc., 347 (1995). Also a LNF Report LNF-95/028(P) (1995).
27. Agnello M. et al., *Nucl. Instr. and Meth.* **A379**, 411 (1996).
28. Benussi L., et al., *Nucl. Instr. and Methods* **A361**, 180 (1995).
29. Hasegawa T., et al., *Phys. Rev. Lett.* **74**, 224 (1995).
30. Baldini R. et al. (The DEAR Collaboration), LNF Report LNF-95/055(IR) (1995).
31. Guaraldo C., et al., *The Kaonic Hydrogen Puzzle*, ref. [2].

CERN's Large Hadron Collider Project

Tom A. Fearnley*

*Niels Bohr Institute, Copenhagen. Currently at CERN - PPE, CH - 1211 Geneva 23, Switzerland

Abstract. The paper gives a brief overview of CERN's Large Hadron Collider (LHC) project. After an outline of the physics motivation, we describe the LHC machine, interaction rates, experimental challenges, and some important physics channels to be studied. Finally we discuss the four experiments planned at the LHC: ATLAS, CMS, ALICE and LHC-B.

INTRODUCTION

The Standard Model (SM) has now been tested at a precision of a few permille, and it has withstood all tests. Experiments at CERN, SLAC and Fermilab, among other laboratories, have resulted in extraordinary agreement between theory and data, effectively validating the Standard Model up to the electroweak (EW) energy scale $\sqrt{s} = 100$ GeV, with one qualification: there is a crucial piece missing in the verification of the Standard Model, which is, of course, the mechanism responsible for the spontaneous electroweak symmetry breaking (of SU(2) $\times$ U(1)), which gives different masses to the electroweak gauge bosons and to the quarks and leptons. Recall that $m_W/m_\gamma > 10^{26}$! In essence, the origin of mass has not been established. In the Standard Model the Higgs mechanism breaks the symmetry and generates mass. A scalar Higgs field permeating vacuum has a non-zero vacuum expectation value. Associated with this field is a neutral scalar boson H^0, whose mass is unknown, basically because m_W and m_t have not been measured with sufficient precision. Nonetheless the mass of H^0 (m_H) is constrained by direct searches: $m_H > 65$ GeV (LEP), and by theory: $m_H < 1000$ GeV. Also, "indirect" estimates of m_H have been made by fitting electroweak data from LEP, SLD, Tevatron and νN data (since these depend logarithmically on m_H). Recent results [1] seem to favour a "light" Higgs, but the errors are very large. At 95 % C.L. $m_H < 550$ GeV.

CP400, *First Latin American Symposium on High Energy Physics/ VII Mexican School of Particles and Fields*, edited by D'Olivo/Klein-Kreisler/Mendez

The search for Higgs is central at the LHC, as it is today at Fermilab's Tevatron and CERN's LEP, where the mass limit can be pushed up to $m_H > 90$ GeV if the Higgs is not discovered. CERN's LHC is a true "discovery machine" in the sense that it will reach unprecedented energy and luminosity (L) in proton-proton collisions: $\sqrt{s} = 14$ TeV and L = 10^{34} $cm^{-2}s^{-1}$.

Such a discovery machine clearly opens up a vast field of SM and possibly non-SM physics. Examples are:
- Top physics: this is an ideal testing ground for non-SM physics, as the Standard Model gives testable predictions for top physics once the top mass has been measured (more precisely than today).
- CP violation: One would like to observe and study CP violation in the B meson system, where the violation may be significant. So far CP violation has only been measured in the $K^0\bar{K^0}$ system. B mesons will be produced copiously at the TeV energy scale. In addition to CP violation, interesting B physics waiting to be done at the TeV scale include mixing of neutral B mesons, lifetimes, and spectroscopy of beauty baryons. A dedicated B physics experiment is being planned at the LHC (LHC-B), to which we will return later.

- Non-SM physics: First, a non-discovery of Higgs at the LHC would be extremely intriguing, and would trigger intense experimental and theoretical searches for alternative mechanisms for electroweak symmetry breaking. Technicolor or other dynamic mechanisms have been proposed. In fact, whether Higgs is discovered or not, the task will be to search for new, non-SM physics. Physicists will not be content with discovering the Higgs, since there are strong *theoretical* (albeit not yet experimental) reasons for believing that the Standard Model is a low-energy approximation of a new and fundamental theory which should manifest itself at the TeV scale. This belief is first and foremost founded in the "naturalness" (or "fine-tuning") problem, which draws our attention to the incredible fine-tuning which is required within the Standard Model in order to keep m_H from exploding at high energy scales due to quadratically divergent quantum corrections. A way out is to assume that the energy scale of the Standard Model has an upper bound (cut-off), beyond which new physics should appear and modify the ultraviolet ($\Lambda \gg \Lambda_{max}$) behaviour of the SM. By making a few simple assumptions one finds that this cut-off is likely to be in the few-TeV regime. Another argument in favour of a more fundamental underlying theory: Although internally self-consistent, SM contains many free parameters (18 if neutrinos are massless, 25 if they have mass). These cannot be computed in the context of the SM. Can a theory with so many free parameters be fundamental ? These problems have led to the development of other, non-SM theories:

- Supersymmetry (SUSY): it transforms bosons to fermions and vice-versa. It "solves" the naturalness problem by exactly cancelling out the problematic loops which cause the divergence of m_H. The mass of the SUSY particles must be of order 1 TeV in order to solve the naturalness problem.
- Minimal Supersymmetric SM (MSSM): it is a supersymmetric extension of the SM, which thus can go beyond the TeV scale without provoking the naturalness problem. MSSM contains five Higgs bosons: scalar $h^0, H^0, H^\pm$ and pseudoscalar A^0. Each SM particle is accompanied by a SUSY partner (squarks, sleptons, and fermionic super-partners to SM bosons, e.g. gluinos), plus charginos and neutralinos. One common objection to SUSY (and MSS-M) is that it generically increases the number of free parameters. So while it solves one problem (naturalness), it hardly solves the problem of the many free parameters ! So far there are no experimental signs of these SUSY particles, and lower limits are being pushed upwards (in the GeV range). Search for SUSY particles has high priority at LHC.
- Technicolor (TC): is a proposed new force (technicolor interaction) which should become strong at energy scale around 500 GeV. It breaks the EW symmetry without any fundamental Higgs boson. Instead, the "Higgs" is a composite particle, similar to the meson in QCD. The experimental signal is the presence of strongly interacting dynamics at the TeV scale, which should produce new resonnances similar to those found in the hadron spectrum at the GeV scale. One of the problems with TC is that it predicts sizable corrections to the Standard Model at LEP energies, not seen in the precision data from LEP. For this and other reasons technicolor is not the "preferred" solution.
- Compositeness: quark and/or lepton substructure, which could explain the EW symmetry breaking dynamically. Quark substructure can be probed by looking for deviations from QCD in the p_t distribution of jets. At LHC the compositeness scale (Λ_c) could be probed up to 15 TeV [2], a factor 10-15 above todays limit of about 1 TeV (CDF).
- Other searches: new Z' or W' heavy vector gauge bosons, leptoquarks, and anomalous couplings in gauge boson pair production (may probe the origin and structure of the Higgs sector).

Other physics:
- Ultra-relativistic heavy ion collisions: at very high energy densities (above 1-3 GeV/fm^3) nuclear matter is expected to undergo a phase transition to quark gluon plasma (deconfinement). LHC will be equipped to accelerate counterrotating heavy ions (up to lead) to 5.5 TeV/nucleon, or 1150 TeV for the lead nucleus, at a luminosity of $10^{27}\ cm^{-2}s^{-1}$ [3]. An energy density of 5-8 GeV/fm^3 will be reached, well above the expected critical density for creation of quark-gluon plasma, and above the density produced by RHIC (3-5 GeV/fm^3).

Electron-proton collisions: CERN is keeping the option open to collide electrons with protons in a modified LHC (re-using LEP equipment). Such a "high-energy HERA" machine would allow unprecedented probing of structure functions and compositeness, as well as searches. The option being studied envisages 60 GeV electrons colliding with 7 TeV protons at $L > 10^{32} cm^{-2}s^{-1}$ (compared with HERA's 30 GeV $\oplus$ 0.82 TeV).

To summarize, the physics motivation and potential for experiments at the LHC in the multi-TeV regime is very strong. It allows us to *i)* search for the Higgs boson, or other mechanisms for electroweak symmetry breaking; *ii)* expand and deepen our knowledge of Standard Model physics, including CP violation and top physics; *iii)* search for new, non-SM physics, for which there are strong theoretical arguments. SUSY is an interesting candidate, but also the unexpected should be expected; and *iv)* study ultra-relativistic heavy ion collisions at extremely high energy densities, where a phase transition to quark-gluon plasma is expected to occur.

THE LHC MACHINE

The project was approved by the CERN Council in December 1994. The LHC machine is a proton-proton (and ion-ion) collider which will be housed in the existing LEP tunnel. As seen in Figure 1, CERN's accelerator complex will be fully employed, from the LINAC (50 MeV), through the Proton Synchrotron (PS) booster (1.4 GeV), the PS (25 GeV), the Super Proton Synchrotron (SPS, 450 GeV), and finally the LHC (7 TeV). As mentioned above, the design specification is a proton-proton centre-of-mass energy of 14 TeV (7 times Tevatron) and a peak luminosity of 10^{34} $cm^{-2}s^{-1}$ (about 1000 times *today's* Tevatron, which however is being upgraded). Table 1 lists some key parameters of the machine. We notice that the bunch interval is only 25 ns, which is short compared with the bunch separation in existing machines. It translates into a very high repetition rate (40 MHz) and an even higher interaction rate ($\sim$ 0.7 GHz), which makes experimentation at LHC extremely challenging (trigger, data acquisition, radiation).

To achieve the specified beam energy, the LHC magnets must produce a magnetic field of 8.28 Tesla, which is significantly higher than the fields in the Tevatron (4.5 T) and HERA (5.5 T). The high field is made possible (and has been exceeded in tests [4] by using a large number of superconducting magnets (around 5000 in total) [5]. The magnets will be cooled by superfluid Helium to 1.9° K. In addition, the coils will be protected against synchrotron radiation by means of a liquid Helium insulation at 4.2° K. The total cold mass is 30000 tons. Cooling this mass down without large thermal gradients is a formidable task! Thanks to the superconductivity, the LHC will deliver

TABLE 1. Main parameters of the LHC machine for proton-proton and lead ion-ion collisions.

Parameter	pp	Pb ions
Operational energy [TeV]	14	1150
Dipole field [T]	8.28	8.28
Luminosity [$cm^{-2}s^{-1}$]	10^{34}	10^{27}
Bunch intervals [m | ns]	7.5 | 25	40.5 | 135
Particles per bunch	10^{11}	10^{8}
Number of bunches	2835	496
Particles per beam	2.8 10^{14}	4.7 10^{10}

FIGURE 1. The CERN accelerator complex. Protons will pass through the linac, the PS booster, the PS and the SPS, before being injected into the LHC.
Figure from [10].

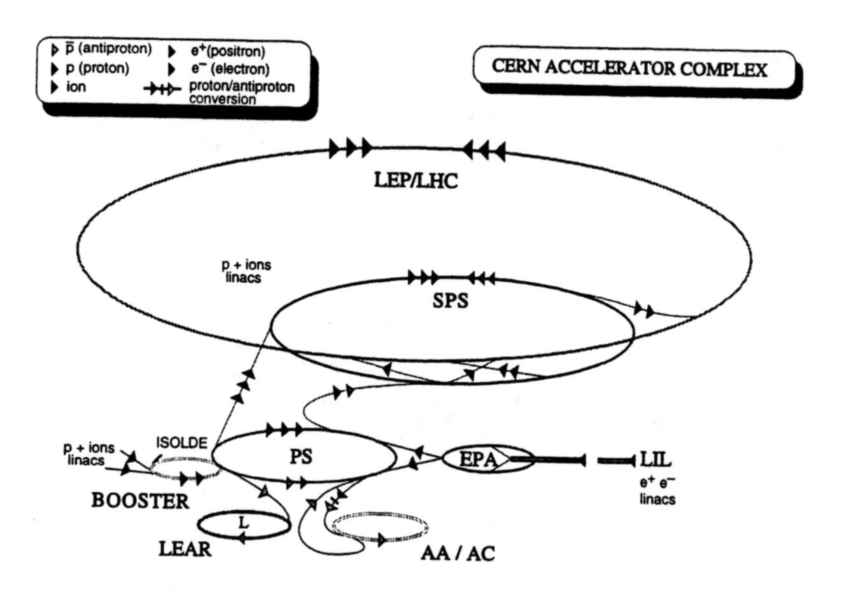

25 times the energy and 10000 times the luminosity of CERN's existing SPS collider at comparable power consumption. Also the RF accelerating cavities are superconducting.

The dipole magnets will use a novel and economical design coined "2-in-1", in which the two dipole windings for the counter-rotating proton beams are embedded in the same iron yoke and cryostat.

INTERACTION RATES AND EXPERIMENTAL CHALLENGES

The following concerns proton-proton collisions in the LHC.
- Rates: At $\sqrt{s}$ = 14 TeV the non-diffractive inelastic proton-proton cross section is expected to be *approximately* 70 mb [7]. At peak LHC luminosity (about 10^{34} $cm^{-2}s^{-1}$), the interaction rate then becomes *approximately* 0.7 GHz. Since the bunch crossing rate is 40 MHz this means that there will be on average about 18 interactions per bunch crossing [7]. Hence, every 25 ns one will observe what will look like one big "event", which in fact is $\sim$ 18 overlapping ("pile-up") events. It will be very difficult to determine the vertex of each of these $\sim$ 18 interactions. Thus, whenever vertexing is crucial, for instance in dedicated B physics runs, it may be necessary to run the LHC below maximum luminosity (by an order of magnitude).
- Trigger: The large bunch crossing rate at peak luminosity (40 MHz) must be reduced to a manageable level ($\sim$ 100 Hz) by a sophisticated 3-level trigger (online event selection). To achieve the necessary data reduction, very complex algorithms for physics selection are needed, which will run on a "farm" of powerful computers with total capacity of the order of 10^6 MIPS [6]. Even at a final 3rd level trigger rate of 100 Hz, the data rate to be recorded reaches $\sim$ 100 MBytes/s.
- Trigger signals: Since an interesting event will typically be embedded in large background (from other interactions in the same bunch crossing), clean signatures are required for the triggering, for instance high-p_t leptons and photons, jets, and missing transverse energy (E_t^{miss}). This sets stringent requirements on the detectors.
- Speed: It is not feasible to make a trigger decision during the 25 ns between bunch crossings. Nonetheless a decision must be made for every crossing ! The solution is to use pipelined trigger processing/memory, in which many bunch crossings are processed concurrently by a chain of processing/memory elements. These pipelines must sit on the detectors. This requires efficient cooling of several million channels which each generate several mW of heat.
- Multiplicity: The expected multiplicity of a typical proton-proton interaction at 14 TeV is of the order of 100. With about 18 interactions per bunch crossing, the average total multiplicity will be almost 2000 particles per (over-

lapping, pile-up) "event", or roughly 80 billion particles per second ! Such high multiplicities require detectors with high granularity and good time resolution to ensure low occupancy, particularly in the central region. In other words, the total number of detector channels must be large ($\sim 10^7$).
- Radiation: The radiation will mainly consist of charged particles and neutrons, the latter coming from the breakup of nuclei in the calorimeters (induced by hadronic showers). The radiation level will vary within a detector, depending on the magnetic field (looping particles), the calorimeter geometry, rapidity, etc. Hence different subdetectors will suffer different amount of radiation. One of the experiments (CMS) estimates [6] that over a 10-year period the absorbed radiation dose will be, at peak luminosity, $1.8\ 10^8/r^2$ rad, where r is distance in cm from the interaction region, in addition to an almost isotropic neutron flux of 10^{14} neutrons/cm^2. Hence, over the lifetime of a typical detector the absorbed dose close to the interaction point will be several Mrad. Clearly, detectors, front-end electronics, optical read-out etc will suffer radiation damage, especially the central trackers and very-forward calorimeters. Radiation-hard electronics will therefore be used. Several radiation hard technologies are being studied, such as silicon-on-insulator (SOI) CMOS, junction field-effect transistor (JFET), SOI-CMOS-JFET, and silicon-on sapphire (SOS) [4].

KEY PROTON-PROTON PHYSICS CHANNELS

Standard Model Higgs boson (H^0):
The dominant production mechanism for $m_H < 700$ GeV is gluon-gluon fusion via a top loop. The optimal detection channel of Higgs depends on its mass, as seen in Figure 2. For $m_H < 90$ GeV the Higgs should be discovered at LEP-200 or Tevatron.

For $80 < m_H < 120$ GeV the most promising channel is $H^0 \rightarrow \gamma\gamma$, which is a rare decay mode. After 1 year running (10^7 sec) at peak luminosity (10^{34} $cm^{-2}s^{-1}$), which corresponds to an integrated luminosity of 10^5 pb^{-1}, one expects around 1000 $H^0 \rightarrow \gamma\gamma$ events. Recall that fits to world EW data now prefer a "light" Higgs around 150 GeV, so that $H^0 \rightarrow \gamma\gamma$ is a potentially important channel to probe and be sensitive to. $\gamma\gamma$ detection requires a good di-photon mass resolution, because of the large background, which demands superb electromagnetic calorimetry. Figure 3 shows an ATLAS simulation [7] of the signal above the prompt $\gamma\gamma$ background, where $m_H = 120$ GeV, 1 % energy resolution and 10^5 pb^{-1} integrated luminosity have been assumed. For lower Higgs masses several years of running is needed, and one considers using the $H^0 \rightarrow b\bar{b}$ channel which may yield a better signal-to-background ratio. In any case this latter channel dominates, but it is considered difficult due to

FIGURE 2. Decay channels of the Standard Model Higgs boson which are experimentally the most promising for SM Higgs search, as a function of the Higgs mass.
Figure from [10].

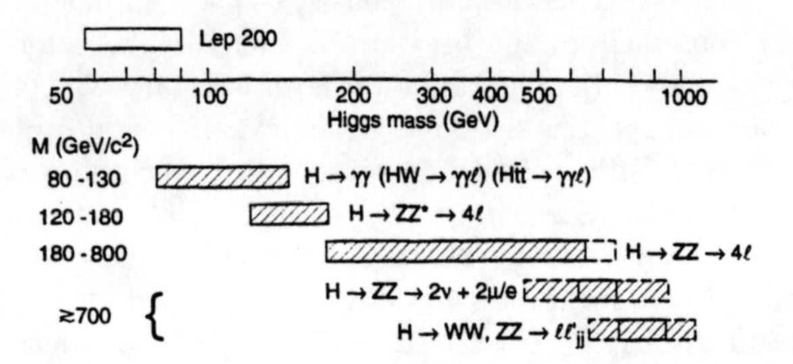

FIGURE 3. Expected $H^0 \to \gamma\gamma$ signal for Standard Model Higgs mass of 120 GeV, above the prompt $\gamma\gamma$ background. An integrated luminosity of 10^5 pb^{-1}, i.e. one year running at peak luminosity, has been assumed in this ATLAS simulation.
Figure from [10].

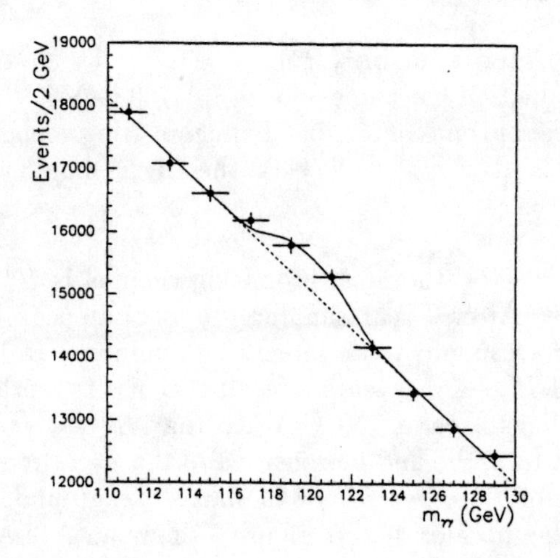

the QCD background and b tagging. Hence, $80 < m_H < 120$ is a difficult but potentially important mass range, and recent studies indicate that both channels $H^0 \to \gamma\gamma, b\bar{b}$ should be explored to improve the signal-to-background.

For $130 < m_H < 180$ GeV ($2m_Z$) the Higgs decay to $b\bar{b}$ again dominates, but it is swamped by QCD background. Instead one will focus on $H^0 \to ZZ^* \to 4l^{\pm}$ (one or both Z are off-shell) [7]. The background is significant and the event rate is small. Nonetheless simulations suggest that observation of Higgs in this mass range through this channel is possible after a few years of running, even below peak luminosity. It requires, however, very good mass resolution of the leptons. For Higgs mass below 130 GeV the expected signal rate drops very rapidly, and several years running at high luminosity would be needed to observe a signal in this channel.

For $180 < m_H < 800$ GeV (above $2m_Z$), the "gold-plated" channel is $H^0 \to ZZ \to 4l^{\pm}$. Expected signal rates are large, background is small and reducible, and for $m_H < 500$ GeV and 10^5 pb^{-1} (1 year at peak luminosity) a signal-to-background ratio of 20 can be achieved. This mass range is also interesting for MSSM Higgs search. For $m_H > 500$ GeV the Higgs width increases rapidly, and the signal from the "gold-plated" channel will be rate limited, but still observable in 1 year (10^5 pb^{-1}) up to $m_H \approx 800$ GeV.

For $800 < m_H < 1000$ GeV the channel $H^0 \to ZZ \to ll\nu\bar{\nu}$ is being considered. The disadvantage is of course that the decay cannot be fully reconstructed because of the escaping neutrinos. Also, for $m_H > 800$ GeV the signal peak is very broad and not easy to distinguish from background. Clearly, an important requirement is large geometric and kinematic acceptance for leptons. One will also consider the channels $H^0 \to ZZ \to lljj$ (j is jet) and $H^0 \to WW \to l\nu jj$, which have high signal rates, but high background as well. Realistically, several years of running at full luminosity may be needed to establish a signal in these channels [7].

Top Physics:
As noted, given a precisely measured top mass, SM gives predictions for top physics such as branching ratios. Top physics is therefore an ideal testing ground for possible non-SM physics. LHC will be a true top "factory", producing $\sim 10^7 t\bar{t}$ per year even below peak luminosity (10^{33} $cm^{-2}s^{-1}$). An obvious goal is to measure the top mass with high precision, using different channels ($t \to jjb, multileptons$). The measured mass is now 175 ± 6 GeV, obtained by CDF and D0 ([1]). Also high on the list is search for rare top decays (in SM the only significant decay mode is $t \to bW$). One would for instance look for $t \to Zc$ (window to new physics), and SUSY decays such as $t \to \tilde{t}\tilde{\gamma}$ and $t \to bH^+$ (MSSM charged Higgs).

SUSY:
SUSY particle decays will have missing energy, since the lightest supersymmetric particle (LSSP) is often assumed to be stable and weakly interacting, thus escaping detection. The most common signature of gluino and squark production is therefore missing transverse energy plus n high-energy jets. Also (depending on the choice of free parameters in MSSM) one may have excess of b jets or multimuons (plus missing energy plus jets) from gluino cascade decay. Such measurements require calorimetry with large pseudorapidity acceptance to avoid forward jets escaping detection and thus mimmick missing energy. (As seen above, a problem is that calorimeters extending close to the beam operate in a very high radiation environment.) The LHC will allow search for squarks and gluinos into the TeV mass range, a factor ~10 higher than todays limits. As for the MSSM Higgs, a wide region of couplings and masses can be explored [7].

Heavy Vector Bosons:
W' and Z' may be detected via their decays into pairs of leptons (ee or $\mu\mu$) or jets, i.e. $Z' \rightarrow l^+l^-, jj$ and $W' \rightarrow e\nu$. The best sensitivity is obtained in $Z' \rightarrow e^+e^-(\mu^+\mu-)$. Such searches require good mass resolution for both the electron and muon channel, i.e. a good electromagnetic calorimeter and muon spectrometer. Simulations show that a $W' \rightarrow e\nu$ signal appears after 1 year running at peak luminosity for W' mass of 4 TeV [7].

CP Violation and B Physics:
Even at "low" luminosity (10^{33} $cm^{-2}s^{-1}$) the LHC can be considered a beauty and top factory. The large $b\bar{b}$ production rate should allow a precise measurement of the parameters characterizing CP violation. Significant CP violation is expected in the $B\bar{B}$ system. The most promising channels which measure the β and α angles of the unitarity triangle are: *i)* $B_d^0 \rightarrow J/\Psi K_s^0$, for measurement of β. This should give a clear signature, manageable background, and relatively high branching ratio. *ii)* $B_d^0 \rightarrow \pi^+\pi^-$, for measurement of α. This channel requires very good di-pion mass resolution to minimize background from other two-body decays of B_d^0, B_s^0 and Λ_b. Moreover, good impact parameter resolution and secondary vertex reconstruction, from a central tracker, will reduce other backgrounds and allow study of B_d^0 and B_s^0 oscillations.

THE FOUR PROPOSED EXPERIMENTS

ATLAS and CMS:
Both are general-purpose pp detectors designed to exploit the full discovery potential of the LHC, at the peak luminosity (10^{34} $cm^{-2}s^{-1}$), *and* at the lower luminosity expected in the first years of LHC running ([7,8]). Both experi-

ments have been approved by CERN, which however has set strict cost ceilings, at CHF 475 million each ($\approx$ USD 325 million). The detectors have:
i) magnetic inner tracking: 2 T solenoidal field in ATLAS, 4 T solenoidal field in CMS;
ii) electromagnetic calorimetry: lead/liquid argon "accordion" in ATLAS, and a new lead tungstate $PbWO_4$ crystal in CMS, which give an estimated mass resolution of $H^0 \rightarrow \gamma\gamma$ of 1.2 GeV and 0.8 GeV in ATLAS and CMS, respectively, for m_H =100 GeV [7,8];
iii) Hadron calorimetry: copper/liquid argon, steel/plastic scintillator tiles and tungsten/liquid argon in ATLAS, and copper/plastic scintillator and iron/parallel plate gas chamber sandwich (or copper/quartz fibre matrix) in CMS;
iv) Muon detection and momentum measurement: a 0.6 T superconducting air core toroid, instrumented with 3 superlayers of drift chambers and cathode strip chambers, for "stand-alone" measurement of muon momentum in ATLAS, and a 4 T solenoidal field surrounded by a 1.8 m thick iron yoke for flux return, absorption and muon chamber support in CMS. Simulations show that these detectors provide a mass resolution of $H^0 \rightarrow ZZ^* \rightarrow 4\mu$ of 1.3 GeV and 0.8 GeV in ATLAS and CMS, respectively, for a 150 GeV Higgs ([7,8]). (In stand-alone mode the resolutions are 1.7 GeV and 3.4 GeV, respectively.)

Current cost estimates of ATLAS and CMS are CHF 475.5 million and 459.0 million, respectively, which are at or just below the cost ceiling. By July 1996 ATLAS members consisted of 148 institutions and 1606 authors, of which non-member states account for 51% of institutions and 42% of authors. There is participation from N/S America, W/E Europe, Russia, Asia and Australia. As for CMS, the Technical Proposal is signed by 132 institutions and 1243 authors, of which non-member states account for about 55% of institutions and some 50% of authors [8]. Participation is from USA, W/E Europe, Russia, China and India (USA participates with 37 institutions!).

ALICE:
Standing for "A Large Ion Collider Experiment", it is a general-purpose heavy ion experiment which is sensitive to hadrons, electrons, muons and photons [3]. ALICE will explore the *expected* phase transition of nuclear matter to quark-gluon plasma (QGP) at extremely high energy densities, as well as QCD and the fundamental question of confinement and chiral symmetry breaking. Heavy ion collisions up to lead-lead will be studied, at a luminosity of 10^{27} $cm^{-2}s^{-1}$. In fact, LHC is the only machine which will reach, and even extend, the energy range probed by most cosmic ray nucleus-nucleus collisions in the atmosphere (except of course the ultra-high energy cosmic ray collisions).
LHC will reach unprecedented energy densities of 5-8 GeV/fm^3, well above the expected critical energy density of quark-gluon deconfinement (plasma)

of 1-3 GeV/fm^3, a phase transition which probably took place, in reverse, a fraction of a second after the Big Bang.

ALICE must deal with huge particle multiplicities. In lead-lead collisions at the LHC energy, $dN_{ch}/dy \sim 8000$ (y is rapidity). In addition to heavy ions (e.g. lead-lead), ALICE will study collisions of lower-mass ions in order to vary the energy density. Also proton-proton and proton-ion collisions will be studied as a reference for the ion-ion data.

In order to detect quark-gluon plasma, the flavor content and phase-space distribution of a large number of soft particles will be measured, event by event. If QGP is formed, all heavy quark bound states, except $\Upsilon(1s)$, are suppressed by colour screening. Hence one will measure $\mu\mu$ decay rates in the Υ family and the suppression of $\Upsilon(2s)$ and $\Upsilon(3s)$ relative to $\Upsilon(1s)$, with different ion species, and relative to pp collisions. One will also study jet production, since jet quenching is expected when QGP is formed [3]. Notice also that ALICE has the potential to study $\gamma\gamma$ collisions and large-cross-section physics (diffraction, elastic scattering etc.), although there are no firm plans for this yet. ALICE plans to run in ion-ion mode for about 10% of the LHC year ($\sim 10^6$ sec), which will allow collection of some 10^7 central events for offline analysis.

The cost estimate for ALICE is currently CHF 116-123 million (about USD 80 million). Participation is again global: 63 institutions and 560 authors from W/E Europe, Russia, Israel, USA, Mexico, China and India have signed the Technical Proposal [3].

LHC-B:
This is a collider experiment dedicated to B physics and in particular CP violation [9], with participation from 39 institutions in W/E Europe, USA, Russia and China. It will utilize the full $\sqrt{s} = 14$ TeV collision energy with its high production rate of B mesons. At this energy the $b\bar{b}$ production cross section will be of the order of 500 μb. LHC-B will preferably operate at "low" luminosity ($\sim 10^{33} cm^{-2} s^{-1}$), where pile-up is small.

When LHC-B becomes operational in 2005, initial measurements of CP violation in the B system will already have been made by other dedicated B physics experiments (HERA-B, BABAR, BELLE, maybe BTEV). LHC-B will therefore be a 2nd generation experiment, its main advantage being the high production rate of B mesons.

In studying CP violation one will measure the angles α, β and γ in the unitarity triangle. Channels of special interest are therefore $B_d^0 \to \pi^+\pi^-$ for the

α angle, $B_d^0 \rightarrow J/\Psi K_s^0$ for β, and $B_s^0 \rightarrow J/\Psi\Phi$ for γ. With an integrated luminosity of $10^4\ pb^{-1}$ (one year running at L = $10^{33}cm^{-2}s^{-1}$), the expected sensitivities are $sin2\beta > 0.03$ and $sin2\alpha > 0.04$ [7]. The Standard Model prediction of very small CP asymmetry in $B_s^0 \rightarrow J/\Psi\Phi$ can be tested.

Another important measurement will be that of $B_s^0\bar{B}_s^0$ oscillations. The mixing parameter x_s can be measured for values up to around 40, covering the statistically preferred range of SM predictions. Other B physics studies include rare decays, such as $B_d^0 \rightarrow \mu^+\mu^-$ and $B_s^0 \rightarrow \mu^+\mu^-$, which are FCNC forbidden at the tree level and therefore have very small branching ratios in the Standard Model. To test these branching ratio predictions at the 2σ level will require about 2 years running at L = $10^{33}cm^{-2}s^{-1}$. This would constrain the CKM matrix elements and probe FCNC processes and perhaps new physics. Also planned are measurements of B baryons and rare B mesons.

LHC SCHEDULE

CERN aims to have LHC and its detectors ready for physics at 14 TeV in 2005. This plan is made possible by significant financial contributions from non-member states. Such contributions have already been negotiated with USA, Canada, Japan, Russia, India and Israel. These and many other non-member states will be heavily involved in the four experiments. Current plans envisage running LEP-200 until the end of 1999. Civil engineering for the LHC should begin in 1998. Removal of LEP equipment should start around 2001, and the first injection tests for LHC are foreseen in 2003. The total cost of the LHC project (machine and experiments) is estimated to about CHF 3 billion (USD 2.1 billion).

CONCLUSIONS

The subject matter is vast, and this short paper has of course only scratched the surface of all the issues involved. The main message we want to transmit is the following:
The physics potential of the Large Hadron Collider is enormous. The four detectors being planned for LHC promise unprecedented potential for:
- Search for the Higgs boson or another mechanism for spontaneous symmetry breaking (origin of mass);
- Top quark physics;
- B-meson physics and CP violation;
- Other Standard Model physics (such as QCD, soft processes, $\gamma\gamma$ physics);
- New non-Standard Model physics (such as supersymmetry, compositeness,

heavy vector gauge bosons, leptoquarks);
- Ultra-relativistic heavy ion physics and the formation of quark gluon plasma;
- The *possibility* to do electron-proton physics if this option should be chosen, which would enable us to probe an order of magnitude lower in x and higher in Q^2 than we can at present machines.

An intense *R&D* programme is in progress, both for the detectors and the machine. The LHC programme is the largest and most challenging basic science project ever undertaken. The project will have truly global participation, as non-member states are heavily involved. When the LHC is switched on at full energy (14 TeV) in 2005, it will be the only *multi-TeV* laboratory in the world. There are compelling theoretical arguments for exploring this energy regime, where new physics beyond the Standard Model may unfold.

Acknowledgements
I thank CINVESTAV, and in particular professor Miguel Angel Perez, for inviting me to this stimulating conference in Merida. I also thank professor Jørn Dines Hansen for useful comments on this paper.

REFERENCES

1. The LEP Electroweak Working Group and the SLD Heavy Flavor Group, "A Combination of Preliminary LEP and SLD Electroweak Measurements and Constraints on the Standard Model", Internal Note prepared from contributions of the LEP and SLD experiments to the 1996 summer conferences, CERN LEPEWWG/ 96-02 and SLAC SLD Physics Note 1996/52 (1996).
2. *Proceedings of the Large Hadron Collider Workshop*, Aachen, edited by G. Jarlskog and D. Rein, CERN 90-10/ECFA 90-133 (1990).
3. *ALICE Technical Proposal*, CERN/LHCC 95-71 LHCC/P3 (1995).
4. *CERN LHC News* no. 7, CERN, September 1995.
5. The LHC Study Group, "The Large Hadron Collider Project", CERN AC/93-03 (1993).
6. Virdee, T.S., "Physics at the LHC", CMS TN/95-168, CERN (1995), and *Proceedings of the Int. Europhysics Conf. on High Energy Physics*, Marseille, France, 1993.
7. *ATLAS Technical Proposal*, CERN/LHCC/94-43 LHCC P2 (1994).
8. *CMS Technical Proposal*, CERN/LHCC/94-38 LHCC P1 (1994).
9. *LHC-B Letter of Intent*, CERN/LHCC/95-5 (1995).
10. Fabjan, C.W., *Proceedings of the CAM-94 Physics Meeting*, Cancun, Mexico, 1994. AIP Conference Proceedings 342, edited by A. Zepeda.

Spin in Semi-Inclusive DIS Processes

D. de Florian*, C. García Canal†, and R. Sassot††

* *Theoretical Physics Division, CERN, CH 1211 Geneva 23, Switzerland*
† *Departamento de Física, Universidad de La Plata, C.C.67 (1900) La Plata, Argentina*
†† *Departamento de Física, Universidad de Buenos Aires, Ciudad Universitaria, Pab.1 (1428) Bs.As., Argentina*

Abstract. We review the most recent developments regarding the perturvative QCD picture for semi-inclusive spin dependent deep inelastic scattering processes and their applications to current and programmed experiments.

INTRODUCTION

Ever since the first release of data on polarized deep inelastic scattering produced by the EMC Collaboration [1], almost nine years ago, the issue of spin in hard processes has been an object of considerable theoretical and experimental interest [2,3].

Almost immediately after the presentation of these data, which suggested that only a very small fraction of the proton spin was carried by its quarks, in opposition to what it is expected from the most naive quark model picture, an exciting theoretical discussion was launched on the most appropriate interpretation of the quantities measured (spin dependent asymmetries and through them, spin dependent structure functions), the role of the gluon anomaly in these observables and in parton model sum rules, their scale dependence, and other features related to QCD corrections, that were not taken into account in the original analysis [2,3].

At the same time, the controversial results triggered new experimental programmes [4–7], designed to corroborate the EMC results and also extend them in order to analyze the spin structure of the neutron and also test sacrosant current algebra sum rules such as the Bjorken sum rule [8].

In recent years both lines of work, the experimental and the theoretical programmes, have been actively pursued producing more and more precise data on spin dependent structure functions, and also clarifying the issue of QCD corrections and extending their computation to higher orders, respectively. Today we have then access to next to leading order global fits for polarized

CP400, *First Latin American Symposium on High Energy Physics/ VII Mexican School of Particles and Fields*, edited by D'Olivo/Klein-Kreisler/Mendez

parton distributions [9–11], similar to those performed in the unpolarized case, which summaryze our knowledge of the QCD structure of nucleons and allow us to make predictions for other expermients exploiting universality. However, the accuracy in the polarized case is far to be as good as in the unpolarized one, essentially because we are only dealing with only two observables of the same kind, totally inclusive polarized deep inelastic scaterring.

One of the consequences of this situation is, for example, that it is not possible yet to constrain the polarization carried by the gluons or by sea quarks without making non trivial and extremely model dependent assumptions. As a way out from these limitations, the attention has been driven from totally inclusive polarized deep inelastic experiments to those in which a given particle (a pion, kaon, nucleon, or eventually a Λ) is identified in the final state. Last year, for example, the SMC collaboration [12] presented a first set of semi-inclusive data, and at present there are ongoing programmes by the Compass [6] and Hermes collaborations [13] (at CERN and HERA, respectively) aimed at measuring semi-inclusive observables with high precision.

Semi-inclusive processes not only may help the extraction of polarized parton distributions in global QCD fits but also involve other issues of theoretical and phenomenological interest, such as the way to implement QCD corrections to them and the role of fracture functions in a perturbative description of hadronization proceses [14–16]. Rather than reviewing the latest developments in totally inclusive experiments, in this presentation we would like to concentrate in what has been done most recently regarding semi-inclusive spin dependent processes.

SEMI-INCLUSIVE DIS

Let us first remind how semi-inclusive DIS processes are described in a perturbative framework and what is the role of the so called fragmentation and fracture functions. We are dealing typically with events where a lepton scatters off a proton and certain hadron is identified in the final state. The traditional partonic description is that where the struck parton hadronizes (fig.1) leading to a cross section which schematically reads [17]

$$\sigma_c \propto f_{i/p} \otimes \sigma^i_{hard} \otimes D^h_i \tag{1}$$

where $f_{i/p}$ is the parton density for the flavour i, σ^i_{hard} is the hard cross section involving the parton i, D^h_i is the fragmentation function, i.e. the probability for parton i to hadronize into a hadron h, and $\otimes$ denotes the convolution. Typically, the structure functions are obtained from totally inclusive DIS, the fragmentation functions from $e^+e^- \rightarrow h + X$, and the cross section is calculated perturbatively. Both structure and fracture functions contain essentially nonperturbative information and that is why they have to be extracted from experiments. More recently, Veneziano and Trentadue introduced the so

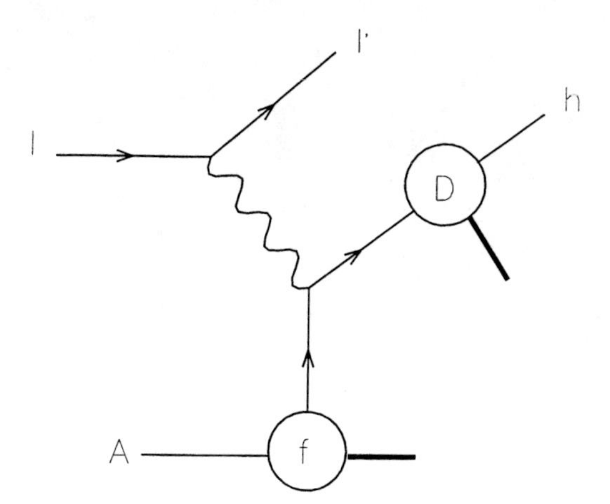

FIGURE 1. Current Fragmentation

called fracture functions [18], which take into account the possibility of target fragmentation (fig.2) for events where the hadron is produced in the target fragmentation region.

For these, the cross sections look like

$$\sigma_t \propto M^h_{i/p} \otimes \sigma^i_{hard} \tag{2}$$

where $M^h_{i/p}$ is the fracture function, the probability of finding a parton i in a fragmented proton. These functions are not a mere refinement of the partonic picture but are essential ingredients when higher order QCD corrections are applied to the cross section. Specifically, they make possible a consistent factorization of collinear singularities which arise in semi-inclusive processes [19]. Taking into account now the helicity states of the particles, the differential cross section for the interaction between a lepton of momentum l and helicity λ_l and a nucleon A of momentum P and helicity λ_A leading to the production of a hadron h with energy $E_h = z\,E_A(1-x)$ and helicity λ_h (with n partons in the final state) can be written as [14,16]

$$\begin{aligned}\frac{d\sigma^{\lambda_l\lambda_A\lambda_h}}{dx\,dy\,dz} &= \int \frac{du}{u} \sum_n \sum_{j=q,\bar{q},g} \int dPS^{(n)} \frac{\alpha^2}{S_H x} \frac{1}{e^2(2\pi)^{2d}} \\ &\times \left[Y_M(-g^{\mu\nu}) + Y_L \frac{4x^2}{Q^2} P_\mu P_\nu + \lambda_l Y_P \frac{x}{Q^2}\, i\epsilon^{\mu\nu qP} \right] \\ &\times \sum_{\lambda_1,\lambda_2=\pm 1} H_{\mu\nu}(\lambda_1,\lambda_2) \left\{ M_{j,h/A}\left(\frac{x}{u}, \frac{E_h}{E_A}, \frac{\lambda_1}{\lambda_A}, \frac{\lambda_h}{\lambda_A}\right)(1-x) \right. \\ &\left. + f_{j/A}\left(\frac{x}{u}, \frac{\lambda_1}{\lambda_A}\right) \sum_{i_\alpha=q,\bar{q},g} D_{h/i_\alpha}\left(\frac{z}{\rho}, \frac{\lambda_h}{\lambda_2}\right) \frac{1}{\rho} \right\}\end{aligned} \tag{3}$$

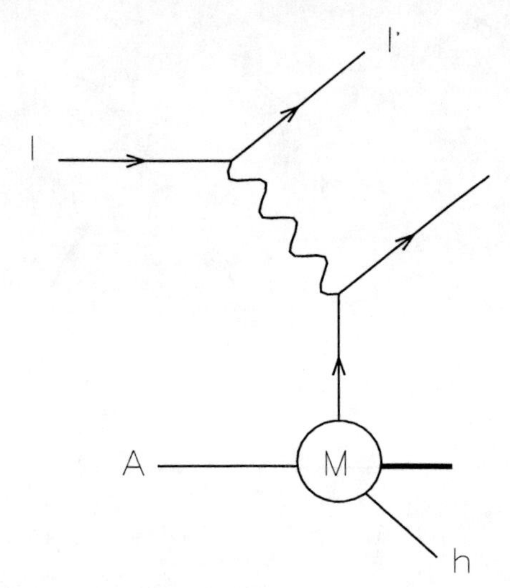

FIGURE 2. Target Fragmentation

where

$$x = \frac{Q^2}{2\,P.q}, \quad y = \frac{P.q}{P.l}, \quad Q^2 = -q^2, \quad S_H = (P+l)^2 \tag{4}$$

and $\rho = E_\alpha / E_A$. The helicity dependent partonic tensor is defined by

$$H_{\mu\nu}(\lambda_1, \lambda_2) = M_\mu(\lambda_1, \lambda_2) M_\nu^\dagger(\lambda_1, \lambda_2) \tag{5}$$

where M_μ is the parton-photon matrix element with the photon polarization vector substracted. Here λ_1 and λ_2 are the helicities of the initial and final state partons respectively. Due to symmetry considerations, the distributions depend only on relative helicities.

Taking different linear combinations of cross sections for targets and final state hadrons with equal or opposite helicities, it is possible to isolate parton distributions, fragmentation functions and fracture functions of different kind and also their QCD corrections. For example, taking the sum over all the helicity states

$$\sigma = \sigma^{\lambda_l++} + \sigma^{\lambda_l+-} + \sigma^{\lambda_l-+} + \sigma^{\lambda_l--} \tag{6}$$

the result is proportional to a convolution of unpolarized parton distributions, fragmentation and fracture functions [19]. Taking the difference between cross sections with opposite target helicities,

$$\Delta\sigma = (\sigma^{\lambda_l++} + \sigma^{\lambda_l+-}) - (\sigma^{\lambda_l-+} + \sigma^{\lambda_l--}) \tag{7}$$

the result contains polarized parton distributions, single polarized fracture functions (single initial state polarization), and unpolarized fragmentation functions. In reference [14] we have shown the results for this cross section computed up to next to leading order in QCD and the details for the factorization of divergencies. This cross section is the relevant for the analysis of semi-inclusive production spin-less objects like scalar mesons and is the one to be used in the analysis of SMC semi-inclusive data. Very recently we have extended the computation to the two remaining possibilities [16], which represent the case where only the final state hadron polarization is put in evidence (single final state polarization) at cross section level,

$$\sigma\Delta = (\sigma^{\lambda_l++} - \sigma^{\lambda_l+-}) + (\sigma^{\lambda_l-+} - \sigma^{\lambda_l--}) \tag{8}$$

and that where both polarizations are relevant (double polarization)

$$\Delta\sigma\Delta = (\sigma^{\lambda_l++} - \sigma^{\lambda_l+-}) - (\sigma^{\lambda_l-+} - \sigma^{\lambda_l--}) \tag{9}$$

It is straightforward to show that the single (final state) polarization cross section $\sigma\Delta$ is proportional to the convolutions

$$\begin{aligned}\sigma\Delta \propto\ & f_{j/A}(x/u) \otimes H\Delta(u,\rho) \otimes \Delta D_{h/i_\alpha}(z/\rho) \\ & +(1-x)\, M\Delta_{j,h/A}(x/u,(1-x)z) \otimes \Delta H'(u)\end{aligned} \tag{10}$$

where the first term corresponds to current fragmentation processes and the second one to fragmentation of the target. In the former, $f_{j/A}(x/u)$ are just the unpolarized parton distributions, $\Delta D_{h/i_\alpha}(z/\rho)$ are the polarized fragmentation functions, and $H\Delta = \sum_{\lambda_1,\lambda_2} \lambda_2\, H(\lambda_1,\lambda_2)$ is the partonic tensor for polarized final states. The convolutions are those defined by equation (3). In the second term of equation (10) we have defined the final state polarized fracture function as

$$M\Delta = M_{j,h/A}\,(+,+) + M_{j,h/A}\,(-,-) - M_{j,h/A}\,(+,-) - M_{j,h/A}\,(-,+) \tag{11}$$

where we have omitted the two first kinematical arguments in the fracture functions, and the single polarized tensor $\Delta H' = \sum_{\lambda_1,\lambda_2} \lambda_1\, H(\lambda_1,\lambda_2)$ (integrated in ρ).

A similar analysis yields for $\Delta\sigma\Delta$ the following convolutions

$$\begin{aligned}\Delta\sigma\Delta \propto\ & \Delta f_{j/A}(x/u) \otimes \Delta H\Delta(u,\rho) \otimes \Delta D_{h/i_\alpha}(z/\rho) \\ & +(1-x)\, \Delta M\Delta_{j,h/A}(x/u,(1-x)z) \otimes H'(u)\end{aligned} \tag{12}$$

where again the first term corresponds to current fragmentation processes and the second to fragmentation of the target. In the former, $\Delta f_{j/A}(x/u)$ are now polarized parton distributions, $\Delta D_{h/i_\alpha}(z/\rho)$ are polarized fragmentation functions, and

$$\Delta M \Delta = M_{j,h/A}(+,+) - M_{j,h/A}(-,-) - M_{j,h/A}(+,-) + M_{j,h/A}(-,+) \quad (13)$$

The details of how QCD corrections can be obtained for these functions are not relevant in the present context, for the moment it is enough to say that the functions defined here develop a dependence on the energy scales and obey certain kind of Altarelli-Parisi equations. For a exhaustive treatment see references [14,16]

APPLICATIONS

Having produced the theoretical framework for NLO QCD corrections for semi-inclusive DIS processes, it is now possible to perform a combined analysis of totally inclusive and semi-inclusive data, and of course to evaluate the impact of these corrections in the phenomenological analysis of the available data.

Up to now, the analyses [12] of semi-inclusive spin asymmetries have been performed in the naive quark-parton model, neglecting higher order corrections and trying to avoid contributions coming from the target fragmentation region imposing kinematical cuts. However, higher order corrections can be non-negligible if gluons are polarized in the proton. In fact, we have found [15] that higher order corrections to semi-inclusive spin asymmetries are not negligible and can be treated quantitatively using a sensible model for fracture functions. Polarized fracture functions are a relatively new concept and have not been measured yet, so there are not parametrisations available for them. However, taking into account that these functions measure the probability for finding a hadron and a struck parton in a target nucleon, one can approximate them as a simple convolution products between known distributions [15].

A second application of the scheme presented here is related to the production of polarized Λ particles in DIS. In fact, in recent years, a considerable degree of attention has been paid to the semi-inclusive production of Λ hyperons and their polarization. The approaches include experimental programmes to measure polarized fragmentation functions in $e^+e^- \rightarrow \Lambda + X$ [20], semi-inclusive lepton proton deep inelastic scattering (DIS) [6,21], both in the current fragmentation region, and also models for the production of these hyperons in present $p\bar{p}$ and DIS experiments, but in the target fragmentation region. Many of these studies are closely related to the interpretation of the proton spin structure [2,3], due to the possibility of reconstructing it from the observed Λ polarization.

Given the increasing experimental interest in spin dependent semi-inclusive Λ physics, and also the important phenomenological insights related to the production of these hyperons in different processes, it is worthwhile to analyze QCD corrections to such cross sections and their phenomenological implications. These corrections, have been shown to be crucial for the interpretation of totally inclusive polarized deep inelastic experiments related to the spin structure of the proton, and certainly would have a non-negligible role in polarized Λ production. On the other hand, one would naturally expect any parton model-inspired phenomenological description of the processes to mold into the more formal QCD improved description, at least in some adequate limit.

CONCLUSIONS

Spin dependent hard processes have been an exciting field of theoretical and experimental research in recent years, and have yielded more interesting results than it was originally expected. Semi-inclusive experiments open now new phenomenological perspectives and serve as a benchmark for some new ideas such as fracture functions and, in general, hadronization phenomena.

REFERENCES

1. J.Ashman et al., *Phys. Lett.* **B206**, 364 (1988).
2. M. Anselmino, A. Efremov and E. Leader, *Phys. Rep.* **261**, 1 (1995).
3. J. Ellis, M. Karliner, CERN-TH/95-334, hep-ph/9601280 (1995).
4. P.L. Anthony et al, *Phys. Rev. Lett.* **71**, 959 (1993);
   K. Abe et al, *Phys. Rev. Lett.* **74**, 346 (1995).
5. D. Adeva et al, *Phys. Lett.* **B302**, 533, (1993).
6. Compass Proposal, CERN SPSLC 96-14, SPSC/P297, (1996).
7. The Hermes Collaboration, Technical Design Report, DESY-PRC93/06 (1993).
8. J. D. Bjorken, *Phys. Rev.* **D148**, 1467 (1966).
9. M. Gluck et al., *Phys. Rev.* **D53**, 4775 (1996).
10. T. Gehrmann, W. J. Stirling, *Phys. Rev.* **D53**, 6100 (1996).
11. R. D. Ball, S. Forte, G. Ridolfi, *Phys. Lett.* **B378**, 255 (1996).
12. B. Adeva et al, *Phys. Lett.* **B369**, 93 (1995).
13. E. E. W. Bruins, hep-ex/9611010 (1996).
14. D. de Florian, C.A. García Canal and R. Sassot, *Nucl. Phys.* **B470**, 195 (1996).
15. D. de Florian et al., hep-ph/9603302 (to be published in *Phys.Lett.* **B**).
16. D. de Florian, R. Sassot, hep-ph/9610362 (to be published *Nucl. Phys.* **B**).
17. E. Leader, E. Predazzi, *An Introduction to Gauge Theories and Modern Particle Physics.* Cambridge University Press 1996.
18. L. Trentadue, G. Veneziano, *Phys. Lett.* **B323**, 201 (1994).
19. D. Graudenz, *Nucl. Phys.* **B432**, 351 (1994).
20. M. Burkhardt, R.L. Jaffe, *Phys. Rev. Lett.* **70**, 234 (1993).
21. R.L. Jaffe, MIT-CTP-2534, hep-ph/9605456 (1996);
    J. Ellis, D. Kharzeev and A. Kotzinian, *Z. Phys.* **C69**, 467 (1996).

The Pierre Auger Observatory

Carlos Hojvat

Fermilab
Batavia Illinois 60510 USA

Abstract. The Pierre Auger Observatory is an international collaboration for the detailed study of the highest energy cosmic rays. The Observatory will operate at two similar sites, one in the northern hemisphere and one in the southern hemisphere. Important contributors to this effort are institutions in latin american countries. The Observatory is designed to collect a statistically significant data set of events with energies greater than 10^{19} eV and with equal exposures for the northern and southern skies.

INTRODUCTION

The Pierre Auger Observatory (PAO) is an international collaboration of over 200 astro-physicists and elementary particle physicists from 18 different countries, representing more than 20 institutions. It has strong participation from both the southern and northern hemispheres and from all continents [1].

The concept of PAO was developed at workshops in Paris 1992, Adelaide 1993, Tokyo 1993 and Fermilab 1995. Professors Jim Cronin and Alan Watson are the spokespersons. A design report has been written, a new version of which will be available February 1997. During an organizational meeting in Paris, November 1995, the southern site location was selected in Argentina. At the first collaboration meeting in San Rafel, Argentina, September 1996, the northern site was selected to be in the USA.

Since the first evidence of their existence, in balloon flights by V. Hess in 1912, cosmic rays (CR) have been extensively studied. In 1938 Pierre Auger observed for the first time extensive air showers (EAS) by observing CR coincidences with detectors separated from each other. Estimates of the energies involved in the first EAS observed was already in excess of 10^{15} eV.

Enrico Fermi proposed in 1949 a possible method of accelerating particles that could explained the presence of such extremely high energies. In effect,

CP400, *First Latin American Symposium on High Energy Physics/ VII Mexican School of Particles and Fields,*
edited by D'Olivo/Klein-Kreisler/Mendez

the CR energy spectrum has now been measured over more than 10 orders of magnitude in energy.

COSMIC RAY FLUXES

The measured flux of CR falls with a power of the energy of the order of 2.5 to 3. For each decade of increasing energy, the flux decreases three orders of magnitude. This logarithmic dependence of the flux with energy is predicted by different acceleration models.

The slope of the flux versus energy shows some structure near 10^{16} eV and also around 10^{19} eV, the so called "knee" and "ankle" features. No "toe" has been found [2]. At 10^{20} eV the measured flux is as low as approximately one event per Km^2 per steradian per century per decade of energy. The combined design acceptance for both sites of the the PAO is 14,000 Km^2 sr, yielding an estimated data collection rate of the order of 140 events per year, at the highest observed energies.

Only about 10 events with energies around 10^{20} eV have been detected to date. In 1991, at the Fly's Eye detector (Utah, USA) an event of 3.2 x 10^{20} eV was detected; an energy of 50 Joules! This is the highest energy event ever recorded. In 1993, at the Agasa detector (Akeno, Japan) an event of 2.0 x 10^{20} eV was detected, 10^8 times the energy of protons accelerated in the highest energy ground based accelerator, the Fermilab Tevatron.

The measurement of the 2.7 °K Microwave Background Radiation (MBR), by A. Penzias and R. Wilson in 1966, had an important consequence; the realization that CR interactions with the MBR will cause an energy loss mechanism. Nucleons above a certain energy threshold produce pions in collisions with photons. This process continues until the energy falls below threshold. For long distances, particles originally with an energy distribution above threshold, will continue to propagate with the "same" energy just below threshold. Therefore, for CR of energies in excess of 10^{19} eV, the observed energy at an earth-based detector will depend on the distance traversed by the primary particle between the source and the detector. The energy spectrum is then expected to be "limited" by this threshold effect to around 10^{20} eV for particles travelling intergalactic distances. This is the so called GZK cut-off, from the calculations of K. Greisen, V. Kuzmin and G. Zatsepin. These calculations have been recently updated by M. T. Dova, L. Anchordoqui, L. Epele and J. Swain of the Auger Collaboration. The energy loss takes place within distances of the order of 100 Mpc. For these distances, depending on the actual distribution of primary sources and their flux, an enhancement could appear in the energy spectrum between 10^{19} and 10^{20} eV due to a lower detected energy than the original at the source. Particles detected with energies above the enhancement must originate within 20 or 30 Mpc from the detector.

Primary high energy CR nuclei undergo a loss of nucleons through the intergalactic space of about 4 nucleons per Mpc. Therefore, for distances longer than 50 Mpc, only nucleons should be observed. The Fly's Eye detector has evidence for a change of composition of CR with energy, mainly nuclei at lower energies changing to nucleons at higher ones.

The transverse momentum exchange in the interaction with the MBR is expected to be very low, therefore the CR direction is not significantly affected. Observation of an enhancement in the energy spectrum could be used for separating the data in "near" and "far" samples when tracing back the arrival directions. The apparent arrival direction of the observed CR may not point back to the source due to the possible magnetic fields in their trajectory. For punctual sources emitting at different energies, and given sufficient statistics, one could observe a distribution of arrival directions, converging to the source as the detected energy increases. The magnetic field in our galaxy of the order of 2 to 3 μ gauss over some kpc, will cause only a small deflection for CR of the order of 10^{20} eV, so that they could not have originated from the center of our galaxy, and therefore are most likely of extra-galactic origin.

The large increase in data collection to be provided by the Auger Observatory will allow a detailed study of the energy spectrum, arrival directions and possible assymetries of CR near and above the GZK energy cut-off. This information could help to identify their source and acceleration mechanisms. Possible candidates are: powerful radio sources, Active Galactic Nuclei by shock acceleration, association with Gamma Ray Bursts, annihilation of topological defects, etc. No proposed mechanism can currently explain acceleration beyond 10^{20} eV. If the energy spectrum extends significantly beyond the GZK cut-off one may need to consider the possibility of some new astrophysics or particle physics phenomena.

THE PIERRE AUGER OBSERVATORY

An EAS develops when a high energy primary particle (e.g. a nucleon, a heavy ion, or a neutrino) collides with the atmosphere and produces a number of secondaries that interact further with the atmosphere, giving rise to succesive generation of particles. The details of the EAS depend on the type of primary particle and its energy. A typical 10^{19} eV proton shower when fully developed could contain as many as 10^{11} individual low energy particles, 90% gammas and 10% electrons of MeV energies and 1% muons with 1 GeV energies.

An EAS can be characterized by experimentally determining the position of shower maximum, X_{max}, its radial particle density distribution (e.g. at the ground), its arrival time and front shape and two angles defining its direction. Of equal importance is to determine the species of the primary particle, that can be infered by the muon content of the EAS and the time characteristics of its front.

The atmosphere is a good calorimeter with a vertical depth of approximately 1,000 gm cm^{-2}. Shower maximum for primary charged particles occurs at about a depth of 850 gm cm^{-2}, or 1500 meters above sea level. The atmospheric vertical thickness represents approximately 26 radiation lengths and 11 interaction lengths, sufficient for good calorimetric measurements. For horizontal showers the atmospheric depth is about 36 times thicker, becoming a good "beam dump". This mass is enough to open the possibilty of utilizing the PAO for the detection of neutrino induced horizontal EAS.

The PAO will consist of two detectors, one in each hemisphere, with a total acceptance of 14,000 Km^2 sr. Each site will occupy a surface of 3,000 Km^2 with an acceptance from the zenith down to 60°. Each detector will combine two well developed CR calorimetric detection techniques, ground array (GA) detection and atmospheric fluorescence (FD) detection.

The GA samples the EAS at ground level with capabilities for the measurement of the radial particle density distribution, the timing characteristics of the shower front, and for distinguishing muons from electromagnetic components. Timing information gives the shower direction as well as contributes to the muon separation. Both southern and northern hemisphere sites are situated at 1400 m altitude, or very close to shower maximum. The ground level sampling is then done close to the peak of the shower development. The GA will have a 100% duty cycle and full sky coverage for anisotropy studies. Each site will consist of more than 1500 water Cherenkov detectors located 1.5 Km apart. Each detector will be circular, 10 m^2 of surface and 1.2 m high. It will be instrumented with 3 photomultipliers, 22 cm in diameter (or an equivalent total photocathode area), looking down into the water. The detectors will be arranged in an hexagon, or in a three fold symmetrical shape, of 3,000 Km^2 total area.

The FD measures the longitudinal development of the EAS by detecting the nitrogen fluorescence from charged particle excitation of the nitrogen atoms; it therefore tracks the longitudinal development of the EAS, giving the position of X_{max} and information on its direction by fitting the longitudinal distribution. Due to the requirement of a dark sky, the FD can only operate for about 10% of the total time. This fraction may be increased (to 20% ?) as techniques are developed for triggering the FD with the GA. Two proposed designs exist for the FD. They differ in the number of "eyes" or locations. More than one eye is required because of the attenuation length of the fluorescence light in the atmosphere, of the order of 17 Km. The minimum number is three "eyes", symmetrically placed within the GA surface area in its three fold symmetrical configuration. Each "eye" looks at the sky above the GA from 3° to 30° from the horizontal and 360° on the horizontal plane. Pixel sizes are 1.5° by 1.5 ° each. The alternative design is for 6 "eyes", one in each corner of the hexagon plus one "eye" in the center of the array. Both designs require the same total number of pixels, or individual photomultipliers.

Two sets of data will be available, the hybrid set with GA and FD information and the GA only data. Because of the following advantages, the hybrid data set is one of the important characteristics of the PAO,

- independent measurement techniques allow control of systematics
- FD and GA give a more reliable energy and measurements of angles
- FD calibrates the energy determination of the GA
- EAS front timing, core position and surface particle density improve the energy and angle determination for the FD
- the two techniques measure the primary mass in complementary ways

MEXICO AND THE PIERRE AUGER OBSERVATORY

The following Mexican institutions participate in the PAO:
CINVESTAV
Benemerita Universidad Autónoma de Puebla
Universidad Nacional Autónoma de Mexico
Universidad Michoacana de San Nicolas de Hidalgo
Universidad de Guadalajara
Universidad Autónoma de San Luis Potosí
Instituto Nacional de Astrofísica Óptica y Electrónica
Centro de Investigaciones en Materiales Avanzados

The Mexican Auger group is working on a variety of topics for the project. Front End Electronics using Switch Capacitors Array or Flash ADC is being developed at Puebla. Surface detector simulations are underway at CINVESTAV IPN and at Puebla. Water Cherenkov tank prototypes made of high density polyethylene and reflective TYVEK^{MR} material on the internal walls is being devloped jointly by Puebla and Michoacan. An Optical design has been proposed for the FD jointly by Puebla and INAOE.

COST AND SCHEDULE

The total cost for both PAO sites is estimated to be of the order of 100M $us. Of this, about 50M $us is expected to be the cost of the GA and about 30M $us the cost of the FD. The remainder of the estimated cost is for infrastructure, the central stations, site preparation, engineering costs, etc. To date no real breakdown of the sources of support exists. Expectations are for 25% of the support to come from each of the North American, South American, European and Asian continents. Most of the contributions are expected to be "in kind" with only a fraction of them required for a "common fund" for joint expenditures.

In parallel with the completion of the PAO design, most of 1997 will be dedicated to the submission of proposals to the funding agencies of each country by the collaborating institutions. During this year a number of scientific and design reviews are expected to take place. Expectations are for funding for the construction of PAO to start late in 1998. The collaboration has a 5 year construction and deployment plan. After the second year of construction is completed, PAO is expected to have the largest operating ground detector array in existence. As the PAO is completed it will continue to increase in size, to be more than double in size on the following 3 years. Early in the 21st century the Pierre Auger Observatory should have a catalogued hundreds of Cosmic Rays with energies in the 10^{20} eV range.

REFERENCES

1. Due to the limited space for further information and references please contact the author at "hojvat@fnal.gov".
2. The presence of a "toe" would indicated the end of the energy spectrum, Alvaro de Rujula, private comunication.

Supersymmetric Scalar Masses, Z', and E(6)

Ernest Ma

Department of Physics, University of California
Riverside, CA 92521, USA

Abstract. Assuming the existence of a supersymmetric U(1) gauge factor at the TeV energy scale (motivated either by the superstring-inspired E_6 model or low-energy electroweak phenomenology), several important consequences are presented. The two-doublet Higgs structure at the 100 GeV energy scale is shown to be different from that of the Minimal Supersymmetric Standard Model (MSSM). A new neutral gauge boson Z' corresponding to the extra U(1) mixes with the Z. The supersymmetric scalar quarks and leptons receive new contributions to their masses from the spontaneous breaking of this extra U(1). The assumption of universal soft supersymmetry breaking terms at the grand-unification energy scale implies a connection between the U(1) breaking scale and the ratio of the vacuum expectation values of the two electroweak Higgs doublets.

INTRODUCTION

Consider the sequential reduction in rank of the symmmetry group E_6:

$$E_6 \to SO(10)\ [\times U(1)_\psi] \tag{1}$$
$$\to SU(5)\ [\times U(1)_\chi] \tag{2}$$
$$\to SU(3)_C \times SU(2)_L\ [\times U(1)_Y]. \tag{3}$$

At each step, a U(1) gauge factor may or may not appear, depending on the details of the symmetry breaking. If E_6 is indeed the grand-unification group, it is often assumed that a single U(1) survives [1] down to the TeV energy range, given by

$$U(1)_\psi \times U(1)_\chi \to U(1)_\alpha. \tag{4}$$

This talk is concerned mainly with the phenomenological consequences of extending the MSSM to include this $U(1)_\alpha$.

CP400, *First Latin American Symposium on High Energy Physics/ VII Mexican School of Particles and Fields*, edited by D'Olivo/Klein-Kreisler/Mendez

Under the maximal subgroup $SU(3)_C \times SU(3)_L \times SU(3)_R$, the fundamental representation of E_6 is given by

$$\mathbf{27} = (3,3,1) + (3^*,1,3^*) + (1,3^*,3). \tag{5}$$

Under the subgroup $SU(5) \times U(1)_\psi \times U(1)_\chi$, we then have

$$\begin{aligned}\mathbf{27} = &(10;1,-1)\ [(u,d),u^c,e^c] \\ &+ (5^*;1,3)\ [d^c,(\nu_e,e)] \\ &+ (1;1,-5)\ [N] \\ &+ (5;-2,2)\ [h,(E^c,N_E^c)] \\ &+ (5^*;-2,-2)\ [h^c,(\nu_E,E)] \\ &+ (1;4,0)\ [S],\end{aligned} \tag{6}$$

where the U(1) charges refer to $2\sqrt{6}Q_\psi$ and $2\sqrt{10}Q_\chi$. Note that the known quarks and leptons are contained in $(10;1,-1)$ and $(5^*;1,3)$, and the two Higgs scalar doublets are represented by (ν_E,E) and (E^c,N_E^c). Let

$$Q_\alpha = Q_\psi \cos\alpha - Q_\chi \sin\alpha, \tag{7}$$

then the so-called η-model [1,2] is obtained with $\tan\alpha = \sqrt{3/5}$ and we have

$$\begin{aligned}\mathbf{27} = &(10;2) + (5^*;-1) + (1;5) \\ &+ (5;-4) + (5^*;-1) + (1;5),\end{aligned} \tag{8}$$

where $2\sqrt{15}Q_\eta$ is denoted; and the N-model [3] is obtained with $\tan\alpha = -1/\sqrt{15}$ resulting in

$$\begin{aligned}\mathbf{27} = &(10;1) + (5^*;2) + (1;0) \\ &+ (5;-2) + (5^*;-3) + (1;5),\end{aligned} \tag{9}$$

where $2\sqrt{10}Q_N$ is denoted. The η-model is theoretically attractive because it is obtained if the symmetry breaking of E_6 occurs via only the adjoint **78** representation which is what the superstring flux mechanism may do [1]. It is also phenomenologically interesting because it allows for an explanation of the experimental R_b excess [2]. The N-model is so called because N has $Q_N = 0$. It allows S to be a naturally light singlet neutrino and is ideally suited to explain the totality of all neutrino-oscillation experiments [3]. It is also a natural consequence of an alternative $SO(10)$ decomposition [4] of E_6, *i.e.*

$$\mathbf{16} = [(u,d),u^c,e^c;h^c,(\nu_E,E);S], \tag{10}$$

$$\mathbf{10} = [h,(E^c,N_E^c);d^c,(\nu_e,e)], \tag{11}$$

$$\mathbf{1} = [N], \tag{12}$$

which differs from the conventional assignment by how the $SU(5)$ multiplets are embedded.

HIGGS SECTOR

The Higgs sector of the $U(1)_\alpha$-extended supersymmetric model consists of two doublets and a singlet. They transform under $SU(3)_C \times SU(2)_L \times U(1)_Y \times U(1)_\alpha$ as follows.

$$\tilde{\Phi}_1 \equiv \begin{pmatrix} \bar{\phi}_1^0 \\ -\phi_1^- \end{pmatrix} \equiv \begin{pmatrix} \tilde{\nu}_E \\ \tilde{E} \end{pmatrix} \sim \left(1, 2, -\frac{1}{2}; -\frac{1}{\sqrt{6}} \cos\alpha + \frac{1}{\sqrt{10}} \sin\alpha\right), \quad (13)$$

$$\Phi_2 \equiv \begin{pmatrix} \phi_2^+ \\ \phi_2^0 \end{pmatrix} \equiv \begin{pmatrix} \tilde{E}^c \\ \tilde{N}_E^c \end{pmatrix} \sim \left(1, 2, \frac{1}{2}; -\frac{1}{\sqrt{6}} \cos\alpha - \frac{1}{\sqrt{10}} \sin\alpha\right), \quad (14)$$

$$\chi \equiv \tilde{S} \sim \left(1, 1, 0; \sqrt{\frac{2}{3}} \cos\alpha\right). \quad (15)$$

Hence the Higgs potential has the contribution

$$V_F = f^2[(\Phi_1^\dagger \Phi_2)(\Phi_2^\dagger \Phi_1) + (\Phi_1^\dagger \Phi_1 + \Phi_2^\dagger \Phi_2)(\bar{\chi}\chi)], \quad (16)$$

where f is the Yukawa coupling of the $\tilde{\Phi}_1 \Phi_2 \chi$ term in the superpotential. From the gauge interactions, we have the additional contribution

$$V_D = \frac{1}{8} g_2^2[(\Phi_1^\dagger \Phi_1)^2 + (\Phi_2^\dagger \Phi_2)^2 + 2(\Phi_1^\dagger \Phi_1)(\Phi_2^\dagger \Phi_2) - 4(\Phi_1^\dagger \Phi_2)(\Phi_2^\dagger \Phi_1)]$$
$$+ \frac{1}{2} g_1^2[-\frac{1}{2}\Phi_1^\dagger \Phi_1 + \frac{1}{2}\Phi_2^\dagger \Phi_2]^2 + \frac{1}{2} g_\alpha^2[\left(-\frac{1}{\sqrt{6}} \cos\alpha + \frac{1}{\sqrt{10}} \sin\alpha\right) \Phi_1^\dagger \Phi_1$$
$$+ \left(-\frac{1}{\sqrt{6}} \cos\alpha - \frac{1}{\sqrt{10}} \sin\alpha\right) \Phi_2^\dagger \Phi_2 + \sqrt{\frac{2}{3}} \cos\alpha \ \bar{\chi}\chi]^2. \quad (17)$$

Let $\langle \chi \rangle = u$, then $\sqrt{2} Re\chi$ is a physical scalar boson with

$$M^2 = \frac{4}{3} \cos^2\alpha \ g_\alpha^2 u^2, \quad (18)$$

and the $(\Phi_1^\dagger \Phi_1)\sqrt{2} Re\chi$ coupling is

$$F = \sqrt{2} u \left[f^2 + g_\alpha^2 \sqrt{\frac{2}{3}} \cos\alpha \left(-\frac{1}{\sqrt{6}} \cos\alpha + \frac{1}{\sqrt{10}} \sin\alpha\right)\right]. \quad (19)$$

The effective $(\Phi_1^\dagger \Phi_1)^2$ coupling λ_1 is thus given by [5–7]

$$\lambda_1 = \frac{1}{4}(g_1^2+g_2^2) + g_\alpha^2\left(-\frac{1}{\sqrt{6}}\cos\alpha + \frac{1}{\sqrt{10}}\sin\alpha\right)^2 - \frac{F^2}{M^2}$$
$$= \frac{1}{4}(g_1^2+g_2^2) + \left(1-\sqrt{\frac{3}{5}}\tan\alpha\right)f^2 - \frac{3f^4}{2\cos^2\alpha\, g_\alpha^2}. \quad (20)$$

Similarly,

$$\lambda_2 = \frac{1}{4}(g_1^2+g_2^2) + \left(1+\sqrt{\frac{3}{5}}\tan\alpha\right)f^2 - \frac{3f^4}{2\cos^2\alpha\, g_\alpha^2}, \quad (21)$$

$$\lambda_3 = -\frac{1}{4}g_1^2 + \frac{1}{4}g_2^2 + f^2 - \frac{3f^4}{2\cos^2\alpha\, g_\alpha^2}, \quad (22)$$

$$\lambda_4 = -\frac{1}{2}g_2^2 + f^2, \quad (23)$$

where the effective two-doublet Higgs potential has the generic form

$$V = m_1^2\Phi_1^\dagger\Phi_1 + m_2^2\Phi_2^\dagger\Phi_2 + m_{12}^2(\Phi_1^\dagger\Phi_2 + \Phi_2^\dagger\Phi_1)$$
$$+ \frac{1}{2}\lambda_1(\Phi_1^\dagger\Phi_1)^2 + \frac{1}{2}(\Phi_2^\dagger\Phi_2)^2 + \lambda_3(\Phi_1^\dagger\Phi_1)(\Phi_2^\dagger\Phi_2) + \lambda_4(\Phi_1^\dagger\Phi_2)(\Phi_2^\dagger\Phi_1). \quad (24)$$

¿From Eqs. (20) to (23), it is clear that the MSSM is recovered in the limit of $f = 0$. Let $\langle\phi_{1,2}^0\rangle \equiv v_{1,2}$, $\tan\beta \equiv v_2/v_1$, and $v^2 \equiv v_1^2 + v_2^2$, then this V has an upper bound on the lighter of the two neutral scalar bosons given by

$$(m_h^2)_{max} = 2v^2[\lambda_1\cos^4\beta + \lambda_2\sin^4\beta + 2(\lambda_3+\lambda_4)\sin^2\beta\cos^2\beta] + \epsilon, \quad (25)$$

where we have added the radiative correction due to the t quark and its supersymmetric scalar partners, *i.e.*

$$\epsilon = \frac{3g_2^2 m_t^4}{8\pi^2 M_W^2}\ln\left(1+\frac{\tilde{m}^2}{m_t^2}\right). \quad (26)$$

Using Eqs. (20) to (23), we obtain

$$(m_h^2)_{max} = M_Z^2\cos^2 2\beta + \epsilon$$
$$+ \frac{1}{\sqrt{2}G_F}\left[f^2\left(\frac{3}{2} - \sqrt{\frac{3}{5}}\tan\alpha\cos 2\beta - \frac{1}{2}\cos^2 2\beta\right) - \frac{3f^4}{2\cos^2\alpha\, g_\alpha^2}\right]. \quad (27)$$

Hence the MSSM bound can be exceeded for a wide range of values of α and β. Normalizing $U(1)_Y$ and $U(1)_\alpha$ at the grand-unification energy scale, we find it to be a very good approximation [8] to have $g_\alpha^2 = (5/3)g_1^2$. We use this and vary f^2 in Eq. (27) subject to the condition that V be bounded from below. We find the largest numerical value of m_h to be about 142 GeV, as compared to 128 GeV in the MSSM, and this is achieved with

$$\tan\alpha = -\frac{2\sqrt{3/5}\cos 2\beta}{3-\cos^2 2\beta}, \quad (28)$$

which is possible in the η-model.

Z - Z' SECTOR

The new Z' of this model mixes with the standard Z so that the experimentally observed Z is actually

$$Z_1 = Z \cos\theta + Z' \sin\theta, \tag{29}$$

where

$$\theta \simeq -\frac{1}{2}\sqrt{\frac{3}{2}}\frac{1}{\cos\alpha}\frac{g_Z}{g_\alpha}\left(\sin^2\beta - \frac{1}{2} + \frac{1}{2}\sqrt{\frac{3}{5}}\tan\alpha\right)\frac{v^2}{u^2}, \tag{30}$$

resulting in a slight shift of its mass from that predicted by the standard model, as well as a slight change in its couplings to the usual quarks and leptons. These deviations can be formulated in terms of the oblique parameters [6]:

$$\epsilon_1 = \left[\sin^4\beta - \frac{1}{4}\left(1 - \sqrt{\frac{3}{5}}\tan\alpha\right)^2\right]\frac{v^2}{u^2} \simeq \alpha T, \tag{31}$$

$$\epsilon_2 = \frac{1}{4}(3 - \sqrt{15}\tan\alpha)\left[\sin^2\beta - \frac{1}{2}\left(1 - \sqrt{\frac{3}{5}}\tan\alpha\right)\right]\frac{v^2}{u^2} \simeq -\frac{\alpha U}{4\sin^2\theta_W}, \tag{32}$$

$$\begin{aligned}\epsilon_3 = {} & \frac{1}{4}\left[1 - 3\sqrt{\frac{3}{5}}\tan\alpha + \frac{1}{2\sin^2\theta_W}\left(1 + \sqrt{\frac{3}{5}}\tan\alpha\right)\right] \\ & \times \left[\sin^2\beta - \frac{1}{2}\left(1 - \sqrt{\frac{3}{5}}\tan\alpha\right)\right]\frac{v^2}{u^2} \simeq \frac{\alpha S}{4\sin^2\theta_W}.\end{aligned} \tag{33}$$

Note that for $\sin^2\beta$ near $(1/2)(1 - \sqrt{3/5}\tan\alpha)$, $\epsilon_{1,2,3}$ are all suppressed. In any case, the experimental errors on these quantities are fractions of a percent, hence $u \sim$ TeV is allowed.

The mass of Z' is approximately equal to that of $\sqrt{2}Re\chi$, *i.e.* M of Eq. (18). Its interactions are of course determined by $U(1)_\alpha$. In particular, in the N-model, two S's are light singlet neutrinos, hence the ratio of the decay rates of Z' to $\nu\bar{\nu} + S\bar{S}$ over Z' to $\ell^-\ell^+$ is 62/15, instead of 4/5 without the S's. This would be a great experimental signature.

SUPERSYMMETRIC SCALAR MASSES

Consider the masses of the supersymmetric scalar partners of the quarks and leptons:

$$m_B^2 = m_0^2 + m_R^2 + m_F^2 + m_D^2, \tag{34}$$

where m_0 is a universal soft supersymmetry breaking mass at the grand-unification scale, m_R^2 is a correction generated by the renormalization-group equations running from the grand-unification scale down to the TeV scale, m_F is the explicit mass of the fermion partner, and m_D^2 is a term induced by gauge-boson masses. In the MSSM, m_D^2 is of order M_Z^2 and does not change m_B significantly. In the $U(1)_\alpha$-extended model, m_D^2 is of order $M_{Z'}^2$ and will affect m_B in a nontrivial way. For example, for the ordinary quarks and leptons,

$$\Delta m_D^2(10; 1, -1) = \frac{1}{8} M_{Z'}^2 \left(1 + \sqrt{\frac{3}{5}} \tan\alpha \right), \tag{35}$$

$$\Delta m_D^2(5^*; 1, 3) = \frac{1}{8} M_{Z'}^2 \left(1 - 3\sqrt{\frac{3}{5}} \tan\alpha \right). \tag{36}$$

This would have important consequences on the experimental search of supersymmetric particles. In fact, depending on m_F, it is possible for exotic scalars to be lighter than the usual scalar quarks and leptons.

Another important outcome of Eq. (34) is that the $U(1)_\alpha$ and electroweak symmetry breakings are related [9]. To see this, go back to the two-doublet Higgs potential V of Eq. (24). Using Eqs. (20) to (23), we can express the parameters m_{12}^2, m_1^2, and m_2^2 in terms of the mass of the pseudoscalar boson, m_A, and $\tan\beta$.

$$m_{12}^2 = -m_A^2 \sin\beta\cos\beta, \tag{37}$$

$$\begin{aligned} m_1^2 &= m_A^2 \sin^2\beta - \frac{1}{2} M_Z^2 \cos 2\beta \\ &\quad - \frac{2f^2}{g_Z^2} M_Z^2 \left[2\sin^2\beta + \left(1 - \sqrt{\frac{3}{5}} \tan\alpha \right) \cos^2\beta - \frac{3f^2}{2\cos^2\alpha\, g_\alpha^2} \right], \end{aligned} \tag{38}$$

$$\begin{aligned} m_2^2 &= m_A^2 \cos^2\beta + \frac{1}{2} M_Z^2 \cos 2\beta \\ &\quad - \frac{2f^2}{g_Z^2} M_Z^2 \left[2\cos^2\beta + \left(1 + \sqrt{\frac{3}{5}} \tan\alpha \right) \sin^2\beta - \frac{3f^2}{2\cos^2\alpha\, g_\alpha^2} \right]. \end{aligned} \tag{39}$$

On the other hand, using Eq. (34), we have

$$m_{12}^2 = f A_f u, \tag{40}$$

$$m_1^2 = m_0^2 + m_R^2(\tilde{g}, f) + f^2 u^2 - \frac{1}{4} \left(1 - \sqrt{\frac{3}{5}} \tan\alpha \right) M_{Z'}^2, \tag{41}$$

$$m_2^2 = m_0^2 + m_R^2(\tilde{g}, f) + f^2 u^2 - \frac{1}{4} \left(1 + \sqrt{\frac{3}{5}} \tan\alpha \right) M_{Z'}^2 + m_R^2(\lambda_t), \tag{42}$$

where fA_f is the coupling of the soft supersymmetry breaking $\tilde{\Phi}_1\Phi_2\chi$ scalar term, $\tilde{g}$ is the gluino, and λ_t is the Yukawa coupling of Φ_2 to the t quark. Matching Eqs. (37) to (39) with Eqs. (40) to (42) allows us to determine u and $\tan\beta$ as a function of f for a given set of parameters at the grand-unification scale.

In the MSSM assuming Eq. (34),

$$m_1^2 - m_2^2 = -m_R^2(\lambda_t) = -(m_A^2 + M_Z^2)\cos 2\beta. \tag{43}$$

Since $m_R^2(\lambda_t) < 0$, we must have $\tan\beta > 1$. In the $U(1)_\alpha$-extended model, because of the extra D-term contribution, $\tan\beta < 1$ becomes possible. Another consequence is that because of Eq. (35), a light scalar t quark is not possible unless $\tan\alpha < -\sqrt{5/3}$.

CONCLUSIONS

(1) Supersymmetric $U(1)_\alpha$ from E_6 is a good possiblity at the TeV scale. (2) The two-doublet Higgs structure at around 100 GeV will be different from that of the MSSM. (3) Supersymmetric scalar masses depend crucially on $U(1)_\alpha$. (4) The $U(1)_\alpha$ breaking scale and $\tan\beta$ are closely related.

ACKNOWLEDGMENTS

I thank Juan Carlos D'Olivo, Miguel Perez, and Rodrigo Huerta for their great hospitality and a stimulating symposium. This work was supported in part by the U. S. Department of Energy under Grant No. DE-FG03-94ER40837.

REFERENCES

1. J. L. Hewett and T. G. Rizzo, Phys. Rept. **183**, 193 (1989); M. Cvetic and P. Langacker, Phys. Rev. **D54**, 3570 (1996); Mod. Phys. Lett. **A11**, 1247 (1996).
2. K. S. Babu *et al.*, Phys. Rev. **D54**, 4635 (1996).
3. E. Ma, Phys. Lett. **B380**, 286 (1996).
4. E. Ma, Phys. Rev. **D36**, 274 (1987).
5. E. Ma and D. Ng, Phys. Rev. **D49**, 6164 (1994); T. V. Duong and E. Ma, Phys. Lett. **B316**, 307 (1993); J. Phys. **G21**, 159 (1995).
6. E. Keith and E. Ma, Phys. Rev. **D54**, 3587 (1996).
7. X. Li and E. Ma, hep-ph/9608398.
8. E. Keith, E. Ma, and B. Mukhopadhyaya, hep-ph/9607488, Phys. Rev. **D55**, in press (1997).
9. E. Keith, E. Ma, and B. Mukhopadhyaya, in preparation.

Asymmetries and Correlations in Charm Production

Marleigh Sheaff[1]

Departamento de Fisica, CINVESTAV-IPN, Apdo. Postal 14-740, 07000, Mexico, D. F., MEXICO

Abstract. The high-statistics samples of charm hadrons produced in recent Fermilab fixed target experiments allow tests of calculations made in next-to-leading-order (NLO) quantum chromodynamics (QCD) or of predictions made using QCD-motivated Monte Carlo models. The large asymmetries measured in the production of charged D mesons can not be explained in either of these frameworks in the absence of fragmentation effects. The PYTHIA/JETSET Monte Carlo including LUND string fragmentation gives a good fit to the asymmetry data with theoretically acceptable input parameters. Charge correlations in production between a meson containing a heavy quark and the charged pion closest to it in phase space are expected in the LUND string model of fragmentation. This effect has important consequences, since it provides a means of self-tagging of neutral B decays for measurements of CP violation in the B system. These correlations have been observed for the first time in charm production by Fermilab E791.

INTRODUCTION

Recent high-statistics measurements of charm production properties allow more precise comparison to theoretical models than has been possible in the past. There have been significant advances on the theoretical side as well. Full next-to-leading-order (NLO) calculations in quantum chromodynamics (QCD) have been performed [1] [2] [3] [4] [5]. Also available for comparison to data are the predictions of a QCD-motivated Monte Carlo, PYTHIA/JETSET [6] which incorporates the effects of fragmentation using the LUND string model [7].

One property of charm production that has been studied for a number of years is the asymmetry, which is a difference in the production of a charm particle (p) and its anti-particle (ap),

$$A = \frac{\sigma_p - \sigma_{ap}}{\sigma_p + \sigma_{ap}}. \tag{1}$$

[1] Work supported by the Consejo Nacional de Ciencia y Tecnologia and the National Science Foundation

CP400, *First Latin American Symposium on High Energy Physics/ VII Mexican School of Particles and Fields,* edited by D'Olivo/Klein-Kreisler/Mendez

This can be an overall asymmetry or an asymmetry over some range in a kinematic variable. One example of the latter type of asymmetry is the "leading particle effect". This is the enhancement in the production of a charm (anti-charm) particle relative to the anti-charm (charm) particle seen at high x_f when the charm (anti-charm) particle has a valence quark in common with the incident beam particle but the anti-charm (charm) particle does not. E.g., for a π^- ($\bar{u}$d) beam, D^- ($\bar{c}$d) is leading and D^+ (c$\bar{d}$) is non-leading. Enhanced production of D^- relative to D^+ in the forward direction, which grows larger with increasing x_f, has been observed experimentally [8] [9] [10].

Another production effect that has been investigated recently is a possible charge correlation between a meson containing a heavy quark and the charged pion nearest to it in phase space. This is expected in the string model of fragmentation and has been proposed as a means of "self-tagging" for neutral B decays to be used in measurements of CP violation in the B system [11]. The first experimental evidence for these charge correlations in charm production has been observed recently by Fermilab E791 [12].

ASYMMETRIES

The charm quark production process is factorizable, providing the charm quark mass is "heavy enough" relative to Λ_{QCD}. Assuming this is the case, it can be calculated by convoluting the parton level cross sections calculated using perturbation theory with the parton distribution functions determined from previous experiments, typically from deep inelastic scattering, and summing over all processes that contribute. For hadroproduction at leading order, i.e., order α_s^2, these processes include anti-quark-quark annihilation ($\bar{q}$q $\rightarrow \bar{Q}$Q) and gluon-gluon fusion (gg $\rightarrow \bar{Q}$Q). The latter dominates at Fermilab fixed target energies and is responsible for over 90% of the cross section. These leading order calculations do not predict any asymmetries when comparing the c and $\bar{c}$ distributions. However, there is a small asymmetry predicted in NLO QCD [1]. The cross sections for charm states containing $\bar{c}$ quarks are expected to be higher than those for states containing c quarks at high x_f when produced by a π^- beam. This is an interference effect that arises mostly from the $\bar{q}$q annihilation subprocess. It was first observed as an asymmetry due to radiative corrections in $e^+e^- \rightarrow \mu^+\mu^-$. The predicted enhancement would have to be multiplied by a factor of 20 in order to explain the large charged D production asymmetry, measured first by WA82 [8] at CERN and then confirmed by Fermilab E769 [9]. This is demonstrated in Figure 1 [9], which shows the WA82 and E769 measured asymmetries versus x_f. The dashed curve is the NLO QCD prediction. The dot-dash curve is the prediction from the PYTHIA/JETSET Monte Carlo using the default parameters and including fragmentation effects. The latter shows the correct trend but has too large an overall asymmetry.

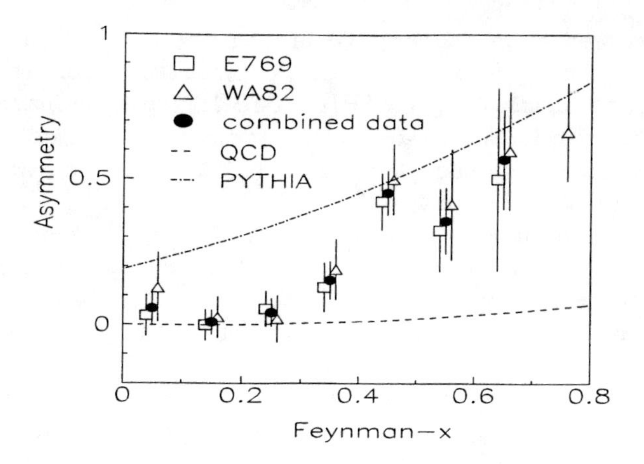

FIGURE 1. Asymmetry versus x_f in the production of charged D mesons. WA82 data come from π^- beam only and are for D^- (Leading) relative to D^+ (Non-Leading) production. E769 data include pion beams of both signs and D^* as well as D production.

Recently, forward asymmetries have been measured by E769 for kaon and proton beams incident [13]. These are included with the E769 π^- beam results in Table 1. The charged D asymmetry is seen to be the same for π^- or p beams incident, as one might expect, since each species contains one d-quark. The D_s asymmetry for the K beams is in statistical agreement with being the same as well. Thus, the shared valence s-quark leads to an asymmetry of similar size to the asymmetry that results from the shared valence d-quark. This represents the first reported asymmetry measurement for D_s produced by an incident kaon beam. Directly produced D^o's are leading in a π^- beam while the D^o's that are the daughters of D^{*+}'s, which constitute a large fraction of the D^o sample, are non-leading. This may explain why the D^o's show no statistically significant asymmetry. However, for protons incident, $\bar{D}^o$'s are leading whether they are directly or indirectly produced. The measured asymmetry is still small. This is not yet understood.

TABLE 1. E769 "leading particle" production asymmetries. Inequalities indicate 90% confidence level lower limits.

Beam	Particle	Anti-particle	A (defined in Eqn. 1)
π^-	D^-	D^+	0.18 ± 0.06
	D^o	$\bar{D}^o$	-0.06 ± 0.07
	D^{*-}	D^{*+}	0.09 ± 0.06
K^-, K^+	D_s^-, D_s^+	D_s^+, D_s^-	0.25 ± 0.11
	$\bar{\Lambda}_c, \Lambda_c$	$\Lambda_c, \bar{\Lambda}_c$	> 0.6
p	D^-	D^+	0.18 ± 0.05
	$\bar{D}^o$	D^o	0.06 ± 0.06
	D^{*-}	D^{*+}	0.36 ± 0.13
	Λ_c	$\bar{\Lambda}_c$	> 0.6

Following $\bar{c}$c production, neither the remnant π^- beam particle nor the remnant struck nucleon are in color singlet states. Since the hadrons that emerge from the interaction must be, PYTHIA/JETSET starts the hadronization process by forming color singlet strings from the available quarks. The struck nucleon is split into a di-quark, qq_N and quark, q_N, allowing two possible string combinations; either (c,$\bar{u}_\pi$), ($\bar{c}$,q_N), and (d_π,qq_N), or ($\bar{c}$,d_π), (c,qq_N), and ($\bar{u}_\pi$,q_N). As the quarks in each string move apart and therefore the energy increases, $\bar{q}$q pairs are created along the string. This is shown schematically in Figure 2 for the two possible diagrams in which the target is a proton which splits into a ud di-quark and a u quark. Some of the time, when the $\bar{c}$ and d quark shown in the lower diagram are close together in phase space, they coalesce to produce a D^-. Since the d tends to be moving forward in the center of mass, this effect will be expected to occur at high x_f [14]. It is therefore known as "beam dragging" [15]. It is also a "leading particle effect", since the d valence quark from the beam π^- coalesces with the produced $\bar{c}$ to form the D^-. This is one possible explanation [10] for the charged D asymmetry shown in Figure 1. Another proposed cause of this asymmetry is for the initial state

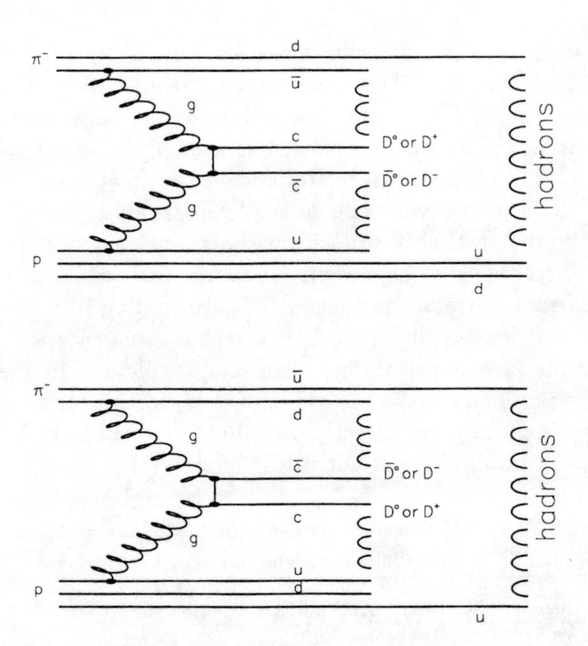

FIGURE 2. Two of the eight possible string combinations that can be formed for πN production of charm in the LUND fragmentation model used by PYTHIA/JETSET.

to fluctuate into a Fock state containing a $\bar{c}$c pair as well as the $\bar{u}$d. This process [16] is known as "intrinsic charm". The $\bar{c}$c pair is then kicked on-shell by the interaction. Since the $\bar{c}$ and c quarks are co-moving with the $\bar{u}$ and d, there is likely to be coalescence and therefore enhanced production of D^- (and D^o) in the forward direction.

Figure 3 [10] shows the measured charged D asymmetry for the sample of 40364±252 D^- and 34104±237 D^+ decays to the $K\pi\pi$ final state that have been fully reconstructed from E791 data. These data are compared to the default PYTHIA prediction (dashed curve) in which m_c is assumed to be 1.35 GeV/c^2 and $< k_t^2 >$ of the gluons is set to 0.44 GeV/c^2. These parameters are changed to 1.7 GeV/c^2 and 1.0 GeV/c^2 for the tuned PYTHIA prediction (solid curve) shown on the figure, which gives a good fit to the data. The Vogt-Brodsky prediction (dot-dash curve) shows a similar trend to the data, but assumes equal numbers of D^- and D^+ are produced at $x_f > 0$ and therefore does not demonstrate the overall asymmetry of the observed distribution. Note that if, in the lower diagram shown in Figure 2, the c quark were to coalesce with the qq_N pair to form a charm baryon a large fraction of the time, more D^- than D^+ would be produced overall. This asymmetry would result from a deficit of D^+ in the target fragmentation region rather than from an enhancement of D^- in the beam fragmentation region.

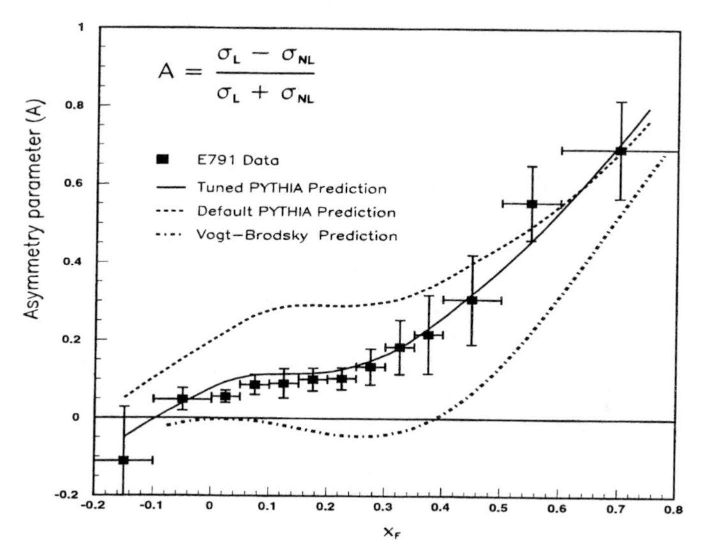

FIGURE 3. Asymmetry versus x_f for D^- (Leading) relative to D^+ (Non-Leading) production from π^-N interactions in E791.

A key prediction of the Vogt-Brodsky model is that the asymmetry at high x_f and low p_t^2 will increase with decreasing p_t^2. Figure 4 [10] shows the data for $0.4 < x_f < 0.8$ along with comparisons to the same theoretical models as shown in the previous figure (curves coded as in that figure). Both the default PYTHIA and the tuned PYTHIA give adequate fits to the data. The Vogt-Brodsky model disagrees as to the overall asymmetry even in this limited region of x_f. The trend toward increasing asymmetry with decreasing p_t^2 predicted in this model is not seen in the data, although the statistics are not good enough to totally rule it out providing the normalization difference can be explained.

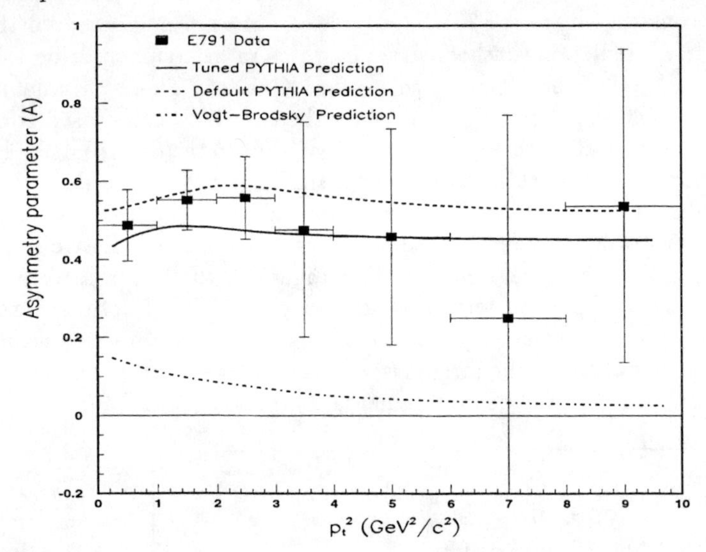

FIGURE 4. Asymmetry versus p_t^2 for D^- (Leading) relative to D^+ (Non-Leading) production for E791 data shown in the previous figure for $0.4 < x_f < 0.8$.

The Fermilab charm photoproduction experiment E687 has recently completed an analysis of production asymmetries [17] for charm mesons and baryons within their region of good acceptance, which is most of forward x_f. Their data are in good agreement with the default PYTHIA for charm photoproduction, including LUND string fragmentation. Two models were examined, one in which the quark called q_N in the above discussion receives an energy fraction

$$\frac{dN}{d\chi} \sim \frac{(1-\chi)^3}{\chi} \tag{2}$$

and the other in which q_N receives an energy fraction

$$\frac{dN}{d\chi} \sim 2(1-\chi) \tag{3}$$

of the remnant nucleon momentum. The latter model, which gives 1/3 of the nucleon momentum to q_N on average, gives a better fit to the measured distributions than the former, which gives only a small fraction of the nucleon momentum to q_N. Since there are no quarks in the incident beam, there is only one set of strings that can be formed in photoproduction, $(\bar{c}, q_N)$ and (c, qq_N).

The beam fragmentation part of the lower diagram shown in Figure 2 is shown under an imaginary looking glass in Figure 5, both for the usual case where some number of $\bar{q}$q pairs are produced along the string (upper diagram) and for the case of coalescence (lower diagram). Looking at the upper diagram, since a $\bar{d}$d pair must be produced to provide the d quark to create a D^-, there will be an available $\bar{d}$ nearby. Thus, if the nearest pion to the D^- in phase space is a charged pion, one would expect that it would be a π^+. The nearest pion must come from a different string in the case shown in the lower diagram, so that the sign of the nearest charged pion can not be predicted in the same way, Since the D^-'s that result from this "beam dragging" diagram are produced at high x_f, $D^-\pi^+$ charge correlations, if observed, would be expected to become smaller with increasing x_f. The analog to the upper diagram for the D^+ is the beam fragmentation part of the upper diagram shown in Figure 2. In this case, in order to form a D^+, a $\bar{d}$d must be produced along the string, since there is no $\bar{d}$ valence quark in the incident π^-.. The available d quark that remains nearby after formation of the D^+ leads to the expectation that if the nearest pion in phase space to the D^+ is charged, it will be a π^-. Since there is no analog to the "beam dragging" diagram for the D^+, $D^+\pi^-$ correlations would not be expected to diminish with increasing x_f as for the $D^-\pi^+$.

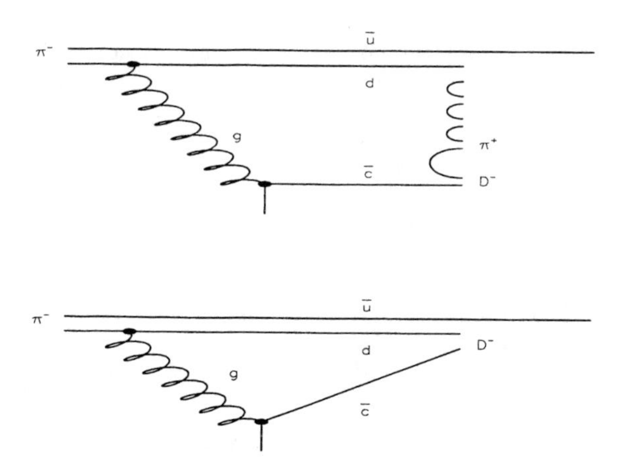

FIGURE 5. The two diagrams in which a D^- is formed in the beam fragmentation region in the LUND fragmentation model used by PYTHIA/JETSET. See the text for a detailed discussion.

These effects have been studied in E791 [12] using the fully reconstructed charm decays shown in Table 2. For each of the D decays listed, all tracks which originated at the primary and which were well-identified as pions by

TABLE 2. E791 decay samples used for charge correlation analysis.

	$D^o \to K\pi$	$D^+ \to K\pi\pi$	$D^{*+} \to D^o\pi$
Particle	22587±210	24569±204	4997±84
Anti-particle	24237±216	29649±238	6048±93

the two threshold Cerenkov counters were then combined one at a time with the reconstructed D to form the Dπ invariant mass. The combination with the smallest Δm, where

$$\Delta m = M(D\pi) - M(D) \tag{4}$$

was selected for further analysis, since small Δm corresponds to small relative velocity between the D and the pion. The Δm for the selected combination is called Δm_{min} and the pion, the "partner pion" in the discussion below.

The PYTHIA/JETSET Monte Carlo was used to correct the observed distributions in Δm_{min} for acceptance, including incorrect assignment of the partner pion. The overall acceptance ranged from 2 to 4%, depending on the Dπ combination. The primary tracks generated by the Monte Carlo were compared before and after reconstruction by matching the slopes, intercepts, and momenta. The wrong choice of the partner pion came about for three reasons,

- Ghost Track - reconstructed partner pion does not match any pion generated. This ranged from 10-15% depending on the decay mode.
- Smearing - wrong pion selected, but pion has same sign as the generated closest pion.
- Dilution - wrong pion selected, and pion has opposite sign from the generated closest pion.

A correlation parameter, α, is then defined for the acceptance-corrected data. E.g., for the D^- and D^+:

$$\alpha(D^-) = \frac{N(D^-\pi^+) - N(D^-\pi^-)}{N(D^-\pi^+) + N(D^-\pi^-)} \tag{5}$$

$$\alpha(D^+) = \frac{N(D^+\pi^-) - N(D^+\pi^+)}{N(D^+\pi^-) + N(D^+\pi^+)} \tag{6}$$

Figure 6(a) shows $\alpha(D^+)$ and $\alpha(D^-)$ versus x_f for all events for which Δm_{min} is less than 0.74 GeV/c^2. The expected correlations are seen for the D^+. They do not show a strong dependence on x_f. The correlations are also seen for the D^- at low x_f, although they are somewhat smaller than for the D^+. The decrease in α with increasing x_f expected for the D^- because of the relative importance of the "beam-dragging" diagram in that kinematic regime is also seen. In fact the correlation for the D^- even changes sign above x_f of about 0.2. Since it is the d from the incoming π^- which becomes part of the D^- in the "beam-dragging" diagram, the quark closest to the D^- in phase space in this case will be the co-moving $\bar{u}$. If this is to fragment into a charged pion, a $d\bar{d}$ must be produced closest to it along its string and the produced pion will be a π^-. This could explain the negative correlation at high x_f. The effects are similar, although not as pronounced, for the D^*'s as shown in Figure 6(b).

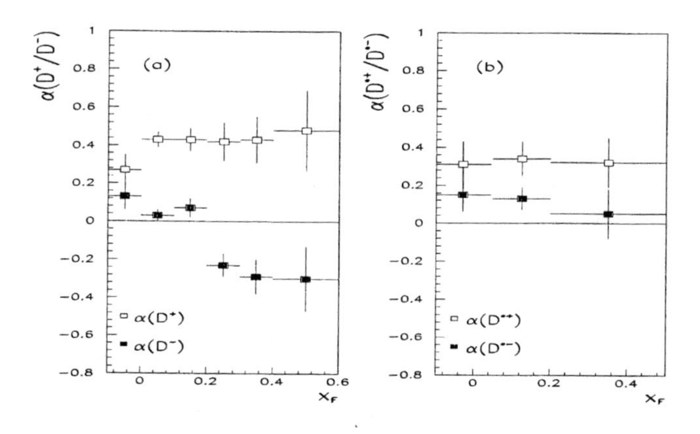

FIGURE 6. Correlation parameter versus x_f for $D^\pm$ and $D^{*\pm}$.

CONCLUSIONS

Recent high-statistics measurements of charm production properties in both hadron and photon beams at Fermilab show good agreement with the predictions of the PYTHIA/JETSET Monte Carlo including LUND string fragmentation. Further tests of the model are planned using the decays of charm mesons and baryons containing different light quarks. These include measurements of D_s and D^o asymmetries and of correlations between the D and $\bar{D}$ meson produced in the same event.

ACKNOWLEDGEMENTS

The author wishes to thank the many dedicated physicists working on fixed target charm experiments at Fermilab for their contributions to the work presented in this article. The support of a number of funding agencies, both in the U. S. and in other countries around the world, has made these experiments possible. This support is gratefully acknowledged. Special thanks are due to the staff of Fermilab, who keep the accelerator, beamlines, computers, etc., alive and well in support of these activities.

REFERENCES

1. P. Nason, S. Dawson, and K. Ellis, Nuc. Phys. **B327**, 49 (1989).
2. M. Mangano, P. Nason, and G. Ridolfi, Nuc. Phys. **B405**, 507 (1993).
3. S. Frixione,M. Mangano, P. Nason, and G. Ridolfi, Nuc. Phys. **B412**, 225 (1994).
4. S. Frixione,M. Mangano, P. Nason, and G. Ridolfi, Nuc. Phys. **B431**, 453 (1994).
5. P. Nason, S. Frixione, and G. Ridolfi, *Proceedings of the XV International Conference on Physics in Collision*, Cracow, Poland, edited by M. Rozanka and K. Rybicki, World Scientific, Singapore, 1995.
6. T. Sjostrand, Computer Phys. Comm. **82**, 74 (1994).
7. B. Andersson, G. Gustafson, G. Ingelman, and T. Sjostrand, Physics Reports **97**, 31 (1983).
8. WA82 Collaboration, M. Adamovich *et al.*, Phys. Lett. **B305**, 402 (1993).
9. G.A. Alves *et al.*, Phys. Rev. Lett. **72**, 812 (1994).
10. E. M. Aitala *et al.*, Phys. Lett. **B371**, 157 (1996). Figures 3 and 4 are reprinted from pages 160 and 161 of this article entitled "Asymmetries between the production of D^+ and D^- mesons from 500 GeV/c π^--nucleus interactions as a function of x_F and p_t^2" with the kind permission of Elsevier Science - NL, Sara Burgerhartstraat 25, 1055 KV Amsterdam, The Netherlands.
11. M. Gronau, A. Nippe, and J. L. Rosner, Phys. Rev. **D47**, 1988 (1992).
12. E. M. Aitala *et al.*, submitted for publication in Phys. Lett. **B**.
13. G.A. Alves *et al.*, Phys. Rev. Lett. **77**, 2388 (1996).
14. J. Leslie, PhD. thesis, University of California at Santa Cruz, 1996, to be published.
15. T. Sjostrand, Int. Journal of Modern Physics A, Vol. 3, 751 (1988).
16. R. Vogt and S. J. Brodsky, Nuc. Phys. **B438**, 261 (1995).
17. P. L. Frabetti *et al.*, Phys. Lett. **B370**, 222 (1996).

Gauge Invariance and the Unstable Particle

Robin G. Stuart

Randall Laboratory of Physics
Ann Arbor, Michigan 48109-1120
USA

Abstract. It is shown how to construct exactly gauge-invariant S-matrix elements for processes involving unstable gauge particles such as the Z^0 boson. The results are applied to derive a physically meaningful expression for the cross-section $\sigma(e^+e^- \to Z^0Z^0)$ and thereby provide a solution to the long-standing *problem of the unstable particle.*

I INTRODUCTION

A resonance is fundamentally a non-perturbative object and is thus not amenable to the methods of standard perturbation theory. In order to describe physics at the Z^0 resonance one is forced, therefore, to employ some sort of non-perturbative procedure. Such a procedure is Dyson summation that sums strings of one-particle irreducible (1PI) self-energy diagrams as a geometric series to all orders in the coupling constant, α and effectively replaces the tree-level Z^0 propagator by a dressed propagator,

$$\frac{1}{s-M_Z^2} \to \frac{1}{s-M_Z^2}\sum_n \left(\frac{\Pi_{ZZ}(s)}{s-M_Z^2}\right)^n = \frac{1}{s-M_Z^2-\Pi_{ZZ}^{(1)}(s)} \tag{1}$$

where $\Pi_{ZZ}(s)$ is the one-loop Z^0 self-energy. The problem here is that electroweak physics is described by a gauge theory. Results of calculations of physical processes must be exactly gauge-invariant but this comes about through delicate cancellations between many different Feynman diagrams each of which is separately gauge-dependent. The cancellation happens at each order in α when all diagrams of a given order are combined. The Z^0 self-energy, $\Pi_{ZZ}(s)$, is gauge-dependent at $\mathcal{O}(\alpha)$ and hence the rhs of eq.(1) is gauge-dependent at all orders in α.

CP400, *First Latin American Symposium on High Energy Physics/ VII Mexican School of Particles and Fields,*
edited by D'Olivo/Klein-Kreisler/Mendez

If the dressed propagator is used in a finite-order calculation the result will be gauge-dependent at some order because the will be no diagrams available to cancel the gauge-dependence beyond the order being calculated. This gauge-dependence should be viewed as an indicator that the approximation scheme being used is inconsistent or does not represent a physical observable.

In constructing amplitudes that represent physical observables, one must take care to respect the requirements that are laid down by analytic S-matrix theory [1]. These conditions are derived from general considerations such as energy conservation and causality and it will be found that appealing to them leads to a procedure for generating gauge-invariant amplitudes with no flexibility in the final result. The procedure will also make it possible to give an answer to the *problem of the unstable particle* that was put forward by Peierls in the early fifties [2]. Much of what appears here can be found in ref.s [3–5].

II GAUGE-INVARIANT S-MATRIX ELEMENTS

We will begin by reviewing what is known about S-matrix elements for processes containing unstable particles. An unstable particle is associated with a pole, s_p, lying on the second Riemann sheet below the real s axis. The scattering amplitude, $A(s)$ for a process which contains an intermediate unstable particle can then be written in the form,

$$A(s) = \frac{R}{s - s_p} + B(s) \tag{2}$$

where R and s_p are complex constants and $B(s)$ is regular at $s = s_p$. The first term on the rhs of eq.(2) will be called the resonant term and the second is the non-resonant background term. It is known that s_p is process-independent in the sense that any process that contains a given unstable particle as an intermediate state will have its pole at the same position. It can also be shown from Fredholm theory that the residue factorizes as, $R = R_i \cdot R_f$, into pieces that depend separately on the initial- and final-state. One can prove, by very simple arguments, that s_p, R and $B(s)$ are separately and exactly gauge-invariant [3].

¿From analytic S-matrix theory it is known that production thresholds are associated with a branch cut. Branch points for stable particles lie on the real s-axis and those for unstable particles, such as W bosons, lie significantly below it. Provided there are no nearby thresholds the amplitude, $A(s)$, can be adequately described in the resonance region by a Laurent expansion about the pole, s_p. It should be emphasized that $A(s)$ can always be written in the form (2) even when thresholds are present. In that case $B(s)$ will be an analytic function containing a branch point and the rhs of (2) continues to be an exact representation of $A(s)$. Laurent expansion refers to how the resonant term is identified. The aim is only to separate $A(s)$ into gauge-invariant resonant and background pieces and there is no necessity to perform a Laurent expansion beyond its leading term although this can provide a useful way for parameterizing electroweak data [6].

Let us consider the production process for a massless fermion pair in $e^+e^- \rightarrow f\bar{f}$. Away from the Z^0 resonance it is known how to calculate the amplitude to arbitrary accuracy in a gauge-invariant manner. Near resonance we must perform a Dyson summation and then separate the amplitude into its resonant and background pieces. Doing so allows expressions for the gauge-invariant quantities, s_p, R and $B(s)$ to be identified in terms of the 1PI functions that occur in perturbation theory. The scattering amplitude for $e^+e^- \rightarrow f\bar{f}$ is

$$A(s,t) = \frac{R_{iZ}(s_p)R_{Zf}(s_p)}{s-s_p} + \frac{R_{iZ}(s)R_{Zf}(s) - R_{iZ}(s_p)R_{Zf}(s_p)}{s-s_p} + \frac{V_{i\gamma}(s)V_{\gamma f}(s)}{s-\Pi_{\gamma\gamma}(s)} + B(s,t) \qquad (3)$$

in which

$$R_{iZ}(s) = \left[V_{i\gamma}(s)\frac{\Pi_{\gamma Z}(s)}{s-\Pi_{\gamma\gamma}(s)} + V_{iZ}(s)\right] F_{ZZ}^{\frac{1}{2}}(s), \qquad (4)$$

$$R_{Zf}(s) = F_{ZZ}^{\frac{1}{2}}(s)\left[V_{Zf}(s) + \frac{\Pi_{Z\gamma}(s)}{s-\Pi_{\gamma\gamma}(s)}V_{\gamma f}(s)\right]. \qquad (5)$$

The pole position s_p is a solution of the equation

$$s - M_Z^2 - \Pi_{ZZ}(s) - \frac{\Pi_{Z\gamma}^2(s)}{s-\Pi_{\gamma\gamma}(s)} = 0 \qquad (6)$$

and $F_{ZZ}(s)$ is defined through the relation

$$s - M_Z^2 - \Pi_{ZZ}(s) - \frac{\Pi_{Z\gamma}^2(s)}{s-\Pi_{\gamma\gamma}(s)} = \frac{1}{F_{ZZ}(s)}(s-s_p). \qquad (7)$$

It should be emphasized that eq.(3) is exact and valid anywhere on the complex s-plane. The effect of Z^0-γ mixing has been included. The quantity $\Pi_{\gamma\gamma}(s)$ and $\Pi_{Z\gamma}(s)$ are the photon self-energy and the Z-γ mixing respectively. $V_{iZ}(q^2)$, $V_{Zf}(q^2)$ are the initial- and final-state Z^0 vertices, into which the external wavefunctions have been absorbed, and $V_{i\gamma}(q^2)$, $V_{\gamma f}(q^2)$ are the corresponding vertices for the photon. Here $B(s,t)$ denotes 1PI corrections to the matrix element that include things like as box diagrams. The first term on the rhs of eq.(3) is the resonant part of $A(s,t)$ and the three terms on the second line taken together are form the non-resonant background.

Calculations are most conveniently performed in terms of the real renormalized parameters of the theory such as the renormalized mass, M_Z. Eq.(6) can be solved iteratively in terms of M_Z to give

$$s_p = M_Z^2 + \Pi_{ZZ}(M_Z^2) + ... \tag{8}$$

The rhs of eq.(8) may be substituted for s_p where it appears in eq.(3)–(5). Taylor series expansion can then be used to obtain perturbative expressions for s_p, R and $B(s)$ in terms of Greens functions with real arguments up to any desired order. At any given order these perturbative expressions will be exactly gauge-invariant as will scattering amplitudes constructed from them.

A couple of points should be noted here. The appearance of Greens functions with complex arguments in eq.(3) is a natural consequence of the analyticity of the S-matrix. The S-matrix itself is never evaluated with complex s. The arguments of $A(s,t)$ on the lhs of eq.(3) is real as is the 's' in the denominator of the first term on the rhs.

In the procedure described above, one starts by extracting the resonant term in a scattering amplitude by Laurent expansion about the exact pole position s_p and then specializes to lower orders by further expanding about the renormalized mass. Other authors [7,8] have attempted to apply the techniques described above by first expanded about the renormalized mass and then added a finite width in the denominator of the resonant part by hand. That procedure cannot be justified and leads to problems when one treats processes like $e^+e^- \rightarrow W^+W^-$. It gives rise to spurious *threshold singularities* or complex scattering angles due the production threshold's being incorrectly located on the real axis. In section IV the process $e^+e^- \rightarrow Z^0Z^0$ will be treated and no threshold singularities or complex scattering angles will arise.

III THE PROBLEM OF THE UNSTABLE PARTICLE

We have thus succeeded in our goal of producing exactly gauge-invariant scattering amplitudes to arbitrary order. One might ask whether what has been done is just a mathematical trick, in which case the gauge-invariance is accidental, or does it have some physical interpretation. In this section it will be shown that the latter is true.

Recall that the coordinate space dressed propagator for a scalar particle has an integral representation

$$\Delta(x'-x) = \int \frac{d^4k}{(2\pi)^4} \frac{e^{-ik\cdot(x'-x)}}{k^2 - m^2 - \Pi(k^2) + i\epsilon} \tag{9}$$

The integrand has a pole at $k^2 = s_p$ where s_p is a solution of the equation $s - m^2 + \Pi(s) = 0$ and as in the previous section we define $F(s)$ via the relation $s - m^2 + \Pi(s) = (s - s_p)/F(s)$. The dressed propagator can then be written as

$$\Delta(x'-x) = \int \frac{d^4k}{(2\pi)^4} e^{-ik\cdot(x'-x)} \left[\frac{F(s_p)}{k^2 - s_p} + \frac{F(k^2) - F(s_p)}{k^2 - s_p} \right] \tag{10}$$

that separates resonant and non-resonant pieces. Performing the k_0 integration resonant gives

$$\begin{aligned}\Delta(x'-x) = &-i\int\frac{d^3k}{(2\pi)^3 2k_0}e^{-ik\cdot(x'-x)}\theta(t'-t)F(s_p)\\ &+\int\frac{d^4k}{(2\pi)^4}\frac{F(k^2)-F(s_p)}{k^2-s_p}\\ &-i\int\frac{d^3k}{(2\pi)^3 2k_0}e^{ik\cdot(x'-x)}\theta(t-t')F(s_p)\end{aligned}\tag{11}$$

where $k_0 = \sqrt{\vec{k}^2 + s_p}$. The non-resonant term contributes only for $t = t'$ and so represents a contact interaction. The resonant part spits into two terms that contribute when $t > t'$ or $t < t'$ and therefore connects points x and x' that are separated by a finite distance in space-time.

The problem of the unstable particle [2] is may be roughly stated as follows: S-matrix theory deals with asymptotic in-states and out-states that propagate from and to infinity. Unstable particles cannot exist as asymptotic states because they decay a finite distance from the interaction region. Indeed it is known [9] that the S-matrix is unitary on the Hilbert space spanned by stable particle states and hence there is not even any room to accommodate unstable particles as external states. How can one use the S-matrix to calculate, say, the production cross-section for an unstable particle when it cannot exist as an asymptotic state?

In the first part of this section it was shown that the resonant part of the dressed propagator connected points with a finite space-time separation. When a similar analysis is applied to a physical matrix element, such as eq.(3), one concludes that the resonant part describes a process in which there is a finite space-time separation between the initial-state vertex, V_i, and the final-state vertex, V_f. In other words, the resonant term describes the finite propagation of a physical Z^0 boson. The non-resonant background represents prompt production of the final state. As these two possibilities are, in principle, physically distinguishable, they must be separately gauge-invariant.

We can thus use finite propagation as a tag for identifying unstable particles without requiring that they appear in the final state. This is, after all, the way b-quarks are identified in vertex detectors. A production cross-section for an unstable particle is obtained by extracting the resonant part of the matrix element for a process containing that particle in an intermediate state and summing over all possible decay modes.

IV THE PROCESS $e^+e^- \to Z^0Z^0$

In this section we will calculate the cross-section for $e^+e^- \to Z^0Z^0$. This is of both theoretical and practical importance. On the theoretical side it represents an example of a calculation of the production cross-section for unstable particles. On the practical side, at high energies $e^+e^- \to Z^0Z^0$ will be a dominant source of fermion pairs $(f_1\bar{f}_1)$ and $(f_2\bar{f}_2)$ due to its double resonant enhancement and hence $\sigma(e^+e^- \to Z^0Z^0)$ is an excellent approximation to the cross-section for 4-fermion pair production. If the experimental situation warrants it, background terms can also be included without difficulty.

In the case of $e^+e^- \to f\bar{f}$, dealt with in section II, the invariant mass squared of the Z^0, s, is fixed by the momenta of the incoming e^+e^-. For the process $e^+e^- \to Z^0Z^0$ the invariant mass of the produced Z^0's is not constant and must be somehow included in phase space integrations. It is not immediately clear how to do this and without further guidance from S-matrix theory there would seem to be considerable flexibility in how to proceed. A new ingredient is required and that is to realize that an expression for an S-matrix element can always be divided into a part that is a Lorentz-invariant function of the kinematic invariants of the problem and Lorentz-covariant objects, such as $\not{p}$ etc. The latter are known as *standard covariants* [10–13]. It is the Lorentz invariant part that satisfies the requirements of analytic S-matrix theory and from which the resonant and non-resonant background parts are extracted while the Lorentz covariant part is untouched.

To calculate $\sigma(e^+e^- \to Z^0Z^0)$, we begin by constructing the cross-section for $e^+e^- \to Z^0Z^0 \to (f_1\bar{f}_1)(f_2\bar{f}_2)$, and will eventually sum over all fermion species. The part of the full matrix element that can give rise to doubly resonant contributions can be written as

$$\begin{aligned}\mathcal{M} = &\sum_i [\bar{v}_{e^+} T^i_{\mu\nu} u_{e^-}] M_i(t, u, p_1^2, p_2^2) \\ &\times \frac{1}{p_1^2 - M_Z^2 - \Pi_{ZZ}(p_1^2)} [\bar{u}_{f_1} \gamma^\mu (V_{Zf_L}(p_1^2)\gamma_L + V_{Zf_R}(p_1^2)\gamma_R) v_{\bar{f}_1}] \\ &\times \frac{1}{p_2^2 - M_Z^2 - \Pi_{ZZ}(p_2^2)} [\bar{u}_{f_2} \gamma^\nu (V_{Zf_L}(p_2^2)\gamma_L + V_{Zf_R}(p_2^2)\gamma_R) v_{\bar{f}_2}]\end{aligned} \quad (12)$$

where $T^i_{\mu\nu}$ are Lorentz covariant tensors that span the tensor structure of the matrix element and γ_L, γ_R are the usual helicity projection operators. The squared invariant masses of the $f_1\bar{f}_1$ and $f_2\bar{f}_2$ pairs are p_1^2 and p_2^2. The M_i, Π_{ZZ} and V_{Zf} are Lorentz scalars that are analytic functions of the independent kinematic Lorentz invariants of the problem.

To extract the piece of the matrix element that corresponds to finite propagation of both Z^0's we extract the leading term in a Laurent expansion in p_1^2 and p_2^2 of the analytic Lorentz-invariant part of eq.(12) leaving the Lorentz-covariant part untouched. This is the doubly-resonant term and is given by

$$\begin{aligned}\mathcal{M} = &\sum_i [\bar{v}_{e^+} T^i_{\mu\nu} u_{e^-}] M_i(t, u, s_p, s_p) \\ &\times \frac{F_{ZZ}(s_p)}{p_1^2 - s_p} [\bar{u}_{f_1} \gamma^\mu (V_{Zf_L}(s_p)\gamma_L + V_{Zf_R}(s_p)\gamma_R) v_{\bar{f}_1}]\end{aligned} \quad (13)$$

$$\times \frac{F_{ZZ}(s_p)}{p_2^2 - s_p}[\bar{u}_{f_2}\gamma^\nu (V_{Zf_L}(s_p)\gamma_L + V_{Zf_R}(s_p)\gamma_R)v_{\bar{f}_2}]$$

where F_{ZZ} defined by a relation like (7). It should be emphasized that eq.(13) is the exact form of the doubly-resonant matrix element to all orders in perturbation theory that we will now specialize to leading order. It is free of threshold singularities noted that were found by other authors [8]. In lowest order eq.(13) becomes, up to overall multiplicative factors,

$$\begin{aligned}\mathcal{M} = &\sum_{i=1}^{2}[\bar{v}_{e^+}T^i_{\mu\nu}u_{e^-}]M_i \\ &\times \frac{1}{p_1^2 - s_p}[\bar{u}_{f_1}\gamma^\mu (V_{Zf_L}\gamma_L + V_{Zf_R}\gamma_R)v_{\bar{f}_1}] \\ &\times \frac{1}{p_2^2 - s_p}[\bar{u}_{f_2}\gamma^\nu (V_{Zf_L}\gamma_L + V_{Zf_R}\gamma_R)v_{\bar{f}_2}].\end{aligned} \tag{14}$$

where $T^1_{\mu\nu} = \gamma_\mu(\not{p}_{e^-} - \not{p}_1)\gamma_\nu$, $M_1 = t^{-1}$; $T^2_{\mu\nu} = \gamma_\nu(\not{p}_{e^-} - \not{p}_2)\gamma_\mu$, $M_2 = u^{-1}$ and the final state vertex corrections take the form $V_{Zf_L} = ie\beta^f_L\gamma_L$ and $V_{Zf_R} = ie\beta^f_R\gamma_R$, The left- and right-handed couplings of the Z^0 to a fermion f are

$$\beta^f_L = \frac{t^f_3 - \sin^2\theta_W Q^f}{\sin\theta_W\cos\theta_W}, \qquad \beta^f_R = -\frac{\sin\theta_W Q^f}{\cos\theta_W}.$$

Squaring the matrix element and integrating over the final state momenta for fixed p_1^2 and p_2^2 gives

$$\begin{aligned}\frac{\partial^3\sigma}{\partial t\,\partial p_1^2\,\partial p_2^2} = &\frac{\pi\alpha^2}{s^2}(|\beta^e_L|^4 + |\beta^e_R|^4)\rho(p_1^2)\ \rho(p_2^2) \\ &\times\left\{\frac{t}{u} + \frac{u}{t} + \frac{2(p_1^2 + p_2^2)^2}{ut} - p_1^2p_2^2\left(\frac{1}{t^2} + \frac{1}{u^2}\right)\right\}\end{aligned} \tag{15}$$

with

$$\begin{aligned}\rho(p^2) &= \frac{\alpha}{6\pi}\sum_f(|\beta^f_L|^2 + |\beta^f_R|^2)\frac{p^2}{|p^2 - s_p|^2}\theta(p_0)\theta(p^2) \\ &\approx \frac{1}{\pi}\cdot\frac{p^2(\Gamma_Z/M_Z)}{(p^2 - M_Z^2)^2 + \Gamma_Z^2M_Z^2}\theta(p_0)\theta(p^2)\end{aligned}$$

where the sum is over fermion species. Note that $\rho(p^2) \to \delta(p^2 - M_Z^2)\theta(p_0)$ as $\mathrm{Im}(s_p) \to 0$ which is the result obtained by cutting a free propagator. The variables s, t, u, p_1^2 and p_2 in eq.(15) arise from products of standard covariants and external wave functions and therefore take real values dictated by the kinematics.

Integrating over t, p_1^2 and p_2^2 leads to

$$\sigma(s) = \int_0^s dp_1^2 \int_0^{(\sqrt{s}-\sqrt{p_1^2})^2} dp_2^2 \sigma(s;p_1^2,p_2^2)\ \rho(p_1^2)\ \rho(p_2^2), \tag{16}$$

where

$$\begin{aligned}\sigma(s;p_1^2,p_2^2) = & \frac{2\pi\alpha^2}{s^2}(|\beta_L^e|^4 + |\beta_R^e|^4) \\ & \times \left\{ \left(\frac{1+(p_1^2+p_2^2)^2/s^2}{1-(p_1^2+p_2^2)/s} \right) \ln \left(\frac{-s+p_1^2+p_2^2+\lambda}{-s+p_1^2+p_2^2-\lambda} \right) - \frac{\lambda}{s} \right\}\end{aligned}$$

and $\lambda = \sqrt{s^2+p_1^4+p_2^4-2sp_1^2-2sp_2^2-2p_1^2p_2^2}$. For $p_1^2 = p_2^2 = M_Z^2$ this agrees with known results [14].

REFERENCES

1. R. J. Eden, P. V. Landshoff, D. I. Olive and J. C. Polkinghorne, *The Analytic S-Matrix*, Cambridge University Press, Cambridge (1966).
2. R. E. Peierls, *Proceedings of the 1954 Glasgow Conference on Nuclear and Meson Physics*, Pergamon Press, New York, (1955) 296.
3. R. G. Stuart, *Phys. Lett.* **B 262** (1991) 113.
4. R. G. Stuart, *Phys. Rev. Lett.* **70** (1993) 3193.
5. R. G. Stuart, hep-ph/9504215.
6. R. G. Stuart, hep-ph/9602300.
7. A. Aeppli, F. Cuypers and G. J. van Oldenborgh, *Phys. Lett.* **B 314** (1993) 413;
8. A. Aeppli, G. J. van Oldenborgh and D. Wyler, *Nucl. Phys.* **B 428** (1994) 126.
9. M. Veltman, *Physica* **29** (1963) 186.
10. A. C. Hearn, *Nuovo Cimento* **21** (1961) 333.
11. K. Hepp, *Helv. Phys. Acta* **36** (1963) 355.
12. D. N. Williams, preprint UCRL-11113 (1963).
13. K. Hepp, *Helv. Phys. Acta* **37** (1964) 11.
14. R. W. Brown and K. O. Mikaelian, *Phys. Rev.* **D 19** (1979) 922.

Electron Beam Detector for the Experiment E831 at Fermilab

J.C. Anjos†, A.F. Barbosa†, N. Barros de Oliveira‡, I.M. Pepe‡, and F.R.A. Simão†

†CBPF, Rua Dr. Xavier Sigaud 150, Rio de Janeiro, R.J., 22290-180, Brazil
‡Inst. Física, UFBa, Rua Caetano Moura 123, Salvador, Bahia, 40210-340, Brazil

Abstract. We describe the construction and testing of Proportional Wire Chambers for the experiment E831 at Fermilab. Two chambers with a total of 6 wire planes and the associated electronics have been built at CBPF as part of the contribution to the experiment. The chambers will be placed in the electron-positron beam, just before the photon radiator, and will allow a better knowledge of the beam divergence.

INTRODUCTION

E831 is a fixed target photoproduction experiment located at the wideband photon beam laboratory at Fermilab. The spectrometer used in experiment E687 [1] was upgraded to enable it to accumulate 10^6 fully reconstructed charm particles. This will allow very high precision studies of the production and decays of charm hadrons [2].

The aim of the Proportional Wire Chambers built at CBPF for E831 is to provide the X and Y coordinates of electrons and positrons of the primary beam just before they hit the photon radiator to generate the photon beam through the Bremsstrahlung process. The main feature of these detectors is the fact that they should cope with an average count rate as high as 3×10^7 events per second. Contributions of these detectors to the experiment include: enhancing photon energy resolution, use in localization algorithms for primary vertex reconstruction, beam tagging in events with more than one electron per RF bucket, and better knowledge of beam divergence, improving resolution in transverse momentum for the inclusive charm production.

Each detector includes three wire plane electrodes, one anode and two cathodes. The wires diameter is 10 μm for anodes and 20 μm for cathodes, except for the guard wires at the sides of each wire plane, whose diameters are 30 μm, 50 μm and 100 μm. The wires pitch is 1 mm and the gap between electrodes

CP400, *First Latin American Symposium on High Energy Physics/ VII Mexican School of Particles and Fields,* edited by D'Olivo/Klein-Kreisler/Mendez

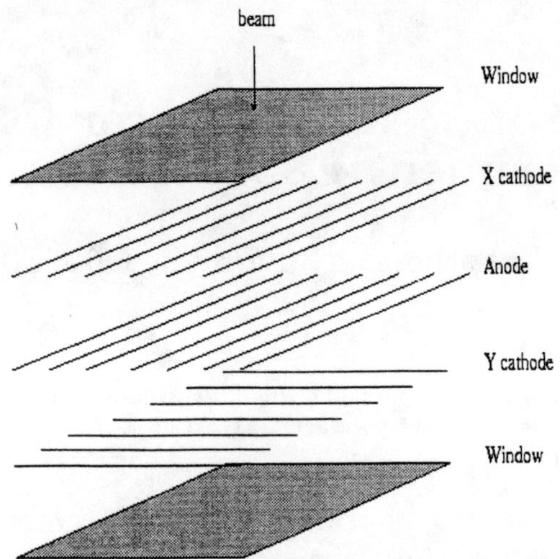

FIGURE 1. Electrodes arrangement in the gas chamber of the detectors

is 3.2 mm. As illustrated in Figure 1, 280 μm thick carbon fiber windows are placed 3.2 mm apart from the wire planes at each (input and output) side of the detectors, so that incident particles cross at least a 12.8 mm gas thickness. The energy loss by collisions of 300 GeV electrons or positrons in this gas thickness (argon, normal pressure) is in the range of 10 KeV, which implies the generation of about 10^2 electron-ion pairs. The windows surface is $80 \times 80\, mm^2$. The active volume is $70 \times 70 \times 12.8\, mm^3$.

The energy loss due to radiation as electrons or positrons cross the active volume is negligible, since the radiation length is approximately 120 m for argon and 0.3 m for carbon. The windows thickness of both detectors is $4 \times 0.28\, mm$ carbon fiber, which corresponds to less than 1% of a radiation length. In order to provide the required count rate, each anode wire works as an independent detector. There are 70 wires in each wire plane, 6 of these are guard wires. Therefore, 64 wires are active counters in each anode plane. Cathode wires only sample avalanche induced charges. The average flux over each anode wire is close to 5×10^5 particles per second. However, according to simulations, the flux is somewhat concentrated in about $4 \times 2\, cm^2$ of the window. As a consequence, individual counters are rather expected to operate under a flux of about 10^6 particles/s.

POSITION MEASUREMENT

The anode planes of the detectors are orthogonal to each other. The cathode planes inside each of them are also orthogonal. When a particle crosses the detector, the deposited electric charge is driven to an anode wire, where the avalanche process takes place, since it operates in the proportional mode. The X and Y coordinates of a particle are therefore related to the position of the anode wires registering the avalanche electric pulse in each detector. In the pre-amplifiers these pulses are shapedby an adequate choice of the differentiating time constant, so that the output pulse duration is less than 100 ns.

The avalanches induce charge distribution on the cathode wire planes. Since the distribution is spread over several wires, the amplitude of cathode wire signals is lower than that observed at anode wires. As shown in Figure 2, every active wire is connected to an electronic channel including bias supply, pre-amplification and discrimination. The state of the wires is latched in a register when an external triggering signal is provided. The information is then read by a data acquisition algorithm.

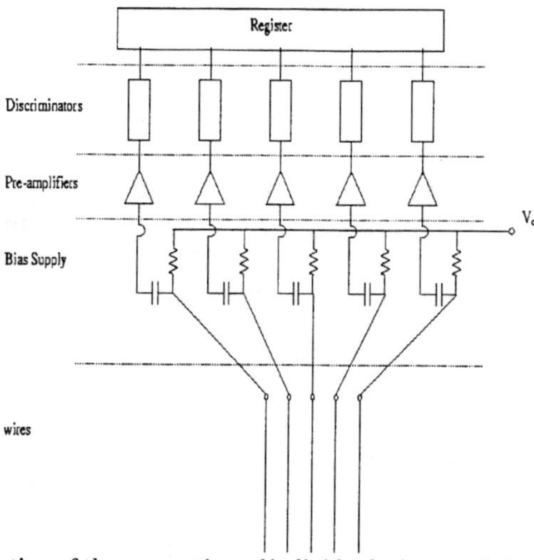

FIGURE 2. Illustration of the connection of individual wires to their electronic processing channels

It should be noted that the $X\&Y$ coordinates may be read from the anodes of both detectors or from the cathodes of each of them. Six coordinates are available - 02 anodes, 04 cathodes - and these are useful to lift out ambiguities in cases where more than one event reach the same wire within a short time interval. For this purpose, one of the detectors may eventually be rotated relatively to the other, so that the electrodes of each of them provide complementary information.

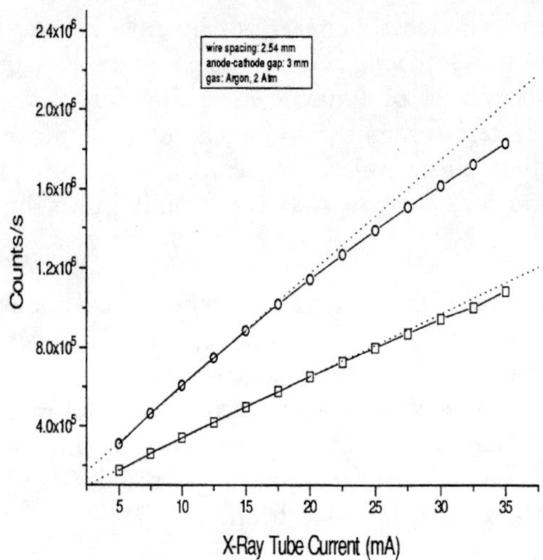

FIGURE 3. Counting rate measurements with a two-wire prototype. Data points correspond to the counts in each of the wires; dashed lines are linear extrapolations of the low count rate points.

FIRST RESULTS

A prototype with two active wires has been mounted and tested in order to estimate the count rate capability. As shown in Figure 3, up to 1 MHz count rate has been measured with negligible loss in the incident *versus* measured linearity relation. The final version of the detectors and the corresponding 384 electronic channels are presently under comissionning at the E831 experiment hall. Installation on the beam line should take place in March 1997.

ACKNOWLEDGMENTS

One of us, J.C. Anjos, would like to thank the organizers for financial support to attend the VII Mexican School of Particles and Fields and the I Latin American Symposium on High Energy Physics.

REFERENCES

1. P.L. Frabetti et al., *Nucl. Inst. and Meth.* **A 320** (1992) 519 .
2. " *A High Statistics Study of States Containing Heavy Quarks Using the Wideband Photon Beam and the E687 Multiparticle Spectrometer*", Fermilab Proposal P831 (1992).

Coulomb Effects on Cold Kaons from Vlasov Dynamics

Alejandro Ayala

Department of Physics, University of Illinois at Urbana-Champaign, Urbana IL 61801-3080, USA

Abstract. We compute the influence of Coulomb effects on kaon distribution at low momentum in high energy nucleus–nucleus collisions. This is accomplished by solving the Vlasov equation in the presence of an expanding, highly charged, fireball.

In recent years, much experimental effort has been devoted to the production of matter at high temperatures and/or densities in relativistic and ultrarelativistic heavy–ion experiments. One way of studying the properties of such matter is to compare the momentum distribution of particles produced in these kind of collisions to those found in nucleon–nucleon experiments at similar energies. One expects that the many–body medium should be most influential on the low momentum (long deBroglie wavelength) part of the spectra and least influential on the high momentum. On the same footing, the large electric charge present in a heavy–ion reaction can, in principle, significantly modify the distribution of the observed (charged) particles, particularly at low momentum. The question we address here is how the enormous electrical charge in a central gold–gold collision ($Z = 158$) modifies the low momentum spectra of charged hadrons.

Recall that the most widely used approach to incorporate Coulomb effects into the description of the charged particle distributions is by means of the Gamow factor [1]. The original motivation for its introduction was the similarity of the problem to barrier penetration in fission. This formula reproduces and extends a perturbative calculation to first order in $Z\alpha$ for the Coulomb interaction of charged particles at low momenta. But for heavy ion reactions, where the total charge Z is such that $Z\alpha \sim 1$, a first order perturbative approach to Coulomb effects is likely to fail and so will an approach that mimics such. In addition, this approach does not take into account that in a high energy heavy ion collision the matter blows apart, and in fact it is the low energy detected charged particle which is left behind.

CP400, *First Latin American Symposium on High Energy Physics/ VII Mexican School of Particles and Fields*, edited by D'Olivo/Klein-Kreisler/Mendez

Therefore it is desirable to look for an alternative approach to the description of Coulomb effects. We explore the influence of a time–dependent electric field produced by an expanding charged fireball on the low momentum spectrum of a test charge particle. We take charged kaons as the test particles.

For simplicity and definiteness, we assume a thermal momentum distribution at freezout and dynamically evolve this distribution by means of the Vlasov equation. The particles that produce the electric field are moving away from the collision region and thus the external field is time dependent. We describe the motion of non–relativistic test particles and finally, we model the expanding fireball as a uniformly charged sphere, as viewed from the center-of-mass of the colliding nuclei, with a radius growing linearly with time. This expansion model is motivated by the similarity solution to a hydrodynamically expanding fireball [2]. This model turns out to be solvable analytically albeit in terms of Bessel functions. A serious comparison with experiment may require a simulation based on an event generator like ARC [3] or RQMD [4].

The sphere has a total charge Ze and radius R_0 at time t_0 which increases with time at a constant speed v_s. The fireball parameters are not independent but are related by $R_0 = v_s t_0$. We first concentrate on the interior ($r \leq v_s t$) potential. If $f^{\pm}(\mathbf{r},\mathbf{p},t)$ represents the $\pm e$ charge particle phase space distribution then, when ignoring particle collisions after decoupling, its dynamics is governed by Vlasov's equation. In the interior region

$$\left[\frac{\partial}{\partial t} + \frac{\mathbf{p}}{m}\cdot\frac{\partial}{\partial \mathbf{r}} \pm \frac{m}{4}\frac{t_s}{t^3}\mathbf{r}\cdot\frac{\partial}{\partial \mathbf{p}}\right] f^{\pm}(\mathbf{r},\mathbf{p},t) = 0\,, \tag{1}$$

where

$$t_s \equiv \frac{4Z\alpha}{m v_s^3}\,, \tag{2}$$

$\alpha = e^2/4\pi$, m is the kaon's mass, and $t \geq t_0$.

The solution to eq. (1) is found by the method of characteristics. This involves solving the classical equations of motion and using the solutions to evolve the initial distribution $f^{\pm}(\mathbf{r},\mathbf{p},t_0)$ forward in time.

For the initial distribution we take a Maxwell–Boltzmann equilibrium distribution at temperature T with the constraint that the kaons lie within the initial sphere.

The solution to eq. (1) incorporating the initial condition is

$$\begin{aligned} f^{\pm}(\mathbf{r},\mathbf{p},t) = \mathcal{N} \exp&\left\{-\frac{m}{2T}\left(\left[\frac{c_1(t)\mathbf{r}}{t_s} - c_2(t)\mathbf{v}\right]^2 \pm \frac{t_s}{4t_0^3}\left[c_3(t)\mathbf{r} - c_4(t)t_s\mathbf{v}\right]^2\right)\right\} \\ &\times\Theta(R_0 - |c_3(t)\mathbf{r} - c_4(t)t_s\mathbf{v}|)\,. \end{aligned} \tag{3}$$

The normalization constant $\mathcal{N}$ is chosen so that the distribution is normalized to unity at the initial time t_0, this ensures that the distribution is normalized to unity for all later times, even when account is taken of the fact that the solution differs outside the expanding sphere.

For positive (negative) kaons the dimensionless functions of time $c_i(t)$ are combinations of Bessel functions $K_{1,2}$, $I_{1,2}$ ($N_{1,2}$, $J_{1,2}$) of $\sqrt{t_s/t}$. We now concentrate on positive kaons, since its asymptotic distribution, contrary to the negative kaons, can be obtained from analytic expressions over the whole momentum domain. The treatment of the negative kaon distribution will be given elsewhere [5]. The asymptotic momentum distribution is obtained by integrating f^+, given by eq. (3), over all space and taking the limit $t \to \infty$. Due to spherical symmetry, the distribution is a function only of the magnitude of the three dimensional momentum vector. The coordinate integration can be performed analytically. It is important to remember that the above treatment is valid only so long as the test particle remains within the expanding sphere. This will be the case if the asymptotic speed is less than v_s. To compute the distribution for asymptotic velocities greater than v_s is straightforward due to the fact that once a kaon has crossed the boundary of the expanding sphere, it sees a time independent potential so that its energy is thereafter conserved. Since a positive kaon will see a positive potential, its kinetic energy will not decrease and neither will its velocity. Once a positive kaon has crossed the surface, it will not come back.

The energy distribution is obtained by multiplying the flux of particles crossing the surface of the sphere at time t by the surface area, inserting an energy conserving δ function, and integrating over time.

$$dN^+/dE = \int_{t_0}^{\infty} dt\, v_s^2 t^2 \int d\Omega_{\mathbf{r}} \int d^3p\, \mathbf{v}\cdot\hat{\mathbf{r}} f^+(\mathbf{r},\mathbf{p},t)|_{r=v_s t}$$
$$\times\Theta(\mathbf{p}\cdot\mathbf{r})\Theta(\mathbf{v}\cdot\hat{\mathbf{r}}-v_s)\delta(E-p^2/2m-Ze^2/v_s t) \qquad (4)$$

The asymptotic momentum distribution is readily obtained from the energy distribution. Fig. 1 shows plots of the positive kaons momentum distribution d^3N/d^3p as a function of the non–relativistic energy $E = p^2/2m$ for different fireball surface velocities and parameters $Z = 158$, $R_0 = 8$ fm, $T = 120$ MeV. Notice that the distributions remain finite at zero energy, since the charged fireball blows apart and leaves a finite number of low energy kaons behind. This cannot happen with a static fireball where a classical turning point exists. In the limit where the fireball expansion velocity is large compared to the thermal velocity of the kaons, the kaon distribution is hardly changed from the simple exponential form, the fireball expands so fast that it leaves the kaon distribution frozen to its initial shape. In the limit that the expansion velocity goes to zero, the static fireball result is recovered.

In conclusion, we have shown the importance of Coulomb effects on charged kaons in central collisions of large nuclei at high energies. We have shown, in a quantitative way, how the expansion of the fireball affects the distribution. Comparison with data at kinetic energies greater than about 100 MeV requires

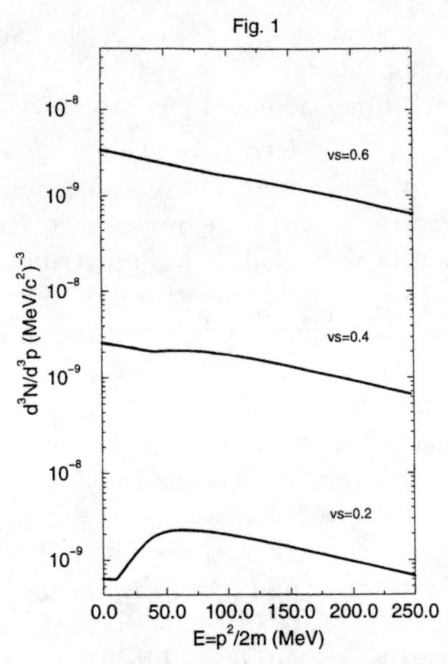

FIGURE 1. Positive kaon momentum distribution d^3N/d^3p for surface fireball velocities $v_s/c = 0.2, 0.4, 0.6$ and parameters $Z = 158$, $R_0 = 8$ fm, $T = 120$ MeV

a relativistic treatment. Also, by changing the initial phase–space distribution or by including a collective flow component, one should be able to extract quantitative information on the dynamics of kaons in heavy–ion collisions. This issues will be explored in a upcoming report [6].

This work was supported by the US National Science Foundation under grant NSF PHY94-21309 and by the US Department of Energy under grant DE-FG02-87ER40328.

REFERENCES

1. M. Gyulassy and S. K. Kauffmann, Nucl. Phys. **A362**, 503 (1981).
2. L. P. Csernai and J. Kapusta, Phys. Rev. D **46**, 1379 (1992).
3. Y. Pang, T. J. Schlagel and S. H. Kahana, Nucl. Phys. **A544**, 435c (1992).
4. H. Sorge, H.Stöcker and W. Greiner, Nucl. Phys. **A498**, 507c (1989).
5. A. Ayala and J. Kapusta, to appear in Phys. Rev. C.
6. A. Ayala and J. Kapusta, work in preparation.

Non-resonant Contribution in Non-leptonic Charm Meson Decays

I. Bediaga*, C. Göbel* and R. Méndez-Galain†

*Centro Brasileiro de Pesquisas Físicas, R. Dr. Xavier Sigaud 150
22290 - 180 - Rio de Janeiro, RJ, Brazil
†Instituto de Física, Facultad de Ingeniería,
CC 30, CP 11000 Montevideo, Uruguay

Abstract. We claim that the non-resonant contribution to non-leptonic charm meson decays cannot be considered constant in the phase space of the reaction as it usually is. We discuss as an example the decay $D^+ \to K^-\pi^+\pi^+$.

Non-leptonic charm meson decays have been extensively studied both theoretically and experimentally. The high diversity and low multiplicity of decay channels provide important information on both weak and strong interactions. These decays have contributions from resonances in intermediate states, as well from the direct non-resonant (NR) decay. The understanding of the decay pattern of charm mesons as a whole, and therefore the extraction of the decay partial widths for all contributing states, is essential in addressing many open problems in charm physics.

The Dalitz plot analysis is a powerful technique widely used in the study of resonance substructures on charmed meson decays. The plot represents the phase space of the decay and it is weighted by the squared amplitude of the reaction. Therefore, it contains information on both the kinematics and the dynamics. Within this technique, intermediate resonant and non-resonant contributions are fitted to get the respective amplitudes and phases. The corresponding partial decay widths can then be obtained.

Data on non-leptonic decays of the D meson has been fitted [1–5] using Breit-Wigner functions [6] to represent the various resonances (with the respective angular distribution) and a constant function to describe the NR contribution [7].

Although the above parameterization is widely used, a very poor fit has been reported [2,5], suggesting that it may not be adequate to describe these decays. These poor results do not improve with higher statistics or considering a larger number of resonances [5]. Moreover, this problem appears in all the

CP400, *First Latin American Symposium on High Energy Physics/ VII Mexican School of Particles and Fields*, edited by D'Olivo/Klein-Kreisler/Mendez

$D \to K\pi\pi$ decay channels already measured.

A possible explanation for these discrepancies is the incorrect use of a constant amplitude for the NR contribution. An incorrect parameterization will certainly influence the fit of the resonances and consequently the extracted values of amplitudes and phases.

Here, we claim that NR charm meson decays cannot be simply considered with a flat dynamics. In weak interactions between quarks and leptons helicity plays an important role. Consequently, one expects a significant dependence of the weak amplitudes on the momenta of the interacting particles. Thus, the dynamics of these reactions vary from point to point of the phase space and the significance of this variation depends on the specific physical reaction.

This should be particularly important in weak decays of charm mesons. The large value of the charm quark mass allows for a quasi perturbative treatment of QCD. Furthermore, charm quark decays into light quarks and this enhances the importance of helicity. Nevertheless, due to the hadronization process of the partons after their weak interaction, one has to take into account non-perturbative QCD effects involved in the final hadronic state formation.

In order to make an estimate of the effect of the dynamics in the Dalitz plot, we use an approximate method to describe hadronic decays. The method is based on both the factorization technique [8] and an effective Hamiltonian [9,10] for the partonic interaction and has been successfully used to describe heavy meson decays [8,11].

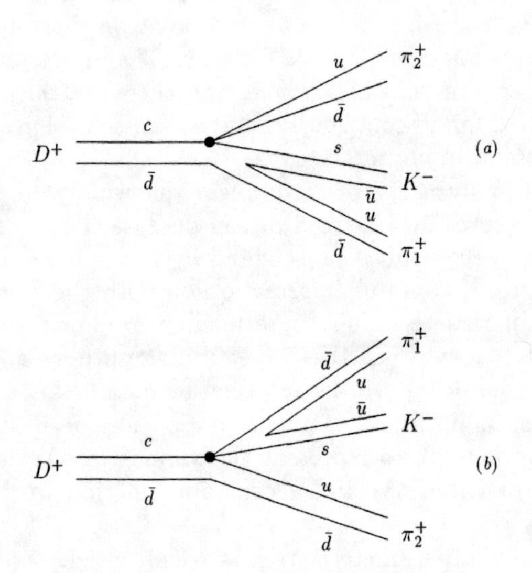

FIGURE 1. The two diagrams contributing to the decay $D^+ \to K^-\pi^+\pi^+$.

As we are interested in the NR contributions, we analyse the channel

$D^+ \to K^-\pi^+\pi^+$, which has a very large NR branching ratio. The diagrams contributing to the decay $D^+ \to K^-\pi^+\pi^+$ are shown in Figure (1). We obtain [12] the following decomposition for the hadronic amplitude

$$\mathcal{M}_{D^+\to K^-\pi^+\pi^+} = (\frac{G_F}{\sqrt{2}})\cos^2\theta_c[a_1\langle K^-\pi_1^+|\bar{s}c|D^+\rangle\langle\pi_2^+|\bar{u}d|0\rangle$$
$$+a_2\langle K^-\pi_1^+|\bar{s}d|0\rangle\langle\pi_2^+|\bar{u}c|D^+\rangle \; + \; (\pi_1^+ \leftrightarrow \pi_2^+)] \; . \qquad (1)$$

where $(\bar{q}q')$ is a short-hand notation for $\bar{q}\gamma^\mu(1-\gamma_5)q'$. The coefficients a_1 and a_2 characterize the contribution of the effective charged and neutral currents respectively, which include short-distance QCD effects; their values have been fitted in charm meson two-body decays [9].

We have calculated the first term of eq. (1) to be

$$\langle K^-\pi_1^+|\bar{s}c|D^+\rangle\langle\pi_2^+|\bar{u}d|0\rangle = (p_{\pi_2\;\mu}\,F_4)\,(if_\pi p_{\pi_2}{}^\mu) = if_\pi m_\pi^2 F_4 \; . \qquad (2)$$

while the second is

$$\langle\pi^+|\bar{u}c|D^+\rangle\langle K^-\pi^+|\bar{s}d|0\rangle = F_{D\pi}^{1^-}(m_1^2)f_+(m_1^2)\,(m_D^2 + m_K^2 + 2m_\pi^2 - 2m_2^2 - m_1^2)$$
$$+ \; [F_{D\pi}^{1^-}(m_1^2)f_+(m_1^2) - F_{D\pi}^{0^+}(m_1^2)f_0(m_1^2)]\,\frac{(m_D^2 - m_\pi^2)(m_K^2 - m_\pi^2)}{m_1^2}$$
$$+ \; (m_1^2 \leftrightarrow m_2^2) \qquad (3)$$

where we have explicitly introduced the Dalitz plot variables $m_1^2 = (p_K + p_{\pi_1})^2$ and $m_2^2 = (p_K + p_{\pi_2})^2$. The five form factors appearing in eqs. above come from the hadronic matrix elements.

The contribution of diagram (1.a), given by equation (2) is proportional to $f_\pi m_\pi^2$. Thus, unless the form factor F_4 is unacceptably large ($F_4 \sim 10^3$), we can safely neglect this contribution in favor of that of diagram (1.b), given by equation (3) which contains m_D^2.

The NR contribution to the amplitude of the decay $D^+ \to K^-\pi^+\pi^+$ can thus be simply written replacing equation (3) in (1), neglecting the contribution of Figure (1.a). The final expression thus depends on the effective coefficient a_2 and the four known [10,13,14] form factors.

In order to check the validity of this calculation scheme, we have evaluated the NR partial decay width $\Gamma(D^+ \to K^-\pi^+\pi^+)_{NR}$ using the expressions above. With the value of a_2 extracted from two body decay [9], we find a branching ratio (BR) of 9% which is close to the reported experimental value [14] $7.3 \pm 1.4\%$ obtained by fitting the NR contribution to a constant.

Figure (2) shows the Dalitz plot for the NR contribution to the decay $D^+ \to K^-\pi^+\pi^+$ as a function of the variables m_1^2 and m_2^2. It has been generated by Monte Carlo simulation with a weight proportional to the square of the amplitude in equation (1), using equation (3). As one can see from equation (3) and Figure (2), according to this calculation the matrix element

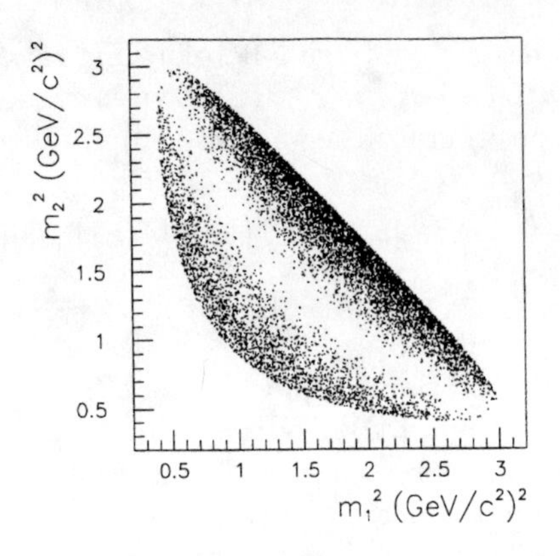

FIGURE 2. The Dalitz plot of the decay $D^+ \to K^-\pi^+\pi^+$, weighted by $|\mathcal{M}_{D^+\to K^-\pi^+\pi^+}|^2$ as in equations (1) and (3), generated via Monte Carlo. The Dalitz plot variables are $m_1^2 \equiv (p_K + p_{\pi_1})^2$ and $m_2^2 \equiv (p_K + p_{\pi_2})^2$.

describing the dynamics of the NR contribution to the decay $D^+ \to K^-\pi^+\pi^+$ significantly varies along the phase space of the reaction. Its shape remains almost unchanged when the parameters defining the form factors vary within the experimental allowed values.

However, the result presented in Figure (2) has been obtained using an approximate method. Non-perturbative effects, present in this decay through the exchange of soft gluons, or final state interactions could change the structure shown in this figure. In the extreme case where non-perturbative effects completely dominate the decay, the structure will be washed out because of the dispersive nature of these effects, therefore obtaining the flat contribution predicted by the pure hadronic decay of a zero spin particle. Comparison between the distribution shown in Figure (2) and experimental data will thus be a test for the validity of the factorization method.

In summary, we have shown that the natural parameterization for the non-resonant part of charm decays – based in the spin amplitude of the hadronic decay – could significantly change due to the fundamental weak interaction between quarks. The appearance of these structures in the plot could be responsible for the problems of the various experimental teams with the convergence of their fits. To clarify this point, it is important in future analyses to use a parameterization for the non-resonant contribution going beyond the simple constant.

REFERENCES

1. R.H. Schindler *et al.*, MarkII Collab., Phys. Rev. **D24**, 78 (1981).
2. J. Adler *et al.*, MarkIII Collab., Phys. Lett. **B196**, 107 (1987).
3. J.C. Anjos *et al.*, E691 Collab., Phys. Rev. **D48**, 56 (1993).
4. H. Albrecht *et al.*, ARGUS Collab., Phys. Lett. **B308**, 435 (1993).
5. P.L. Frabetti *et al.*, E687 Collab., Phys. Lett. **B331**, 217 (1994).
6. J.D. Jackson, Nuovo Cimento **34**, 1644 (1964).
7. In fact, Mark III tried to fit using a different parameterization for the NR amplitude, since they were having problems with their fit.
8. D. Fakirov and B. Stech, Nucl. Phys. **B133**, 315 (1978); N. Cabibbo and L. Maiani, Phys. Lett. **B73**, 418 (1978).
9. M. Bauer, B. Stech and M. Wirbel, Z. Phys. **C34**, 103 (1987).
10. A.J. Buras, J.-M. Gérard, and R. Rückl, Nucl. Phys. **B268**, 16 (1986).
11. J.Bjorken, Nucl. Phys. **B11** (Proc. Suppl.) 325 (1989); H. Yamamoto, hep-ph/9402269.
12. Details of this calculation can be found in I. Bediaga, C. Göbel and R. Méndez-Galain; Phys. Rev. Lett. **78**, 22 (1997).
13. L.B. Okun, *Leptons and Quarks* (North-Holland, Amsterdam, 1982).
14. Particle Data Group; Review of Particle Properties, Phys. Rev. **D54** (1996).

Algebraic model of baryon resonances

R. Bijker[1)] and A. Leviatan[2)]

[1)] *Instituto de Ciencias Nucleares, Universidad Nacional Autónoma de México, A.P. 70-543, 04510 México D.F., México*
[2)] *Racah Institute of Physics, The Hebrew University, Jerusalem 91904, Israel*

Abstract. We discuss recent calculations of electromagnetic form factors and strong decay widths of nucleon and delta resonances. The calculations are done in a collective constituent model of the nucleon, in which the baryons are interpreted as rotations and vibrations of an oblate top.

I INTRODUCTION

The study of the properties of baryon resonances is entering a new era with the forthcoming new and more accurate data from new facilities, such as Jefferson Lab., MAMI, ELSA and Brookhaven. Effective models of baryons which are based on three constituents (qqq) share a common spin-flavor-color structure but differ in their assumptions on the spatial dynamics. For example, quark potential models in nonrelativistic [1] or relativized [2] forms emphasize the single-particle aspects of quark dynamics for which only a few low-lying configurations in the confining potential contribute significantly to the eigenstates of the Hamiltonian. On the other hand, some regularities in the observed spectra, such as linear Regge trajectories and parity doubling, hint that an alternative, collective type of dynamics may play a role in the structure of baryons.

In this contribution we discuss a collective model within the context of an algebraic approach [3] and present some results for electromagnetic form factors [4] and strong decay widths [5].

II COLLECTIVE MODEL OF BARYONS

We consider a collective model in which the baryon resonances are interpreted in terms of rotations and vibrations of the string configuration in Fig. 1. A fit to the 3 and 4 star nucleon and delta resonances gives a r.m.s. deviation of 39 MeV [3]. The corresponding oblate top wave functions are spread

CP400, *First Latin American Symposium on High Energy Physics/ VII Mexican School of Particles and Fields*, edited by D'Olivo/Klein-Kreisler/Mendez

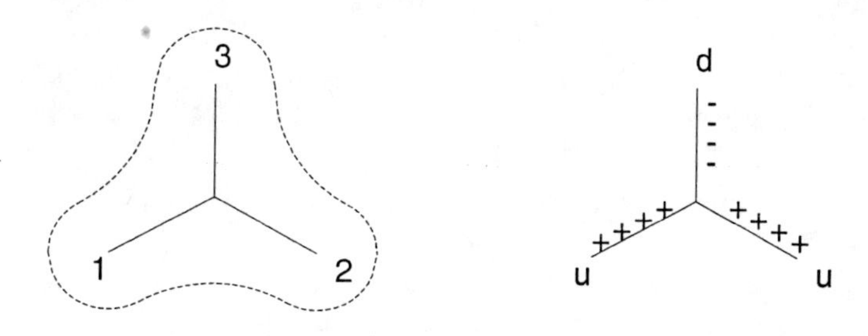

FIGURE 1. Collective model of baryons (the charge distribution of the proton is shown as an example).

over many oscillator shells and hence are truely collective. The baryon wave functions have the form

$$\left| {}^{2S+1}\dim\{SU_f(3)\}_J \, [\dim\{SU_{sf}(6)\}, L^P]_{(v_1,v_2);K} \right\rangle \; . \tag{1}$$

The spin-flavor part has the usual $SU_{sf}(6)$ classification and determines the permutation symmetry of the state. The spatial part is characterized by the labels: $(v_1, v_2); K, L^P$, where (v_1, v_2) denotes the vibrations (stretching and bending) of the string configuration in Fig. 1; K denotes the projection of the rotational angular momentum L on the body-fixed symmetry-axis and P the parity. Finally, S and J are the spin and the total angular momentum $\vec{J} = \vec{L} + \vec{S}$. In this contribution we focus on the nucleon resonances which are interpreted as rotational excitations of the $(v_1, v_2) = (0, 0)$ vibrational ground state.

The electromagnetic (strong) coupling is assumed to involve the absorption or emission of a photon (elementary meson) from a single constituent. The collective form factors and decay widths are obtained by folding with a probability distribution for the charge and magnetization along the strings of Fig. 1

$$g(\beta) = \beta^2 \, \mathrm{e}^{-\beta/a}/2a^3 \; , \tag{2}$$

where β is a radial coordinate and a is a scale parameter. In the algebraic approach these form factors and decay widths can be obtained in closed analytic form (in the limit of a large model space) which allows us to do a straightforward and systematic analysis of the experimental data.

The ansatz of Eq. (2) for the probability distribution is made to obtain a dipole form for the proton electric form factor

$$G_E^p(k) = \frac{1}{(1 + k^2 a^2)^2} \; . \tag{3}$$

The same distribution is used to calculate transition form factors and decay widths. As a result, all collective form factors are found to drop as powers of the momentum transfer [4]. This property is well-known experimentally and is in contrast with harmonic oscillator based quark models in which all form factors fall off exponentially.

III ELECTROPRODUCTION

Electromagnetic inelastic form factors can be measured in electroproduction of baryon resonances. They are expressed in terms of helicity amplitudes A^N_ν, where ν indicates the helicity and N represents proton (p) or neutron (n) couplings. In a string-like model of hadrons one expects [6] on the basis of QCD that strings will elongate (hadrons swell) as their energy increases. This effect can be easily included in the present analysis by making the scale parameters of the strings energy- dependent. We use here the simple ansatz

$$a = a_0\left(1 + \xi\,\frac{W-M}{M}\right) , \tag{4}$$

where M is the nucleon mass and W the resonance mass. This ansatz introduces a new parameter (ξ), the stretchability of the string.

In Fig. 2 we show the transverse helicity amplitudes A^p_ν for the $N(1520)D_{13}$ resonance calculated in the Breit frame (a factor of $+i$ is suppressed). The scale parameter a is determined in a simultaneous fit to the nucleon charge radii and the nucleon elastic form factors: $a = 0.232$ fm [4]. The effect of stretching on the helicity amplitudes is sizeable (especially if one takes the value $\xi \approx 1$ which is suggested by QCD arguments [6] and the Regge behavior of nucleon resonances). The data for $N(1520)D_{13}$ (and also for $N(1680)F_{15}$) show a clear indication that the form factors are dropping faster than expected on the basis of the dipole form. The disagreement at low momentum transfer may be due to the neglect of the meson cloud.

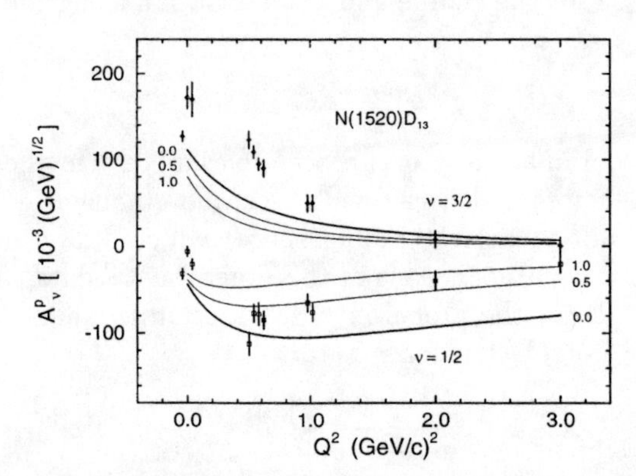

FIGURE 2. Effect of hadron swelling for excitation of $N(1520)D_{13}$. The curves are labelled by the stretching parameter ξ of Eq. (4).

IV STRONG DECAY WIDTHS

In addition to electromagnetic couplings, strong decays provide an important, complementary, tool to study the structure of baryons. We consider decays with emission of π and η. The experimental data [7] are shown in Table 1, where they are compared with the results of our calculation. The calculated values depend on two parameters g and h in the transition operator and on the scale parameter a of Eq. (2). These parameters are determined from a least-square fit to the $N\pi$ partial widths (which are relatively well known) with the exclusion of the S_{11} resonances whose assignments are not clear due to possible mixing between the $N(1535)$ and $N(1650)$ and/or the possible presence of a third S_{11} resonance [8].

The calculation of $N\pi$ decay widths is found to be in fair agreement with experiment (see Table 1). These results are to a large extent a consequence of spin-flavor symmetry. The calculated widths for the $N\eta$ channel are systematically small. We emphasize here that, since the transition operator was determined from the $N\pi$ decays, the η decays are calculated without introducing any further parameters. The results of this analysis suggest that the large η width for the $N(1535)S_{11}$ is not due to a conventional q^3 state. One possible explanation is the presence of another state in the same mass region, *e.g.* a quasi-bound meson-baryon S wave resonance just below or above threshold, for example $N\eta$, $K\Sigma$ or $K\Lambda$ [9]. Another possibility is an exotic configuration of four quarks and one antiquark ($q^4\bar{q}$).

TABLE 1. $N\pi$ and $N\eta$ decay widths of 3 and 4 star nucleon resonances in MeV.

State	Mass	Resonance	$\Gamma(N\pi)$		$\Gamma(N\eta)$	
			th	exp	th	exp
S_{11}	$N(1535)$	${}^28_{1/2}[70,1^-]_{(0,0);1}$	85	79 ± 38	0.1	74 ± 39
S_{11}	$N(1650)$	${}^48_{1/2}[70,1^-]_{(0,0);1}$	35	130 ± 27	8	11 ± 6
P_{13}	$N(1720)$	${}^28_{3/2}[56,2^+]_{(0,0);0}$	31	22 ± 11	0.2	
D_{13}	$N(1520)$	${}^28_{3/2}[70,1^-]_{(0,0);1}$	115	67 ± 9	0.6	
D_{13}	$N(1700)$	${}^48_{3/2}[70,1^-]_{(0,0);1}$	5	10 ± 7	4	
D_{15}	$N(1675)$	${}^48_{5/2}[70,1^-]_{(0,0);1}$	31	72 ± 12	17	
F_{15}	$N(1680)$	${}^28_{5/2}[56,2^+]_{(0,0);0}$	41	84 ± 9	0.5	
G_{17}	$N(2190)$	${}^28_{7/2}[70,3^-]_{(0,0);1}$	34	67 ± 27	11	
G_{19}	$N(2250)$	${}^48_{9/2}[70,3^-]_{(0,0);1}$	7	38 ± 21	9	
H_{19}	$N(2220)$	${}^28_{9/2}[56,4^+]_{(0,0);0}$	15	65 ± 28	0.7	
$I_{1,11}$	$N(2600)$	${}^28_{11/2}[70,5^-]_{(0,0);1}$	9	49 ± 20	3	

V SUMMARY AND CONCLUSIONS

We have analyzed simultaneously electromagnetic form factors and strong decay widths in a collective model of the nucleon. The helicity amplitudes are folded with a probability distribution for the charge and magnetization, which is determined from the well-established dipole form of the nucleon electromagnetic form factors. The same distribution function is used to calculate the transition form factors for the excited baryons. As a result, all form factors drop as powers of the momentum transfer.

For electromagnetic couplings we find that the inclusion of the stretching of baryons improves the calculation of the helicity amplitudes for large values of the momentum transfer. The disagreement for small values of the momentum transfer $0 \leq Q^2 \leq 1$ $(\text{GeV/c})^2$ may be due to coupling of the photon to the meson cloud, *i.e.* configurations of the type $q^3 - q\bar{q}$. Since such configurations have much larger spatial extent than q^3, their effects are expected to drop faster with momentum transfer than the constituent form factors.

An analysis of the strong decay widths into the $N\pi$ and $N\eta$ channels shows that the π decays follow the expected pattern. Our calculations do not show any indication for a large η width, as is observed for the $N(1535)S_{11}$ resonance. The observed large η width indicates the presence of another configuration, which is outside the present model space.

The results reported in this article are based on work done in collaboration with F. Iachello (Yale). The work is supported in part by grant No. 94-00059 from the United States-Israel Binational Science Foundation (BSF), Jerusalem, Israel (A.L.) and by CONACyT, México under project 400340-5-3401E and DGAPA-UNAM under project IN105194 (R.B.).

REFERENCES

1. N. Isgur and G. Karl, Phys. Rev. D **18**, 4187 (1978); **19**, 2653 (1979); **20**, 1191 (1979).
2. S. Capstick and N. Isgur, Phys. Rev. D **34**, 2809 (1986).
3. R. Bijker, F. Iachello and A. Leviatan, Ann. Phys. (N.Y.) **236**, 69 (1994).
4. R. Bijker, F. Iachello and A. Leviatan, Phys. Rev. C **54**, 1935 (1996).
5. R. Bijker, F. Iachello and A. Leviatan, preprint nucl-th/9608057, Phys. Rev. D, in press.
6. K. Johnson and C.B. Thorn, Phys. Rev. D **13**, 1934 (1974); I. Bars and H.J. Hanson, Phys. Rev. D **13**, 1744 (1974).
7. Particle Data Group, Phys. Rev. D **54**, 1 (1996).
8. Z. Li and R. Workman, Phys. Rev. C **53**, R549 (1996).
9. N. Kaiser, P.B. Siegel and W. Weise, Phys. Lett. **B362**, 23 (1995).

New particles and FCNC[1]

Umberto Cotti[2] and Arnulfo Zepeda

Departamento de Física, Centro de Investigación y de Estudios Avanzados del IPN, A.P. 14-740, 07001 México D.F., México.

Abstract. We discuss the case of simultaneous mixing of gauge bosons and mixing of fermions in a model independent way and for a variety of extra-fermion representations. In this context we analyze a class of lepton family violating processes, namely $Z \to e\bar{\tau}$, $Z \to \mu\bar{\tau}$, $Z \to e\bar{\mu}$, $\mu \to ee\bar{e}$, $\tau \to ee\bar{e}$ and $\tau \to \mu\mu\bar{\mu}$. in the presence of one extra neutral gauge boson, Z', with universal, non-universal or family changing couplings. We derive bounds on the combined effect of Z–Z' mixing and ordinary–exotic lepton mixing. We demonstrate that for some classes of exotic fermions, such as mirror fermions, is not possible to constraint individually the mixing parameters by *family changing neutral currents* (FCNC) bounds. This situation can suggest some class of new mechanism which cancels FCNC effects in a variety of extensions of the *standard model* (SM)

I INTRODUCTION

Tree level *family changing neutral current* (FCNC) interactions arise in extended models from three possible sources: (i) the exchange of family changing neutral gauge boson, (ii) the mixing between exotic and ordinary fermions and (iii) the existence of neutral scalars in the Higgs sector with family violating couplings.

In previous works an extensive research has been performed in the context of FCNC produced by the mixing of the standard neutral gauge boson with one which does not couple universally to fermion generations [1], or by the mixing between exotic and ordinary fermions [2,3]. We have shown [4], how this phenomenon arises in the general case of simultaneous mixing of neutral gauge bosons and mixing of ordinary fermions with exotic ones. Here we apply our model independent formalism to several lepton family violating processes in the e–μ, μ–τ and e–τ sectors considering several possible exotic fermionic representations. We limit our discussion to the case of just one additional

1) This work was partially supported by CONACyT in Mexico.

2) Present address: Instituto de Física, Universidad Autónoma de Puebla, A.P. J-48, 72570 Puebla Pue., México. e-mail: ucotti@fis.cinvestav.mx .

CP400, *First Latin American Symposium on High Energy Physics/ VII Mexican School of Particles and Fields,* edited by D'Olivo/Klein-Kreisler/Mendez

massive neutral gauge boson Z′. We obtain bounds for the mixing parameters including the possibility that the contribution of the neutral gauge boson mixing and that of the fermion mixing are of the same order[3].

II MIXING EFFECTS: THE GENERAL CASE

Denoting by Θ the Z-Z′ mixing angle and by

$$\mathsf{U}_a = \begin{pmatrix} \mathsf{A}_a & \mathsf{E}_a \\ \mathsf{F}_a & \mathsf{G}_a \end{pmatrix}, \tag{1}$$

the mixing between light and heavy fermions of a given helicity a and a given electric charge, the neutral-current lagrangian in the light–light sector has the following expression:

$$-\mathcal{L}^{\rm nc} = \frac{e}{s_{\theta_{\rm w}} c_{\theta_{\rm w}}} \sum_{a={\rm L,R}} \bar{\psi}_{la} \gamma^\mu \left(\mathsf{K}_a,\ \mathsf{K}'_a\right) \psi_{la} \begin{pmatrix} \mathrm{Z} \\ \mathrm{Z}' \end{pmatrix}_\mu , \tag{2}$$

where

$$\begin{aligned} \mathsf{K}_{\rm L} &= \left(\mathbf{\Lambda}_{\rm L} + \mathrm{t}_{3{\rm OL}} - \mathrm{Q}s^2_{\theta_{\rm w}}\right) \cos\Theta + \mathbf{\Xi}_{\rm L} \sin\Theta, \\ \mathsf{K}_{\rm R} &= \left(\mathbf{\Lambda}_{\rm R} - \mathrm{Q}s^2_{\theta_{\rm w}}\right) \cos\Theta + \mathbf{\Xi}_{\rm R} \sin\Theta, \\ \mathbf{\Lambda}_{\rm L} = \left(\mathsf{F}^\dagger \mathsf{F}\right)_{\rm L} (t_{3{\rm EL}} - t_{3{\rm OL}}), &\qquad \mathbf{\Lambda}_{\rm R} = \left(\mathsf{F}^\dagger \mathsf{F}\right)_{\rm R} t_{3{\rm ER}}, \qquad \mathbf{\Xi}_a = \left(\mathsf{H}_{ll}\right)_a , \end{aligned} \tag{3}$$

and H_{ll} contains the couplings of the extra gauge boson, Z′, to the light sector.

III APPLICATIONS TO THE LEPTONIC SECTOR

A Constraints from $\mathrm{Z} \to l_i \bar{l}_j$

With the approximation $M_{\rm Z} \gg m_{l_i}, m_{l_j}$, and for $i \neq j$

$$\mathrm{B}(\mathrm{Z} \to l_i \bar{l}_j + \bar{l}_i l_j) \simeq 4 \frac{\mathrm{B}(\mathrm{Z} \to l\bar{l})}{|g_{\rm V}|^2 + |g_{\rm A}|^2} \left(\left|\Lambda^{ij}_{\rm L} + \Xi^{ij}_{\rm L}\Theta\right|^2 + \left|\Lambda^{ij}_{\rm R} + \Xi^{ij}_{\rm R}\Theta\right|^2 \right) + \mathrm{O}\left(\Theta^2\right).$$

It then follows that

$$\left|\Lambda^{ij}_{\rm L} + \Xi^{ij}_{\rm L}\Theta\right|^2 + \left|\Lambda^{ij}_{\rm R} + \Xi^{ij}_{\rm R}\Theta\right|^2 < c\widetilde{\mathrm{B}}_{l_i \bar{l}_j}, \tag{4}$$

---

3) this should be the manifestation of some new symmetry which can cancel FCNC contributions

where $c^{-1} = \left(4\frac{\mathrm{B}(\mathrm{Z}\to l\bar{l})}{|g_V|^2+|g_A|^2}\right) = 0.536$ [4] and where

$$\mathrm{B}_{l_i\bar{l}_j} \equiv \mathrm{B}(\mathrm{Z}\to l_i\bar{l}_j + \bar{l}_i l_j) = \begin{cases} \mathrm{B}_{e\bar{\mu}} & < \; 1.7\times 10^{-6} \; \equiv \; \widetilde{\mathrm{B}}_{e\bar{\mu}} \\ \mathrm{B}_{e\bar{\tau}} & < \; 7.3\times 10^{-6} \; \equiv \; \widetilde{\mathrm{B}}_{e\bar{\tau}} \\ \mathrm{B}_{\mu\bar{\tau}} & < \; 1.0\times 10^{-5} \; \equiv \; \widetilde{\mathrm{B}}_{\mu\bar{\tau}}, \end{cases} \tag{5}$$

according to the experimental limits [5,6]. This means that the fermion mixing parameters Λ_a^{ij} are bounded to lie in a circular region centered at $(-\Xi_{\mathrm{L}}^{ij}\Theta,$ $-\Xi_{\mathrm{R}}^{ij}\Theta\,)$ and of radius $\sim 10^{-3}$.

B Constraints from $l_i \to l_j l_j \bar{l}_j$

Assuming $m_{l_i} \gg m_{l_j}$ and ignoring possible contributions from scalars, the branching ratio $\mathrm{B}(l_i \to l_j l_j \bar{l}_j)$ for $i \neq j$ is

$$\frac{\mathrm{B}(l_i\to l_j l_j\bar{l}_j)}{\mathrm{B}(l_i\to l_j\bar{\nu}_{l_j}\nu_{l_i})} \simeq 4\Bigg[\left(2\left|-\tfrac{1}{2}+\mathrm{s}_{\theta_\mathrm{w}}^2\right|^2+\left|\mathrm{s}_{\theta_\mathrm{w}}^2\right|^2\right)\left|\Lambda_{\mathrm{L}}^{ij}+\Xi_{\mathrm{L}}^{ij}\Theta\right|^2 + \left(\left|-\tfrac{1}{2}+\mathrm{s}_{\theta_\mathrm{w}}^2\right|^2+2\left|\mathrm{s}_{\theta_\mathrm{w}}^2\right|^2\right)\left|\Lambda_{\mathrm{R}}^{ij}+\Xi_{\mathrm{R}}^{ij}\Theta\right|^2\Bigg] + \mathrm{O}\left(\Theta^2\right),$$

$$\mathrm{B}_{l_i l_j l_j \bar{l}_j} \equiv \mathrm{B}(l_i\to l_j l_j\bar{l}_j) = \begin{cases} \mathrm{B}_{\mu e e\bar{e}} & < \; 1.0\times 10^{-12} \; \equiv \; \widetilde{\mathrm{B}}_{\mu e e\bar{e}} \\ \mathrm{B}_{\tau e e\bar{e}} & < \; 3.3\times 10^{-6} \; \equiv \; \widetilde{\mathrm{B}}_{\tau e e\bar{e}} \\ \mathrm{B}_{\tau\mu\mu\bar{\mu}} & < \; 1.9\times 10^{-6} \; \equiv \; \widetilde{\mathrm{B}}_{\tau\mu\mu\bar{\mu}} \end{cases} \tag{6}$$

and $\mathrm{s}_{\theta_\mathrm{w}}^2 = 0.2237$, the constraints on the mixing parameters are

$$0.203\left|\Lambda_{\mathrm{L}}^{ij}+\Xi_{\mathrm{L}}^{ij}\Theta\right|^2 + 0.176\left|\Lambda_{\mathrm{R}}^{ij}+\Xi_{\mathrm{R}}^{ij}\Theta\right|^2 < c_{l_i}\widetilde{\mathrm{B}}_{l_i l_j l_j \bar{l}_j}, \tag{7}$$

where $c_{l_i} = \left(4\mathrm{B}(l_i\to l_j\bar{\nu}_{l_j}\nu_{l_i})\right)^{-1}$ and

$$\mathrm{B}(l_i\to l_j\bar{\nu}_{l_j}\nu_{l_i}) = \begin{cases} \mathrm{B}_{\mu e\bar{\nu}_e\nu_\mu} & \approx \; 1.00 \\ \mathrm{B}_{\tau\mu\bar{\nu}_\mu\nu_\tau} & = \; 0.1735\pm 0.0014 \\ \mathrm{B}_{\tau e\bar{\nu}_e\nu_\tau} & = \; 0.1783\pm 0.0008. \end{cases} \tag{8}$$

[4] using the conventional SM branching ratio 0.0337 for $\mathrm{B}_{l\bar{l}}$ and the standard values for g_V and g_A

IV SOME $SU(2)_L$ REPRESENTATIONS FOR ADDITIONAL FERMIONS

Some improvement on the above derived bounds for the mixing parameters may be obtained with information about the $SU(2)_L$ transformation properties of the additional fermions and for this reason we analyze here a few simple $SU(2)_L$ representations in which new additional charged leptons may appear. In this analysis we will not consider any particular case for the Ξ_a^{ij} parameters, but we will assume that they are of O(1). What follows is valid for one or more additional families, independently of whether the extra families are fundamental or excited leptons in the context of composite models.

a *No additional fermions.* When FCNC exists, Eqs. (4) and (7) read:

$$\left|\Xi_L^{ij}\Theta\right|^2 + \left|\Xi_R^{ij}\Theta\right|^2 < c\widetilde{B}_{l_i\bar{l}_j} = \begin{cases} c\widetilde{B}_{e\bar{\mu}} & = & 3.2\times10^{-6} \\ c\widetilde{B}_{e\bar{\tau}} & = & 1.4\times10^{-5} \\ c\widetilde{B}_{\mu\bar{\tau}} & = & 1.9\times10^{-5} \end{cases} \tag{9}$$

$$0.203\left|\Xi_L^{ij}\Theta\right|^2 + 0.176\left|\Xi_R^{ij}\Theta\right|^2 = c_{l_i}B_{l_il_jl_j\bar{l}_j} < \begin{cases} c_\mu\widetilde{B}_{\mu ee\bar{e}} & = & 0.25\times10^{-12} \\ c_\tau\widetilde{B}_{\tau ee\bar{e}} & = & 4.8\times10^{-6} \\ c_\tau\widetilde{B}_{\tau\mu\mu\bar{\mu}} & = & 2.7\times10^{-6} \end{cases} \tag{10}$$

respectively. All the constraints are for the product $\Xi_a^{ij}\Theta$ of the couplings of the light fermions to the Z′ and the Z-Z′ mixing angle. In particular $\Xi_a^{e\mu}\Theta < 10^{-6}$.

b *Sequential fermions.* The situation is the same as that of no additional fermions. Eqs. (9) and (10) hold. As in the previous case the strongest constraint is for $\Xi_a^{e\mu}\Theta < 10^{-6}$.

c *Vector singlets.* Eqs. (4) and (7) now read

$$\left|\Lambda_L^{ij} + \Xi_L^{ij}\Theta\right|^2 + \left|\Xi_R^{ij}\Theta\right|^2 < c\widetilde{B}_{l_i\bar{l}_j} \tag{11}$$

$$0.203\left|\Lambda_L^{ij} + \Xi_L^{ij}\Theta\right|^2 + 0.176\left|\Xi_R^{ij}\Theta\right|^2 < c_{l_i}\widetilde{B}_{l_il_jl_j\bar{l}_j} \tag{12}$$

respectively. The contribution to FCNC from the ordinary–exotic fermion mixing is only left handed. If $\Xi_a^{ij} \sim O(1)$, then the stringent bounds on Θ, consequence of $\Xi_R^{e\mu}\Theta < 10^{-6}$, imply an equally stringent bound on $\Lambda_L^{e\mu}$.

d *Vector doublets (homodoublets).* Eqs. (4) and (7) now read

$$\left|\Xi_L^{ij}\Theta\right|^2 + \left|\Lambda_R^{ij} + \Xi_R^{ij}\Theta\right|^2 < c\widetilde{B}_{l_i\bar{l}_j} \tag{13}$$

$$0.203\left|\Xi_L^{ij}\Theta\right|^2 + 0.176\left|\Lambda_R^{ij} + \Xi_R^{ij}\Theta\right|^2 < c_{l_i}\widetilde{B}_{l_il_jl_j\bar{l}_j}. \tag{14}$$

The contribution to FCNC from the ordinary–exotic fermions mixing is only right handed. The conclusions are the same as in the previous case with $L \leftrightarrow R$.

e Mirror fermions and self conjugated triplets. Eqs. (4) and (7) are unchanged. The contribution to FCNC from the ordinary–exotic fermions mixing is both left and right handed. As a consequence there are no stringent bounds on Θ and the limits on Λ_a^{ij} and Θ are strongly correlated. Is there some higer symmetry relating boson and fermion mixing parameters which can explain the cancellation of FCNC effects?

V CONCLUSIONS

In a model independent way we obtained bounds for the strength of the FCNC, $(\mathbf{\Lambda} + \mathbf{\Xi}\Theta)_a$, in the ordinary charged–leptons sector, produced both by the ordinary–exotic fermion mixing, Λ_a^{ij}, and by the Z–Z′ mixing, Θ. There may be a strong correlation between Θ and Λ_a^{ij} and then it is not safe to take the limit $\Theta \to 0$. In the same way, if one consider specific extended models, e.g. [1,7–15], some additional statements may be drawn on the Ξ_a^{ij}. In this work we have concentrated our attention to *lepton family violation* (LFV) in decay processes. On the other hand, there may be LFV processes of a different type [16] which will certainly put additional constraints on the LFV parameters.

REFERENCES

1. T. Kuo and N. Nakagawa, Phys. Rev. **D32**, 306 (1985).
2. P. Langacker and D. London, Phys. Rev. **D38**, 886 (1988).
3. E. Nardi, E. Roulet, and D. Tommasini, Nucl. Phys. **B386**, 239 (1992).
4. U. Cotti and A. Zepeda, Phys. Rev. **D55**, 2998 (1997).
5. R. Akers et al., Z. Phys. **C67**, 555 (1995), OPAL Collaboration.
6. L3-Note-1798, (1995), The L3 Collaboration.
7. J. Bernabéu, E. Nardi, and D. Tommasini, Nucl. Phys. **B409**, 69 (1993).
8. A. Ilakovac and A. Pilaftsis, Nucl. Phys. **B437**, 491 (1995).
9. J. Bernabeu, A. Santamaria, J. Vidal, A. Mendez, and J. W. F. Valle, Phys.Lett. **187B**, 303 (1987).
10. G. Eilam and G. Rizzo, Phys. Lett. **188B**, 91 (1987).
11. J. Bernabeu and A. Santamaria, Phys. Lett. **197B**, 418 (1987).
12. J. W. F. Valle, Phys.Lett. **199B**, 432 (1987).
13. F. Gabbiani, E. Gabrielli, A. Masiero, and L. Silvestrini, Nucl. Phys. **B477**, 321 (1996).
14. E. Gabrielli, A. Masiero, and L. Silvestrini, Phys. Lett. **B374**, 80 (1996).
15. R. Gaitán-Lozano, A. Hernández-Galeana, S. A. Tomás, W. A. Ponce, and A. Zepeda, Phys. Rev. **D51**, 6474 (1995).
16. F. Sciulli and S. Yang, Lepton flavor violation searches, in *Future Physics at HERA*, edited by G. Ingelman, A. D. Roeck, and R. Klanner, page 260, DESY, Hamburg, 1996.

The Charm of the Proton and the Λ_c^+ Production [1]

J. dos Anjos, G. Herrera[2], J. Magnin and F.R.A. Simão

CBPF, Rua Dr. Xavier Sigaud 150, CEP 22290-180 - Urca RJ, Rio de Janeiro, Brazil

Abstract. We propose a two component model for charmed baryon production in pp collisions consisting of the conventional parton fusion mechanism and fragmentation plus quarks recombination in which a ud valence diquark from the proton recombines with a c-sea quark to produce a Λ_c^+. Our two-component model is compared with the intrinsic charm two-component model and experimental data.

INTRODUCTION

The production mechanism of hadrons containing heavy quarks is not well understood. Although the fusion reactions $gg \to Q\bar{Q}$ and $q\bar{q} \to Q\bar{Q}$ are supposed to be the dominant processes, they fail to explain important features of heavy quark hadro-production like the leading particle effects observed in $D^{\pm}$ produced in $\pi^- p$ collisions [1], Λ_c^+ production in pp interactions [2] [3] and in others baryons containing heavy quarks [4], the J/Ψ cross section at large x_F observed in πp collisions [5], etc.

The above mentioned effects have been explained using a two-component model [6] consisting of the parton fusion mechanism calculable in perturbative QCD plus the coalescence of intrinsic charm [7].

In hadron-hadron collisions the recombination of valence spectator quarks with c-quarks present in the sea of the initial hadron is a possible mechanism for charmed hadron production. Here we explore that possibility for the Λ_c^+'s production in pp interactions. We will assume that in addition to the usual parton fusion processes, a ud diquark recombines with c-sea quark both from the incident proton.

---

1) This work was partially supported by Centro Latino Americano de Física (CLAF).

2) Permanent address: CINVESTAV, Apdo Postal 14-740 México DF,México

CP400, *First Latin American Symposium on High Energy Physics/ VII Mexican School of Particles and Fields,*
edited by D'Olivo/Klein-Kreisler/Mendez

We compare our results with those of the intrinsic charm two-component model and the experimental data available.

Λ_C^+ PRODUCTION *VIA* PARTON FUSION

In the parton fusion mechanism the Λ_c^+ is produced via the subprocesses $q\bar{q}(gg) \to c\bar{c}$ with the subsequent fragmentation of the c quark. The inclusive x_F distribution of the Λ_c^+ in pp collisions is given by [8] [9]

$$\frac{d\sigma^{pf}}{dx_F} = \frac{1}{2}\sqrt{s} \int H_{ab}(x_a, x_b, Q^2) \frac{1}{E} \frac{D_{\Lambda_c/c}(z)}{z} dz dp_T^2 dy \,, \tag{1}$$

where

$$\begin{aligned} H_{ab}(x_a, x_b, Q^2) = \Sigma_{a,b} \Big(& q_a(x_a, Q^2)\bar{q}_b(x_b, Q^2) \\ & + \bar{q}_a(x_a, Q^2) q_b(x_b, Q^2) \Big) \frac{d\hat{\sigma}}{d\hat{t}} |_{q\bar{q}} \\ & + g_a(x_a, Q^2) g_b(x_b, Q^2) \frac{d\hat{\sigma}}{d\hat{t}} |_{gg} \end{aligned} \tag{2}$$

with x_a and x_b being the parton momentum fractions, $q(x, Q^2)$ and $g(x, Q^2)$ the quark and gluon distribution in the proton, E the energy of the produced c-quark and $D_{\Lambda_c/c}(z)$ the fragmentation function. In eq. 1, p_T^2 is the squared transverse momentum of the produced c-quark, y is the rapidity of the $\bar{c}$ quark and $z = x_F/x_c$ is the momentum fraction of the charm quark carried by the Λ_c^+. The sum in eq. 2 runs over $a, b = u, \bar{u}, d, \bar{d}, s, \bar{s}$.

We use the LO results for the elementary cross-sections $\frac{d\hat{\sigma}}{d\hat{t}} |_{q\bar{q}}$ and $\frac{d\hat{\sigma}}{d\hat{t}} |_{gg}$ [8].

$$\frac{d\hat{\sigma}}{d\hat{t}} |_{q\bar{q}} = \frac{\pi\alpha_s^2(Q^2)}{9\hat{m}_c^4} \frac{cosh\,(\Delta y) + m_c^2/\hat{m}_c^2}{[1 + cosh\,(\Delta y)]^3} \tag{3}$$

$$\frac{d\hat{\sigma}}{d\hat{t}} |_{gg} = \frac{\pi\alpha_s^2(Q^2)}{96\hat{m}_c^4} \frac{8cosh\,(\Delta y) - 1}{[1 + cosh\,(\Delta y)]^3} \left[cosh\,(\Delta y) + \frac{2m_c^2}{\hat{m}_c^2} + \frac{2m_c^4}{\hat{m}_c^4} \right] \tag{4}$$

where Δy is the rapidity gap between the produced c and $\bar{c}$ quarks and $\hat{m}_c^2 = m_c^2 + p_T^2$.

In order to be consistent with the LO calculation of the elementary cross sections, we use the GRV-LO parton distribution functions [10], allowing by a global factor $K \sim 2 - 3$ in eq. 1 to take into account NLO contributions [6].

We take $m_c = 1.5\,GeV$ for the c-quark mass and fix the scale of the interaction at $Q^2 = 2m_c^2$ [8]. Following [6], we use two fragmentation functions to describe the hadronization of the charm quark;

$$D_{\Lambda_c/c}(z) = \delta(1-z) \tag{5}$$

and the Peterson fragmentation function [11]

$$D_{\Lambda_c/c}(z) = \frac{N}{z\left[1 - 1/z - \epsilon_c/(1-z)\right]^2} \tag{6}$$

with $\epsilon_c = 0.06$ and the normalization defined by $\sum_H \int D_{H/c}(z)dz = 1$.

Λ_C^+ PRODUCTION *VIA* RECOMBINATION

The production of leading mesons at low p_T by recombination of quarks was proposed long time ago [12]. The method introduced by Das and Hwa for mesons was extended by Ranft [13] to describe single particle distributions of leading baryons in pp collisions.

In recombination models one assumes that the outgoing hadron is produced in the beam fragmentation region through the recombination of the maximun number of valence and the minimun number of sea quarks coming from the projectile according to the flavor content of the final hadron. Thus, *e.g.* Λ_c^+'s produced in pp collisions are formed by the ud valence diquark and a c-quark from the sea of the incident proton. One ignores other type of contributions involving more than one sea flavor recombination.
The invariant inclusive x_F distribution for leading baryons is given by

$$\frac{2E}{\sqrt{S}\sigma}\frac{d\sigma^{rec}}{dx_F} = \int_0^{x_F} \frac{dx_1}{x_1}\frac{dx_2}{x_2}\frac{dx_3}{x_3} F_3\left(x_1, x_2, x_3\right) R_3\left(x_1, x_2, x_3, x_F\right) \tag{7}$$

where x_i, $i = 1, 2, 3$, is the momentum fraction of the ith quark, $F_3\left(x_1, x_2, x_3\right)$ is the three-quark distribution function in the incident hadron and $R_3\left(x_1, x_2, x_3, x_F\right)$ is the three-quark recombination function.
We use a parametrization containing explicitly the single quark distributions for the three-quark distribution function

$$F_3\left(x_1, x_2, x_3\right) = \beta F_{u,val}\left(x_1\right) F_{d,val}\left(x_2\right) F_{c,sea}\left(x_3\right)\left(1 - x_1 - x_2 - x_3\right)^\gamma \tag{8}$$

with $F_q\left(x_i\right) = x_i q\left(x_i\right)$ and F_u normalized to one valence u quark. The parameters β and γ are constants fixed by the consistency condition

$$F_q\left(x_i\right) = \int_0^{1-x_i} dx_j \int_0^{1-x_i-x_j} dx_k\, F_3\left(x_1, x_2, x_3\right), \quad i, j, k = 1, 2, 3 \tag{9}$$

for the valence quarks of the incoming proton as in ref. [13].

We use the GRV-LO parametrization for the single quark distributions in eq. 8. It must be noted that since the GRV-LO distributions are functions of x and Q^2, then our $F_3(x_1, x_2, x_3)$ also depends on Q^2.

In contrast with the parton fusion calculation, in which the scale Q^2 of the interaction is fixed at the vertices of the appropriated Feynman diagrams, in recombination there is not clear way to fix the value of the parameter Q^2, which in this case is not properly a scale parameter and should be used to give adequately the content of the recombining quarks in the initial hadron.

Since the charm content in the proton sea increases rapidly for Q^2 growing from m_c^2 to Q^2 of the order of some m_c^2's when it become approximately constant, we take $Q^2 = 4m_c^2$, a conservative value, but sufficiently far from the charm threshold in order to avoid a highly depressed charm sea which surely does not represent the real charm content of the proton. At this value of Q^2 we found that the condition of eq. 9 is fulfilled approximately with $\gamma = -0.1$ and $\beta = 75$. We have verified that the recombination cross section does not change appreciably at higher values of Q^2.

For the three-quark recombination function for Λ_c^+ production we take the simple form [13]

$$R_3(x_u, x_d, x_c) = \alpha \frac{x_u x_d x_c}{x_F^2} \delta(x_u + x_d + x_c - x_F) \tag{10}$$

with α fixed by the condition $\int_0^1 dx_F (1/\sigma) d\sigma^{rec}/dx_F = 1$, then σ is the cross section for Λ_c^+'s inclusively produced in pp collisions. From eqs 7 and 8, the invariant x_F distribution for Λ_c is

$$\begin{aligned} \frac{2E}{\sqrt{s}\sigma} \frac{d\sigma^{rec}_{\Lambda_c^+}}{dx_F} &= 75\alpha \frac{(1-x_F)^{-0.1}}{x_F^2} \int_0^{x_F} dx_1 F_{u,val}(x_1) \\ &\times \int_0^{x_F - x_1} dx_2 F_{d,val}(x_2) F_{c,sea}(x_F - x_1 - x_2) \end{aligned} \tag{11}$$

where we already integrated over x_3. The parameter σ will be fixed with experimental data.

The inclusive production cross section of the Λ_c^+ is obtained by adding the contribution of recombination eq. 11 to the QCD processes of eq. 7, then

$$\frac{d\sigma^{tot}}{dx_F} = \frac{d\sigma^{pf}}{dx_F} + \frac{d\sigma^{rec}}{dx_F}. \tag{12}$$

The resulting inclusive Λ_c^+ production cross section $d\sigma^{tot}/dx_F$ is plotted in fig. 1 using the two fragmentation function of eqs. 5 and 6 and compared

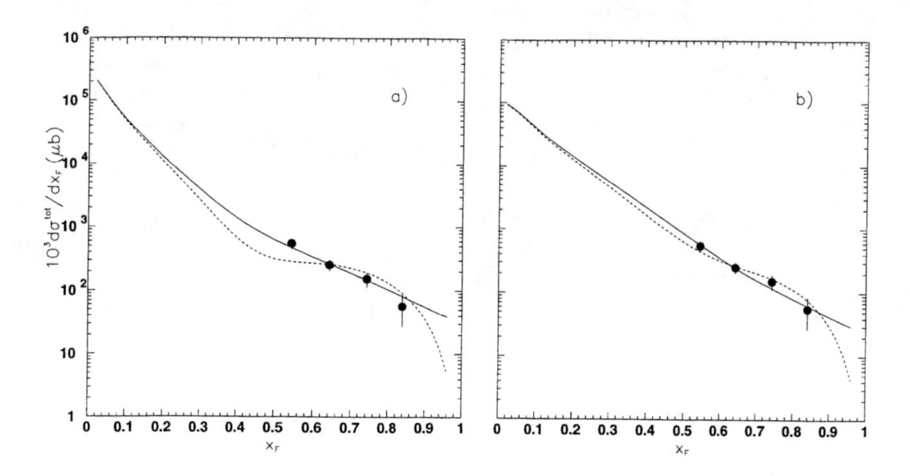

FIGURE 1. x_F distribution predicted by parton fusion plus recombination (full line) and parton fusion plus IC coalescence (dashed line) for Peterson fragmentation (a) and delta fragmentation function (b). Experimental data (black dots) are taken from ref. 3.

with experimental data in *pp* collisions from the ISR [3]. As we can see, the shape of the experimental data is very well described by our model. We use a factor $\sigma = 0.92(0.72)\mu bar$ for Peterson (delta) fragmentation respectively.

In a similar approach R. Vogt *et al.* [6] calculated the Λ_c^+ production in *pp* and πp collisions. The two component model used by them consists of a parton fusion mechanism plus coalescence of the intrinsic charm in the proton. Their results are shown in fig.1. The normalization however has been modified to make a proper comparison to our result.

CONCLUSIONS

We have studied the Λ_c^+ production in *pp* collisions with a two component model. We show that both the intrinsic charm model and the conventional recombination of quarks can describe the shape of the x_F distribution for Λ_c^+'s produced in *pp* collisions. None of them, however, can describe the abnormally high normalization of the ISR data quoted in ref. [3]. This discrepancy between theory and experiment does not exist for charmed meson production, which is well described both in shape and normalization with the parton fusion mechanisn plus intrinsic charm coalescence [9] and with the conventional recombination as proposed here [14].

An interesting test to rule out one of the two models would come from a measurement of the Λ_c polarization as proposed in [15].

ACKNOWLEDGMENTS

We would like to thank the organizers for financial support to participate in the I Simposium Latino Americano de Física de Altas Energías and for the kind hospitality extended to us during the event.

REFERENCES

1. E791 Collaboration (E.M. Aitala *et al.*), Phys. Lett. **B 371**,157 (1996).
2. A.N. Aleev *et al.*, Z. Phys. **C 23**, 333 (1984), G. Bari *et al.*, Nuovo Cimento, **104 A**, 57 (1991).
3. P. Chauvat *et al.* Phys. Lett. **B 199**, 304 (1987).
4. S.F. Biagi *et al.*, Z. Phys. **C 28**, 175 (1985), G. Bari *et al.*, Nuovo Cimento **104 A**, 787 (1991), R. Werding, WA89 Collaboration, in proceeding of ICHEP94, Glasgow.
5. NA3 Collaboration (J. Badier *et al.*), Z. Phys. **C 20**, 101 (1983).
6. R. Vogt and S.J. Brodsky, Nucl. Phys. **B 478**, 311 (1996).
7. S.J. Brodsky, P. Hoyer, C. Peterson and N. Sakai, Phys Lett. **B 93**, 451 (1980) and S.J. Brodsky, C. Peterson and N. Sakai, Phys. Rev. **D 23**, 2745 (1981).
8. J. Babcock, D. Sivers and S. Wolfram, Phys. Rev. **D 18**, 162 (1978), B.L. Combridge, Nucl. Phys. **B 151**, 429 (1979), P. Nason, S. Dawson and R.K. Ellis, Nucl. Phys. **B 327**, 49 (1989), R.K. Ellis, Fermilab-Conf-89/168-T (198 9), I. Inchiliffe, Lectures at the 1989 SLAC Summer Institute, LBL-28468 (1989).
9. R. Vogt, S.J. Brodsky and P. Hoyer, Nucl. Phys **B 383**, 643 (1992).
10. M. Glük, E. Reya and A. Vogt, Z. Phys. **C 53**, 127 (1992).
11. C. Peterson, D. Schlatter, J. Schmitt and P. Zerwas, Phys. Rev. **D 27**, 105 (1983).
12. K.P. Das and R.C. Hwa, Phys. Lett. **B 68**, 459 (1977).
13. J. Ranft, Phys. Rev. **D 18**, 1491 (1978).
14. E. Cuautle, G. Herrera and J. Magnin in preparation.
15. G. Herrera and L. Montano, *Phys. Lett.* **B 381** 337(1996), J. dos Anjos, G. Herrera, J. Magnin and F.R.A. Simão, in preparation.

Quark Mixing Possibilities in Extensions of the Standard Model and the Electromagnetic Current

A. García[a], R. Huerta[b], and G. Sánchez-Colón[b,1]

[a]*Departamento de Física.*
Centro de Investigación y de Estudios Avanzados del IPN.
A.P. 14-740, C.P. 07000, México, D.F., MEXICO.

[b]*Departamento de Física Aplicada.*
Centro de Investigación y de Estudios Avanzados del IPN. Unidad Mérida.
A.P. 73, Cordemex C.P. 97310, Mérida, Yucatán, MEXICO.

ABSTRACT

The existence of new quarks may lead to their mixing with ordinary quarks. We are particularly interested in studying the consequences such mixings could have upon the electromagnetic current. We find that new strong flavor and parity violating electromagnetic interactions may exist. To be specific, for our analysis we work with the model $SU_2 \otimes U_1 \otimes \hat{SU}_2 \otimes \hat{U}_1$, which doubles the standard electroweak sector, both in gauge bosons and in quarks.

INTRODUCTION

It is well know that if new or exotic fermions exist their mixings with the ordinary fermions will lead in general to flavor changing neutral current interactions [1]. But as far as we know, it has not been discussed in detail if mixings of that nature may lead to flavor and parity violating electromagnetic interactions. At first glance, one might think that this is not the case, but a careful analysis may lead to the opposite point of view, as we shall show in what follows. Put in rather reversed terms, one may ask what are the conditions for the elimination of flavor and parity violating contributions to

---

[1)] Also at Physics Department, University of California Riverside. Riverside, CA 92521-0413, U.S.A.

CP400, *First Latin American Symposium on High Energy Physics/ VII Mexican School of Particles and Fields*, edited by D'Olivo/Klein-Kreisler/Mendez

electromagnetic current and if such conditions can always be met with certainty. We shall see that, if those mixings exist then the latter contributions must also exist. The only way to eliminate them is to prohibit such mixings.

For the sake of definiteness, we shall work with the extension of the minimal electroweak model $SU_2 \otimes U_1$ given [2] by $SU_2 \otimes U_1 \otimes \hat{SU}_2 \otimes \hat{U}_1$, which doubles the gauge boson sector, and with new so-called mirror quarks which double the quark content. We shall assume manifest left-right symmetry, which means that the gauge constants of the SU_2's and U_1's obey $g = \hat{g}$ and $g' = \hat{g}'$, respectively. To keep the algebra simple we shall work with two families of ordinary quarks, u, d, s, and c and with two families of "mirror" counter-parts which we will denote with hats, $\hat{u}$, $\hat{d}$, $\hat{s}$, and $\hat{c}$.

The higgs content frequently discussed consists of the doublet $\varphi(\frac{1}{2}, Y_\varphi, 0, 0)$ and a mirror doublet $\hat{\varphi}(0, 0, \hat{\frac{1}{2}}, \hat{Y}_{\hat{\varphi}})$. These two leave massless one of the mirror neutral gauge bosons, the paraphoton. We are interested specifically in the case in which only one massless neutral gauge bosons exists and which couples to all the quarks present, i.e., the mirror quarks will be assigned electric charges. In these circumstances, the paraphoton should become a very massive neutral boson. To achieve this goal, new higgses must be introduced. A bidoublet $\Phi(\frac{1}{2}, Y_\Phi, \frac{1}{2}^*, -Y_\Phi)$ is a natural choice. Bisinglets $\Psi(0, Y_\Psi, 0, -Y_\Psi)$ are also welcome. The vacuum expectation values (VEV's) of all these higgses will be denoted by v, $\hat{v}$, k_1, k_2 and V. These VEV's will give masses to all the gauge bosons, but the photon. A detailed analysis of this sector shows [3] that in order to have two light (the already observed ones) bosons, one charged and one neutral, it is necessary that $V^2 \gg \hat{v}^2 \gg v^2 \sim k_1^2, k_2^2$ or that $\hat{v}^2 \gg V^2 \gg v^2 \sim k_1^2, k_2^2$. In both options the bisinglet is required to exist.

QUARK MASSES AND MIXINGS

Our frame work is now set up. Let us next concentrate in the mixings of quarks. The masses of the quarks will be generated by spontaneous symmetry-breaking (SSB) at the several Yukawa quark-higgs couplings. Straight-forward calculations lead to (a similar expression is found for u and c type quarks)

$$(\bar{q})_L(m_{ij})(q)_R + \text{c.c.} = (\bar{q})_L \begin{pmatrix} (g'_{ij}) & (\ell'_{ij}) \\ (j'_{ij}) & (\hat{g}'_{ij}) \end{pmatrix} (q)_R + \text{c.c.}, \tag{1}$$

where, restricting ourselves to two families $(q)_{L,R} = (d, s, \hat{d}, \hat{s})_{L,R}$. Thus (m_{ij}) is a 4×4 and (g'_{ij}), (ℓ'_{ij}), (j'_{ij}), and $(\hat{g}'_{ij})$ are 2×2 matrices. The primes on the latter stand for the absorption of the VEV's into the corresponding Yukawa coupling constants. These four last matrices are generated by the quark couplings with φ, Φ, Ψ, and $\hat{\varphi}$, respectively. Left-right exchange symmetry,

$\varphi \leftrightarrow \hat{\varphi}$, $\Phi \leftrightarrow \Phi^\dagger$, $\Psi \leftrightarrow \Psi^\dagger$, $q_L \leftrightarrow \hat{q}_R$, $q_R \leftrightarrow \hat{q}_L$, requires $(g_{ij}) = (\hat{g}_{ij})^\dagger$, $(j_{ij}) = (j_{ij})^\dagger$, and $(\ell_{ij}) = (\ell_{ij})^\dagger$. We shall ignore CP violation and accordingly (m_{ij}) may be taken to be real.

The diagonalization of Eq. (1) must go in two steps. First we must extract the large angles and second the small ones. Since one expects mirror matter to be much heavier than ordinary one, we must have $(\hat{g}'_{ij}) \gg (g'_{ij})$. We make a diagonalization of the block-diagonal part of (m_{ij}) containing these two 2×2 submatrices. The diagonalizing matrix will also be block-diagonal and its effect upon $(q)_{L,R}$ will be

$$(q^0)_{L,R} = \begin{pmatrix} (\theta_{L,R}) & 0 \\ 0 & (\hat{\theta}_{L,R}) \end{pmatrix} (q)_{L,R}, \tag{2}$$

where $(\theta_{L,R})$ and $(\hat{\theta}_{L,R})$ are 2×2 rotation matrix and $\theta_{L,R}$ and $\hat{\theta}_{L,R}$ the corresponding rotation angles. These angles will be large and the quarks obtained will the first approximation to the quark mass eigenstates. They can be assigned strong-flavors at this point, the index zero stands for all this. The initial mass matrix becomes

$$(m_{ij}) \underset{\theta,\hat{\theta}}{\longrightarrow} \begin{pmatrix} \begin{matrix} m_d^0 & 0 \\ 0 & m_s^0 \end{matrix} & (\Delta m_{ij}^{(1)}) \\ (\Delta m_{ij}^{(2)}) & \begin{matrix} \hat{m}_d^0 & 0 \\ 0 & \hat{m}_s^0 \end{matrix} \end{pmatrix} \tag{3}$$

where $\hat{m}_d^0$, $\hat{m}_s^0 \gg m_d^0$, m_s^0. The second step, the final rotation, will make (3) completely diagonal and will yield the physical quarks. In order that this rotation does not alter the above mass hierarchy, it is necessary to have $(\hat{g}'_{ij}) \gg (g'_{ij}) \sim (\ell'_{ij}) \sim (j'_{ij})$, which in turn requires

$$\hat{m}_d^0,\ \hat{m}_s^0 \gg m_d^0,\ m_s^0 \sim \Delta m_{ij}^{(1)} \sim \Delta m_{ij}^{(2)}. \tag{4}$$

This means that the angles in the second rotation must be very small and they can be retained to first order. The complete rotation matrices (in 2×2 block form) are

$$R_{L,R} = \begin{pmatrix} I & (\epsilon_{ij}^{L,R}) \\ (-\epsilon_{ij}^{L,R}) & I \end{pmatrix} \begin{pmatrix} (\theta_{L,R}) & 0 \\ 0 & (\hat{\theta}_{L,R}) \end{pmatrix} \tag{5}$$

where $\epsilon_{ij}^{L,R} \ll 1$. The mass matrix is then diagonalized to

$$R_L(m_{ij})R_R^\dagger = \begin{pmatrix} m_d & & & \\ & m_s & & \\ & & \hat{m}_d & \\ & & & \hat{m}_s \end{pmatrix} \tag{6}$$

and $(q^0)_{L,R}$ are replaced by the physical quarks, $(q^{ph})_{L,R}$ which in turn become the new strong-flavor eigenstates. The small angles are determined in terms of the entries of (3). We shall not display their detailed expression here, but it must be mentioned that they are all of the order $m_{d,s}/\hat{m}_{d,s}$ and that it is necessary $m_s - m_d$ not be small.

QUARK ANOMALOUS MAGNETIC MOMENTS AND ELECTROMAGNETIC INTERACTIOS

The next subject to be discussed is the anomalous magnetic moments of quarks. Although quarks are initially assumed to be point-like, QCD will dress them up, so to speak, and they will gain anomalous magnetic moments [4]. How to deal with these QCD-induced anomalous magnetic moments of quarks is the central issue of the present analysis. We shall exploit the fundamental property of the standard model: The commutativity of the electroweak and QCD sectors. Our discussion will be by necessity qualitative, however we intend to keep it as general as possible.

In setting up an effective lagrangian, one usually (i) starts from the standard model basic lagrangian, (ii) applies SSB, (iii) diagonalizes the mass matrices and identifies the physical quarks, and finally (iv) uses QCD. The explicit invariance of operators is lost after step (ii). In contrast, we propose to follow the path (i) start from the basic lagrangian, (ii) apply QCD, (iii) introduce SSB and, (iv) finally, diagonalize the mass matrices and identify physical quarks. We shall call the first path A and the second one B. The advantage of B is that the effective operators induced by QCD must all respect the electroweak gauge symmetry explicitly.

The appearance of QCD-induced anomalous magnetic moment operators of quarks can only take place after SSB, because prior to it one cannot write down electroweak invariant operators of this type. The reason for this is that L and R quarks must be connected to one another. To achieve this connection it is necessary that the higgses intervene, only then can invariant operators of magnetic moment type be written down.

Operators of this type will be induced by SSB upon graphs, for example, like $\varphi + q_L \to q_R + W^0$ or $\Phi + \hat{q}_L \to q_R + \hat{W}_3$. SSB and the necessary rotations will yield operators corresponding to $q_L^{ph} \to q_R^{ph} + A_\mu$ or $\hat{q}_L^{ph} \to q_R^{ph} + A_\mu$, where A_μ represents the photon field. They are anomalous magnetic moment operators. These latter operators are dressed by QCD and, as we have just discussed, intimately related to the higgses. Right after SSB, the anomalous magnetic moment 4×4 matrix will look like (in terms of 2×2 matrices and suppressing Lorentz indices and $\sigma_{\mu\nu}$ matrices)

$$(\bar{q})_L(\mu_{ij})(q)_R + \text{c.c.} = (\bar{q})_L \begin{pmatrix} (\nu_{ij}) & (\nu'_{ij}) \\ (\nu''_{ij}) & (\hat{\nu}_{ij}) \end{pmatrix} (q)_R + \text{c.c.} \tag{7}$$

The rotations with the large angles of Eq. (2) will lead to

$$(\mu_{ij}) \underset{\theta,\hat{\theta}}{\longrightarrow} \begin{pmatrix} \begin{matrix} \mu_d & 0 \\ 0 & \mu_s \end{matrix} & (\Delta\gamma_{ij}) \\ (\Delta\gamma'_{ij}) & \begin{matrix} \hat{\mu}_d & 0 \\ 0 & \hat{\mu}_s \end{matrix} \end{pmatrix}. \tag{8}$$

Our experience with magnetic moments shows that their magnitudes are order of magnitude inversely proportional to the masses for the particles. So one must require that their hierarchy be

$$\mu_d,\ \mu_s \gg \hat{\mu}_d,\ \hat{\mu}_s \simeq \Delta\gamma_{ij},\ \Delta\gamma_{ij'}. \tag{9}$$

The second rotation with the small angles will lead to the matrix (to first order in the small angles)

$$\begin{pmatrix} \mu_d & \mu_d\epsilon_{21}^R + \mu_s\epsilon_{12}^L & \mu_d\epsilon_{31}^R + \Delta\gamma_{11} & \mu_d\epsilon_{41}^R + \Delta\gamma_{21} \\ \mu_d\epsilon_{21}^L + \mu_s\epsilon_{12}^R & \mu_s & \mu_s\epsilon_{32}^R + \Delta\gamma_{21} & \mu_s\epsilon_{42}^R + \Delta\gamma_{22} \\ \mu_d\epsilon_{31}^L + \Delta\gamma'_{11} & \mu_s\epsilon_{32}^L + \Delta\gamma'_{12} & \hat{\mu}_d & 0 \\ \mu_d\epsilon_{41}^L + \Delta\gamma'_{21} & \mu_s\epsilon_{42}^L + \Delta\gamma'_{22} & 0 & \hat{\mu}_s \end{pmatrix}, \tag{10}$$

there will be an analogous matrix for the complex conjugate anomalous magnetic moments. Matrix (10) and its c.c. will be sandwiched between the physical quarks and the electromagnetic field A_μ will appear in the $F_{\mu\nu}$ stress tensor. When all the terms are collected there will appear effective electromagnetic interactions diagonal in the quark fields, like $\mu_d \bar{d}^{ph}\sigma_{\mu\nu}d^{ph}F_{\mu\nu} + \mu_s\bar{s}^{ph}\sigma_{\mu\nu}s^{ph}F_{\mu\nu}$ + (mirror part). These terms will be accompanied by non-diagonal operators, namely, $\bar{d}^{ph}\sigma_{\mu\nu}(g_V^{ds}+g_A^{ds}\gamma_5)s^{ph}+\bar{s}^{ph}\sigma_{\mu\nu}(g_V^{sd}+g_A^{sd}\gamma_5)d^{ph}$ + (mixed ordinary-mirror terms), here there will not appear non-diagonal mirror-mirror terms. The effective tensor and axial-tensor couplings are $g_V^{ds} = g_V^{sd} = (\mu_s - \mu_d)(\epsilon_{12}^L + \epsilon_{12}^R)$ and $g_A^{ds} = -g_A^{sd} = (\mu_s + \mu_d)(\epsilon_{12}^L - \epsilon_{12}^R)$.

From this analysis we may conclude that the existence of mirror matter and its mixing with ordinary matter will lead in general to the appearance of flavor and parity violating terms in the electromagnetic interactions.

DISCUSSION

Let us first discuss what condition must be met for the above flavor and parity violating terms in the electromagnetic current to disappear. We have already assumed that the 2×2 matrices of magnetic moments in the diagonal of Eq. (7) are diagonalized into the 2×2 diagonal blocks of (8), which is what one would ordinarily assume for the minimal electroweak sector. If we now require that (10) be diagonal too, then in its upper 2×2 diagonal block we should have $\mu_d\epsilon_{21}^R + \mu_s\epsilon_{12}^L = 0$ and $\mu_d\epsilon_{21}^L + \mu_s\epsilon_{12}^R = 0$. This is a system of two homogeneous equations for ϵ_{12}^R and ϵ_{12}^L (remember $\epsilon_{21}^R = -\epsilon_{12}^R$ and $\epsilon_{21}^L = -\epsilon_{12}^L$). In order to have non-zero ϵ_{12}^R and ϵ_{12}^L, the determinant of the system must vanish. This means that

$$\mu_d^2 = \mu_s^2. \tag{11}$$

Eq. (11) could only be satisfied in the strong-flavor symmetry limit. Since such symmetries are known to be broken (11) cannot be satisfied and, accordingly, (10) remains non-diagonal.

One way out of this impasse is not to require that the 2×2 blocks in the diagonal of (7) be diagonalized by the first large angle rotations. Detailed calculations show that then one can indeed impose that (10) be completely diagonalized by the second small angle rotations. This means that the price of eliminating electromagnetic flavor and parity violations at the level $SU_2\otimes U_1\otimes \hat{SU}_2\otimes\hat{U}$ is to already accept their existence at the level of the minimal $SU_2\otimes U_1$. This is a rather contradictory point of view. It also means that the limit of the large gauge group (by sending the VEV's of Φ, Ψ, and $\hat{\varphi}$ to infinity) into the minimal one is far from smooth. This would render any attempt to estimate the QCD-induced anomalous magnetic moments of ordinary quarks at level of $SU_2\otimes U_1$ meaningless [4,5]. One would face a new fine-tuning problem which would cast serious doubts on our present understanding of the standard model. This option is not satisfactory either. Another way out would be that (8) be diagonalized very much as (3) was diagonalized into (6). For this to be the case the hierarchy (9) of the magnetic moment matrix elements must be abandoned and, contrastingly, should be required to be analogous to the hierarchy (4) of mass matrix elements. This would lead to the very massive mirror quarks to have enormous anomalous magnetic moments, i.e., a very unphysical situation [5].

One is finally led to the choice either ordinary and mirror quarks do not mix at all and electromagnetic interactions are always flavor and parity conserving or they do mix and then non-vanishing flavor and parity violating electromagnetic interactions appear.

The above analysis shows that the non-existence of flavor and parity violation in electromagnetic interactions is not a question of fundamental principles, but is simply a question of assumption. The possibility of flavor and parity violations in electromagnetic interactions is really an open question, which can be decided upon only by experiment. All we can say as of now is that the ϵ_{ij}^R and ϵ_{ij}^L angles must be very small, in order to comply with the observed intensity of parity and flavor violation in nature.

Our analysis has been qualitative in nature. Nevertheless, we hope it has remained general enough to show that new flavor and parity violating interactions may exist in nature and that they should be considered seriously.

ACKNOWLEDGMENTS

One the authors (A.G.) wishes acknowledge interesting discussions with G. Kane and P. Langacker. This work is supported in part by CONACyT (México).

REFERENCES

1. P. Langacker and D. London, Phys. Rev. **D38**, 886 (1988); and references therein.
2. S. M. Barr, D. Chang, and G. Senjanović, Phys. Rev. Lett. **67**, 2765 (1992). Here can be found references to prior models.
3. O. Miranda, Ph.D. Thesis (CINVESTAV, 1997) and to be submitted for publication.
4. L. Brekke, Ann. Phys. **240**, 400 (1995).
5. H. Georgi, L. Kaplan, D. Morin, and A. Schenk, Phys. Rev. **D51**, 3888 (1995).

Renormalization Group Equations for a Certain Class of Quark-Yukawa Couplings

H. González[1,3], S.R. Juárez W.[2],
P. Kielanowski[1,4] and G. López Castro[1,5]

Abstract. Working in the framework of the CKM parameterization by its eigenvectors we consider the ansatz $y_u \sim y_d^2$ which implies for them a universal diagonalizing matrix. We derive and solve the RGE for the special form of y_d which is compatible with the phenomenological data for CKM.

The mass matrices of quarks and leptons generated by the Higgs mechanism depend on the coupling matrices of the Higgs particle with quarks and leptons and on the Higgs field vacuum expectation value. Some of the properties and symmetries of the couplings of Higgs-quark and Higgs-lepton before symmetry breaking, are not uniquely recovered from the observable quark masses and the parameters of the CKM matrix V_{ckm} at low energies. The experimental data for the masses of quarks and leptons can only be used as constrains on the postulated Higgs-quark couplings. The present approach to the problem of the quark mass and V_{ckm} generation consists of the postulation of the invariant Yukawa Couplings (YC) at the GU scale with some relation between them and then the evolution of these couplings to the scale of m_t (mass of the top quark). The structure of the YC at the GU scale is chosen in such a way as to reproduce the low energy data. The compatibility between our results and the experimental data can be an indication of the validity of new symmetries at the GU scale.

The V_{ckm} matrix relates the quark mass eigenstates and the weak eigenstates for six quarks. By convention, the charged $2e/3$ quarks (u, c, t) are unmixed, and all the mixing is expressed in the V_{ckm} matrix operating on the $-e/3$ quarks (d, s, b). The values of each element of V_{ckm} can in principle

[1] Departamento de Física, CINVESTAV. Ap. Postal 14-740, 07300 México D.F., Mexico.
[2] Escuela Superior de Fís. y Mat., IPN, U.P.-A.L.M. Edif. 9, 07738 México D.F., Mexico.
[3] Programa de Mat. y Fís., Universidad Surcolombiana, Neiva, A.A. 385, Colombia.
[4] Department of Physics, Warsaw University, Poland
[4] Institut de Physique Théorique, Université Catolique de Louvain, Louvain-la-Neuve, Belgium

CP400, *First Latin American Symposium on High Energy Physics/ VII Mexican School of Particles and Fields,* edited by D'Olivo/Klein-Kreisler/Mendez

be determined from weak decays of the relevant quarks Ref. [1]. There are different parameterizations of the CKM matrix in the literature, Refs. [2-5]. We will consider another parameterization in terms of the Eigenvalues and Eigenvectors (EE), Ref. [6]. In this parameterization the V_{ckm} obeys the equation: $V_{ckm}^3 = I$, and the CKM matrix is written in the form $V_{ckm} = \hat{A}D\hat{A}^{\dagger}$. Here D is the diagonal matrix $D = \text{Diag}\left(e^{-2\pi i/3}, \quad e^{2\pi i/3}, \quad 1\right)$. The unitary matrix $\hat{A}$ can be interpreted as a universal matrix that diagonalizes the mass matrices of up and down quarks. Because of the rephasing freedom of the quark fields it has only 4 parameters as the original CKM matrix. When the matrix $\hat{A}$ is real at the grand unification scale (GU) it has the form

$$\hat{A} = \begin{pmatrix} c_1 & -s_1 & 0 \\ s_1 c_2 & c_1 c_2 & -s_2 \\ s_1 s_2 & c_1 s_2 & c_2 \end{pmatrix} \tag{1}$$

where $c_i = \cos\beta_i$, $s_i = \sin\beta_i$. Alternatively the matrix $\hat{A}$ can be written in terms of complex parameters A and λ

$$\hat{A} = \begin{pmatrix} \frac{1}{k_1} & -\frac{\lambda^*}{\sqrt{3}k_1} & 0 \\ \frac{\lambda}{\sqrt{3}k_2} & \frac{1}{k_2} & -\frac{A^*\lambda^{*2}}{\sqrt{3}k_2} \\ \frac{A\lambda^3}{3k_1k_2} & \frac{A\lambda^2}{\sqrt{3}k_1k_2} & \frac{k_1}{k_2} \end{pmatrix}, \tag{2}$$

where $k_1 = \left[1 + \frac{|\lambda^2|}{3}\right]^{1/2}$, and $k_2 = \left[1 + \frac{|\lambda^2|}{3} + \frac{|A\lambda^2|^2}{3}\right]^{1/2}$.

Now we will consider the Renormalization Group Equations (RGE) for the YC in the new scheme

The evolution of the YC in the framework of the new scheme EE is obtained for three models. The models that we have considered are: the standard model (SM), standard model with two Higgs doublets (DHM) and minimal supersymmetric standard model $(MSSM)$. The RGE up to the order λ^4 for the YC, Ref. [7] become modified by a quark field rephasing transformation. Solving the new equations we obtain a different evolution of the matrices y_u and y_d. The RGE for the YC's in the new scenario obtained from Ref. [7] for $\{i, j, k, n = 1, 2, 3\}$ and $t \equiv \ln\left|\frac{\mu}{m_t}\right|$ are:

$$\frac{d\,[\hat{u}]_{ij}}{dt} = \frac{2}{(4\pi)^2}\left[\left(3\,[\hat{u}]_{33} - \sum_n c_n g_n^2\right)[\hat{u}]_{ij} + a_{uu}[\hat{u}]_{i3}[\hat{u}]_{3j}\right] \tag{3}$$

$$\begin{aligned} \frac{d[\hat{d}]_{kj}}{dt} = \frac{1}{(4\pi)^2}\Bigg[2\left(T_d - \sum_n c_n' g_n^2\right)[\hat{d}]_{kj} + \\ a_{du}\left(e^{\frac{2\pi i}{3}k}[\hat{u}]_{k3}[\hat{d}]_{3j} + e^{-\frac{2\pi i}{3}j}[\hat{u}]_{3j}[\hat{d}]_{k3}\right)\Bigg], \end{aligned} \tag{4}$$

Where for convenience we have defined

$$\hat{u} = \widetilde{y_u(t)}\ \widetilde{y_u(t)}^\dagger = \sqrt{D} y_u(t) y_u^\dagger(t) \sqrt{D^*}$$
$$\hat{d} = \widetilde{y_d(t)}\ \widetilde{y_d(t)}^\dagger = \sqrt{D^*} y_d(t) y_d^\dagger(t) \sqrt{D} \qquad (5)$$

using $U_T = \hat{A} D^{\frac{1}{2}}$, $D_T^\dagger = D^{\frac{1}{2}} \hat{A}^\dagger$ which are obtained from $V_{ckm} = U_T D_T^\dagger = \hat{A} D \hat{A}^\dagger$ and $D_{mu} \propto \hat{A} \widetilde{y_u(t)} \hat{A}^\dagger$, $D_{md} \propto \hat{A} \widetilde{y_d(t)} \hat{A}^\dagger$, where

$$D_{mu} = \text{Diag}\,(m_u,\ m_c,\ m_t)\,, \quad D_{md} = \text{Diag}\,(m_d,\ m_s,\ m_b)\,. \qquad (6)$$

The coefficients in Eqs. (3) and (4) are equal to

MODEL	a_{uu}	a_{du}	T_d
SUSY	3	1	$3[\hat{d}]_{33}$
DHM	3/2	1/2	$3[\hat{d}]_{33}$
SM	3/2	-3/2	$3[\hat{u}]_{33}$

(7)

and RGE for the g_i's are

$$\frac{dg_i(t)}{dt} = \frac{1}{8\pi^2} \frac{b_i}{2} g_i^3(t). \qquad (8)$$

Here

$$b_i = \begin{cases} (\frac{33}{5}, 1, -3) \\ (\frac{21}{5}, -3, -7) \\ (\frac{41}{10}, \frac{-19}{6}, -7) \end{cases}, \quad c_i = \begin{cases} (\frac{13}{15}, 3, \frac{16}{3}) \\ (\frac{17}{20}, \frac{9}{4}, 8) \\ (\frac{17}{20}, \frac{9}{4}, 8) \end{cases}, \quad c_i' = \begin{cases} (\frac{7}{15}, 3, \frac{16}{3}) & Susy \\ (\frac{1}{4}, \frac{9}{4}, 8) & DHM \\ (\frac{1}{4}, \frac{9}{4}, 8) & SM \end{cases}. \qquad (9)$$

The solutions of Eqs. (3),(4) give the evolution of the YC's y_u and y_d from the GU to the m_t scale. As an initial condition we considered $y_u(t_i) \propto y_d^2(t_i)$. This assumption is based on the hierarchical structure of the quark masses. The consequences of the analytical solutions of the RGE for the YC on the evolution of V_{ckm} , λ and A are:

$$V_{ckm}(t) = \hat{A}(A, \lambda) D \hat{A}^+(\tilde{A}, \tilde{\lambda}), \qquad (10)$$

where

$$\tilde{\lambda} = \frac{\left(1 + Re^{-2i\pi/3}\right)}{\left(1 + Re^{2i\pi/3}\right)} \lambda, \quad \tilde{A} = \frac{a\left(1 + Re^{2i\pi/3}\right)^3}{\left(1 + Re^{-2i\pi/3}\right)^2} A. \qquad (11)$$

Here R and a are the coefficients that depend on the model that is used. Their form follows from the renormalization group equations and is given in Ref. [8].

The results obtained when adjusting the parameters A and λ to the experimental values for the CKM matrix for the three models are given in the following Tables.

Data	Exp.	SUSY	SM	DHM
$V_{ud}\|$	$.9736 \pm 0.001$	.975164	.97514	.975166
$\|V_{us}\|$	$.2205 \pm 0.0018$	.221411	.221539	.221405
$\|V_{cd}\|$	$.224 \pm 0.016$	.221436	.22149	.221419
$\frac{\|V_{ub}\|}{\|V_{cb}\|}$	$.08 \pm 0.02$	.1380	.1158	.1340
$\|V_{cb}\|$	$.041 \pm 0.003$	.041006	.041004	.041005

TABLE I. Comparison with the experimental data taken from the latest information on the CKM matrix Ref. [1].

	SUSY	SM	DHM
$\|V_{ub}\|$	0.00566	0.00475	0.00550
$\|V_{td}\|$	0.00462	0.00665	0.00490
	$\|V_{ub}\| > \|V_{td}\|$	$\|V_{ub}\| < \|V_{td}\|$	$\|V_{ub}\| > \|V_{td}\|$

TABLE II. Asymmetry of V_{ckm}.

Wolf. Par.	SUSY	SM	DHM
$\|\lambda_w\|$	0.221	0.222	0.221
$\|A_w\|$	0.836	0.835	0.836
$\|\rho\|$	0.565	0.369	0.538
$\|\eta\|$	0.264	0.371	0.279

TABLE III. Wolfenstein Parameters at the m_t scale.

Concluding we see from the Tables I–III that our results are compatible with the experimental data for the CKM matrix. This fact strongly supports the validity of our ansatz $y_u \sim y_d^2$ at the GU scale.

S. R. J. W. gratefully acknowledges partial support by Comisión de Operación y Fomento de Actividades Académicas (Instituto Politécnico Nacional).

REFERENCES

1. Particle Data Group: F. J. Gilman et al., Phys. Rev. D54, 1 (1996).
2. N. Cabbibo, Phys. Rev. Lett. 10, 531 (1963).
3. M. Kobayashi and T. Maskawa, Prog. Theor. Phys. 49, 652 (1973).
4. L. Wolfenstein, Phys. Rev. Lett. 51, 1945 (1983).
5. M. Gronau and J. Schecheter, Phys. Rev. Lett. 54, 385 (1985).
6. P. Kielanowski, Phys. Rev. Lett. 63, 2189 (1989).
7. B. Grzadkowski, M. Lindner and S. Theisen, Phys. Lett.B, 64 (1987).
8. H. González, S. R. Juárez., P. Kielanowski. and G. López. to be published

Sea Contributions to Baryon Semileptonic Decays

V. Gupta*, R. Huerta*, and G. Sánchez-Colón*†.

*Departamento de Física Aplicada.
Centro de Investigación y de Estudios Avanzados del IPN.
Unidad Mérida.
A.P. 73, Cordemex 97310, Mérida, Yucatán, MEXICO.

†Physics Department, University of California Riverside
Riverside, CA 92521-0413, U.S.A.

Abstract. The physical spin 1/2 baryons are treated as a baryon "core" of three valence quarks (as in the standard quark model) surrounded by a "sea" which is specified by its total flavor, spin, and color quantum numbers. In particular we assume the sea to be a flavor octet with spin 0 or 1 but no color. The parameters of the general baryon wavefunction of the model are determined by fitting the magnetic moment data using experimental errors. Preliminarly predictions of this wavefunction for semileptonic decays are given.

INTRODUCTION

The expectations of the standard quark model (SQM) that the three valence quarks give dominant contribution to the low energy properties of the baryons has had limited quantitative success. The valence quarks cannot even account for the proton spin [1]. One must go beyond SQM.

In reality, quarks interact and one expects the physical hadrons to consist of the valence quarks plus a "sea" consisting of gluons and virtual quark-antiquark ($q\bar{q}$) pairs. Different treatments of the sea can be found in the literature [2–7]. We picture the physical baryon as a baryon "core" made of the three valence quarks surrounded by a sea which is specified by its total flavor, spin, and color quantum numbers.

CP400, *First Latin American Symposium on High Energy Physics/ VII Mexican School of Particles and Fields,*
edited by D'Olivo/Klein-Kreisler/Mendez

SPIN 1/2 BARYON WAVEFUNCTION WITH SEA

We assume the core baryon wavefunction is given by the SQM. For the $SU(3)$ flavor octet spin 1/2 baryons we denote this SQM or q^3 wavefunction by $\tilde{B}(\mathbf{8},1/2)$. These octet states are denoted by $\tilde{p}$, $\tilde{\Sigma}^+$, etc. We assume the sea to be a color singlet but with spin and flavor. Its $SU(3)$ flavor singlet and octet wavefunctions are denoted by $S(\mathbf{1})$ and $S(\mathbf{8})$. These can carry spin 0 (wavefunction H_0) or spin 1 (wavefunction H_1). In our model the general sea is described by the four wavefunctions $S(\mathbf{1})H_0$, $S(\mathbf{8})H_0$, $S(\mathbf{1})H_1$, and $S(\mathbf{8})H_1$. We refer to the even parity spin 0 (spin 1) sea as scalar (vector) sea. The spin 0, $SU(3)$ symmetry implicit in SQM is given by $S(\mathbf{1})H_0$. Flavor octet sea $S(\mathbf{8})$ contains states $S(Y;I,I_3)$ which carry hypercharge Y and isospin I. The flavor quantum numbers of these states will combine with those of the $\tilde{B}$ in core to give the desired global quantum numbers for the $J^P = \frac{1}{2}^+$ physical baryon B (p, Σ^+, etc.).

The flavor-spin wavefunction of spin up ($\uparrow$) physical baryon B in our model is (schematically)

$$\begin{aligned} B(1/2\uparrow) = {} & \tilde{B}(\mathbf{8},1/2\uparrow)H_0S(\mathbf{1}) + b_0\left[\tilde{B}(\mathbf{8},1/2)\otimes H_1\right]^{\uparrow} S(\mathbf{1}) \\ & + \sum_N a(N)\left[\tilde{B}(\mathbf{8},1/2\uparrow)H_0 \otimes S(\mathbf{8})\right]_N \\ & + \sum_N b(N)\left\{[\tilde{B}(\mathbf{8},1/2)\otimes H_1]^{\uparrow} \otimes S(\mathbf{8})\right\}_N , \end{aligned} \tag{1}$$

where $N = \mathbf{1}, \mathbf{8_F}, \mathbf{8_D}, \mathbf{10}, \mathbf{\overline{10}}, \mathbf{27}$ from the reduction of $\tilde{B}(\mathbf{8}) \otimes S(\mathbf{8})$. Also, $\left[\tilde{B}(\mathbf{8},1/2)\otimes H_1\right]^{\uparrow}$ represents the appropriate spin 1/2 combination of $\tilde{B}(1/2)$ and H_1 in spin space. The color wavefunctions are not indicated as both the $\tilde{B}$ and sea are color singlets.

The normalization of the physical baryons wavefunction in Eq. (1) can be obtained by using $\langle\tilde{B}(Y,I,I_3)|\tilde{B}(Y',I',I_3')\rangle = \langle S(Y,I,I_3)|S(Y',I',I_3')\rangle = \delta_{YY'}\delta_{II'}\delta_{I_3I_3'}$. Denoting by N_1, N_2, N_3, and N_4, the normalization constants for the (p,n), (Ξ^0,Ξ^-), $(\Sigma^{\pm},\Sigma^0)$, and Λ isospin multiplets, one has $N_1^2 = N_0^2 + a^2(\mathbf{\overline{10}})$, $N_2^2 = N_0^2 + a^2(\mathbf{10})$, $N_3^2 = N_0^2 + \sum_{N=\mathbf{10},\mathbf{\overline{10}}} a^2(N)$, $N_4^2 = N_0^2 + a^2(\mathbf{1})$, where, $N_0^2 = 1 + \sum_{N=\mathbf{8_D},\mathbf{8_F},\mathbf{27}} a^2(N)$. Note, for the physical baryons to have $J^P = 1/2^+$ the scalar sea has $J^P = 0^+$ since $\tilde{B}$ have $J^P = 1/2^+$.

How do the sea wavefunctions with $J^P = 0^+$ or 1^+ and the above flavor properties arise? A sea with flavor $\mathbf{8}$ property can arise from Goldstone bosons (usual $J^P = 0^-$ pseudoscalar mesons, $\pi^{\pm}$, $K^{\pm}$, etc.) and their effect on baryon structure has been considered recently [8]. These can combine with $q\bar{q}$-pairs or gluons to give the total quantum numbers for the sea considered by us.

Our approach can be used to construct wavefunctions for other hadrons incorporating a sea specified by total quantum numbers. Also, since we have an explicit wavefunction we can calculate all relevant physical quantities in terms of the parameters in the wavefunction, namely, b_0, $a(N)$'s, and $b(N)$'s. Since, there is no á priori theoretical knowledge which of these are important, we determine them by confronting the predictions of our wavefunctions with experiment.

APPLICATION TO MAGNETIC MOMENTS

This was done earlier by us and details can be found in Ref. [9] where explicit wavefunctions are given and various fits are discussed.

The baryon magnetic moment operator $\hat{\mu}$ was taken to be $\sum_q (e_q/2m_q)\sigma_z^q$ $(q = u, d, s)$ as in SQM. For the fits μ_q (or m_q) were treated as three additional parameters. Using *experiemntal errors,* we obtained two excellent six parameters fits to the eight magnetic moment data.

Case 1. The values $\mu_u = 2.5007$, $\mu_d = -1.3058$, $\mu_s = -0.8233$, $a(\mathbf{8_F}) = -0.1536$, $a(\mathbf{10}) = 0.5065$, and $b(\mathbf{8_F}) = 0.5272$, gave $\chi^2/\mathrm{DOF} = 1.95/2$.

Case 2. The values $\mu_u = 2.4748$, $\mu_d = -1.3010$, $\mu_s = -0.8243$, $a(\mathbf{8_F}) = -0.1466$, $a(\mathbf{10}) = 0.4941$, and $b_0 = 0.4779$, gave $\chi^2/\mathrm{DOF} = 2.09/2$.

Both cases have similar values for μ_q implying $m_u \approx m_d \approx 0.6 m_s$ and practically the same values for the two parameters $a(\mathbf{8_F})$ and $a(\mathbf{10})$ specifying the scalar sea. The basic difference is that vector sea in Case 1 carries flavor but is flavorless in Case 2.

This, 3-parameter wavefunction determined by the magnetic moment fit can be used to predict other data.

SEMILEPTONIC DECAYS (SLD)

For these we need to calculate $G_{V,A}(B \to B') = \langle B'|J_{V,A}|B\rangle$ using our baryon wavefunction.

Vector Current Operator J_V. For $\Delta s = 0$ decays, $J_V = I_+ = \sum_q I_+^{(q)} + I_+^{(S)}$ is the total isospin raising operator. Superscripts q and S refer to the quark and sea parts.

For $\Delta s = 1$ decays, $J_V = V_- = \sum_q V_-^{(q)} + V_-^{(S)}$ is the total V-spin lowering operator which causes a $s \to u$ transition.

Axial Vector Current Operator J_A. We can write $J_A = J_A^{(q)} + \beta_A J_A^{(S)}$, in general, where the parameter β_A (specifying the strength of the sea current operator $J_A^{(S)}$ relative to the quark current $J_A^{(q)}$) is not known. Further, β_A could be quite different for $\Delta s = 0$ and $\Delta s = 1$ decays. Whatever the values of β_A, $J_A^{(S)}$ can only contribute if the sea has both spin and flavor. For Case 2, the sea won't contribute to G_A.

TABLE 1. Predictions for G_A/G_V for baryon semileptonic decays.

Decay	Data	SQM	Errors for magnetic moment data fit Exptal.	Exptal. + $0.1\mu_N$
$n \to p$	1.2573 ± 0.0028	5/3	1.2463	1.2548
$\Lambda \to p$	0.718 ± 0.015	1	0.7478	0.7529
$\Sigma^- \to n$	-0.34 ± 0.017	$-1/3$	-0.2300	-0.2290
$\Xi^- \to \Lambda$	0.25 ± 0.05	1/3	0.2686	0.2729
χ^2		21,731	61	49

Full numerical analysis of the SLD is not yet complete. Partial results for Case 2 are given in the Table 1 for G_A/G_V for the four measured decays. Column 1 gives the experimental data and Column 2 gives the SQM prediction. Our predictions (using the wavefunction determined by magnetic moment data, Case 2 above) are given in Column 3. These are much better than SQM. The last column gives predictions for SLD when the magnetic moments are fitted by adding a theoretical error of $0.1\mu_N$ in quadrature to the experimental error. This is a popular choice [10]. If we do this we get a $\chi^2/\text{DOF} = 0.22/2$ with $a(\mathbf{8_F}) = -0.1676$, $a(\mathbf{10}) = 0.4912$, and $b_0 = 0.4700$, while $\mu_u = 2.4726$, $\mu_d = -1.2839$, and $\mu_s = -0.8122$. The fits to SLD is much better, but the $\Sigma^- \to n$ prediction is poor in both cases.

It should be noted that the usual phenomenological fit using $SU(3)$ symmetry to SLD is excellent but it has no predictions for the magnetic moments.

Since we have an explicit wavefunction we are at present working on a simultaneous fit to 14 pieces of data (8 magnetic moments, 4 SLD, and spin distributions of p and n). Preliminary results indicate that an excellent fit with 3 or 4 parameters should be possible.

We thank CONACyT (México) for partial support.

REFERENCES

1. J. Ashman *et al.*, Nucl. Phys. **B328**, 1 (1989).
2. J. F. Donoghue and E. Golowich, Phys. Rev. **D15**, 3421 (1977).
3. E. Golowich, E. Haqq and G. Karl, Phys. Rev. **D28**, 160 (1983).
4. He Hanxin, Zhang Xizhen and Zhuo Yizhang, Chinese Physics 4, 359 (1984); J. Franklin, Phys. Rev. **D30**, 1542 (1984).
5. F. E. Close and Z. Li, Phys. Rev. **D42**, 2194 (1994); F. E. Close, Rep. Prog. Phys. **51**, 833(1988); Z. Li, Phys. Rev. **D44**, 2841 (1991); Z. Li and G. Karl, Phys. Rev. **D49**, 2620 (1994).

6. V. Gupta and X. Song, Phys. Rev. **D49**, 2211 (1994). These authors consider a sea with spin and color but no flavor.
7. R. L. Jaffe and H. J. Lipkin, Phys. Lett. **B226**, 458 (1991). These authors consider a sea with spin which is either $SU(2)$ or $SU(3)$ symmetric but has no color.
8. E. J. Eichten, I. Hinchliffe, and C. Quigg, Phys. Rev. **D45**, 2269 (1992); T. P. Cheng and Ling-Fong Li, Phys. Rev. Lett. **74**, 2872 (1995); S. J. Brodsky and Bo-Qiang Ma, Phys. Lett. **B381**, 317 (1996).
9. V. Gupta, R. Huerta, and G. Sánchez-Colón. To be published in Int. J. of Mod. Phys. A (1997).
10. M. Casu and L. M. Sehgal, preprint no. hep-ph/9606264 v3 (1996).

Constraints on New Physics with Effective Lagrangians

J. M. Hernández*, M. A. Pérez*, and J. J. Toscano†

* *Departamento de Física, CINVESTAV-IPN, Apdo. Postal 14-740, 07000, México, D. F., México.*
† *Facultad de Ciencias Físico Matemáticas, Universidad Autónoma de Puebla, Apdo. Postal 1152, 72000, Puebla, Pue., México.*

Abstract. We study possible effects coming from physics beyond the Standard Model by using the effective Lagrangian approach. In particular, we work with the Higgs and Electroweak sectors. We find that the effects coming from new physics may enhance several Standard Model predictions.

I INTRODUCTION

The agreement between the experimental data, from LEP and another accelerators, with the Standard Model (SM) predictions means that physics at the Z scale is described by the $SU(3) \times SU(2) \times U(1)$ gauge structure.

In spite of the remarkable success of the SM, there is a strong feeling among the particle physicists that the SM will not be the ultimate theory of the fundamental interactions. We could use a specific high energy model and work out in detail its consequences at low energy, but it is preferable to use a model and process independent scheme. Henceforth, we need a formalism which give us insight on the next physical scenario of the nature, if it exist at all. This can be achieved by means of the effective Lagrangian approach. The present work will cover some aspects of the effective Lagrangian approach, in the search of physics beyond SM. We will present only a short review on the basis of effective Lagrangians, we do not attempt to describe the formalism in detail here but refer the reader to Ref. [1].

[1)] Work supported by CONACyT.

CP400, *First Latin American Symposium on High Energy Physics/ VII Mexican School of Particles and Fields,* edited by D'Olivo/Klein-Kreisler/Mendez

II EFFECTIVE LAGRANGIANS

To the extent that the high energy theory is weakly coupled[2], the low energy effects of the new physics may be parametrized in terms of an effective Lagrangian [1,2]:

$$\mathcal{L}_{\text{eff}} = \mathcal{L}_o + \sum_{n\geq 5}\left[\sum_{i=1}^{N(n)} \frac{\alpha_i}{\Lambda^{n-5}} (O_i^n + H.c.)\right], \tag{1}$$

where $\mathcal{L}_0$ is the SM Lagrangian, Λ is the onset scale for new physics, O_i^n are monomials built up from SM fields and respect the SM gauge structure; α_i are unknown parameters. As was emphasized in Ref. [3], (1) is the most general Lagrangian we can write, compatible with the symmetries and particle content of the SM. Although we have an infinite tower of increasing dimension operators, it is sufficient to restrict attention to lowest-dimension interactions, since we have an "experimental energy constraint": dimensional analysis would indicate that operators of dimension (mass)^n give contributions of the order:

$$(\frac{E}{\Lambda})^n \ \leq \ \epsilon, \tag{2}$$

where ϵ is the experimental precision of the process which we are interested in. The energy constraint, together with the fact of having all the operators allowed by the SM symmetries in (1), guarantees the predictability of the effective Lagrangian [4]. Since we assume the high energy theory is weakly coupled, there are operators which can be generated at tree-level in a perturbative expansion while others only can be generated at one-loop level or beyond [5]. The one-loop operators will be suppressed by a $(1/16\pi^2)$ factor with respect to the tree-level generated operators, and we do not consider them in the present work.

III NEW PHYSICS EFFECTS ON $H^O \to \bar{f}f, Z \to \bar{f}f$

There are four tree-level-generated operators which can contribute to the decay $Z \to \bar{f}f$, they are given explicitly in Ref. [6]. There are three leptonic operators which can give explicit flavour changing neutral currents (FCNC) [6]. Both classes of operators modify the coefficients of the neutral current Lagrangian but do not change its Lorentz structure. By using the LEP data we may constrain the corresponding effective parameters α_i. Since the current experiments are sensitive to the SM radiative corrections, we have included the QED, QCD and weak one-loop contributions, while only tree-level new physics effects are considered. The corresponding bounds are given in Table 1.

2) The case of strongly coupled high energy theory has been treated elsewhere [3].

	$\epsilon_{\phi f}$	$\epsilon^{(1)}_{\phi f} = \epsilon^{(3)}_{\phi f}$	$\epsilon^{ab}_{\phi f} = \epsilon^{(1)ab}_{\phi f} = \epsilon^{(3)ab}_{\phi f}$
e	$+4.3\ 10^{-3}$	$+4.4\ 10^{-3}$	-
	$-4.3\ 10^{-3}$	$-4.3\ 10^{-3}$	-
μ	$+5.4\ 10^{-3}$	$+5.7\ 10^{-3}$	-
	$-5.5\ 10^{-3}$	$-5.5\ 10^{-3}$	-
τ	$+4.5\ 10^{-3}$	$+6.1\ 10^{-3}$	-
	$-6.1\ 10^{-3}$	$-4.8\ 10^{-3}$	-
U	$+1.1\ 10^{-2}$	$+2.6\ 10^{-3}$	-
	$-1.3\ 10^{-3}$	$-1.9\ 10^{-3}$	-
D	$+1.0\ 10^{-3}$	$+3.5\ 10^{-3}$	-
	$-8.0\ 10^{-3}$	$-4.7\ 10^{-3}$	-
$e^{\pm}\mu^{\mp}$	-	-	$\pm\ 0.949\ 10^{-2}$
$e^{\pm}\tau^{\mp}$	-	-	$\pm\ 0.987\ 10^{-2}$
$\mu^{\pm}\tau^{\mp}$	-	-	$\pm\ 1.19\ 10^{-2}$

TABLE 1. Bounds on the tree-level-generated operators arising from $Z \rightarrow \bar{f}f$

It is important to notice that our bounds for the weak neutral currents induced by new physics effects are stronger than the FCNC ones; this arises because in the former case there is interference between the SM and new physics contributions, while in the latter case there is no SM contribution.

In the SM framework we have $BR(H^o \rightarrow \bar{b}b) \sim 1$, for a Higgs boson mass in the range 60-150 GeV. However, it is possible that the couplings between Higgs and light fermions can be sensitively enhanced by effects of new physics. There are three tree-level-generated effective operators which induce new contributions on the $H^o\bar{f}f$ vertices [6], again without changing its Lorentz structure. In Fig. 1 we display $BR(H^o \rightarrow \bar{f}f)$ as a function of $\epsilon_{f\phi} = (v/\Lambda)^2\ \alpha_{f\phi}$ for the $\bar{e}e$, $\bar{\mu}\mu$, $\bar{\tau}\tau$, $\bar{u}u$, $\bar{s}s$, $\bar{c}c$, and $\bar{b}b$ channels; we take $m_H = 120$ GeV, $\Lambda = 1$ TeV, and assume that all others effective parameters are zero.

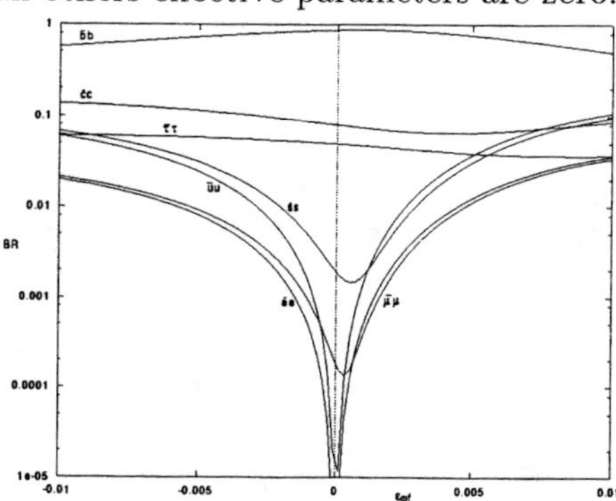

FIGURE 1. Branching Ratio for $H \rightarrow \bar{f}f$. We have set $\Lambda = 1$.

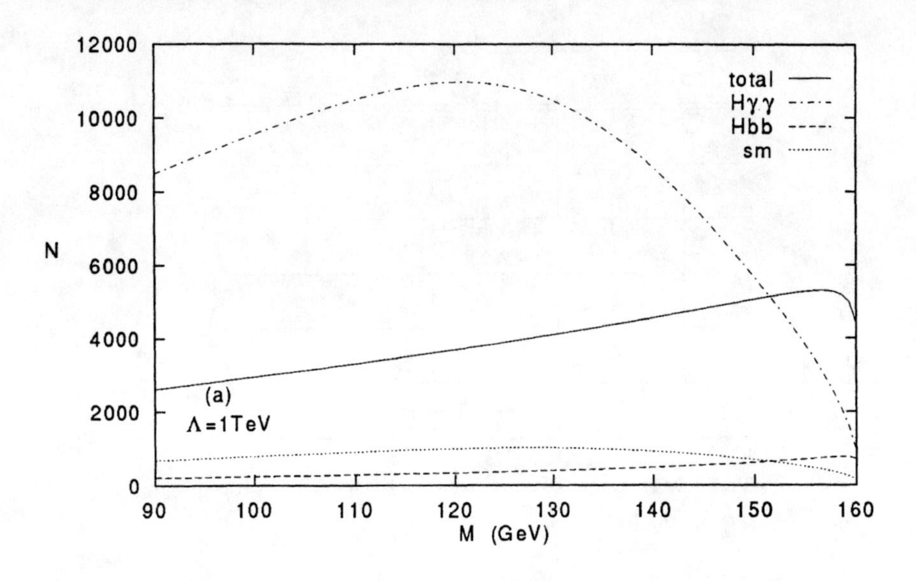

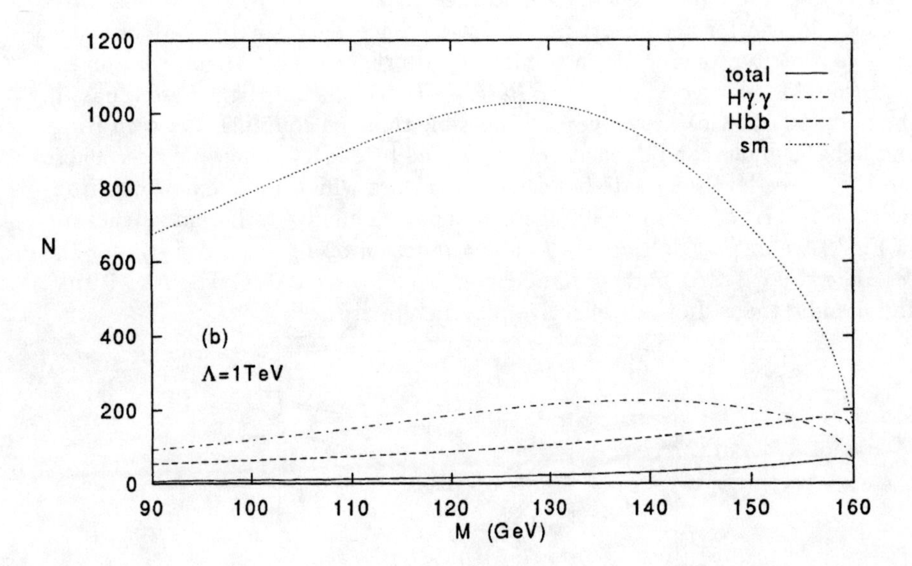

FIGURE 2. Number of events expected for the $\bar{b}b$ final state as a function of the Higgs boson mass and $\Lambda = 1$ TeV, for the scenarious with a maximum (a) and a minimum (b) number of events. In each case, $H^o\gamma\gamma$ (dot-dashed), $H^o\bar{b}b$ (dashed), SM (dotted), and total (solid line) contributions are displayed.

IV AN ENLIGHTING CASE: $\gamma\gamma \to H \to X$

Not only the discovery of a light Higgs boson will enlighten the status of the SM, but also its properties in detail. This will be one of the principal purposes for the $\gamma\gamma$ mode at the next generation of linear e^+e^- colliders [7]. The precise measurement of the Higgs decay width will permit us to discriminate between the SM and new physics predictions.

The process $\gamma\gamma \to H \to X$, where the final states are $\bar{b}b$, WW, ZZ, and $\bar{t}t$, is important because it will show the Higgs boson as a resonance, and it will permit us, with the appropriate collider configuration, an exhaustively search of several Higgs boson properties. We can have effective interactions both in the $H\gamma\gamma$ and $H\bar{f}f$ vertices. While the former vertex is generated by tree-level dimension eight effective operators, the latter can be generated by dimension six ones (also tree-level generated). The calculus is given explicitly in Ref. [8]. In this case, instead of giving to the α_i the values of 1, we search in all the possibilities that will optimize the number of expected events. In figure 2 we present the scenarious with a maximum and minimum number of events for the specific final state $X = \bar{b}b$. In Ref. [8] we found an enhancement by a factor of 5 in the cases $\bar{b}b$, WW, and ZZ, and almost one order of magnitude for the final state $\bar{t}t$.

V CONCLUSIONS

The effective Lagrangian approach provides a systematic way to analyze new physics effects in present and future colliders. The Higgs sector is highly sensitive to new physics contributions. In particular, the Higgs boson production through $\gamma\gamma$ fusion will permit us to discriminate between effects coming from the SM and physics beyond SM.

REFERENCES

1. Wudka, J., Int. J. Mod. Phys. A9 (1994)2301; Feliciano J., et al., Rev. Mex. Fis. 42 (1996) 517.
2. Georgi, H., Ann. Rev.Nucl. Part. Sci. 43 (1995) 209.
3. Dobado, A. and Herrero, M. J., Phys. Lett. B228 (1989) 495; Dobado, A., et al., Phys. Lett. B255 (1995) 405.
4. Gomis, J. and Weinberg, S., Nucl. Phys. B469 (1996) 473.
5. Artz, C., Einhorn, M. B. and Wudka, J., Nucl. Phys. B433 (1995) 41.
6. Hernández, J. M., Sampayo, O. A. and Toscano, J. J., Int. J. Mod. Phys. A11 (1996) 4921.
7. Ginzburg, I. F., et al., Nucl. Instr. and Meth. A219 (1984) 5.
8. Hernández, J. M., Pérez, M. A. and Toscano, J. J., Phys. Lett. B375 (1996) 227.

The Λ_0 Polarization and the Recombination Mechanism[1]

G. Herrera†, J. Magnin‡, Luis M. Montaño†and F.R.A. Simão‡

†CINVESTAV, Apdo postal 14-740 Mexico DF, Mexico
‡CBPF, Rua Dr. Xavier Sigaud 150, CEP 22290-180 - Urca RJ,Rio de Janeiro, Brazil

Abstract. We use the recombination and the Thomas Precession Model to obtain a prediction for the Λ_0 polarization in the $p + p \to \Lambda_0 + X$ reaction. We study the effect of the recombination function on the Λ_0 polarization.

INTRODUCTION

The unexpected discovery of large polarization in inclusive Λ_0 production by unpolarized protons has shown that important spin effects arise in the hadronization process. Several models have been proposed to explain hyperon polarization, being the Thomas Precession Model (TPM) [1] one of the most extensively used to describe polarization in a variety of reactions.

In order to obtain the Λ_0 polarization, we first calculate the momentum fraction of the recombining s-quark in the proton sea using a recombination model [2]. We use two different forms for the recombination function to see their influence on the predicted Λ_0 polarization.

THE Λ_0 POLARIZATION IN THE TPM

In pp collisions the recombining s-quark resides in the sea of the proton and carries a very small fraction $x_s \simeq 0.1$ of the proton momentum. When the s-quark recombines to form a Λ_0, it becomes a valence quark and must carry a large fraction (of the order of $\frac{1}{3}$) of Λ_0's momentum. Then one expects a large increase in the longitudinal momentum of the s-quark as it passes from the proton to the Λ_0,

$$\Delta p \simeq (\frac{1}{3}x_F - x_s)p = \left(\frac{1}{3} - \xi\right) x_F p, \qquad (1)$$

[1] This work was partially supported by Centro Latino Americano de Física (CLAF).

CP400, *First Latin American Symposium on High Energy Physics/ VII Mexican School of Particles and Fields*, edited by D'Olivo/Klein-Kreisler/Mendez

where p is the proton's momentum, $\xi = x_s/x_F$ and $x_F p = p_\Lambda$ is the momentum of the Λ_0 with x_F the Feynman x.

Since the s-quark carries transverse momentum, on the average $p_T(s/p) \sim p_T(s/\Lambda) \sim \frac{1}{2}p_{T\Lambda}$, its velocity vector is not parallel to the change in momentum induced by recombination and it must feel the effect of Thomas precession. Consequently, the Λ_0 is produced with net polarization perpendicular to the plane of the reaction.

According to the TPM the Λ_0 polarization in the reaction $p + p \rightarrow \Lambda_0 + X$ is given by [1]

$$P(p \rightarrow \Lambda) = -\frac{3}{M^2 \Delta x} \frac{(1 - 3\xi)}{\left(\frac{1+3\xi}{2}\right)^2} p_{T\Lambda}, \tag{2}$$

where $M^2 = \left[\frac{m_D^2 + p_{TD}^2}{1-\xi} + \frac{m_s^2 + p_{Ts}^2}{\xi} - m_\Lambda^2 - p_{T\Lambda}^2\right]$ and $\xi = \frac{1}{3}(1 - x_F) + 0.1 x_F$ as was assumed in ref. [1]. $\Delta x = 0.5$ GeV is a characteristic recombination scale and m_D, p_{TD}, m_s, p_{Ts}, m_Λ and $p_{T\Lambda}$ are respectively the masses and transverse momentum of the diquark, the s-quark and the Λ_0.

THE $\xi\,(X_F)$ PARAMETRIZATION IN THE RECOMBINATION MODEL

We use the recombination model proposed in ref. [3], which has been extended to take into account baryon production [4], to obtain a parametrization for ξ as a function of x_F [2]. The inclusive x_F distribution for Λ_0's in pp collisions is

$$\frac{d\sigma}{dx_F} = \int \frac{dx_u}{x_u} \frac{dx_d}{x_d} \frac{dx_s}{x_s} F\left(x_u, x_d, x_s\right) R\left(x_F, x_u, x_d, x_s\right), \tag{3}$$

where $F(x_u, x_d, x_s)$ and $R(x_F, x_u, x_d, x_s)$ are the three quark distribution and recombination functions respectively.

For the three quark distribution function we use the factorized form

$$F\left(x_u, x_d, x_s\right) = \beta F_{u,val}\left(x_u\right) F_{d,val}\left(x_d\right) F_{s,sea}\left(x_s\right) \left(1 - x_u - x_d - x_s\right)^\gamma \tag{4}$$

with $\gamma = -0.3$ as has been proposed in ref. [4] and $\beta = 0.75$. We used the Field and Feynman [5] parametrizations for the single quark distribution.

In order to see how the shape of the recombination function affects the prediction for the Λ_0 polarization, we use two different forms for $R(x_u, x_d, x_s)$:

$$R_1\left(x_u, x_d, x_s\right) = \kappa_1 \frac{x_u x_d x_s}{\left(x_F\right)^3} \delta\left(\frac{x_u + x_d + x_s}{x_F} - 1\right) \tag{5}$$

as in ref. [4] and

$$R_2(x_u, x_d, x_s) = \kappa_2 \left(\frac{x_u x_d}{x_F^2}\right)^a \left(\frac{x_s}{x_F}\right)^b \delta\left(\frac{x_u + x_d + x_s}{x_F} - 1\right), \tag{6}$$

which is inspired in the three valons recombination model proposed by R.C. Hwa [6]. In R_2, unlike R_1, the light quarks are considered with different weight than the more massive s quark introducing two distinct exponents a and b. Indeed, in the recombination model proposed in ref. [6], a recombination function for hyperons is derived and a ratio $\frac{a}{b} = \frac{2}{3}$ is used. We choose $a = 1$, $b = \frac{3}{2}$ by fitting experimental data. κ_1 and κ_2 are normalization constants.

The probability for Λ_0 production at x_F with an $s - quark$ from the sea of the proton at momentum fraction x_s is

$$\frac{d\sigma_i}{dx_s dx_F} = \int \frac{dx_u}{x_u} \frac{dx_d}{x_d} \frac{1}{x_s} F(x_u, x_d, x_s) R_i(x_F, x_u, x_d, x_s) \tag{7}$$

with $i = 1, 2$. The average value of x_s is therefore [2]

$$\langle x_s \rangle_i = \left[\int dx_s x_s \frac{d\sigma_i}{dx_s dx_F}\right] / \frac{d\sigma_i}{dx_F}. \tag{8}$$

We have taken $m_D = \frac{2}{3}$ GeV, $m_s = \frac{1}{2}$ GeV and $\langle p_T^2 \rangle_{s,D} = \frac{1}{4} p_{T\Lambda}^2 + \langle k_T^2 \rangle$ with $\langle k_T^2 \rangle = 0.25$ GeV ^{2} [1]. Figure 1 shows the Λ_0 polarization for the three different parametrizations of $\xi(x_F)$ at $p_T = 0.5$ GeV/c.

CONCLUSIONS

The two forms for ξ obtained with the two different recombination functions of eqs. 5 and 6 are very similar in shape for large x_F. For small x_F however, the difference grows slightly and $\xi_1(x_F = 0) = \frac{1}{3}$ while $\xi_2(x_F = 0) \neq \frac{1}{3}$.

The parametrizations for $\xi(x_F)$ obtained from the recombination model are different to the simple form proposed in ref. [1]. Our calculation of $\xi(x_F)$ shows that, for $x_F \to 1$, $\xi(x_F) \to 0.15$ approximately for both recombination functions. This is consistent with our actual knowledge of the sea structure functions in the proton.

We have seen that for small $p_{T\Lambda}$ our fit gives a good description of experimental data. This is reasonable since recombination models work better for small p_T.

Within the precision of experimental data [3][7], it would be hard to decide which recombination function better describe Λ_0's production. A more accurate measurement of polarization at low p_T and low x_F can help to clarify the right form of the recombination function. It is interesting to note that, although the shape of the recombination function is not important for cross section calculations, it does make a difference when applied to polarization. In this sense, polarization measurements can help to understand the underlying mechanisms in hadroproduction.

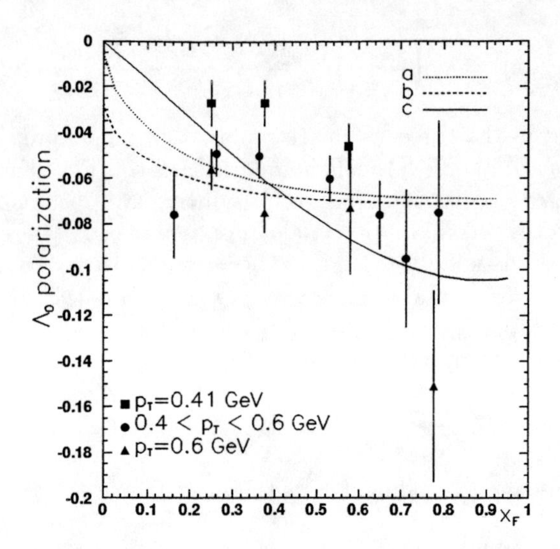

FIGURE 1. Λ_0 polarization at $p_T = 0.5GeV/c$ obtained with $\xi(x_F)$ determined with the recombination functions R_1 (a), and R_2 (b). (c) is the polarization prediction of ref. [1]. Experimental data are taken from refs. [1] and [7].

ACKNOWLEDGMENTS

We would like to thank the organizers for financial support to attend the I Simposium Latino Americano de Física de Altas Energías.

REFERENCES

1. T. De Grand and H. Miettinen,*Phys. Rev.* **D 23** (1981) 1227, *Phys. Rev.* **D 24** (1981) 2419, T. Fujita and T. Matsuyana, *Phys. Rev.***D 38** (1988) 401, T. de Grand, *ib.* 403.
2. G. Herrera, J. Magnin, L.M. Montaño and F.R.A. Simão, *Phys. Lett.***B 382** (1996) 201, G. Herrera and L. Montano, *Phys. Lett.***B 381** (1996) 337.
3. K.P. Das and R.C. Hwa, *Phys. Lett.***B 68** (1977) 459.
4. J. Ranft,*Phys. Rev.***D 18** (1978) 1491.
5. R.D. Field and R.P. Feynman, *Phys. Rev.* **D 15** (1977) 2590.
6. R.C. Hwa, *Phys. Rev.***D 22** (1980) 1593.
7. A. M. Smith et al. ,*Phys. Lett.***B 185** (1987) 209. S. Herhan et al. ,*Phys. Lett.***B 82** (1979) 301.

Top Quark Interactions and the Nonlinear Chiral Lagrangian

F. Larios[†*] and C.P.–Yuan[†]

[†] *Department of Physics and Astronomy, Michigan State University East Lansing, MI 48824*
[*] *Departamento de Física, CINVESTAV-IPN, Apdo. Postal 14-740, 07000 México, D.F., México.*

Abstract. In the context of the $SU(2)_L \times U(1)_Y$ non-linearly realized chiral Lagrangian, we study the effects from dimension 5 operators to $V_L V_L \to t\bar{t}$ amplitudes as compared to the contribution from the dimension 4 Lagrangian. At high energies, there is an equivalence between the longitudinal weak bosons and the would-be Goldstone bosons; thus, these operators contain information from the electroweak symmetry breaking sector. Considering the production of $t\bar{t}$ pairs from $V_L V_L$ fusion processes at the LC and the LHC, if no significant deviation from the SM prediction is found, the size of their coefficients is at most of order 10^{-1} (based on naive dimensional analysis). This constraint is one order of magnitude more stringent than the bounds for the next-to-leading order bosonic operators commonly studied in $V_L V_L \to V_L V_L$ scatterings.

THE NON-LINEAR CHIRAL LAGRANGIAN

The starting point of our analysis is the most general gauge invariant chiral Lagrangian that involves the electroweak couplings of the top quark up to dimension five [1].

$$\begin{aligned}\mathcal{L} = {} & i\bar{t}\gamma^\mu\left(\partial_\mu + i\frac{2s_w^2}{3}\mathcal{A}_\mu\right)t + i\bar{b}\gamma^\mu\left(\partial_\mu - i\frac{s_w^2}{3}\mathcal{A}_\mu\right)b - m_t\bar{t}t - m_b\bar{b}b \\ & -\frac{1}{2}\left(1 - \frac{4s_w^2}{3} + \kappa_L^{NC}\right)\overline{t_L}\gamma^\mu t_L\mathcal{Z}_\mu - \frac{1}{2}\left(\frac{-4s_w^2}{3} + \kappa_R^{NC}\right)\overline{t_R}\gamma^\mu t_R\mathcal{Z}_\mu \\ & -\frac{1}{2}\left(-1 + \frac{2s_w^2}{3}\right)\overline{b_L}\gamma^\mu b_L\mathcal{Z}_\mu - \frac{s_w^2}{3}\overline{b_R}\gamma^\mu b_R\mathcal{Z}_\mu \\ & -\frac{1}{\sqrt{2}}\left(1 + \kappa_L^{CC}\right)\overline{t_L}\gamma^\mu b_L\mathcal{W}_\mu^+ - \frac{1}{\sqrt{2}}\left(1 + \kappa_L^{CC\dagger}\right)\overline{b_L}\gamma^\mu t_L\mathcal{W}_\mu^- + h.c.\end{aligned}$$

CP400, *First Latin American Symposium on High Energy Physics/ VII Mexican School of Particles and Fields*, edited by D'Olivo/Klein-Kreisler/Mendez

$$-\frac{1}{4g^2}\mathcal{W}^a_{\mu\nu}\mathcal{W}^{a\,\mu\nu} - \frac{1}{4g'^2}\mathcal{B}_{\mu\nu}\mathcal{B}^{\mu\nu} + \frac{v^2}{4}\mathcal{W}^+_\mu \mathcal{W}^{-\,\mu} + \frac{v^2}{8}\mathcal{Z}_\mu \mathcal{Z}^\mu + \mathcal{L}^{(5)} \,, \tag{1}$$

where the coefficients $\kappa_L^{\rm NC}$, $\kappa_R^{\rm NC}$, $\kappa_L^{\rm CC}$, and $\kappa_R^{\rm CC}$ parametrize possible deviations from the SM couplings [2]. The term $\mathcal{L}^{(5)}$ contains the dimension five interactions of the top quark. A study of the dimension 4 deviations (κ's) can be made at the energy scale of m_t, whereas for the dimension 5 couplings it is necessary to probe physics at much higher scales.

The fields $\mathcal{W}^\pm_\mu$, $\mathcal{Z}_\mu$ and $\mathcal{A}_\mu$ are defined in terms of the *non-linear* parametrization of the Goldstone boson sector associated with the electroweak symmetry breaking (EWSB), $\Sigma = \exp\left(i\frac{\phi^a\tau^a}{v}\right)$ (cf. Ref. [1]), with $v = 246$ GeV being the EWSB scale. In the unitary gauge ($\Sigma = 1$)

$$\mathcal{W}^\pm_\mu = -gW^\pm_\mu \,, \;\; \mathcal{Z}_\mu = -\frac{g}{c_w}Z_\mu \,, \;\; \mathcal{A}_\mu = -\frac{e}{s_w^2}A_\mu \,, \tag{2}$$

where $W^\pm_\mu$ and Z_μ are the usual weak charged and neutral vector bosons; A_μ is the photon field.

Dimension five operators

Let us study the effects from some of the possible dimension 5 couplings listed in Ref. [1], to the amplitudes $Z_L Z_L \to t\bar{t}$ and $W_L^+ W_L^- \to t\bar{t}$. These operators are:

$$\frac{a_{zz1}}{\Lambda}\bar{t}t\mathcal{Z}_\mu\mathcal{Z}^\mu \;\;, \qquad \frac{a_{ww1}}{\Lambda}\bar{t}t\mathcal{W}^+_\mu\mathcal{W}^{-\mu} \;\;, \qquad \frac{a_m}{\Lambda}\bar{t}\sigma^{\mu\nu}t\mathcal{A}_{\mu\nu} \,, \tag{3}$$

where Λ is either the lowest new heavy mass scale, or something around $4\pi v \sim 3.1$ TeV if no new resonances exist this value [3]. (In this study, we assume there are no new resonances below the 3 TeV scale.)

AMPLITUDES

The leading contributions, in powers of the CM energy $E = \sqrt{s}$ of the $V_L V_L$ system, to the $Z_L Z_L \to t\bar{t}$ helicity amplitudes are:

$$\begin{aligned} T_{zz++} = -T_{zz--} &= \frac{m_t\,E}{v^2}(\kappa_L^{NC} - \kappa_R^{NC} + 1)^2 - \frac{2E^3}{v^2}\frac{a_{zz1}}{\Lambda} \,, \\ T_{zz+-} = T_{zz-+} &= \frac{2\,m_t^2\cot\theta}{v^2}(\kappa_L^{NC} - \kappa_R^{NC} + 1)^2 + 0 \,, \end{aligned} \tag{4}$$

where, for example, T_{zz+-} is the amplitude that corresponds to the production of a right handed top and a left handed anti-top. In the range of energies of interest, $E \gg m_t$, the dominant contribution will come from the dimension

5 coupling a_{zz1}, even in the presence of the possible effects from the κ's. For the other fusion process $W_L W_L$ the contribution from the κ's can in principle be significant. To simplify our analysis we will consider the κ's to be zero. As mentioned before, by the time the higher scales of energy begin to be probed at the colliders, sufficient information about the κ's should already be available.

The leading contributions to the $W_L^+ W_L^- \to t\bar{t}$ helicity amplitudes are:

$$
\begin{aligned}
T_{ww++} = -T_{ww--} \quad &= \frac{m_t E}{v^2}\Big[1 + (1 + c_\theta)(2\kappa_L^{CC} + (\kappa_L^{CC})^2 + (\kappa_R^{CC})^2) \\
&\quad -c_\theta(\kappa_L^{NC} + \kappa_R^{NC})\Big] - \frac{2E^3}{v^2}\frac{(a_{ww1} + 2a_m c_\theta)}{\Lambda}, \\
T_{ww+-} = \frac{2m_t^2 s_\theta}{(1 - c_\theta)v^2} &+ \frac{8E^2}{v^2}s_\theta\left(\frac{m_t}{\Lambda}a_m + \frac{\kappa_R^{NC} - (\kappa_R^{CC})^2}{8}\right), \\
T_{ww-+} = \frac{8E^2}{v^2}s_\theta &\left(\frac{m_t}{\Lambda}a_m + \frac{\kappa_L^{NC} - \kappa_L^{CC}(2 + \kappa_L^{CC})}{8}\right),
\end{aligned}
\tag{5}
$$

where $s_\theta = \sin\theta$; θ is the angle between the top quark and the W_L^+ boson in the $W_L^+ W_L^-$ CM system.

In Figs. 1 and 2, we show the number of $t\bar{t}$ pairs produced from $Z_L Z_L$ and $W_L^+ W_L^-$ fusion processes, respectively. Each Figure contains the number of events at the LC (an $e^- e^+$ collider with $\sqrt{s} = 1.5$ TeV and 200 fb^{-1} of integrated luminosity) and the LHC (a pp collider with $\sqrt{s} = 14$ TeV and 100 fb^{-1} of integrated luminosity). The rapidity of t is required to be within 2 ($|y^{t,b}| \le 2$), and its transverse momentum to be at least 20 GeV. We also require the invariant mass M_{VV} to be larger than 500 GeV so that we can apply the effective-W approximation [4]. For the LHC calculations we used the CTEQ3L parton distribution function with the factorization scale equal to the mass of the W boson [5].

As shown in the Figures, if the coefficients a's are of order 10^{-1} then a significant deviation from the SM prediction will be observed. On the other hand, if no significant deviation is observed we can set a constraint on a. For example, the SM event rate from $W_L^+ W_L^-$ fusion at the LC is equal to 400 $t\bar{t}$ pairs (cf. Fig. 2, $a = 0$); for $a_{ww1} = \pm 0.05$ the rate number is outside the range 400 ± 40 (i.e. a 2σ deviation). This means that if the observed production rate is consistent with the SM prediction at the 95% C.L. we can constrain $| a_{ww1} | < 5 \times 10^{-2}$. This bound is one order of magnitude more stringent than the bounds for the next-to-leading order bosonic operators commonly studied in $V_L V_L \to V_L V_L$ scatterings [6]. Therefore, the production of top quarks via $V_L V_L$ fusions at the LHC and the LC should be carefully studied when data is available because it can be sensitive to the electroweak symmetry breaking mechanism, even more than the commonly studied $V_L V_L \to V_L V_L$ processes in some models of strong dynamics.

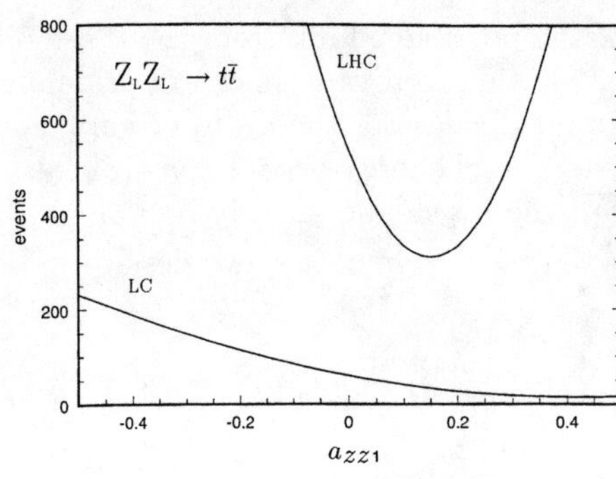

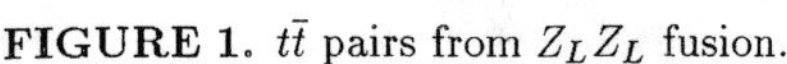

FIGURE 1. $t\bar{t}$ pairs from $Z_L Z_L$ fusion.

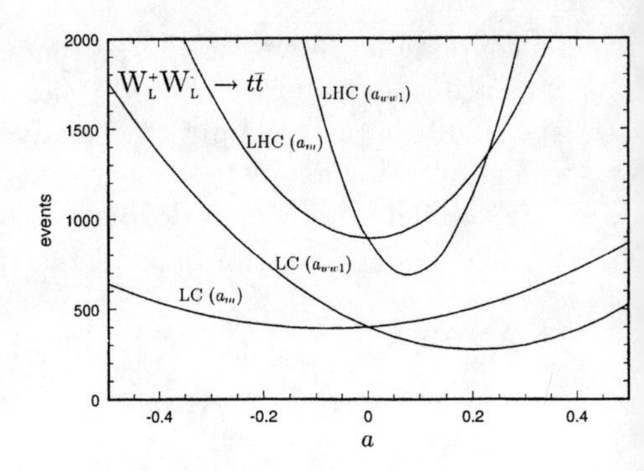

FIGURE 2. $t\bar{t}$ pairs from $W_L^+ W_L^-$ fusion

ACKNOWLEDGMENTS

F. Larios was supported in part by the Organization of American States, and by the Sistema Nacional de Investigadores. CPY was supported in part by the NSF grant No. PHY-9507683. We want to thank C. Balazs for helping with the figures.

REFERENCES

1. F. Larios and C.P.-Yuan, MSUHEP-60620, hep-ph/9606397, to appear on Phys. Rev. **D**, (1997).
2. C.–P. Yuan, published in Proc. of Workshops on Particles and Fields and Phenomenology of Fundamental Interactions, Puebla, Mexico, Nov. 1995;
   F. Larios, E. Malkawi and C.–P. Yuan, hep-ph/9609482, to appear on Acta Phys. Pol.
3. H. Georgi, "Effective Field Theory", in Annu. Rev. Nucl. Part. Sci. **43**, 209 (1993);
   A. Manohar and H. Georgi, Nucl. Phys. **B234**, 189 (1984).
4. S. Dawson, Phys. Lett. **B217**, 347 (1989);
   S. Cortese and R. Petronzio, Phys. Lett. **B276**, 203 (1992);
   I. Kuss and H. Spiesberger, Phys. Rev. **D53**, 6078 (1996).
5. H.L. Lai, J. Botts, J. Huston, J.G. Morfin, J.F. Owens, J.W. Qiu, W.K. Tung, H. Weerts, Phys. Rev. **D51**, 4763 (1995).
6. J. Bagger, V. Barger, K. Cheung, J. Gunion, T. Han, G.A. Ladinsky, R. Rosenfeld, and C.–P. Yuan, Phys. Rev. **D49**, 1246 (1994); and **D52**, 3878 (1995);
   V. Barger, J. F. Beacom, K. Cheung, T. Han, Phys. Rev. **D50**, 6704 (1993);
   H.-J. He, Y.-P. Kuang and C.-P. Yuan, Phys. Lett. **B382**, 149 (1996); and hep-ph/9611316, to appear in Phys. Rev. **D**, (1997).

VMD Approach to $\tau^- \to (\omega, \phi)\pi^-\nu_\tau$ Decays

D. A. López Falcón* and G. López Castro†

*Departamento de Física Aplicada, Centro de Investigación y Estudios Avanzados del IPN, Unidad Mérida, Apdo. Postal 73, Cordemex, 97310 Mérida, Yucatán, MEXICO.
†Departamento de Física, Centro de Investigación y Estudios Avanzados del IPN, Apdo. Postal 14-740, 07000 México, D.F., MEXICO.

Abstract. We give a description of $\tau^- \to (\omega, \phi)\pi^-\nu_\tau$ decays in the approach of the vector-meson dominance (VMD) model and compare our results with the conserved vector current (CVC) predictions and experimental data.

From a pure phenomenological point of view, the decays $\tau^- \to (\omega, \phi)\pi^-\nu_\tau$ are interesting because they can provide a test for the anomalous vertices $VV'P$ in an energy region larger than tested in light meson decays. At a more theoretical level, these decays are interesting because they could provide a test for second class axial currents [1] and violation of the OZI rule [2].

While the $\tau^- \to \omega\pi^-\nu_\tau$ decay has been measured in several experiments [3,4,5], only some upper limits are available for the $\tau^- \to \phi\pi^-\nu_\tau$ [3,6] decay mode. Here we present the results obtained using the VMD model [7,8] and give a comparison with the predictions based on the CVC hypothesis and present experimental data.

The squared mass distribution of the $V\pi$ hadronic system in $\tau^- \to (\omega, \phi)\pi^-\nu_\tau$ decays can be written as [8]

$$\frac{d\Gamma}{ds} = \frac{G_F^2|V_{ud}|^2}{1536\pi^3 m^3}|F_V(s)|^2\frac{(m^2-s)^2}{s^2}(m^2+2s)(s^2-2s\Sigma^2+\Delta^4)^{3/2}, \qquad (1)$$

where $\Sigma^2 \equiv m_V^2 + m_\pi^2$ and $\Delta^2 \equiv m_V^2 - m_\pi^2$. The masses of τ, the vector-meson and π^- are denoted by m, m_V and m_π, respectively. The form factor describing the weak vertex of the hadronic system is defined as

$$< V\pi|\bar{u}\gamma_\mu d|0 >= iF_V(s)\epsilon_{\mu\alpha\beta\gamma}\varepsilon^{*\alpha}q_1^\beta q_2^\gamma, \qquad (2)$$

where $q_{1,2}$ denote the four-momenta of V and π^-; ε^* is the polarization four-vector of V, and $s = (q_1+q_2)^2$.

CP400, *First Latin American Symposium on High Energy Physics/ VII Mexican School of Particles and Fields*, edited by D'Olivo/Klein-Kreisler/Mendez

Another observable associated to $\tau^- \to (\omega, \phi)\pi^-\nu_\tau$ decays is the spectral function $v(s)$ which is defined as,

$$v(s) = \frac{32\pi^2 m^3}{G_F^2|V_{ud}|^2(m^2-s)^2(m^2+2s)}\frac{d\Gamma}{ds}. \tag{3}$$

In the VMD model, the form factor is given by

$$F_V(s) = \Sigma_{V'=\rho,\ldots}\frac{g_{V'}g_{V'V\pi^-}}{m_{V'}^2 - s - im_{V'}\Gamma_{V'}}, \tag{4}$$

where $g_{V'}$ denotes the coupling constant describing the $W^- - V'$ junction and $g_{V'V\pi^-}$ is the $V'V\pi^-$ coupling constant.

As is shown in [4], the experimental data on the spectral function of $\tau^- \to \omega\pi^-\nu_\tau$ requires the presence of at least one ρ' in addition to the $\rho(770)$. Including these two resonances $(\rho+\rho')$, the form factor can be written as follows

$$F_V(s) = \frac{g_{\rho^-}g_{\rho V\pi^-}}{m_\rho^2 - s - im_\rho\Gamma_\rho}\left\{1 + \alpha_V\frac{m_\rho^2 - s - im_\rho\Gamma_\rho}{m_{\rho'}^2 - s - im_{\rho'}\Gamma_{\rho'}}\right\}, \tag{5}$$

where $\alpha_V \equiv g_{\rho'}g_{\rho'V\pi}/g_\rho g_{\rho V\pi}$.

The different parameters entering the previous equation can be fixed in the case of $\tau^- \to \omega\pi^-\nu_\tau$ by using the available data on the spectral function and light mesons decays. For this purpose we use $g_{\rho^-} = \sqrt{2}m_\rho^2/\gamma_\rho = (0.166 \pm 0.005)$ GeV^2 [7,9], and $g_{\rho^+\omega\pi^-} = (13.5 \pm 2.5)$ GeV^{-1}, which cover the range of values for this coupling as extracted from $\rho^{-,0} \to \pi^{-,0}\gamma$, $\omega \to \pi^0\gamma$ and $\pi^0 \to \gamma\gamma$ using the isospin symmetry relation $g_{\rho^-\omega\pi^-} = g_{\rho^0\omega\pi^0}$. Using these couplings, we have fitted the data of Ref. [4] using the $\rho(1450)$ or the $\rho(1700)$ in addition to the $\rho(770)$, leaving α_ω as a free parameter and allowing an overall normalization factor N of the spectral function.

The fit to the data of Ref. [4] in this model is better if the $\rho(770) + \rho(1450)$ combination is chosen. In this case the best fit gives

$$\alpha_\omega = -0.37 \pm 0.14 \tag{6}$$

$$N = 1.22 \pm 0.34\ . \tag{7}$$

Using the above results into Eqs. (5) and (1) we obtain [8]

$$B(\tau^- \to \omega\pi^-\nu_\tau) = \begin{cases} (1.22 \pm 0.56)\%\ if\ N = 1,\ and \\ (1.49 \pm 0.80)\%\ if\ N = 1.22 \end{cases} \tag{8}$$

where the uncertainty reflects the errors in $g_{\rho\omega\pi}$, g_ρ, α_ω and N. The large uncertainty above is dominated by the error we have attributed to $g_{\rho\omega\pi}$.

TABLE 1. Experimental data and predictions based on VMD and CVC

DECAY	BR_{EXP} (%)	BR_{CVC} [10] (%)	BR_{VMD} (%)
$\tau^- \to \omega\pi^-\nu_\tau$	1.91 ± 0.09 [3]	1.75 ± 0.19	1.49 ± 0.80
$\tau^- \to \phi\pi^-\nu_\tau$	$< (1.2 \sim 2.0) \times 10^{-2}$ [6]	$< 3 \times 10^{-2}$	$(1.20 \pm 0.48) \times 10^{-3}$

In order to predict the branching ratio for $\tau^- \to \phi\pi^-\nu_\tau$, we can rely on the same model for $F_\phi(s)$ but in this case we will assume that $\alpha_\phi = \alpha_\omega$ [8], which would mean that

$$\frac{g_{\rho\phi\pi}}{g_{\rho\omega\pi}} = \frac{g_{\rho'\phi\pi}}{g_{\rho'\omega\pi}}, \tag{9}$$

because the spectral function of $\tau^- \to \phi\pi^-\nu_\tau$ is very difficult to measure given the low branching ratio for this decay. Note that the previous equation can be obtained if we assume a $U(3)$ invariant coupling for the $V'VP$ vertex and we replace the 1^3S_1 nonet of vector mesons (the ρ) by the 2^3S_1 nonet of vector mesons (the ρ') [8] (we use the spectroscopic notation $n^{2S+1}L_J$).

On the other hand, the $\rho^+\phi\pi^-$ coupling can be extracted from $\phi \to \rho\pi$. This gives [8]:

$$g_{\rho^+\phi\pi^-} = (1.10 \pm 0.03)\ \text{GeV}^{-1}. \tag{10}$$

Using the preceding value into Eqs. (5) and (1) and relying on the above assumptions, we obtain:

$$B(\tau^- \to \phi\pi^-\nu_\tau) = (1.20 \pm 0.48) \times 10^{-5}, \tag{11}$$

where the large error is dominated by the error in α_ϕ.

The results for the branching ratios obtained in the VMD model, Eqs. (8) and (11), are compared in Table 1 with recent measurements and the predictions based on the CVC hypothesis by using the $e^+e^- \to V\pi^0$ experimental data [10]. We observe that both, the predictions based on CVC and on the VMD model, are in agreement (within errors) with present experimental data, althought central values differ substantially for the $\tau^- \to \omega\pi^-\nu_\tau$ case. Since our results relies strongly on the measured spectral function of $\tau^- \to \omega\pi^-\nu_\tau$, the difference in central values between the VMD model and the CVC predictions would indicate an inconsistency (in normalization) between the results of Ref. [4] and the data on $e^+e^- \to \omega\pi^0$ used to get the CVC prediction.

On the other hand, the VMD model allows to derive an absolute value for the $\tau^- \to \phi\pi^-\nu_\tau$ branching ratio while only an upper limit can be quoted from the CVC hypothesis mainly because of large uncertainties in the measurements of the $e^+e^- \to \phi\pi^0$ cross section.

REFERENCES

1. C. Leroy and J. Pestieau: Phys. Lett. **B72**, (1978) 398;
   S. Weinberg: Phys. Rev. **112**, (1958) 1375.
2. S. Okubo, Phys. Lett. **5**, (1963) 165; G. Zweig, CERN Report no. S419/TH, (1964) 412; J. Iizuka, Prog. Theor. Phys. Suppl. **37-8**, (1966) 21.
3. R. Barnet *et al.*, *Particle Data Group*, Phys. Rev. **D54**, Part I, (1996).
4. H. Albrecht *et al.*, ARGUS Collaboration, Phys. Lett. **B185**, (1987) 223.
5. R. Ballest *et al.*, CLEO Collaboration, Phys. Rev. Lett. **75**, (1995) 3809;
   D. Buskulic *et al.*, ALEPH Collaboration, preprint CERN-PPE-96-103 (1996).
6. R. D. Kass, talk given at the IV International Workshop on Tau lepton physics, 16-19 september 1996, Colorado USA.
7. R. Decker, Z. Phys. **C36**, (1987) 487.
8. G. López Castro and D.A. López Falcón, Phys. Rev. **D54** (1996) 4400.
9. L. Okun, *Lepton and quarks*, (North Holland, Amsterdam, 1982), p. 109.
10. S. I. Eidelman and V. I. Ivanchenko, talk given at the IV International Workshop on Tau lepton physics, 16-19 september 1996, Colorado USA.

Chiral Sum Rules and Hadronic Models

Carlos A. Ramirez

Depto. de Física,
Universidad Industrial de Santander,
A. A. 678, Bucaramanga, Colombia

Abstract. Several Models (mainly two: Vector Meson Dominance and Quark Models) are presented. Their predictions for the Spectral Functions, and the corresponding Sum Rules are obtained. Comparison with the experimental data and QCD constrains is done.

I INTRODUCTION

The two point functions are defined as [1]

$$\delta_{ab}\Pi_{\rm A}^{\mu\nu}(q^2) = \delta_{ab}\left[(q_\mu q_\nu - q^2 g_{\mu\nu})\Pi_A(q^2) - q_\mu q_\nu \frac{F_\pi^2}{q^2+i\epsilon}\right]$$

$$= \delta_{ab}\left[(q^\mu q^\nu - q^2 g^{\mu\nu})\int_0^\infty \mathrm{d}s\frac{\rho_A(s)}{s-q^2-i\epsilon} - \frac{q^\mu q^\nu}{q^2+i\epsilon}F_\pi^2\right] \qquad (1)$$

and similarly for Π_V without the F_π^2 term. $V_\mu^a = \bar{q}T_a\gamma_\mu q$, $A_\mu^a = \bar{q}T_a\gamma_\mu\gamma_5 q$, with $2T_a = \tau_a$ the Pauli matrices and q are the quark doublets of Isospin. We have that $\Pi_{V,A}(q^2) = \int_0^\infty \mathrm{d}s\rho_{V,A}(s)/(s-q^2-i\epsilon)$ and $\rho_{V,A} = \Im\Pi_{V,A}(s)/\pi$, in the spectral representation.

There are several theoretical constrains on the spectral functions. Its asymptotic behavior can be extracted from QCD, in the high and in the low energy limits. For the first case we have that [2] $\rho_{V,A}(s) \to [1+\frac{\alpha_s(s)}{\pi}]/8\pi$ and $\rho_V(s) - \rho_A(s) \to (0.18\mathrm{GeV})^6/s^3$. At low energy the QCD predictions can be obtained by using the Chiral Lagrangian techniques [3]:

$$\rho_V(s) \to \frac{1}{48\pi^2}\left(1-\frac{m_\pi^2}{s}\right)^{3/2}(1+4L_9^r s/F_\pi^2)^2\theta(s-4m_\pi^2)$$

$$\rho_A(s) \to \frac{s}{96\pi^2(4\pi F_\pi)^2}\cdot\sqrt{(1-3m_\pi^2/s)^2 - 4m_\pi^2/s}\cdot\theta(s-9m_\pi^2) \qquad (2)$$

CP400, *First Latin American Symposium on High Energy Physics/ VII Mexican School of Particles and Fields,*
edited by D'Olivo/Klein-Kreisler/Mendez

Besides the former limits the spectral functions have to obey the following sum rules [2,4]:

$$\text{W0}: \int_0^\infty ds \frac{\rho_V(s) - \rho_A(s)}{s} = -4L_{10} + \mathcal{O}(m_\pi^2)$$

$$\text{W1}: \int_0^\infty ds[\rho_V(s) - \rho_A(s)] = F_\pi^2 + \mathcal{O}(m_\pi^4)$$

$$\text{W2}: \int_0^\infty ds\ s[\rho_V(s) - \rho_A(s)] = 0 + \mathcal{O}(m_\pi^2)$$

$$\text{W3}: \int_0^\infty ds\ s\log(s/\Lambda)[\rho_V(s) - \rho_A(s)] = \frac{4\pi F_\pi^2}{3\alpha}(m_{\pi^0}^2 - m_{\pi^+}^2) + \mathcal{O}(m_\pi^2) \quad (3)$$

The experimental data for the spectral functions has been analyzed by Peccei and Sola first and then by Donoghue and Golowich [5].

Many models have been used to describe them: Vector Meson Resonance (VMD) models [6], Quark Models (Free Constituent Quark, or Chiral Quark Models) [7], Nonrelativistic quark models [8], NJL [9], etc. In the present work we want to review the first two, in several versions.

II VECTOR MESON DOMINANCE MODELS

VMD [6] models provide a very useful fitting for the data at the resonance region. The simplest case is the narrow resonance approximation, where the spectral functions are given by $\rho_V = \sum_i F_{V_i}^2 \delta(s - m_{V_i}^2)$ and similarly for ρ_A. For this case the sum rules can be satisfied but the spectral functions are incorrect. The constants $F_V = (154 \pm 3)$ MeV (for the $\rho(770)$) and $F_A = (123 \pm 25)$ MeV (for the $a_1(1230)$) can be obtained from the partial widths [10]. There are not experimental data for the higher resonances so $F_n = F_0\sqrt{m_0/M_n}$ [11] is going to be used. This relation is expected to be valid for heavy quarks and maybe for light ones too.

Here we want to analyze three cases for the narrow approximation (see table): The first one is studied in the old literature where the KSRF relation [12] $F_V^2 = 2F_\pi^2$ is used together with W1 and W2 to obtain $L_{10} = -(3/8) \cdot (F_\pi/m_V)^2$ and $m_{\pi^+}^2 - m_{\pi^0}^2 = (3\alpha/2\pi) \cdot m_\rho^2 \log(2)$. For the second case we take into account the first two resonances (ρ and a_1) together with the experimental values for F_V and F_A. And for the third one we take into account the first four resonances . In this case the values for the unknown constants are taken from the NRQM (see next section)

A more realistic model is obtained by the Breit-Wigner parametrization:

$$\rho_V = \frac{1}{\pi} \frac{F_V^2 m_V \Gamma_V}{(s - m_V)^2 + (m_V \Gamma_V)^2} \to F_V^2 \delta(s - m_V^2) \quad (4)$$

and similarly for ρ_A. In this case the spectral functions can be described at least in an approximate sense, as it is the case case of the $\rho(770)$. For the a_1 it does not work because the shape of the resonance is not symmetric and the Self-Energies have to be included. The results for the sum rules are not very different from those of narrow approximation case.

Coeficient	$\rho(770)$, $a_1(1230)$ and KSRF relat.	$\rho(770)$, $a_1(1230)$	4 Resonan. NRQM	Exper.
-$L_{10}[10^{-3}]$	5.6	7.5	7.9	6.8 ± 0.3
F_π [MeV]	-	92	95	94
W2[MeV^4]	-	$-(308)^4$	$-(344)^4$	0
Δm_π [MeV]	5.23	9.82	7.97	4.593(3)

Table: Results for the Sum Rules in the narrow resonance approximation, for the cases described in the text.

III QUARK MODELS

In the simplest model [7] the spectral functions are described by a quark loop and the result is

$$\rho_V = \frac{1}{8\pi^2}\frac{N_C}{3}\left(1 + \frac{2m_q^2}{s}\right)\sqrt{1 - \frac{4m_q^2}{s}}$$

$$\rho_A = \frac{1}{8\pi^2}\frac{N_C}{3}\left(1 - \frac{4m_q^2}{s}\right)^{3/2} \tag{5}$$

This crude result does not predict any resonance behavior but it describes the average behavior. The predictions for the sum rules are: $L_{10} = -N_C N_D/96\pi^2 = -12.7 \cdot 10^{-3}$, $F_\pi = (N_C N_D/4\pi)m_q^2 \log(\Lambda^2/m_q^2) \simeq 80$ MeV, (with $\Lambda \simeq \Lambda_{\text{ChPT}} \simeq 1$ GeV. For the other two sum rules the predictions depend strongly on m_q and Λ. Finally the large s behavior agrees with the QCD prediction for ρ_V, $\rho_A \to 1/8\pi^2$ but not for $\rho_V - \rho_A$.

One step further consists into include the quark-quark interaction. According to the Nonrelativistic Quark Model [8] $F_V^2 = 8\pi N_C|R(0)|^2/m_V$ and $F_A^2 = 72 N_C |R_p'(0)|^2/3\pi m_A^3$, where $R(r)$ is the radial part of the wave function satisfying the Schrödinger equation

$$\left(\nabla^2 - m_q[V(r) - E]\right)\psi(\vec{r}) = 0 \tag{6}$$

with $V(r) = \int \frac{d^3p}{(2\pi)^3} V(p) e^{ip\cdot r}$, $V(p) = 4\pi\alpha_s(s)/\vec{p}^2$.

This model can predict the spectra, and the vector and axial couplings. The spectra for the V-s and A-s states can be obtained by using the potential $V(r) = -Fr + V_0$ The best fitting for the spectra [10] is obtained for $V_0 + 2m_q \simeq$ 210 MeV and $(F^2/m_q)^{1/3} \simeq 286$ MeV. The third data needed can be obtained from F_V. The parameters thus obtained are $F \simeq 0.1$ GeV, $m_q \simeq 0.4$ MeV and $V_0 = -390$ MeV that are close to the accepted values [8]. Finally $F_a = 134$ MeV is predicted and agrees with the experimental value.

To conclude it can be said that the VMD and NRQM predictions for the spectral functions and sum rules agree with the experimental data, however the NRQM is able to predict the spectra and the Vector and Axial coupling constants (but is not clear what are the predictions for the widths). Besides the NRQM is closer to QCD and is obtained from it at least in some approximate sense. The VMD does not have the same prediction power but it is only a parametrization of the physics near the resonance region.

REFERENCES

1. Barton, G., *Introduction to Dispersion Techniques in Field Theory*, W. A. Benjamin 1965.
2. Shifman M., Vainshtein A., and Zakarov V., Nucl. Phys. **B147**, 385, 448 and 519 (1979). Narison, S., *QCD Spectral Sum Rules*, World Scientific 1989.
3. Gasser J., and Leutwyler H., Ann. Phys. (N. Y.) **158**, 142 (1984); Nucl. Phys. **B250**, 495 (1985).
4. Weinberg, S., Phys. Rev. Lett **18**, 188 and 507 (1967). Das T., et al., Phys. Rev. Lett. **18**, 759 (1967). Das T., et al., **18**, 761 (1967); **19**, 859 (1967).
5. Peccei R., and Solá J., Nucl. Phys. **B281**, 1 (1987). Donoghue J., and Golowich E., Phys. Rev. **D49**, 1513 (1994). Donoghue J., and Holstein B., Phys. Rev. **D46**, 4076 (1992).
6. Sakurai, J., *Currents and Mesons*, U. of Chicago Press 1969. Meissner, V., Phys. Rep. **161**, 213 (1988).
7. Manohar A., and Georgi H., Nucl. Phys. **B234**, 189 (1984). Ball, R., *Chiral Gauge Theory*, Phys. Rep. **182**, 1 (1989).
8. Van Royen R., and Weisskopt V., Il Nouv. Cim. **L A**, 617 (1967). Le Yaouanc A., et al., *Hadron Transitions in the Quark Model*, Gordon Breach 1988.
9. Nambu Y., Jona-Lasinio G., Phys. Rev. **122**, 345 (1961); **124**, 246 (1961).
10. Particle Data Group, *Review of Particle Properties*, Phys. Rev. **D54**, 1 (1996).
11. Donoghue, J., UMHEP-355 (University of Massachusetts, Amherst preprint)
12. Kawarabayashi K., and Suzuki M., Phys. Rev. Lett. **16**, 255 (1966). Riazuddin, and Fayyazuddin, Phys. Rev. **147**, 1071 (1966).

S- and P-wave Kaon-Pion Phase Shifts from Chiral Perturbation Theory

J. Sá Borges[1], J. Soares Barbosa[2], and V. Oguri[3]

Instituto de Física, Universidade do Estado do Rio de Janeiro
Rua São Francisco Xavier 524, Maracanã
Rio de Janeiro. Brasil

Abstract. We fit, by adjusting the parameters L_1^r,L_2^r and L_3 of chiral perturbation theory (ChPT) amplitude, the low energy kaon-pion S- and P-wave experimental phase-shifts. We get the K* resonance, positive and large isospin 1/2 S-wave and a negative and small 3/2 S-wave phase shifts, compatible with experimental data. We propose the present method as the best way to fix ChPT parameters. The unitarization program of current algebra is also discussed.

I INTRODUCTION

The strong interactions involving mesons were described, in the early sixties, by current algebra. In 1979, Weinberg [1] suggested that it is possible to summarize these previous results in a phenomenological Lagrangian that incorporates all the constraints coming from chiral symmetry of the underlying theory, namelly massless Quantum Chromodynamics (QCD). Following these ideas, Gasser and Leutwyller [2] have developed Chiral Perturbation Theory (ChPT), that allows one to compute many different Green functions involving low energy mesons as functions of lowest powers of their momenta, their masses and a few undetermined parameters. These parameters have to be obtained phenomenologically. In the present work, we adjust three of them in order to fit experimental kaon-pion phase shifts.

In two previous works, [3,4] we have related the unitarization procedure, formulated in connection with the Ward identity method of current algebra, to the recent approach of ChPT for meson scattering. As both methods give the same analytical structure for meson amplitudes, we concluded that ChPT

1) E-mail: saborges@vmesa.uerj.br
2) E-mail: jsoares@uerj.br
3) E-mail: oguri@vax.fis.uerj.br

CP400, *First Latin American Symposium on High Energy Physics/ VII Mexican School of Particles and Fields,*
edited by D'Olivo/Klein-Kreisler/Mendez

can be considered as a field theoretical implementation of unitarity on current algebra low energy meson-meson scattering amplitudes.

Chiral Perturbation Theory and the Unitarization Program of Current Algebra lead to amplitudes with few free parameters. In the unitarization program of current algebra, the free parameters were fixed by successfully fitting experimental phase-shifts data [5,6] whereas, in ChPT, they were previously determined [2] from pion-pion scattering length and others low energy experimental parameters.

In a recent paper, [7] we have shown how to fix the relevant parameters $\bar{\ell}_1$ and $\bar{\ell}_2$ of ChPT by fitting low energy pion-pion phase shifts. By using a suitable definition for phase shifts, we have been able to reproduce the ρ resonance with just one parameter and also to obtain a good adjust of experimental data for isospin zero S- wave with the remaining parameter. The isospin two S-wave has no free parameter. We have also analyzed, in pion-pion scattering, the sensitivity of the theoretical results by small variation of the parameters and we have concluded that the method can be considered as an interesting alternative to fix ChPT parameters.

In the present paper, we will follow the same procedure for kaon-pion scattering. We start with ChPT amplitude, that depends on seven parameters. We choose some of them to be zero and we adjust L_1^r,L_2^r and L_3 in order to fit low energy kaon-pion experimental phase-shifts. We propose this method as an interesting way to fix ChPT parameters.

II CHPT KAON-PION SCATTERING AMPLITUDE

We can express the ChPT isospin 3/2 $K\pi$ total amplitude obtained by Bernard et al. [8] in the following form: [4]:

$$T_{3/2}(s,t,u) = \tfrac{1}{2F^2}(M^2 - s) + \tfrac{1}{4F^4}(s - M^2)^2\bar{J}_{\pi K}(s)+$$

$$\tfrac{1}{24F^4}\left[(u - s)(t - 4m_\pi^2) + 3t(2t - m_\pi^2)\right]\bar{J}_{\pi\pi}(t) + \tfrac{m_\pi^2}{8F^4}(t - \tfrac{8}{9}m_K^2)\bar{J}_{\eta\eta}(t)+$$

$$\tfrac{1}{48F^4}\left[(u - s)(t - 4m_K^2) + 9t^2\right]\bar{J}_{KK}(t) + \tfrac{1}{32F^4}\Big[(t - s + \tfrac{m^4}{u})(u - 2M^2 + \tfrac{m^4}{u})-$$

$$(10u - 4M^2 - 3\tfrac{m^4}{u})\tfrac{m^4}{u} + 11u^2 - 12M^2u + 4M^4\Big]\bar{J}_{\pi K}(u)+$$

$$\tfrac{2}{F^4}(4L_1^r + L_3^r)(t - 2m_K^2)(t - 2m_\pi^2) + \tfrac{1}{F^4}4L_2^r(s - M^2)^2 + \tfrac{2}{F^4}(2L_2^r + L_3^r)(u - M^2)^2+$$

$$\tfrac{1}{32F^4}\Big[(t - s + \tfrac{m^4}{u})(u - 2\Sigma^2 + \tfrac{\Delta^4}{u}) + 3\Big(u^2 - m^2\Delta^2(2 - \tfrac{m^2\Delta^2}{u^2})\Big) -$$

$$4M^2(u - \tfrac{m^2\Delta^2}{u} - \tfrac{1}{3}M^2)\Big]\bar{J}_{K\eta}(u) + \tfrac{1}{4F^4}\tfrac{1}{32\pi^2}\Big[\tfrac{1}{6}tu + \tfrac{1}{6}st - \tfrac{1}{3}su + \tfrac{1}{3}m^4-$$

$$\tfrac{7}{2}t^2 + \tfrac{8}{9}m_\pi^2 m_K^2 - (t - s + \tfrac{m^4}{u})\Big(\tfrac{m_K^2 m_\eta^2}{\Delta^2}ln\tfrac{m_\eta^2}{m_K^2} + \tfrac{m_\pi^2 m_K^2}{m^2}ln\tfrac{m_K^2}{m_\pi^2} + \tfrac{1}{2}(M^2 + \Sigma^2)\Big)\Big] +$$

$$\frac{1}{4F^2}\left[\frac{(\mu_K-\mu_\pi)}{4m^2}\left(tu - su + m^4 + 8(s-M^2)^2 + 11u^2 - 12uM^2 + 4M^4\right)+\right.$$

$$\left.\frac{(\mu_\eta-\mu_K)}{4\Delta^2}\left(tu - su + m^4 + 3(u - \tfrac{2}{3}M^2)^2\right)\right]+$$

$$\frac{1}{16F^2}\left[\mu_\pi(10s - 10M^2 + 3m^2) - \mu_K(4s - 4M^2 + 2m^2) - \mu_\eta(6s - 6M^2 + m^2)\right]+$$

$$\frac{1}{4F^2}\left[\frac{\mu_\pi}{m_\pi^2}\left(\tfrac{1}{3}t(s-u) - t(2t - m_\pi^2)\right) + \frac{\mu_K}{m_K^2}\left(\tfrac{1}{6}t(s-u) - \tfrac{3}{2}t^2\right) + \frac{\mu_\eta}{m_\eta^2}m_\pi^2(\tfrac{8}{9}m_K^2 - t)\right]+$$

$$\frac{1}{F^4}\left[8L_4^r M^2 t - 2L_5^r(s-m^2)(M^2+m^2) + 4(2L_6^r + L_8^r - 2L_4^r)(M^4 - m^4)\right] \quad (II.1).$$

The combinations of the meson masses are:

$$m_\pi^2 + m_K^2 = M^2, \quad m_\pi^2 - m_K^2 = m^2, \quad m_K^2 + m_\eta^2 = \Sigma^2, \quad \text{and} \quad m_K^2 - m_\eta^2 = \Delta^2,$$

$$\text{and} \quad 16\pi^2 \bar{J}_{ab}(x) = 1 + \left(\frac{m_a^2 - m_b^2}{x} - \frac{m_a^2 + m_b^2}{m_a^2 - m_b^2}\right)\ln\frac{m_b}{m_a} - \frac{\nu(x)}{x}\ln - \frac{(x - m_a^2 + m_b^2 + \sqrt{\nu(x)})}{2m_a m_b},$$

$$\nu_{a,b}^2(x) = x^2 - 2(m_a^2 + m_b^2)x + (m_a^2 - m_b^2)^2, \quad \mu_P = \frac{1}{16\pi^2}\frac{m_P^2}{F_P^2}\ln\frac{m_P}{\mu}, \quad F_\pi = F_K = F.$$

At this point, we would like to emphasize that ChPT amplitude has the same structure as the first-order quasi-unitarized amplitude (QU1) presented in the appendix of ref. (6). As QU1 fits experimental data, we have decided to fix three ChPT parameters, directly related to the QU1 ones, by fitting the same experimental data.

From now on, we use (II.1) for isospin three-half amplitude and we obtain the isospin one-half amplitude from:

$$T_{1/2}(s,t,u) = \frac{3}{2}T_{3/2}(u,t,s) - \frac{1}{2}T_{3/2}(s,t,u)$$

We can re-obtain the so-called soft meson total isospin I Weinberg amplitudes, T^{ca} [9], by neglecting loop-corrections and fourth order terms in the original ChPT Lagrangian, namely:

$$T^{ca}_{I=3/2} = -\frac{1}{2F^2}(s - M^2) \quad \text{and} \qquad (II.2a)$$

$$T^{ca}_{I=1/2} = \frac{1}{4F^2}(t + 2s - 2u), \qquad (II.2b)$$

The usual definition of S- ($\ell = 0$) and P- ($\ell = 1$) partial wave are:

$$T_{I\ell} = \frac{1}{2}\int_{-1}^{1} T_I(s,t,u)P_\ell(x)dx, \quad t = \frac{\nu^2(s)}{2s}(x-1), \quad \text{and} \quad s + t + u = 2M^2. \qquad (II.3)$$

One can check that one loop ChPT S- and P- waves, in the physical region and for s < Σ^2, satisfy approximate unitarity relation:

$$ImT_{I\ell}(s) = \frac{1}{16\pi}\frac{\nu(s)}{s}T^{ca}_{I\ell}(s)^2,$$

where $T^{ca}_{I\ell}$ is the soft meson current algebra ℓ-partial-wave, with isospin I Weinberg amplitude obtained from eqs. (II.2):

In the next section we will fit the experimental phase shifts by adjusting ChPT parameters.

III PHASE-SHIFTS AND LOW ENERGY PARAMETERS

We would like to comment about phase-shift definition. As elastic unitarity is not exactly satisfied, the definition of partial-wave phase-shifts is arbitrary. Nevertheless, one can make suitable definitions. We have adopted the definition

$$\tan\delta_{I\ \ell} = \frac{Im\ T_{I\ell}}{Re\ T_{I\ell}} \qquad (II.4)$$

the authors of ref. (8) preferred adopting the definition $\delta_{\ell I} = \frac{\rho}{16\pi}\ Re\ T_{\ell I}$, which is valid for small δ.

The strategy to fit experimental phase shifts was the following. First, we have chosen L^r_4,L^r_5,L^r_6 and L^r_8 to be zero; this was because QU1 parameters are related only to L^r_1,L^r_2, and to L_3. Next, we have fitted the K* resonance with two parameters. The polynomial part of $P_{1/2}$ can be written as:

$$T^P_{1/2} = \frac{\nu^2(s)}{6s^2F^4}\left(\lambda_1(s^2+m^4)+\lambda_2(s^2-m^4)\right)$$

where λ_1 and λ_2 are the combinations:

$$\lambda_1 = -8L^r_1 - 2L_3 \quad \text{and} \quad \lambda_2 = 4L^r_2 - L_3$$

We adjust L^r_2 to fit the experimental data for $S_{1/2}$. We obtain the $S_{3/2}$ amplitude without any free parameter. We present in the figures (1) and (2) the fit of the experimental data corresponding to the following values of the parameters:

$$L^r_1 = -.004963, \qquad L^r_2 = -.0034646 \qquad \text{and} \qquad L_3 = .010255.$$

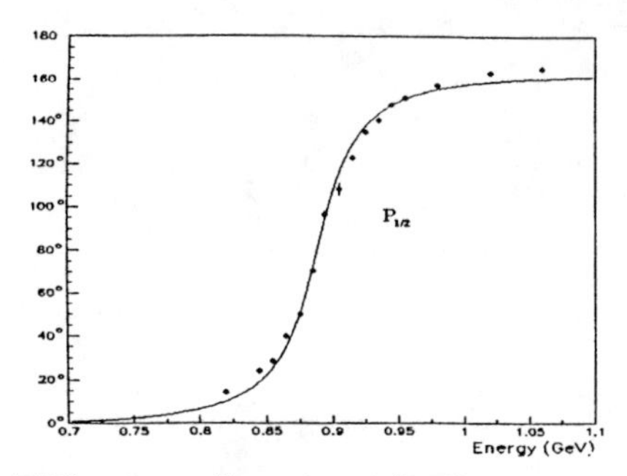

FIG. 1. Isospin 1/2 P-wave phase-shifts. The experimental points are from ref. 10. The parameters used in order to fit the K* (892) resonance are: $\lambda_1 = .0192$, $\lambda_2 = -.0241$

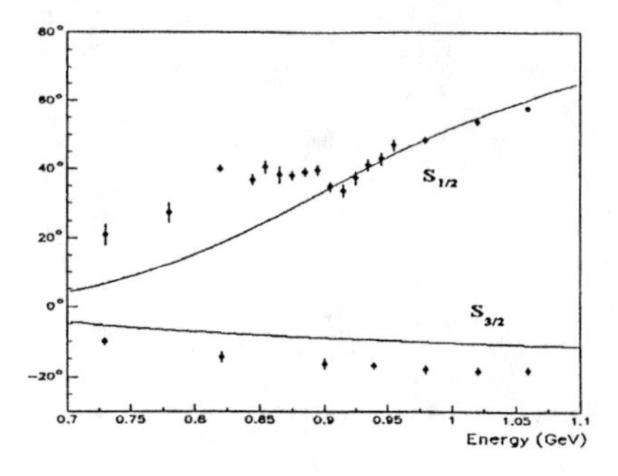

FIG. 2. Isospin 1/2 and 3/2 S-wave phase-shifts. The experimental data are from ref. 10. The parameter used in order to fit the $S_{1/2}$ is L_2^r =-.00346

IV CONCLUSION

In this paper, following the same method used [7] to fix ChPT parameters $\bar{\ell}_1$ and $\bar{\ell}_2$, we have fixed three ChPT parameters L_1^r, L_2^r and L_3 by using kaon-pion S- and P- wave phase-shifts. As a result we obtain the K* resonance, large and positive $S_{1/2}$ wave and negative and small $S_{3/2}$ phase-shifts. This shows that ChPT can handle the main feature of $K - \pi$ scattering up to the resonance region, what appeared to be a difficult task [8,11]. The application of the present method to other processes will finally give a coherent set of parameters.

REFERENCES

1. S. Weinberg, Physica A 96 (1979)327;
2. J.Gasser and H. Leutwyler, Ann. Phys. 158 (1984)142; Nucl. Phys. B250(1985)465;
3. J. Sá Borges, Phys. Lett. B 149, (1984) 21;
4. J. Sá Borges and F.R. Simão, Phys. Rev. D 53 (1996) 4806.
5. J. Sá Borges, Nucl. Phys. B 51(1973)189;
6. J. Sá Borges, Nucl. Phys. B 109 (1976) 357;
7. J. Sá Borges, J. Soares Barbosa and V. Oguri, preprint IF-UERJ-22/96, to appear in Phys. Lett. B.
8. V. Bernard, N. Kaiser and U.G. Meißner, Nucl. Phys. B357 (1991) 131.
9. S. Weinberg, Phys. Rev. Lett. 17 (1966) 616;
10. P. Estrabrooks and A. D. Martin, Nucl. Phys. B 79 (1974) 301;
11. L. Beldjoudi, T. N. Truong, Phys. Lett B 351 (1995) 357.

Limits on Trilinear Gauge Boson Couplings at DØ

A. Sánchez-Hernández

(Representing the DØ Collaboration)
Depto. de Física, CINVESTAV
Apdo. postal 14-740, 07000 México, D.F.

Abstract.
Measurements of trilinear gauge boson couplings are reported based on the direct observation of diboson final states produced in $p\bar{p}$ collisions at $\sqrt{s} = 1.8$ TeV with the DØ detector at Fermilab. The limits on the anomalous couplings were obtained at 95% CL from the following processes: $p\bar{p} \to WW/WZ + X \to e\nu jj + X$, $p\bar{p} \to W\gamma + X \to l\nu\gamma + X$ $(l = e, \mu)$, and $p\bar{p} \to Z\gamma + X \to l\bar{l}\gamma + X$ $(l = e, \nu)$.

INTRODUCTION

The self-couplings of the gauge bosons are completely fixed by the $SU(2) \times U(1)$ symmetry of the Standard Model (SM). The trilinear couplings appear as the three gauge boson vertices and can be measured by studying the gauge boson pair production processes. The measurement of the coupling parameters is one of the few remaining crucial test of the SM. Deviations of the couplings from the SM values signal new physics.

The WWV ($V = \gamma$ or Z) vertices are described by a general effective Lagrangian [1] with two overall couplings ($g_{WW\gamma} = -e$ and $g_{WWZ} = -e \cdot \cot\theta_W$) and six dimensionless coupling parameters g_1^V, κ_V and λ_V, where $V = \gamma$ or Z, after imposing C, P and CP invariance. g_1^γ is restricted to unity by electromagnetic gauge invariance. The genaral Lagrangian is reduced to the SM Lagrangian by setting $g_1^\gamma = g_1^Z$, $\kappa_V = 1$ ($\Delta\kappa_V \equiv \kappa_V - 1 = 0$) and $\lambda_V = 0$. The cross section with the non-SM coupling parameters grows with $\hat{s}$. In order to avoid unitarity violation, the coupling parameters are modified by form factors with a scale Λ; $\lambda_V(\hat{s}) = \frac{\lambda_V}{(1+\hat{s}/\Lambda^2)^2}$ and $\Delta\kappa_V(\hat{s}) = \frac{\Delta\kappa_V}{(1+\hat{s}/\Lambda^2)^2}$.

The $Z\gamma V$ ($V = \gamma$ or Z) vertices are described by a general vertex function [2] with eight dimensionless coupling parameters h_i^V ($i = 1, 4; V = \gamma$ or Z). In the SM, all h_i^V's are zero. The form factors for these vertices, which are required to

CP400, *First Latin American Symposium on High Energy Physics/ VII Mexican School of Particles and Fields,*
edited by D'Olivo/Klein-Kreisler/Mendez

constrain the cross sections within the unitarity limit, are $h_i^V(\hat{s}) = \frac{h_{i0}^V}{(1+\hat{s}/\Lambda^2)^n}$, where $n = 3$ for $i = 1, 3$ and $n = 4$ for $i = 2, 4$.

The data presented here were collected with the DØ detector during the Tevatron collider runs 1A (1992-1993) and 1B (1993-1995) at Fermilab. Limits on the anomalous coupling parameters were obtained at a 95% CL from the following processes: $p\bar{p} \to WW/WZ + X \to e\nu jj + X$, $p\bar{p} \to W\gamma + X \to l\nu\gamma + X$ ($l = e, \mu$), and $p\bar{p} \to Z\gamma + X \to l\bar{l}\gamma + X$ ($l = e, \nu$) at $\sqrt{s} = 1.8$ TeV.

WW/WZ ANALYSIS

The WW/WZ candidates were selected by searching for events containing an isolated electron with high E_T ($E_T > 25$ GeV), large missing transverse energy $\not{E}_T$ ($\not{E}_T > 25$ GeV) and two high E_T jets from ~ 76 pb^{-1} of data taken during 1993-1995 Tevatron collider run. The transverse mass of the electron and neutrino system was requiered to be consistent with a W boson decay ($M_T > 40$ GeV/c^2). The invariant mass of the two jet system was required to be $50 < m_{jj} < 110$ GeV/c^2, as expected for a W or Z decay. It was also required that the p_T of the two gauge bosons was balanced ($|p_T(jj) - p_T(e\nu)| < 40$ GeV/c) as expected for WW/WZ production. The number of events that satisfied all of the requirements was 399. There were two major sources of background for this process, QCD multijet events with a jet misidentified as an electron and W boson production with two associated jets. Total number of background events was estimated to be 387±38. The SM predicts 16±3 events for the above requirements and thus no significant deviation from the SM prediction was seen. A maximum likelihood fit to the p_T^W spectrum, calculated from the E_T of electron and missing E_T, was performed to set limits on the anomalous couplings. Assuming the WWZ couplings and the $WW\gamma$ couplings are equal and using $\Lambda = 1.5$ TeV, the following limits at 95% confidence level (CL) were obtained, as shown in Fig. 1a: $-0.56 < \Delta\kappa < 0.75$ (with $\lambda = 0$) and $-0.42 < \lambda < 0.44$ (with $\Delta\kappa = 0$).

Different assumptions for the relationship between the WWZ couplings and the $WW\gamma$ couplings were examined. The limits in Fig. 1b were obtained using the HISZ [3] relations. In Fig. 1c limits on the WWZ couplings are shown under the assumption that the $WW\gamma$ couplings take the SM values. In Fig. 1d limits on the $WW\gamma$ couplings are shown with the assumption that the WWZ couplings take the SM values. These plots indicate that this analysis is more sensitive to WWZ couplings.

The limits presented here have improved by a factor of two in comparison to our published results [4]. They are even tighter than our combined fit from Run 1A $WW\gamma/Z$ couplings measurements [5].

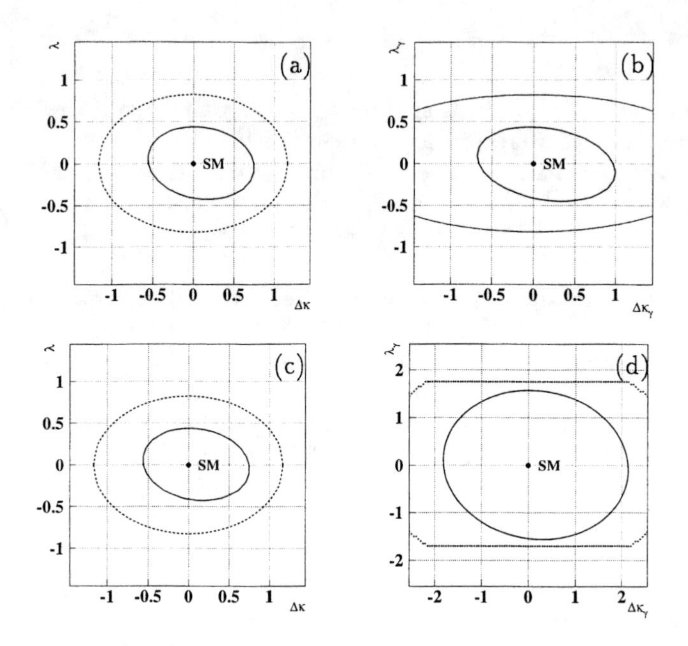

FIGURE 1. Countour limits on anomalous couplings at the 95% CL (inner curves) and limits from S-matrix unitarity (outer curves), assuming (a) $\Delta\kappa \equiv \Delta\kappa_\gamma = \Delta\kappa_Z$, $\lambda \equiv \lambda_\gamma = \lambda_Z$; (b) HISZ realtions; (c) SM $WW\gamma$ couplings, and (d) SM WWZ couplings. $\Lambda = 1.5$ TeV is used.

$W\gamma$ ANALYSIS

$W\gamma$ events from two decay modes of the W boson were studied: $W \to e\nu$ and $W \to \mu\nu$. In each case the photon was required to have a minimum transverse momentum of 10 GeV/c and to be spatially separated from the charged lepton by at least 0.7 units of $\mathcal{R}_{l\gamma}$, $\mathcal{R} \equiv \sqrt{(\Delta\eta)^2 + (\Delta\phi)^2}$ ($\eta = -\log\tan(\theta/2)$). We have observed $84.4^{+12.3}_{-11.3} \pm 8.7$ signal events from $\sim$ 89pb^{-1} of data taken during 1992-1993 and 1993-1995 Tevatron collider runs. The asymmetrical error is the 1σ uncertainty due to Poisson statistics, and the second error is due to the uncertainties in the background estimates.

¿From this observation we calculated the $W\gamma$ cross section times branching ratio of W bosons to leptons, for our photon requirements, to be: $\sigma(p\bar{p} \to W\gamma + X) \times$ BR$(W \to l\nu) = 11.8^{+1.7}_{-1.6} \pm 1.6(syst) \pm 1.2(lum)$ pb. This is in agreement with the SM prediction of 12.5 ± 1.0 pb.

A combined likelihood analysis of the p_T^γ spectra from the individual $W(e\nu)$ and $W(\mu\nu)$ analyses allowed us to set 95% CL limits on the anomalous $WW\gamma$ coupling parameters of $-0.98 < \Delta\kappa < 1.01$ and $-0.33 < \lambda < 0.31$. These are the 95% CL limits when only one of the couplings is allowed to vary at a time.

$Z\gamma$ ANALYSIS

Measurements of $Z\gamma$ production through the $ee\gamma$ and $\mu\mu\gamma$ decay channels with the DØ detector were previously reported [6]. Here we present a new measurement of $ee\gamma$ channel based on 89 pb^{-1} of data collected in 1993-1995 Tevatron collider run. We also present the first measurement of the $\nu\nu\gamma$ production at hadron colliders based on 13.5 pb^{-1} of data collected in the 1992-1993 run.

Event selection for the $ee\gamma$ analysis required two electrons with $E_T > 25$ GeV and a photon with $E_T > 10$ GeV. We additionally required that the photon was separated from either electron by at least 0.7 units in $\eta - \phi$ space. The above selection criteria yielded 14 candidate events with an estimated background of 1.6 ± 0.5 events, dominated by $Z + j$ and multijet production with jets faking the photon or electrons. This background was derived from data. The results agree well with the SM predictions of 12.1 ± 1.2 signal events. The following 95% CL limits on anomalous couplings (with assumption of one coupling being non-zero at a time) were obtained by a fit of the E_T^γ spectrum: $|h_{10,30}^V| < 1.8$, $|h_{20,40}^V| < 0.38$ for the form-factor scale of $\Lambda = 500$ GeV.

For the $\nu\nu\gamma$ analysis, we required a much tighter cut on the photon energy: $E_T > 40$ GeV which was forced by a dominant background from $W \rightarrow e\nu$ decays with the electron being misidentified for a photon due to inefficiency of the central tracker. Additional cuts were applied to the shape of the photon EM shower in transverse and longitudinal directions to ensure that it was consistent with a photon originating from a real vertex. The residual background, which had roughly equal contributions from $W \rightarrow e\nu$ decays and *bremsstrahlung* photons from cosmic and beam halo muons, was derived from data.

We observed 4 candidate events with an expected background of 5.8 ± 1.0. This is consistent with the SM prediction of 1.8 ± 0.2 events. Limits on anomalous couplings were set at 95% CL by the E_T^γ fit: $|h_{10,30}^V| < 0.87$, $|h_{20,40}^Z| < 0.21$, and $|h_{20,40}^\gamma| < 0.22$. This represents a factor of two improvement compared to the combined $ee\gamma$ and $\mu\mu\gamma$ limits [6] based on the same data.

REFERENCES

1. K. Hagiwara, R.D. Peccei, D. Zeppenfeld and K. Hikasa, *Nucl. Phys.* **B 282**, 253 (1987).
2. U. Baur and E. L. Berger, *Phys. Rev.* **D 41**, 1476 (1990).
3. K. Hagiwara, S. Ishihara, R. Szalapski and D. Zeppenfeld, *Phys. Rev.* **D 48**, 2182 (1993).
4. S. Abachi, *et.al.*, (DØ Collaboration), *Phys. Rev. Lett.* **77**, 3303 (1996).
5. T. Yasuda, "Double Boson Production at DØ", FERMILAB-Conf-96/239-E.
6. S. Abachi, *et.al.*, (DØ Collaboration), *Phys. Rev. Lett.* **75**, 1034 (1995).

Polarizabilities and Electromagnetic Decays of Hyperons in the Bound State Soliton Model

Carlos L. Schat[a], Carlo Gobbi[b] and Norberto N. Scoccola[a]

[a] *Physics Department, Comisión Nacional de Energía Atómica, Av.Libertador 8250, (1429) Buenos Aires, Argentina.*
[b] *University of Pavia, and INFN, Sezione di Pavia, via Bassi 6, I-27100 Pavia, Italy.*

Abstract. The radiative decays of hyperons [1] and their electric and magnetic static polarizabilities [2] are studied in the framework of the bound state soliton model. Detailed predictions for the total decay widths and the E2/M1 ratios corresponding to decuplet–to–octet electromagnetic transitions are presented in relation to future planned experiments at CEBAF and Fermilab. The results are compared to those obtained in quark based models.

Experiments at CEBAF and Fermilab will provide new and precise information about the electromagnetic decays of excited hyperons. The experimental situation for the electromagnetic polarizabilities, which are quantities of fundamental interest in the understanding of hadron structure, is also likely to change. With the advent of hyperon beams at FNAL and CERN hyperon polarizabilities will be measured soon. All these up coming experiments [5] have triggered a number of theoretical investigations in different hadron models. As an example, let us only mention two of those models and the problems they face when trying to describe the hyperon polarizabilities. Within the non relativistic quark model (NRQM) the large diamagnetic contribution to the magnetic polarizability is rather difficult to understand. In the case of heavy baryon chiral perturbation theory (HBCPT) predictions are not expected to be very accurate unless the contributions due to P-wave excitations (Δ-like), which are of higher order in the chiral expansion, are included. It is, therefore, interesting to attempt a description based on a completely different point of view, like the one given by the topological Skyrme soliton model.

In the bound state soliton model one starts with an effective $SU(3)$ chiral action

CP400, *First Latin American Symposium on High Energy Physics/ VII Mexican School of Particles and Fields*, edited by D'Olivo/Klein-Kreisler/Mendez

$$\Gamma = \int d^4x \Big\{ -\frac{f_\pi^2}{4}\,\mathrm{Tr}(L_\mu L^\mu) + \frac{1}{32\epsilon^2}\,\mathrm{Tr}[L_\mu, L_\nu]^2 \Big\} + \Gamma_{\mathrm{WZ}} + \Gamma_{\mathrm{sb}}\,. \tag{1}$$

Here Γ_{WZ} is the non–local Wess-Zumino action and Γ_{sb} is a symmetry breaking term. Their explicit form can be found elsewhere. In Eq.(1) the left current L_μ is expressed in terms of the chiral field U as $L_\mu = U^\dagger \partial_\mu U$.

Next, the Callan–Klebanov ansatz [3] is introduced: $U = \sqrt{U_\pi}\, U_K \sqrt{U_\pi}$, where

$$U_K = \exp\left[i \frac{\sqrt{2}}{f_K} \begin{pmatrix} 0 & K \\ K^\dagger & 0 \end{pmatrix} \right] , \quad K = \begin{pmatrix} K^+ \\ K^0 \end{pmatrix} , \tag{2}$$

and U_π is the soliton background field written as a direct extension to $SU(3)$ of the $SU(2)$ field u_π, i.e.,

$$U_\pi = \begin{pmatrix} u_\pi & 0 \\ 0 & 1 \end{pmatrix} , \tag{3}$$

with u_π being the conventional hedgehog solution $u_\pi = \exp[i\vec{\tau}\cdot\hat{r}F(r)]$.

According to the usual procedure, one expands up to second order in the kaon field. The Lagrangian density can therefore be rewritten as the sum of a pure $SU(2)$ Lagrangian and an effective Lagrangian describing the interaction between the soliton and the kaon fields. *In this picture strange hyperons arise as bound states of kaons in the background field of the soliton.* The octet and decuplet hyperons are obtained by populating the lowest kaon bound state. The gauged effective action can be more conveniently written as

$$\Gamma = \Gamma^{strong} + \Gamma^{lin} + \Gamma^{quad} , \tag{4}$$

where we have singled out the contributions linear and quadratic in the e.m. field A_μ:

$$\Gamma^{lin} = \int d^4x \; e \; A_\mu J^\mu \quad , \quad \Gamma^{quad} = -\int d^4x \; e^2 \; A_\mu \; G^{\mu\nu} \; A_\nu \, .$$

In this work we are interested in the radiative decays of the decuplet hyperons, $\Sigma^* \to \Lambda\,\gamma, \Sigma^* \to \Sigma\,\gamma, \Xi^* \to \Xi\,\gamma$. For all these processes both M1 and E2 transitions are allowed. The electric (α) and magnetic (β) static polarizabilities are defined through the shift in the particle's energy due to the presence of an external constant electric and magnetic field as:

$$\delta H = -\frac{1}{2}\,\alpha\, E^2 - \frac{1}{2}\,\beta\, B^2 \, . \tag{5}$$

As it is clear from the form of the interaction (4), there will be in principle two contributions to the static polarizabilities, one coming from the term quadratic in A^μ, known as the *seagull contribution* and one coming from second order

TABLE 1. Electric and magnetic polarizabilities (in $10^{-4} fm^3$) for the low lying octet hyperons.

particle	SET III			
	α_{tot}	β_s	β_d	β_{tot}
N	17.3	-8.3	5.6	-2.7
Λ	18.1	-8.7	12.1	3.4
Σ^0	18.1	-8.7	-4.0	-12.7
Σ^+	18.8	-9.1	10.4	1.3
Σ^-	17.4	-8.4	0.48	-7.9
Ξ^0	19.9	-9.6	14.0	4.4
Ξ^-	18.0	-8.7	1.5	-7.2

TABLE 2. Hyperon radiative decay widths (in keV) as calculated in the bound state soliton model. Also listed are the predictions of the non-relativistic quark model (NRQM) and bag model (BM).

decay	This work		NRQM	BM	EMP
	SET I	E2/M1 (in %)			
$\Sigma_0 \rightarrow \Lambda\gamma$	8	0	8.5	4.6	8.9 ± 0.7
$\Sigma_0^* \rightarrow \Lambda\gamma$	243	-4.56	273	152	
$\Sigma_0^* \rightarrow \Sigma_0\gamma$	19	0	18	15	
$\Sigma_+^* \rightarrow \Sigma_+\gamma$	91	-4.84	110		
$\Sigma_-^* \rightarrow \Sigma_-\gamma$	1	-57.7	2.5		
$\Xi_0^* \rightarrow \Xi_0\gamma$	148	-3.13	135		
$\Xi_-^* \rightarrow \Xi_-\gamma$	5	-17.8	3.2		

perturbation theory applied to the term linear in A^μ, the so called *dispersive contribution.*

We performed the numerical calculations adopting different sets of parameters, namely

$$\text{SET I} : f_\pi = 54 MeV,\ \epsilon = 4.84\ , \text{SET III} : f_\pi = 93 MeV,\ \epsilon = 4.26$$

We always take the empirical values $m_K = 495\ MeV$, $m_\pi = 138\ MeV$ and $f_K/f_\pi = 1.22$. For SET I the pion decay constant f_π and the Skyrme parameter ϵ are taken in order to fit the mass of the nucleon and the Δ. In the case of SET III we take the empirical value of f_π and adjust the Skyrme parameter ϵ to reproduce the empirical $\Delta - N$ mass difference. In Table 1 we report our results for the hyperon polarizabilities. In the electric case the seagull contribution is dominant, while in the magnetic case both seagull and dispersive contributions are relevant. The seagull contributions are always dominated by the purely solitonic terms and are responsible for the general pattern of the electric polarizabilities, where we obtain small splittings within the same set of parameters. The structure is richer in the magnetic polarizability case because of the interplay between a large negative seagull part with the relevant dispersive contribution. The results obtained by using baryon chiral perturbation are rather different from ours. In should be noticed that such a calculation does not include P-wave excitations (Δ-like) since they are of higher order in the chiral expansion. Results for the total decay widths are given in Table 2. We observe a good agreement between our results and those obtained using the (NRQM) and the bag model (BM). This overall agreement between different models contrasts with the situation for the $\Lambda(1405)$ decay widths [4] where the NRQM prediction is much larger than the ones obtained with other models. This can be considered as an indication that contrary to other low-lying hyperons the $\Lambda(1405)$ can not be simply understood as a 3-quark state. Another interesting feature of our results is the strong suppression of the $\Sigma^*_- \to \Sigma_- \gamma$ and $\Xi^*_- \to \Xi_- \gamma$ in agreement with the well-known $SU(3)$ U-spin selection rule. Presently the only radiative hyperon decay for which the empirical decay width is accurately known is $\Sigma_0 \to \Lambda\gamma$. As we see, in this case the soliton model prediction is reasonable good.

Experimental verification of these predictions as well as those on the e.m. decays of the other excited hyperons will certainly help to improve our understanding on the structure of the hyperons.

REFERENCES

1. C. L. Schat, C. Gobbi and N. N. Scoccola, Phys. Lett. **B356** (1995) 1.
2. C. Gobbi, C. L. Schat and N. N. Scoccola, Nucl. Phys. **A598** (1996) 318.
3. C. G. Callan and I. Klebanov, Nucl.Phys. **B262** (1985) 365.
4. C. L. Schat, N. N. Scoccola and C. Gobbi, Nucl. Phys. **A585** (1995) 627.
5. R. A. Schumacher, Nucl. Phys. **A585** (1995) 63c; J. S. Russ, Nucl. Phys. **A585** (1995) 39c; M. A. Moinester, Proc. of Workshop on Chiral Dynamics, MIT, July 1994, eds. A. Bernstein and B. Holstein, Los Alamos archive *hep-ph/9409463*; J. Russ, spokesman, FNAL E781 Collaboration; J. Russ, Proc. of the CHARM2000 Workshop, Fermilab, June 1994, eds. D. M. Kaplan and S. Kwan, Fermilab-Conf-94/190, p.111 (1994).

PART II.

COSMOLOGY AND QUANTUM FIELD THEORY

Symmetries and the Antibracket: The Batalin-Vilkovisky Method

Jorge Alfaro

Facultad de Física, Universidad Católica de Chile, Casilla 306, Santiago 22, CHILE

I INTRODUCTION

The most general quantization prescription available today is provided by the Batalin-Vilkovisky (BV) method [1,2]. It includes all the results of the BRST method, and also permits a systematic quantization of systems with an open algebra of constraints. Moreover it possesses a rich algebraic structure, that have been considered in the description of String Field Theory [3,4].

Because of the many advantages of the BV method, it is useful to understand more deeply the nature and reasons for the different prescriptions that appear there. This we have done in a series of papers [5–8]: Impossing that the most general Schwinger -Dyson equations of a theory should come from a symmetry principle(SD-BRST symmetry), we have been able to derive the BV method from standard BRST Lagrangian quantization. In so doing, we have understood the meaning of the antifields, the appearence of the graded canonical structure generated by the antibracket and ultimately, the underlying reason for the existence of a master equation. From this pesrpective, it is quite natural to generalize the BV structure, first to non-abelian invariances of the path integral measure [6], later to the most general open algebra [7] [8].

In these lectures, we review these recent developments: In section II, we study the derivation of BV from Schwinger-Dyson-BRST symmetry, in section III, we consider some generalizations of the BV structure suggested by our approach. In section IV, we present a connection between the Poisson bracket and the antibracket and use it to relate the Nambu bracketc [23] with the generalized n-brackets of section III. Finally in section V, we draw some conclusions.

CP400, *First Latin American Symposium on High Energy Physics/ VII Mexican School of Particles and Fields,* edited by D'Olivo/Klein-Kreisler/Mendez

II DERIVATION OF THE BATALIN-VILKOVISKY METHOD FROM SCHWINGER-DYSON BRST SYMMETRY

In this section, we review the derivation of the BV method from Schwinger-Dyson BRST symmetry done in [5].

Let us recapitulate the basic ingredients of the Batalin-Vilkovisky(BV) method [1,2].

Begin with a set of fields $\phi^A(x)$ of given Grassmann parity (statistics) $\epsilon(\phi^A) = \epsilon_A$, and then introduce for each field a corresponding antifield ϕ^*_A of opposite Grassmann parity $\epsilon(\phi^*_A) = \epsilon_A + 1$. The fields and antifields are taken to be canonically conjugate,

$$(\phi^A, \phi^*_B) = \delta^A_B \ , \quad (\phi^A, \phi^B) = (\phi^*_A, \phi^*_B) \ = \ 0, \tag{1}$$

within a certain graded bracket structure $(\cdot,\cdot)$, the antibracket:

$$(F, G) = \frac{\delta^r F}{\delta \phi^A} \frac{\delta^l G}{\delta \phi^*_A} - \frac{\delta^r F}{\delta \phi^*_A} \frac{\delta^l G}{\delta \phi^A} \ . \tag{2}$$

The subscripts l and r denote left and right differentiation, respectively. The summation over indices A includes an integration over continuous variables such as space-time x, when required.

This antibracket is statistics-changing in the sense that

$$\epsilon[(F,G)] = \epsilon(F) + \epsilon(G) + 1, \tag{3}$$

and satisfies the following exchange relation:

$$(F,G) \ = \ -(-1)^{(\epsilon(F)+1)(\epsilon(G)+1)}(G,F) \ . \tag{4}$$

Furthermore, one may verify that the antibracket acts as a derivation of the kind

$$\begin{aligned} (F,GH) &= (F,G)H + (-1)^{\epsilon(G)(\epsilon(F)+1)}G(F,H) \\ (FG,H) &= F(G,H) + (-1)^{\epsilon(G)(\epsilon(H)+1)}(F,H)G \ , \end{aligned} \tag{5}$$

and satisfies a Jacobi identity of the form

$$(-1)^{(\epsilon(F)+1)(\epsilon(H)+1)}(F,(G,H)) + \text{cyclic perm.} = 0 \ . \tag{6}$$

Some simple consequences of these relations are that $(F,F) = 0$ for any Grassmann odd F, and $(F,(F,F)) = ((F,F),F) = 0$ for any F.

The antifields ϕ^*_A are also given definite ghost numbers $gh(\phi^*_A)$, related to those of the fields ϕ^A:

$$gh(\phi_A^*) = -gh(\phi^A) - 1 \, . \tag{7}$$

The absolute value of the ghost number can be fixed by requiring that the action carries ghost number zero.

The Batalin-Vilkovisky quantization prescription can now be formulated as follows. First solve the equation

$$\frac{1}{2}(W, W) \;=\; i\hbar\Delta W \tag{8}$$

where

$$\Delta \;=\; (-1)^{\epsilon_A+1}\frac{\delta^r}{\delta\phi^A}\frac{\delta^r}{\delta\phi_A^*} \, . \tag{9}$$

This W will be the "quantum action", presumed expandable in powers of $\hbar$:

$$W = S + \sum_{n=0}^{\infty}\hbar^n M_n, \tag{10}$$

and a boundary condition is that S in eq. (10) should coincide with the classical action when all antifields are removed, *i.e.*, after setting $\phi_A^* = 0$. One can solve for the additional M_n-terms through a recursive procedure, order-by-order in an $\hbar$-expansion. To lowest order in $\hbar$ this is the Master Equation:

$$(S, S) \;=\; 0 \, . \tag{11}$$

Similarly, one can view eq. (8) as the full "quantum Master Equation".

A correct path integral prescription for the quantization of the classical theory $S[\phi^A, \phi_A^* = 0]$ is that one should find an appropriate "gauge fermion" Ψ such that the partition function is given by

$$\mathcal{Z} \;=\; \int [d\phi^A][d\phi_A^*]\delta(\phi_A^* - \frac{\delta^r\Psi}{\delta\phi^A})\exp\left[\frac{i}{\hbar}W\right] \, . \tag{12}$$

This prescription guarantees gauge-independence of the S-matrix of the theory. The "extended action" $S \equiv S_{ext}[\phi^A, \phi_A^*]$ is a solution of the Master Equation (11), and can been given by an expansion in powers of antifields [1]. After the elimination of the antifields by the δ-function constraint in eq. (12), one can verify that the action is invariant under the usual BRST symmetry, which we will here denote by δ.

In the following, we will uncover and derive in a simple manner the Batalin-Vilkovisky formalism starting from two basic ingredients: A) Standard BRST Lagrangian quantization, and B) The requirement that the most general Schwinger-Dyson equations of the full quantum theory follow from a symmetry principle. We shall throughout, unless otherwise stated, assume that when ultraviolet regularization is required, a suitable regulator which preserves the relevant BRST symmetry exists.

Schwinger-Dyson equations will thus play a crucial rôle in this analysis. The idea is to enlarge the usual BRST symmetry in precisely such a way that both the usual gauge-symmetry Ward Identities *and* the most general Schwinger-Dyson equations both follow from the same BRST Ward Identities. The way to do this is known [10]; it is a special case of collective field transformations that can be used to gauge arbitrary symmetries [11].

A No gauge Symmetries

Consider a quantum field theory based on an action $S[\phi^A]$ without any internal gauge symmetries. Such a quantum theory can be described by a path integral.

$$\mathcal{Z} = \int [d\phi^A] \exp\left[\frac{i}{\hbar} S[\phi^A]\right], \tag{13}$$

and the associated generating functional.

Equivalently, such a quantum field theory can be entirely described by the solution of the corresponding Schwinger-Dyson equations, once appropriate boundary conditions have been imposed. At the path integral level they follow from invariances of the measure. Let us for simplicity consider the case of a flat measure which is invariant under arbitrary local shifts, $\phi^A(x) \to \phi^A(x)+\varepsilon^A(x)$. We can gauge this symmetry by means of collective fields $\varphi^A(x)$: Suppose we transform the original field as

$$\phi^A(x) \to \phi^A(x) - \varphi^A(x), \tag{14}$$

then the transformed action $S[\phi^A - \varphi^A]$ is trivially invariant under the local gauge symmetry

$$\delta\phi^A(x) = \Theta(x) , \quad \delta\varphi^A(x) = \Theta(x), \tag{15}$$

and the measure for ϕ^A in eq. (13) is also invariant. The gauge invariant functions are of the form $F[\phi^A - \varphi^A]$, for any F.

We next integrate over the collective field in the transformed path integral, using the same flat measure. The integration is of course very formal since it will include the whole volume of the gauge group.[1] To cure this problem, we gauge-fix in the standard BRST Lagrangian manner [9]. That is, we add to the transformed Lagrangian a BRST-exact term in such a way that the local gauge symmetry is broken. In this case an obvious BRST multiplet consists of a ghost-antighost pair $c^A(x), \phi^*_A(x)$, and a Nakanishi-Lautrup field $B_A(x)$:

---

1) This situation is no different from usual path integral manipulations of gauge theories.

$$
\begin{aligned}
\delta\phi^A(x) &= c^A(x) \\
\delta\varphi^A(x) &= c^A(x) \\
\delta c^A(x) &= 0 \\
\delta\phi^*_A(x) &= B_A(x) \\
\delta B_A(x) &= 0.
\end{aligned}
\tag{16}
$$

No assumptions will be made as to whether ϕ^A are of odd or even Grassmann parity. We assign the usual ghost numbers to the new fields,

$$
gh(c^A) = 1 \;, \quad gh(\phi^*_A) = -1 \;, \quad gh(B_A) = 0, \tag{17}
$$

and the operation δ is statistics-changing. The rules for operating with δ are given in the Appendix.

Let us choose to gauge-fix the transformed action by adding to the Lagrangian a term of the form

$$
-\delta[\phi^*_A(x)\varphi^A(x)] = (-1)^{\epsilon(A)+1}B_A(x)\varphi^A(x) - \phi^*_A(x)c^A(x). \tag{18}
$$

The partition function is now again well-defined:

$$
\begin{aligned}
\mathcal{Z} = \int [d\phi][d\varphi][d\phi^*][dc][dB] \exp\Big[\frac{i}{\hbar}\Big(S[\phi-\varphi] \\
- \int dx\{(-1)^{\epsilon(A)}B_A(x)\varphi^A(x) + \phi^*_A(x)c^A(x)\}\Big)\Big].
\end{aligned}
\tag{19}
$$

Since the collective field has just been gauge fixed to zero, it may appear useful to integrate both it and the field $B_A(x)$ out. We are then left with

$$
\begin{aligned}
\mathcal{Z} &= \int [d\phi^A][d\phi^*_A][dc^A] \exp\left[\frac{i}{\hbar}S_{ext}\right] \\
S_{ext} &= S[\phi^A] - \int dx \phi^*_A(x)c^A(x) \;,
\end{aligned}
\tag{20}
$$

which obviously coincides with the original expression (13) apart from the trivially decoupled ghosts. But the remnant BRST symmetry is still non-trivial: We find it in the usual way by substituting for $B_A(x)$ its equation of motion. This gives

$$
\begin{aligned}
\delta\phi^A(x) &= c^A(x) \\
\delta c^A(x) &= 0 \\
\delta\phi^*_A(x) &= -\frac{\delta^l S}{\delta\phi^A(x)}.
\end{aligned}
\tag{21}
$$

The functional measure is also invariant under this symmetry, according to our assumption about the measure for ϕ^A, and assuming a flat measure for ϕ^*_A as well. The Ward Identities following from this symmetry are the seeked-for Schwinger-Dyson equations:

$$0 = \langle \delta\{\phi^*_A(x) F[\phi^A]\}\rangle \ ,$$

where we have chosen F to depend only on ϕ^A just to ensure that the whole object carries overall ghost number zero. After integrating over both ghosts c^A and antighosts ϕ^*_A, this Ward Identity can be written

$$\langle \frac{\delta^l F}{\delta\phi^A(x)} + \left(\frac{i}{\hbar}\right) \frac{\delta^l S}{\delta\phi^A(x)} F[\phi^A]\rangle = 0 \ , \tag{22}$$

that is, precisely the most general Schwinger-Dyson equations for this theory. The symmetry (21) can be viewed as the BRST Schwinger-Dyson algebra.

Consider now the equation that expresses BRST invariance of the extended action S_{ext}:

$$\begin{aligned} 0 = \delta S_{ext} &= \int dx \frac{\delta^r S_{ext}}{\delta\phi^A(x)} c^A(x) - \int dx \frac{\delta^r S_{ext}}{\delta\phi^*_A(x)} \frac{\delta^l S}{\delta\phi^A(x)} \\ &= \int dx \frac{\delta^r S_{ext}}{\delta\phi^A(x)} c^A(x) - \int dx \frac{\delta^r S_{ext}}{\delta\phi^*_A(x)} \frac{\delta^l S_{ext}}{\delta\phi^A(x)} \ . \end{aligned} \tag{23}$$

In the last line we have used the fact that S differs from S_{ext} by a term independent of ϕ^A. Using the notation of the antibracket (2), this is seen to correspond to a Master Equation of the form

$$\frac{1}{2}(S_{ext}, S_{ext}) = -\int dx \frac{\delta^r S_{ext}}{\delta\phi^A(x)} c^A(x) \ . \tag{24}$$

The ghosts c^A play the rôle of spectator fields in the antibracket. But their appearance on the r.h.s. of the Master Equation ensures that the solution S_{ext} will contain these fields.

The extended action of Batalin and Vilkovisky does however not coincide with S_{ext} as defined above. But suppose we integrate *only* over these ghosts $c^A(x)$, without integrating over the corresponding antighosts $\phi^*_A(x)$. Then the partition function reads

$$\mathcal{Z} = \int [d\phi^A][d\phi^*_A]\delta(\phi^*_A) \exp\left[\frac{i}{\hbar} S[\phi^A]\right] . \tag{25}$$

What has happened to the BRST algebra? For the present case of a ghost field c^A appearing linearly in the action before being integrated out, we can derive the correct substitution rule as follows. First, we should really phrase the question in a more precise manner. What we need to know is how to replace c inside the path integral, *i.e.*, inside Green functions. This will automatically give us the correct transformation rules for those fields that are not integrated out. Consider the identity

$$\int [dc] F(c^B(y)) \exp\left[-\frac{i}{\hbar}\int dx \phi^*_A(x) c^A(x)\right] =$$
$$F\left(i\hbar \frac{\delta^l}{\delta\phi^*_B(y)}\right) \exp\left[-\frac{i}{\hbar}\int dx \phi^*_A(x) c^A(x)\right] . \tag{26}$$

Eq. (26) teaches us that it is not enough to replace c by its equation of motion ($c(x) = 0$); a "quantum correction" in the form of the operator $\hbar\delta/\delta\phi^*$ must be added as well. The appearance of this operator is the final step towards unravelling the canonical structure in the formalism of Batalin and Vilkovisky. It also shows that even in this trivial case we have to include "quantum corrections" to BRST symmetries if we insist on integrating out only one ghost field, while keeping its antighost.

It is important that the operator $\hbar\delta^l/\delta\phi^*$ in eq. (26) always acts on the integral (really a δ-function) to its right.

We now make this replacement, having always in mind that it is only meaningful inside the path integral. For the BRST transformation itself we get, upon one partial integration

$$\delta\phi^A(x) = i\hbar(-1)^{\epsilon_A}\frac{\delta^r}{\delta\phi^*_A(x)}$$
$$\delta\phi^*_A(x) = -\frac{\delta^l S}{\delta\phi^A(x)} . \tag{27}$$

We know from our derivation that this transformation leaves at least the combination of measure and action invariant. As a check, if we consider the same Ward Identity as above, based on $0 = \langle\delta\{\phi^*_A(x)F[\phi^A]\}\rangle$, we recover the Schwinger-Dyson equation (22).

The original $S[\phi^A]$ can in this case be identified with the extended action of Batalin and Vilkovisky, and the antighost ϕ^*_A is the antifield corresponding to ϕ^A. Because there are no internal gauge symmetries, the extended action turns out to be independent of the antifields. Although our S_{ext} of eq. (20) cannot be identified with the extended action of Batalin-Vilkovisky, the Master Equation derived in eq. (23) is of course very similar to their corresponding Master Equation. Writing eq.(23) in terms of the antibracket is a little forced. It is done in order to facilitate the comparison.

Finally, it only remains to be seen what has happened to this bracket structure after having integrated out the ghost. We will keep the same notation as before, so that in this case the "extended" action is trivially equal to the original action: $S_{ext} = S[\phi^A]$. Let us then again consider the variation of an arbitrary functional G, this time only a function of ϕ^A and ϕ^*_A. Inside the path integral (and only there!) we can represent the variation of G as:

$$\delta G[\phi^A, \phi^*_A] = \int dx \frac{\delta^r G}{\delta\phi^A(x)}\left[\frac{\delta^l S_{ext}}{\delta\phi^*_A(x)} + (i\hbar)(-1)^{\epsilon_A}\frac{\delta^r}{\delta\phi^*_A(x)}\right] - \int dx \frac{\delta^r G}{\delta\phi^*_A(x)}\frac{\delta^l S_{ext}}{\delta\phi^A(x)} , \tag{28}$$

where the derivative operator no longer acts on the δ-function of ϕ^*_A. We have kept the term proportional to $\delta^l S_{ext}/\delta\phi^*_A$, even though S_{ext} in this simple case is independent of ϕ^*_A (and that term therefore vanishes). It comes from the partial integration with respect to the operator δ^l/δ^*_A, and is there in general when S_{ext} depends on ϕ^*_A.

This equation precisely describes the "quantum deformation" of the classical BRST charge, as it occurs in the Batalin-Vilkovisky framework:

$$\delta G = (G, S_{ext}) - i\hbar\Delta G \; . \tag{29}$$

The BRST operator in this form is often denoted by σ.

We have here introduced

$$\Delta \equiv (-1)^{\epsilon_A+1} \frac{\delta^r \delta^r}{\delta\phi^*_A(x)\delta\phi^A(x)} \tag{30}$$

which is identical to the operator (9) of Batalin and Vilkovisky. Again, this term arises as a consequence of the partial integration which allows us to expose the operator $\delta/\delta\phi^*_A(x)$ that otherwise acts only on the δ-functional $\delta(\phi^*_A)$. In other words, the δ-function constraint on $\phi^*_A(x)$ is now considered as part of the functional measure for ϕ^*_A.

Precisely which properties of the partially-integrated extended action are then responsible for the canonical structure behind the Batalin-Vilkovisky formalism? As we have seen, the crucial ingredients come from integrating out the Nakanishi-Lautrup fields B_A and the ghosts c^A. Integrating out B_A changes the BRST variation of the antighosts ϕ^*_A into $-\delta^l S_{ext}/\delta\phi^A$. *This is an inevitable consequence of introducing collective fields as shifts of the original fields (and hence enforcing Schwinger-Dyson equations), and then gauge-fixing them to zero.* But this only provides half of the canonical structure, making, loosely speaking, ϕ^A canonically conjugate to ϕ^*_A, but not vice versa. The rest is provided by integrating over the ghosts c^A; at the linear level it changes the BRST variation of the fields ϕ^A themselves into $\delta^l S_{ext}/\delta\phi^*_A$. *This in turn is again an inevitable consequence of having introduced the collective fields as shifts, and then having gauge-fixed them to zero.*[2]

The latter operation is, however, complicated by the fact that the fields ϕ^*_A which are fixed in the process of integrating out c^A are chosen to remain in the path integral. This makes it impossible to discard the "quantum correction" to $\delta S_{ext}/\delta\phi^*_A$. So, in fact, if one insists on keeping these antighosts ϕ^*_A, now seen as canonically conjugate partners of ϕ^A, the simple canonical structure is in this sense never truly realized.

---

[2] Gauge-fixing the collective fields to zero implies linear couplings to the auxiliary fields and ghosts, respectively. This is one of the central properties of the extended action that leads to the canonical structure, and to the fact that the extended action S_{ext} itself is the (classical) BRST generator. However, this does not exclude the possibility that different gauge fixings of the shift symmetries could produce a more general formalism.

We have seen these features only in what is the trivial case of no internal gauge symmetries. But as we shall show in the following section, they hold in greater generality.

B Gauge Theories: Yang-Mills

Let us illustrate our approach with another simple example: Yang-Mills theory.

We start with the pure Yang-Mills action $S[A_\mu]$, and define a covariant derivative with respect to A_μ as

$$D_\mu^{(A)} \equiv \partial_\mu - [A_\mu, \ ]. \tag{31}$$

Next, we introduce the collective field a_μ which will enforce Schwinger-Dyson equations as a result of the BRST algebra:

$$A_\mu(x) \to A_\mu(x) - a_\mu(x). \tag{32}$$

In comparison with the previous example, the only new aspect here is that the transformed action $S[A_\mu - a_\mu]$ actually has two independent gauge symmetries. Because of the redundancy introduced by the collective field, we can write the two symmetries in different ways. To make contact with the Batalin-Vilkovisky formalism, we will choose the following version,

$$\begin{aligned} \delta A_\mu(x) &= \Theta_\mu(x) \\ \delta a_\mu(x) &= \Theta_\mu(x) - D_\mu^{(A-a)}\varepsilon(x), \end{aligned} \tag{33}$$

which also shows the need for being careful in defining what we mean by a covariant derivative. (The general principle is that we choose the original gauge symmetry of the original field to be carried entirely by the collective field; the transformation of the original gauge field is then always just a shift.) Although $\Theta(x)$ includes arbitrary deformations, it only leaves the transformed *field* invariant, while of course the action is also invariant under Yang-Mills gauge transformations of this transformed field itself. Hence the need for including two independent gauge transformations.

We now gauge-fix these two gauge symmetries, one at a time, in the standard BRST manner. As a start, we introduce a suitable multiplet of ghosts and auxiliary fields. We need one Lorentz vector ghost $\psi_\mu(x)$ for the shift symmetry of A_μ, and one Yang-Mills ghost $c(x)$. These are of course Grassmann odd, and both carry the same ghost number

$$gh(\psi_\mu) = gh(c) = 1. \tag{34}$$

Next, we gauge-fix the shift symmetry of A_μ by removing the collective field a_μ. This leads us to introduce a corresponding antighost $A_\mu^*(x)$, Grassmann odd, and an auxiliary field $b_\mu(x)$, Grassmann even. They have the usual ghost number assignments,

$$gh(A^*_\mu) = -1 , \quad gh(b_\mu) = 0 , \tag{35}$$

and we now have the nilpotent BRST algebra

$$\begin{aligned}
\delta A_\mu(x) &= \psi_\mu(x) \\
\delta a_\mu(x) &= \psi_\mu(x) - D_\mu^{(A-a)} c(x) \\
\delta c(x) &= -\frac{1}{2}[c(x), c(x)] \\
\delta \psi_\mu(x) &= 0 \\
\delta A^*_\mu(x) &= b_\mu(x) \\
\delta b_\mu(x) &= 0.
\end{aligned} \tag{36}$$

Fixing $a_\mu(x)$ to zero is achieved by adding a term

$$-\delta[A^*_\mu(x) a^\mu(x)] = -b_\mu(x) a^\mu(x) - A^*_\mu(x)\{\psi^\mu - D^\mu_{(A-a)} c(x)\} \tag{37}$$

to the Lagrangian. We shall follow the usual rule of only starting to integrate over *pairs* of ghost-antighosts in the partition function. With this rule we shall still keep $c(x)$ unintegrated (since we have not yet introduced its corresponding antighost), but we can now integrate over both $\psi_\mu(x)$ and $A^*_\mu(x)$. This leads to the following extended, but not yet fully gauge-fixed, action S_{ext}:

$$\begin{aligned}
\mathcal{Z} &= \int [dA_\mu][da_\mu][d\psi_\mu][dA^*_\mu][db_\mu] \exp\left[\frac{i}{\hbar} S_{ext}\right] \\
S_{ext} &= S[A_\mu - a_\mu] - \int dx \{b_\mu(x) a_\mu(x) + A^*_\mu(x)[\psi^\mu(x) - D^\mu_{(A-a)} c(x)]\} .
\end{aligned} \tag{38}$$

This extended action is invariant under the BRST transformation (36). The full integration measure is also invariant. Of course, the expression above is still formal, since we have not yet gauge fixed ordinary Yang-Mills invariance.

Furthermore, we are eventually going to integrate over the ghost c, which already now appears in the extended action. If we insist that the Schwinger-Dyson equations involving this field, *i.e*, equations of the form

$$0 = \int [dc] \frac{\delta^l}{\delta c(x)} \left[F e^{\frac{i}{\hbar}[\text{Action}]}\right] \tag{39}$$

are to be satisfied automatically (for reasonable choices of functionals F) by means of the full unbroken BRST algebra, we must introduce yet one more collective field. This new collective field, call it $\tilde{c}(x)$, is Grassmann odd, and has $gh(\tilde{c}) = 1$. Now shift the Yang-Mills ghost:

$$c(x) \to c(x) - \tilde{c}(x). \tag{40}$$

To fix the associated fermionic gauge symmetry, we introduce a new BRST multiplet of a ghost-antighost pair and an auxiliary field. We follow the same rule as before and let the transformation of the new collective field $\tilde{c}$ carry the (BRST) transformation of the original ghost c, *viz.*,

$$\begin{aligned}
&\delta c(x) = C(x) \\
&\delta\tilde{c}(x) = C(x) + \frac{1}{2}[c(x) - \tilde{c}(x), c(x) - \tilde{c}(x)] \\
&\delta C(x) = 0 \\
&\delta c^*(x) = B(x) \\
&\delta B(x) = 0.
\end{aligned} \tag{41}$$

It follows from (41) that ghost number assignments should be as follows:

$$gh(C) = 2\ , \quad gh(c^*) = -2\ , \quad gh(B) = -1, \tag{42}$$

and B is a fermionic Nakanishi-Lautrup field. Next, gauge-fix $\tilde{c}(x)$ to zero by adding a term

$$-\delta[c^*(x)\tilde{c}(x)] = B(x)\tilde{c}(x) - c^*(x)\{C(x) + \frac{1}{2}[c(x) - \tilde{c}(x), c(x) - \tilde{c}(x)]\} \tag{43}$$

to the Lagrangian. This gives us the fully extended action,

$$\begin{aligned}
S_{ext} = S[A_\mu - a_\mu] - \int dx\{&b_\mu(x)a_\mu(x) + A^*_\mu(x)[\psi^\mu(x) - D^\mu_{(A-a)}\{c(x) - \tilde{c}(x)\}] \\
&- B(x)\tilde{c}(x) + c^*(x)(C(x) + \frac{1}{2}[c(x) - \tilde{c}(x), c(x) - \tilde{c}(x)])\},
\end{aligned} \tag{44}$$

with the partition function so far being integrated over all fields appearing above, except for c (whose antighost $\bar{c}$ still has to be introduced when we gauge-fix the original Yang-Mills symmetry).

The extended action *and* the functional measure is formally invariant under the following set of transformations:

$$\begin{array}{lcl lcl}
\delta A_\mu(x) & = & \psi_\mu(x), & \delta\psi_\mu(x) & = & 0, \\
\delta a_\mu(x) & = & \psi_\mu(x) - D^{(A-a)}_\mu[c(x) - \tilde{c}(x)], & \delta c(x) & = & C(x), \\
\delta A^*_\mu(x) & = & b_\mu(x) & \delta b_\mu(x), & = & 0, \\
\delta\tilde{c}(x), & = & C(x) + \frac{1}{2}[c(x) - \tilde{c}(x), c(x) - \tilde{c}(x)] & \delta C(x) & = & 0, \\
\delta c^*(x) & = & B(x) & \delta B(x) & = & 0\ .
\end{array}$$

As the notation indicates, the fields $A^*_\mu(x)$ and $c^*(x)$ can be identified with the Batalin-Vilkovisky antifields of $A_\mu(x)$ and $c(x)$, respectively. *These antifields are the usual antighosts of the collective fields enforcing Schwinger-Dyson equations through shift symmetries.*

Note that the general rule of assigning ghost number and Grassmann parity to the antifields,

$$gh(\phi_A^*) = -(gh(\phi^A) + 1) , \qquad \epsilon(\phi_A^*) = \epsilon(\phi^A) + 1 , \tag{45}$$

arises in a completely straightforward manner. Here, it is a simple consequence of the fact that the BRST operator raises ghost number by one unit (and changes statistics), supplemented with the usual rule that antighosts have opposite ghost number of the ghosts.

The extended action (44) above has more fields than the extended action of Batalin and Vilkovisky, and the transformation laws of what we identify as antifields do not match those of ref. [1]. In the form we have presented it, the BRST symmetry is nilpotent also off-shell. If we integrate over the auxiliary fields b_μ and B (and subsequently over a_μ and $\tilde{c}$) the BRST symmetry becomes nilpotent only on-shell. The extended action then takes the following form:[3]

$$S_{ext} = S[A_\mu] - \int dx \{A_\mu^*(x)[\psi^\mu(x) - D^\mu c(x)] + c^*(x)(C(x) + \frac{1}{2}[c(x), c(x)])\}. \tag{46}$$

This differs from the extended action of Batalin and Vilkovisky by the terms involving the ghost fields $\psi_\mu(x)$ and $c^*(x)$. Comparing with the case of no gauge symmetries, this is exactly what we should expect. These ghost fields ψ_μ and c^* ensure the correct Schwinger-Dyson equations for A_μ and c, respectively.

To find the corresponding BRST symmetry we use the equations of motion for the auxiliary fields b_μ and B, and use the δ-function constraints on a_μ and $\tilde{c}$. This gives

$$\begin{aligned} \delta A_\mu(x) &= \psi_\mu(x) \\ \delta \psi_\mu(x) &= 0 \\ \delta c(x) &= C(x) \\ \delta C(x) &= 0 \\ \delta A_\mu^*(x) &= -\frac{\delta^l S_{ext}}{\delta A_\mu(x)} \\ \delta c^*(x) &= -\frac{\delta^l S_{ext}}{\delta c(x)} \end{aligned} \tag{47}$$

BRST invariance of the extended action (46) immediately implies that it satisfies a Master Equation we can write as

$$\int dx \frac{\delta^r S_{ext}}{\delta A_\mu^*(x)} \frac{\delta^l S_{ext}}{\delta A^\mu(x)} + \int dx \frac{\delta^r S_{ext}}{\delta c^*(x)} \frac{\delta^l S_{ext}}{\delta c(x)} = \int dx \frac{\delta^r S_{ext}}{\delta A_\mu(x)} \psi_\mu(x) + \int dx \frac{\delta^r S_{ext}}{\delta c(x)} C(x) . \tag{48}$$

---

[3] We are dropping the subscript on the covariant derivative since no confusion can arise at this point.

Note that this Master Equation precisely is of the form

$$\frac{1}{2}(S_{ext}, S_{ext}) = -\int dx \frac{\delta S_{ext}}{\delta \phi^A(x)} c^A(x) \ , \tag{49}$$

with two ghost fields c^A that are just ψ_μ and $C(x)$.

To finally make contact with the Batalin-Vilkovisky formalism, let us integrate out the ghost ψ_μ. As in eq.(26), we shall use an identity of the form

$$\begin{aligned}
&\int [d\psi_\mu] F[\psi^\mu(y)] \exp\left[-\frac{i}{\hbar}\int dx A^*_\mu(x)\{\psi^\mu(x) - D^\mu c(x)\}\right] \\
&= F\left[D^\mu c(y) + (i\hbar)\frac{\delta^l}{\delta A^*_\mu(y)}\right] \\
&\quad \times \int [d\psi_\mu] \exp\left[-\frac{i}{\hbar}\int dx A^*_\mu(x)\{\psi^\mu(x) - D^\mu c(x)\}\right] \\
&= \exp\left[\frac{i}{\hbar}\int dx A^*_\mu(x) D^\mu c(x)\right] F\left[(i\hbar)\frac{\delta^l}{\delta A^*_\mu(y)}\right]\delta(A^*_\mu)
\end{aligned} \tag{50}$$

This shows that we should replace $\psi_\mu(x)$ by its equation of motion, plus the shown quantum correction of $\mathcal{O}(\hbar)$ which then acts on the rest of the integral, or equivalently, by just the derivative operator (which then acts solely on the functional δ-function). To get a useful representation of $\psi^\mu(y)$ we integrate the latter version of the identity by parts, thus letting the derivative operator act on everything except $\delta(A^*_\mu)$. This automatically brings down the equations of motion for ψ^μ.

Having in this manner integrated out ψ^μ and C, the partition function reads

$$\begin{aligned}
\mathcal{Z} &= \int [dA_\mu][dA^*_\mu][dc^*]\delta(A^*_\mu)\delta(c^*) \exp\left[\frac{i}{\hbar}S_{ext}\right] \\
S_{ext} &= S[A_\mu] + \int dx \{A^*_\mu(x) D^\mu c(x) - \frac{1}{2}c^*[c(x), c(x)]\},
\end{aligned} \tag{51}$$

and the *classical* BRST symmetry follows as discussed above by substituting only the equations of motion for ψ_μ and C:

$$\begin{aligned}
\delta A_\mu(x) &= \frac{\delta^l S_{ext}}{\delta A^*_\mu(x)} = D_\mu c(x) \\
\delta c(x) &= \frac{\delta^l S_{ext}}{\delta c^*(x)} - \frac{1}{2}[c(x), c(x)] \\
\delta A^*_\mu(x) &= -\frac{\delta^l S_{ext}}{\delta A_\mu(x)} \\
\delta c^*(x) &= -\frac{\delta^l S_{ext}}{\delta c(x)} \ .
\end{aligned} \tag{52}$$

This is the usual extended action of Batalin and Vilkovisky and the corresponding classical BRST symmetry. Of course, in the partition function the integrals over A^*_μ and c^* are trivial. This is as it should be, because by integrating out these antighosts we should finally recover the starting measure and the still not gauge-fixed Yang-Mills action. What we have provided here is thus only a very precise functional derivation of the extended action. It shows that we *can* understand the extended action in the usual path integral framework, and that the integration measures for A^*_μ and c^* (with the accompagnying δ-function constraints) are provided automatically.

Obviously, the extended action is not yet very useful from the point of view of ordinary BRST gauge fixing. To unravel some of the mechanisms behind the Batalin-Vilkovisky scheme, it is nevertheless advantageous to keep the antifields. In fact, it is even more useful to return to the formulation in eq. (46), where there is yet no split into a classical and a quantum part of the symmetry. Let us therefore take (46) as the starting action, and now just gauge-fix in a standard manner the Yang-Mills symmetry. Choosing, *e.g.*, a covariant gauge, we therefore finally extend the BRST multiplet to include a Yang-Mills antighost $\bar{c}$ and a Nakanishi-Lautrup scalar b. For there to be no doubt, let us also note that these fields have

$$gh(\bar{c}) = -1 \ , \qquad gh(b) = 0 \ . \tag{53}$$

The BRST transformations are the usual $\delta\bar{c}(x) = b(x)$ and $\delta b(x) = 0$. Gauge fixing to a covariant α-gauge can be achieved by adding a term

$$\delta[\bar{c}(x)\{\partial_\mu A^\mu(x) - \frac{1}{2\alpha}b(x)\}] = b(x)\partial_\mu A^\mu(x) + \bar{c}(x)\partial_\mu\psi^\mu(x) - \frac{1}{2\alpha}b(x)^2 \tag{54}$$

to the Lagrangian. The corresponding completely gauge-fixed extended action then reads

$$\begin{aligned} S_{ext} = S[A_\mu] - \int dx\{&A^*_\mu(x)[\psi^\mu(x) - D^\mu c(x)] \\ &+c^*(x)(C(x) + \frac{1}{2}[c(x), c(x)]) \\ &-b(x)\partial_\mu A^\mu(x) - \bar{c}(x)\partial_\mu\psi^\mu(x) + \frac{1}{2\alpha}b(x)^2\}. \end{aligned} \tag{55}$$

Now integrate out ψ^μ and C. The result is indeed a partition function of the form introduced by the formalism of Batalin and Vilkovisky:

$$\begin{aligned} \mathcal{Z} &= \int [dA_\mu][dA^*_\mu][d\bar{c}][dc][dc^*][db]\delta(A^*_\mu + \partial_\mu\bar{c})\delta(c^*)e^{[\frac{i}{\hbar}S_{ext}]} \\ S_{ext} &= S[A_\mu] + \int dx\{A^*_\mu(x)D^\mu c(x) - \frac{1}{2}c^*(x)[c(x), c(x)] \\ &\quad +b(x)\partial_\mu A^\mu(x) - \frac{1}{2\alpha}b(x)^2\}. \end{aligned} \tag{56}$$

Note that by adding the Yang-Mills gauge-fixing terms, the δ-function constraint on the antifield A^*_μ has been shifted. Thus when doing the A^*_μ-integral, we are in effect substituting not $A^*_\mu(x) = 0$ but

$$A^*_\mu(x) = -\partial_\mu \bar{c}(x) = \frac{\delta^r \Psi}{\delta A^\mu(x)} , \tag{57}$$

where Ψ is defined as the term whose BRST variation is added to the action, *i.e.*, in this particular case,

$$\Psi = \int dx \{\bar{c}(x)(\partial^\mu A_\mu(x) - \frac{1}{2\alpha} b(x))\} . \tag{58}$$

Upon doing the A^*_μ and c^* integrals, one recovers the standard covariantly gauge-fixed Yang-Mills theory

$$S = S[A_\mu] + \int dx \{\bar{c}(x)\partial^\mu D_\mu c(x) + b(x)\partial^\mu A_\mu(x) - \frac{1}{2\alpha} b(x)^2\}. \tag{59}$$

The identification (57) strongly suggests that one can see gauge fixing as a particular canonical transformation involving new fields (the antighosts $\bar{c}$). But there are terms in eq. (59) (those involving $b(x)$) which do not immediately follow from this perspective. In the Batalin-Vilkovisky framework, this is resolved by noting that one can always add terms of new fields and antifields with trivial antibrackets. In this version of the gauge-fixing procedure, one returns to the ("minimally") extended action of eq. (56) and extends it in a "non-minimal" way. In the Yang-Mills case, this includes an additional term in the action of the form

$$S_{nm} = \int dx \bar{c}^*(x) b(x) , \tag{60}$$

with $\bar{c}^*$ and b having the same ghost number and Grassmann parity:

$$gh(\bar{c}^*) = gh(b) = 0 ; \qquad \epsilon(\bar{c}^*) = \epsilon(b) = 0 . \tag{61}$$

As the notation indicates, these new fields $\bar{c}^*$ and b are indeed just the antifield of $\bar{c}$, and the usual Nakanishi-Lautrup field, respectively.

Gauge fixing to the same gauge as above can then be achieved by the same gauge fermion (58) which now affects both A^*_μ and $\bar{c}^*$ within the antibracket. It can thus be seen as the canonical transformation that shifts A^*_μ and $\bar{c}^*$ from zero to

$$A^*_\mu(x) = \frac{\delta^r \Psi}{\delta A^\mu(x)} , \qquad \bar{c}^*(x) = \frac{\delta^r \Psi}{\delta \bar{c}(x)} . \tag{62}$$

Since Ψ does not depend on the antifields, this canonical transformation leaves all *fields* A_μ, c and $\bar{c}$ unchanged.

Can we understand the non-minimally extended action from our point of view too? Consider the stage at which we introduce the antighost $\bar{c}$. This field does not yet appear in the action, but we can of course still introduce a corresponding collective "shift" field $\bar{c}'$ for $\bar{c}$ as well. The corresponding BRST multiplet consists of a new "shift-antighost" $\lambda(x)$, an "anti-antighost" $\bar{c}^*(x)$, and the associated auxiliary field $B'(x)$:

$$\begin{aligned}
\delta\bar{c}(x) &= \lambda(x) \\
\delta\bar{c}'(x) &= \lambda(x) - b(x) \\
\delta\lambda(x) &= 0 \\
\delta b(x) &= 0 \\
\delta\bar{c}^*(x) &= B'(x) \\
\delta B'(x) &= 0 \ .
\end{aligned} \tag{63}$$

The assignments will then have to be exactly as in eq. (61), supplemented with $gh(B') = \epsilon(B') = 1$. We are again dealing with *two* symmetries, because the shifted field $\bar{c}(x) - \bar{c}'(x)$ itself can still be shifted by the usual Nakanishi-Lautrup field. Let us now gauge-fix this large symmetry. We do it in the most simple manner by adding a term

$$-\delta[\bar{c}^*(x)\bar{c}'(x)] = B'(x)\bar{c}'(x) - \bar{c}^*(x)(\lambda(x) - b(x)) \tag{64}$$

to the Lagrangian. The integrals over B' and $\bar{c}'$ are of course trivial and we are left with the non-minimally extended action for this theory plus, as expected, the corresponding term with the new ghost λ. The final gauge-fixing of the Yang-Mills symmetry will now consist in adding, instead of eq. (54),

$$\delta[\bar{c}(x)\{\partial_\mu A^\mu(x) - \frac{1}{2\alpha}b(x)\}] = \lambda(x)\partial_\mu A^\mu(x) + \bar{c}(x)\partial_\mu\psi^\mu(x) - \frac{1}{2\alpha}\lambda(x)b(x) \ . \tag{65}$$

Before Yang-Mills gauge fixing, the integral over $\lambda(x)$ just gave a factor of $\delta(\bar{c}^*)$. After adding the gauge-fixing term, this is modified:

$$\delta(\bar{c}^*) \rightarrow \delta(\bar{c}^*(x) - \partial^\mu A_\mu(x) + \frac{1}{2\alpha}b(x)) \ . \tag{66}$$

Substituting this back into the extended action, we recover the result (59). Note that this indeed can be viewed as a canonical transformation within the antibracket. All the correct δ-function constraints are provided by the collective fields and their ghosts. Since the functional Ψ has been chosen to depend only on the fundamental fields, and not on the antifields, the fields A_μ, c and $\bar{c}$ are all left untouched by this canonical transformation. Extending the action from the minimal to the non-minimal case is equivalent to demanding that also Schwinger-Dyson equations for $\bar{c}(x)$ follow as Ward identities of the BRST symmetry. Since the antighost $\bar{c}$ remains in the path integral after gauge fixing, it would indeed be very unnatural not to demand that correct Schwinger-Dyson equations for this field follow as well. As shown, this requirement automatically leads to the *non-minimally* extended action.

As for the functional measure, we have stressed earlier that we always assume the existence of a suitable regulator that preserves the pertinent BRST symmetry. We can make this statement a little more explicit by detailing the required symmetries of the measure in this Yang-Mills case. Before integrating out any fields, the measures for A_μ, c and $\bar{c}$ should all be invariant under local shifts. For A_μ this corresponds to the usual euclidean measure (and it is very difficult to imagine this shift symmetry being broken by any reasonable regulator), while for c and $\bar{c}$ this is consistent with the usual rules of Berezin integration. The measures for the three collective fields should in addition be invariant under what corresponds to usual Yang-Mills BRST transformations, a property that indeed holds formally. Finally, the measures for all antifields are only required to be invariant under local shifts. After having integrated out the auxiliary fields B_A, invariance of these measures of the antifields is now non-trivial but one can check explicitly that it is formally satisfied. This should indeed be the case, because it is straightforward to check that the action remains invariant. Since at least the *combination* of measure and action must remain invariant after integrating out some of the fields, invariance of the measure is in this case formally guaranteed.

Let us finally point out that once the antighost $\bar{c}$ is being treated on equal footing with A_μ and c, a Master Equation of the form

$$\frac{1}{2}(S_{ext}, S_{ext}) = -\int dx \frac{\delta^r S_{ext}}{\delta\phi^A(x)} c^A(x) \tag{67}$$

now holds with ϕ^A denoting *all* the fields that finally remain in the path integral: A_μ, c and $\bar{c}$. Similarly, the BRST algebra becomes, upon integrating out the collective fields φ^A and the auxiliary fields B_A, of the very simple form (21) we encountered already in the case of no internal gauge symmetries.

The general case of arbitrary gauge symmetries is considered in [5].

C Quantum Master Equations

We have seen from the collective field method that for closed irreducible gauge algebras we get an extended action that can be split into a part independent of the new ghosts c^A, and a simple quadratic term of the form $\phi^*_A c^A$. Let us, for reasons that will become evident shortly, denote the part which is independent of c^A by $S^{(BV)}$, *i.e.*:

$$S_{ext}[\phi, \phi^*, c] = S^{(BV)}[\phi, \phi^*] - \phi^*_A c^A \ . \tag{68}$$

This action is invariant under the transformations

$$\delta\phi^A = c^A$$
$$\delta c^A = 0$$

$$\delta\phi_A^* = -\frac{\delta^l S_{ext}}{\delta\phi^A} . \tag{69}$$

Moreover, the functional measure is also formally guaranteed to be invariant in this case. It follows that in this case the Ward Identities of the kind $0 = \langle\delta[\phi_A^* F[\phi]\rangle$ are the most general Schwinger-Dyson equations for the quantum theory defined by the classical action $S[\phi]$.

But demanding that *both* the action S_{ext} *and* the functional measure be invariant under the BRST Schwinger-Dyson symmetry above is not the most general condition. To derive the correct Ward Identities we only need that just the *combination* of action and measure is invariant. In this subsection we want to discuss the more general case in which the set of transformations (69) still generate a symmetry of the combination of measure and action, but not of each individually. If we insist on a solution of the form (68), then the other property that is required, $\langle c^A\phi_B^*\rangle = -i\hbar\delta_B^A$, follows automatically.

It thus remains to be found under what conditions the combination of the action and the measure remain invariant under the BRST Schwinger-Dyson symmetry. With an action S_{ext} of the form (68), we get

$$\begin{aligned}
\delta S_{ext} &= \frac{\delta^r S^{(BV)}}{\delta\phi^A}c^A + \frac{\delta^r S^{(BV)}}{\delta\phi_A^*}\left(-\frac{\delta^l S_{ext}}{\delta\phi^A}\right) - \frac{\delta^r(\phi_A^* c^A)}{\delta\phi_B^*}\left(-\frac{\delta^l S_{ext}}{\delta\phi^A}\right)\\
&= \frac{\delta^r S^{(BV)}}{\delta\phi^A}c^A + \frac{\delta^r S^{(BV)}}{\delta\phi_A^*}\left(-\frac{\delta^l S^{(BV)}}{\delta\phi^A}\right) - (-1)^{\epsilon_A+1}c^A\left(-\frac{\delta^l S^{(BV)}}{\delta\phi^A}\right)\\
&= -\frac{\delta^r S^{(BV)}}{\delta\phi_A^*}\frac{\delta^l S^{(BV)}}{\delta\phi^A} = \frac{1}{2}(S^{(BV)}, S^{(BV)}) .
\end{aligned} \tag{70}$$

We will still assume that we are integrating over a flat euclidean measure for the fundamental field ϕ^A. This measure is formally invariant under the transformation (69). However, for a corresponding flat euclidean measure for ϕ_A^*, the Jacobian of the transformation (69) will in general be different from unity. As we already discussed above, the Jacobian equals

$$J = 1 - \frac{\delta^r}{\delta\phi_A^*}\left(\frac{\delta^l S_{ext}}{\delta\phi^A(x)}\mu\right) . \tag{71}$$

Thus to demand that the combination of measure and action remains invariant, we must in general require that

$$\frac{1}{2}(S_{ext}, S_{ext}) = -\frac{\delta^r S_{ext}}{\delta\phi^A}c^A + i\hbar\Delta S_{ext} , \tag{72}$$

which, assuming the form (68) – since we know that this is sufficient to guarantee the correct Schwinger-Dyson equations – reduces to the quantum Master Equation of Batalin and Vilkovisky:

$$\frac{1}{2}(S^{(BV)}, S^{(BV)}) = i\hbar\Delta S^{(BV)} . \tag{73}$$

Let us emphasize that this equation follows even *before* possible gauge fixings. It is required in order that the general Schwinger-Dyson equations for the fundamental fields are satisfied, and is not postulated on only the requirement that the final functional integral be independent of the gauge-fixing function. However, gauge independence of the functional integral upon the addition of a term of the form $\delta\Psi[\phi]$ now follows straightforwardly, since for a functional Ψ that depends only on the fields ϕ, we have $\delta^2\Psi[\phi] = 0$.

D Quantum BRST

In subsection A we noted that the usual BRST Schwinger-Dyson symmetry acquires a "quantum correction" if one insists on using the formalism where the new ghost fields c^A have been integrated out of the path integral. As we saw already in the case of no gauge symmetries, this deforms the BRST operator:

$$\delta \rightarrow \sigma = \delta - i\hbar\Delta . \tag{74}$$

The notation is not entirely precise, because the operator δ on the right hand side of this equation of course equals the operator δ on the left hand side only modulo those changes incurred by integrating out the ghosts c^A. But we keep it like this to avoid complications in the notation. After having integrated out the ghosts c^A, the BRST operator δ will become identical to the variation within the antibracket.

Since this quantum deformation involves the same operator $\hbar\Delta$ that in certain specific cases may modify the classical Master Equation, one might be led to believe that these two issues are related, *i.e.*, that the "quantum BRST" operator should only be applied when there are, (or as a consequence of having) quantum corrections in the full gauge-fixed action. This is actually not the case, and we therefore find it useful to return briefly to the meaning of the quantum BRST operator, here denoted by σ.

Let us again choose the simplest solution to the Master Equation of the form (68). We emphasize that it is immaterial whether this extended action S_{ext} satisfies the classical or quantum Master Equations. Since we are interested in seeing the effect of integrating out the ghosts c^A, consider, as in subsection A, the expectation value of the BRST variation of an arbitrary functional $G = G[\phi^A, \phi^*_A]$:

$$\langle \delta G[\phi, \phi^*] \rangle = \mathcal{Z}^{-1} \int [d\phi][d\phi^*][dc] \delta G[\phi, \phi^*] \exp\left[\frac{i}{\hbar}\left(S^{(BV)} - \phi^*_A c^A\right)\right]$$

$$
\begin{aligned}
&= \mathcal{Z}^{-1} \int [d\phi][d\phi^*][dc] \left\{ \frac{\delta^r G}{\delta \phi^A} c^A + \frac{\delta^r G}{\delta \phi^*_A} \left(- \frac{\delta^l S_{ext}}{\delta \phi^A} \right) \right\} \\
&\times \exp \left[\frac{i}{\hbar} \left(S^{(BV)} - \phi^*_A c^A \right) \right] \\
&= \mathcal{Z}^{-1} \int [d\phi][d\phi^*] \left\{ \frac{\delta^r G}{\delta \phi^A} (i\hbar) \frac{\delta^l}{\delta \phi^*_A} \delta(\phi^*) - \frac{\delta^r G}{\delta \phi^*_A} \frac{\delta^l S^{(BV)}}{\delta \phi^A} \delta(\phi^*) \right\} \\
&\times \exp \left[\frac{i}{\hbar} S^{(BV)} \right] \\
&= \mathcal{Z}^{-1} \int [d\phi][d\phi^*] \delta(\phi^*) \left\{ \frac{\delta^r G}{\delta \phi^A} \frac{\delta^l S^{(BV)}}{\delta \phi *_A} \right. \\
&+ (i\hbar)(-1)^{\epsilon_A} \frac{\delta^r \delta^r G}{\delta \phi *_A \delta \phi^A} - \left. \frac{\delta^r G}{\delta \phi^*_A} \frac{\delta^l S^{(BV)}}{\delta \phi^A} \right\} \exp \left[\frac{i}{\hbar} S^{(BV)} \right] \\
&= \langle (G, S^{(BV)}) - i\hbar \Delta G \rangle \ .
\end{aligned} \tag{75}
$$

The derivation given here corresponds to the path integral before gauge fixing, but it goes through in entirely the same manner in the gauge-fixed case. (The only difference is that the relevant δ-function reads $\delta(\phi^* - \delta^r \Psi / \delta \phi)$ instead of $\delta(\phi^*)$; this does not affect the manipulations above).

The emergence of the "quantum correction" in the BRST operator is thus completely independent of the particular solution $S^{(BV)}[\phi, \phi^*]$; it must always be included when one uses the formalism in which the ghosts c^A have been integrated out. The quantum BRST operator σ is unusual, because it appears only after functional manipulations inside the path integral.

Since by construction the partition function is invariant under δ (when keeping the ghosts c^A) and σ (after having integrated out these ghosts), it follows that all expectation values involving these operators vanish:

$$
\langle \delta G[\phi, \phi^*] \rangle = 0 \tag{76}
$$

when keeping c^A, and

$$
\langle \sigma G[\phi, \phi^*] \rangle = 0 \tag{77}
$$

when the c^A have been integrated out.

This of course holds for the action as well:

$$
\langle \delta S_{ext} \rangle = 0 \ ; \quad \langle \sigma S^{(BV)} \rangle = 0 \ . \tag{78}
$$

The first of these equations is trivially satisfied when S_{ext} satisfies the classical Master Equation, because then the variation δS_{ext} itself vanishes. This equation is then only non-trivially satisfied when $\Delta S_{ext} \neq 0$.

Since the two operations δ and σ are equivalent in the precise sense given above, the same considerations should apply to the second equation. Indeed it does: When $S^{(BV)}$ satisfies the classical Master Equation, $\sigma S^{(BV)} = 0$ at the operator level, while that equation is satisfied only in terms of expectation values when $\Delta S^{(BV)} \neq 0$.

Note that when $\Delta S_{ext} \neq 0$ (or $\Delta S^{(BV)} \neq 0$), the quantum action is neither invariant under δ nor σ. The action precisely has to remain non-invariant in order to cancel the non-trivial contribution from the measure in that case. This is the origin of the factor 1/2 difference between the quantum Master Equation

$$\frac{1}{2}(S^{(BV)}, S^{(BV)}) - i\hbar\Delta S^{(BV)} = 0 \tag{79}$$

and the operator σ (when acting on $S^{(BV)}$):

$$\langle (S^{(BV)}, S^{(BV)}) - i\hbar\Delta S^{(BV)} \rangle = 0 \ . \tag{80}$$

The combination of these two equations yields the new identities

$$\langle \Delta S^{(BV)} \rangle = 0 \ , \quad \langle (S^{(BV)}, S^{(BV)}) \rangle = 0 \tag{81}$$

which can also be verified directly using the path integral.

The operator δ defines a BRST cohomology only on the subspace of fields ϕ^A; it is only nilpotent on that subspace. The operator σ is nilpotent in general: $\sigma^2 = 0$ (a consequence of having performed partial integrations in deriving it). However, the two operators share the same physical content.

III GENERALIZATIONS OF THE BATALIN-VILKOVISKY FORMALISM

In order to see how the conventional antibracket formalism of Batalin and Vilkovisky can be generalized, it is important to have a fundamental principle from which this formalism can be derived. As has been discussed in the previous sections, this principle is that Schwinger-Dyson BRST symmetry must be imposed on the full path integral.

Schwinger-Dyson BRST symmetry can be derived from the local symmetries of the given path integral measure. When the measure is flat, the relevant symmetry is that of local shifts, and the resulting Schwinger-Dyson BRST symmetry leads directly to a quantum Master Equation on the action S which is exponentiated inside the path integral. This action depends on two new sets of ghosts and antighosts, c^A and ϕ^*_A [5]. The conventional Batalin-Vilkovisky formalism for an action S^{BV} follows if one substitutes $S[\phi, \phi^*, c] = S^{BV}[\phi, \phi^*] - \phi^*_A c^A$ and integrates out the ghosts c^A. The so-called "antifields" of the Batalin-Vilkovisky formalism are simply the Schwinger-Dyson BRST antighosts ϕ^*_A [5].

It is of interest to see what happens if one abandons[4] the assumption of flat measures for the fields ϕ^A, and if one does not restrict oneself to local transformations that leave the functional measure invariant. Some steps in this direction were recently taken in ref. [6]. One here exploits the reparametrization invariance encoded in the path integral by performing field transformations $\phi^A = g^A(\phi', a)$ depending on new fields a^i. It is natural to assume that these transformations form a group, or more precisely, a quasigroup [13]. The objects

$$u_i^A \equiv \left.\frac{\delta^r g^A}{\delta a^i}\right|_{a=0} \tag{82}$$

are gauge generators of this group. They satisfy

$$\frac{\delta^r u_i^A}{\delta \phi^B} u_j^B - (-1)^{\epsilon_i \epsilon_j} \frac{\delta^r u_j^A}{\delta \phi^B} u_i^B = -u_k^A U_{ij}^k , \tag{83}$$

where the U_{ij}^k are structure "coefficients" of the group. They are supernumbers with the property

$$U_{ij}^k = -(-1)^{\epsilon_i \epsilon_j} U_{ji}^k . \tag{84}$$

In ref. [6], specializing to compact supergroups for which $(-1)^{\epsilon_i} U_{ij}^i = 0$, the following Δ-operator was derived:

$$\Delta G \equiv (-1)^{\epsilon_i} \left[\frac{\delta^r}{\delta \phi^A} \frac{\delta^r}{\delta \phi_i^*} G\right] u_i^A + \frac{1}{2}(-1)^{\epsilon_i + 1} \left[\frac{\delta^r}{\delta \phi_j^*} \frac{\delta^r}{\delta \phi_i^*} G\right] \phi_k^* U_{ji}^k . \tag{85}$$

When the coefficients U_{ij}^k are constant, this Δ-operator is nilpotent: $\Delta^2 = 0$. As noted by Koszul [14], and rediscovered by Witten [15], one can define an antibracket (F, G) by the rule

$$\Delta(FG) = F(\Delta G) + (-1)^{\epsilon_G}(\Delta F)G + (-1)^{\epsilon_G}(F, G) . \tag{86}$$

Explicitly, for the case above, this leads to the following new antibracket [6]:

$$(F, G) \equiv (-1)^{\epsilon_i(\epsilon_A + 1)} \frac{\delta^r F}{\delta \phi_i^*} u_i^A \frac{\delta^l G}{\delta \phi^A} - \frac{\delta^r F}{\delta \phi^A} u_i^A \frac{\delta^l G}{\delta \phi_i^*} + \frac{\delta^r F}{\delta \phi_i^*} \phi_k^* U_{ij}^k \frac{\delta^l G}{\delta \phi_j^*} . \tag{87}$$

This antibracket is statistics-changing, $\epsilon((F, G)) = \epsilon(F) + \epsilon(G) + 1$, and has the following properties:

---

4) See the 2nd reference in [6]. This is related to the covariant formulations of the antibracket formalism [12].

$$(F,G) = (-1)^{\epsilon_F\epsilon_G+\epsilon_F+\epsilon_G}(G,F) \tag{88}$$

$$\begin{aligned}(F,GH) &= (F,G)H + (-1)^{\epsilon_G(\epsilon_F+1)}G(F,H)\\ (FG,H) &= F(G,H) + (-1)^{\epsilon_G(\epsilon_H+1)}(F,H)G\end{aligned} \tag{89}$$

$$0 = (-1)^{(\epsilon_F+1)(\epsilon_H+1)}(F,(G,H)) + \text{cyclic perm.}\ . \tag{90}$$

Furthermore,

$$\Delta(F,G) = (F,\Delta G) - (-1)^{\epsilon_G}(\Delta F,G)\ . \tag{91}$$

The Δ given in eq. (85) is clearly a non-Abelian generalization of the conventional Δ-operator of the Batalin-Vilkovisky formalism.

We shall now show how to extend this construction to the general case of non-linear and open algebras. Recently, interest in more complicated algebras such as strongly homotopy Lie algebras [16] has arisen in the context of string field theory [17].

Consider the quantized Hamiltonian BRST operator Ω for first-class constraints with an arbitrary, possibly open, gauge algebra [18].[5] Apart from a set of phase space operators Q^i and P_i, introduce a ghost pair $\eta^i, \mathcal{P}_i$. They have Grassmann parities $\epsilon(\eta^i) = \epsilon(\mathcal{P}_i) = \epsilon(Q^i)+1 \equiv \epsilon_i+1$, and are canonically conjugate with respect to the usual graded commutator:

$$[\eta^i,\mathcal{P}_j] = \eta^i\mathcal{P}_j - (-1)^{(\epsilon_i+1)(\epsilon_j+1)}\mathcal{P}_j\eta^i = i\delta^i_j\ . \tag{92}$$

In addition $[\eta^i,\eta^j] = [\mathcal{P}_i,\mathcal{P}_j] = 0$. The quantum mechanical BRST operator can then be written in the form of a $\mathcal{P}\eta$ normal-ordered expansion in powers of the $\mathcal{P}$'s [18]:

$$\Omega = G_i\eta^i + \sum_{n=1}^{\infty}\mathcal{P}_{i_n}\cdots\mathcal{P}_{i_1}U^{i_1\cdots i_n}\ . \tag{93}$$

Here

$$U^{i_1\cdots i_n} = \frac{(-1)^{\epsilon^{i_1\cdots i_{n-1}}_{j_1\cdots j_n}}}{(n+1)!}U^{i_1\cdots i_n}_{j_1\cdots j_{n+1}}\eta^{j_{n+1}}\cdots\eta^{j_1}\ , \tag{94}$$

and the sign factor is defined by:

$$\epsilon^{i_1\cdots i_{n-1}}_{j_1\cdots j_n} = \sum_{k=1}^{n-1}\sum_{l=1}^{k}\epsilon_{i_l} + \sum_{k=1}^{n}\sum_{l=1}^{k}\epsilon_{j_l}\ . \tag{95}$$

---

5) For a comprehensive review of the classical Hamiltonian BRST formalism, see, *e.g.*, ref. [19].

The $U^{i_1 \cdots i_1}_{j_n \cdots j_{n+1}}$'s are generalized structure coefficients. For rank-1 theories the expansion ends with the 2nd term, involving the usual Lie algebra structure coefficients U^k_{ij}. The number of terms that must be included in the expansion of eq. (93) increases with the rank. By construction $\Omega^2 = 0$.

The functions G_i appearing in eq. (93) are the constraints. In the quantum case they satisfy the constraint algebra

$$[G_i, G_j] \;=\; iG_k U^k_{ij} \;. \tag{96}$$

We choose these to be the ones associated with motion on the supergroup manifold defined by the transformation $\phi^A = g^A(\phi', a)$.

When considering representations of the (super) Heisenberg algebra (92), one normally chooses the operators to act to the right. Thus, in the ghost coordinate representation we could take

$$\mathcal{P}_j = i(-1)^{\epsilon_j} \frac{\delta^l}{\delta \eta^j} \;, \tag{97}$$

and similarly in the ghost momentum representation we could take

$$\eta^j = i \frac{\delta^l}{\delta \mathcal{P}_j} \;. \tag{98}$$

On the other hand, the most convenient representation of the constraint G_j is [13]

$$\overleftarrow{G}_j = -i \frac{\overleftarrow{\delta}^r}{\delta \phi^A} u^A_j \;, \tag{99}$$

which involves a right-derivative *acting to the left.* Using eq. (83), $\overleftarrow{G}_j$ is seen to satisfy

$$[\overleftarrow{G}_i, \overleftarrow{G}_j] = i\overleftarrow{G}_k U^k_{ij} \;. \tag{100}$$

Since we wish Ω of eq. (93) to act in a definite way, we choose representations of the (super) Heisenberg algebra (92) that involve operators acting to the left as well. These are

$$\overleftarrow{\mathcal{P}}_j = i \frac{\overleftarrow{\delta}^r}{\delta \eta^j} \tag{101}$$

in the ghost coordinate representation, and

$$\overleftarrow{\eta}_j = i(-1)^{\epsilon_j}\frac{\overleftarrow{\delta^r}}{\delta\mathcal{P}_j} \tag{102}$$

in the ghost momentum representation. Inserting these operators into eq. (93) will give the corresponding BRST operator $\overleftarrow{\Omega}$ acting to the left. We now identify the ghost momentum $\mathcal{P}_j$ with the Lagrangian antighost ("antifield") ϕ^*_j.

As a special case, consider the operator $\overleftarrow{\Omega}$ in the case of an ordinary rank-1 super Lie algebra for which $(-1)^{\epsilon_i}U^i_{ij} = 0$. In the ghost momentum representation $\overleftarrow{\Omega}$ takes the form

$$\overleftarrow{\Omega} = (-1)^{\epsilon_i}\frac{\overleftarrow{\delta^r}}{\delta\phi^A}u^A_i\frac{\overleftarrow{\delta^r}}{\delta\phi^*_i} - \frac{1}{2}(-1)^{\epsilon_j}\phi^*_k U^k_{ij}\frac{\overleftarrow{\delta^r}}{\delta\phi^*_j}\frac{\overleftarrow{\delta^r}}{\delta\phi^*_i} \, . \tag{103}$$

One notices that the $\overleftarrow{\Omega}$ of the above equation coincides with our non-Abelian Δ-operator of eq. (85). In detail:

$$\Delta F \;\equiv\; F\overleftarrow{\Omega} \, . \tag{104}$$

For a rank-0 algebra – the Abelian case – we get, with the same identification,

$$\overleftarrow{\Omega} = (-1)^{\epsilon_A}\frac{\overleftarrow{\delta^r}}{\delta\phi^A}\frac{\overleftarrow{\delta^r}}{\delta\phi^*_A} \, . \tag{105}$$

The associated Δ-operator, defined through eq. (104) is seen to agree with the Δ of the conventional Batalin-Vilkovisky formalism eq.(9).

We define the general Δ-operator through the identification (104) and the complete expansion

$$\overleftarrow{\Omega} = (-1)^{i}\frac{\overleftarrow{\delta^r}}{\delta\phi^A}u^A_i\frac{\overleftarrow{\delta^r}}{\delta\phi^*_i} + \sum_{n=1}^{\infty}\phi^*_{i_n}\cdots\phi^*_{i_1}\overleftarrow{U}^{i_1\cdots i_n} \, . \tag{106}$$

Here

$$\overleftarrow{U}^{i_1\cdots i_n} = \frac{(-1)^{\epsilon^{i_1\cdots i_{n-1}}_{j_1\cdots j_n}}}{(n+1)!}(i)^{n+1}(-1)^{\epsilon_{j_1}+\cdots+\epsilon_{j_{n+1}}}U^{i_1\cdots i_n}_{j_1\cdots j_{n+1}}\frac{\overleftarrow{\delta^r}}{\delta\phi^*_{j_{n+1}}}\cdots\frac{\overleftarrow{\delta^r}}{\delta\phi^*_{j_1}} \, . \tag{107}$$

By construction we then have $\Delta^2 = 0$.

It is quite remarkable that the above derivation, based on Hamiltonian BRST theory in the operator language, has a direct counterpart in the Lagrangian path integral. The two simplest cases, that of rank-0 and rank-1 algebras have been derived in detail in the Lagrangian formalism in ref. [6]. It is intriguing that completely different manipulations (integrating out the corresponding ghosts c^i, and partial integrations inside the functional integral) in the Lagrangian framework leads to these quantized Hamiltonian BRST operators. The rank-0 case, that of the conventional Batalin-Vilkovisky formalism, corresponds to the gauge generators

$$\overleftarrow{G}_A = -i\frac{\overleftarrow{\delta}^r}{\delta\phi^A} . \tag{108}$$

These are generators of translations: when the functional measure is flat, the Schwinger-Dyson BRST symmetry is generated by local translations. The non-Abelian generalizations correspond to imposing different symmetries as BRST symmetries in the path integral [6].

These non-Abelian BRST operators $\overleftarrow{\Omega}$ can be Abelianized by canonical transformations involving the ghosts [20], but the significance of this in the present context is not clear. Since in general the number of "antifields" ϕ^*_i will differ from that of the fields ϕ^A, it is obvious that u^A_i in general will be non-invertible. Even when the number of antifields matches that of fields, the associated matrix u^A_B may be non-invertible ("degenerate").[6]

Having the general Δ-operator available, the next step consists in extracting the associated antibracket. By the definition (86), this antibracket measures the failure of Δ to be a derivation. When Δ is a second-order operator, the antibracket so defined will itself obey the derivation rule (89). For higher-order Δ-operators this is no longer the case. The antibracket will then in all generality only obey the much weaker relation

$$(F, GH) = (F, G)H - (-1)^{\epsilon_G}F(G, H) + (-1)^{\epsilon_G}(FG, H) . \tag{109}$$

The relation (91) also holds in all generality. When the Δ-operator is of order three or higher, the antibracket defined by (86) will not only fail to be a derivation, but will also violate the Jacobi identity (90).

For higher-order Δ-operators one can, as explained by Koszul [14], use the failure of the antibracket to be a derivation to define *higher antibrackets*. These are Grassmann-odd analogues of Nambu brackets [23,24]. The construction is

---

6) In the special case where u^A_B is invertible, the transformation $\phi^*_A \rightarrow \phi^*_B(u^{-1})^B_A$ makes the corresponding Δ-operator Abelian [11], but we are not interested in that case here. See also refs. [21,22].

most conveniently done in an iterative procedure, starting with the Δ-operator itself [14,25]. To this end, introduce objects Φ^n_Δ which are defined as follows:[7]

$$\begin{aligned}
\Phi^1_\Delta(A) &= (-1)^{\epsilon_A}\Delta A \\
\Phi^2_\Delta(A,B) &= \Phi^1_\Delta(AB) - \Phi^1_\Delta(A)B - (-1)^{\epsilon_A}A\Phi^1_\Delta(B) \\
\Phi^3_\Delta(A,B,C) &= \Phi^2_\Delta(A,BC) - \Phi^2_\Delta(A,B)C - (-1)^{\epsilon_B(\epsilon_A+1)}B\Phi^2_\Delta(A,C) \\
&\vdots \\
\Phi^{n+1}_\Delta(A_1,\ldots,A_{n+1}) &= \Phi^n_\Delta(A_1,\ldots,A_nA_{n+1}) - \Phi^n_\Delta(A_1,\ldots,A_n)A_{n+1} \\
&\quad -(-1)^{\epsilon_{A_n}(\epsilon_{A_1}+\cdots+\epsilon_{A_{n-1}}+1)}A_n\Phi^n_\Delta(A_1,\ldots,A_{n-1},A_{n+1}) \, .
\end{aligned} \tag{110}$$

The Φ^n_Δ's define the higher antibrackets. For example, the usual antibracket is given by

$$(A,B) \equiv (-1)^{\epsilon_A}\Phi^2_\Delta(A,B) \, . \tag{111}$$

The iterative procedure clearly stops at the first bracket that acts like a derivation. For example, the "three-antibracket" defined by $\Phi^3_\Delta(A,B,C)$ directly measures the failure of Φ^2_Δ to act like a derivation. But more importantly, it also measures the failure of the usual antibracket to satisfy the graded Jacobi identity:

$$\begin{aligned}
&\sum_{\text{cycl.}} (-1)^{(\epsilon_A+1)(\epsilon_C+1)}(A,(B,C)) \\
&\quad = (-1)^{\epsilon_A(\epsilon_C+1)+\epsilon_B+\epsilon_C}\Phi^1_\Delta(\Phi^3_\Delta(A,B,C)) \\
&\quad + \sum_{\text{cycl.}} (-1)^{\epsilon_A(\epsilon_C+1)+\epsilon_B+\epsilon_C}\Phi^3_\Delta(\Phi^1_\Delta(A),B,C) \, ,
\end{aligned} \tag{112}$$

and so on for the higher brackets.

When there is an infinite number of higher antibrackets, the associated algebraic structure is analogous to a strongly homotopy Lie algebra L_∞. The L_1 algebra is then given by the nilpotent Δ-operator, the L_2 algebra is given by Δ and the usual antibracket, the L_3 algebra by these two and the additional "three-antibracket", etc. The set of higher antibrackets defined above seems natural in closed string field theory [17], the corresponding Δ-operator being given by the string field BRST operator Q.

---

7) Note that our definitions differ slightly from ref. [14,25] due to our Δ-operators being based on right-derivatives, while those of ref. [14,25] are based on left-derivatives.

Having constructed the Δ-operator (and its associated hierarchy of antibrackets), it is natural to consider a quantum Master Equation of the form

$$\Delta \exp\left[\frac{i}{\hbar}S(\phi,\phi^*)\right] = 0 \ . \tag{113}$$

Using the properties of the Φ^n's defined above, we can write this Master Equation as a series in the higher antibrackets,

$$\sum_{k=1}^{\infty}\left(\frac{i}{\hbar}\right)^k \frac{\Phi^k(S,S,\ldots,S)}{k!} = 0 \ , \tag{114}$$

where each of the higher antibrackets $\Phi^k(S,S,\ldots,S)$ has k entries. The series terminates at a finite order if the associated BRST operator terminates at a finite order. For example, in the Abelian case of shift symmetry the general equation (114) reduces to $i\hbar\Delta S - \frac{1}{2}(S,S) = 0$, the Master Equation of the conventional Batalin-Vilkovisky formalism.

A solution S to the general Master Equation (114) is invariant under deformations

$$\delta S = \sum_{k=1}^{\infty}\left(\frac{i}{\hbar}\right)^{k-1} \frac{\Phi^k(\epsilon,S,S,\ldots,S)}{(k-1)!} \ , \tag{115}$$

where again each Φ^k has k entries, and ϵ is Grassmann-odd. One can view this as the possibility of adding a BRST variation

$$\sigma\epsilon = \sum_{k=1}^{\infty}\left(\frac{i}{\hbar}\right)^{k-1} \frac{\Phi^k(\epsilon,S,S,\ldots,S)}{(k-1)!} \tag{116}$$

to the action. Here σ is the appropriately generalized "quantum BRST operator".[8] In the case of the Abelian shift symmetry, the above σ-operator becomes $\sigma\epsilon = \Delta\epsilon + (i/\hbar)(\epsilon,S)$, which precisely equals ($(i\hbar)^{-1}$ times) the quantum BRST operator of the conventional Batalin-Vilkovisky formalism.

We note that the general Master Equation (114) and the BRST symmetry (115) has the same relation to closed string field theory [17,27] that the conventional Batalin-Vilkovisky Master Equation and BRST symmetry has to open string field theory [15]. The rôle of the action S is then played by the string field Ψ, and the Master Equation (114) is the analogue of the closed string field equations. The symmetry (115) is then the analogue of the closed string field theory gauge transformations.

---

[8] For finite order, a rearrangement in terms of increasing rather than decreasing orders of $\hbar$ may be more convenient.

The present definition of higher antibrackets suggests the existence of an analogous hierarchy of Grassmann-even brackets based on a supermanifold and a non-Abelian open algebra – a natural generalization of Poisson-Lie brackets. It is also interesting to investigate the Poisson brackets and Nambu brackets generated by the generalized antibrackets and suitable vector fields V anticommuting with the generalized Δ-operator (and in particular certain Hamiltonian vector fields within the antibrackets), as described in the case of the usual antibracket in ref. [26]. We explore this in the next section.

IV POISSON BRACKET AND ANTIBRACKET

In this section we want to make an explicit connection between Poisson bracket and the antibracket.

For this purpose, consider the canonical algebra:

$$[x^i, x^j] = 0 \tag{117}$$

$$[p_i, p_j] = 0 \tag{118}$$

$$[x^i, p^j] = i\delta_{ij}1 \tag{119}$$

Now consider the non-abelian antibracket corresponding to a Lie algebra (see the third term of equation (87)):

$$(A, B) = \frac{\partial_r A}{\partial z_i^*} z_k^* U_{ij}^k \frac{\partial_l B}{\partial z_j^*} \tag{120}$$

We choose U_{ij}^k corresponding to the structure constants of the canonical algebra. For each generator of the algebra we include an antifield, z_i^* (z_0^* is the antifield associated to the generator 1).

Now we introduce the operator d ("exterior derivative"). For functions of z^i alone (i.e they do not depend on the antifields), it is:

$$dA = A_{,i}\, z_i^* \tag{121}$$

$$dB = B_{,i}\, z_i^* \tag{122}$$

For the canonical algebra, we get:

$$(dA, dB) = \{A, B\} z_0^* \tag{123}$$

$$\{A, B\} = \frac{\partial A}{\partial x^i}\frac{\partial B}{\partial p_i} - \frac{\partial A}{\partial p_i}\frac{\partial B}{\partial x^i} \tag{124}$$

We see that $\{,\}$ is the usual Poisson bracket.

A Nambu bracket and generalizations

In [23,24] it is considered a possible generalization of the canonical formalism of classical mechanics, where the evolution equation contains two (or more) Hamiltonians:

$$\frac{dF}{dt} = [F, H_1, H_2] \tag{125}$$

$$[G_1, G_2, G_3] = \epsilon_{ijk} \frac{\partial G_1}{\partial x^i} \frac{\partial G_2}{\partial x^j} \frac{\partial G_3}{\partial x^k}, \qquad i, j, k = 1, 2, 3 \tag{126}$$

$[G_1, G_2, G_3]$ is called a Nambu bracket.

Now we show how to obtain it (and its generalizations) from the higher antibrackets of [7]

Introduce:

$$\Delta = \lambda \epsilon_{ijk} \frac{\partial}{\partial x_i^*} \frac{\partial}{\partial x_j^*} \frac{\partial}{\partial x_k^*}, \qquad i, j, k = 1, 2, 3 \tag{127}$$

$$\Delta^2 = 0 \tag{128}$$

For the 3-bracket we obtain:

$$\Phi_3(A, B, C) = 6\lambda(-1)^{\epsilon(B)} \epsilon_{ijk} \frac{\partial A}{\partial x_i^*} \frac{\partial B}{\partial x_j^*} \frac{\partial C}{\partial x_k^*} \tag{129}$$

Now, as in the previous subsection, choose:

$$dG_1 = G_{1,i} x_i^* \tag{130}$$

$$dG_2 = G_{2,i} x_i^* \tag{131}$$

$$dG_3 = G_{3,i} x_i^* \tag{132}$$

the G_k are functions of x^j alone (they do not depend on the antifields).

Choosing λ appropiately, we get:

$$\Phi_3(dG_1, dG_2, dG_3) = [G_1, G_2, G_3] \tag{133}$$

It is clear that the identities satisfied by $\Phi_3(A, B, C)$ imply identities for $[G_1, G_2, G_3]$ [24]. They, in turn, are a direct consequence of Δ being nilpotent and of third order in the derivatives.

The most general Nambu bracket and its properties are obtained in a similar way, starting from a suitable nilpotent Δ-operator and the identification eq.(132) and eq.(133.

V CONCLUSIONS

In these lectures we have studied the Batalin-Vilkovisky(BV) method of quantization from the perspective of the BRST-Schwinger-Dyson symmetry. This has led us directly to a generalization of the algebraic structure behind BV, first to non-abelian Lie algebras and later on, to general open algebras. New Δ operators emerge quite naturally. From them, using the Koszul procedure, we have derived a tower of n-brackets, a master equation and a corresponding invariance of the master equation.

As a simple application of the new formalism, we have exhibit a connection between the non-abelian antibracket and the Poisson bracket. In addition to this, we have explained how to get the Nambu brackets from a suitable Δ operator combined with the Koszul procedure.

ACKNOWLEDGEMENTS

The author wants to express his gratitude to the organizers of the *VII Mexican School of Particles and Fields* and *I Latin American Symposium on High Energy Physics* for a very pleasant stay at Mérida. He also wants to thank the hospitality of L.F. Urrutia at Universidad Nacional Autónoma de México.

His work has been partially supported by Fondecyt # 1950809 and a collaboration Conacyt(México)-Conicyt(Chile).

Appendix

In this appendix we give some additional conventions, and list some useful identities.

The Leibniz rules for derivations of the left and right kind read

$$\frac{\delta^l(F\cdot G)}{\delta A}=\frac{\delta^l F}{\delta A}G+(-1)^{\epsilon_F\cdot\epsilon_A}F\frac{\delta^l G}{\delta A} \tag{134}$$

and

$$\frac{\delta^r(F\cdot G)}{\delta A}=F\frac{\delta^r G}{\delta A}+(-1)^{\epsilon_G\cdot\epsilon_A}\frac{\delta^r F}{\delta A}G\ , \tag{135}$$

where A denotes a field (or antifield) of arbitrary Grassmann parity ϵ_A. Similarly, ϵ_F and ϵ_G are the Grassmann parities of the functionals F and G.

Actual variations, let us denote them by $\bar{\delta}$ in contrast to the BRST transformations δ of the paper, are defined as follows:

$$F[A+\bar{\delta}A] - F[A] \equiv \bar{\delta}F \equiv \bar{\delta}A\frac{\delta^l F}{\delta A} \equiv \frac{\delta^r F}{\delta A}\bar{\delta}A \ . \tag{136}$$

The commutation rule of two arbitrary fields is

$$A \cdot B = (-1)^{\epsilon_A \cdot \epsilon_B} B \cdot A \ , \tag{137}$$

and for actual variations one has the simple rule that

$$\bar{\delta}(F \cdot G) = (\bar{\delta}F)G + F(\bar{\delta}G) \tag{138}$$

independent of the Grassmann parities ϵ_F and ϵ_G. The rules (122) and (123) in conjunction lead to the useful identity

$$\frac{\delta^l F}{\delta A} = (-1)^{\epsilon_A(\epsilon_F+1)}\frac{\delta^r F}{\delta A} \ . \tag{139}$$

The BRST variations we have worked with in this paper correspond to right derivation rules. This is of course not imposed upon us, but it is convenient if we wish to compare our expressions with those of Batalin and Vilkovisky. It follows from requiring the actual variations to be related to the BRST transformations by multiplication of an anticommuting parameter μ *from the right*. This then provides us with very helpful operational rules for the BRST transformations δ. In particular,

$$\bar{\delta}F \equiv (\delta F)\mu = \frac{\delta^r F}{\delta A}\bar{\delta}A \ . \tag{140}$$

Now, since

$$\delta F \equiv \frac{\delta^r \bar{\delta}F}{\delta \mu} \ , \tag{141}$$

it follows that

$$\delta F = \frac{\delta^r F}{\delta A}\delta A \ . \tag{142}$$

From this it also follows directly that the BRST transformations act as right derivations:

$$\delta(F \cdot G) = F(\delta G) + (-1)^{\epsilon_G}(\delta F)G \ . \tag{143}$$

These are the basic rules that are needed for the manipulations in the main text.

REFERENCES

1. I.A. Batalin and G.A. Vilkovisky, Phys. Lett. **B102** (1981) 27; Phys. Rev. **D28** (1983) 2567 [E: **D30** (1984) 508]; Nucl. Phys. **B234** (1984) 106; J. Math. Phys. **26** (1985) 172.
2. M. Henneaux, Nucl. Phys. B (Proc. Suppl.) **18A** (1990) 47.
   J.M.L. Fisch, Univ. Brux. preprint ULB TH2/90-01.
   M. Henneaux and C. Teitelboim, "Quantization of Gauge Systems", Princeton University Press, Princeton, New Jersey (1992).
3. W. Siegel, Phys. Lett. **151B** (1985) 391; 396.
   C. Thorn, Nucl. Phys. **B287** (1987) 61.
   M. Bochicchio, Phys. Lett. **B193** (1987) 31.
   H. Hata, Nucl. Phys. **B329** (1990) 698.
   B. Zwiebach, Nucl.Phys.B390:33-152,1993.
   E. Witten, Phys.Rev.D46:5467-5473,1992.
   H. Hata and B. Zwiebach,Ann.Phys.229:177-216,1994.
4. E. Verlinde, Nucl. Phys. **B381** (1992) 141.
5. J. Alfaro and P.H. Damgaard,Nucl. Phys. B404(1993)751.
6. J. Alfaro and P.H. Damgaard, Nucl. Phys. B455(1995)409; Phys. Lett. B334(1994)369.
7. J. Alfaro and P.H. Damgaard, Phys. Lett. B369(1996)289.
8. K. Bering, P. H. Damgaard and J. Alfaro, Nucl.Phys.B478(1996)459.
9. T. Kugo and S. Uehara, Nucl. Phys. **B197** (1982) 378.
   L. Baulieu, Phys. Rep. **129** (1985) 1.
10. J. Alfaro and P.H. Damgaard, Phys. Lett. **B222** (1989) 425.
    J. Alfaro, P.H. Damgaard, J. Latorre and D. Montano, Phys. Lett. **B233** (1989) 153.
11. J. Alfaro and P.H. Damgaard, Ann. Phys.(N.Y.) **202** (1990) 398.
12. O.M. Khudaverdian, J. Math. Phys. **32** (1991) 1934.
    O.M. Khudaverdian and A.P. Nersessian, Mod. Phys. Lett. **A8** (1993) 2377.
    A. Schwarz, Commun. Math. Phys. **155** (1993) 249; **158** (1993) 373.
    I.A. Batalin and I.V. Tyutin, Int. J. Mod. Phys. **A8** (1993) 2333; Mod. Phys. Lett. **A8** (1993) 3673.
    H. Hata and B. Zwiebach, Ann. Phys. (NY) **229** (1994) 177.
13. I.A. Batalin, J. Math. Phys. **22** (1981) 1837.
14. J.-L. Koszul, Astérisque, hors serie (1985) 257.
15. E. Witten, Mod. Phys. Lett. **A5** (1990) 487.
16. T. Lada and J. Stasheff, Int. J. Theor. Phys. **32** (1993) 1087.
17. E. Witten and B. Zwiebach, Nucl. Phys. **B377** (1992) 55.
    B. Zwiebach, Nucl. Phys. **B390** (1993) 33.
18. I.A. Batalin and E.S. Fradkin, Phys. Lett. **128B** (1983) 303.
19. M. Henneaux, Phys. Rep. **126** (1985) 1.
20. I.A. Batalin and E.S. Fradkin, J. Math. Phys. **25** (1984) 2426.
21. I.A. Batalin and I.V. Tyutin, Int. J. Mod. Phys. **A9** (1994) 517.
22. A. Nersessian, hep-th/9305181.

23. Y. Nambu, Phys. Rev. **D7** (1973) 2405.
24. L. Takhtajan, Commun. Math. Phys. **160** (1994) 295.
25. F. Akman, q-alg/9506027.
26. A. Nersessian and P.H. Damgaard, Phys. Lett. **B355** (1995) 150.
27. T. Kugo and K. Suehiro, Nucl. Phys. **B337** (1990) 434.

Searching for a theory of mass

P. Binétruy

LPTHE, Université Paris-Sud
F-91405 Orsay Cedex, France

Abstract. Recent attempts to understand the diversity of the masses and mixing angles in the observed spectrum of fundamental particles is reviewed.

I INTRODUCTION

Ever since the days of Galilei, the problem of mass has been nagging physicists [1]. We certainly have learnt a great deal, in particular that all the observed masses seem to appear through the spontaneous breaking of the electroweak symmetry. But there remains a lot to understand: (i) if there are heavier mass scales yet to be observed, why is not the electroweak scale destabilized by the associated quantum fluctuations ? (ii) why are not all observed masses of the same order? in other words, why is there such a variety of masses? (iii) what governs the mixing of quarks and leptons? The first problem is often referred to as the *gauge hierarchy problem.* And the second as the *fermion mass hierarchy problem.* The third one is somewhat connected with the latter.

Ever since the days of the successes of QED, the electron had been considered as the prototype of an elementary particle. And questions such as 'why are there quarks as heavy as the charm or bottom quarks' were raised. But the discovery of the top quark around 180 GeV [2] has somewhat changed the perspective. It was realized that the top quark mass falls precisely in the range of mass scale which is typical of the spontaneous breaking of the electroweak symmetry: $(G_F\sqrt{2})^{-1/2} \sim 250$ GeV, which provides the vacuum expectation value of the Higgs field in the Standard Model. Then why is the electron so much lighter? Even the charm and bottom quarks look light by comparison to the top.

This new perspective may lead to new ways of answering the problem of fermion mass hierarchies, and there has been therefore a flurry of activity in recent years. In what follows, I will try to present the various attempts. The very fact that there are so many different attempts is of course a sign that

CP400, *First Latin American Symposium on High Energy Physics/ VII Mexican School of Particles and Fields,* edited by D'Olivo/Klein-Kreisler/Mendez

none of them is completely satisfactory. This is, I believe, what makes the interest of such an enumeration. It is probable that, if a successful model of fermion masses appears in the near future, it will incorporate some of the ideas which are presently scattered among different proposals. It is, in this respect, encouraging to see that the borderlines between the different theories beyond the Standard Model (say supersymmetry, technicolor and/or compositeness) have started to fade with the recent advances on dynamical symmetry breaking.

II YUKAWA COUPLINGS IN THE STANDARD MODEL

It is well-known that in the Standard Model the spontaneous breaking of the electroweak gauge symmetry is the origin of all non-zero masses in the gauge *and* fermion sector. The vacuum expectation value of the scalar field $< H^0 >= v/\sqrt{2}$ is fixed by the Fermi constant: $v = (G_F\sqrt{2})^{-1/2} = 246$ GeV. This fixes the characteristic mass scale of the theory.

Masses in the fermion sector are obtained through the couplings of fermions to the scalar sector, known as Yukawa couplings. More precisely, they read

$$\begin{aligned} \mathcal{L}_Y = &-\lambda_d \bar{\psi}_Q H d_R - \lambda_u \bar{\psi}_Q \tilde{H} u_R \\ &-\lambda_e \bar{\psi}_L H e_R - \lambda_\nu \bar{\psi}_L \tilde{H} \nu_R + \text{h.c.} \end{aligned} \tag{1}$$

where

$$\psi_Q = \begin{pmatrix} u_L \\ d_L \end{pmatrix} \in (\mathbf{2}_L, Y = \frac{1}{3}), H = \begin{pmatrix} H^+ \\ H^0 \end{pmatrix} \in (\mathbf{2}_L, Y = 1) \tag{2}$$

$$\psi_L = \begin{pmatrix} \nu_L \\ e_L \end{pmatrix} \in (\mathbf{2}_L, Y = -1), \tilde{H} = i\tau_2 H^* = \begin{pmatrix} H^0 \\ -H^- \end{pmatrix} \in (\mathbf{2}_L, Y = -1),$$

and we have introduced a right-handed neutrino $\nu_R \in (1_L, Y = 0)$ which does not couple to the gauge fields and does not appear in the Standard Model.

After spontaneous breaking of the gauge symmetry, fermion masses are given in terms of the Yukawa couplings

$$\begin{aligned} m_d &= \lambda_d \frac{v}{\sqrt{2}}, m_u = \lambda_u \frac{v}{\sqrt{2}} \\ m_e &= \lambda_e \frac{v}{\sqrt{2}}, m_\nu = \lambda_\nu \frac{v}{\sqrt{2}}, \end{aligned} \tag{3}$$

and the diversity of quark and lepton masses is reflected in the diversity of Yukawa couplings. Indeed, whereas a top mass around 180 GeV corresponds to a top Yukawa coupling of order one, the electron mass yields an electron

Yukawa coupling some six orders of magnitude below. Not to mention neutrinos for which, in case they are massive, one need introduce even smaller couplings.

The Standard Model gives no clue as to why one encounters such a large span of orders of magnitude for the Yukawa couplings. Indeed, neither the intergeneration $(m_t/m_u, m_b/m_d, m_\tau/m_e, \cdots)$ nor the intrageneration $(m_t : m_b : m_\tau, m_u : m_d : m_e, \cdots)$ hierarchies are explained in this context. One therefore has to call for physics beyond the Standard Model to look for clues.

We have until now suppressed the family indices. When one restores them, there appears mixing among the fermion fields. These mixings also have a hierarchical structure whose origin is also to be found in new physics.

Indeed, let us introduce family indices in (1) and restrict our attention to the couplings to the neutral scalar :

$$\begin{aligned}\mathcal{L}_Y &= -\left[\Lambda_d\right]_{ij} \bar{d}_{Li} H^0 d_{Rj} - \left[\Lambda_u\right]_{ij} \bar{u}_{Li} H^0 u_{Rj} \\ &= -\left[\Lambda_e\right]_{ij} \bar{e}_{Li} H^0 e_{Rj} - \left[\Lambda_\nu\right]_{ij} \bar{\nu}_{Li} H^0 \nu_{Rj}.\end{aligned} \tag{4}$$

In all generality, each Yukawa matrix Λ is a complex matrix. Thus, after the spontaneous breaking of the gauge symmetry the resulting mass matrices are diagonalized with independent rotations of the left-handed and of the right-handed fields. For example,

$$M_d \equiv \Lambda_d < H^0 >= V_L^{d\dagger} D_d V_R^d \tag{5}$$

where V_L^d and V_R^d are unitary matrices and D_d is the diagonal mass matrix: $D_d = \text{Diag}(m_d, m_s, m_b)$. And similarly for Λ_u, Λ_e and Λ_ν.

Therefore the charge eigenstates, *i.e.* the fields $u_{L,Ri}$, $d_{L,Ri}$, $e_{L,Ri}$ and $\nu_{L,Ri}$ which appear in the multiplets of the $SU(2) \times U(1)$ gauge symmetry, are different from the mass eigenstates:

$$\begin{aligned}\hat{u}_{Li} &= (V_L^u)_{ij}\, u_{Lj}, \hat{d}_{Li} = \left(V_L^d\right)_{ij} d_{Lj} \\ \hat{u}_{Ri} &= (V_R^u)_{ij}\, u_{Rj}, \hat{d}_{Ri} = \left(V_R^d\right)_{ij} d_{Rj}, \text{ etc.}\end{aligned} \tag{6}$$

Note that, in the absence of right-handed neutrinos, the rotation matrix V_L^ν can be chosen arbitrarily: massless neutrinos are degenerate.

The corresponding mixings are observed in interactions with the gauge bosons. Let us start with charged currents, *i.e.* couplings to the charged intermediate gauge bosons:

$$\begin{aligned}\mathcal{L}_{cc} = -g/\sqrt{2} \sum_i & \left[\bar{u}_{Li}\gamma^\mu(1-\gamma_5)\, d_{Li}\right] W_\mu^+ \\ & + \left[\bar{\nu}_{Li}\gamma^\mu(1-\gamma_5)\, e_{Li}\right] W_\mu^+ + \text{h.c.}\end{aligned}$$

$$= -g/\sqrt{2}\sum_{k,j}\Big(\bar{\hat{u}}_{Lk}\left[V_L^u V_L^{d\dagger}\right]_{kj}\gamma^\mu(1-\gamma_5)\,\hat{d}_{Lj}W_\mu^+$$

$$+\bar{\hat{\nu}}_{Lk}\left[V_L^\nu V_L^{e\dagger}\right]_{kj}\gamma^\mu(1-\gamma_5)\,\hat{e}_{Lj}W_\mu^+\Big) + \text{h.c.} \tag{7}$$

One recognizes the Cabibbo-Kobayashi-Maskawa matrix

$$V_{CKM} = V_L^u V_L^{d\dagger}, \tag{8}$$

and the neutrino mixing matrix

$$V_\nu = V_L^\nu V_L^{e\dagger} \tag{9}$$

Experimental data shows that the former matrix has a hierarchical structure. It is best expressed in the Wolfenstein parametrization [3]:

$$V_{CKM} \sim \begin{pmatrix} 1-\frac{\lambda^2}{2} & \lambda & A\lambda^3(\rho-i\eta) \\ -\lambda & 1-\frac{\lambda^2}{2} & A\lambda^2 \\ A\lambda^3(\rho-i\eta) & -A\lambda^2 & 1 \end{pmatrix} \tag{10}$$

where λ is the sine of the Cabibbo angle:

$$\lambda = \sin\theta_c \sim 0.22, \tag{11}$$

and $A \sim 0.78$, $\sqrt{\rho^2+\eta^2} \sim 0.36$. Such a structure needs to be explained on the same grounds as the mass hierarchies discussed above.

As is well-known, the situation is quite different for neutral currents. Indeed

$$\mathcal{L}_{nc} = -\frac{g}{2\cos\theta_W}\sum_i\left(\bar{u}_{Li}\gamma^\mu\left[\frac{1-\gamma_5}{2}-\frac{4}{3}\sin^2\theta_W\right]u_{Li}+\cdots\right)Z_\mu$$

$$= -\frac{g}{2\cos\theta_W}\sum_{ijk}\left(\bar{\hat{u}}_{Lk}\,(V_L^u)_{ki}\,\gamma^\mu\left[\frac{1-\gamma_5}{2}-\frac{4}{3}\sin^2\theta_W\right]\left(V_L^{u\dagger}\right)_{ij}\hat{u}_{Lj}+\cdots\right)Z_\mu$$

$$= -\frac{g}{2\cos\theta_W}\sum_i\left(\bar{\hat{u}}_{Li}\gamma^\mu\left[\frac{1-\gamma_5}{2}-\frac{4}{3}\sin^2\theta_W\right]\hat{u}_{Li}+\cdots\right)Z_\mu. \tag{12}$$

Hence at tree level neutral currents are flavor diagonal in the Standard Model.

It is a remarkable property of the Standard Model that this property remains approximately true even at one loop. This provides stringent constraints on extensions of the Model and the amount of flavor changing neutral currents has proven to be the stumbling block of many theories.

Because of its importance I will therefore describe in more details the origin of the one-loop suppression in the Standard Model. Let us consider the standard Feynman diagram which contributes to the $K_L - K_S$ mass difference: a box diagram made with two internal lines of W and two internal lines of up-type quarks.

The corresponding amplitude is found to be of the form

$$\sum_{ij} \left(V^{\dagger}\right)_{di} (V)_{is} \left(V^{\dagger}\right)_{dj} (V)_{js} f\left(\frac{m_{u_i}^2}{M_W^2}, \frac{m_{u_j}^2}{M_W^2}\right), \tag{13}$$

where V is the Cabibbo-Kobayashi-Maskawa mixing matrix and f is a function of the masses of the quarks running in the loop.

If all quark masses were equal, this function would factorize out of the sum, and the amplitude would be proportional to $V_{di}^{\dagger} V_{is} = \left(V^{\dagger} V\right)_{ds} = 0$. Restricting our attention to the first two families, one therefore expects an answer proportional to $m_c^2 - m_u^2 \sim m_c^2$. Indeed, one finds:

$$\frac{\Delta m_K}{m_K} \sim \frac{\alpha}{\pi} G_F m_c^2 \sin^2 \theta_c. \tag{14}$$

This strong suppression of the Flavor Changing Neutral Currents (FCNC) at one loop level in the Standard Model is difficult to achieve in its extensions. As we will see, such theories introduce new fields coupled to the quarks and leptons. If we take as an example the process considered above, these new fields tend to run in the loop diagram and give a contribution which may be too large.

III COMPOSITENESS AND THE PROBLEM OF MASS

A Composite fermions

If one follows the trend of XXth century physics towards an ever more fundamental theory (molecule-atom-nucleus-nucleon-quark), the natural step beyond the Standard Model should be composite quarks and leptons. There are however some indications that the story might not repeat itself this time.

Let us first consider the last step on the substructure road: the proton is made of quarks u and d. Indirect measurements based on the breaking of chiral symmetry tend to indicate that the intrinsic masses of these quarks are at most a few MeV. This shows that the mass of the proton, of order 1 GeV, is mostly binding energy. Indeed, this is the typical scale of the binding force: the strong interaction.

Suppose now that quarks are made of some more fundamental preon fields, bound by a new interaction. Since this interaction has not been detected yet, its typical scale must be larger that the range of energy presently available, say larger than 1 TeV. On the hand, the quarks that we consider have masses of order of a few MeV. How can one ever reconcile this mass scale with a binding energy which we could have naively expected to be of the order of the

TeV? No theory seems to have come up with a satisfactory solution to this problem.

Let us note however that the problem is less severe if one considers the top quark since its mass is around 180 GeV. The top quark might therefore be a candidate for a composite fermion.

B Technicolor

An attractive alternative is to make the Higgs composite: we have stressed earlier its central role in the generation of mass. This is the path followed by technicolor theories.

In this class of models [4], a new gauge interaction –technicolor– becomes strong at a scale $\Lambda_{TC} \sim 1$ TeV. Massless fermions charged under technicolor are introduced: the technifermions F. The associated chiral symmetry is broken at Λ_{TC} much in the same way as the chiral symmetry associated with massless quarks is spontaneously broken by strong interactions. The characteristic scale of this breaking is given by the technipion decay constant F_Π, again in analogy with QCD.

This breaking induces the breaking of the electroweak symmetry because the technifermions are assumed to carry $SU(2) \times U(1)$ quantum numbers. Hence the relation between F_Π and the scale v of spontaneous breaking of $SU(2) \times U(1)$:

$$F_\Pi = v = 246 \text{ GeV}. \tag{15}$$

The technifermion condensates have a typical value at the technicolor scale:

$$< \bar{F}F > |_{\mu=\Lambda_{TC}} \sim \Lambda_{TC}^3. \tag{16}$$

The problems arise when one tries to generate ordinary fermion masses. One needs to introduce a new gauge interaction –known as Extended Technicolor or ETC– broken at a large scale M_{ETC} [5]. This new interaction couples ordinary fermions to technifermions and induces at low energies a 4-fermion interaction (to be more precise, it involves 2 fermions and 2 technifermions) which plays the role of the Yukawa coupling in the Standard Model (the 2 technifermions condense much in the way of the former scalar field). The diagram contributing to the fermion masses involves a loop made with a technifermion and an ETC gauge boson.

One thus obtains typically for quark and lepton masses:

$$m_f = \frac{g_{ETC}^2}{M_{ETC}^2} < \bar{F}F > |_{\mu=M_{ETC}}; \tag{17}$$

If technicolor behaves as QCD,

$$< \bar{F}F > |_{\mu=M_{ETC}} \sim < \bar{F}F > |_{\mu=\Lambda_{TC}} \sim \Lambda_{TC}^3. \quad (18)$$

Hence

$$m_f = \frac{g_{ETC}^2}{M_{ETC}^2} \Lambda_{TC}^3. \quad (19)$$

We see that such a mechanism explains masses of order 1 GeV if the ETC scale is $M_{ETC}/g_{ETC} \sim 30$ TeV. On the other hand, the top mass would require $M_{ETC}/g_{ETC} \sim 2$ TeV, which is a value much too close to the technicolor scale for the whole picture to make sense: the top mass is a real problem for ETC models.

In any case, FCNC are the source of even more serious problems for this class of models [6]. They arise through the tree-level exchange of new boson fields. These bosons could be new ETC gauge bosons which couple an ordinary fermion f with another ordinary fermion f' (their presence is required to close the algebra since the successive couplings of two ETC gauge bosons coupling an ordinary fermion with a technifermion generate this type of coupling). They could also be pseudo-Goldstone fields, that is the Goldstone bosons which appear after the breaking of the chiral technifermion symmetry and which receive a mass through explicit breaking terms (arising from standard or ETC gauge interactions for example).

Constraints on processes such as $K^0-\bar{K}^0$ or $B_d^0-\bar{B}_d^0$ mixing impose to choose M_{ETC}/g_{ETC} larger than a few hundred TeV. It can be inferred from (19) that this in turn makes the ETC interaction useless to generate the masses of the Standard Model. Hence the standard ETC models are excluded.

Ways out of this contradiction have been found. In walking technicolor scenarios [7], one assumes that the coupling α_{TC} runs more slowly than a QCD-type coupling would (this needs many technifermion fields to achieve). Then

$$\frac{< \bar{F}F > |_{\mu=M_{ETC}}}{< \bar{F}F > |_{\mu=\Lambda_{TC}}} \gg 1. \quad (20)$$

This allows to reconcile the FCNC constraints on the ETC scale with fermion masses of order 1 GeV, but not with masses of order of a few hundred GeV like the top. Hence technicolor models fail to explain the top mass.

This is why in recent years, technicolor aficionados have turned to topcolor-assisted technicolor (TC 2) [8]: the top quark gets a dynamical mass from a new strong interaction –topcolor– much in the way of top condensate models [9]; the bottom and light quarks get their masses through ETC interactions.

IV SUPERSYMMETRIC MODELS

In supersymmetric models, the Yukawa couplings as well as the scalar potential are determined by the superpotential which is an analytic function of the fields. This analyticity property is in some sense a reminiscence in the scalar sector of the chirality of the fermions associated by supersymmetry.

When describing the supermultiplets involved in the supersymmetric description of the Standard model, it is easier to use fermions fields of the same chirality. Hence for example, the right-handed u-quark u_R is present through its charge conjugate u_L^c. This understood, the minimal content is given by the superfields (fermion-sfermion):

$$
\begin{aligned}
Q_i &= \begin{pmatrix} U_i \\ D_i \end{pmatrix} \in (\mathbf{3}_c, \mathbf{2}_L, Y = 1/3) \\
L_i &= \begin{pmatrix} N_i \\ E_i \end{pmatrix} \in (\mathbf{1}_c, \mathbf{2}_L, Y = -1) \\
U_i^c &\in (\bar{\mathbf{3}}_c, \mathbf{1}_L, Y = -4/3) \\
D_i^c &\in (\bar{\mathbf{3}}_c, \mathbf{1}_L, Y = 2/3) \\
E_i^c &\in (\bar{\mathbf{1}}_c, \mathbf{1}_L, Y = 2)
\end{aligned}
$$

where i is a family index, and two (Higgs-Higgsino) superfields:

$$
\begin{aligned}
H_d &= \begin{pmatrix} H_d^+ \\ H_d^0 \end{pmatrix} \in (\mathbf{1}_c, \mathbf{2}_L, Y = 1) \\
H_u &= \begin{pmatrix} H_u^0 \\ H_u^- \end{pmatrix} \in (\mathbf{1}_c, \mathbf{2}_L, Y = -1).
\end{aligned}
$$

One notes $< H_d^0 > \equiv v_1$ and $< H_u^0 > \equiv v_2$ and $\tan\beta \equiv v_2/v_1$.

Let us consider the field content of the superfields U and U^c: U contains the left-handed quark u_L and the corresponding squark $\tilde{u}_L$ whereas U^c contains the left-handed squark u_L^c and the squark $\tilde{u}_R^*$, which are the charge conjugates of the right-handed quark u_R and the associated squark $\tilde{u}_R$. Therefore, $\tilde{u}_L$ and $\tilde{u}_R$ are totally independent scalar fields (although, as we will see below, they might have mixed mass terms).

Yukawa couplings are obtained through second derivatives of the superpotential:

$$
\mathcal{L}_Y = \frac{1}{2}\frac{\partial^2 W}{\partial\phi_i \partial\phi_j}\bar{\psi}_i\psi_j + \text{h.c.} \tag{21}
$$

where ψ_i is the fermionic partner of the scalar ϕ_i under supersymmetry.

Terms in the superpotential which yield the Yukawa couplings present in the Standard Model are:

$$W^{(3)} = \Lambda_{D_{ij}} Q_i \cdot H_d D_j^c + \Lambda_{U_{ij}} Q_i \cdot H_u U_j^c + \Lambda_{E_{ij}} L_i \cdot H_d E_j^c, \tag{22}$$

where $Q \cdot H \equiv \epsilon_{ab} Q_a H_b$ with a, b $SU(2)_L$ indices ($\epsilon_{12} = -\epsilon_{21} = 1$) such that each term is a $SU(2)_L$ singlet.

There are actually other terms compatible with the gauge symmetry of the Standard Model. They are:

$$\Lambda_{ijk}^{[LLE^c]} L_i \cdot L_j E_k^c, \ \Lambda_{ijk}^{[QLD^c]} Q_i \cdot L_j D_k^c, \ \Lambda_{ijk}^{[U^cD^cD^c]} U_i^c D_j^c D_k^c. \tag{23}$$

The first two violate lepton number whereas the last one violates baryon number. The presence of all three simultaneously in the superpotential with couplings of order one would therefore lead to a fast decay of the proton. One thus needs a rationale as to why some of these couplings are either extremely small or vanishing: one possibility is to impose a discrete symmetry known as R-parity [10].

R-parity

One of the original motivations of R-parity was to avoid the appearance of new interactions involving the exchange of a single squark or slepton between ordinary matter fields. For it not to occur, the existence of a new discrete symmetry was postulated: R-parity under which ordinary matter fields have charge $+1$ and supersymmetric partners have charge -1.

It can be checked that such a parity can be expressed in terms of the baryon B, lepton number L and spin S of the field considered:

$$R_p = (-1)^{3B+L+2S} \tag{24}$$

An immediate consequence of R-parity is that supersymmetric partners are pair-produced from ordinary matter. For example only a slepton pair, which has R-parity $(-1)^2 = 1$, can be produced in an e^+e^- collision. If supersymmetric partners are heavy, they are thus all the more difficult to produce.

One infers the well-known result that the Lightest Supersymmetric Particle (LSP) is stable: having $R_p = -1$ it cannot decay into ordinary matter. Supersymmetric theories therefore provide a suitable candidate for dark matter.

But although R-parity plays an important role in supersymmetric models, there is nothing sacred to it. The interaction associated with squark (or slepton) exchange may be too weak to have been detected yet. Indeed the corresponding amplitude is proportional to $\lambda_{\not R_p}^2 / m_{\tilde{q}}^2$ where $\lambda_{\not R_p}$ is a R-parity-violating coupling. It can be small in much the same way as Higgs exchange leads to an interaction of order $(m_f/v)^2/m_H^2$ (m_f/v is the order of magnitude of the Yukawa coupling of a fermion of mass m_f) much smaller than the Fermi constant $G_F = 1/v^2$ which characterizes weak interaction. Hence it is possible to include R-parity violation at the condition that the couplings $\lambda_{\not R_p}$ be small enough.

Let us stress again the connection between R-parity and baryon and lepton number: the violation of R-parity is necessarily associated with B or L violations. But there is no problem with proton decay if there is not simultaneous violation of B and L. Indeed, the decay modes of the proton ($p \to \pi^0 e^+$, $p \to K^+ \nu$,...) are forbidden either by B or L conservation.

To conclude on R-parity, we stress that a supersymmetric theory of Yukawa couplings must:

- either explain R-parity,
- or account for the order of magnitude of R-parity violating couplings, that is explain the hierarchies $\Lambda_{\not R_p}/\Lambda_{R_p}$ or $\Lambda^{[LLE^c]} : \Lambda^{[QLD^c]} : \Lambda^{[U^cD^cD^c]}$ for the couplings introduced in (23).

The mu-term

Returning back to the superpotential, one must include terms quadratic in the fields. Assuming R-parity, one finds only one term which involves the Higgs fields:

$$W^{(2)} = \mu H_u \cdot H_d \qquad (25)$$

This term raises a problem, known as the μ-problem. Indeed, the parameter μ must be nonzero in order to have $< H_u > \neq 0$ and $< H_d > \neq 0$ at the minimum of the scalar potential. But, since μ is the only mass scale in the superpotential, it sets the scale for the electroweak symmetry breaking and it must be of the order of 100 GeV. But how to generate such a scale in a theory where the fundamental scale is presumably the grand unification scale or the Planck scale? A possible solution is that the μ parameter arises through supersymmetry breaking and a number of proposals have been made in this direction [11].

Although the mu-term involves only the two Higgs superfields, we will see that it plays a role in issues connected with flavor hierarchies.

If one does not impose R-parity, another term quadratic in the fields may be written:

$$W^{(2)}_{\not R_p} = \mu_{\not R_p} L \cdot H_u. \qquad (26)$$

This term is not completely independent of the previous R-parity violating couplings written in (23). Indeed, a redefinition of H_d ($H'_d = H_d + (\mu_{\not R_p}/\mu) L$) cancels this term but creates new terms of the form (23) out of the Yukawa couplings.

Squarks and sleptons. The issue of flavor changing neutral currents.

One might wonder whether problems associated with the masses of fermions have anything to do with the corresponding sfermion spectrum. At first, since scalars acquire a mass through supersymmetry breaking whereas fermions do not, the question of sfermion masses seems to be related with the issue of supersymmetry breaking rather than with the spectrum of ordinary fermions. We will see that this is not so because of the constraints imposed on Flavor Changing Neutral Currents (FCNC). Let us first discuss squark and slepton masses.

The mass matrix of squarks and sleptons receives two types of contributions:

- supersymmetric, as they appear from the scalar potential which is expressed as a sum of F-terms associated with the superpotential and D-terms associated with gauge interactions:

$$V = \sum_i \left| \frac{\partial W}{\partial \phi_i} \right|^2 + \frac{1J}{2} \sum_a \left| g \phi_i^+ T^a_{ij} \phi_j \right|^2 . \tag{27}$$

- arising from the breaking of supersymmetry; if this breaking is soft (i.e. it does not generate quadratic divergences), it may only induce mass terms for scalars and gauginos and trilinear analytic terms in the potential, known as A-terms.

Let us illustrate this on a single squark field. The mass term has the following structure:

$$(\tilde{q}_L^* \ \ \tilde{q}_R^*) \begin{pmatrix} m_q^2 + m_D^2 + m_{\not{S}}^2 & Am_q + \mu m_q \cot\beta \\ Am_q + \mu m_q \cot\beta & m_q^2 + m_D^2 + m_{\not{S}}^2 \end{pmatrix} \begin{pmatrix} \tilde{q}_L \\ \tilde{q}_R \end{pmatrix} . \tag{28}$$

Among the supersymmetric terms, one finds:

(i) mass terms arising from the superpotential after gauge symmetry breaking. Such terms yield on the diagonal the corresponding fermion mass m_q, and off the diagonal a mass term which arises through the mu-term; for example since $\mu H_u H_d + \lambda_u Q H_u U^c \in W$, we have $|\partial W / \partial H_u|^2 = |\mu H_d + \lambda_u Q U^c|^2$ which gives $\mu v_1 \lambda_u \tilde{u}_L \tilde{u}_R^* +$ h.c. $= \mu m_u \cot\beta \ \tilde{u}_L \tilde{u}_R^* +$ h.c. If we consider all three families, such terms have the same family structure as the quark matrix.

(ii) terms arising from the D-terms in (27); let us consider for example the generator T^3 of $SU(2)$. Since

$$g \ \phi_i^+ T^a_{ij} \phi_j = g \ (\tilde{u}_L^* \frac{1}{2} \tilde{u}_L - \tilde{d}_L^* \frac{1}{2} \tilde{d}_L + H_d^{0*} \frac{1}{2} H_d^0 - H_u^{0*} \frac{1}{2} H_u^0 + \cdots),$$

the corresponding term in the potential yields in particular a mass term $(g^2/2)(v_1^2 - v_2^2)(|\tilde{u}_L J|^2 - |\tilde{d}_L|^2)$. Such a contribution has no family dependence because the standard gauge interactions have no such dependence.

As for the contributions arising from supersymmetry breaking, the soft supersymmetry breaking terms include:

(i) a mass term for scalar fields which we denote generically by $m_{\not{S}}$; the family structure of such terms $(m_{\not{S}})_{ij}$ depends obviously on the way supersymmetry is broken.

(ii) trilinear A-terms; they also contribute to the mass matrix through the coupling:

$$A(\lambda_u Q H_u U^c + \text{h.c.})$$

which yields $A(m_u \tilde{u}_L \tilde{u}_R^* + \text{ h.c.})$. Again the family structure (A_{ijk}) depends on the way supersymmetry is broken.

Thus supersymmetry-breaking contributions a priori introduce a new family dependence. In the process of identifying the mass eigenstates, quarks and leptons undergo different rotations: this is a source of potentially dangerous FCNC [12].

Indeed, let us consider the quark-squark gluino coupling:

$$\bar{q}\tilde{g}\tilde{q} = \bar{\hat{q}} V^q \tilde{g} \tilde{V}^{q\dagger} \hat{\tilde{q}}, \tag{29}$$

where we have adopted notations similar to those introduced for fermions in (6): $\hat{\tilde{q}}$ are the squark mass eigenstates and $\tilde{V}^q$ is the corresponding rotation matrix. The mismatch between quark and squark rotations leads to the mixing matrix

$$W = V^q \tilde{V}^{q\dagger}. \tag{30}$$

This may lead to untolerable FCNC.

Let us consider a diagram which is very similar to the box diagram considered for the Standard Model in sect.1, with internal quarks and W replaced by squarks and gluinos. The corresponding contribution is typically:

$$\frac{\Delta m_K}{m_K} \sim \alpha_S^2 \sum_{i,j} W_{di}^\dagger W_{is} W_{dj}^\dagger W_{js} f\left(\frac{\tilde{m}_i^2}{m_{\tilde{g}}^2}, \frac{\tilde{m}_i^2}{m_{\tilde{g}}^2}\right), \tag{31}$$

where f is a function of the masses $\tilde{m}_i, \tilde{m}_j$ running in the loop.

The data is very constraining:

$$\frac{\Delta m_K}{m_K} = 7 \times 10^{-15}, \tag{32}$$

and similarly for other flavors:

$$\frac{\Delta m_B}{m_B} = 7 \times 10^{-14}, \quad \frac{\Delta m_D}{m_D} \leq 7 \times 10^{-14}, \quad \text{Br}(B \to X_s \gamma) \leq 5.4 \times 10^{-4}. \tag{33}$$

To translate these results into constraints on the parameters, let us write the general squark mass matrix as:

$<\delta_{12}^d>$	$(\delta_{MM}^d)_{12}$	$(\delta_{LR}^d)_{12}$	$<\delta_{13}^d>$	$(\delta_{MM}^d)_{13}$
0.006	0.05	0.008	0.04	0.1
$(\delta_{LR}^d)_{13}$	$(\delta_{LR}^d)_{23}$	$<\delta_{12}^u>$	$(\delta_{MM}^u)_{12}$	$(\delta_{LR}^u)_{12}$
0.06	0.04	0.04	0.1	0.06

TABLE 1. Upper bound on squark mass parameters from FCNC (M stands for L or R).

$$\tilde{M}^{q2} = \begin{pmatrix} \tilde{M}_{LL}^{q2} & \tilde{M}_{LR}^{q2} \\ \tilde{M}_{LR}^{q2} & \tilde{M}_{RR}^{q2} \end{pmatrix}, \tag{34}$$

where each entry is a 3×3 matrix in family space. One then defines [13]:

$$(\delta_{MN}^q)_{ij} \equiv \left(V_M^q \tilde{M}_{MN}^{q2} V_N^{q\dagger} \right)_{ij} / \tilde{m}^2, \tag{35}$$

where M and N run over $\{L, R\}$, V_M^q is the corresponding quark rotation matrix and $\tilde{m}$ is an average squark mass. If V_M^q was equal to $\tilde{V}_M^q$, the squark rotation matrix, the matrix δ_{MN}^q would be diagonal.

The bounds on the matrix elements of δ_{MN}^q coming from the data given in (33) is presented in Table 1.

The most stringent constraints are on the parameter:

$$<\delta_{ij}^q> \equiv \sqrt{(\delta_{LL}^q)_{ij} (\delta_{RR}^q)_{ij}} \tag{36}$$

In general, contributions such as (31) are too large to satisfy the constraints of Table 1. Possible solutions out of this are [13]:

- squark mass degeneracy

  In this case, the function f of the squark masses factorizes in (31) and $\Delta m_K / m_K$ is proportional to $(W^\dagger W)_{ds} = \mathbf{1}_{ds} = 0$. More generally, one concludes that

  $$\frac{\Delta m_K}{m_K}\Big|_{\text{squark exchange}} \sim \frac{\Delta \tilde{m}^2}{\tilde{m}^2} \tag{37}$$

  where $\Delta \tilde{m}^2$ is a typical squark mass difference.

  This solution is often referred to as universality. But universality is often assumed at a grand unified or a string scale close to the Planck scale. One has to renormalize the squark masses down to a low energy scale; this introduces violations to universality ($\Delta \tilde{m}^2$) which may be too large.

- quark-squark alignment

  If the same matrix diagonalizes the quark and squark mass matrices, *i.e.* $V^q \sim \tilde{V}^q$, then $W = V^q \tilde{V}^{q\dagger} \sim 1$ and $W_{ij} \sim 0$ for $i \neq j$. Thus using (31), one finds that the contribution of squark exchange to $\Delta m_K / m_K$ vanishes.

To conclude on this point, we see that, in general, *a supersymmetric theory of fermion masses has to be a theory of fermion* **and** *sfermion masses.* Otherwise, one is likely to run into undesirable FCNC.

V THE INFRARED FIXED POINT APPROACH

Let us turn to Grand Unified Theories (GUTs) which unify some of the Yukawa couplings at the scale of gauge coupling unification M_U. For example in the case of the minimal supersymmetic $SU(5)$ model, since both chiralities of the quark b and lepton τ^c are in the same representations (**10** and **5**), we have the following relation[1]:

$$\lambda_b(M_U) = \lambda_\tau(M_U) \tag{38}$$

The ratio λ_b/λ_τ is equal to m_b/m_τ since both fields receive their masses through their couplings to the Higgs field H_d.

We have to renormalize the relation (38) down to low energies using the Renormalisation Group Equations (RGE) for the Yukawa couplings. In particular, the equation for λ_b reads:

$$\frac{\mu}{\lambda_b}\frac{d\lambda_b}{d\mu} = \frac{1}{8\pi^2}\left(3\lambda_b^2 + \frac{1}{2}\lambda_t^2 - \frac{8}{3}g_s^2 + \cdots\right) \tag{39}$$

where g_s is the QCD gauge coupling and we have neglected the other gauge interactions. It turns out that, starting at the common value (38) which can be evaluated from the measured τ mass, the term in g_s^2 tends to make $\lambda_b(m_b)$ too large. In order to compensate for this, we need to start with a large enough value for the top coupling $\lambda_t(M_U)$, which has the opposite effect on λ_b in (39).

Let us indeed suppose that we start with a large value for λ_t at the unification scale. The RGE for this coupling is similar to the one for b:

$$\frac{\mu}{\lambda_t}\frac{d\lambda_t}{d\mu} = \frac{1}{8\pi^2}\left(3\lambda_t^2 + \frac{1}{2}\lambda_b^2 - \frac{8}{3}g_s^2 + \cdots\right) \tag{40}$$

The term in λ_t^2 dominates at high scales and makes the value of $\lambda(\mu)$ drop rapidly with decreasing μ, until it reaches a value of order g_s where the gauge coupling term has a stabilizing effect [14]. The surprise is that the value obtained this way is very close to the experimental value given by the top quark mass. Let us indeed look more closely at the dynamics.

Quasi-infrared fixed point.

We neglect here λ_b and the $SU(2) \times U(1)$ gauge couplings for simplicity. Then (40) simply reads

---

[1] Similar relations for the first two families are more questionable.

$$\frac{\mu}{\lambda_t^2}\frac{d\lambda_t^2}{d\mu} = \frac{1}{8\pi^2}\left(6\lambda_t^2 - \frac{16}{3}g_s^2\right) \tag{41}$$

and using the RGE for the strong gauge coupling:

$$\frac{\mu}{g_s^2}\frac{dg_s^2}{d\mu} = \frac{1}{8\pi^2}\left(-3g_s^2\right) \tag{42}$$

one obtains the following equation

$$\mu\frac{d\ln[\lambda_t^2/g_s^2]}{d\mu} = \frac{1}{8\pi^2}\left(6\lambda_t^2 - \frac{7}{3}g_s^2\right). \tag{43}$$

This equation obviously has an infrared fixed point at the value:

$$\left(\frac{\lambda_t^2}{g_s^2}\right)^* = \frac{7}{18}. \tag{44}$$

In fact, the 14 orders of magnitude between $\mu_0 = M_U$ and M_Z are not sufficient to reach the fixed point and the value of the couplings at low energies goes like:

$$\left(\frac{\lambda_t^2}{g_s^2}\right)(\mu) \sim \frac{(\lambda_t^2/g_s^2)^*}{1 - \left(\frac{\alpha_s(\mu)}{\alpha_s(\mu_0)}\right)^{-7/9}}. \tag{45}$$

Since $(\alpha_s(\mu)/\alpha_s(\mu_0))^{-7/9}$ at $\mu = M_Z$, this ratio is approximately 0.7 at this scale. One refers to this value as a quasi-infrared fixed point (QIRFP) [15].

The top mass is then

$$m_t \sim v_2\lambda_t|_{QIRFP} = v\sin\beta\lambda_t|_{QIRFP}. \tag{46}$$

Including all corrections, one finds [16]

$$m_t \sim 200 \text{ GeV} \sin\beta \sim 200 \text{ GeV}\frac{\tan\beta}{\sqrt{1+\tan^2\beta}}. \tag{47}$$

Thus, as a function of $\tan\beta$, m_t rises linearly before reaching a plateau at 200 GeV. For $\tan\beta \geq 30$, the bottom Yukawa coupling $\lambda_b = \lambda_t(m_b/m_t)\tan\beta$ cannot be neglected and the previous analysis must be changed: it turns out that after $\tan\beta$ values of order 30, m_t is decreasing with $\tan\beta$.

This is obviously the situation encountered in the case of top-bottom-tau unification:

$$\lambda_b(M_U) = \lambda_t(M_U) = \lambda_\tau(M_U) \tag{48}$$

such as in $SO(10)$ unification. Indeed in this case,

$$\tan\beta \sim \frac{\lambda_t v_2}{\lambda_b v_1} \sim \frac{m_t}{m_b} \gg 1. \tag{49}$$

To summarize, we see that, in the supersymmetric Standard Model, the presence of an infrared fixed point in the RGE allows the determination of the top Yukawa coupling using only low energy information:

- gauge interaction
- multiplet content

and ignoring the nature of the more fundamental theory. Could this be applied to determine more Yukawa couplings?

Dynamical determination of Yukawa couplings through infrared fixed points

Such a determination seems hopeless in the context of the low energy theory [17]. For example, one obtains from (39) and (40)

$$\mu\frac{d\ln(\lambda_b/\lambda_t)}{d\mu} = \frac{5}{16\pi^2}(\lambda_b^2 - \lambda_t^2) \tag{50}$$

which signals the presence of a fixed point $\lambda_b = \lambda_t$. Such a fixed point is obviously only compatible with a large $\tan\beta$ scenario. Note that this is characteristic of fermions which carry the same quantum numbers (here under $SU(3)_c$).

But what about the physics between M_{GUT} and M_{Pl}? A priori, the situation seems hopeless since there are only 2 or 3 orders of magnitude in scale. But let us reconsider for one moment the previous analysis. We found:

$$\left(\frac{\lambda_t^2}{g_s^2}\right)(\mu) \sim \frac{(\lambda_t^2/g_s^2)^*}{1-\left(\frac{\alpha_s(\mu)}{\alpha_s(\mu_0)}\right)^B} \tag{51}$$

In the case considered then, there was a sizable correction to the fixed point value $(\lambda_t^2/g_s^2)^*$ because:

- $\alpha_s(\mu)$ runs slowly which ensures that $\alpha_s(\mu)/\alpha_s(\mu_0)$ is of order one.
- $B = -7/9$ is also of order one.

But if there exists a new interaction for which:

- α runs quickly
- B is large

one may fall very quickly into the infrared fixed point.

This line of reasoning has been pushed in recent years by G. Ross and his collaborators [18]. For the purpose of illustration, let us consider as an example the unified gauge group $SU(3)\times SU(3)\times SU(3)$ with identical gauge couplings. The field content consists of 3 copies of the following representations: $(\mathbf{1},\mathbf{3},\bar{\mathbf{3}})$, $(\bar{\mathbf{3}},\mathbf{1},\mathbf{3})$ and $(\mathbf{3},\bar{\mathbf{3}},\mathbf{1})$. We suppose that the gauge symmetry is valid between the scales $\mu = M_c \sim 10^{18}$ GeV and $\mu = M_U \sim 10^{16}$ GeV where it is broken down to the Standard Model symmetry. When we go from M_c down to M_U, the gauge coupling decreases (because of the number of fields, the theory is not asymptotically free) as well as the Yukawa couplings. They obey an equation similar to (51) with $(\lambda_t^2/g_s^2)^* = 22/9$ and $B = 11/3$. This large value of B is responsible for the fact that the couplings fall much more quickly to their fixed point value. Indeed, one finds $(\alpha(M_U)/\alpha(M_c))^B \sim 0.48$; in other words, the 2 orders of magnitude between M_U and M_c in this theory are as efficient as the 14 orders of magnitude in the case of the minimal supersymmetric model discussed earlier.

This strategy of infrared fixed points might also be attractive for the squark mass matrix. The goal would be the presence of an infrared fixed point such that the soft scalar mass terms are degenerate. For this purpose, one may consider the new symmetry above M_U to be family independent. Then, since the squarks $\tilde{u}$, $\tilde{c}$, $\tilde{t}$ carry the same quantum numbers under this symmetry, their masses obey:

$$(\tilde{m}_u^2 - \tilde{m}_c^2)^* = 0, \quad (\tilde{m}_c^2 - \tilde{m}_t^2)^* = 0, \quad (\tilde{m}_u^2 - \tilde{m}_t^2)^* = 0. \tag{52}$$

VI FAMILY SYMMETRIES

General strategy

A hierarchical pattern is observed among the masses of the quarks and the charged leptons. It is tempting to attribute such a pattern to a family symmetry and the goal of this section is to try and identify the properties of such a symmetry. For example, when one renormalizes the quark and charged lepton masses up to the scale of grand unification, one observes the following hierarchical structure:

$$\begin{aligned} m_u : m_c : m_t &\sim \lambda^8 : \lambda^4 : 1 \\ m_d : m_s : m_b &\sim \lambda^4 : \lambda^2 : 1 \\ m_e : m_\mu : m_\tau &\sim \lambda^4 : \lambda^2 : 1 \end{aligned} \tag{53}$$

where $\lambda \equiv \sin\theta_c \sim 0.22$ and only orders of magnitude are given[2]. One obtains in particular from (53) that

---

[2] In other words, constants of order one are not written explicitly; because the value of λ

$$\frac{m_d m_s m_b}{m_e m_\mu m_\tau} \sim O(1). \tag{54}$$

A similar hierarchical pattern is observed among the mixing angles. For example, the Cabibbo-Kobayashi-Maskawa matrix is written in the Wolfenstein parametrization [3]:

$$V_{CKM} \sim \begin{pmatrix} 1-\frac{\lambda^2}{2} & \lambda & A\lambda^3(\rho - i\eta) \\ -\lambda & 1-\frac{\lambda^2}{2} & A\lambda^2 \\ A\lambda^3(\rho - i\eta) & -A\lambda^2 & 1 \end{pmatrix} \tag{55}$$

where $A \sim 0.78$ and $\sqrt{\rho^2 + \eta^2} \sim 0.36$.

The obvious question is whether one can explain such structures with the help of a family or horizontal symmetry.

The Froggatt-Nielsen proposal

How such a symmetry would work has been explained by C. Froggatt and H. Nielsen almost some 20 years ago [17], when they proposed an illustrative example which remains the prototype of such models. They assume the existence of a symmetry which requires some quark and lepton masses to be zero and generate a finite mass at some order in a symmetry breaking interaction.

To be more specific they choose the example of a continuous symmetry: an abelian gauge symmetry which is spontaneously broken by the vacuum expectation value of a scalar field θ charged under the symmetry. Hierarchies appear because the low energy Yukawa couplings originate from the underlying theory through the couplings of Fig. 5 which involve the couplings of heavy fermions of mass M to the symmetry-breaking scalar θ; after the spontaneous breaking of the family symmetry, the effective Yukawa couplings include factors of the ratio $< \theta > /M$ which is assumed to be small. In the model proposed by Froggatt and Nielsen, the mass M is obtained through the couplings of the heavy fermions to a scalar (neutral under the symmetry) which acquires a vacuum expectation value.

This line of research has been pursued by many authors since [19]- [27]. We will illustrate it on an example.

An illustrative example using a horizontal abelian gauge symmetry

Let us consider an abelian gauge symmetry $U(1)_X$ which forbids any renormalisable coupling of the Yukawa type except the top quark coupling. Hence the matrix Λ_U in (22) has the form:

---

is not very small compared to one, the exponents in (53) are to be understood up to one unit.

	Q_1	Q_2	Q_3
$X = a_0 +$	$a_8 + a_3$	$a_8 - a_3$	$-2a_8$
	U^c	C^c	T^c
$X = b_0 +$	$b_8 + b_3$	$b_8 - b_3$	$-2b_8$
	D^c	S^c	B^c
$X = c_0 +$	$c_8 + c_3$	$c_8 - c_3$	$-2c_8$

TABLE 2. $U(1)_X$ charges for the standard supermultiplets

$$\Lambda_U = \begin{pmatrix} 0 & 0 & 0 \\ 0 & 0 & 0 \\ 0 & 0 & 1 \end{pmatrix} \tag{56}$$

where 1 in the last entry means a matrix element of order one. The presence of such a non-zero entry means that the charges under $U(1)_X$ obey the relation:

$$X_{Q_3} + X_{U_3^c} + X_{H_u} = 0 \tag{57}$$

whereas similar combinations for the other field are non-vanishing and prevent the presence of a non-zero entry elsewhere in the matrix Λ_U.

We assume that this symmetry is spontaneously broken through the vacuum expectation value of a field θ of charge X_θ normalized to -1: $< \theta > \neq 0$. The presence of non-renormalizable terms of the form $Q_i U_j^c H_u (\theta/M)^{n_{ij}}$ induces in the effective theory below the scale of $U(1)_X$ breaking an effective Yukawa matrix of the form:

$$\Lambda_U = \begin{pmatrix} \lambda^{n_{11}} & \lambda^{n_{12}} & \lambda^{n_{13}} \\ \lambda^{n_{21}} & \lambda^{n_{22}} & \lambda^{n_{23}} \\ \lambda^{n_{31}} & \lambda^{n_{32}} & 1 \end{pmatrix} \tag{58}$$

where $\lambda =< \theta > /M$ and

$$n_{ij} = X_{Q_i} + X_{U_j^c} + X_{H_u} \tag{59}$$

($n_{33} = 0$). Such nonrenormalizable interactions may arise through integrating out heavy fermions of mass M as in the Froggatt-Nielsen model or appear if the underlying theory incorporates gravity, *e.g.* in string theories, in which case the scale M is the Planck scale M_P.

Let us note for example the X-charges of the standard supermultiplets as given in Table 2 (we assume that $3a_8 + b_8 > a_3 + b_3 > 0$ and $3a_8 + b_8 > a_3 + b_3 > 0$).

Then the CKM matrix reads

$$V = \begin{pmatrix} 1 & \lambda^{2a_3} & \lambda^{3a_8+a_3} \\ \lambda^{2a_3} & 1 & \lambda^{3a_8-a_3} \\ \lambda^{3a_8+a_3} & \lambda^{3a_8-a_3} & 1 \end{pmatrix} \tag{60}$$

One has in particular $V_{us}V_{cb} = V_{ub}$. As for the mass ratios, one finds

$$\begin{aligned} \frac{m_u}{m_t} &\sim \lambda^{3(a_8+b_8)+a_3+b_3}, \frac{m_c}{m_t} \sim \lambda^{3(a_8+b_8)-a_3-b_3} \\ \frac{m_u}{m_t} &\sim \lambda^{3(a_8+c_8)+a_3+c_3}, \frac{m_u}{m_t} \sim \lambda^{3(a_8+c_8)-a_3-c_3} \end{aligned} \tag{61}$$

For example, if $a_3 = c_3$, one obtains

$$V_{us} \sim \lambda^{2a_3} \sim \sqrt{\frac{m_d}{m_s}}, \tag{62}$$

which is a classical relation [28].

It might seem on this example that, by choosing the charges of the different fields, one may accomodate any observed pattern of masses. There are however constraints on the symmetry: in particular those coming from the cancellation of anomalies. This is indeed one of the reasons to choose a local gauge symmetry. We will see that this gives very interesting constraints on the model.

Anomalies

In order to make sense of the horizontal symmetry, one must make sure that the anomalies are cancelled: not only the anomaly C_X corresponding to the triangle diagram with three $U(1)_X$ gauge bosons, but also the mixed anomalies C_i corresponding with triangle diagram with one $U(1)_X$ gauge boson and two gauge bosons of the Standard Model gauge symmetry: $U(1)_Y$, $SU(2)$ and $SU(3)$ for $i \in \{1,2,3\}$ respectively.

Since they are linear in the $U(1)_X$ charges the coefficients C_i depend only on the family independent part of the quantum numbers, respectively a_0, b_0, c_0, d_0 and e_0 for Q_i, U_i^c, D_i^c, L_i and E_i^c. They read explicitly

$$\begin{aligned} C_1 &= a_0 + 8b_0 + 2c_0 + 3d_0 + 6e_0 + h_u + h_d \\ C_2 &= 3(3a_0 + d_0) + h_u + h_d \\ C_3 &= 3(2a_0 + b_0 + c_0) \end{aligned} \tag{63}$$

where h_u and h_d are the X-charges of H_u and H_d.

On the other hand, mass ratios are also given by the X-charges. In the example given above, which is fairly general, one finds:

$$m_u m_c m_t = \lambda^{3(a_0+b_0+h_u)} \tag{64}$$

$$m_d m_s m_b = \lambda^{3(a_0+c_0+h_d)} \tag{65}$$

$$m_e m_\mu m_\tau = \lambda^{3(d_0+e_0+h_d)} \tag{66}$$

Anomaly cancellation would require $C_1 = C_2 = C_3 = \cdots = 0$ which gives $a_0 + b_0 = a_0 + c_0 = 0$ and $3(d_0 + e_0) = -(h_u + h_d)$. Then comparing (64) and (65) with the data (53) yields $h_u = 4$ and $h_d = 2$. The last equation (66) is then incompatible with the same data. This shows that the observed pattern of masses is incompatible with a nonanomalous family symmetry: the $U(1)_X$ symmetry must be anomalous [19,23]. Is this the end of the story?

Before we address this question, let us derive from the mass spectrum (53) a relation among the anomaly coefficients. Since

$$\frac{m_d m_s m_b}{m_e m_\mu m_\tau} = \lambda^{3(a_0+c_0-d_0-e_0)} = \lambda^{h_u+h_d-(C_1+C_2-\frac{8}{3}C_3)/2}. \tag{67}$$

Assuming the presence of a mu-term (25) imposes that $h_u + h_d = 0$. Since the data (54) imposes this ratio of masses to be of order one, one finds:

$$C_1 + C_2 - \frac{8}{3}C_3 = 0. \tag{68}$$

There is one instance where one finds a seemingly anomalous symmetry: in superstring models, there is a $U(1)$ symmetry whose anomaly is compensated by the 4-dimensional version [29] of the Green-Schwarz mechanism [30]. This is possible trough the couplings of the gauge fields to a dilaton-axion-dilatino supermultiplet:

$$\mathcal{L} = -\frac{1}{4}s(x)\sum_i k_i F_{i\mu\nu}F^i_{\mu\nu} + \frac{1}{4}a(x)\sum_i k_i F_{i\mu\nu}\tilde{F}^i_{\mu\nu} \tag{69}$$

where $s(x)$ and $a(x)$ are the dilaton and the axion fields, and k_i is the Kac-Moody level of the corresponding gauge group G_i. If one compares with the ordinary Lagrangian of gauge theories, one realizes that the gauge coupling of G_i at the string scale M is given by

$$\frac{1}{g_i^2(M)} = k_i < s(x) > . \tag{70}$$

Thus, since there is a unique dilaton in string theories

$$k_1 g_1^2(M) = k_2 g_2^2(M) = k_3 g_3^2(M) = k_X g_X^2(M) \tag{71}$$

which is the stringy version of the unification of gauge couplings (we have included here the family group $U(1)_X$).

Performing a $U(1)_X$ gauge transformation: $A^X_\mu(x) \to A^X_\mu(x) + \partial_\mu\theta(x)$ yields

$$\delta\mathcal{L} = \sum_i C_i\theta(x)F^{i\mu\nu}\tilde{F}^i_{\mu\nu}; \tag{72}$$

We can complement this with a Peccei-Quinn transformation of the axion: $a(x) \to a(x) - 4\theta(x)\delta_{GS}$ where δ_{GS} is a number,

$$\delta'\mathcal{L} = -\delta_{GS}\theta \sum_i k_i F^{i\mu\nu}\tilde{F}^i_{\mu\nu}. \tag{73}$$

The total transformation is an invariance of the Lagrangian if $C_i = \delta_{GS}k_i$. Hence the necessary condition for the cancellation of anomalies *à la* Green-Schwarz is

$$\frac{C_1}{k_1} = \frac{C_2}{k_2} = \frac{C_3}{k_3} = \frac{C_X}{k_X} = \delta_{GS}. \tag{74}$$

This set of relations allowed L. Ibáñez [31] to relate the value of the Weinberg angle to the mixed anomaly coefficients of the anomalous $U(1)$. Indeed, using (71) and (74), one finds:

$$\tan^2\theta_W(M) = \frac{g_1^2(M)}{g_2^2(M)} = \frac{k_2}{k_1} = \frac{C_2}{C_1} \tag{75}$$

But we have seen in (68) that, if this anomalous $U(1)$ is to be interpreted as a family symmetry, the observed pattern of mass hierarchies imposes a relation among these anomaly coefficients. In the context of string models where all the nonabelian symmetries appear at the same Kac-Moody level, $k_2 = k_3$ and thus $C_2 = C_3$ We thus find:

$$C_1 = \frac{5}{3}C_2 \tag{76}$$

and

$$\sin^2\theta_W(M) = \frac{3}{8}. \tag{77}$$

Thus, if the horizontal abelian symmetry is precisely the anomalous $U(1)_X$, observed hierarchies of fermion masses are compatible with the standard value of $\sin^2\theta_W$ at gauge coupling unification. And this result is obtained without ever making reference to a grand unified gauge group (which is rarely present in superstring models).

There is another advantage of working in the context of string models. Indeed, in this case, the properties of the anomalous $U(1)_X$ are constrained. In particular, its absolute normalisation is fixed [32] and

$$\lambda^2 = \frac{<\theta>^2}{M^2} = \frac{g^2}{192\pi^2}\mathrm{Tr}\ X \sim 10^{-2} \text{ to } 10^{-1}. \tag{78}$$

Hence, one naturally obtains the small parameter (the Cabibbo angle) that was necessary for the whole picture to make sense.

Let us stress however two potential problems of the approach:

(i) in all the preceding formulas we have neglected factors of order one and discussed only the orders of magnitude as powers of the small parameter

$\lambda \sim \sin\theta_c \sim 1/5$. But since this is not such a small parameter, the actual value of the factor of order one introduces some uncertainties: for example $\lambda^n/2 \sim 3\lambda^{n+1}$. G. Ross [18] proposed to combine this approach with the fixed point method in order to determine these factors of order one. Indeed, we saw that the fixed point approach has difficulties in dealing with large ratios of couplings but can easily account for ratios of coupling constants of order one.

(ii) we already stressed that a theory of mass should also discuss sfermion masses. It turns out that sfermion masses are also constrained by the symmetry $U(1)_X$ [20,13].

Let us consider the LL squark mass matrix. If $X_{Q_i} > X_{Q_j}$, the following mass term is allowed:

$$\tilde{m}\tilde{Q}_i\tilde{Q}_j^* \left(\frac{\theta}{M}\right)^{X_{Q_i}-X_{Q_j}} + \text{h.c.} \tag{79}$$

where $\tilde{m}$ is an overall supersymmetry-breaking scale, and if $X_{Q_j} > X_{Q_i}$,

$$\tilde{m}\tilde{Q}_i\tilde{Q}_j^* \left(\frac{\theta^\dagger}{M}\right)^{X_{Q_j}-X_{Q_i}} + \text{h.c.} \tag{80}$$

(the use of hermitian conjugates of fields is allowed since this is a supersymmetry-breaking contribution).

Thus, after $U(1)_X$ breaking, this yields $\tilde{m}\tilde{Q}_i\tilde{Q}_j^*\lambda^{|X_{Q_i}-X_{Q_j}|}$ + h.c., where $\lambda = <\theta>/M = <\theta^\dagger>/M$ and the scalar mass matrix reads:

$$\tilde{M}_{LL}^{q2} = \begin{pmatrix} 1 & \lambda^{|X_{Q_1}-X_{Q_2}|} & \lambda^{|X_{Q_1}-X_{Q_3}|} \\ \lambda^{|X_{Q_2}-X_{Q_1}|} & 1 & \lambda^{|X_{Q_2}-X_{Q_3}|} \\ \lambda^{|X_{Q_3}-X_{Q_1}|} & \lambda^{|X_{Q_3}-X_{Q_2}|} & 1 \end{pmatrix} \tag{81}$$

Thus the squark mass matrices are already approximately diagonal: there is a partial alignment due to the family symmetry [13].

One finds for the parameters introduced in (35) and (36)

$$(\delta_{LR}^q)_{ij} \sim \frac{\sqrt{m_{q_i}m_{q_j}}}{\tilde{m}} \ll 1, \tag{82}$$

and more constraining conditions for δ_{LL} and δ_{RR}. Typically [27],

$$\left(\delta_{12}^d\right)^2 = (X_{Q_1} - X_{Q_2})(X_{D_1^c} - X_{D_2^c})\left(V_L^Q\right)_{12}\left(V_L^{D^c}\right)_{12}\frac{m_d}{m_s}. \tag{83}$$

where the matrix elements on the right-hand side are to be renormalized down to low energies where they tend to decrease. This still gives some nontrivial constraints on the models.

VII NEUTRINO MASSES

Neutrinos provide a sector where one can still make predictions: one may try to gather enough information on the candidate horizontal symmetry from the quark and charged lepton sector and use it to predict neutrino masses and mixings. In this section, we will see how far this program goes.

Let us recall the two kinds of masses that one may consider for neutral leptons. In order to write a standard neutrino mass, we need to introduce a right-handed neutrino N_R (precisely the degree of freedom which is missing in the Standard Model). Then a Dirac mass term reads

$$\mathcal{L}_{Dirac} = \; - \; m_{\nu D} \bar{\nu}_L N_R + \text{ h.c.} \tag{84}$$

On the other hand, for the Majorana mass term, one does not need the introduction of a right-handed neutrino. Indeed, a Majorana mass term reads:

$$\mathcal{L}_{Majorana} = \; - \; m_{\nu M} \overline{(\nu^c)_R} \nu_L + \text{ h.c.} \tag{85}$$

The seesaw model [33], which represents the prototype of neutrino mass models in all theories which involve large scales such as grand unified or superstring theories, includes both Dirac and Majorana mass terms:

$$\mathcal{L} = -\frac{1}{2} (\overline{\nu}_L \; \overline{(N^c)_L}) \begin{pmatrix} 0 & m_D \\ m_D & M \end{pmatrix} \begin{pmatrix} (\nu^c)_R \\ N_R \end{pmatrix}. \tag{86}$$

In the case where $M \gg m_D$, the eigenvalues are respectively:

$$m_1 \sim \frac{m_D^2}{M}, \quad m_2 \sim M \tag{87}$$

and, due to the presence of a zero in the matrix, the mixing angle is given in terms of mass ratios:

$$\tan\theta \sim \sqrt{\frac{m_1}{m_2}} = \frac{m_D}{M}. \tag{88}$$

We have discussed the case of one family but the discussion easily generalises to the 3-family case with a mass matrix of the form:

$$\mathcal{M} = \begin{pmatrix} 0 & \mathcal{M}_D \\ \mathcal{M}_D & \mathcal{M}_M \end{pmatrix}, \tag{89}$$

where the Dirac and Majorana mass matrices, respectively $\mathcal{M}_D$ and $\mathcal{M}_M$, are 3x3 matrices. Then the light neutrino mass matrix reads:

$$\mathcal{M}_\nu = -\mathcal{M}_D \mathcal{M}_M^{-1} \mathcal{M}_D^T. \tag{90}$$

Of course, given the freedom we have on each of the specific entries in $\mathcal{M}_D$ and $\mathcal{M}_M$, seesaw models really form a class of models and one has to go to specifics in order to discuss their phenomenology. This is precisely what a family symmetry provides us with [34]- [39], [26,40].

In order to discuss neutrino masses in this context, we introduce one right-handed neutrino for each family: $\bar{N}_i$, $i \in \{1,2,3\}$. The neutrino Dirac mass term is generated from the non-renormalisable couplings:

$$L_i H_u \bar{N}_j \left(\frac{\theta}{M}\right)^{p_{ij}} \tag{91}$$

where L_i is the left-handed lepton doublet and $p_{ij} = X_{L_i} + X_{\bar{N}_j} + X_{H_u}$ is assumed to be positive[3]. This yields a Dirac mass matrix:

$$(\mathcal{M}_D)_{ij} \sim < H_u > \left(\frac{< \theta >}{M}\right)^{p_{ij}} . \tag{92}$$

The entries of the Majorana matrix $\mathcal{M}_M$ are generated in the same way, with non-renormalizable interactions of the form:

$$M \bar{N}_i \bar{N}_j \left(\frac{\theta}{M}\right)^{q_{ij}} \tag{93}$$

giving rise to effective Majorana masses

$$(\mathcal{M}_M)_{ij} \sim M \left(\frac{< \theta >}{M}\right)^{q_{ij}} \tag{94}$$

provided that $q_{ij} = X_{\bar{N}_i} + X_{\bar{N}_j}$ is a positive integer[2].

The order of magnitude of the entries of the light neutrino mass matrix $\mathcal{M}_\nu = -\mathcal{M}_D \mathcal{M}_M^{-1} \mathcal{M}_D^T$ (cf. (90)) are therefore fixed by the $U(1)_X$ symmetry.

Let us suppose that all entries of $\mathcal{M}_D$ and $\mathcal{M}_M$ are nonzero (i.e. $p_{ij}, q_{ij} \geq 0$); one obtains [26]:

$$(\mathcal{M}_\nu)_{ij} \sim \frac{< H_u >^2}{M} \lambda^{X_{L_i} + X_{L_j} + 2X_{H_u}} \tag{95}$$

which leads to the following light neutrino masses and lepton mixing matrix:

$$m_{\nu_i} \sim \frac{< H_u >^2}{M} \lambda^{2X_{L_i} + 2X_{H_u}} U_{ij} \sim \lambda^{|X_{L_i} - X_{L_j}|} \tag{96}$$

One therefore finds that the neutrino spectrum is hierarchical. Moreover, the structure of the lepton mixing matrix is very similar to the Cabibbo-Kobayashi-Maskawa matrix (60). And there is a relation between neutrino mass ratios and mixing angles [37]:

$$U_{ij}^2 \sim \frac{m_{\nu_i}}{m_{\nu_j}} m_{\nu_i} < m_{\nu_j} \tag{97}$$

---

3) Otherwise, this coupling is absent, which leads to a supersymmetric zero in the mass matrix.

A Neutrino degeneracies

It would be wrong however to consider that a hierarchical spectrum is a generic feature of this type of models. Indeed, one can work out models which allow for degeneracies in the light neutrino spectrum [40]. It is easy to see (for example from (97)) that such degeneracies are associated in these models with large mixings. And large mixing angles may be necessary to solve some of the neutrino problems: the solar neutrino deficit in the large angle branch of the MSW interpretation [41] or the present atmospheric neutrino data [42]. Indeed, approximate mass degeneracy between two or even all three neutrinos has been advocated to explain the different neutrino puzzles [43–45].

In the case of family symmetries, one may first note that the situation in the neutrino sector might be very different from the quark and charged lepton sector where one Yukawa coupling (the top) dominates over all the rest, thus providing a clear starting point for the $U(1)_X$ symmetric situation, summarized in (56). In the case of neutrinos, the $U(1)_X$ symmetry, even though it is abelian, may induce some degeneracies. Consider for example the following matrix [46,47]:

$$\mathcal{M}_\nu = \begin{pmatrix} 0 & 0 & 0 \\ 0 & 0 & a \\ 0 & a & 0 \end{pmatrix} \tag{98}$$

where a is a number of order one. This pattern corresponds to the conservation of a combination of lepton number *à la* Zeldovich-Konopinsky-Mahmoud [48]: indeed $\mathbf{L_e}$ and $\mathbf{L}_\mu + \mathbf{L}_\tau$ are separately conserved. It has two degenerate eigenvalues and the corresponding diagonalizing matrix R_ν has one large mixing angle:

$$D_\nu = \begin{pmatrix} 0 & 0 & 0 \\ 0 & -a & 0 \\ 0 & 0 & a \end{pmatrix} \quad R_\nu = \begin{pmatrix} 1 & 0 & 0 \\ 0 & \frac{1}{\sqrt{2}} & \frac{1}{\sqrt{2}} \\ 0 & -\frac{1}{\sqrt{2}} & \frac{1}{\sqrt{2}} \end{pmatrix} \tag{99}$$

where $D_\nu = R_\nu^T \mathcal{M}_\nu R_\nu$.

Of course, as soon as the $U(1)_X$ symmetry is broken, the vanishing entries in (98) are filled by powers of the small parameter λ. This lifts the degeneracy at a level which is fixed by the charges under the $U(1)_X$ symmetry. In one model considered in Ref. [40], where the lepton doublet charges follow the pattern $0 < X_{L_2} = -X_{L_3} \equiv l < X_{L_1} \equiv l'$, one obtains:

$$\mathcal{M}_\nu \simeq \begin{pmatrix} b\,\lambda^{2l'} & c\,\lambda^{l'+l} & e\,\lambda^{l'-l} \\ c\,\lambda^{l'+l} & d\,\lambda^{2l} & a \\ e\,\lambda^{l'-l} & a & f\,\lambda^{2l} \end{pmatrix} \tag{100}$$

The eigenvalues are found to be $m_{\nu_1} \simeq \lambda^{2l'}$, $-m_{\nu_2} \simeq m_{\nu_3} \simeq a$ to leading order but $m^2_{\nu_3} - m^2_{\nu_2} \simeq \lambda^{2l}$. Also the lepton mixing matrix (obtained by multiplying R_ν with the charged lepton diagonalisation matrix) reads, to leading order,

$$U \simeq \begin{pmatrix} 1 & A\,\lambda^{l'-l} & A\,\lambda^{l'-l} \\ -\sqrt{2}\,A^*\,\lambda^{l'-l} & \frac{1}{\sqrt{2}} & \frac{1}{\sqrt{2}} \\ B\,\lambda^{l'+l} & -\frac{1}{\sqrt{2}} & \frac{1}{\sqrt{2}} \end{pmatrix} \tag{101}$$

where A and B are constant of order one and one recognizes the large mixing angles in the (2,3) sector.

As stressed above, approximate degeneracies among neutrinos may be welcome to solve some of the problems encountered in neutrino phenomenology. If, as in the models discussed above, degeneracies are related with the presence of large mixing angles, one may envisage the following mass patterns:

1. $m_{\nu_{\tau(e)}} \ll m_{\nu_{e(\tau)}} \ll m_{\nu_\mu}$ (MSW non-adiabatic and LSND effect)
2. $m_{\nu_e} \simeq m_{\nu_\tau} \ll m_{\nu_\mu}$ (MSW adiabatic and LSND effect)
3. $m_{\nu_\mu} \ll m_{\nu_e} \simeq m_{\nu_\tau}$ (MSW adiabatic and LSND effect)
4. $m_{\nu_e} \ll m_{\nu_\mu} \simeq m_{\nu_\tau}$ (atmospheric ν problem and LSND effect)
5. $m_{\nu_e} \simeq m_{\nu_\mu} \simeq m_{\nu_\tau}$ (MSW adiabatic and atmospheric ν problem)
6. $m_{\nu_\tau}, m_{\nu_\mu} \ll m_{\nu_e}$ (atmospheric ν problem and LSND effect)

where pattern 3, 4 and 5 can account for the hot dark matter of the Universe [49].

For example, if we wish to use the model described above to understand the atmospheric neutrino problem through $\nu_\mu - \nu_\tau$ oscillations, one finds [40] that $\bar{\nu}_\mu - \bar{\nu}_e$ oscillations may be compatible with the positive result announced by the LSND collaboration [50]. We are therefore in scenario 4. But $\nu_e - \nu_\tau$ and $\nu_\mu - \nu_\tau$ are below the sensitivity of CHORUS or NOMAD (which is the main difference from the phenomenology of the Zee model discussed above) and neutrinoless double beta decay is out of reach of planned experiments.

Another version of the model may explain the solar neutrino deficit through adiabatic MSW and leads to scenario 3, which has an inverted mass hierarchy and has also been advocated in [44].

VIII YUKAWA COUPLINGS IN STRING MODELS

In this section, we will discuss the generic features of Yukawa couplings in string models and the different lines of attack chosen to tackle the problem of hierarchies among the Yukawa couplings.

It is a general property of superstring models that if a Yukawa coupling λ_Y is non-zero, it is typically of order of a gauge coupling g

$$\lambda_Y \sim g. \tag{102}$$

Such a relation should be understood at the string scale and, as we will discuss below, gets renormalized at lower scales. The relation (102) is best understood in the context of 10-dimensional string models compactified to 4 dimensions. Indeed the gauge supermultiplet of the 10-dimensional theory which consists of gauge fields $A_M, M = 0, \cdots, 9$ (8 physical degrees of freedom) and gauginos $\chi_A, A = 1, \cdots, 8$ gives rise, through compactification on a 6-dimensional compact manifold $\mathcal{M}$, to:

(i) gauge supermultiplets (A_μ, λ_α) in 4-dimensional Minkowski space ($\mu = 0, \cdots, 3$);

(ii) chiral supermultiplets (A_I, ψ_α) in 4 dimensions: $I = 4, \cdots, 9$ is a compact index, therefore A_I is a scalar field ϕ in Minkowski space.

Thus a Yukawa term of the form $\phi\bar{\psi}\psi$ originates from a term $A_M\bar{\chi}\chi$ in the underlying 10-dimensional theory, *i.e.* the minimal coupling of gaugino to gauge fields. The coupling constant is therefore a gauge coupling constant:

$$\mathcal{L}_{Yukawa} = \int_{\mathcal{M}} d^6y \; \mathrm{g} f_{abc} A_I^a \psi_\eta^b \Gamma^I_{\eta\zeta} \psi_\zeta^c \tag{103}$$

where Γ^I is a gamma function of the 10-dimensional theory with compact index I.

We stress that the relation (102) is valid at the string unification scale M_S and should therefore be renormalized down to low energies. The standard procedure makes use of the renormalisation group equations. We have seen that such equations have quasi-infrared fixed points; they play an important role for the type of couplings that we consider here. Because the fixed point lies close to the generic value (102) at the string scale, it is most probable that it will be reached at low energies. As we have seen, it corresponds to a typical value for the top quark mass of $m_t \sim m_0 \sin\beta$ where m_0 is around 190 to 200 GeV. The boundary condition therefore yields a natural explanation as to why the top is heavy. But it seems more difficult to understand the origin of the smaller Yukawa couplings: $\lambda_Y \ll g$.

But before we proceed to this question, let us see an example of a model with a large top mass due to a generic value of the corresponding Yukawa coupling; we consider a flipped $SU(5) \times U(1)$ model [51] where the low energy fermions are in representations $(\mathbf{10}, 1/2)$ noted F, $(\bar{\mathbf{5}}, -3/2)$ noted $\bar{f}$ and the singlet representation noted L^c. The superpotential reads

$$W = g \sum_{ijk} \alpha_{ijk} F_i F_j h_k + \beta_{ijk} F_i \bar{f}_j h_k + \gamma_{ijk} \bar{f}_i L_j^c h_k, \tag{104}$$

where i, j, k are family indices and $\alpha_{ijk}, \beta_{ijk}, \gamma_{ijk}$ are numbers equal to 0 or 1. At this stage, massive fermions such as the top are obtained from a mixing of massive states (of mass $g < h >$) and massless states [52]. Hence

$$\lambda_t \sim g \cos\theta \tag{105}$$

where θ is some mixing angle. Thus $\lambda_t < g$ and one recovers bounds for the top mass of the order of the infrared fixed point discussed above. One should note that in this model one also finds $\lambda_b = \lambda_\tau = g \cos\theta'$ where θ' is some other mixing angle. This allows to obtain at low energies $m_b \sim 3m_\tau \sim 5\ GeV$. The hierarchy is thus realized through some mixing to the massless states which must be such that $\cos\theta \gg \cos\theta'$.

Why should Yukawas be small ($\lambda_Y \ll g$)?

Let us recall how one obtains Yukawa couplings in string theories: they are given by the correlation functions of the corresponding vertex operators V computed on the 2-dimensional world-sheet, described by the string in its motion through spacetime. In other words,

$$\mathcal{L}_{Yukawa} = < \mathbf{V}(\phi)\mathbf{V}(\bar{\psi})\mathbf{V}(\psi)V(T_i)\cdots V(\tilde{\phi}_j)\cdots > . \tag{106}$$

In this formula extra fields appear. Indeed Yukawa couplings depend in general on the vacuum expectation values (vevs) of fields such as:

(i) moduli fields T_i, known as Kähler moduli, whose vevs describe radii and shapes of the six-dimensional compact manifold.

(ii) extra singlet and non-singlet fields $\tilde{\phi}_j$: complex structure moduli, etc.

It is this dependence on extra fields which allows to generate some hierarchies among the Yukawa couplings. One can distinguish two different origins for this hierarchy.

- non-perturbative contributions

  A simple non-renormalisation theorem states that the superpotential cannot depend on the moduli T [53]. The idea behind this result goes as follows.

  In the case of a single (Kähler) modulus T, its real part can be interpreted as the radius-squared R^2 of the compact manifold $\mathcal{M}$ whereas its imaginary part is related to the antisymmetric tensor B_{MN} present in the 10-dimensional massless spectrum (M and N are here 6-dimensional compact indices). Similar interpretations apply to the other Kähler moduli. Because of its origin, $Im\ T$ behaves in a way very similar to an axion and only has derivative couplings. Hence the superpotential cannot depend on $Im\ T$, and being analytic in the fields, cannot depend on T as a whole.

  This, however, is a perturbative statement. As a matter of fact, T can be interpreted as $1/\gamma^2$, where γ is the coupling constant of the 2-dimensional sigma-model that characterizes the string model considered on the world-sheet. And the above non-renormalisation theorem can be interpreted as stating the absence of quantum corrections to the superpotential, to all finite orders in the sigma model perturbation. However, non-perturbative contributions such as world-sheet instantons may appear: they are typically of order $\exp(-1/\gamma^2) \sim \exp(-T) \sim \exp(-R^2)$. Hence we may have Yukawa couplings with an exponential suppression [54]:

$$\lambda_Y = g e^{-R^2/\alpha'} \tag{107}$$

where the right dimensions have been restored by using the dimensionful string coupling α'. If $R \gg \sqrt{\alpha'}$, this generates small Yukawa couplings, as looked for.

- family symmetries Effective superstring models generally include many abelian $U(1)$ symmetries, one of which is often anomalous. We have seen in the previous section how such symmetries may account for the fermion mass hierarchies.

The string unification threshold

Nonrenormalisation theorems can also be seen from the point of view of string perturbation theory [55]. One consequence is that the superpotential is not renormalized, to all orders of string perturbation theory (to be explicit, the one loop contribution in this expansion corresponds to world sheets with one handle, *i.e.* with the topology of a torus). At the field theory level, the string coupling is expressed in terms of the vev of the dilaton field ϕ:

$$g_{string} = e^{2\phi} \tag{108}$$

Through supersymmetry, ϕ is related to the antisymmetric tensor $B_{\mu\nu}$ (this time with 4-dimensional indices). Together with a fermion, the dilatino, they form what is known as a linear supermultiplet L, which is real. The superpotential, being analytic in the fields, cannot depend on L. This is related to the gauge invariance associated with the antisymmetric tensor. This in turn ensures that W cannot depend on ϕ and hence is not renormalized, to all orders of string perturbation theory.

This does not mean however that Yukawa couplings do not get renormalised. In fact, because there is wave function renormalisation, fields do get renormalized and so do the Yukawas, in order to cancel the renormalisation effect in the superpotential.

Take for example a one-loop diagram contributing to the renormalisation of a fermion field. At the string level, it consists of a torus with the two vertex operators of the external fermions (of say momentum Q) attached to it. Such a contribution involves two distinct classes of one-loop field theory diagrams: the ones involving massless string states and the ones involving massive string states. The first class of diagrams contributes, through infrared divergent logarithmic contributions $\log Q^2$, to the external fermion anomalous dimension. On the other hand, since the massive states have moduli-dependent (*i.e.* radii-dependent) masses, the second class gives some moduli-dependent contributions known as threshold corrections.

Indeed, these corrections make the boundary conditions (102) at the string scale moduli-dependent [56]. This is similar to what happens in the case of gauge couplings [57], which is not surprising since we have seen earlier the strong connection between generic Yukawa couplings and gauge couplings. Indeed, in the case of the untwisted fields of an orbifold model, it has been shown [56] that, when one expresses the Yukawa couplings in terms of the string couplings, the moduli-dependent corrections of the two types of couplings cancel again each other:

$$\lambda_{klm}(M_S) = g(M_S)\lambda^0_{klm}(1 + g^2(M_S)\alpha_{klm})^{-1/2} \tag{109}$$

where λ^0_{klm} and α_{klm} are moduli-independent constants.

IX DYNAMICAL DETERMINATION IN THE CONTEXT OF SUPERSTRINGS

Finally, I would like to describe a new mechanism proposed by Nambu [58] for generating hierarchies among Yukawa couplings. It was originally devised within the Standard Model but remarkably adapts to the context of strings [59]- [64].

Nambu's idea is to minimize the vacuum energy with respect to the Yukawa couplings subject to a quadratic constraint. Minimization in the absence of the constraints would lead to instabilities (the Yukawa couplings being either zero or infinite). The role of the constraint is precisely to restrain the Yukawas to a finite region of the parameter space. Minimization then tends to favor the boundary regions where one coupling is of order one whereas the others are exponentially smaller.

The quadratic constraint chosen by Nambu originates from the cancellation of the quadratic divergence, typically a Veltman condition [65] but we will see in what follows that other types of constraints may arise in string models [62].

Nambu proposes a toy model with two Yukawa couplings in which the Veltman condition simply reads

$$\lambda_1^2 + \lambda_2^2 = a^2 \tag{110}$$

with a a constant. This constrains both Yukawas to the region $[0, a]$. Such a condition cancels the order Λ^2 contributions to the vacuum energy, where Λ is the cut-off. One is left in the scalar potential with the $O(\ln \Lambda)$ contributions which read:

$$< V_1 >= -A(\lambda_1^4 + \lambda_2^4) + B(\lambda_1^2 \ln \lambda_1^2 + \lambda_2^2 \ln \lambda_2^2). \tag{111}$$

If one disregards for a moment the logarithmic corrections proportional to B, this type of potential favors, in the case of $A > 0$, large hierarchies of couplings: $(\lambda_1, \lambda_2) = (a, 0)$ or $(0, a)$. The general effect of the logarithmic corrections is to generate a non-zero value ($\sim a \exp -(2aA/B)$) for the smaller Yukawa coupling.

A nice property of the Nambu mechanism is that in the case of more than two Yukawa couplings, one coupling is large whereas all the others are small (and of a similar magnitude).

The procedure of minimizing with respect to the Yukawas is certainly meaningful in the case of strings. Indeed we have seen that Yukawa couplings depend on the moduli fields. In this context, the minimization amounts to varying with respect to the underlying moduli. According to the non-renormalisation theorems discussed above, these moduli correspond to flat directions of the potential. The associated degeneracy may be lifted once supersymmetry is broken. For our approach to make sense, one needs to suppose that, at the low energies that we consider, some of the moduli still remain undetermined so that we can minimize with respect to them.

The question of the constraint may receive a different treatment in string models. If one considers the quadratic divergences, one is led to compute the supertrace of the squared mass

$$STrM^2 = \sum_J (-)^J (2J+1) m_J^2 \tag{112}$$

where the sum runs over all fields and all spins J. After supersymmetry breaking, the leading terms in $STrM^2$ are of order $m_{3/2}^2$, where $m_{3/2}$ is the gravitino mass. These terms must cancel since otherwise they would lead to instabilities by driving $m_{3/2}$ to zero or infinity [66]. The next order $O(m_{3/2}^4/M_{Pl}^2)$ should also cancel or at least be bounded when we consider the Yukawa couplings to be dynamical (and minimize with respect to them): the reason again is that otherwise they would lead to instabilities *i.e.* Yukawa couplings being zero or infinite. This gives a condition which is quadratic in the Yukawa couplings [61]. A similar condition could be obtained by minimising with respect to $m_{3/2}$ [60,63] but trying to obtain by minimisation the constraint that confines the Yukawas to a finite region might lead to instabilities. Finally, it is possible that the geometry of the moduli themselves leads to some constraints on the Yukawa couplings: such constraints are in this sense more geometrical than dynamical, which is probably a welcome feature. Let us illustrate this approach on a simple example [62] with two moduli fields T_1 and T_2 and two superfields ϕ_1 and ϕ_2 with moduli-independent Yukawa couplings λ_1 and λ_2. The model is defined by

$$\begin{aligned} K &= -\frac{3}{2}\ln(T_1 + T_1^\dagger - |\phi_1|^2) - \frac{3}{2}\ln(T_2 + T_2^\dagger - |\phi_2|^2), \\ W &= \frac{1}{3}\lambda_1\phi_1^3 + \frac{1}{3}\lambda_2\phi_2^3. \end{aligned} \tag{113}$$

In the low-energy limit, the fields have to be normalized and the scalar potential reads in terms of the normalized fields $\hat{\phi}_i$:

$$V = \hat{\lambda}_1^2 |\hat{\phi}_1|^2 + \hat{\lambda}_2^2 |\hat{\phi}_2|^2 \tag{114}$$

with

$$\hat{\lambda}_1^2 =\sim \left(\frac{T_1 + T_1^\dagger}{T_2 + T_2^\dagger} \right)^{3/2} \lambda_1^2, \; \hat{\lambda}_2^2 =\sim \left(\frac{T_2 + T_2^\dagger}{T_1 + T_1^\dagger} \right)^{3/2} \lambda_2^2. \tag{115}$$

The constraint in this case is therefore multiplicative:

$$\hat{\lambda}_1 \hat{\lambda}_2 = \lambda_1 \lambda_2 \tag{116}$$

which appears to be typical of this type of relation arising from the geometry of moduli.

Minimization at a low-energy scale of the vacuum energy with respect to the top and bottom Yukawa couplings subject to a constraint of the type (116) was studied in detail in [62]. It was shown there that, qualitatively, the ratio of the two couplings behaves as

$$\left(\frac{\lambda_b}{\lambda_t} \right) (1 \text{ TeV}) \sim g^4(M_P) \frac{\mu}{M_{SUSY}}, \tag{117}$$

where g is the gauge coupling, μ is the mu-parameter discussed in Section 3 and M_{SUSY} is the typical mass splitting between superpartners.

It turns out [64] that this type of approach might not be after all so different from the $U(1)$ family symmetry discussed above. One can indeed account for the intergenerational hierarchies in a very similar way: the small parameter $\lambda =< \theta > /M$ of the $U(1)_X$ approach is replaced here by $\lambda = (T_1 + T_1)^\dagger/(T_2 + T_2)^\dagger$ and the $U(1)_X$ charges by the modular weights[4]. The anomalies discussed then are replaced by one-loop modular anomalies.

Given the huge number of different string models, the task of describing the properties of Yukawa couplings might seem overwhelming. It turns out that, barring the technicalities and intricacies of the task of computing the Yukawa couplings in a specific model, the general features are remarkably simple and general.

Generically, the Yukawa couplings are of the order of the gauge couplings at the string unification. In order to obtain smaller couplings, one needs to consider extra fields, either moduli or extra gauge non-singlet fields associated

---

[4] The modular weights are the exponents $n_i^{(\alpha)}$ defined in an expansion of the Kähler potential in terms of the $|\phi_i|^2$: $K = K_0(T_\alpha, T_\alpha^\dagger) + \sum_i (T_\alpha + T_\alpha^\dagger)^{n_i^{(\alpha)}} |\phi_i|^2 + \cdots$ Hence, in the example (113) above, the non-zero modular weights are all equal to -1.

with the breaking of an extra gauge symmetry which provides approximate selection rules. Finally, the dynamical scenario might provide some new ways of getting a hierarchy; it also invoves the underlying moduli structure.

Acknowledgements

I wish to thank the organizers of the VII Mexican School of Particles and Fields for providing such a warm welcome and a stimulating atmosphere during the school.

REFERENCES

1. For a concise and thought-provoking account of the first four hundred years, see L. Okun, Proc. of Erice 1991 "Physics at the highest energy and luminosity", Erice, July 1422, 1991, p.1.
2. F. Abe *et al.*, CDF collaboration, Phys. Rev. Lett. 74 (1995) 2626; S. Abachi *et al.*, D0 collaboration, Phys. Rev. Lett. 74 (1995) 2632.
3. L. Wolfenstein, Phys. Rev. Lett. 51 (1983) 1945.
4. S. Weinberg, Phys. Rev. D19 (1979) 1277; L. Susskind, Phys. Rev. D20 (1979) 2619.
5. S. Dimopoulos and L. Susskind, Nucl. Phys. B155 (1979) 237; E. Eichten and K. Lane, Phys. Lett. B90 (1980) 125.
6. S. Dimopoulos and J. Ellis, Nucl. Phys. B182 (1981) 505.
7. B. Holdom, Phys. Rev. D24 (1981) 1441, Phys. Lett. B150 (1985) 301; T. Appelquist, D. Karabali and L.C.R. Wijewardhana, Phys. Rev. Lett. 57 (1986) 957; T. Appelquist and L.C.R. Wijewardhana, Phys. Rev. D36 (1987) 568; K. Yamawaki, M. Bando and K. Matumoto, Phys. Rev. Lett. 56 (1986) 1335; T. Akiba and T. Yanagida, Phys. Lett. B169 (1986) 432.
8. C.T. Hill, Phys. Lett. B345 (1995) 483.
9. Y. Nambu, Proc. of the XI International Symposium on Elementary Particle Physics, Kazimierz, Poland, 1988, ed. by Z. Adjuk, S. Pokorski and A. Trautman; Enrico Fermi Institute report EFI 89-08 (unpublished); V.A. Miransky, M. Tanabashi and K. Yamawaki, Phys. Lett. B221 (1989) 171, Mod. Phys. Lett. A4 (1989) 1043; W.A. Bardeen, C.T. Hill and M. Lindner, Phys. Rev. D41 (1990) 1647.
10. G. Farrar and P. Fayet, Phys. Lett. B76 (1978) 575.
11. G.F. Giudice and A. Masiero, Phys. Lett. B206 (1988) 480; V.S. Kaplunovsky and J. Louis, Phys. Lett. B306 (1993) 269.
12. H. P. Nilles, Phys. Rep. 110 (1984) 1; L.J. Hall, V.A. Kostelecky and S. Raby, Nucl. Phys. B 267 (1986) 415; F. Gabbiani and A. Masiero, Nucl. Phys. B322 (1989) 235; M. Dine, A. Kagan and R. Leigh, Phys. Rev. D48 (1993) 4269.
13. Y. Nir and N. Seiberg, Phys. Lett. B 309 (1993) 337.
14. B. Pendelton and G.G. Ross, Phys. Lett. 98B (1981) 291.

15. C.T. Hill, Phys. Rev. D24 (1981) 691; C.T. Hill, C.N. Leung and S. Rao, Nucl. Phys. B262 (1985) 517.
16. M. Carena, S. Pokorski and C. Wagner, Nucl. Phys. B406 (1993) 59; W.A. Bardeen, M. Carena, S. Pokorski and C. Wagner, Phys. Lett. B320 (1994) 110; M. Carena, M. Olechowski, S. Pokorski and C. Wagner, Nucl. Phys. B419 (1994) 213, B426 (1994) 269.
17. C.D. Froggatt and H.B. Nielsen, Nucl. Phys. B147 (1979) 277 and B164 (1979) 114.
18. M. Lanzagorta and G.G. Ross, Phys. Lett. B349 (1995) 319 and B364 (1995) 163; G.G. Ross, Phys. Lett. B364 (1995) 216.
19. J. Bijnens and C. Wetterich, Nucl. Phys. B283 (1987) 237.
20. M. Leurer, Y. Nir and N. Seiberg, Nucl. Phys. B398 (1993) 319, B420 (1994) 468.
21. L. Ibáñez and G.G. Ross, Phys. Lett. B332 (1994) 100.
22. V. Jain and R. Shrock, Phys. Lett. B352 (1995) 83.
23. P. Binétruy and P. Ramond, Phys. Lett. B350 (1995) 49.
24. E. Dudas, S. Pokorski and C.A. Savoy, Phys. Lett. B356 (1995) 45.
25. Y. Nir, Phys. Lett. B354 (1995) 107.
26. P. Binétruy, S. Lavignac and P. Ramond, Nucl. Phys. B477 (1996) 353.
27. E. Dudas, S. Pokorski and C.A. Savoy, Phys. Lett. B369 (1996) 255; E. Dudas, C. Grojean, S. Pokorski and C.A. Savoy, Nucl. Phys. B481 (1996) 85.
28. R. Gatto, G. Sartori and M. Tonin, Phys. Lett. B28 (1968) 128.
29. M. Dine, N. Seiberg and E. Witten, Nucl. Phys. B289 (1987) 317.
30. M. Green and J. Schwarz, Phys. Lett. B149 (1984) 117.
31. L. Ibáñez, Phys. Lett. B303 (1993) 55.
32. J. Atick, L. Dixon and A. Sen, Nucl. Phys. B292 (1987) 109.
33. M. Gell-Mann, P. Ramond, and R. Slansky in Sanibel Talk, CALT-68-709, Feb 1979, and in *Supergravity* (North Holland, Amsterdam 1979). T. Yanagida, in *Proceedings of the Workshop on Unified Theory and Baryon Number of the Universe*, KEK, Japan, 1979.
34. A. Rasin and J.P. Silva, Phys. Rev. D49 (1994) R20.
35. H. Dreiner, G.K. Leontaris, S. Lola, G.G. Ross and C. Scheich, Nucl. Phys. B436 (1995) 461; G.K. Leontaris, S. Lola, C. Scheich and J.D. Vergados, Phys. Rev. D53 (1996) 6381.
36. E. Papageorgiu, Z. Phys. C64 (1994) 509, Phys. Lett. B343 (1995) 263, hep-ph/9510352.
37. Y. Grossman and Y. Nir, Nucl. Phys. B448 (1995) 30.
38. P. Ramond, 25th Anniversary Volume of the Centre de Recherches Mathématiques de l'Université de Montréal, hep-ph/9506319.
39. E.J. Chun and A. Lukas, Phys. Lett. B387 (1996) 99.
40. P. Binétruy, S. Lavignac, S. Petcov and P. Ramond, preprint hep-ph/9610481 (to be published in Nuclear Physics).
41. S.P. Mikheyev and A.Yu. Smirnov, Yad. Fiz. **42**, 1441 (1985) [Sov. J. Nucl. Phys. **42**, 913 (1985)]; Il Nuovo Cimento C **9**, 17 (1986); L. Wolfenstein, Phys. Rev. D **17**, 2369 (1978); Phys. Rev. D **20**, 2634 (1979).

42. Y. Fukuda et al., Phys. Lett. B335 (1994) 237; R. Becker-Szendy et al., Nucl. Phys. B (Proc. Suppl.) 38 (1995) 331; M. Goodman et al., Nucl. Phys. B (Proc. Suppl.) 38 (1995) 337; T.K. Gaisser, F. Halzen and T. Stanev, Phys. Rep. 258 (1995) 173.
43. S.T. Petcov and A.Yu. Smirnov, Phys. Lett. B322 (1994) 109.
44. D.O. Caldwell and R.N. Mohapatra, Phys. Lett. B354 (1995) 371; G. Raffelt and J. Silk, Phys. Lett. B366 (1996) 429.
45. D.O. Caldwell and R.N. Mohapatra, Phys. Rev. D48 (1993) 3259 and D50 (1994) 3477.
46. S.T. Petcov, Phys. Lett. 110B (1982) 245; C.N. Leung and S.T. Petcov, Phys. Lett. 125B (1983) 461.
47. D. Wyler and L. Wolfenstein, Nucl. Phys. B218 (1983) 205.
48. Ya.B. Zeldovich, DAN SSSR 86 (1952) 505; E.J. Konopinsky and H. Mahmoud, Phys. Rev. 92 (1953) 1045.
49. J.R. Primack et al., Phys. Rev. Lett. 74 (1995) 2160.
50. C. Athanassopoulos et al., Phys. Rev. Lett. 75 (1995) 2650, and Los Alamos Nat. Lab. report LA–UR–96–1326, April 24, 1996; see also J.E. Hill, Phys. Rev. Lett. 75 (1995) 2654.
51. S.M. Barr, Phys. Lett. 112B (1982) 219; J.P. Derendinger, J.E. Kim and D.V. Nanopoulos, Phys. Lett. 139B (1984) 170; I. Antoniadis, J. Ellis, J. Hagelin and D.V. Nanopoulos, Phys. Lett. 194B (1987) 231.
52. I. Antoniadis, J. Ellis, J. Hagelin and D.V. Nanopoulos, Phys. Lett. 205B (1988) 459, 208B (1988) 209, 231B (1989) 65; G. Leontaris, J. Rizos and K. Tamvakis, Phys. Lett. 251B (1990) 83.
53. E. Witten, Nucl. Phys. B268 (1986) 79.
54. M. Dine, N. Seiberg, X.G. Wen and E. Witten, Nucl. Phys. B289 (1987) 319.
55. E. Martinec, Phys. Lett. 171B (1986) 189; M. Dine and N. Seiberg, Phys. Rev. Lett. 57 (1986) 2625.
56. I. Antoniadis, E. Gava, K.S. Narain and T.R. Taylor, Nucl. Phys. B407 (1993) 706.
57. V. Kaplunovsky, Nucl. Phys. B307 (1988) 145; L. Dixon, V. Kaplunovsky and J. Louis, Nucl. Phys. B355 (1991) 649.
58. Y. Nambu, *Proceedings of the International Conference on Fluid Mechanics and Theoretical Physics in honor of Professor Pei-Yuan Chou's 90th anniversary*, Beijing, 1992; preprint EFI 92-37.
59. P. Binétruy, Proceedings of the Yukawa workshop, Gainesville, February 1994.
60. C. Kounnas, I. Pavel and F. Zwirner, Phys. Lett. B335 (1994) 403.
61. P. Binétruy and E. Dudas, Phys. Lett. B338 (1994) 23.
62. P. Binétruy and E. Dudas, Nucl. Phys. B442 (1995) 21.
63. C. Kounnas, I. Pavel, G. Ridolfi and F. Zwirner, Phys. Lett. B354 (1995) 322.
64. P. Binétruy and E. Dudas, Nucl. Phys. B451 (1995) 31.
65. M. Veltman, *Acta Phys. Pol.* B12 (1981) 437
66. S. Ferrara, C. Kounnas and F. Zwirner, Nucl. Phys. B429 (1994) 589.

Inflation, Topological Defects and Baryogenesis: Selected Topics at the Interface between Particles & Fields and Cosmology[1]

Robert H. Brandenberger

Brown University Physics Department
Providence, R.I. 02912, USA

Abstract. Modern cosmology provides a connection between the physics of particles and fields and observational cosmology. Making use of this link, a wealth of new observational data can be utilized to explore and constrain theories of fundamental physics. Inflationary Universe models and topological defect theories are the most popular current paradigms for explaining the origin of structure in the Universe. In these lectures, I discuss various aspects of inflation and topological defects in which there has been interesting recent progress or in which there are outstanding problems. Particular emphasis is given to how baryogenesis scenarios can be influenced by inflation and topological defects.

INTRODUCTION AND OUTLINE

Most aspects of high energy physics beyond the standard model can only be tested by going to energies far greater than those which present accelerators can provide. Fortunately, the marriage between particle physics and cosmology has provided a way to "experimentally" test the new theories of fundamental forces.

The key realization, discovered both in the context of the scenario of inflation [1] and of topological defects models [2] is that physics of the very early Universe may explain the origin of the structure which is observed. It now appears that a rich set of data concerning the nonrandom distribution of matter on a wide range of cosmological scales, and on the anisotropies in the cosmic microwave background (CMB), may potentially be explained by high energy physics. In addition, studying the consequences of particle physics

1) Invited lectures at the VII Mexican School of Particles and Fields and the I Latin American Symposium on High Energy Physics, Merida, Mexico, 10/29 - 11/6 1996.

CP400, *First Latin American Symposium on High Energy Physics/ VII Mexican School of Particles and Fields*, edited by D'Olivo/Klein-Kreisler/Mendez

models in the context of cosmology may lead to severe constraints on new microscopic theories. Finally, particle physics and field theory may provide explanations of some deep cosmological puzzles, e.g. why the Universe at the present time appears so homogeneous, so close to being spatially flat, and why it contains the observed small net baryon to entropy ratio.

In these lectures, I focus on three important aspects of modern cosmology. The first concerns some fundamental problems of inflationary cosmology. In particular, some recent progress in the understanding of "reheating" in inflation will be reviewed.

The second topic concerns topological defect models of structure formation (due to lack of space and time I focus mostly on cosmic strings). Although at the moment defect theories do not explain some of the basic problems of standard cosmology which inflation does, defects do provide conceptually straightforward and in principle quite predictive theories of structure formation. I will review the main points of the cosmic string model, focusing on the predictions with which defect models and inflation-based structure formation theories can be distinguished.

The third main topic is baryogenesis. Recent progress on electroweak baryogenesis will be reviewed, with particular attention to the role which topological defects may play.

The specific outline is as follows:

1. **Introduction and Outline**

2. **Inflationary Universe: Progress and Problems**
   2.A Problems of Standard Cosmology
   2.B Inflationary Universe Scenario
   2.C Problems of Inflation
   2.D Reheating in Inflationary Cosmology
   2.E Summary

3. **Topological Defects and Structure Formation**
   3.A Quantifying Data on Large-Scale Structure
   3.B Topological Defects
   3.C Formation of Defects in Cosmological Phase Transitions
   3.D Evolution of Strings and Scaling
   3.E Cosmic Strings and Structure Formation
   3.F Specific Predictions

4. **Topological Defects and Baryogenesis**
   4.A Principles of Baryogenesis
   4.B GUT Baryogenesis and Topological Defects
   4.C Electroweak Baryogenesis and Topological Defects
   4.D Summary

In the Merida lectures I also discussed a further important topic, the classical and quantum theory of cosmological perturbations, which has become the main tool of modern cosmology. A general relativistic and quantum mechanical analysis of the generation and evolution of linearized fluctuations is essential in order to be able to accurately calculate the amplitude of density perturbations and CMB anisotropies. However, due to lack of space I refer the readers interested in this topic to other recent conference contributions [3,4] and to a detailed review article [5].

Unless otherwise specified, units in which $\hbar = c = k_B = 1$ will be used. Distances are expressed in Mpc (1pc $\simeq$ 3.06 light years). Following the usual convention, h indicates the expansion rate of the Universe in units of 100 km s^{-1} Mpc^{-1}, $\Omega = \rho/\rho_c$ is the ratio of the energy density ρ to the critical density ρ_c (the density which yields a spatially flat Universe), G is Newton's constant and m_{pl} is the Planck mass.

INFLATIONARY UNIVERSE: PROGRESS AND PROBLEMS

The hypothesis that the Universe underwent a period of exponential expansion at very early times has become the most popular theory of the early Universe. Not only does it solve some of the problems of standard big bang cosmology, but it also provides a causal theory for the origin of inhomogeneities in the Universe which is predictive and in reasonable agreement with current observational results. Nevertheless, there are several problems of principle which merit further study.

Problems of Standard Cosmology

The standard big bang cosmology rests on three theoretical pillars: the cosmological principle, Einstein's general theory of relativity and a perfect fluid description of matter.

The cosmological principle states that on large distance scales the Universe is homogeneous. This implies that the metric of space-time can be written in Friedmann-Robertson-Walker (FRW) form:

$$ds^2 = a(t)^2 \left[\frac{dr^2}{1 - kr^2} + r^2(d\vartheta^2 + \sin^2 \vartheta d\varphi^2) \right] , \tag{1}$$

where the constant k determines the topology of the spatial sections. In the following, we shall usually set $k = 0$, i.e. consider a spatially closed Universe. In this case, we can without loss of generality take the scale factor $a(t)$ to be equal to 1 at the present time t_0, i.e. $a(t_0) = 1$. The coordinates r, ϑ and φ are comoving spherical coordinates. World lines with constant comoving coordinates are geodesics corresponding to particles at rest. If the Universe is expanding, i.e. $a(t)$ is increasing, then the physical distance $\Delta x_p(t)$ between two points at rest with fixed comoving distance Δx_c grows:

$$\Delta x_p = a(t)\Delta x_c \,. \tag{2}$$

The dynamics of an expanding Universe is determined by the Einstein equations, which relate the expansion rate to the matter content, specifically to the energy density ρ and pressure p. For a homogeneous and isotropic Universe, they reduce to the Friedmann-Robertson-Walker (FRW) equations

$$\left(\frac{\dot{a}}{a}\right)^2 - \frac{k}{a^2} = \frac{8\pi G}{3}\rho \tag{3}$$

$$\frac{\ddot{a}}{a} = -\frac{4\pi G}{3}(\rho + 3p) \,. \tag{4}$$

These equations can be combined to yield the continuity equation (with Hubble constant $H = \dot{a}/a$)

$$\dot{\rho} = -3H(\rho + p) \,. \tag{5}$$

The third key assumption of standard cosmology is that matter is described by an ideal gas with an equation of state

$$p = w\rho \,. \tag{6}$$

For cold matter, pressure is negligible and hence $w = 0$. From (5) it follows that

$$\rho_m(t) \sim a^{-3}(t) \,, \tag{7}$$

where ρ_m is the energy density in cold matter. For radiation we have $w = 1/3$ and hence it follows from (5) that

$$\rho_r(t) \sim a^{-4}(t) \,, \tag{8}$$

$\rho_r(t)$ being the energy density in radiation.

The three classic observational pillars of standard cosmology are Hubble's law, the existence and black body nature of the nearly isotropic CMB, and the abundances of light elements (nucleosynthesis). These successes are discussed in detail in many textbooks on cosmology, and will therefore not be reviewed here.

It is, however, important to recall two important aspects concerning the thermal history of the early Universe. Since the energy density in radiation redshifts faster than the matter energy density, it follows by working backwards in time from the present data that although the energy density of the

Universe is now mostly in cold matter, it was initially dominated by radiation. The transition occurred at a time denoted by t_{eq}, the "time of equal matter and radiation", which is also the time when perturbations can start to grow by gravitational clustering. The second important time is t_{rec}, the "time of recombination" when photons fell out of equilibrium. The photons of the CMB have travelled without scattering from t_{rec}. Their spatial distribution is predicted to be a black body since the cosmological redshift preserves the black body nature of the initial spectrum (simply redshifting the temperature) which was in turn determined by thermal equilibrium. CMB anisotropies probe the density fluctuations at t_{rec}. Note that for the usual values of the cosmological parameters, $t_{eq} < t_{rec}$.

Standard Big Bang cosmology is faced with several important problems. Only one of these, the age problem, is a potential conflict with observations. The three problems which are most often discussed in the context of inflation - the homogeneity, flatness and formation of structure problems (see e.g. [1]) - are questions which have no answers within the standard theory but which can be successfully addressed in the context of inflationary cosmology.

The horizon problem is the fact that, within the context of standard cosmology, the comoving region $\ell_p(t_{rec})$ over which the CMB is observed to be homogeneous to better than one part in 10^4 is much larger than the comoving forward light cone $\ell_f(t_{rec})$ at t_{rec}, which is the maximal distance over which microphysical forces could have caused the homogeneity. Hence, standard cosmology cannot explain the observed isotropy of the CMB.

In standard cosmology and in an expanding Universe, $\Omega = 1$ is an unstable fixed point. As the temperature decreases, $\Omega - 1$ increases. In fact, in order to explain the present small value of $\Omega - 1 \sim \mathcal{O}(1)$, the initial energy density had to be extremely close to critical density. For example, at $T = 10^{15}$ GeV, we require $\Omega - 1 \sim 10^{-50}$. What is the origin of these fine tuned initial conditions? This is the flatness problem of standard cosmology.

The third of the classic problems of standard cosmological model is the "formation of structure problem." Observations indicate that galaxies and even clusters of galaxies have nonrandom correlations on scales larger than 50 Mpc (see e.g. [6,7]). This scale is comparable to the comoving horizon at t_{eq}. Thus, if the initial density perturbations were produced much before t_{eq}, the correlations cannot be explained by a causal mechanism. Gravity alone is, in general, too weak to build up correlations on the scale of clusters after t_{eq} (see, however, the explosion scenario of [8]). Hence, the two questions of what generates the primordial density perturbations and what causes the observed correlations, do not have an answer in the context of standard cosmology. This problem is illustrated in Fig. 1.

There are other serious problems of standard cosmology, e.g. the age and the cosmological constant problems. However, to date modern cosmology does not shed any light on these problems, and I will therefore not address them here.

Inflationary Universe Scenario

The idea of inflation [1] is very simple (for some early reviews of inflation see e.g. [9–12]). We assume there is a time interval beginning at t_i and ending at t_R (the "reheating time") during which the Universe is exponentially expanding, i.e.,

$$a(t) \sim e^{Ht}, \quad t \epsilon \, [t_i, \, t_R] \tag{9}$$

with constant Hubble expansion parameter H. Such a period is called "de Sitter" or "inflationary." The success of Big Bang nucleosynthesis sets an upper limit to the time of reheating, namely the time of nucleosynthesis.

The phases of an inflationary Universe are sketched in Fig. 2. Before the onset of inflation there are no constraints on the state of the Universe. In some models a classical space-time emerges immediately in an inflationary state, in others there is an initial radiation dominated FRW period. Our sketch applies to the second case. After t_R, the Universe is very hot and dense, and the subsequent evolution is as in standard cosmology. During the inflationary phase, the number density of any particles initially in thermal equilibrium at $t = t_i$ decays exponentially. Hence, the matter temperature $T_m(t)$ also decays exponentially. At $t = t_R$, all of the energy which is responsible for inflation (see later) is released as thermal energy. This is a nonadiabatic process during which the entropy increases by a large factor.

Inflation can easily solve the homogeneity problem. Let $\Delta t = t_R - t_i$ denote the period of inflation. During inflation, the forward light cone increases

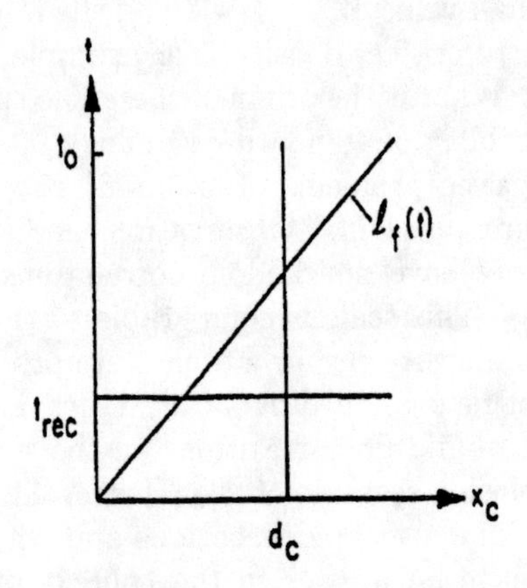

FIGURE 1. A sketch (conformal separation vs. time) of the formation of structure problem: the comoving separation d_c between two clusters is larger than the forward light cone at time t_{eq}.

exponentially compared to a model without inflation, whereas the past light cone is not affected for $t \geq t_R$. Hence, provided Δt is sufficiently large, $\ell_f(t_R)$ will be greater than $\ell_p(t_R)$.

Inflation also can solve the flatness problem [13,1] The key point is that the entropy density s is no longer constant. As will be explained later, the temperatures at t_i and t_R are essentially equal. Hence, the entropy increases during inflation by a factor $\exp(3H\Delta t)$. Thus, ϵ decreases by a factor of $\exp(-2H\Delta t)$. Hence, $(\rho - \rho_c)/\rho$ can be of order 1 both at t_i and at the present time. In fact, if inflation occurs at all, then rather generically, the theory predicts that at the present time $\Omega = 1$ to a high accuracy (now $\Omega < 1$ requires special initial conditions or rather special models [14]).

Most importantly, inflation provides a mechanism which in a causal way generates the primordial perturbations required for galaxies, clusters and even larger objects. In inflationary Universe models, the Hubble radius ("apparent" horizon), $3t$, and the "actual" horizon (the forward light cone) do not coincide at late times. Provided that the duration of inflation is sufficiently long, then (as sketched in Fig. 3) all scales within our apparent horizon were inside the actual horizon since t_i. Thus, it is in principle possible to have a casual generation mechanism for perturbations [15–18].

The generation of perturbations is due to a causal microphysical process. Such a process can only act coherently on length scales smaller than the Hubble radius $\ell_H(t)$ where $\ell_H(t) = H^{-1}(t)$. A heuristic way to understand the meaning of $\ell_H(t)$ is to realize that it is the distance which light (and hence the maximal distance any causal effects) can propagate in one expansion time.

The density perturbations produced during inflation are due to quantum fluctuations in the matter and gravitational fields [16,17]. The amplitude of these inhomogeneities corresponds to a temperature T_H whose value is

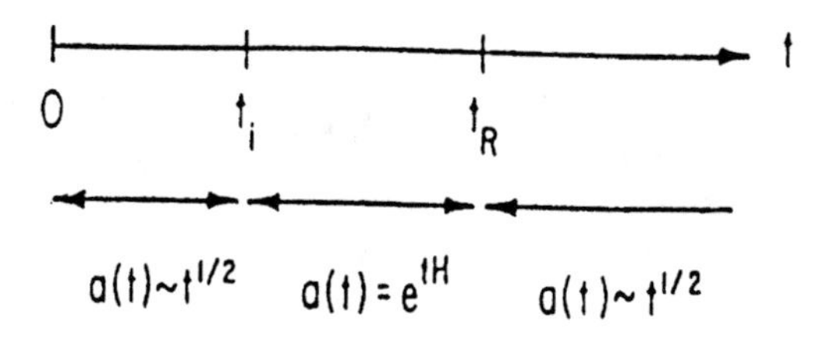

FIGURE 2. The phases of an inflationary Universe. The times t_i and t_R denote the beginning and end of inflation, respectively. In some models of inflation, there is no initial radiation dominated FRW period. Rather, the classical space-time emerges directly in an inflationary state from some initial quantum gravity state.

$T_H \sim H$, the Hawking temperature of the de Sitter phase. This implies that at all times t during inflation, perturbations with a fixed physical wavelength $\sim H^{-1}$ will be produced. Subsequently, the length of the waves is stretched with the expansion of space, and soon becomes larger than the Hubble radius. The phases of the inhomogeneities are random. Thus, the inflationary Universe scenario predicts perturbations on all scales ranging from the comoving Hubble radius at the beginning of inflation to the corresponding quantity at the time of reheating. In particular, provided that inflation lasts sufficiently long, perturbations on scales of galaxies and beyond will be generated. Note, however, that it is very dangerous to interpret de Sitter Hawking radiation as thermal radiation. In fact, the equation of state of this "radiation" is not thermal [19].

Obviously, the key question is how to obtain inflation. From the FRW equations, it follows that in order to get exponential increase of the scale factor, the equation of state of matter must be

$$p = -\rho \tag{10}$$

This is where the connection with particle physics comes in. The energy

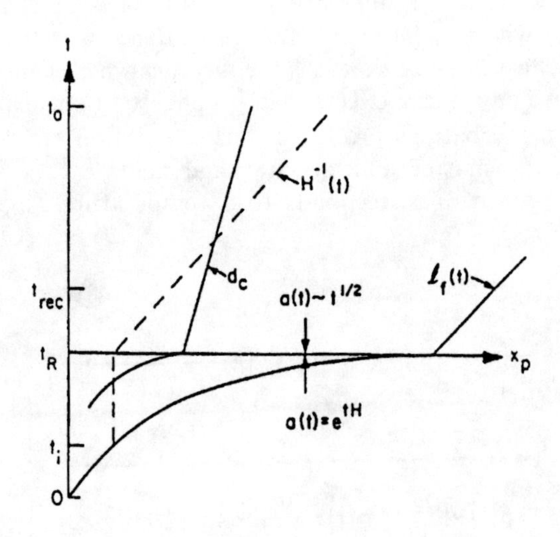

FIGURE 3. A sketch (physical coordinates vs. time) of the solution of the formation of structure problem. Provided that the period of inflation is sufficiently long, the separation d_c between two galaxy clusters is at all times smaller than the forward light cone. The dashed line indicates the Hubble radius. Note that d_c starts out smaller than the Hubble radius, crosses it during the de Sitter period, and then reenters it at late times.

density and pressure of a scalar quantum field φ are given by

$$\rho(\varphi) = \frac{1}{2}\dot{\varphi}^2 + \frac{1}{2}(\nabla\varphi)^2 + V(\varphi) \tag{11}$$

$$p(\varphi) = \frac{1}{2}\dot{\varphi}^2 - \frac{1}{6}(\nabla\varphi)^2 - V(\varphi)\,. \tag{12}$$

Thus, provided that at some initial time t_i

$$|\dot{\varphi}(\underline{x}, t_i)|, |\nabla\varphi(\underline{x}_i\, t_i)| \ll V(\varphi(\underline{x}_i, t_i))\,, \tag{13}$$

the equation of state of matter will be (10).

The next question is how to realize the required initial conditions (13) and to maintain the dominance of potential over kinetic and gradient energy for sufficiently long. Various ways of realizing these conditions were put forward, and they gave rise to different models of inflation. I will focus on "old inflation," "new inflation"" and "chaotic inflation." There are many other attempts at producing an inflationary scenario, but there is as of now no convincing realization.

Old Inflation

The old inflationary Universe model [1,20] is based on a scalar field theory which undergoes a first order phase transition. As a toy model, consider a scalar field theory with the potential $V(\varphi)$ of Fig. 4. This potential has a metastable "false" vacuum at $\varphi = 0$, whereas the lowest energy state (the "true" vacuum) is $\varphi = a$. Finite temperature effects [21] lead to extra terms in the finite temperature effective potential which are proportional to $\varphi^2 T^2$ (the resulting finite temperature effective potential is also depicted in Fig. 4). Thus, at high temperatures, the energetically preferred state is the false vacuum state. Note that this is only true if φ is in thermal equilibrium with the other fields in the system.

For fairly general initial conditions, $\varphi(x)$ is trapped in the metastable state $\varphi = 0$ as the Universe cools below the critical temperature T_c. As the Universe expands further, all contributions to the energy-momentum tensor $T_{\mu\nu}$ except for the contribution

$$T_{\mu\nu} \sim V(\varphi) g_{\mu\nu} \tag{14}$$

redshift. Hence, provided that the potential $V(\varphi)$ is shifted upwards such that $V(a) = 0$, then the equation of state in the false vacuum approaches $p = -\rho$, and inflation sets in. After a period Γ^{-1}, where Γ is the tunnelling rate, bubbles of $\varphi = a$ begin to nucleate [22] in a sea of false vacuum $\varphi = 0$. Inflation lasts until the false vacuum decays. During inflation, the Hubble constant is given by

$$H^2 = \frac{8\pi G}{3} V(0)\,. \tag{15}$$

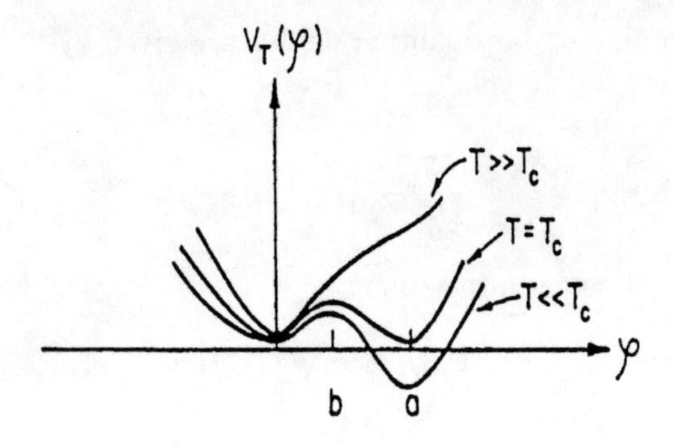

FIGURE 4. The finite temperature effective potential in a theory with a first order phase transition.

Note that the condition $V(a) = 0$, which looks rather unnatural, is required to avoid a large cosmological constant today (none of the present inflationary Universe models manages to circumvent or solve the cosmological constant problem).

It was immediately realized that old inflation has a serious "graceful exit" problem [1,23]. The bubbles nucleate after inflation with radius $r \ll 2t_R$ and would today be much smaller than our apparent horizon. Thus, unless bubbles percolate, the model predicts extremely large inhomogeneities inside the Hubble radius, in contradiction with the observed isotropy of the microwave background radiation.

For bubbles to percolate, a sufficiently large number must be produced so that they collide and homogenize over a scale larger than the present Hubble radius. However, with exponential expansion, the volume between bubbles expands exponentially whereas the volume inside bubbles expands only with a low power. This prevents percolation.

New Inflation

Because of the graceful exit problem, old inflation never was considered to be a viable cosmological model. However, soon after the seminal paper by Guth, Linde [24] and independently Albrecht and Steinhardt [25] put forwards a modified scenario, the New Inflationary Universe.

The starting point is a scalar field theory with a double well potential which undergoes a second order phase transition (Fig. 5). $V(\varphi)$ is symmetric and $\varphi = 0$ is a local maximum of the zero temperature potential. Once again, it was argued that finite temperature effects confine $\varphi(x)$ to values near $\varphi = 0$ at temperatures $T \geq T_c$. For $T < T_c$, thermal fluctuations trigger the instability of $\varphi(x) = 0$ and $\varphi(x)$ evolves towards either of the global minima at $\varphi = \pm\sigma$ by the classical equation of motion

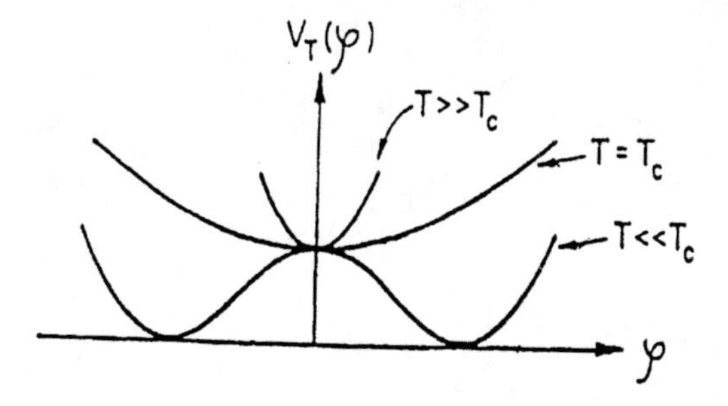

FIGURE 5. The finite temperature effective potential in a theory with a second order phase transition.

$$\ddot{\varphi} + 3H\dot{\varphi} - a^{-2}\,\nabla^2\,\varphi = -V'(\varphi)\,. \tag{16}$$

Within a fluctuation region, $\varphi(x)$ will be homogeneous. In such a region, we can neglect the spatial gradient terms in Eq. (16). Then, from (11) and (12) we can read off the induced equation of state. The condition for inflation is

$$\dot{\varphi}^2 \ll V(\varphi)\,, \tag{17}$$

i.e. slow rolling. Often, the "slow rolling" approximation is made to find solutions of (16). This consists of dropping the $\ddot{\varphi}$ term.

There is no graceful exit problem in the new inflationary Universe. Since the fluctuation domains are established before the onset of inflation, any boundary walls will be inflated outside the present Hubble radius.

Let us, for the moment, return to the general features of the new inflationary Universe scenario. At the time t_c of the phase transition, $\varphi(t)$ will start to move from near $\varphi = 0$ towards either $\pm\sigma$ as described by the classical equation of motion, i.e. (16). At or soon after t_c, the energy-momentum tensor of the Universe will start to be dominated by $V(\varphi)$, and inflation will commence. t_i shall denote the time of the onset of inflation. Eventually, $\phi(t)$ will reach large values for which nonlinear effects become important. The time at which this occurs is t_B. For $t > t_B$, $\varphi(t)$ rapidly accelerates, reaches $\pm\sigma$, overshoots and starts oscillating about the global minimum of $V(\varphi)$. The amplitude of this oscillation is damped by the expansion of the Universe and (predominantly) by the coupling of φ to other fields. At time t_R, the energy in φ drops below the energy of the thermal bath of particles produced during the period of oscillation.

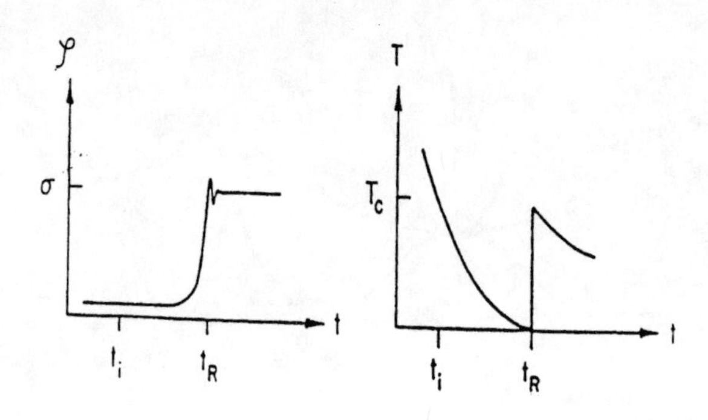

FIGURE 6. Evolution of $\varphi(t)$ and $T(t)$ in the new inflationary Universe.

The evolution of $\varphi(t)$ is sketched in Fig. 6. The time period between t_B and t_R is called the reheating period and is usually short compared to the Hubble expansion time. For $t > t_R$, the Universe is again radiation dominated.

In order to obtain inflation, the potential $V(\varphi)$ must be very flat near the false vacuum at $\varphi = 0$. This can only be the case if all of the coupling constants appearing in the potential are small. However, this implies that the φ cannot be in thermal equilibrium at early times, which would be required to localize φ in the false vacuum. In the absence of thermal equilibrium, the initial conditions for φ are only constrained by requiring that the total energy density in φ not exceed the total energy density of the Universe. Most of the phase space of these initial conditions lies at values of $|\varphi| >> \sigma$. This leads to the "chaotic" inflation scenario [26].

Chaotic Inflation

Consider a region in space where at the initial time $\varphi(x)$ is very large, homogeneous and static. In this case, the energy-momentum tensor will be immediately dominated by the large potential energy term and induce an equation of state $p \simeq -\rho$ which leads to inflation. Due to the large Hubble damping term in the scalar field equation of motion, $\varphi(x)$ will only roll very slowly towards $\varphi = 0$. The kinetic energy contribution to $T_{\mu\nu}$ will remain small, the spatial gradient contribution will be exponentially suppressed due to the expansion of the Universe, and thus inflation persists. Note that in contrast to old and new inflation, no initial thermal bath is required. Note also that the precise form of $V(\varphi)$ is irrelevant to the mechanism. In particular, $V(\varphi)$ need not be a double well potential. This is a significant advantage, since for scalar fields other than Higgs fields used for spontaneous symmetry breaking, there is no particle physics motivation for assuming a double well potential, and since the inflaton (the field which gives rise to inflation) cannot be a conventional Higgs field due to the severe fine tuning constraints.

The field and temperature evolution in a chaotic inflation model is similar to what is depicted in Fig. 6, except that φ is rolling towards the true vacuum at $\varphi = \sigma$ from the direction of large field values.

Chaotic inflation is a much more radical departure from standard cosmology than old and new inflation. In the latter, the inflationary phase can be viewed as a short phase of exponential expansion bounded at both ends by phases of radiation domination. In chaotic inflation, a piece of the Universe emerges with an inflationary equation of state immediately after the quantum gravity (or string) epoch.

The chaotic inflationary Universe scenario has been developed in great detail (see e.g. [27] for a recent review). One important addition is the inclusion of stochastic noise [28] in the equation of motion for φ in order to take into account the effects of quantum fluctuations. It can in fact be shown that for sufficiently large values of $|\varphi|$, the stochastic force terms are more important than the classical relaxation force $V'(\varphi)$. There is equal probability for the quantum fluctuations to lead to an increase or decrease of $|\varphi|$. Hence, in a substantial fraction of comoving volume, the field φ will climb up the potential. This leads to the conclusion that chaotic inflation is eternal. At all times, a large fraction of the physical space will be inflating. Another consequence of including stochastic terms is that on large scales (much larger than the present Hubble radius), the Universe will look extremely inhomogeneous.

Problems of Inflationary Cosmology

In spite of its great success at resolving some of the problems of standard cosmology and of providing a causal, predictive theory of structure formation, there are several important unresolved conceptual problems in inflationary cosmology. I will focus on three of these problems, the cosmological constant mystery, the fluctuation problem, and the dynamics of reheating.

Cosmological Constant Problem

Since the cosmological constant acts as an effective energy density, its value is bounded from above by the present energy density of the Universe. In Planck units, the constraint on the effective cosmological constant Λ_{eff} is (see e.g. [29])

$$\frac{\Lambda_{eff}}{m_{pl}^4} \leq 10^{-122}\,. \tag{18}$$

This constraint applies both to the bare cosmological constant and to any matter contribution which acts as an effective cosmological constant.

The true vacuum value of the potential $V(\varphi)$ acts as an effective cosmological constant. Its value is not constrained by any particle physics requirements (in the absence of special symmetries). The cosmological constant problem is thus even more accute in inflationary cosmology than it usually is. The same unknown mechanism which must act to shift the potential (see Fig. 4) such that inflation occurs in the false vacuum must also adjust the potential to vanish in the true vacuum.

Supersymmetric theories may provide a resolution of this problem, since unbroken supersymmetry forces $V(\varphi) = 0$ in the supersymmetric vacuum. However, supersymmetry breaking will induce a nonvanishing $V(\varphi)$ in the true vacuum after supersymmetry breaking.

We may therefore be forced to look for realizations of inflation which do not make use of scalar fields. There are several possibilities. It is possible to obtain inflation in higher derivative gravity theories. In fact, the first model with exponential expansion of the Universe was obtained [30] in an R^2 gravity theory. The extra degrees of freedom associated with the higher derivative terms act as scalar fields with a potential which automatically vanishes in the true vacuum. For some recent work on higher derivative gravity inflation see also [31].

Another way to obtain inflation is by making use of condensates (see [32] and [33] for different approaches to this problem). An additional motivation for following this route to inflation is that the symmetry breaking mechanisms observed in nature (in condensed matter systems) are induced by the formation of condensates such as Cooper pairs. Again, in a model of condensates there is no freedom to add a constant to the effective potential.

The main problem when studying the possibility of obtaining inflation using condensates is that the quantum effects which determine the theory are highly nonperturbative. In particular, the effective potential written in terms of a condensate $\langle\varphi\rangle$ does not correspond to a renormalizable theory and will in general [34] contain terms of arbitrary power in $\langle\varphi\rangle$. However (see [35]), one may make progress by assuming certain general properties of the effective potential.

Let us [35] consider a theory in which at some time t_i a condensate $\langle\varphi\rangle$ forms, i.e. $\langle\varphi\rangle = 0$ for $t < t_i$ and $\langle\varphi\rangle \neq 0$ for $t > t_i$. The expectation value of the Hamiltonian H written in terms of the condensate $\langle\varphi\rangle$ contains terms of arbitrary powers of $\langle\varphi\rangle$:

$$\langle H\rangle = \sum_n (-1)^n n! a_n \langle\varphi\rangle^n \,. \tag{19}$$

We summarize our ignorance of the nonperturbative physics in the assumption that the resulting series is asymptotic, and in particular Borel summable, with coefficients $a_n \propto 1$. In this case, we can resum the series to obtain [35]

$$\langle H\rangle = \int_0^\infty \frac{f(t)dt}{t(tm_{pl} + \langle\varphi\rangle)} e^{-1/t} \,, \tag{20}$$

where the function $f(t)$ is related to the coefficients a_n via

$$a_n = \frac{1}{n!}\int_0^\infty dt f(t) t^{-n-2} e^{-1/t} \,. \tag{21}$$

The expectation value of the Hamiltonian $\langle H \rangle$ can be interpreted as the effective potential V_{eff} of this theory. The question is under which conditions this potential gives rise to inflation. If we regard $\langle \varphi \rangle$ as a classical field (i.e. neglect the ultraviolet and infrared divergences of the theory), then the dynamics of the model can be read off directly from (20), with initial conditions for $\langle \varphi \rangle$ at the time t_i close to $\langle \varphi \rangle = 0$. It is easy to check that rather generically, the conditions required to have slow rolling of φ, namely

$$V' m_{pl} << \sqrt{48\pi} V \tag{22}$$

$$V'' m_{pl}^2 << 24\pi V \,, \tag{23}$$

are satisfied. However, since the potential decays only slowly at large values of $\langle \varphi \rangle$ and since there is no true vacuum state at finite values of $\langle \varphi \rangle$, the slow rolling conditions are satisfied for all times. In this case, inflation would never end - an obvious cosmological disaster.

However, $\langle \varphi \rangle$ is not a classical scalar field but the expectation value of a condensate operator. Thus, we have to worry about diverging contributions to this expectation value. In particular, in a theory with symmetry breaking there will often be massless excitations which will give rise to infrared divergences. It is necessary to introduce an infrared cutoff energy ε whose value is determined in the context of cosmology by the Hubble expansion rate. Note in particular that this cutoff is time-dependent. Effectively, we thus have a theory of two scalar fields $\langle \varphi \rangle$ and ε. In this case, the first of the slow rolling conditions becomes (if ε is expressed in Planck units)

$$\dot{\varepsilon}^2 m_{pl}^2 + \dot{\varphi}^2 << 2V \,. \tag{24}$$

The infrared cutoff changes the form of the effective potential. We assume that this change can be modelled by replacing $\langle \varphi \rangle$ by $\langle \varphi \rangle / \varepsilon$. If we (following [36]) take the infrared cutoff to be

$$\varepsilon(t) = \frac{H(0)}{m_{pl}} [1 - a(Ht)^p] \,, \tag{25}$$

where $0 < a << 1$ and p is an integer and the time at the beginning of the rolling has been set to $t = 0$, then it can be shown [35] that an period of inflation with a graceful exit is realized. After the condensate $\langle \varphi \rangle$ starts rolling at $\langle \varphi \rangle \sim 0$, inflation will commence. As inflation proceeds, $\varepsilon(t)$ will slowly grow and will eventually dominate the energy functional, signaling an end of the inflationary period. From (25) it follows that inflation lasts until $a^{1/p} Ht = 1$.

This analysis demonstrates that it is in principle possible to obtain inflation from condensates. However, the model must be studied in much more detail before we can determine whether it gives a realization of inflation which is free of problems.

Fluctuation Problem

A generic problem for all realizations of inflation studied up to now concerns the amplitude of the density perturbations which are induced by quantum fluctuations during the period of exponential expansion. From the amplitude of CMB anisotropies measured by COBE, and from the present amplitude of density inhomogeneities on scales of clusters of galaxies, it follows that the amplitude of the mass fluctuations $\delta M/M$ on a length scale given by the comoving wavenumber k at the time $t_H(k)$ when that scale crosses the Hubble radius in the FRW period is

$$\frac{\delta M}{M}(k, t_H(k)) \propto 10^{-5} . \tag{26}$$

The perturbations originate during inflation as quantum excitations (see e.g. [5] for a comprehensive review). Their amplitude at the time $t_i(k)$ when the scale k leaves the Hubble radius during inflation is given by

$$\frac{\delta M}{M}(k, t_i(k)) \simeq \frac{V'\delta\varphi}{\rho}|_{t_i(k)} , \tag{27}$$

where $\delta\varphi$ is the amplitude of the quantum fluctuation of $\delta\varphi(k)$ (note that this is a momentum space quantity). While the scale k is larger than the Hubble radius, the fluctuation amplitude grows by general relativistic gravitational effects. The amplitudes at $t_i(k)$ and $t_H(k)$ are related by

$$\frac{\delta M}{M}(k, t_H(k)) \simeq \frac{1}{1+p/\rho}|_{t_i(k)} \frac{\delta M}{M}(k, t_i(k)) \tag{28}$$

(see e.g. [37,38,5]). Combining (27) and (28) and working out the result for the potential

$$V(\varphi) = \lambda\varphi^4 \tag{29}$$

we obtain the result [39,40,37]

$$\frac{\delta M}{M}(k, t_H(k)) \simeq 10^2\lambda^{1/2} . \tag{30}$$

Thus, in order to agree with the observed value (26), the coupling constant λ must be extremely small:

$$\lambda \leq 10^{-12} . \tag{31}$$

It has been shown in [41] that the above conclusion is generic, at least for models in which inflation is driven by a scalar field. In order that inflation does not produce a too large amplitude of the spectrum of perturbations, a dimensionless number appearing in the potential must be set to a very small value. Models in which inflation is NOT driven by a scalar field but realized in some unified theory of all fundamental forces might avoid the fluctuation problem, in particular if there is some principle such as asymptotic freedom during the period of inflation [42] which suppresses scalar perturbations.

Reheating Problem

A question which has recently received a lot of attention and will be discussed in greater detail shortly is the issue of reheating in inflationary cosmology. The question concerns the energy transfer between the inflaton and matter fields which is supposed to take place at the end of inflation (see Fig. 6).

According to either new inflation or chaotic inflation, the dynamics of the inflaton leads first to a transfer of energy from potential energy of the inflaton to kinetic energy. After the period of slow rolling, the inflaton φ begins to oscillate about the true minimum of $V(\varphi)$. Quantum mechanically, the state of homogeneous oscillation corresponds to a coherent state. Any coupling of φ to other fields (and even self coupling terms of φ) will lead to a decay of this state. This corresponds to the particle production. The produced particles will be relativistic, and thus at the conclusion of the reheating period a radiation dominated Universe will emerge.

The key questions are by what mechanism and how fast the decay of the coherent state takes place. It is important to determine the temperature of the produced particles at the end of the reheating period. The answers are relevant to many important questions regarding the post-inflationary evolution. For example, it is important to know whether the temperature after reheating is high enough to allow GUT baryogenesis and the production of GUT-scale topological defects. In supersymmetric models, the answer determines the predicted abundance of gravitinos and other moduli fields.

Recently, there has been a complete change in our understanding of reheating. This topic will be discussed in detail below.

Reheating in Inflationary Cosmology

Reheating is an important stage in inflationary cosmology. It determines the state of the Universe after inflation and has consequences for baryogenesis, defect formation, and, as will be shown below, maybe even for the composition of the dark matter of the Universe.

After slow rolling, the inflaton field begins to oscillate uniformly in space about the true vacuum state. Quantum mechanically, this corresponds to a coherent state of $k = 0$ inflaton particles. Due to interactions of the inflaton with itself and with other fields, the coherent state will decay into quanta of elementary particles. This corresponds to post-inflationary particle production.

Reheating is usually studied using simple scalar field toy models. The one we will adopt here consists of two real scalar fields, the inflaton φ with Lagrangian

$$\mathcal{L}_o = \frac{1}{2}\partial_\mu\varphi\partial^\mu\varphi - \frac{1}{4}\lambda(\varphi^2 - \sigma^2)^2 \tag{32}$$

interacting with a massless scalar field χ representing ordinary matter. The interaction Lagrangian is taken to be

$$\mathcal{L}_I = \frac{1}{2}g^2\varphi^2\chi^2 . \tag{33}$$

Self interactions of χ are neglected.

By a change of variables

$$\varphi = \tilde{\varphi} + \sigma , \tag{34}$$

the interaction Lagrangian can be written as

$$\mathcal{L}_I = g^2\sigma\tilde{\varphi}\chi^2 + \frac{1}{2}g^2\tilde{\varphi}^2\chi^2 . \tag{35}$$

During the phase of coherent oscillations, the field $\tilde{\varphi}$ oscillates with a frequency

$$\omega = m_\varphi = \lambda^{1/2}\sigma \tag{36}$$

(neglecting the expansion of the Universe which can be taken into account as in [43,44]).

Elementary Theory of Reheating

According to the elementary theory of reheating (see e.g. [45] and [46]), the decay of the inflaton is calculated using first order perturbations theory. According to the Feynman rules, the decay rate Γ_B of φ (calculated assuming that the cubic coupling term dominates) is given by

$$\Gamma_B = \frac{g^2\sigma^2}{8\pi m_\phi} . \tag{37}$$

The decay leads to a decrease in the amplitude of φ (from now on we will drop the tilde sign) which can be approximated by adding an extra damping term to the equation of motion for φ:

$$\ddot{\varphi} + 3H\dot{\varphi} + \Gamma_B\dot{\varphi} = -V'(\varphi) . \tag{38}$$

From the above equation it follows that as long as $H > \Gamma_B$, particle production is negligible. During the phase of coherent oscillation of φ, the energy density and hence H are decreasing. Thus, eventually $H = \Gamma_B$, and at that point reheating occurs (the remaining energy density in φ is very quickly transferred to χ particles).

The temperature T_R at the completion of reheating can be estimated by computing the temperature of radiation corresponding to the value of H at which $H = \Gamma_B$. From the FRW equations it follows that

$$T_R \sim (\Gamma_B m_{pl})^{1/2} . \tag{39}$$

If we now use the "naturalness" constraint[2]

$$g^2 \sim \lambda \tag{40}$$

in conjunction with the constraint on the value of λ from (31), it follows that for $\sigma < m_{pl}$,

$$T_R < 10^{10}\text{GeV} . \tag{41}$$

This would imply no GUT baryogenesis, no GUT-scale defect production, and no gravitino problems in supersymmetric models with $m_{3/2} > T_R$, where $m_{3/2}$ is the gravitino mass. As we shall see, these conclusions change radically if we adopt an improved analysis of reheating.

Modern Theory of Reheating

However, as was first realized in [47], the above analysis misses an essential point. To see this, we focus on the equation of motion for the matter field χ coupled to the inflaton φ via the interaction Lagrangian $\mathcal{L}_I$ of (35). Taking into account for the moment only the cubic interaction term, the equation of motion becomes

$$\ddot{\chi} + 3H\dot{\chi} - ((\frac{\nabla}{a})^2 - m_\chi^2 - 2g^2\sigma\varphi)\chi = 0 . \tag{42}$$

Since the equation is linear in χ, the equations for the Fourier modes χ_k decouple:

$$\ddot{\chi}_k + 3H\dot{\chi}_k + (k_p^2 + m_\chi^2 + 2g^2\sigma\varphi)\chi_k = 0, \tag{43}$$

where k_p is the time-dependent physical wavenumber.

Let us for the moment neglect the expansion of the Universe. In this case, the friction term in (43) drops out and k_p is time-independent, and Equation (43) becomes a harmonic oscillator equation with a time-dependent mass determined by the dynamics of φ. In the reheating phase, φ is undergoing oscillations. Thus, the mass in (43) is varying periodically. In the mathematics

---

[2] At one loop order, the cubic interaction term will contribute to λ by an amout $\Delta\lambda \sim g^2$. A renormalized value of λ smaller than g^2 needs to be finely tuned at each order in perturbation theory, which is "unnatural".

literature, this equation is called the Hill equation (or the Mathieu equation in the special case of an oscillating perturbation). It is well known that there is an instability. In physics, the effect is known as **parametric resonance** (see e.g. [48]). At frequencies ω_n corresponding to half integer multiples of the frequency ω of the variation of the mass, i.e.

$$\omega_k^2 = k_p^2 + m_\chi^2 = (\frac{n}{2}\omega)^2 \quad n = 1, 2, ..., \tag{44}$$

there are instability bands with widths $\Delta\omega_n$. For values of ω_k within the instability band, the value of χ_k increases exponentially:

$$\chi_k \sim e^{\mu t} \text{ with } \mu \sim \frac{g^2\sigma\varphi_0}{\omega}, \tag{45}$$

with φ_0 being the amplitude of the oscillation of φ. Since the widths of the instability bands decrease as a power of the (small) coupling constant g^2 with increasing n, for practical purposes only the lowest instability band is important. Its width is

$$\Delta\omega_k \sim g\sigma^{1/2}\varphi_0^{1/2}. \tag{46}$$

Note, in particular, that there is no ultraviolet divergence in computing the total energy transfer from the φ to the χ field due to parametric resonance.

It is easy to include the effects of the expansion of the Universe (see e.g. [47,43,44]). The main effect is that the value of ω_k becomes time-dependent. Thus, (for the model with a bare inflaton mass which we are considering) a mode slowly enters and leaves the resonance bands. As a consequence, any mode lies in the resonance band for only a finite time. This implies that the calculation of energy transfer is perfectly well-behaved. No infinite time divergences arise.[3]

It is now possible to estimate the rate of energy transfer, whose order of magnitude is given by the phase space volume of the lowest instability band multiplied by the rate of growth of the mode function χ_k. Using as an initial condition for χ_k the value $\chi_k \sim H$ given by the magnitude of the expected quantum fluctuations, we obtain

$$\dot{\rho} \sim \mu(\frac{\omega}{2})^2\Delta\omega_k H e^{\mu t}. \tag{47}$$

From (47) it follows that provided that the condition

$$\mu\Delta t >> 1 \tag{48}$$

[3] Note, however, that even without expansion, scattering of the produced particles leads to a cutoff of the instability after some finite time (see e.g. [49] for a recent numerical study).

is satisfied, where $\Delta t < H^{-1}$ is the time a mode spends in the instability band, then the energy transfer will procede fast on the time scale of the expansion of the Universe. In this case, there will be explosive particle production, and the energy density in matter at the end of reheating will be given by the energy density at the end of inflation.

The above is a summary of the main physics of the modern theory of reheating. The actual analysis can be refined in many ways (see e.g. [43,44]). First of all, it is easy to take the expansion of the Universe into account explicitly (by means of a transformation of variables), to employ an exact solution of the background model and to reduce the mode equation for χ_k to a Hill equation.

The next improvement consists of treating the χ field quantum mechanically (keeping φ as a classical background field). At this point, the techniques of quantum field theory in a curved background can be applied. There is no need to impose artificial classical initial conditions for χ_k. Instead, we may assume that χ starts in its initial vacuum state (excitation of an initial thermal state has been studied in [50]), and the Bogoliubov mode mixing technique (see e.g. [51]) can be used to compute the number of particles at late times.

Using this improved analysis, we recover the result (47). Thus, provided that the condition (48) is satisfied, reheating will be explosive. Working out the time Δt that a mode remains in the instability band for our model, expressing H in terms of φ_0 and m_{pl}, and ω in terms of σ, and using the naturalness relation $g^2 \sim \lambda$, the condition for explosive particle production becomes

$$\frac{\varphi_0 m_{pl}}{\sigma^2} >> 1\,, \tag{49}$$

which is satisfied for all chaotic inflation models with $\sigma < m_{pl}$ (recall that slow rolling ends when $\varphi \sim m_{pl}$ and that therefore the initial amplitude φ_0 of oscillation is of the order m_{pl}).

We conclude that rather generically, reheating in chaotic inflation models will be explosive. This implies that the energy density after reheating will be approximately equal to the energy density at the end of the slow rolling period. Therefore, as suggested in [52,53] and [54], respectively, GUT scale defects may be produced after reheating and GUT-scale baryogenesis scenarios may be realized, provided that the GUT energy scale is lower than the energy scale at the end of slow rolling.

Note, however, that the state of χ after parametric resonance is **not** a thermal state. The spectrum consists of high peaks in distinct wave bands. An important question which remains to be studied is how this state thermalizes. For some interesting work on this issue see [55,49]. As emphasized in [52] and [53], the large peaks in the spectrum may lead to symmetry restoration and to the efficient production of topological defects (for a differing view on this issue see [56,57]). Since the state after explosive particle production is not a thermal state, it is useful to follow [43] and call this process "preheating" instead of reheating.

A futher interesting conjecture which emerges from the parametric resonance analysis of preheating [43,44] is that the dark matter in the Universe may consist of remnant coherent oscillations of the inflaton field. In fact, it can easily be checked from (49) that the condition for efficient transfer of energy eventually breaks down when φ_0 has decreased to a sufficiently small value. For the model considered here, an order of magnitude calculation shows that the remnant oscillations may well contribute significantly to the present value of Ω.

Note that the details of the analysis of preheating are quite model-dependent. In fact [43], in many models one does not get the kind of "narrow-band" resonance discussed here, but "wide-band" resonance. In this case, the energy transfer is even more efficient.

There has recently been a lot of work on various aspects of reheating (see e.g. [58–61] for different approaches). Many important questions, e.g. concerning thermalization and back-reaction effects during and after preheating (or parametric resonance) remain to be fully analyzed.

Summary

The inflationary Universe is an attractive scenario for early Universe cosmology. It can resolve some of the problems of standard cosmology, and in addition gives rise to a predictive theory of structure formation (see e.g. [62] for a recent review).

However, important unsolved problems of principle remain. Rather generically, the predicted amplitude of perturbations is too large (the spectral shape, however, is in quite good agreement with the observations). The present realizations of inflation based on scalar field also make the cosmological constant problem more accute. In addition, there are no convincing particle-physics based realizations of inflation. Many models of inflation resort to introducing a new matter sector. It is important to search for a better connection between modern particle physics / field theory and inflation. String cosmology and dilaton gravity (see e.g. the recent reviews in [63]) may provide an interesting new approach to the unification of inflation and fundamental physics.

Recently, there has been much progress in the understanding of the energy transfer at the end of inflation between the inflaton field and matter. It appears that resonance phenomena such as parametric resonance play a crucial role. These new reheating scenarios lead to a high reheating temperature, although much more work remains to be done before one can reach a final conclusion on this issue.

TOPOLOGICAL DEFECTS AND STRUCTURE FORMATION

Quantifying Data on Large-Scale Structure

It is length scales corresponding to galaxies and larger which are of greatest interest in cosmology when attempting to find an imprint of the primordial fluctuations produced by particle physics. On these scales, gravitational effects are assumed to be dominant, and the fluctuations are not too far from the linear regime. On smaller scales, nonlinear gravitational and hydrodynamical effects determine the final state and mask the initial perturbations.

To set the scales, consider the mean separation of galaxies, which is about $5h^{-1}$ Mpc [64], and that of Abell clusters which is around $25h^{-1}$ Mpc [65]. The largest coherent structures seen in current redshift surveys have a length of about $100h^{-1}$ Mpc [6,7], the recent detections of CMB anisotropies probe the density field on length scales of about 10^3h^{-1} Mpc, and the present horizon corresponds to a distance of about $3 \cdot 10^3 h^{-1}$ Mpc.

Galaxies are gravitationally bound systems containing billions of stars. They are non-randomly distributed in space. A quantitative measure of this non-randomness is the "two-point correlation function" $\xi_2(r)$ which gives the excess probability of finding a galaxy at a distance r from a given galaxy:

$$\xi_2(r) = < \frac{n(r) - n_0}{n_0} > . \tag{50}$$

Here, n_0 is the average number density of galaxies, and $n(r)$ is the density of galaxies a distance r from a given one. The pointed braces stand for ensemble averaging.

Recent observational results from a various galaxy redshift surveys yield reasonable agreement [66] with a form

$$\xi_2(r) \simeq \left(\frac{r_0}{r}\right)^\gamma \tag{51}$$

with scaling length $r_0 \simeq 5h^{-1}$ Mpc and power $\gamma \simeq 1.8$. A theory of structure formation must explain both the amplitude and the slope of this correlation function.

On scales larger than galaxies, a better way to quantify structure is by means of large-scale systematic redshift surveys. Such surveys have discovered coherent planar structures and voids on scales of up to $100h^{-1}$ Mpc. Fig. 7 is a sketch of redshift z versus angle α in the sky for one 6° slice of the sky [6]. The second direction in the sky has been projected onto the $\alpha - z$ plane. The most prominent feature is the band of galaxies at a distance of about $100h^{-1}$ Mpc. This band also appears in neighboring slices and is therefore presumably part of a planar density enhancement of comoving planar size of

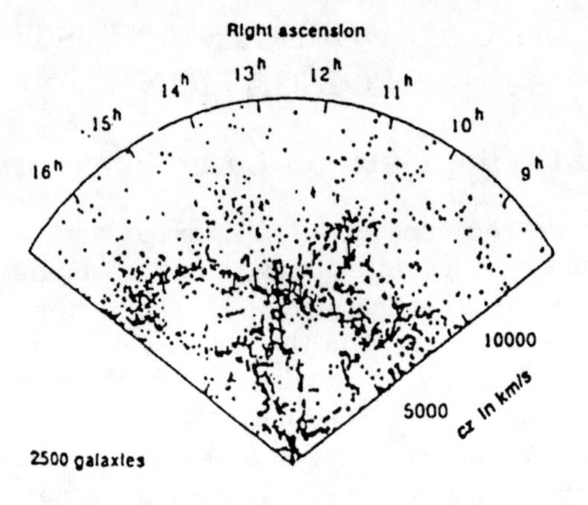

FIGURE 7. Results from the CFA redshift survey. Radial distance gives the redshift of galaxies, the angular distance corresponds to right ascension. The results from several slices of the sky (at different declinations) have been projected into the same cone.

at least $(50 \times 100) \times h^{-2}$ Mpc^2. This structure is often called the "Great Wall." The challenge for theories of structure formation is not only to explain the fact that galaxies are nonrandomly distributed, but also to predict both the observed scale and topology of the galaxy distribution. Topological defect models of structure formation attempt to address these questions.

Until 1992 there was little evidence for any convergence of the galaxy distribution towards homogeneity. Each new survey led to the discovery of new coherent structures in the Universe on a scale comparable to that of the survey. In 1996, results of a much deeper redshift survey were published [7] which for the first time find no coherent structures on the scale of the entire survey. In fact, no coherent structures on scales larger than $100h^{-1}$ Mpc are seen. This is the first direct evidence for the cosmological principle from optical surveys (the isotropy of the CMB has for a long time been a strong point in its support).

In summary, a lot of data from optical and infrared galaxies are currently available, and new data are being collected at a rapid rate. The observational constraints on theories of structure formation are becoming tighter.

Toplogical Defects

According to particle physics theories, matter at high energies and temperatures must be described in terms of fields. Gauge symmetries have proved to be extremely useful in describing the standard model of particle physics, according to which at high energies the laws of nature are invariant under a nonabelian group G of internal symmetry transformations $G = \mathrm{SU}(3)_c \times \mathrm{SU}(2)_L \times U(1)_Y$ which at a temperature of about 200 MeV is spontaneously broken down to $G' = \mathrm{SU}(3)_c \times \mathrm{U}(1)$. The subscript on the SU(3) subgroup indicates that it is the color symmetry group of the strong interactions, $\mathrm{SU}(2)_L\times \mathrm{U}(1)_Y$ is the Glashow-Weinberg-Salam (WS) model of weak and electromagnetic interactions, the subscripts L and Y denoting left handedness and hypercharge respectively. At low energies, the WS model spontaneously breaks to the U(1) subgroup of electromagnetism.

Spontaneous symmetry breaking is induced by an order parameter φ taking on a nontrivial expectation value $< \varphi >$ below a certain temperture T_c. In some particle physics models, φ is a fundamental scalar field in a nontrivial representation of the gauge group G which is broken. However, φ could also be a fermion condensate, as in the BCS theory of superconductivity.

Earlier we have seen that symmetry breaking phase transitions in gauge field theories do not, in general, lead to inflation. In most models, the coupling constants which arise in the effective potential for the scalar field φ driving the phase transition are too large to generate a period of slow rolling which lasts more than one Hubble time $H^{-1}(t)$. Nevertheless, there are interesting remnants of the phase transition: topological defects.

Consider a single component real scalar field with a typical symmetry breaking potential

$$V(\varphi) = \frac{1}{4}\lambda(\varphi^2 - \eta^2)^2 \tag{52}$$

Unless $\lambda \ll 1$ there will be no inflation. The phase transition will take place on a short time scale $\tau < H^{-1}$, and will lead to correlation regions of radius $\xi < t$ inside of which φ is approximately constant, but outside of which φ ranges randomly over the vacuum manifold $\mathcal{M}$, the set of values of φ which minimizes $V(\varphi)$ – in our example $\varphi = \pm\eta$. The correlation regions are separated by domain walls, regions in space where φ leaves the vacuum manifold $\mathcal{M}$ and where, therefore, potential energy is localized. Via the usual gravitational force, this energy density can act as a seed for structure.

Topological defects are familiar from solid state and condensed matter systems. Crystal defects, for example, form when water freezes or when a metal crystallizes [67]. Point defects, line defects and planar defects are possible. Defects are also common in liquid crystals [68]. They arise in a temperature quench from the disordered to the ordered phase. Vortices in ^{4}He are analogs of global cosmic strings. Vortices and other defects are also produced [69] during a quench below the critical temperature in ^{3}He. Finally, vortex lines may play an important role in the theory of superconductivity [70].

The analogies between defects in particle physics and condensed matter physics are quite deep. Defects form for the same reason: the vacuum manifold is topologically nontrivial. The arguments [71] which say that in a the-

TABLE 1. Classification of cosmologically allowed (v) and forbidden (x) defects

defect name	n	local defect	global defect
domain wall	1	x	x
cosmic string	2	v	v
monopole	3	x	v
texture	4	-	v

ory which admits defects, such defects will inevitably form, are applicable both in cosmology and in condensed matter physics. Different, however, is the defect dynamics. The motion of defects in condensed matter systems is friction-dominated, whereas the defects in cosmology obey relativistic equations, second order in time derivatives, since they come from a relativistic field theory.

After these general comments we turn to a classification of topological defects [71]. We consider theories with an n-component order parameter φ and with a potential energy function (free energy density) of the form (6.1) with $\varphi^2 = \sum_{i=1}^{n} \varphi_i^2$.

There are various types of local and global topological defects (regions of trapped energy density) depending on the number n of components of φ (see e.g. [72] for a comprehensive survey of topological defect models). The more rigorous mathematical definition refers to the homotopy of $\mathcal{M}$. The words "local" and "global" refer to whether the symmetry which is broken is a gauge or global symmetry. In the case of local symmetries, the topological defects have a well defined core outside of which φ contains no energy density in spite of nonvanishing gradients $\nabla\varphi$: the gauge fields A_μ can absorb the gradient, *i.e.*, $D_\mu\varphi = 0$ when $\partial_\mu\varphi \neq 0$, where the covariant derivative D_μ is defined by $D_\mu = \partial_\mu + ie\,A_\mu$, e being the gauge coupling constant. Global topological defects, however, have long range density fields and forces.

Table 1 contains a list of topological defects with their topological characteristics. A "v" marks acceptable theories, a "x" theories which are in conflict with observations (for $\eta \sim 10^{16}$ GeV).

Theories with domain walls are ruled out [73] since a single domain wall stretching across the Universe today would overclose the Universe. Local monopoles are also ruled out [74] since they would overclose the Universe. Local textures are ineffective at producing structures because there is no traped potential energy.

From now on we will focus on one type of defects: cosmic strings (see e.g. [72,75,76] for recent reviews, and [77] for a classic review paper). These arise in theories with a complex order parameter ($n = 2$). In this case the vacuum manifold of the model is

$$\mathcal{M} = S^1\,, \tag{53}$$

which has nonvanishing first homotopy group:

$$\Pi_1(\mathcal{M}) = Z \neq 1\,. \tag{54}$$

A cosmic string is a line of trapped energy density which arises whenever the field $\varphi(x)$ circles $\mathcal{M}$ along a closed path in space (*e.g.*, along a circle). In this case, continuity of φ implies that there must be a point with $\varphi = 0$ on any disk whose boundary is the closed path. The points on different sheets connect up to form a line overdensity of field energy.

To construct a field configuration with a string along the z axis [78], take $\varphi(x)$ to cover $\mathcal{M}$ along a circle with radius r about the point $(x, y) = (0, 0)$:

$$\varphi(r, \vartheta) \simeq \eta e^{i\vartheta}\,,\; r \gg \eta^{-1}\,. \tag{55}$$

This configuration has winding number 1, *i.e.*, it covers $\mathcal{M}$ exactly once. Maintaining cylindrical symmetry, we can extend (55) to arbitrary r

$$\varphi(r,\, \vartheta) = f(r)e^{i\vartheta}\,, \tag{56}$$

where $f(0) = 0$ and $f(r)$ tends to η for large r. The width w can be found by balancing potential and tension energy. The result is

$$w \sim \lambda^{-1/2}\eta^{-1}\,. \tag{57}$$

For local cosmic strings, *i.e.*, strings arising due to the spontaneous breaking of a gauge symmetry, the energy density decays exponentially for $r \gg w$. In this case, the energy μ per unit length of a string is finite and depends only on the symmetry breaking scale η

$$\mu \sim \eta^2 \tag{58}$$

(independent of the coupling λ). The value of μ is the only free parameter in a cosmic string model.

Formation of Defects in Cosmological Phase Transitions

The symmetry breaking phase transition takes place at $T = T_c$ (called the critical temperature). From condensed matter physics it is well known that in many cases topological defects form during phase transitions, particularly if the transition rate is fast on a scale compared to the system size. When cooling a metal, defects in the crystal configuration will be frozen in; during a temperature quench of ^{4}He, thin vortex tubes of the normal phase are trapped in the superfluid; and analogously in a temperature quench of a superconductor, flux lines are trapped in a surrounding sea of the superconducting Meissner phase.

In cosmology, the rate at which the phase transition proceeds is given by the expansion rate of the Universe. Hence, topological defects will inevitably be produced in a cosmological phase transition [71], provided the underlying particle physics model allows such defects.

The argument which ensures that in theories which admit topological or semitopological defects, such defects will be produced during a phase transition in the very early Universe is called the Kibble mechanism [71]. To illustrate the physics, consider a mechanical toy model, first introduced by Mazenko, Unruh and Wald [79]. Take (see Fig. 8) a lattice of points on a flat table. At each point, a pencil is pivoted. It is free to rotate and oscillate. The tips of nearest neighbor pencils are connected with springs (to mimic the spatial gradient terms in the scalar field Lagrangean). Newtonian gravity creates a potential energy $V(\varphi)$ for each pencil (φ is the angle relative to the vertical direction). $V(\varphi)$ is minimized for $|\varphi| = \eta$ (in our toy model $\eta = \pi/2$). Hence, the Lagrangean of this pencil model is analogous to that of a scalar field with symmetry breaking potential (52).

At high temperatures $T \gg T_c$, all pencils undergo large amplitude high frequency oscillations. However, by causality, the phases of oscillation of pencils with large separation s are uncorrelated. For a system in thermal equilibrium, the length s beyond which phases are random is the correlation length $\xi(t)$. However, since the system is quenched rapidly, there is a causality bound on ξ:

$$\xi(t) < t \,, \tag{59}$$

where t is the causal horizon.

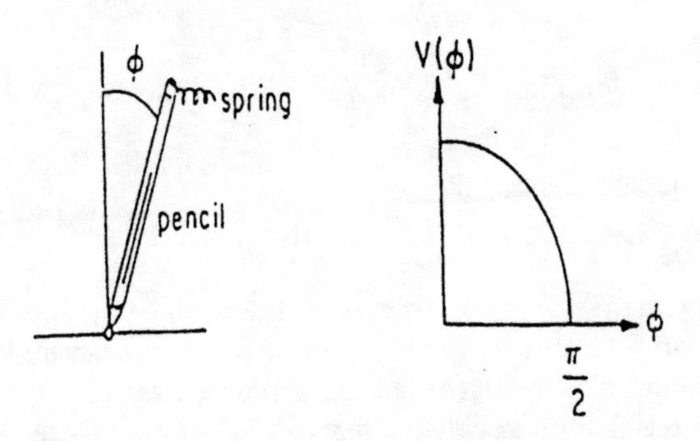

FIGURE 8. The pencil model: the potential energy of a simple pencil has the same form as that of scalar fields used for spontaneous symmetry breaking. The springs connecting nearest neighbor pencils give rise to contributions to the energy which mimic spatial gradient terms in field theory.

The critical temperature T_c is the temperature at which the thermal energy is equal to the energy a pencil needs to jump from horizontal to vertical position. For $T < T_c$, all pencils want to lie flat on the table. However, their orientations are random beyond a distance $\xi(t)$ determined by equating the free energy gained by symmetry breaking (a volume effect) with the gradient energy lost (a surface effect). As expected, $\xi(T)$ diverges at T_c. Very close to T_c, the thermal energy T is larger than the volume energy gain E_{corr} in a correlation volume. Hence, these domains are unstable to thermal fluctuations. As T decreases, the thermal energy decreases more rapidly than E_{corr}. Below the "Ginsburg temperature" T_G, there is insufficient thermal energy to excite a correlation volume into the state $\varphi = 0$. Domains of size

$$\xi(t_G) \sim \lambda^{-1}\eta^{-1} \tag{60}$$

freeze out [71,80]. The boundaries between these domains become topological defects. An improved version of this argument has recently been given by Zurek [81] (see also [82]).

We conclude that in a theory in which a symmetry breaking phase transitions satisfies the topological criteria for the existence of a fixed type of defect, a network of such defects will form during the phase transition and will freeze out at the Ginsburg temperature. The correlation length is initially given by (60), if the field φ is in thermal equilibrium before the transition. Independent of this last assumption, the causality bound implies that $\xi(t_G) < t_G$.

For times $t > t_G$ the evolution of the network of defects may be complicated (as for cosmic strings) or trivial (as for textures). In any case (see the caveats of [83]), the causality bound persists at late times and states that even at late times, the mean separation and length scale of defects is bounded by $\xi(t) \leq t$.

Applied to cosmic strings, the Kibble mechanism implies that at the time of the phase transition, a network of cosmic strings with typical step length $\xi(t_G)$ will form. According to numerical simulations [84], about 80% of the initial energy is in infinite strings (strings with curvature radius larger than the Hubble radius) and 20% in closed loops.

Evolution of Strings and Scaling

The evolution of the cosmic string network for $t > t_G$ is complicated. The key processes are loop production by intersections of infinite strings (see Fig. 9) and loop shrinking by gravitational radiation. These two processes combine to create a mechanism by which the infinite string network loses energy (and length as measured in comoving coordinates). It can be shown (see e.g. [77]) that, as a consequence, the correlation length of the string network is always proportional to its causality limit

$$\xi(t) \sim t\,. \tag{61}$$

Hence, the energy density $\rho_\infty(t)$ in long strings is a fixed fraction of the background energy density $\rho_c(t)$

$$\rho_\infty(t) \sim \mu\xi(t)^{-2} \sim \mu t^{-2} \tag{62}$$

or

$$\frac{\rho_\infty(t)}{\rho_c(t)} \sim G\mu\,. \tag{63}$$

We conclude that the cosmic string network approaches a "scaling solution" in which the statistical properties of the network are time independent if all distances are scaled to the horizon distance.

Cosmic Strings and Structure Formation

The starting point of the structure formation scenario in the cosmic string theory is the scaling solution for the cosmic string network, according to which at all times t (in particular at t_{eq}, the time when perturbations can start to grow) there will be a few long strings crossing each Hubble volume, plus a distribution of loops of radius $R \ll t$ (see Fig. 10).

The cosmic string model admits three mechanisms for structure formation: loops, filaments, and wakes. Cosmic string loops have the same time averaged field as a point source with mass [85] $M(R) = \beta R\mu$, R being the loop radius and $\beta \sim 2\pi$. Hence, loops will be seeds for spherical accretion of dust and radiation.

For loops with $R \leq t_{eq}$, growth of perturbations in a model dominated by cold dark matter starts at t_{eq}. Hence, the mass at the present time will be

$$M(R,\, t_0) = z(t_{eq})\beta\, R\mu\,. \tag{64}$$

In the original cosmic string model [2,86] it was assumed that loops dominate over wakes. However, according to the newer cosmic string evolution

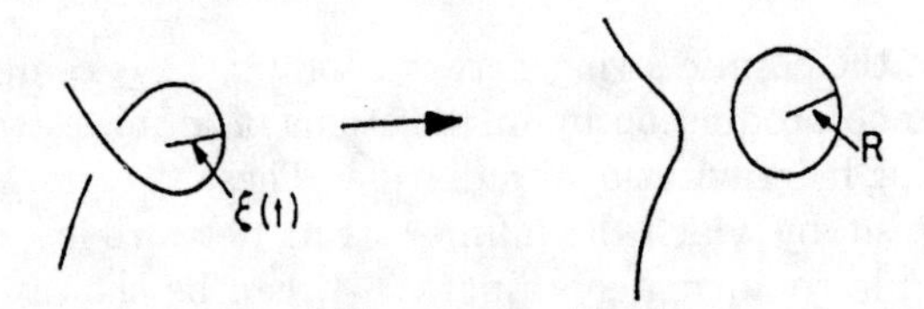

FIGURE 9. Formation of a loop by a self intersection of an infinite string. According to the original cosmic string scenario, loops form with a radius R determined by the instantaneous coherence length of the infinite string network.

simulations [87], most of the energy in strings is in the long strings, and hence the loop accretion mechanism is subdominant.

The second mechanism involves long strings moving with relativistic speed in their normal plane which give rise to velocity perturbations in their wake [88]. The mechanism is illustrated in Fig. 11: space normal to the string is a cone with deficit angle [89]

$$\alpha = 8\pi G\mu\,. \tag{65}$$

If the string is moving with normal velocity v through a bath of dark matter, a velocity perturbation

$$\delta v = 4\pi G\mu v\gamma \tag{66}$$

[with $\gamma = (1-v^2)^{-1/2}$] towards the plane behind the string results. At times after t_{eq}, this induces planar overdensities, the most prominent (*i.e.*, thickest at the present time) and numerous of which were created at t_{eq}, the time of equal matter and radiation [90–92]. The corresponding planar dimensions are (in comoving coordinates)

$$t_{eq}z(t_{eq}) \times t_{eq}z(t_{eq})v \sim (40 \times 40v)\,\mathrm{Mpc}^2\,. \tag{67}$$

The thickness d of these wakes can be calculated using the Zel'dovich approximation [93]. The result is (for $G\mu = 10^{-6}$)

$$d \simeq G\mu v\gamma(v)z(t_{eq})^2\,t_{eq} \simeq 4v\,\mathrm{Mpc}\,. \tag{68}$$

Wakes arise if there is little small scale structure on the string. In this case, the string tension equals the mass density, the string moves at relativistic speeds, and there is no local gravitational attraction towards the string.

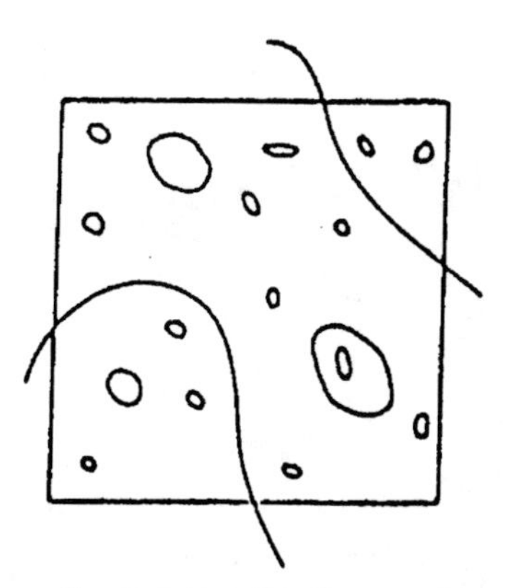

FIGURE 10. Sketch of the scaling solution for the cosmic string network. The box corresponds to one Hubble volume at arbitrary time t.

In contrast, if there is small scale structure on strings, then the coarse-grained string tension T is smaller [94] than the mass per unit length μ , and thus there is a gravitational force towards the string which gives rise to cylindrical accretion, producing filaments [95].

Which of the mechanisms – filaments or wakes – dominates is determined by the competition between the velocity induced by the Newtonian gravitational potential of the strings and the velocity perturbation of the wake.

The cosmic string model predicts a scale-invariant spectrum of density perturbations, exactly like inflationary Universe models but for a rather different reason. Consider the *r.m.s.* mass fluctuations on a length scale $2\pi k^{-1}$ at the time $t_H(k)$ when this scale enters the Hubble radius. From the cosmic string scaling solution it follows that a fixed (*i.e.*, $t_H(k)$ independent) number $\tilde{v}$ of strings of length of the order $t_H(k)$ contribute to the mass excess $\delta M(k,\, t_H(k))$. Thus

$$\frac{\delta M}{M}(k,\, t_H(k)) \sim \frac{\tilde{v}\mu t_H(k)}{G^{-1} t_H^{-2}(k) t_H^3(k)} \sim \tilde{v}\, G\mu\,. \tag{69}$$

Note that the above argument predicting a scale invariant spectrum will hold for all topological defect models which have a scaling solution, in particular also for global monopoles and textures.

The amplitude of the *r.m.s.* mass fluctuations (equivalently: of the power spectrum) can be used to normalize $G\mu$. Since today on galaxy cluster scales

$$\frac{\delta M}{M}(k,\, t_0) \sim 1\,, \tag{70}$$

the growth rate of fluctuations linear in $a(t)$ yields

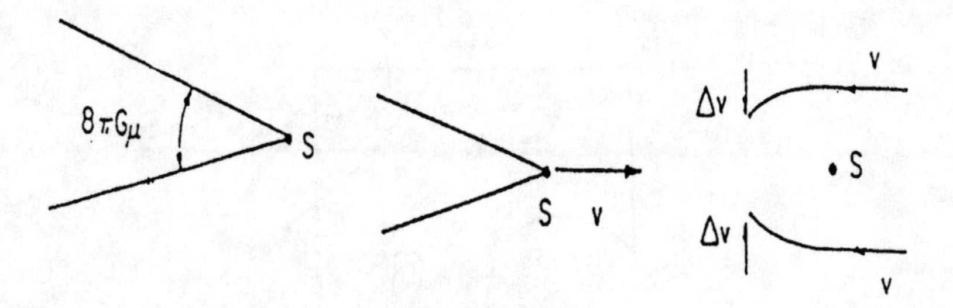

FIGURE 11. Sketch of the mechanism by which a long straight cosmic string S moving with velocity v in transverse direction through a plasma induces a velocity perturbation Δv towards the wake. Shown on the left is the deficit angle, in the center is a sketch of the string moving in the plasma, and on the right is the sketch of how the plasma moves in the frame in which the string is at rest.

$$\frac{\delta M}{M}(k,\, t_{eq}) \sim 10^{-4}\,, \tag{71}$$

and therefore, using $\tilde{v} \sim 10$,

$$G\mu \sim 10^{-5}\,. \tag{72}$$

A similar value is obtained by normalizing the model to the COBE amplitude of CMB anisotropies on large angular scales [96,97] (the normalizations from COBE and from the power spectrum of density perturbations on large scales agree to within a factor of 2). Thus, if cosmic strings are to be relevant for structure formation, they must arise due to a symmetry breaking at energy scale $\eta \simeq 10^{16}$GeV. This scale happens to be the scale of unification (GUT) of weak, strong and electromagnetic interactions. It is tantalizing to speculate that cosmology is telling us that there indeed was new physics at the GUT scale.

A big advantage of the cosmic string model over inflationary Universe models is that HDM is a viable dark matter candidate. Cosmic string loops survive free streaming, as discussed in Section 3.B, and can generate nonlinear structures on galactic scales, as discussed in detail in [98,99]. Accretion of hot dark matter by a string wake was studied in [92]. In this case, nonlinear perturbations develop only late. At some time t_{nl}, all scales up to a distance $q_{\max}$ from the wake center go nonlinear. Here

$$q_{\max} \sim G\mu v\gamma(v)z(t_{eq})^2 t_{eq} \sim 4v\,\mathrm{Mpc}\,, \tag{73}$$

and it is the comoving thickness of the wake at t_{nl}. Demanding that t_{nl} corresponds to a redshift greater than 1 leads to the constraint

$$G\mu > 5 \cdot 10^{-7}\,. \tag{74}$$

Note that in a cosmic string and hot dark matter model, wakes form nonlinear structures only very recently. Accretion onto loops and small scale structure on the long strings provide two mechanisms which may lead to high redshift objects such as quasars and high redshift galaxies. The first mechanism has recently been studied in [100], the second in [101,102].

The power spectrum of density fluctuations in a cosmic string model with HDM has recently been studied numerically by Mähönen [103], based on previous work of [104] (see also [105] for an earlier semi-analytical study). The spectral shape agrees quite well with observations, and a bias factor of less than 2 is required to give the best-fit amplitude for a COBE normalized model. Note, however, that the results depend quite sensitively on the details of the string scaling solution which are at present not well understood.

Due to lack of space, I will not discuss the global monopole [106] and global texture [107] models of structure formation. There has been a lot of work on the texture model, and the reader is referred to [108,109] for recent review articles.

Specific Signatures

The cosmic string theory of structure formation makes several distinctive predictions, both in terms of the galaxy distribution and in terms of CMB anisotropies. On large scales (corresponding to the comoving Hubble radius at t_{eq} and larger, structure is predicted to be dominated either by planar [90–92] or filamentary [95] galaxy concentrations. For models in which the strings have no local gravity, the resulting nonlinear structures will look very different from the nonlinear structures in models in which local gravity is the dominant force. As discovered and discussed recently in [110], a baryon number excess is predicted in the nonlinear wakes. This may explain the "cluster baryon crisis" [111], the fact that the ratio of baryons to dark matter in rich clusters is larger than what is compatible with the nucleosynthesis constraints in a spatially flat Universe.

As described in the previous subsection, space perpendicular to a long straight cosmic string is conical with deficit angle given by (65). Consider now CMB radiation approaching an observer in a direction normal to the plane spanned by the string and its velocity vector (see Fig. 12). Photons arriving at the observer having passed on different sides of the string will obtain a relative Doppler shift which translates into a temperature discontinuity of amplitude [112]

$$\frac{\delta T}{T} = 4\pi G\mu v\gamma(v)\,, \tag{75}$$

where v is the velocity of the string. Thus, the distinctive signature for cosmic strings in the microwave sky are line discontinuities in T of the above

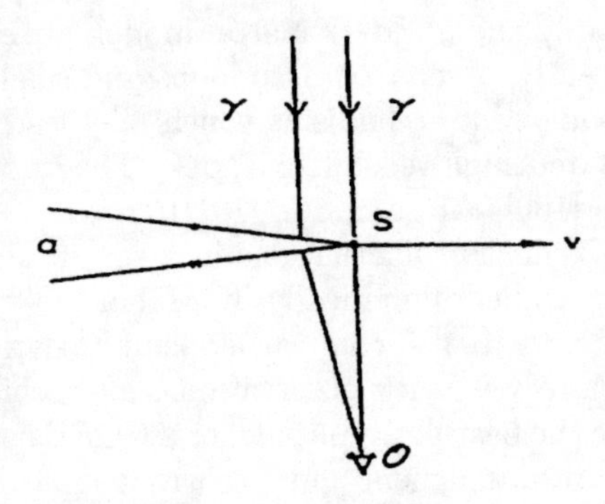

FIGURE 12. Sketch of the Kaiser-Stebbins effect by which cosmic strings produce linear discontinuities in the CMB. Photons γ passing on different sides of a moving string S (velocity v) towards the observer $\mathcal{O}$ receive a relative Doppler shift due to the conical nature of space perpendicular to the string (deficit angle α).

magnitude.

Given ideal maps of the CMB sky it would be easy to detect strings. However, real experiments have finite beam width. Taking into account averaging over a scale corresponding to the beam width will smear out the discontinuities, and it turns out to be surprisingly hard to distinguish the predictions of the cosmic string model from that of inflation-based theories using quantitative statistics which are easy to evaluate analytically, such as the kurtosis of the spatial gradient map of the CMB [113]. There may be ways to distinguish between string and inflationary models by looking at the angular power spectrum of CMB anisotropies. Work on this subject, however, is still controversial [114–116].

Global textures also produce distinctive non-Gaussian signatures [117] in CMB maps. In fact, these signatures are more pronounced and on larger scales than the signatures in the cosmic string model.

TOPOLOGICAL DEFECTS AND BARYOGENESIS

Principles of Baryogenesis

Baryogenesis is another area where particle physics and cosmology connect in a very deep way. The goal is to explain the observed asymmetry between matter and antimatter in the Universe. In particular, the objective is to be able to explain the observed value of the net baryon to entropy ratio at the present time

$$\frac{\Delta n_B}{s}(t_0) \sim 10^{-10} \tag{76}$$

starting from initial conditions in the very early Universe when this ratio vanishes. Here, Δn_B is the net baryon number density and s the entropy density.

As pointed out by Sakharov [118], three basic criteria must be satisfied in order to have a chance at explaining the data:

1. The theory describing the microphysics must contain baryon number violating processes.

2. These processes must be C and CP violating.

3. The baryon number violating processes must occur out of thermal equilibrium.

As was discovered in the 1970's [119], all three criteria can be satisfied in GUT theories. In these models, baryon number violating processes are mediated by superheavy Higgs and gauge particles. The baryon number violation is visible in the Lagrangian, and occurs in perturbation theory (and is therefore in principle easy to calculate). In addition to standard model CP violation, there are typically many new sources of CP violation in the GUT sector. The third Sakharov condition can also be realized: After the GUT symmetry-breaking phase transition, the superheavy particles may fall out of thermal equilibrium. The out-of-equilibrium decay of these particles can thus generate a nonvanishing baryon to entropy ratio.

The magnitude of the predicted n_B/s depends on the asymmetry ε per decay, on the coupling constant λ of the n_B violating processes, and on the ratio n_X/s of the number density n_X of superheavy Higgs and gauge particles to the number density of photons, evaluated at the time t_d when the baryon number violating processes fall out of thermal equilibrium, and assuming that this time occurs after the phase transition. The quantity ε is proportional to the CP-violation parameter in the model. In a GUT theory, this CP violation parameter can be large (order 1), whereas in the standard electroweak theory it is given by the CP violating phases in the CKM mass matrix and is very small. As shown in [119] it is easily possible to construct models which give the right n_B/s ratio after the GUT phase transition (for recent reviews of baryogenesis see [120] and [121]).

GUT Baryogenesis and Topological Defects

The ratio n_B/s, however, does not only depend on ε, but also on $n_X/s(t_d)$. If the temperature T_d at the time t_d is greater than the mass m_X of the superheavy particles, then it follows from the thermal history in standard cosmology that $n_X \sim s$. However, if $T_d < m_X$, then the number density of X particles is diluted exponentially in the time interval between when $T = m_X$ and when $T = T_d$. Thus, the predicted baryon to entropy ratio is exponentially suppressed:

$$\frac{n_B}{s} \sim \frac{1}{g^*}\lambda^2 \varepsilon e^{-m_X/T_d}\,, \tag{77}$$

where g^* is the number of spin degrees of freedom in thermal equilibrium at the time of the phase transition. In this case, the standard GUT baryogenesis mechanism is ineffective.

However, topological defects may come to the rescue [122]. As we have seen in the previous section, topological defects will inevitably be produced in the symmetry breaking GUT transition provided they are topologically allowed in that symmetry breaking scheme. The topological defects provide an alternative mechanism of GUT baryogenesis.

Inside of topological defects, the GUT symmetry is restored. In fact, the defects can be viewed as solitonic configurations of X particles. The continuous decay of defects at times after t_d provides an alternative way to generate a nonvanishing baryon to entropy ratio. The defects constitute out of equilibrium configurations, and hence their decay can produce a nonvanishing n_B/s in the same way as the decay of free X quanta.

The way to compute the estimate n_B/s ratio is as follows: The defect scaling solution gives the energy density in defects at all times. Taking the time derivative of this density, and taking into account the expansion of the Universe, we obtain the loss of energy attributed to defect decay. By energetics, we can estimate the number of decays of individual quanta which the defect decay corresponds to. We can then use the usual perturbative results to compute the resulting net baryon number.

Provided that $m_X < T_d$, then at the time when the baryon number violating processes fall out of equilibrium (when we start generating a nonvanishing n_B) the energy density in free X quanta is much larger than the defect density, and hence the defect-driven baryogenesis mechanism is subdominant. However, if $m_X > T_d$, then as indicated in (77), the energy density in free quanta decays exponentially. In constrast, the density in defects only decreases as a power of time, and hence soon dominates baryogenesis.

One of the most important ingredients in the calculation is the time dependence of $\xi(t)$, the separation between defects. Immediately after the phase transition at the time t_f of the formation of the defect network, the separation is $\xi(t_f) \sim \lambda^{-1}\eta^{-1}$. In the time period immediately following, the time period of relevance for baryogenesis, $\xi(t)$ approaches the Hubble radius according to the equation [80]

$$\xi(t) \simeq \xi(t_f)(\frac{t}{t_f})^{5/4} . \tag{78}$$

Using this result to calculate the defect density, we obtain after some algebra

$$\frac{n_B}{s}|_{\text{defect}} \sim \lambda^2 \frac{T_d}{\eta} \frac{n_B}{s}|_0 , \tag{79}$$

where $n_B/s|_0$ is the unsuppressed value of n_B/s which can be obtained using the standard GUT baryogenesis mechanism. We see from (79) that even for low values of T_d, the magnitude of n_B/s which is obtained via the defect mechanism is only suppressed by a power of T_d. However, the maximum strength of the defect channel is smaller than the maximum strength of the usual mechanism by a geometrical suppression factor λ^2 which expresses the fact that even at the time of defect formation, the defect network only occupies a small volume.

Electroweak Baryogenesis and Topological Defects

It has been known for some time that there are baryon number violating processes even in the standard electroweak theory. These processes are, however, nonperturbative. They are connected with the t'Hooft anomaly [123], which in turn is due to the fact that the gauge theory vacuum is degenerate, and that the different degenerate vacuum states have different quantum numbers (Chern-Simons numbers). In theories with fermions, this implies different baryon number. Configurations such as sphalerons [124] which interpolate between two such vacuum states thus correspond to baryon number violating processes.

As pointed out in [125], the anomalous baryon number violating processes are in thermal equilibrium above the electroweak symmetry breaking scale. Therefore, any net baryon to entropy ratio generated at a higher scale will be erased, unless this ratio is protected by an additional quantum number such as a nonvanishing $B - L$ which is conserved by electroweak processes.

However, as first suggested in [126] and discussed in detail in many recent papers (see [127] for reviews of the literature), it is possible to regenerate a nonvanishing n_B/s below the electroweak symmetry breaking scale. Since there are n_B violating processes and both C and CP violation in the standard model, Sakharov's conditions are satisfied provided that one can realize an out-of-equilibrium state after the phase transition. Standard model CP violation is extremely weak. Thus, it appears necessary to add some sector with extra CP violation to the standard model in order to obtain an appreciable n_B/s ratio. A simple possibility which has been invoked often is to add a second Higgs doublet to the theory, with CP violating relative phases.

The standard way to obtain out-of-equilibrium baryon number violating processes immediately after the electroweak phase transition is [127] to assume that the transition is strongly first order and proceeds by the nucleation of bubbles (note that these are two assumptions, the second being stronger than the first!).

Bubbles are out-of-equilibrium configurations. Outside of the bubble (in the false vacuum), the baryon number violating processes are unsuppressed, inside they are exponentially suppressed. In the bubble wall, the Higgs fields have a nontrivial profile, and hence (in models with additional CP violation in the Higgs sector) there is enhanced CP violation in the bubble wall. In order to obtain net baryon production, one may either use fermion scattering off bubble walls [128] (because of the CP violation in the scattering, this generates a lepton asymmetry outside the bubble which converts via sphalerons to a baryon asymmetry) or sphaleron processes in the bubble wall itself [129,130]. It has been shown that, using optimistic parameters (in particular a large CP violating phase $\Delta\theta_{CP}$ in the Higgs sector) it is possible to generate the observed n_B/s ratio. The resulting baryon to entropy ratio is of the order

$$\frac{n_B}{s} \sim \alpha_W^2 (g^*)^{-1} (\frac{m_t}{T})^2 \Delta\theta_{CP} , \tag{80}$$

where α_W refers to the electroweak interaction strength, g^* is the number of spin degrees of freedom in thermal quilibrium at the time of the phase transition, and m_t is the top quark mass. The dependence on the top quark mass enters because net baryogenesis only appears at the one-loop level.

However, analytical and numerical studies show that, for the large Higgs masses which are indicated by the current experimental bounds, the electroweak phase transition will unlikely be sufficiently strongly first order to proceed by bubble nucleation. In addition, there are some concerns as to whether it will proceed by bubble nucleation at all (see e.g. [131]).

Once again, topological defects come to the rescue. In models which admit defects, such defects will inevitably be produced in a phase transition independent of its order. Moving topological defects can play the same role in baryogenesis as nucleating bubbles. In the defect core, the electroweak symmetry is unbroken and hence sphaleron processes are unsuppressed [132]. In the defect walls there is enhanced CP violation for the same reason as in bubble walls. Hence, at a fixed point in space, a nonvanishing baryon number will be produced when a topological defect passes by.

Defect-mediated electroweak baryogenesis has been worked out in detail in [133] (see [134] for previous work) in the case of cosmic strings. The scenario is as follows: at a particular point x in space, antibaryons are produced when the front side of the defect passes by. While x is in the defect core, partial equilibration of n_B takes place via sphaleron processes. As the back side of the defect passes by, the same number of baryons are produced as the number of antibaryons when the front side of the defect passes by. Thus, at the end a positive number of baryons are left behind.

As in the case of defect-mediated GUT baryogenesis, the strength of defect-mediated electroweak baryogenesis is suppressed by the ratio SF of the volume which is passed by defects divided by the total volume, i.e.

$$\frac{n_B}{s} \sim \mathrm{SF}\frac{n_B}{s}|_0 \,, \tag{81}$$

where $(n_B/s)|_0$ is the result of (80) obtained in the bubble nucleation mechanism.

A big caveat for defect-mediated electroweak baryogenesis is that the standard electroweak theory does not admit topological defects. However, in a theory with additional physics just above the electroweak scale it is possible to obtain defects (see e.g. [135] for some specific models). The closer the scale η of the new physics is to the electroweak scale η_{EW}, the larger the volume in defects and the more efficient defect-mediated electroweak baryogenesis. Using the result of (78) for the separation of defects, we obtain (for non-superconducting strings)

$$\mathrm{SF} \sim \lambda(\frac{\eta_{EW}}{\eta})^{3/2} v_D \,. \tag{82}$$

where v_D is the mean defect velocity.

Obviously, the advantage of the defect-mediated baryongenesis scenario is that it does not depend on the order and on the detailed dynamics of the electroweak phase transition.

Summary

As we have seen, topological defects may play an important role in cosmology. Defects are inevitably produced during symmetry breaking phase transitions in the early Universe in all theories in which defects are topologically stable. Theories giving rise to domain walls or local monopoles are ruled out by cosmological constraints. Those producing cosmic strings, global monopoles and textures are quite attractive.

If the scale of symmetry breaking at which the defects are produced is about 10^{16} GeV, then defects can act as the seeds for galaxy formation. Defect theories of structure formation predict a roughly scale-invariant spectrum of density perturbations, similar to inflation-based models. However, the phases in the density field are distributed in a non-Gaussian manner. Thus, the predictions of defect models can be distinguished from those of inflationary models. In addition, the predictions of different defect models can be distinguished from eachother.

As shown in this section, topological defects may also play a crucial role in baryogenesis. This applies both to GUT and electroweak baryogenesis. The crucial point is that defects constitute out-of-equilibrium configurations, and may therefore be the sites of net baryon production.

Acknowledgements

I wish to thank the organizers of the school and symposium for inviting me to speak in Merida. I am grateful to all of my research collaborators, on whose work I have freely drawn. Partial financial support for the preparation of this manuscript has been provided at Brown by the US Department of Energy under Grant DE-FG0291ER40688,

REFERENCES

1. A. Guth, *Phys. Rev.* **D23**, 347 (1981).
2. Ya.B. Zel'dovich, *Mon. Not. R. astron. Soc.* **192**, 663 (1980);
   A. Vilenkin, *Phys. Rev. Lett.* **46**, 1169 (1981).
3. R. Brandenberger, 'Particle Physics Aspects of Modern Cosmology', in "Field Theoretic Methods in Fundamental Physics" (Mineumsa Co Ltd., Seoul, 1997), hep-ph/9701262 (1997).
4. R. Brandenberger, H. Feldman, V. Mukhanov and T. Prokopec, 'Gauge Invariant Cosmological Perturbations: Theory and Applications,' publ. in "The

Origin of Structure in the Universe," eds. E. Gunzig and P. Nardone (Kluwer, Dordrecht, 1993).

5. V. Mukhanov, H. Feldman and R. Brandenberger, *Phys. Rep.* **215**, 203 (1992).
6. V. de Lapparent, M. Geller and J. Huchra, *Ap. J. (Lett)* **302**, L1 (1986).
7. S. Landy, S. Shectman, H. Lin, R. Kirschner, A. Oemler and D. Tucker, *Ap. J. (Lett.)* **456**, 1L (1996);
   H. Lin, R. Kirshner, S. Shectman, S. Landy, A. Oemler, D. Tucker and P. Schechter, *Ap. J. (Lett.)* **464**, 60L (1996);
   S. Shectman, S. Landy, A. Oemler, D. Tucker, H. Lin, R. Kirshner and P. Schechter, *Ap. J. (Suppl.)* **470**, 172S (1996).
8. J. Ostriker and L. Cowie, *Ap. J. (Lett.)* **243**, L127 (1981).
9. A. Linde, 'Particle Physics and Inflationary Cosmology' (Harwood, Chur, 1990).
10. S. Blau and A. Guth, 'Inflationary Cosmology,' in '300 Years of Gravitation' ed. by S. Hawking and W. Israel (Cambridge Univ. Press, Cambridge, 1987).
11. K. Olive, *Phys. Rep.* **190**, 307 (1990).
12. R. Brandenberger, *Rev. Mod. Phys.* **57**, 1 (1985).
13. D. Kazanas, *Ap. J.* **241**, L59 (1980).
14. J. Gott III and T. Statler, *Phys. Lett.* **136B**, 157 (1984);
    M. Bucher, A. Goldhaber and N. Turok, *Phys. Rev.* **D52**, 3314 (1995);
    A. Linde, *Phys. Lett.* **B351**, 99 (1995).
15. W. Press, *Phys. Scr.* **21**, 702 (1980).
16. G. Chibisov and V. Mukhanov, 'Galaxy Formation and Phonons,' Lebedev Physical Institute Preprint No. 162 (1980);
    G. Chibisov and V. Mukhanov, *Mon. Not. R. Astron. Soc.* **200**, 535 (1982).
17. V. Lukash, *Pis'ma Zh. Eksp. Teor. Fiz.* **31**, 631 (1980).
18. K. Sato, *Mon. Not. R. Astron. Soc.* **195**, 467 (1981).
19. R. Brandenberger, *Phys. Lett.* **129B**, 397 (1983).
20. A. Guth and S.-H. Tye, *Phys. Rev. Lett.* **44**, 631 (1980).
21. D. Kirzhnits and A. Linde, *Pis'ma Zh. Eksp. Teor. Fiz.* **15**, 745 (1972);
    D. Kirzhnits and A. Linde, *Zh. Eksp. Teor. Fiz.* **67**, 1263 (1974);
    C. Bernard, *Phys. Rev.* **D9**, 3313 (1974);
    L. Dolan and R. Jackiw, *Phys. Rev.* **D9**, 3320 (1974);
    S. Weinberg, *Phys. Rev.* **D9**, 3357 (1974).
22. S. Coleman, *Phys. Rev.* **D15**, 2929 (1977);
    C. Callan and S. Coleman, *Phys. Rev.* **D16**, 1762 (1977);
    M. Voloshin, Yu. Kobzarev and L. Okun, *Sov. J. Nucl. Phys.* **20**, 644 (1975);
    M. Stone, *Phys. Rev.* **D14**, 3568 (1976);
    M. Stone, *Phys. Lett.* **67B**, 186 (1977);
    P. Frampton, *Phys. Rev. Lett.*, **37**, 1380 (1976);
    S. Coleman, in 'The Whys of Subnuclear Physics' (Erice 1977), ed by A. Zichichi (Plenum, New York, 1979).
23. A. Guth and E. Weinberg, *Nucl. Phys.* **B212**, 321 (1983).
24. A. Linde, *Phys. Lett.* **108B**, 389 (1982).
25. A. Albrecht and P. Steinhardt, *Phys. Rev. Lett.* **48**, 1220 (1982).

26. A. Linde, *Phys. Lett.* **129B**, 177 (1983).
27. A. Linde, D. Linde and A. Mezhlumian, *Phys. Rev.* **D49**, 1783 (1994);
A. Linde, 'Lectures on Inflationary Cosmology', Stanford preprint SU-ITP-94-36, hep-th/9410082 (1994).
28. A. Starobinsky, in 'Current Trends in Field Theory, Quantum Gravity, and Strings', Lecture Notes in Physics, ed. by H. de Vega and N. Sanchez (Springer, Heidelberg, 1986).
29. S. Weinberg, *Rev. Mod. Phys.* **61**, 1 (1989);
S. Carroll, W. Press and E. Turner, *Ann. Rev. Astron. Astrophys.* **30**, 499 (1992).
30. A. Starobinsky, *Phys. Lett.* **91B**, 99 (1980).
31. V. Mukhanov and R. Brandenberger, *Phys. Rev. Lett.* **68**, 1969 (1992).
32. R. Ball and A. Matheson, *Phys. Rev.* **D45**, 2647 (1992).
33. L. Parker and Y. Zhang, *Phys. Rev.* **D47**, 416 (1993).
34. A. Zhitnitsky, "Effective Field Theories as Asymptotic Series: From QCD to Cosmology", hep-ph/9601348 (1996).
35. R. Brandenberger and A. Zhitnitsky, "Can Asymptotic Series Resolve the Problems of Inflation?", Brown preprint BROWN-HET-1035, hep-ph/9604407 (1996), *Phys. Rev. D*, in press (1997).
36. N. C. Tsamis and R. P. Woodard, *Phys. Lett.* **B301**, 351 (1993); *Ann. Phys.* **238**, 1 (1995); "Quantum Gravity Slows Inflation", hep-ph/9602315 (1996).
37. J. Bardeen, P. Steinhardt and M. Turner, *Phys. Rev.* **D28**, 1809 (1983).
38. R. Brandenberger and R. Kahn, *Phys. Rev.* **D28**, 2172 (1984).
39. V. Mukhanov and G. Chibisov, *JETP Lett.* **33**, 532 (1981).
40. A. Guth and S.-Y. Pi, *Phys. Rev. Lett.* **49**, 110 (1982);
S. Hawking, *Phys. Lett.* **115B**, 295 (1982);
A. Starobinskii, *Phys. Lett.* **117B**, 175 (1982);
V. Mukhanov, *JETP Lett.* **41**, 493 (1985).
41. F. Adams, K. Freese and A. Guth, *Phys. Rev.* **D43**, 965 (1991).
42. R. Brandenberger, V. Mukhanov and A. Sornborger, *Phys. Rev.* **D48**, 1629 (1993).
43. L. Kofman, A. Linde and A. Starobinski, *Phys. Rev. Lett.* **73**, 3195 (1994).
44. Y. Shtanov, J. Traschen and R. Brandenberger, *Phys. Rev.* **D51**, 5438 (1995).
45. A. Dolgov and A. Linde, *Phys. Lett.* **116B**, 329 (1982).
46. L. Abbott, E. Farhi and M. Wise, *Phys. Lett.* **117B**, 29 (1982).
47. J. Traschen and R. Brandenberger, *Phys. Rev.* **D42**, 2491 (1990).
48. L. Landau and E. Lifshitz, 'Mechanics' (Pergamon, Oxford, 1960);
V. Arnold, 'Mathematical Methods of Classical Mechanics' (Springer, New York, 1978).
49. T. Prokopec and T. Roos, "Lattice Study of Classical Inflaton Decay", hep-ph/9610400.
50. M. Hotta, I. Joichi, S. Matsumoto and M. Yoshimura, "Quantum System under Periodic Perturbation: Effect of Environment", hep-ph/9608374 (1996).
51. N. Birrell and P. Davies, 'Quantum Fields in Curved Space' (Cambridge Univ. Press, Cambridge, 1982).

52. L. Kofman, A. Linde and A. Starobinski, *Phys. Rev. Lett.* **76**, 1011 (1996).
53. I. Tkachev, *Phys. Lett.* **B376**, 35 (1996).
54. E. Kolb, A. Linde and A. Riotto, *Phys. Rev. Lett.* **77**, 4290 (1996).
55. S. Khlebnikov and I. Tkachev, *Phys. Rev. Lett.* **77**, 219 (1996);
S. Khlebnikov and I. Tkachev, "The Universe after Inflation: the Wide Resonance Case", hep-ph/9608458 (1996).
56. R. Allahverdi and B. Campbell, "Cosmological Reheating and Selfinteracting Final State Bosons", hep-ph/9606463 (1996).
57. D. Boyanovsky, H. de Vega, R. Holman and J. Salgado, *Phys. Rev.* **D54**, 7570 (1996);
D. Boyanovsky, D. Cormier, H. de Vega, R. Holman, A. Singh and M. Srednicki, "Preheating in FRW Universes", hep-ph/9609527 (1996).
58. M. Yoshimura, *Prog. Theor. Phys.* **94**, 873 (1995);
H. Fujisaki, K. Kumekawa, M. Yamaguchi and M. Yoshimura, *Phys. Rev.* **D53**, 6805 (1996).
59. D. Boyanovsky, H. de Vega, R. Holman, D.-S. Lee and A. Singh, *Phys. Rev.* **D51**, 4419 (1995);
D. Boyanovsky, M. D'Attanasio, H. de Vega, R. Holman and D.-S. Lee, *Phys. Rev.* **D52**, 6805 (1995).
60. D. Kaiser, *Phys. Rev.* **D53**, 1776 (1996).
61. G. Anderson, A. Linde and A. Riotto, *Phys. Rev. Lett.* **77**, 3716 (1996).
62. A. Liddle and D. Lyth, *Phys. Rep.* **231**, 1 (1993).
63. M. Gasperini, "Status of String Cosmology: Phenomenological Aspects", hep-th/9509127 (1995);
G. Veneziano, "String Cosmology: Concepts and Consequences", hep-th/9512091 (1995).
64. M. Davis and J. Huchra, *Ap. J.* **254**, 437 (1982).
65. N. Bahcall and R. Soneira, *Ap. J.* **270**, 20 (1983);
A. Klypin and A. Kopylov, *Sov. Astr. Lett.* **9**, 41 (1983).
66. M. Strauss et al., *Ap. J.* **385**, 421 (1992).
67. D. Mermin, *Rev. Mod. Phys.* **51**, 591 (1979).
68. P. de Gennes, 'The Physics of Liquid Crystals' (Clarendon Press, Oxford, 1974);
I. Chuang, R. Durrer, N. Turok and B. Yurke, *Science* **251**, 1336 (1991);
M. Bowick, L. Chandar, E. Schiff and A. Srivastava, *Science* **263**, 943 (1994).
69. M. Salomaa and G. Volovik, *Rev. Mod. Phys.* **59**, 533 (1987).
70. A. Abrikosov, *JETP* **5**, 1174 (1957).
71. T.W.B. Kibble, *J. Phys.* **A9**, 1387 (1976).
72. A. Vilenkin and E.P.S. Shellard, 'Strings and Other Topological Defects' (Cambridge Univ. Press, Cambridge, 1994).
73. Ya.B. Zel'dovich, I. Kobzarev and L. Okun, *Zh. Eksp. Teor. Fiz.* **67**, 3 (1974).
74. Ya.B. Zel'dovich and M. Khlopov, *Phys. Lett.* **79B**, 239 (1978);
J. Preskill, *Phys. Rev. Lett.* **43**, 1365 (1979).
75. M. Hindmarsh and T.W.B. Kibble, *Rept. Prog. Phys.* **58**, 477 (1995).
76. R. Brandenberger, *Int. J. Mod. Phys.* **A9**, 2117 (1994).

77. A. Vilenkin, *Phys. Rep.* **121**, 263 (1985).
78. H. Nielsen and P. Olesen, *Nucl. Phys.* **B61**, 45 (1973).
79. G. Mazenko, W. Unruh and R. Wald, *Phys. Rev.* **D31**, 273 (1985).
80. T.W.B. Kibble, *Acta Physica Polonica* **B13**, 723 (1982).
81. W. Zurek, *Acta Phys. Pol.* **B24**, 1301 (1993);
W. Zurek, "Cosmological Experiments in Condensed Matter Systems", cond-mat/9607135 (1996), *Phys. Rep.* (in press).
82. M. Hindmarsh, A.-C. Davis and R. Brandenberger, *Phys. Rev.* **D49**, 1944 (1994);
R. Brandenberger and A.-C. Davis, *Phys. Lett.* **B332**, 305 (1994).
83. T.W.B. Kibble and E. Weinberg, *Phys. Rev.* **D43**, 3188 (1991).
84. T. Vachaspati and A. Vilenkin, *Phys. Rev.* **D30**, 2036 (1984).
85. N. Turok, *Nucl. Phys.* **B242**, 520 (1984).
86. N. Turok and R. Brandenberger, *Phys. Rev.* **D33**, 2175 (1986);
A. Stebbins, *Ap. J. (Lett.)* **303**, L21 (1986);
H. Sato, *Prog. Theor. Phys.* **75**, 1342 (1986).
87. D. Bennett and F. Bouchet, *Phys. Rev. Lett.* **60**, 257 (1988);
B. Allen and E.P.S. Shellard, *Phys. Rev. Lett.* **64**, 119 (1990);
A. Albrecht and N. Turok, *Phys. Rev.* **D40**, 973 (1989).
88. J. Silk and A. Vilenkin, *Phys. Rev. Lett.* **53**, 1700 (1984).
89. A. Vilenkin, *Phys. Rev.* **D23**, 852 (1981);
J. Gott, *Ap. J.* **288**, 422 (1985);
W. Hiscock, *Phys. Rev.* **D31**, 3288 (1985);
B. Linet, *Gen. Rel. Grav.* **17**, 1109 (1985);
D. Garfinkle, *Phys. Rev.* **D32**, 1323 (1985);
R. Gregory, *Phys. Rev. Lett.* **59**, 740 (1987).
90. T. Vachaspati, *Phys. Rev. Lett.* **57**, 1655 (1986).
91. A. Stebbins, S. Veeraraghavan, R. Brandenberger, J. Silk and N. Turok, *Ap. J.* **322**, 1 (1987).
92. R. Brandenberger, L. Perivolaropoulos and A. Stebbins, *Int. J. of Mod. Phys.* **A5**, 1633 (1990);
L. Perivolarapoulos, R. Brandenberger and A. Stebbins, *Phys. Rev.* **D41**, 1764 (1990);
R. Brandenberger, *Phys. Scripta* **T36**, 114 (1991).
93. Ya. B. Zeldovich, *Astr. Astrophys.* **5**, 84 (1970).
94. B. Carter, *Phys. Rev.* **D41**, 3869 (1990).
95. T. Vachaspati and A. Vilenkin, *Phys. Rev. Lett.* **67**, 1057 (1991);
T. Vachaspati, *Phys. Rev.* **D45**, 3487 (1992);
D. Vollick, *Phys. Rev.* **D45**, 1884 (1992).
96. D. Bennett, A. Stebbins and F. Bouchet, *Ap. J. (Lett.)* **399**, L5 (1992).
97. L. Perivolaropoulos, *Phys. Lett.* **298B**, 305 (1993).
98. R. Brandenberger, N. Kaiser, D. Schramm and N. Turok, *Phys. Rev. Lett.* **59**, 2371 (1987);
R. Brandenberger, N. Kaiser and N. Turok, *Phys. Rev.* **D36**, 2242 (1987).
99. R. Scherrer, A. Melott and E. Bertschinger, *Phys. Rev. Lett.* **62**, 379 (1989).

100. R. Moessner and R. Brandenberger, *Mon. Not. R. astr. Soc.* **280**, 797 (1996), astro-ph/9510141.
101. A. Aguirre and R. Brandenberger, *Int. J. Mod. Phys.* **D4**, 711 (1995), astro-ph/9505031.
102. V. Zanchin, J.A.S. Lima and R. Brandenberger, *Phys. Rev.* **D54**, 7129 (1996), astro-ph/9607062.
103. P. Mähönen, *Ap. J. (Lett.)* **459**, L45 (1996).
104. T. Hara and S. Miyoshi, *Prog. Theor. Phys.* **81**, 1187 (1989);
T. Hara, S. Morioka and S. Miyoshi, *Prog. Theor. Phys.* **84**, 867 (1990);
T. Hara et al., *Ap. J.* **428**, 51 (1994).
105. A. Albrecht and A. Stebbins, *Phys. Rev. Lett.* **69**, 2615 (1992).
106. M. Barriola and A. Vilenkin, *Phys. Rev. Lett.* **63**, 341 (1989);
S. Rhie and D. Bennett, *Phys. Rev. Lett.* **65**, 1709 (1990).
107. N. Turok, *Phys. Rev. Lett.* **63**, 2625 (1989).
108. N. Turok, *Phys. Scripta* **T36**, 135 (1991).
109. R. Durrer and Z.H. Zhou, *Phys. Rev.* **D53**, 5394 (1996).
110. A. Sornborger, R. Brandenberger, B. Fryxell and K. Olson, "The Structure of Cosmic String Wakes", astro-ph/9608020 (1996), *Ap. J.* (in press).
111. S. White, J. Navarro, A. Evrard and C. Frenk, *Nature* **366**, 429 (1993).
112. N. Kaiser and A. Stebbins, *Nature* **310**, 391 (1984).
113. R. Moessner, L. Perivolaropoulos and R. Brandenberger, *Ap. J.* **425**, 365 (1994), astro-ph/9310001.
114. J. Magueijo, A. Albrecht, P. Ferreira and D. Coulson, *Phys. Rev.* **D54**, 3727 (1996), astro-ph/9605047.
115. W. Hu and M. White, *Phys. Rev. Lett.* **77**, 1687 (1996), astro-ph/9602020.
116. N. Turok, *Phys. Rev. Lett.* **77**, 4138 (1996), astro-ph/9607109.
117. N. Turok and D. Spergel, *Phys. Rev. Lett.* **64**, 2736 (1990).
118. A. Sakharov, *Pisma Zh. Eksp. Teor. Fiz.* **5**, 32 (1967).
119. M. Yoshimura, *Phys. Rev. Lett.* **41**, 281 (1978);
A. Ignatiev, N. Krasnikov, V. Kuzmin and A. Tavkhelidze, *Phys. Lett.* **76B**, 436 (1978);
S. Dimopoulos and L. Susskind, *Phys. Rev.* **D18**, 4500 (1978);
S. Weinberg, *Phys. Rev. Lett.* **42**, 850 (1979);
D. Toussaint, S. Trieman, F. Wilczek and A. Zee, *Phys. Rev.* **D19**, 1036 (1979).
120. A. Dolgov, *Phys. Rep.* **222**, 309 (1992).
121. V. Rubakov and M. Shaposhnikov, "Electroweak Baryon Number Nonconservation in the Early Universe and in High-Energy Collisions", hep-ph/9603208 (1996).
122. R. Brandenberger, A.-C. Davis and M. Hindmarsh, *Phys. Lett.* **B263**, 239 (1991).
123. G. t'Hooft, *Phys. Rev. Lett.* **37**, 8 (1976).
124. N. Manton, *Phys. Rev.* **D28**, 2019 (1983);
F. Klinkhamer and N. Manton, *Phys. Rev.* **D30**, 2212 (1984).
125. V. Kuzmin, V. Rubakov and M. Shaposhnikov, *Phys. Lett.* **B155**, 36 (1985);

P. Arnold and L. McLerran, *Phys. Rev.* **D36**, 581 (1987).

126. M. Shaposhnikov, *JETP Lett.* **44**, 465 (1986);
M. Shaposhnikov, *Nucl. Phys.* **B287**, 757 (1987);
L. McLerran, *Phys. Rev. Lett.* **62**, 1075 (1989).
127. N. Turok, in 'Perspectives on Higgs Physics', ed. G. Kane (World Scientific, Singapore, 1992);
A. Cohen, D. Kaplan and A. Nelson, *Ann. Rev. Nucl. Part. Sci.* **43**, 27 (1993).
128. A. Cohen, D. Kaplan and A. Nelson, *Phys. Lett.* **B245**, 561 (1990);
A. Cohen, D. Kaplan and A. Nelson, *Nucl. Phys.* **B349**, 727 (1991);
A. Nelson, D. Kaplan and A. Cohen, *Nucl. Phys.* **B373**, 453 (1992);
M. Joyce, T. Prokopec and N. Turok, *Phys. Lett.* **B338**, 269 (1994);
M. Joyce, T. Prokopec and N. Turok, *Phys. Rev. Lett.* **75**, 1695 (1995).
129. N. Turok and T. Zadrozny, *Phys. Rev. Lett.* **65**, 2331 (1990);
N. Turok and J. Zadrozny, *Nucl. Phys.* **B358**, 471 (1991);
L. McLerran, M. Shaposhnikov, N. Turok and M. Voloshin, *Phys. Lett.* **B256**, 451 (1991);
M. Dine, P. Huet, R. Singleton and L. Susskind, *Phys. Lett.* **B257**, 351 (1991).
130. A. Cohen, D. Kaplan and A. Nelson, *Phys. Lett.* **B263**, 86 (1991);
M. Joyce, T. Prokopec and N. Turok, *Phys. Rev.* **D53**, 2958 (1996).
131. M. Gleiser, *Phys. Rev. Lett.* **73**, 3495 (1994);
J. Borrill and M. Gleiser, *Phys. Rev.* **D51**, 4121 (1995).
132. W. Perkins, *Nucl. Phys.* **B449**, 265 (1995).
133. R. Brandenberger, A.-C. Davis, T. Prokopec and M. Trodden, *Phys. Rev.* **D53**, 4257 (1996).
134. R. Brandenberger, A.-C. Davis and M. Trodden, *Phys. Lett.* **B332**, 305 (1994);
R. Brandenberger and A.-C. Davis, *Phys. Lett.* **B308**, 79 (1993).
135. M. Trodden, A.-C. Davis and R. Brandenberger, *Phys. Lett.* **B349**, 131 (1995).

Finite Temperature Field Theory

H. Arthur Weldon

Department of Physics, West Virginia University
Morgantown, WV 26506-6315 USA

Abstract. This paper is a pedagogical review for non-experts of some topics in finite temperature quantum field theory. It begins with a review of thermodynamics and a summary of the lattice results for phase transitions in the electroweak and QCD theories. The main topic is the perturbative computation of thermal Green functions and the five standard formulations in imaginary time and real time. A summary of nonrelativistic plasma physics emphasizes the TEM and longitudinal waves. The last section reviews chirally-invariant gauge theories at high temperature, which have the same modes of oscillation. It demonstrates in simple situations the breakdown of perturbation theory and the Braaten-Pisarski resummation program that solves it.

I INTRODUCTION

The first introduction of temperature into relativistic quantum field theory occured in 1974 [1] and was motivated by the idea that a spontaneously broken symmetry would be unbroken at sufficently high temperature. Since then there has been great progress in finite temperature field theory. An indication of the growing interest in this subject is the recent appearance of three excellent books on finite temperature field theory by Joseph Kapusta [2], Michel Le Bellac [3], and Ashok Das [4]. Much of what is said here can be found in more detail in those books, which also contain more extensive references to the literature.

Statistical mechanics textbooks often say that quantum mechanics is only important at extremely low temperatures, e.g. 10^{-4} eV for Helium, and that high temperature physics is classical. The reason is that at high temperature, particles get farther apart and quantum mechanics becomes less and less important. However this argument breaks down when the temperature becomes much higher than the mass of the particles: $T \gg m$. In this regime the entropy increases from the production of additional particle-antiparticle pairs. The new particles make the typical separation smaller and quantum mechanics is again important. At $T \gg m$ the number of particles and antiparticles

CP400, *First Latin American Symposium on High Energy Physics/ VII Mexican School of Particles and Fields*, edited by D'Olivo/Klein-Kreisler/Mendez

mechanics is again important. At $T \gg m$ the number of particles and antiparticles in a volume V is $N \sim VT^3$. The typical separation between particles is $d \sim (V/N)^{1/3} \sim 1/T$, which decreases with T. The characteristic deBroglie wavelength of such particles is $\lambda \sim 1/\langle p \rangle \sim 1/T$. Because the quantum mechanical wave length is comparable to the particle separation, $\lambda \sim d$, the system is always quantum mechanical. It is this regime that is of interest to particle physicists.

This paper is organized as follows. Sec II reviews thermodynamics with particular emphasis on the phase transitions expected in electroweak theory and QCD. Sec III presents five different but equivalent ways to formulate finite temperature field theory and compute the Green functions. Sec. IV summarizes classical, nonrelativistic plasma physics in which there are two modes of propagation: transverse electromagnetic waves and longitudinal plasma oscillations. Sec. V discusses chirally invariant gauge theories at high temperature and shows that these same oscillations occur but with a different dispersion relation. It demonstrates in simple situations the breakdown of naive perturbation theory that was recognized, diagosed, and solved by Braaten and Pisarski. Each section may be read independently.

II THERMODYNAMICS

For some physicists finite temperature field theory serves only to calculate the partition function of particular field theories, viz. electroweak theory and QCD. This section will briefly summarize some of those results. It is important to recognize that thermodynamics deals only with bulk properties of a system. Typically these include the equation of state, the phase structure as a function of various parameters, and various coefficients of thermal expansion and compressibility. All thermodynamic properties may be computed the partition function $Z \equiv Tr[e^{-\beta H}]$. For a system at fixed termperature and volume the appropriate thermodynamic potential is the Helmholtz free energy $F(T,V) = -T\ln[Z]$. From this one can directly compute the pressure, entropy, and energy density:

$$P = -\frac{\partial F}{\partial V} \qquad S = -\frac{\partial F}{\partial T} \qquad U = F + TS. \tag{1}$$

By far the most interesting and most dramatic thermodynamic property of a system are its phase transitions. The most familiar system that undergoes phase transitions is water, whose phase diagram is shown in Figure 1. There is a first order phase transition across the lines, a second order transition at the critical point, and no phase transition beyond the critical point.

First Order Transition. A first order transition allows two phases, such as ice and water, to coexist in a mixed phase precisely at the transition point. To produce a transition, latent heat must be supplied or removed.

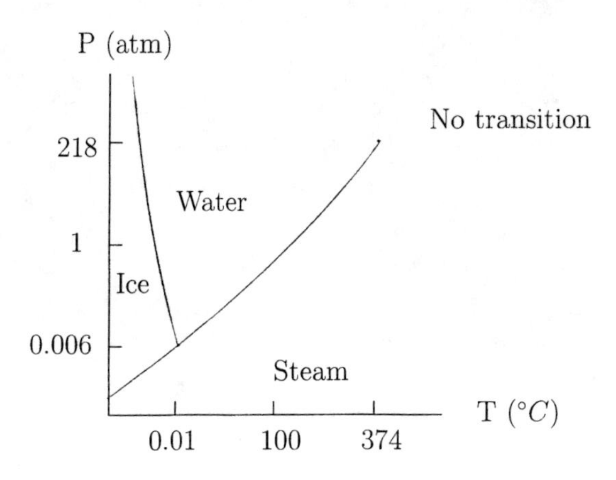

FIGURE 1. Phase diagram of H_2O.

For example, to melt one gram of ice at 1 atmosphere of pressure requires 80 calories of heat. The latent heat is related to the discontinuity in the entropy at the transition: $L = T_c \Delta S$. The transition occurs locally in space by nucleation and growth of new-phase bubbles in the old phase. Completing the transition requires a finite time; it is not instantaneous.

Second Order Transition. In a second order transition the system is either in one phase or the other; there is no coexistence. The transition occurs instantaneously. There is no latent heat associated with the transition. The entropy is continuous at the transition, but the derivative $\partial S/\partial T$ is discontinuous.

No Transition. For H_2O at $T > 374°C$ and $P > 218$ atmospheres there is no precise difference between "water" and "steam". One can qualitatively distinguish "water" from "steam" by its smaller correlation length. Another system with no true phase transition is a partially ionized gas, i.e. a nonrelativistic plasma. One can qualitatively distinguish the phases by the fraction of atoms that are ionized. When there is no phase transition the entropy and all its derivatives are continuous. However there can still be a very rapid change in the entropy as occurs when a gas is ionized or when passing around the critical point in Figure 1. The behavior of both the electroweak theory and of QCD appear to be of this type, i.e. no true phase transition.

A Electroweak Transition

The phase structure of electroweak gauge theory is controlled by the vector and scalar bosons. Quarks and leptons play a negligible role. It is customary to let $\phi(x)$ denote a complex SU(2) doublet of spinless fields (i.e. four real fields) that enjoy a self-interaction through the potential

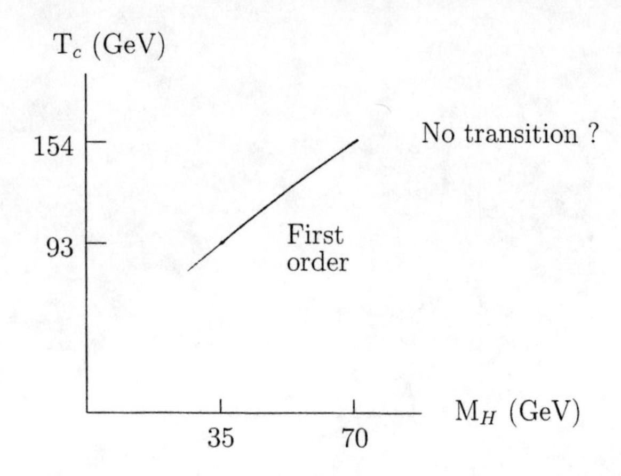

FIGURE 2. Qualitative dependence of the electroweak transition temperature on the Higgs mass.

$$V(\phi) = -\mu^2\phi^\dagger\phi + \lambda(\phi^\dagger\phi)^2. \tag{2}$$

The lagrangian for the boson sector is

$$\mathcal{L} = |(\partial_\mu + igT_iW^i_\mu + \frac{1}{2}ig'YB_\mu)\phi|^2 - V(\phi) - \frac{1}{4}W^i_{\mu\nu}W^{\mu\nu}_i - \frac{1}{4}B_{\mu\nu}B^{\mu\nu} \tag{3}$$

The minimum of the classical potential occurs at $\phi_0 = (\mu^2/2\lambda)^{1/2} \equiv v/\sqrt{2}$, which breaks three of the four local symmetries of $SU(2) \times U(1)$. The $W^\pm$ acquires a mass $M_W = gv/2$; the Z^0 acquires a mass $M_Z = (g^2 + g'^2)^{1/2}v/2$. The remaining massless combination is the photon which has a coupling $e = gg'(g^2 + g'^2)^{-1/2}$. Experimental measurements of M_W, M_Z, and e fix the values of g, g', and v. The remaining neutral component of ϕ is the Higgs field H^0 with a mass $M_H = \mu\sqrt{2}$ that is unknown. Experimental searches for the Higgs have bounded it from below $M_H > 60$ GeV with 95% confidence level.

At finite temperature the one-loop corrections to the potential give a transition that is second order, but this is altered in higher orders. The lattice simulations of Kajantie et al [5] show a first order transtion with $T_c = 93, 138, 154$ GeV for $M_H = 35, 60, 70$, respectively as shown in Figure 2. The latent heat gradually decreases with increasing Higgs mass: $L/T_c^4 = 0.26, 0.041, 0.027$ for $M_H = 35, 60, 70$, respectively. If the quartic coulpling λ is increased further, it is not known whether the transition remains first order or terminates at a critical point (i.e. second order transition). There are theoretical arguments that at large Higgs masses, $M_H > M_W = 80$ GeV, the transition disappears [6].

B QCD Transition

Quantum chromodynamics also has a finite-temperature phase transition. In contrast to the electroweak transition, in QCD the local gauge symmetry is not broken and the nature of the transition does depend on the fermions. In the full QCD there are six flavors of quarks (u,d,s,c,b,t) and eight color gluon fields A_μ^A with a lagrangian

$$\mathcal{L} = \overline{\psi}(i\gamma^\mu\partial_\mu - \frac{1}{2}g\gamma^\mu\lambda_A A_\mu^A - m)\psi - \frac{1}{4}F_{\mu\nu}^A F_A^{\mu\nu}, \tag{4}$$

where m is the mass matrix for the six quarks. From electroweak symmetry breaking all the quarks acquire some mass. The lightest are $m_u \approx 5$ MeV and $m_d \approx 10$ MeV, values which are derived from current algebra. Isospin would be an exact global symmetry if m_u and m_d were equal. The isospin current $J^{\mu a} = \overline{\psi}t^a\gamma^\mu\psi$, where $a = 1,2,3$, would be exactly conserved and the three charges would generate isospin rotations: $[Q^a, \psi_i] = t_{ij}^a\psi_j$, where $i = 1,2$ correspond to u and d.

Chiral Symmetry Breaking. If m_u and m_d were zero, there would be three additional conserved axial currents, $J_5^{\mu a} = \overline{\psi}\gamma_5 t^a\gamma^\mu\psi$, and three time-independent axial charges that generate axial rotations: $[Q_5^a, \psi_i] = t_{ij}^a\gamma_5\psi_j$. Note that the isospin charges Q^a do not commute with the axial charges Q_5^a. However the left chiral charges $Q_L^a = (Q^a - Q_5^a)/2$ and right chiral charges $Q_R^a = (Q^a + Q_5^a)/2$ do commute with each other: $[Q_L^a, Q_5^b] = 0$. The name chiral symmetry refers to $SU(2)_L \times SU(2)_R$. If chiral symmetry is broken then only the vector isospin symmetry remains $SU(2)_V$. Although the lagrangian is very nearly invariant under axial rotations, the spectrum of hadrons is not. For example, chiral symmetry predicts that the vector mesons ρ and the axial vector mesons a_1 should have the same mass because $[Q_5^a, \rho^b(x)] = \epsilon^{abc}a_1^c(x)$. However, the ρ mass is 770 MeV and the a_1 mass is 1260 MeV. Another incorrect prediction is that the positive parity nucleon doublet (p,n) should be degenerate with a negative parity doublet (p',n'). The nucleon mass is 940 MeV but its negative parity partner has a mass of 1535 MeV. The source of chiral symmetry breaking is in nonperturbative effects such as instantons. It follows from Goldstone's theorem that the nonperturbative breaking of the approximate global symmetry means that there must be three low mass, spinless mesons with the same quantum numbers as the broken generators Q_5^a. These are, of course, the three pi mesons.

Order Parameter for Chiral Symmetry. The order parameter for chiral symmetry is $\overline{\psi}_u\psi_u + \overline{\psi}_d\psi_d$. This is easy to see from the commutation relation

$$2\delta^{ab}\overline{\psi}_\ell\psi_\ell = [Q_5^a, \overline{\psi}_i t_{ij}^b\psi_j]. \tag{5}$$

Take the vacuum expectation value of this relation. If chiral symmetry were unbroken, the vacuum $|0\rangle$ would be an eigenstate of Q_5^a in which case

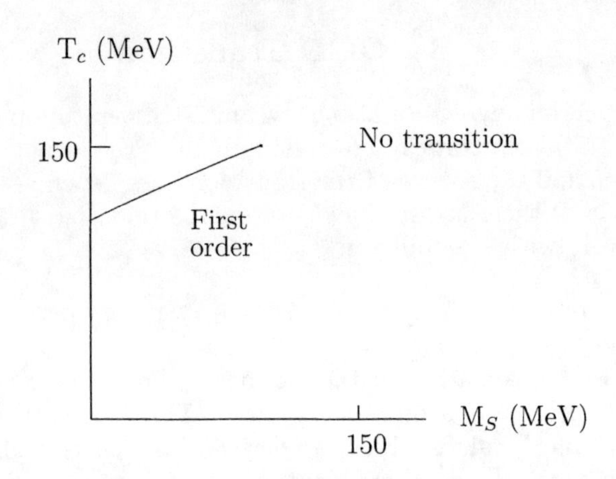

FIGURE 3. Qualitative dependence of the QCD chiral transition temperature on the strange quark mass.

$\langle 0|\overline{\psi}\psi|0\rangle$ would vanish. With broken chiral symmetry, $Q_5^a|0\rangle$ gives some new, odd-parity state and therefore $\langle 0|\overline{\psi}\psi|0\rangle \neq 0$. It is easy to obtain the order parameter at finite temperature. Multiply (5) by $\exp(-\beta H)$ and take the trace over energy eigenstates $|N\rangle$:

$$\mathrm{Tr}(e^{-\beta H}\overline{\psi}\psi) = \sum_N e^{-\beta E_N}\langle N|Q_5^a\,\overline{\psi}t^a\gamma_5\psi|N\rangle - \sum_N e^{-\beta E_N}\langle N|\overline{\psi}t^a\gamma_5\psi\,Q_5^a|N\rangle. \quad (6)$$

If chiral symmetry were unbroken then ρ and a_1 would have the same energy. Then states such as $|N\rangle = (|\rho\rangle \pm |a_1\rangle)/\sqrt{2}$ would be eigenstates of H and of Q_5^a. All hadronic states would be in such parity doubled multiplets and the right hand side would vanish. Thus $\mathrm{Tr}[\exp(-\beta H)\overline{\psi}\psi]$ vanishes if the spectrum is chirally invariant but is non-vanishing if chiral symmetry is broken. Thus it is an order parameter for chiral symmetry.

Lattice simulations of finite temperature QCD have shown that if $m_u = m_d = 0$ the chiral transition is first order for m_s up to about 100 MeV and second order for heavier values of m_s. This behavior is peculiar to the case of u and d massless. The Columbia group has shown that if m_u and m_d are are non-zero and smaller than 50 or 100 MeV then the chiral transition is first order for m_s up to about 100 MeV, terminates at a critical point, and disappears for heavier m_s [7]. This is show in Figure 3. Thus at the physical value of m_s it appears that there is no true phase transition. However there is a very rapid change in the value of $\langle\overline{\psi}\psi\rangle$ with T. It is also important to realize that there is no order parameter that signals deconfinement. However there is a very rapid increase in the entropy, suggestive of liberated quarks and gluons, that occurs in a temperture interval of less than 10 MeV.

III THERMAL GREEN FUNCTIONS

Thermodynamics is able to provide only a few bulk properties of any finite-temperature system. It is the thermal Green functions that provide dynamical information about the microscopic behavior for eventual comparison with experiment. The propagator is the basic Green function. The poles in the propagator give the elementary excitations of the system. For example, in condensed matter physics typical elementary excitations are electron holes (vacancies), phonons (quantized sound waves), polarons (lattice distortions), or magnons (spin waves in ferromagnets). The real part of the pole energy gives the dispersion relation $E(k)$ of the excitation. The imaginary part gives the damping rate $\gamma(k)$, which is the rate at which a fluctuation in the number of elementary excitations will be damped back to the equilibrium Bose or Fermi distribution.

One of the major obstacles for beginners in finite temperature field theory is that there are a variety of different formulations. This section will present the five standard formulations. These five are equivalent, but each is useful in its own way. The discussion will focus on a real scalar field $\phi(x)$ for clarity. More details on these formulations can be found in [2–4,8].

Why Complex Time is Important. Even though finite temperature field theory can be developed entirely with real time variables (i.e. Minkowski space-time), the results inevitably have special properties when the time variable is made complex. The reason for this is essentially that the time evolution operator $\exp(-iHt)$ becomes the density operator $\exp(-\beta H)$ when $t \rightarrow -i\beta$. This leads to a specific constraint on the correlation functions that is originally due to Kubo, Martin, and Schwinger [9]. The simplest correlation function is

$$D_{>}(\vec{x},t) = \mathrm{Tr}[e^{-\beta H}\phi(\vec{x},t)\phi(0)]\big/Z, \tag{7}$$

where $Z = \mathrm{Tr}[\exp(-\beta H)]$ is the partition function. Using the time dependence of the Heisenberg fields, $\phi(\vec{x},t+i\alpha) = \exp(-\alpha H)\phi(\vec{x},t)\exp(\alpha H)$, gives

$$D_{>}(\vec{x},t+i\alpha) = \mathrm{Tr}[e^{-(\alpha+\beta)H}\phi(\vec{x},t)e^{\alpha H}\phi(0)]\big/Z. \tag{8}$$

The spectrum of the Hamiltonian H is bounded from below but will always have arbitrarily large positive eigenvalues. These infinitely large energies will not spoil (8) if α is such that the exponents are always negative. In other words, (8) is analytic for $-\beta \leq \alpha \leq 0$. In the same manner the correlation function

$$D_{<}(\vec{x},t) = \mathrm{Tr}[e^{-\beta H}\phi(0)\phi(\vec{x},t)]\big/Z, \tag{9}$$

has the property that

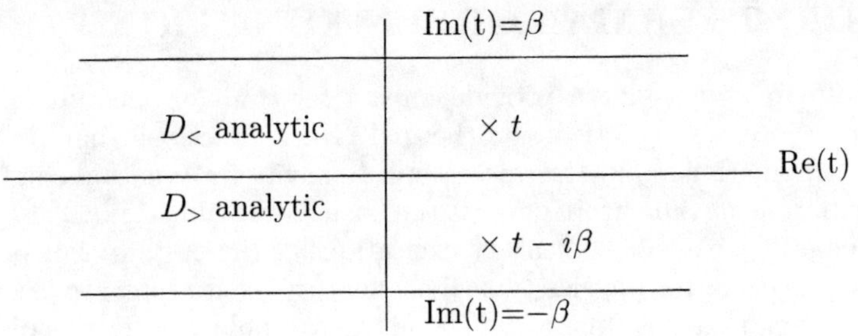

Im(t)=β

$D_<$ analytic

$\times\, t$

Re(t)

$D_>$ analytic

$\times\, t - i\beta$

Im(t)=$-\beta$

FIGURE 4. The KMS condition (10) relates the functions $D_>$ and $D_<$ in complex t.

$$D_<(\vec{x}, t + i\alpha') = \mathrm{Tr}[e^{(\alpha'-\beta)H}\phi(0)e^{-\alpha' H}\phi(\vec{x}, t)]\big/Z, \tag{10}$$

which is analytic for $0 \leq \alpha' \leq \beta$. The two functions are related by choosing $\alpha = -\beta$ in (8) to obtain

$$D_>(\vec{x}, t - i\beta) = D_<(\vec{x}, t). \tag{11}$$

This is the Kubo, Martin, Schwinger relation [9]. It holds for any complex time t as shown in Figure 4.

A Canonical Quantization in Imaginary Time

The formal similarity between inverse temperature and imaginary time was first discussed by Felix Bloch [10]. The idea was not fully developed until the 1955 work of Matsubara [11]. It is described in many textbooks [12,13]. The starting point is to canonically quantize on a finite interval of imaginary time, namely $t = -i\tau$ where $0 \leq \tau \leq \beta$.It will be convenient to denote the Euclidean space-time coordinate by $x_E = (\vec{x}, -i\tau)$. Then

$$\phi(x_E) = \phi(\vec{x}, -i\tau) = e^{H\tau}\phi(\vec{x}, 0)e^{-H\tau} \tag{12}$$

The Hamiltonian can be split into a free part H_0 and an interaction H_I, i.e. $H = H_0 + H_I$. The interaction picture operator satisfies the free field equations and is given explcitly by

$$\phi^{ip}(x_E) = \sum_{\vec{k}} \frac{1}{\sqrt{2EV}}[a_{\vec{k}} e^{i\vec{k}\cdot\vec{x} - E\tau} + a^{\dagger}_{\vec{k}} e^{-i\vec{k}\cdot\vec{x} + E\tau}]. \tag{13}$$

Note that this is not hermitian but does satisfy $[\phi^{ip}(x_E)]^* = \phi^{ip}(x_E^*)$. The interaction picture field is related to the the exact Heisenberg field by

$$\phi(x_E) = V(0,\tau)\phi^{ip}(x_E)V(\tau,0), \tag{14}$$

where $V(\tau,\tau') = \exp(H_0\tau)\exp(-H(\tau-\tau'))\exp(-H_0\tau')$. To compute a thermal correlation such as

$$\mathrm{Tr}[e^{-\beta H}\phi(x_E)\phi(x'_E)], \tag{15}$$

rewrite it in terms of interaction picture fields:

$$\mathrm{Tr}[e^{-\beta H_0}V(\beta,\tau)\phi^{ip}(x_E)V(\tau,\tau')\phi^{ip}(x'_E)V(\tau',0)]. \tag{16}$$

If this product is ordered with respect to the times τ and τ', it becomes the exact propagator

$$\mathrm{Tr}\Big[e^{-\beta H}T_\tau[\phi(x_E)\phi(x'_E)]\Big] = \mathrm{Tr}\Big[e^{-\beta H_0}T_\tau[\phi^{ip}(x_E)\phi^{ip}(x'_E)V(\beta,0)]\Big]. \tag{17}$$

Since $\partial V(\tau,0)/\partial\tau = -H_I^{ip}(\tau)V(\tau,0)$, it has the usual perturbative expansion

$$V(\beta,0) = 1 - \int_0^\beta d\tau_1 H_I^{ip}(\tau_1) + \frac{1}{2!}\int_0^\beta d\tau_1 \int_0^{\tau_1} d\tau_2 H_I^{ip}(\tau_1)H_I^{ip}(\tau_2) + \cdots. \tag{18}$$

The perturbative expansion of the exact propagator (17) and other n-point functions becomes a combinatoric problem in the free creation and annihilation operators $a^\dagger$ and a. Wick's theorem gives a Feynman diagram expansion in imaginary time.

Free Propagator. The first term in the perturbative expansion of (17) is the free propagator

$$D(x_E) = -\frac{i}{Z_0}\mathrm{Tr}[e^{-\beta H_0}T_\tau[\phi^{ip}(x_E)\phi^{ip}(0)]]. \tag{19}$$

Using the expansion (13) leads to the thermal average of bilinears:

$$\frac{1}{Z_0}\mathrm{Tr}[e^{-\beta H_0}a^\dagger_{\vec{k}}a_{\vec{k}}] = \frac{1}{\exp(\beta E)-1} \equiv n. \tag{20}$$

Obiously n is the Bose-Einstein distribution function. The free propagator is

$$D(x_E) = \frac{1}{i}\int \frac{d^3k}{2E}\frac{e^{i\vec{k}\cdot\vec{x}}}{(2\pi)^3}[e^{-E|\tau|} + n(e^{-E\tau} + e^{E\tau})]. \tag{21}$$

The time argument of the field operator is $t = -i\tau$ where $0 \le \tau \le \beta$. Since the time argument of the propagator is the difference of two such times, the variable τ in (21) has the range $-\beta \le \tau \le \beta$. Therefore for positive τ

$$0 \le \tau \le \beta: \quad D_>(\vec{x},-i\tau) = \int \frac{d^3k}{2E}\frac{e^{i\vec{k}\cdot\vec{x}}}{(2\pi)^3}[(1+n)e^{-E\tau} + ne^{E\tau}]. \tag{22}$$

This is analytic in the region shown in Figure 4. This arises because τ must be positive for $1 \times \exp(-E\tau)$ to be bounded at infinite E and must be less than β for $n \exp(E\tau)$ to be bounded. For τ in this range, let $\tau = \tau' + \beta$ where $-\beta \leq \tau' \leq 0$. Then using the identity $(1+n)\exp(-\beta E) = n$ gives

$$-\beta \leq \tau' \leq 0 : \quad D_>(\vec{x}, -i\tau' - i\beta) = \int \frac{d^3k}{2E} \frac{e^{i\vec{k}\cdot\vec{x}}}{(2\pi)^3} [ne^{-E\tau'} + (1+n)e^{E\tau'}]. \quad (23)$$

The right hand side is $D_<$ so that

$$-\beta \leq \tau' \leq 0 : \quad D_>(\vec{x}, -i\tau' - i\beta) = D_<(\vec{x}, -i\tau'), \quad (24)$$

which verifies the KMS condition (11) for the free propagator.

Example. To perform perturbative calculations it usually convenient to transform to momentum space. The τ-dependence of the free propagator (21) may be written as a Fourier sum:

$$D(x_E) = iT \sum_{\ell=-\infty}^{\infty} \int \frac{d^3k}{(2\pi)^3} e^{i\vec{k}\cdot\vec{x} - i(2\pi\ell T)\tau} \frac{-1}{(2\pi\ell T)^2 + \vec{k}^2 + m^2}. \quad (25)$$

Using the identification $(2\pi\ell)\tau = k^0 t$ and $t = -i\tau$ implies that $k^0 = i2\pi\ell T$. Thus the momentum-space propagator is

$$D(\vec{k}, k^0) = \frac{1}{(k_0)^2 - \vec{k}^2 - m^2} = \frac{-1}{(2\pi\ell T)^2 + \vec{k}^2 + m^2}. \quad (26)$$

As an example, let the interaction lagrangian be $\mathcal{L}_I = \lambda\phi^3/3!$. The one-loop proper self-energy is

$$\Pi(\vec{k}, i2\pi\ell T) = -\lambda^2 T \sum_{n=-\infty}^{\infty} \int \frac{d^3p}{(2\pi)^3} D(\vec{p}, i2\pi nT) D(\vec{k} - \vec{p}, i2\pi(\ell - n)T). \quad (27)$$

The sum on n can be calculated. The resulting self-energy is then known at discrete points $k^0 = i2\pi\ell T$. To determine the self-energy at other values of k^0 and particularly at real k^0 is not entirely trivial because there are an infinite number of functions that have the correct value at the discrete points $k^0 = i2\pi\ell T$. One must require that the self-energy be analytic in k^0 and vanish as $|k^0| \to \infty$ in order to obtain the correct value on the real k^0 axis. This example is worked out in [4].

B Path Integral Quantization in Imaginary Time

An alternative to canonical quantization is provided by the path integral form of quantum field theory. This is described in the book by Kapusta [2]. The starting point is the Feynman, Matthews, Salam formula [14]. Let

$\Phi_a(\vec{x})$ and $\Phi_b(\vec{x})$ be any static field configurations. Choose quantum states that are eigenstates of the field operator: $\phi(\vec{x},0)|\Phi_a\rangle = |\Phi_a\rangle\Phi_a(\vec{x})$. Then the general FMS formula is

$$\langle\Phi_b|e^{-iH(t_f-t_i)}|\Phi_a\rangle = \int_{\phi(t_i)=\Phi_a}^{\phi(t_f)=\Phi_b} d\phi \, \exp\,[i\int_{t_i}^{t_f} dt \int d^3x \, \mathcal{L}]. \tag{28}$$

In the functional integral $\Phi_a(\vec{x})$ and $\Phi_b(\vec{x})$ specify the initial and final values of the field integration over $\phi(\vec{x},t)$. To apply this to a finite temperature problem requires two steps: (i) Let $t_f - t_i = -i\beta$; (ii) Set $\Phi_a = \Phi_b$ and sum over all Φ_a in order to produce a trace. Thus the partition function is

$$\mathrm{Tr}[e^{-\beta H}] = \int_{\text{periodic}} d\phi \, \exp[i\int_{t_i}^{t_i-i\beta} dt \int d^3x \, \mathcal{L}], \tag{29}$$

where the integration is over all fields satisfying $\phi(\vec{x},t_i-i\beta) = \phi(\vec{x},t_i)$. The generating functional for the Green functions requires a source J:

$$Z(J) = \int_{\text{periodic}} d\phi \, \exp[i\int_{t_i}^{t_i-i\beta} dt \int d^3x \, (\mathcal{L} - J\phi)]. \tag{30}$$

The time integration must run from some t_i to $t_i - i\beta$. The simplest choice is to let $t_i = 0$ and let the contour run along the negative, imaginary axis from 0 to $-i\beta$. Thus $t = -i\tau$ where $0 \le \tau \le \beta$:

$$Z(J) = \int_{\text{periodic}} d\phi \, \exp[\int_0^{\beta} d\tau \int d^3x \, (\mathcal{L} - J\phi)]. \tag{31}$$

Integrating the fields over this Euclidean time produces the Euclidean Green functions. The periodic fields may be expressed in Fourier series as

$$\phi(\vec{x},-i\tau) = \left(\frac{\beta}{V}\right)^{1/2} \sum_{\ell=-\infty}^{\infty} \sum_{\vec{k}} e^{i(\vec{K}\cdot\vec{x}-i2\pi\ell T\tau)} \, \phi_\ell(\vec{k}). \tag{32}$$

Note that this periodic c-number field is quite different from the free field operator (13) of the previous section.

Free Propagator. For the free propagator the action is quadratic in the field:

$$S_0 = -\frac{1}{2}\int_0^{\beta} d\tau \int d^3x \Big[\Big(\frac{\partial\phi}{\partial\tau}\Big)^2 + (\vec{\nabla}\phi)^2 + m^2\phi^2\Big]. \tag{33}$$

In terms of the Fourier decomposition this is

$$S_0 = -\frac{\beta^2}{2}\sum_{\ell=-\infty}^{\infty}\sum_{\vec{k}} [(2\pi\ell T)^2 + \vec{k}^{\,2} + m^2] \, |\phi_\ell(\vec{k})|^2. \tag{34}$$

The free propagator is therefore

$$D(\vec{k}, i2\pi\ell T) = \frac{-1}{(2\pi\ell T)^2 + \vec{k}^{\,2} + m^2} \tag{35}$$

Naturally this is the same as (26).

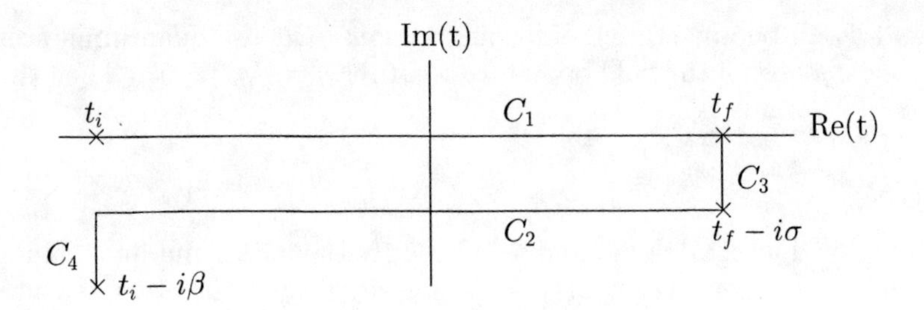

FIGURE 5. The t contour for real-time Feynman rules.

C Path Integral Quantization in Real Time

In the 1960's Schwinger and Keldysh independently developed a path integral formulation in real time to apply to problems in nonequilibrium statistical mechanics [15,16]. In 1984 Niemi and Semenoff adapted the real-time path integral to the context of thermal equilibrium [17]. The starting point is (30):

$$Z(J) = \int_{\text{periodic}} d\phi \, \exp [i \int_{t_i}^{t_i - i\beta} dt \int d^3x \, (\mathcal{L} - J\phi)]. \tag{36}$$

The only way in which temperature enters is that the functional integration is over fields satisfying the periodicity condition $\phi(\vec{x}, t_i - i\beta) = \phi(\vec{x}, t_i)$. One may choose any value of t_i and any contour in the complex time plane that starts at t_i and ends at $t_i - i\beta$. We choose t_i to real, negative, and large and the contour to be the union of four straight-line segments: $C_1 \cup C_2 \cup C_3 \cup C_4$ as shown in Figure 5. C_1 runs along the real time axis from t_i to some real, positive, large time t_f. C_3 drops vertically from t_f to $t_f - i\sigma$. Often σ is chosen to be $\beta/2$ but in general the only constraint is that $0 \leq \sigma \leq \beta$. C_2 runs antiparallel to the real time axis from $t_f - i\sigma$ to $t_i - i\sigma$. C_4 is a vertical secgment from $t_i - i\sigma$ to $t_i - i\beta$. In the limit that $t_i \to -\infty$ and $t_f \to \infty$ the contributions from the contours C_4 and C_3 vanish. The surviving contributions come from the fields along the segments C_1 and C_2. The fields along these two segments are

$$\phi_1(\vec{x}, t) \equiv \phi(\vec{x}, t) \qquad \phi_2(\vec{x}, t) \equiv \phi(\vec{x}, t - i\sigma). \tag{37}$$

These are denoted ϕ_a where $a = 1, 2$. The propagator has four components D_{ab} which are

$$D_{11}(x, x') = \ \text{Tr}[e^{-\beta H} \, T[\phi(\vec{x}, t)\phi(\vec{x}', t')]]/Z \tag{38a}$$

$$D_{12}(x, x') = \text{Tr}[e^{-\beta H} \, \phi(\vec{x}, t)\phi(\vec{x}', t' - i\sigma)]/Z \tag{38b}$$

$$D_{21}(x, x') = \text{Tr}[e^{-\beta H} \, \phi(\vec{x}, t - i\sigma)\phi(\vec{x}', t')]/Z \tag{38c}$$

$$D_{22}(x, x') = \ \text{Tr}[e^{-\beta H} \, \overline{T}[\phi(\vec{x}, t)\phi(\vec{x}', t')]]/Z. \tag{38d}$$

Note that contour ordering along C_2 results in the anti-time-ordering operation $\overline{T}$ in D_{22}.

Free propagator. The field integrations require no temperature-dependent boundary conditions at t_i or t_f. However the temperature does enter via the KMS condition (11). The solutions to $(\partial \cdot \partial + m^2)D_>(x) = 0$ and $(\partial \cdot \partial + m^2)D_<(x) = 0$ subject to the KMS condition are

$$D_>(x) = \frac{1}{i}\int \frac{d^3k}{(2\pi)^3}\frac{e^{i\vec{k}\cdot\vec{x}}}{2E}[e^{-iEt}(1+n) + e^{iEt}n] \tag{39}$$

$$D_<(x) = \frac{1}{i}\int \frac{d^3k}{(2\pi)^3}\frac{e^{i\vec{k}\cdot\vec{x}}}{2E}[e^{iEt}(1+n) + e^{-iEt}n], \tag{40}$$

where n is the Bose-Einstein function (20). The four components of the contour-ordered propagator are

$$D_{11}(x) = \theta(t)D_>(x) + \theta(-t)D_<(x) \tag{41a}$$
$$D_{12}(x) = D_<(\vec{x}, t + i\sigma) \tag{41b}$$
$$D_{21}(x) = D_>(\vec{x}, t - i\sigma) \tag{41c}$$
$$D_{22}(x) = \theta(t)D_<(x) + \theta(-t)D_>(x). \tag{41d}$$

In momentum space the free propagator is

$$D_{11}(K) = \Delta_F - i2\pi\, n\, \delta(K^2 - m^2) \tag{42a}$$
$$D_{12}(K) = e^{\sigma k_0}[\theta(-k_0) + n]2\pi\delta(K^2 - m^2) \tag{42b}$$
$$D_{21}(K) = e^{-\sigma k_0}[\theta(k_0) + n]2\pi\delta(K^2 - m^2) \tag{42c}$$
$$D_{22}(K) = -\Delta_F^* - i2\pi\, n\, \delta(K^2 - m^2), \tag{42d}$$

where $\Delta_F = 1/(K^2 - m^2 + i\eta)$ is the zero-temperature free propagator. It is natural to organize this into a 2×2 matrix:

$$D_{ab}(K) = U_{ac}\begin{pmatrix} \Delta_F & 0 \\ 0 & -\Delta_F^* \end{pmatrix}_{cd} U_{db}, \tag{43}$$

where U is the non-unitary matrix

$$U = \sqrt{n(k_0)}\begin{pmatrix} e^{k_0\beta/2} & e^{k_0(\sigma-\beta/2)} \\ e^{k_0(-\sigma+\beta/2)} & e^{k_0\beta/2} \end{pmatrix}. \tag{44}$$

Exact Propagator. The proper self-energy is also a 2×2 matrix. For example, if $\mathcal{L}_I = \lambda\phi^3/3!$ the one-loop contribution in coordinate space is $\Pi_{ab}(x) = (-1)^{a+b}\lambda^2 D_{ab}(x)D_{ab}(x)$. The four-components are intimately related. In momentum space the proper self-energy always has the form

$$\Pi_{ab}(K) = U_{ac}^{-1}\begin{pmatrix} \Pi_F & 0 \\ 0 & -\Pi_F^* \end{pmatrix}_{cd} U_{db}^{-1}, \tag{45}$$

with the same matrix U as above. The full propagator that results from summing the self-energy insertions has the same form as (43):

$$D'_{ab}(K) = U_{ac} \begin{pmatrix} \Delta'_F & 0 \\ 0 & -\Delta'^*_F \end{pmatrix}_{cd} U_{db}, \tag{46}$$

where $\Delta'_F = 1/(K^2 - m^2 - \Pi_F + i\eta)$.

D Canonical Quantization in Real Time

Real-time thermal field theory does not have to be formulated in terms of the path integral. Nieves showed how canonical quantization produces the same results [18]. The starting point is the relation of the exact Heisenberg field operator $\phi(x)$ to the interaction picture operator:

$$\phi(\vec{x}, t) = U(-\infty, t)\phi^{ip}(\vec{x}, t)U(t, -\infty). \tag{47}$$

As usual this leads to the operator relation

$$T[\phi(x_1)\phi(x_2)] = S^\dagger T[\phi^{ip}(x_1)\phi^{ip}(x_2)S], \tag{48}$$

where S is the scattering operator

$$S = U(\infty, -\infty) = T\exp\left[-i\int d^4x \mathcal{L}_I^{ip}(x)\right]. \tag{49}$$

The zero-temperature propagator is the vacuum expectation value of (48) The $S^\dagger$ operator disappears because $\langle 0|S^\dagger = \langle 0^{ip}|$. Thus the vacuum expectation of (48) requires only the pertubative expansion of S itself. Using Wick's theorem to organize the contractions gives the usual Feynman rules.

At non-zero temperature one must take expectation values of (48) in all energy eigenstates $|N_{\rm in}\rangle$:

$$\langle N_{\rm in}|T[\phi(x_1)\phi(x_2)]|N_{\rm in}\rangle = \langle N_{\rm in}|S^\dagger T[\phi^{ip}(x_1)\phi^{ip}(x_2)S]|N_{\rm in}\rangle. \tag{50}$$

Now the $S^\dagger$ can not be eliminated. The perturbative expansion and Wick's theorem gives time-ordered contractions from S, anti-time-ordered contractions from $S^\dagger$, and non-ordered contractions if one vertex is from S and one vertex from $S^\dagger$. These free propagators are labelled D_{11}, D_{22}, and D_{12} or D_{21} as before. For example, the one-loop correction to $D_{11}(w-z)$ requires integrating $D_{1a}(w-x)[D_{ab}(x-y)]^2 D_{b1}(y-z)$ over x and y. The type 1 vertices come from S; the type 2 vertices, from $S^\dagger$.

E Thermofield Dynamics

The four previous formulations allow calculations of Green functions. They make no change in the operator structure of the theory. Thermofield dynamics changes the operator structure to accomodate finite temperature more directly [4,8,19,20]. The typical thermal trace that must be computed is

$$\sum_N \langle N|T[\phi(x_1)\phi(x_2)]|N\rangle/Z. \tag{51}$$

For every operator $\phi(x)$, TFD introduces an automorphic image $\tilde{\phi}(x)$. Every $\mathcal{O}$ commutes with every $\tilde{\mathcal{O}}$. The hamiltonian H has an image $\tilde{H}$ constructed out of image fields. The vector space is enlarges so that corresponding to every multiparticle state $|N\rangle$ is an image state $|\tilde{N}\rangle$. Every ordinary state $|N_1\rangle$ is orthogonal to every image state $|\tilde{N}_2\rangle$. The general state is a direct product $|N_1, \tilde{N}_2\rangle \equiv |N_1\rangle \otimes |\tilde{N}_2\rangle$. It is helpful to introduce $\hat{H} = H - \tilde{H}$ with the property

$$\hat{H}|N_1, \tilde{N}_2\rangle = (E_{N_1} - E_{N_2})\, |N_1, \tilde{N}_2\rangle. \tag{52}$$

Thus there are an infinite number of generalized states with zero eigenvalue under $\hat{H}$. Because the ordinary field $\phi(x)$ does not operate on the tilde states

$$\langle M_1, \tilde{M}_2|\phi(x)|N_1, \tilde{N}_2\rangle = \langle M_1|\phi(x)|N_1\rangle\, \langle \tilde{M}_2|\tilde{N}_2\rangle, \tag{53}$$

which vanishes if $M_2 \neq N_2$. It therefore follows that

$$\langle M, \tilde{M}|\phi(x)\phi(0)|N, \tilde{N}\rangle = \begin{cases} \langle N|\phi(x)\phi(0)|N\rangle & \text{if } |M\rangle = |N\rangle \\ 0 & \text{if } |M\rangle \neq |N\rangle \end{cases} \tag{54}$$

Thus in the physical subspace only the diagonal matrix element, $\langle N|\cdots|N\rangle$, contributes. This property makes is useful to introduce the thermal superposition of states

$$|O(\beta)\rangle = \frac{1}{\sqrt{Z}} \sum_N |N, \tilde{N}\rangle\, e^{-\beta E_N/2}. \tag{55}$$

This has the marvelous property that expectation values in the larger space are thermal averages in the original space:

$$\langle O(\beta)|T[\phi(x)\phi(0)]|O(\beta)\rangle = \frac{1}{Z}\sum_N \langle N|T[\phi(x)\phi(0)]|N\rangle\, e^{-\beta E_N} = iD_{11}(x). \tag{56}$$

The other three components of the propagator are also expectation values: D_{12} of $\phi(x)\tilde{\phi}(0)$, D_{21} of $\tilde{\phi}(x)\phi(0)$, and D_{22} of $\tilde{\phi}(x)\tilde{\phi}(0)$. Although the thermal ground state has all physical energies E_N it has no generalized energy in the sense that $\hat{H}|O(\beta)\rangle = 0$. Similarly it has no generalized momentum, angular momentum, charge, flavor, or color. It provides a natural treatment of order parameters such as $\langle O(\beta)|\phi|O(\beta)\rangle$ for the electroweak transition and $\langle O(\beta)|\overline{\psi}\psi|O(\beta)\rangle$ for the chiral transition in QCD.

Perturbation Theory. The development of the perturbation expansion closely parallels the $T = 0$ treatment using the interaction picture. Since $\hat{\mathcal{L}}_I = \mathcal{L}_I - \tilde{\mathcal{L}}_I$ the evolution operator is

$$\hat{S} = T \exp[-i \int d^4x \, \hat{\mathcal{L}}_I^{ip}(x)] = S \, \tilde{S}^\dagger. \tag{57}$$

The propagator is

$$D_{11}(x) = -i\langle 0^{ip}(\beta)|T[\phi^{ip}(x)\phi^{ip}(0)S \, \tilde{S}^\dagger]|0^{ip}(\beta)\rangle. \tag{58}$$

Expanding this using Wick's theorem follows the same pattern as canonical quantization and, of course, gives exactly the same results.

IV NONRELATIVISTIC PLASMA PHYSICS

High temperature field theory pertains to quantum phenomena that take place in a thermalized background of particles and antiparticles. Many features of the ultrarelativistic plasma are already present in nonrelativistic classical plasmas consisting of positive ions and negative electrons and it is worth reviewing them. Plasma physics began when Guglielmo Marconi accidentally disovered the ionosphere surrounding the earth. Marconi was born in 1874 and as a teenager followed the work of Heinrich Hertz in Germany on radio waves. Marconi built increasingly powerful radio transmitters so that by the turn of the century he could send and receive signals between ships 100 km from shore. In 1901 he built a transmitter in England for the purpose of sending a 3 MHz signal across the Atlantic Ocean to a receiver in Newfoundland. Several distinguished mathematical physicists stated that that this was clearly impossible since Maxwell's equations show that radio waves travel in straight lines. Because of the Earth's curvature, straight line transmission between very tall towers would only be possible at a distance of less than 200 km.

Marconi was not convinced by these arguments. He built the transmitter and received the signal 3500 km away in Newfoundland in December 1901. The public was impressed but the mathematical physicists were astounded. Within a year Heaviside and Kennedy deduced that the radio waves had reflected off of an ionized layer, i.e. a plasma, in the upper atmosphere produced by ultraviolet radiation from the sun. In 1909 Marconi received the Nobel Prize in physics for his work in radio.

TEM Waves. The radio signals sent by Marconi were transverse electromagnetic waves. In a plasma the wave has a modified dispersion relation however so that $\omega \neq kc$. The dispersion relation is controlled by the plasma frequency $(\omega_{\text{pl}})^2 = ne^2/m_e$. For a typical electron density $n \approx 10^6$ per cm^3 in the iononosphere, the plasma frequency $\omega_{\text{pl}} = 16$ MHz. The dispersion relation is

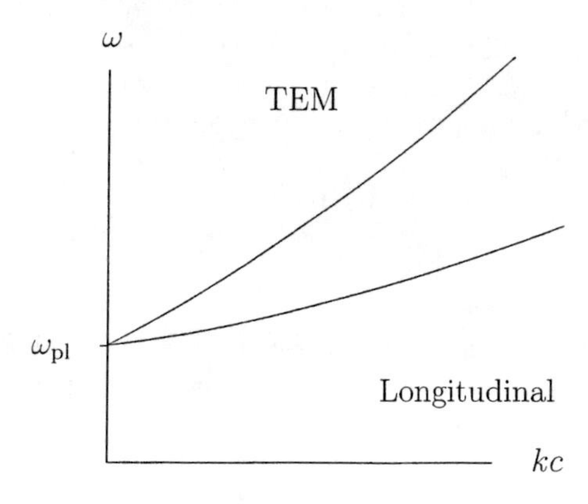

FIGURE 6. Nonrelativistic dispersion relation for TEM and longitudinal plasma waves.

$$\omega^2 = \omega_{\rm pl}^2 + (kc)^2, \tag{59}$$

which is shown in Figure 6. For $\omega > \omega_{\rm pl}$ the wave vector is real and the field configuration for linearly polarized wave propagating along z is

$$E_x = E_0 e^{i(kz-\omega t)} \qquad B_y = \frac{kc}{\omega} E_0 e^{i(kz-\omega t)}. \tag{60}$$

Accompanying the electric field oscillation is a current within the plasma:

$$J_x = i\frac{\omega_{\rm pl}^2}{\omega} E_0 e^{i(kz-\omega t)}. \tag{61}$$

The beam Marconi transmitted from England was at 3 MHz, which is less than $\omega_{\rm pl}$. Thefore the wavevector is pure imaginary $k = i\alpha$ so that (60) and (61) suffer exponential attenuation along the direction of propagation: $\exp(-\alpha z - i\omega t)$. The energy of the TEM wave must be conserved and the boundary conditions for Maxwell's equations show that most of the wave energy is reflected back by the ionosphere. This reflection is what allowed Marconi's signal to reach Newfoundland.

The above discussion has neglected the damping of the TEM waves due to two-body collisions, in particular the Rayleigh process $\gamma + e^- \to \gamma + e^-$. This produces a collision rate γ that is of order e^4. It corresponds to a plasma conductivity $\sigma = i\omega_{\rm pl}^2/(\omega + i\gamma)$ and a plasma current $J_x = \sigma E_x$. In the collisionless regime considered previously, $\omega \gg \gamma$, so that the conduction was inductive: $\sigma \approx i\omega_{\rm pl}^2/\omega$. In the collision-dominated regime appropriate to metals the conduction is resistive: $\sigma \approx \omega_{\rm pl}^2/\gamma$. In general the dispersion relation is

$$\omega^2 = \frac{\omega\,\omega_{\rm pl}^2}{\omega + i\gamma} + (kc)^2. \tag{62}$$

The collisions give ω a small negative imaginary part that slightly damps the wave in time even for $\mathrm{Re}(\omega) > \omega_{\mathrm{pl}}$. From the collisionless dispersion relation (59) the transverse modes appear to be screened in the static limit because $k \to i\omega_{\mathrm{pl}}/c$ as $\omega \to 0$. However, at low freqency collisions are important and the correct limiting wave vector from (62) is

$$\omega \to 0: \qquad k \to (1+i)\frac{\omega_{\mathrm{pl}}}{c}\left(\frac{\omega}{2\gamma}\right)^{1/2}. \tag{63}$$

This is the familiar skin depth effect that screens all frequencies except $\omega = 0$. Perturbative calculations for the quark-gluon plasma show the same screening of the transverse mode in the static limit. In the quark-gluon case it results in infrared divergences.

Longitudinal Plasma Waves. There is another mode of field oscillation distinct from the transverse modes. It was discovered experimentally by Langmuir in 1929 in a cold plasma. J.J. Thomson and his son studied the oscillations at non-negligible electron temperature in 1933. The correct dispersion relation was found by Landau in 1946:

$$\omega^2 = \omega_{\mathrm{pl}}^2 + k^2 v^2, \tag{64}$$

where $v^2 = 3T/m$ is the average thermal velocity of the electrons. This is shown in Figure 6. For a wave propagating along z the electric field is also polarized along z:

$$E_z = E_0 e^{i(kz-\omega t)}. \tag{65}$$

The associated plasma current is along the direction of propagation:

$$J_z = i\omega E_0 e^{i(kz-\omega t)} \qquad J_0 = ikE_0 e^{i(kz-\omega t)}. \tag{66}$$

For $\omega > \omega_{\mathrm{pl}}$ the wave vector k is real; for $\omega < \omega_{\mathrm{pl}}$ the wave vector k is pure imaginary, which results in exponential attenuation. In the static limit,

$$\omega \to 0: \qquad k \to i\omega_{\mathrm{pl}}/v = i\left(\frac{ne^2}{3T}\right)^{1/2}, \tag{67}$$

which is know as Debye screening of the longitudinal waves.

The longitudinal modes also suffers collisional damping from $\gamma + e^- \to \gamma + e^-$, but there is a more important process discovered by Landau. At large k the dispersion relation crosses into the space-like region $kc > \omega$ as shown in Figure 6. In this region the process $\gamma + e^- \to e^-$ can conserve energy and momentum. This allows a plasma electron to absorb energy $\hbar\omega$ and momentum $\hbar k$ from the longitudinal plasma wave. The amplitude is first order in e and the rate is of order e^2. This is called collisionless damping or Landau damping. No two-body collisions are involved. It occurs whenever the wave propagation is space-like. In the quantum system, all virtual bosons that are space-like are damped by this mechanism. Landau damping does not alter the static Debye screening (67).

V HIGH TEMPERATURE GAUGE THEORIES

A Breakdown of Perturbation Theory

It has long been known that interacting scalar fields develop an effective thermal mass proportional to temperature [1]. In a theory of massless scalars with quartic self-interaction $\mathcal{L}_I = \lambda\phi^4$ the one-loop self-energy is momentum-independent and has a value $\Pi = \lambda T^2$. The radiatively corrected scalar propagator is

$$D(K) = \frac{1}{K^2} + \frac{1}{K^2}\lambda T^2 \frac{1}{K^2} + \cdots \tag{68}$$

If the components of K^μ are very small, then the second term is larger than the first term. Specifically, if the components of K^μ are smaller than $\lambda^{1/2}T$, the second term is larger. This is a breakdown of perturbation theory in that for small momenta the perturbative corrections to the propagator are larger than the unperturbed answer. The solution to this particular breakdown is trivial: Sum up the self-energy insertions to obtain the propagator $1/(K^2 - \lambda T^2)$. Although it is not obvious, this completely solves the problem in the scalar theory and there is no further breakdown in any of the n-point Green functions.

In gauge theories the breakdown of naive perturbation theory begins with the propagators but is much more pervasive and the solution is much more interesting. The problem was recognized, diagnosed, and solved in a series of papers by Braaten and Pisarski [21–23]. The breakdown occurs when the temperature is the largest scale in the problem and all masses are negligible. It is simplest to look first at the one-loop self-energy of the gauge bosons. At $\vec{k} = 0$ this self-energy is a function only of ω and T. When $\omega \ll T$ it is only a function of T. A simple computation gives $\Pi^{mn} = -\delta^{mn} m_g^2$ where where m_g is a thermal gluon mass parameter given by

$$m_g^2 = g^2T^2(2N + N_f)/18, \tag{69}$$

for N colors and N_f flavors of four-component massless quarks. The $\vec{k} = 0$ propagator is

$$D^{mn}(0,\omega) = -\delta^{mn}\Big[\frac{1}{\omega^2} + \frac{1}{\omega^2} m_g^2 \frac{1}{\omega^2} + \cdots\Big]. \tag{70}$$

Perturbation theory has broken down because in the region $\vec{k} = 0$ and $\omega \ll gT$ the perturbative correction is larger than the unperturbed term. Summing the geometric series to $1/(\omega^2 - m_g^2)$ solves this particular problem, but it does not eliminate the breakdown. Because of gauge invariance there is a nonabelian Ward identity that relates the three gluon amplitude to the two gluon amplitude. The Ward identity must hold at each order of perturbation theory. Therefore the order T^2 breakdown of perturbation theory in the propagator gurantees that there will be an order T^2 breakdown of perturbation theory in the three-gluon vertex. But it doesn't stop there. Another Ward identity relates the three-gluon vertex to the four-gluon vertex. Thus the four-gluon amplitude will have a breakdown as well. Indeed the one-loop correction to every n-gluon amplitude must contain a contribution of order T^2 that dominates the tree amplitude at momenta smaller than gT.

There is another independent breakdown that occurs in the fermion sector. For massless fermions the one-loop self-energy contains a contribution of order g^2T^2. The one-loop corrected propagator at zero momentum is

$$S(0,\omega) = \gamma_0\omega\Big[\frac{1}{\omega^2} + \frac{1}{\omega^2}m_f^2\frac{1}{\omega^2} + \cdots\Big], \tag{71}$$

where

$$m_f^2 = g^2T^2(N^2-1)/8N, \tag{72}$$

for four-component massless quarks in SU(N). It is customary to call this a thermal mass because it plays that role in kinematics. However the propagator does not break chiral symmetry since $\{S,\gamma_5\} = 0$. As before there is a breakdown of pertubation theory at $\omega \ll gT$. Here the Ward identity that relates the quark self-energy to the quark-gluon vertex is the same as in QED. For zero-momentum quarks it is

$$(\omega-\omega')\Gamma_0(\omega,\omega') = S^{-1}(0,\omega) - S^{-1}(0,\omega'), \tag{73}$$

and therefore the vertex is

$$\Gamma_0(\omega,\omega') = \gamma_0(1 - \frac{m_f^2}{\omega\omega'} + \cdots). \tag{74}$$

Thus when ω and ω' are small, the perturbative corrections are larger than the bare vertex. Because of gauge invariance there is a relation between this quark+ gluon vertex and the quark+two-gluon vertex. Ultimately the breakdown occurs in the one-loop correction to every quark+n-gluon vertex.

The solution to the breakdown of naive perturbation theory was given by Braaten and Pisarski [21–23]. The order g^2T^2 part of the one-loop corrections to all amplitudes (propagators, n-gluon and quark+ n-gluons) must be the starting point of any calculation. These are the effectrive vertices. They define the skeleton diagrams of a reorganized pertubation theory. In the new expansion all pertubative corrections are genuinely smaller than the starting amplitudes so that the new pertubative expansion is completely consistent. The only textbook coverage of this resummation program is by Le Bellac [3].

B One-Loop Thermal Gluon Propagator

To analyze the gluon propagator an non-zero three-momentum one needs the full tensor $\Pi^{\mu\nu}(\vec{k},\omega)$. At finite temperature this generally has the property that $K_\mu \Pi^{\mu\nu} \neq 0$. However, in the region $k \ll T$ and $\omega \ll T$ it does satisfy $K_\mu \Pi^{\mu\nu} = 0$ and this part of $\Pi^{\mu\nu}(K)$ is gauge-invariant. One may express this tensor in terms of $g^{\mu\nu}, K^\mu K^\nu, u^\mu u^\nu$, and $K^\mu u^\nu + u^\mu K^\nu$, where u^μ is the four-velocity of the quark-gluon plasma. In the rest frame of the plasma this means that the tensor is covariant under rotations but there is not manifest covariance under boosts. Note that even the rotational covariance restricts the possible choices of gauge-fixing. The usual choices of covariant, Coulomb, or temporal are fine. Other choices, such as axial gauge $A_3 = 0$, would spoil the rotational invariance and are more awkward. The rotationally-invariant self-energy must be of the form

$$\Pi^{\mu\nu}(K) = \Pi_T(k,\omega)A^{\mu\nu} + \Pi_L(k,\omega)B^{\mu\nu}. \tag{75}$$

In the rest frame of the plasma the two basis tensors are

$$A^{\mu\nu}\Big|_{\text{rest}} = \begin{pmatrix} 0 & 0 \\ 0 & -\delta^{mn} + \hat{k}^m \hat{k}^n \end{pmatrix} \qquad B^{\mu\nu}\Big|_{\text{rest}} = -\frac{1}{K^2}\begin{pmatrix} k^2 & \omega k^n \\ \omega k^m & \omega^2 \hat{k}^m \hat{k}^n \end{pmatrix}. \tag{76}$$

These satisfy $K_\mu A^{\mu\nu} = K_\mu B^{\mu\nu} = 0$. The second tensor may be written more compactly as $B^{\mu\nu} = -\tilde{K}^\mu \tilde{K}^\nu / K^2$ where $\tilde{K}^\mu = (k, \omega\,\hat{k})$. The real-time propagator matrix $D^{\mu\nu}_{ab}$ can be diagonaized as in (50) in terms of $D^{\mu\nu}_F$. For example, the time-ordered part of the propagator is

$$D^{\mu\nu}_{11}(K) = (1+n)D^{\mu\nu}_F(K) - nD^{\mu\nu}_F(K)^*, \tag{77}$$

where $n = 1/(\exp(\beta|\omega|) - 1)$ is the Bose-Einstein function. In covariant gauge the Feynman propagator is

$$D^{\mu\nu}_F(K) = -\frac{A^{\mu\nu}}{K^2 - \Pi_T} + \frac{\tilde{K}^\mu \tilde{K}^\nu}{K^2(K^2 - \Pi_L)} - \xi \frac{K^\mu K^\nu}{K^4}. \tag{78}$$

It is straightforward to compute the order g^2T^2 part of the one-loop self-energy [24,25]:

$$\Pi_T = \frac{3}{2} m_g^2 \Big[\frac{\omega^2}{k^2} + (1 - \frac{\omega^2}{k^2})\frac{\omega}{2k} \ln\Big(\frac{\omega+k}{\omega-k}\Big)\Big] \tag{79}$$

$$\Pi_L = 3m_g^2 \Big[1 - \frac{\omega^2}{k^2}\Big]\Big[1 - \frac{\omega}{2k}\ln\Big(\frac{\omega+k}{\omega-k}\Big)\Big]. \tag{80}$$

These functions are gauge invariant. The solution to the transverse disper-

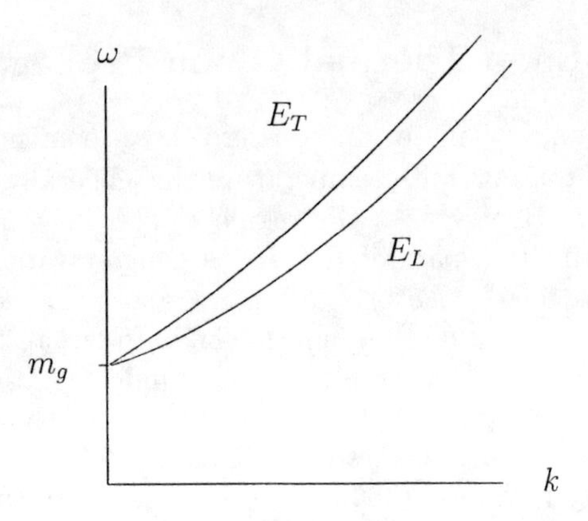

FIGURE 7. Dispersion relation for transverse and longitudinal gluon modes.

sion relation $K^2 = \Pi_T$ are shown in Figure 7. It has the same qualitative features as the nonrelativistic dispersion relation in Figure 6. The nonrelativistic plasma frequency $\omega_{\rm pl} = e(n/m_e)^{1/2}$ is replaced by the thermal gluon mass m_g. The physics of the thermal gluon modes in the chirally symmetric plasma is almost exactly the same as that of photons in the nonrelativistic plasma discussed in Sec. IV.

For the transvervse mode the propagating gluons have exactly the same $\vec{E}$ and $\vec{B}$ as before (omitting the commutator terms) . For $\omega < m_g$ the wave vector is complex and the waves are exponentially attenuated. In the static limit the wave vector is

$$TEM \quad \omega \to 0: \qquad k \to i\Big(\frac{3\pi m_g^2\,\omega}{4}\Big)^{1/3}. \tag{81}$$

This imaginary wave vector vanishes at $\omega = 0$ just as occured with the nonrelativistic skin depth (63). This does not eliminate all infrared divergences but does make them less severe. Typical power divergences become logarithmic and logarithmic divergences become finite.

The longitudinal dispersion relation $K^2 = \Pi_L$ shown in Figure 7 satisfies $\omega > k$ and in this respect differs from the nonrelativistic longitudinal mode shown in Figure 6. The field is however the same, i.e. $E_z \propto \exp(ikz - iE_Lt)$. Waves with $\omega < m_g$ suffer exponential damping in z. In the static limit the imaginary wave vector

$$L \quad \omega \to 0: \qquad k \to i m_g\sqrt{3} \tag{82}$$

provides Debye screening of longitudinal phenomena.

C One-Loop Thermal Quark Propagator

The real-time propagator matrix S_{ab} can be diagonalized in a fashion similar to (50) in terms of a Feynman propagator S_F. For example the time-ordered part is related to S_F by

$$S_{11}(K) = (1 - n_f) S_F(K) + n_f \gamma_0 (S_F(K))^\dagger \gamma_0, \tag{83}$$

where $n_f = 1/(\exp(\beta|\omega|) + 1)$ is the Fermi-Dirac distribution function. Rotational symmetry and chiral invariance require that the fermion self-energy be a linear combination of γ_0 and $\vec{\gamma} \cdot \hat{k}$. Therefore

$$[S_F(\vec{k}, \omega)]^{-1} = \gamma_0 A(k, \omega) - \vec{\gamma} \cdot \hat{k} B(k, \omega), \tag{84}$$

so that the propagator itself is

$$S_F(K) = \frac{1}{2} \frac{\gamma_0 - \vec{\gamma} \cdot \hat{k}}{A - B} + \frac{1}{2} \frac{\gamma_0 + \vec{\gamma} \cdot \hat{k}}{A + B}. \tag{85}$$

For the free propagator $A = \omega$ and $B = k$. The one-loop contibution to A and B contains a part proportional to $g^2 T^2$ [21,26,27]:

$$A(k, \omega) = \omega - \frac{m_f^2}{2k} \ln \left(\frac{\omega + k}{\omega - k} \right) \tag{86}$$

$$B(k, \omega) = k + \frac{m_f^2}{k} \left[1 - \frac{\omega}{2k} \ln \left(\frac{\omega + k}{\omega - k} \right) \right], \tag{87}$$

where m_f is given in (72). These functions are gauge invariant. The poles in the propagator are at ω for which $A^2 = B^2$. The two positive energy solutions are shown in Figure 8. It is quite unusual for there to be two positive energy solutions. (There are automatically two negative energy solutions also.) To interpret the solutions it convenient to use $\vec{\gamma} = \gamma_0 \gamma_5 \vec{\Sigma}$ to write the inverse propagator as

$$[S_F(\vec{k}, \omega)]^{-1} = \gamma_0 (A - \gamma_5 \vec{\Sigma} \cdot \hat{k} B) \tag{88}$$

The spinor wave functions annihilated by this are of two types: either $A = B$ and the chirality equals the helicity; or $A = -B$ and the chirality is opposite the helicity. The chirality operator can be expressed in terms of creation and annihilation operators as

$$Q_5 = \sum_{\vec{k}} \sum_{\lambda=-1}^{1} \lambda [b^\dagger(\vec{k}, \lambda) b(\vec{k}, \lambda) - d^\dagger(\vec{k}, \lambda) d(\vec{k}, \lambda)], \tag{89}$$

where $\lambda = \pm 1$ labels the helicity.

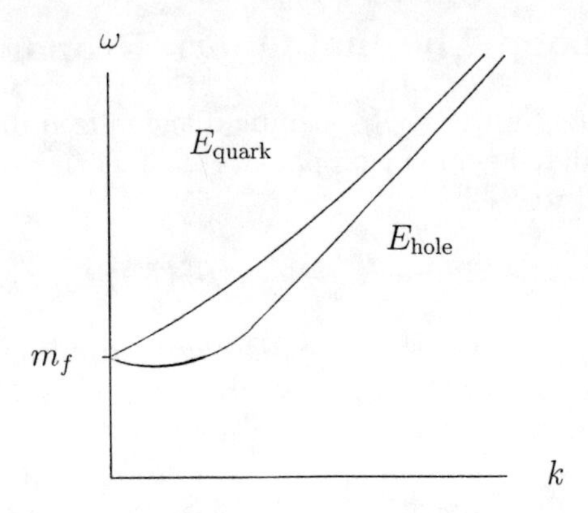

FIGURE 8. Dispersion relation for quarks and holes.

Solutions with $A = B$. Solutions to (88) for which $A = B$ must have chirality equal to helicity. The one quark state $b^\dagger(\vec{k},\lambda)|\mathrm{vac}\rangle$ has chirality equal helicity because applying Q_5 yields the eigenvalue λ. In a thermal problem the background is not the vacuum but rather a superposition of many quantum states $|\Phi\rangle$. One could work in an ensemble in which all states have zero chirality. It is more typical to work in an unconstrained ensemble with the chiral chemical potential set equal to zero. This guarantees that the ensemble average of chirality is zero, but also implies that almost all the states $|\Phi\rangle$ have zero chirality. A one quark excitation on any such state, $b^\dagger(\vec{k},\lambda)|\Phi\rangle$ will still have chirality equal to helicity. These comprise the positive energy solutions to $A = B$ and are inidcated in Figure 8 by $E_{\rm quark}$. Another class of states are those with one missing antiquark, i.e. one antiquark-hole, $d(\vec{k},\lambda)|\Phi\rangle$, which also have chirality equal to helicity. These hole states comprise the negative energy solution to $A = B$.

Solutions with $A = -B$. Wave functions annihilated by (88) that have helicity opposite to chirality must have ω such that $A = -B$. These are of two types: the one antiquark state $d^\dagger(\vec{k},\lambda)|\Phi\rangle$ solves $A = -B$ with negative ω; the one hole state $b(\vec{k},\lambda)|\Phi\rangle$ solves $A = -B$ with positive ω and is shown in Figure 8 by $E_{\rm hole}$.

D The Effective Action

There is a great deal of interesting physics associated with the high temperature propagators for gauge bosons and fermions. However as argued in Sec A, the order T^2 effects also signal a failure of the naive perturbative expansion and imply that there will be one-loop corrections of order T^2 to the higher N-point functions. These corrections were all computed by Braaten and Pisarski. Afterwards they found that the results could be summarized by an effective action that reproduced all the order T^2 corrections.

Some features of the effective action are easy to understand. The temperature-dependent part of any one-loop diagram with n external legs is of the form

$$\int \frac{d^4P\delta(P^2)}{\exp(\beta|p_0|)\pm 1}\, F(P;K_1,K_2,\cdots K_n), \tag{90}$$

where the Feynman amplitude F is a product of propagators and vertex factors. For massless gauge theories F is homogeneous in the loop momentum $P^\mu = (p_0,\vec{p})$ and therefore depends only on P^μ/p^0. After integrating over p_0 the remaining three-dimensional integral factors into an integral over p multiplied by an integral over angles $d\Omega$:

$$\int_0^\infty \frac{pdp}{\exp(\beta p)\pm 1}\int d\Omega\, F(Q;K_1,K_2,\cdots K_n), \tag{91}$$

where $Q^\mu = P^\mu/p_0 = (1,\hat{q})$. The first integral gives a factor T^2 and the angular integrals must be computed.

Fermions. It is simplest to first discuss the fermion sector. The lowest order term should produce the fermion self-energy in Sec C. The order g^2 term is

$$W^{(2)}(K) = -m_f^2\int\frac{d\Omega}{4\pi}\frac{\gamma_\mu Q^\mu}{Q\cdot K} = -m_f^2\int\frac{d\Omega}{4\pi}\frac{\gamma_0-\vec{\gamma}\cdot\vec{k}}{\omega-k\cos\theta}, \tag{92}$$

where $Q^\mu=(1,\hat{q})$ and Ω are the solid angles of the unit vector $\hat{q}$. Evaluating the integral gives

$$W(K)=\gamma_0\Big[\frac{-m_f^2}{2k}\ln\Big(\frac{\omega+k}{\omega-k}\Big)\Big]-\vec{\gamma}\cdot\hat{k}\,\frac{m_f^2}{k}\Big[1-\frac{\omega}{2k}\ln\Big(\frac{\omega+k}{\omega-k}\Big)\Big], \tag{93}$$

which reproduces (86) and (87). The momentum space action to order g^2 is

$$S_f^{(2)} = \int\frac{d^4K}{(2\pi)^4}\overline{\psi}(-K)W^{(2)}(K)\psi(K). \tag{94}$$

In coordinate space the action is

$$S_f^{(2)} = -m_f^2\int d^4x d^4y\overline{\psi}(x)W^{(2)}(x-y)\psi(y) \tag{95}$$

$$W^{(2)}(x-y)=\int\frac{d\Omega}{4\pi}\gamma_\mu Q^\mu\,\langle x|\frac{1}{Q\cdot\partial}|y\rangle, \tag{96}$$

with the notation

$$\langle x|\frac{1}{Q\cdot\partial}|y\rangle = i\int\frac{d^4K}{(2\pi)^4}\,e^{-iK\cdot(x-y)}\frac{1}{Q\cdot K}. \tag{97}$$

It takes considerable effort to determine the higher order terms but the answer is remarkably simple: Just replace the partial derivative ∂^μ by the gauge-covariant derivative $(D^\mu)_{ij} = \delta_{ij}\partial^\mu + ig(L_a)_{ij}A^\mu$. Then $W^{(2)}(x-y)$ is replaced by the full W:

$$W(x-y) = \int\frac{d\Omega}{4\pi}\gamma_\mu Q^\mu\,\langle x|\frac{1}{Q\cdot D}|y\rangle. \tag{98}$$

Expanding this to any order g^n gives the vertex for n gluons interacting with a quark.

Gauge Bosons The effective action that correctly produces all the multi-gluon vertices to order T^2 is

$$S_g = -\frac{3}{4}m_g^2\int d^4xd^4yF_{\lambda\alpha}(x)V^\alpha_\beta(x-y)F^{\beta\lambda}(y) \tag{99}$$

$$V^{\mu\nu}(x-y) = \int\frac{d\Omega}{4\pi}\,Q^\mu Q^\nu\langle x|\frac{1}{(Q\cdot D)(Q\cdot D)}|y\rangle. \tag{100}$$

In momentum space the action is

$$S_g = -\frac{3}{4}m_g^2\int\frac{d^4K}{(2\pi)^4}F_{\lambda\alpha}(-K)V^\alpha_\beta(K)F^{\beta\lambda}(K). \tag{101}$$

The lowest order kernel is

$$V^{(2)\alpha\beta}(K) = \int\frac{d\Omega}{4\pi}\frac{Q^\alpha Q^\beta}{(Q\cdot K)(Q\cdot K)}. \tag{102}$$

The lowest order term in the action is quadratic in the vector potential:

$$S_g^{(2)} = \frac{1}{2}\int\frac{d^4K}{(2\pi)^4}A_\mu(-K)\Pi^{\mu\nu}(K)A_\nu(K), \tag{103}$$

where

$$\Pi^{\mu\nu}(K) = -\frac{3m_g^2}{2}\int\frac{d\Omega}{4\pi}\Big[\frac{K^\mu Q^\nu + Q^\mu K^\nu}{K\cdot Q} - g^{\mu\nu} - K^2\frac{Q^\mu Q^\nu}{(K\cdot Q)^2}\Big]. \tag{104}$$

Computing this integral gives the self-energy tensor (75) as designed. The higher order terms in g give all the multi-gluon vertex corrections.

E Applications

With the effective action $S_f + S_g$ as a starting point it is possible to compute a variety of processes consistently. These include damping rates for gluons, dilepton production from $q\bar{q}$ annihilation, energy loss dE/dx for a propagating quark and a propagating gluon, the viscosity of the plasma, and many others. The book by Le Bellac [3] treats several of these applications and gives references to the original calculations.

Much of the motivation for these calculations comes from the RHIC and LHC projects that are designed to produce a quark-gluon plasma. For over a decade there has been a program of heavy ion experiments at Brookhaven, Fermilab, and CERN colliding a variety of nuclear beams on various fixed targets. The fixed target experiments will be superceded by colliding two beams of heavy nuclear ions. Brookhaven is on schedule to complete the Relativistic Heavy Ion Collider (RHIC) in 1999. It will collide gold nucleii at an energy of $s^{1/2} = 200$ GeV per baryon. CERN will build the Large Hadron Collider (LHC) to be finished in 2005. Part of the LHC program will be collisions heavy nucleii up to lead at $s^{1/2} = 6300$ GeV per baryon. The motivation for the colliders is to produce a small region of quark-gluon plasma in chich chiral symmetry is restored and color is not confined.

Collisions at RHIC and LHC will be much more complicated than normal accelerator experiments. Although the ultimate motiviation is to detect the quark-gluon plasma predicted by finite temperature QCD, the theory employed to understand the full collision involves much more than field theory. There are three stages of a typical collision. The first stage is prelude to local thermal equilibrium. The theoretical techniques for this include hard parton scattering, quantum decoherence, relativistic kinetic theory, and relativistic fluid mechanics. The second stage is the quark-gluon plasma where the topic of these lectures, viz. finite temperature field theory, apply. The third stage is hadronization, in which the plasma cools back into the hadrons that are ulitmately detected. A new book edited by R.C. Hwa [28] contains an excellent collection of articles that review the entire subject of the quark-gluon plasma.

ACKNOWLEDGMENTS

It is a pleasure to thank all the organizers of VII Mexican School of Particles and Fields for their work and especially Juan Carlos D'Olivo for his gracious hospitality. This research was supported in part by the U.S. National Science Foundation grant PHY-9630149.

REFERENCES

1. Dolan, L., and Jackiw, R., Phys. Rev. **D9**, 3320 (1974); Weinberg, S., Phys. Rev.**D9**, 3357 (1974).
2. Kapusta, J., *Finite Temperature Field Theory*, Cambridge: Cambridge Univ. Press, 1989.
3. Le Bellac, M., *Thermal Field Theory*, Cambridge: Cambridge Univ. Press, 1996.
4. Das, A., *Finite Temperature Field Theory*, World Scientific: Singapore, 1997.
5. Kajantie, K., Laine, M., Rummukainen, K., and Shaposhnikov, M., Nucl. Phys. **B466**, 189 (1996) and **B458**, 90 (1996).
6. Banks, T. and Rabinovici, E., Nucl. Phys. **B160**, 349 (1979); Fradkin, E. and Shenker, S., Phys. Rev. **D19**, 3682 (1979).
7. Brown, F.R., Butler, F.P., Chen, H., Christ, N.H., Dong, Z., Schaffer, W., Phys. Rev. Lett. **65**, 2491 (1990).
8. Landsman, N.P., and van Weert, Ch.G., Phys. Rep. **145**, 141 (1987).
9. R. Kubo, J. Phys. Soc. Japan **12**, 570 (1957); Martin, P.C., and Schwinger, J., Phys. Rev. **115**, 1342 (1959).
10. Bloch, F., Z. Phys. **74**, 295 (1932).
11. Matsubara, T., Prog. Theor. Phys. **14**, 351 (1955).
12. Abrikosov, A.A., Gorkov, L.P., and Dzyaloshinski, I.E., *Methods of Quantum Field Theory in Statistical Physics*, Prentice-Hall: Englewood Cliffs, 1963.
13. Mahan, G.D., *Many-Particle Physics*, Plenum: New York, 1990.
14. Feynman, R.P., Phys. Rev. **91**, 1291 (1953); Matthews, P.T., and Salam, A., Nuovo Cim. **2**, 120 (1955).
15. Schwinger, J., J. Math. Phys. **2**, 407 (1961).
16. Keldysh, L.V., Sov. Phys. JETP **20**, 1018 (1964).
17. Niemi, A.J., and Semenoff, G.W., Ann. Phys. (N.Y.) **152**, 105 (1984); Nucl. Phys. **B230**, 181 (1984).
18. Nieves, J.F., Phys. Rev. **D42**, 4123 (1990).
19. Umezawa, H., Matsumoto, H., and Tachiki, M., *Thermofield Dynamics and Condensed States*, North-Holland: Amsterdam, 1982.
20. Ojima, I., Ann. Phys. (N.Y.) **137**, 1 (1981).
21. Pisarski, R.D., Phys. Rev. Lett. **63**, 1129 (1989); Nucl. Phys. **A498**, 423c (1989).
22. Braaten, E. and Pisarski, R.D., Nucl. Phys. **B337**, 569 (1990) and **B339**, 310 (1990).
23. Braaten, E. and Pisarski, R.D., Phys. Rev. **D45**, 1827 (1992).
24. Kalashnikov, O.K., and Klimov, V.V., Phys. Lett. **B95**, 234 (1980).
25. Weldon, H.A., Phys. Rev. **D26**, 1394 (1982).
26. Klimov, V.V., Sov. J. Nucl. Phys. **33**, 934 (1981).
27. Weldon, H.A., Phys. Rev. **26**, 1394 (1982) and **D40**, 2410 (1989).
28. Hwa, R.C., ed *Quark-Gluon Plasma 2*, Singapore: World Scientific, 1995.

Black Hole Electromagnetic Duality

S. Deser

Department of Physics, Brandeis University, Waltham, MA 02254, USA

Abstract. After defining the concept of duality in the context of general n-form abelian gauge fields in $2n$ dimensions, we show by explicit example the difference between apparent but unrealizable duality transformations, namely those in $D = 4k + 2$, and those, in $D = 4k$, that can be implemented by explicit dynamical generators. We then consider duality transformations in Maxwell theory in the presence of gravitation, particularly electrically and magnetically charged black hole geometries. By comparing actions in which both the dynamical variables and the charge parameters are "rotated," we show their equality for equally charged electric and magnetic black holes, thus establishing their equivalence for semiclassical processes which depend on the value of the action itself.

I begin this lecture by paying my respects to the memory of Juan José Giambiagi, to whom this conference is dedicated. Having known him since the early '60's, I have had the opportunity of understanding his importance not only through his physics (universal as many of his ideas have become) but also through the inspiration he provided in the evolution of physics research in Argentina and indeed throughout all Latin America. He was a man of great culture, with both knowledge and perspective across a wide spectrum of human ideas, and a man of great courage as I was able to observe in the dark days around 1970 when he was exiled to La Plata. He was an optimist in spite of his dark insights. We will all miss him.

An earlier important loss to Latin American physics was that of Carlos Aragone of Uruguay and Venezuela, with whom I had the pleasure of a 25 year collaboration. He was another leader of theoretical physics in our far-flung community, who twice helped create fruitful environments – in his original and in his adopted homelands.

Finally, I thank the organizers for inviting me, even though my topic is not in the mainstream of this conference. The work described here was performed in collaboration with M. Henneaux and C. Teitelboim. Indeed it builds on work first done with the latter [1] some 20 years ago! It will appear in Phys. Rev. D early in 1997 [2], and in another paper still in process, from which the general n-form discussion is drawn.

CP400, *First Latin American Symposium on High Energy Physics/ VII Mexican School of Particles and Fields,*
edited by D'Olivo/Klein-Kreisler/Mendez

Our motivation for returning to so old a topic is its relevance to current research. I will have time here to discuss only one aspect of duality, namely its application to black hole physics, particularly that of charged black holes and their semiclassical behavior, that is when the actions themselves ($I/\hbar$) and not just the field equations matter. Since both electrically and magnetically charge black holes can exist, investigating their equivalence in this regime is tantamount to establishing a generalized Maxwell duality in presence of sources, both electric and magnetic, as well as of a gravitational field. We will indeed show (after reviewing the flat space, sourcefree case) that the actions of magnetically and electrically (equally) charged black holes are in fact the same, a conclusion recently reached in [3] by very different means.[1]

Let me begin with some introductory notions about duality in a more general framework to show also what duality is *not*, as there are still a number of misconceptions in the literature. Consider a general (n–1)-form potential and its associated field strength $F_{1..n} \equiv \partial_n A_{1..n-1}$. [All potential and field indices are to be understood to be totally antisymmetrized and suitably normalized; also I use "mostly plus" metric signature.] The dual of a field is always defined to be

$$^*F^{1..n} \equiv \frac{1}{n!}\,\epsilon^{1..n\,n+1..2n}\,F_{n+1..2n} \tag{1}$$

where ϵ is the Levi–Civita symbol (with $\epsilon^{01..} = +1$) in $2n$ dimensions. Clearly only in $2n$ dimensions will n-form fields be of the same rank as their duals so that one can even attempt to speak of duality transformations, let alone invariances. Now the action, field equations, and Bianchi identities for a source-free field are

$$I = -c_n \int d^{2n}x\, F_{1..n}F^{1..n}, \qquad \partial_1 F^{1..n} = 0, \qquad \partial_1\, {}^*F^{1..n} \equiv 0 \tag{2}$$

where $c_1 = 1/2$, $c_2 = 1/4$ etc. The (source-free) field equations and Bianchi identities are of the same form so that formally any linear transformation

$$F \to aF + b\,{}^*F \tag{3}$$

together with its dual, ${}^*F \to a\,{}^*F + b\,{}^{**}F$ also gives F's that obey this pair of equations. Double duality is an operation that depends on whether $n = d/2$ is even or odd, as a little reflection on the ϵ symbol verifies:

$$^{**}F = F, \;\; n = 2k+1, \qquad ^{**}F = -F, \;\; n = 2k \tag{4}$$

---

[1] This is not directly related to the very different question of charge quantization in the *e.g.*, $\sim \hbar$ sense. Also, we will be considering here the fixed charge sectors rather than the complementary case of fixed chemical potential, but the results should carry through to that situation as well.

(this is also the reason self-duality is only realizable in the $n = 2k + 1$ case). Either way, the above formal transformation is compatible with the equations. Is this symmetry shared by other physical quantities of these theories, in particular by their actions (our main interest here) and by their stress-tensors? Although it is only the Poincaré generators that are physical in flat space, the local stress tensor becomes an observable current in presence of gravity. These quantities are bilinear in the fields so they should impose more stringent conditions than the – linear – equations. To see most clearly what restrictions on (3) they impose let us rewrite the bilinears symmetrically in terms of F and $^*F$. Surprisingly, the actions and stress tensors are of the same form in all dimensions, because the scalar identity $F_{\mu..}F^{\mu..} \equiv -^*F^*_{\mu..}F^{\mu..}$ is by (4) dimension-independent. It then follows from (2) that

$$I = -\frac{1}{2}c_n \int d^{2n}x(F^2 - {}^*F^2) \,. \tag{5a}$$

The corresponding stress-tensors are then easily found, by varying with respect to the metric in the usual way:

$$T^\mu_\nu = \frac{1}{2}(F^{\mu..}F_{\nu..} + {}^*F^{\mu..}\,{}^*F_{\nu..}) \,. \tag{5b}$$

In accordance with conformal invariance of the action, $T^\mu_\mu = \frac{1}{2}(F^2 + {}^*F^2) \equiv 0$. In all cases, there is the same "mismatch" between the signs in the action and stresses, so that not both would seem to remain invariant under a duality transformation. The latter must be defined as either a normal rotation or a hyperbolic one rather than the general (3) to even formally keep either a sum or a difference of squares invariant. There is also no help from the fact that cross terms in the form F^*F are total divergences and hence irrelevant to the action (apart from possible topological effects). That is, in $4k$ dimensions $F_{\mu\nu..}\,{}^*F^{\mu\nu..} = \partial_\mu[\epsilon^{\mu..}A\partial A]$ is the divergence of a Chern–Simons structure, while in $4k+2$, F^*F actually vanishes identically, *e.g.*, $F_\mu(\epsilon^{\mu\nu}F_\nu) \equiv 0$. So we have a paradox: the equations and identities in all dimensions are together invariant under any linear variation of F and $^*F$ into each other, while the action and stress tensor can seemingly never both be invariant under any transformation at all. In fact, as we will now show, none of the above considerations is even meaningful and (despite the uniformity in (5a) and (5b)) the correct answer is that Maxwell theory and its $4k$ extensions are perfectly invariant in a precise sense under duality rotations, while duality is not even definable for scalar theory and *its* ($4k+2$) generalizations.

The basis for those statements is the simple remark that in a dynamical theory, only transformations that can be generated by functionals of the canonical variables are even meaningful. Until the latter are given, one cannot even know what (if any) duality change is possible, let alone whether it defines an invariance. Thus, the scalar field in $D=2$,

$$I = -\frac{1}{2}\int d^2x\, F_\mu F^\mu \qquad F_\mu \equiv \partial_\mu \phi \qquad {}^*F^\mu \equiv \epsilon^{\mu\nu}\partial_\nu \phi \tag{6}$$

has Hamiltonian form

$$I = \int d^2x[\pi\dot{\phi} - \frac{1}{2}(\pi^2 + \phi'^2)] \, , \tag{7}$$

the field strength having components $F_0 = \dot{\phi} = \pi$, $F_1 = \phi'$. Now it is clear that there is no generator $G = \int dx \mathcal{G}(\pi, \phi)$ such that its Poisson bracket with π and ϕ' will rotate them into each other (with either sign). For example $[G, \pi(x)] \sim \phi'(x)$ would require $G \sim \int dy\, \phi(y)\phi'(y)$ but that is clearly a total divergence and similarly for $[G, \phi'] \sim \pi$. It is easy to see (by counting signs in ϵ) that this impossibility extends to the general $D = 4k + 2$ case.[2]

Let us turn to $D = 4k$, in particular to electrodynamics in $D = 4$, our main topic. We start with a quick review of the flat space source-free sector [1]. Here the Maxwell action may be written in terms of the reduced first order conjugate variables $(\mathbf{E}, \mathbf{A})$ as

$$I_M[\mathbf{E}, \mathbf{A}] = \int d^4x[-\mathbf{E} \cdot \dot{\mathbf{A}} - \frac{1}{2}(\mathbf{E}^2 + \mathbf{B}^2)] \, , \qquad \boldsymbol{\nabla} \cdot \mathbf{E} = 0 \, , \tag{8}$$

where $\mathbf{B} \equiv \nabla \times \mathbf{A}$. In the absence of sources, the Gauss constraint says that $\mathbf{E}$ is purely transverse,

$$\mathbf{E} \equiv \boldsymbol{\nabla} \times \mathbf{Z} \tag{9}$$

and therefore only the transverse, gauge-invariant, part of $\dot{\mathbf{A}}$ survives in the kinetic term, which may be rewritten as

$$\int d^4x\, \epsilon^{ijk}\partial_j Z_k \dot{A}_i \, . \tag{10}$$

We assert, and it is easy to check, that the above reduced I_M is invariant under the rotation of the 2 dimensional vector with components $V \equiv (\mathbf{Z}, \mathbf{A})$ or its curl $W \equiv (\mathbf{E}, \mathbf{B})$ under the usual 2-dimensional rotation,

$$V' = RV \qquad \text{or} \qquad W' = RW \, , \qquad R \equiv \exp(i\sigma_2 \cos\theta) \, . \tag{11}$$

Equally important is that the generator of this transformation exists and has a very elegant "topological" (metric independent) Chern–Simons form,

$$G = -\frac{1}{2}\int d^3x\, \epsilon^{ijk}[Z_i\partial_j Z_k + A_i\partial_j A_k] \, . \tag{12}$$

---

[2]) Eq. (12) immediately shows that $\epsilon^{ijklm}A_{ij}\partial_k A_{lm}$ is a total divergence for even form potentials represented here by A_{ij}.

The Poisson bracket or commutator of G with V or with W engenders (11) by virtue of the canonical commutation relations $[\mathbf{E}^i, \mathbf{A}'_j] = [\delta^i_j(\mathbf{r}-\mathbf{r}')]^T$ where δ^T is the usual transverse projection of the unit operator. As usual there is some asymptotic falloff to be specified; here and in curved space we take $\mathbf{A} \sim a(\Omega)r^{-1} + \mathcal{O}(r^{-2})$ and $\mathbf{E} \sim \mathbf{e}(\Omega)r^{-2} + \mathcal{O}(r^{-3})$ where $\mathbf{a}, \mathbf{e}$ depend only on solid angle.

We must now generalize the above analysis to include nontrivial geometries and charges. The former is easy: Just write the Maxwell action in the covariant first order form,[3]

$$I_M = -\frac{1}{2}\int d^4x \left[F^{\mu\nu}(\partial_\mu A_\nu - \partial_\nu A_\mu) - \frac{1}{2}\, F^{\mu\nu}F^{\alpha\beta}g_{\mu\alpha}g_{\nu\beta}(-g)^{-1/2}\right] \tag{13}$$

where $F^{\mu\nu}$ is a contravariant tensor density to be varied independently, then insert the usual 3+1 decomposition of the metric into its spatial part g_{ij}, mixed part $g_{0i} \equiv N_i$ and time-time part $g^{00} \equiv -N^{-2}$, so that $\sqrt{-g} = N\sqrt{g}$ where $|g|$ is the 3-metric determinant. Then it immediately follows that I_M can be written as [4]

$$I_M[\mathbf{E},\mathbf{A}] = -\int d^4x[E^i\dot{A}_i + \tfrac{1}{2}\, Ng^{-1/2}g_{ij}(E^iE^j + B^iB^j) - \epsilon_{ijk}N^iE^jB^k] \tag{14}$$

where $F^{0i} \equiv E^i$ is the electric, $B^i \equiv \epsilon^{ijk}\partial_j A_k$ the magnetic, field (both are contravariant three-densities) and all metric operators are in 3-space; we have solved the Gauss constraint (still $\partial_i E^i = 0$) so that both E^i and B^i are identically transverse, $\partial_i E^i = 0 = \partial_i B^i$. Note that although it is on an arbitrary curved background space, (14) is easily seen to be invariant under (11) via the same (metric independent!) generator G of (12) since the canonical variables and kinetic term are unchanged while $(\mathbf{E}^2 + \mathbf{B}^2)$ and $\mathbf{E}\times\mathbf{B}$ are clearly locally invariant under (11).

We now turn to the black hole case and include electric and magnetic sources. To stick to the problem of interest in [3], where only the exterior solution is considered, one can still work with the source-free Maxwell equations but one must allow for non-vanishing electric and magnetic fluxes at infinity. This is possible because the spatial sections Σ have a hole. There are thus two-surfaces that are not contractible to a point, namely, the surfaces surrounding the hole (we assume for simplicity a single black hole but the analysis can straightforwardly be extended to the multi-black hole case).

Let us first dispose of a technicality when varying in presence of electrical sources or fluxes. The variation of the action under changes of E^i,

$$\delta_E I_M = -\int d^4x \delta E^i(\dot{A}_i + Ng^{-1/2}g_{ij}E^j - \epsilon_{ijk}B^jN^k), \tag{15}$$

[3] This can be done in the same way for all form actions and incidentally exhibits their common Weyl invariance.

vanishes for arbitrary variations δE^i subject to the transversality conditions[4] $\partial_i \delta E^i = 0$ and $\delta \oint_{S^2_\infty} E^i dS_i = 0$ if and only if the coefficient of δE^i in (15) fulfills the condition

$$\dot{A}_i + N g^{-1/2} g_{ij} E^j - \epsilon_{ijk} B^j N^k = \partial_i V \tag{16}$$

where V $(\equiv A_0)$ is an arbitrary function which behaves asymptotically as $C + O(r^{-1})$: In that case, $\delta I_M = -\int d^4x \delta E^i \partial_i V = -\oint_{S^2_\infty} \delta E^i V dS_i = -C\delta(\text{electric flux}) = 0$. No special conditions are required, on the other hand, when varying A_i. Thus, (14) is appropriate as it stands, *i.e.*, without "improving" it by adding surface terms to the variational principle in which the competing histories all have the same given electric flux at infinity and thus also the same given electric charge (here equal to zero). As pointed out in [3], it is necessary to allow the temporal component V of the vector potential to approach a non-vanishing constant at infinity since this is what happens in the black hole case if V is required to be regular on the horizon. However, as we have just shown, in order to achieve this while working with this action, it is unnecessary to keep all three components E^i of the electric field fixed at spatial infinity; only the electric flux $\oint_{S^2_\infty} E^i dS_i$ must be kept constant in the variational principle.

In the presence of a non-vanishing magnetic flux, the magnetic field is given by the expression

$$B^i = \epsilon^{ijk} \partial_j A_k + B^i_S \tag{17}$$

where B^i_S is a fixed field that carries the magnetic flux,

$$\oint_{S^2_\infty} B^i_S dS_i = 4\pi\mu, \tag{18}$$

and where $B^i_T = \epsilon^{ijk}\partial_j A_k$ is the transverse part of B^i,

$$\partial_i B^i_T = 0, \oint_{S^2_\infty} B^i_T dS_i = 0. \tag{19}$$

Following Dirac, we can take B^i_S to be entirely localized on a string running from the source-hole to infinity, say along the positive z-axis $\theta = 0$. We shall not need the explicit form of B^i_S in the sequel, but only to remember that for a given magnetic charge μ, B^i_S is completely fixed and hence is not a field to be varied in the action. The only dynamical components of the magnetic field B^i are still the transverse ones, *i.e.*, A_i.

One can also decompose the electric field as

---

4) The condition $\delta \oint_{S^2_\infty} E^i dS_i = 0$ is actually a consequence of $\partial_i \delta E^i = 0$ (and of smoothness) on spatial sections with R^3-topology. We write it separately, however, because this is no longer the case if Σ has holes, as below.

$$E^i = E_T^i + E_L^i \tag{20}$$

where the longitudinal part carries all the electric flux

$$\oint_{S^2_\infty} E_L^i dS_i = 4\pi e \tag{21}$$

and the transverse field obeys

$$\partial_i E_T^i = 0, \ \oint_{S^2_\infty} E_T^i dS_i = 0 \tag{22}$$

and can thus again be written as $E_T^i = \epsilon^{ijk}\partial_j Z_k$ for some Z_k. Given the electric charge e, the longitudinal electric field is completely determined if we impose in addition, say, that it be spherically symmetric. As we have done above, we shall work with a variational principle in which we have solved Gauss's law and in which the competing histories have a fixed electric flux $\oint_{S^2_\infty} E^i dS_i$ at infinity. This means that the longitudinal electric field is completely frozen and that only the tranverse components E_T^i or Z^i are dynamical, as for the magnetic field.

In order to discuss duality, it is convenient to treat the non-dynamical components of E^i and B^i symmetrically. To that end, one may either redefine B_S^i by adding to it an appropriate transverse part so that it shares the spherical symmetry of E_L^i, or one may redefine E_L^i by adding to it an appropriate transverse part so that it is entirely localized on the string. Both choices (or, actually, any other intermediate choice) are acceptable here. For concreteness we may take the first choice; the fields then have no string-singularity.

In the Maxwell action, E^i and B^i are now the *total* electric and magnetic fields. Since E_L^i may be taken to be time-independent (the electric charge is constant), one may replace E^i by E_T^i in the kinetic term of (14), yielding as alternative action

$$I_M^{e,\mu}[\mathbf{E}_T, \mathbf{A}] = -\int d^4x [E_T^i \dot{A}_i + \tfrac{1}{2} N g^{-1/2} g_{ij}(E^i E^j + B^i B^j) - \epsilon_{ijk} N^i E^j B^k]. \tag{23}$$

This amounts to dropping a total time derivative – equal to zero for periodic boundary conditions – and shows explicitly that the kinetic term is purely transverse. Note that there is actually a *different* action (23), hence a distinct variational principle, for each choice of e and μ, as the notation indicates.

Consider now a duality rotation acting on the transverse, dynamical variables A_i (or B_T^i) and E_T^i. Just as in the sourceless case, the kinetic term of is invariant under this transformation: it is the same kinetic term and the transformation law is the same; the surface term at the horizon in the variation vanishes because $\dot{A}_i = 0$ and $\dot{Z}_i = 0$ there.[5] Thus, if we also rotate the (non-dynamical) components of the electric and magnetic fields in the same way, that is, if we relabel the external parameters e, μ by the same 2D rotation, so that the 2-vector $Q \equiv (e, \mu)$, becomes

$$Q' = RQ \tag{24}$$

then the actions $I_M^{e,\mu}$ and $I_M^{e',\mu'}$ are equal since $\mathbf{E}$ and $\mathbf{B}$ enter totally symmetrically in the energy and momentum densities. More explicitly, if we write the longitudinal fields as $B_L^i = \mu V^i$, $E_L^i = eV^i$, then the relevant terms in (23) are just

$$-\int d^4x \{ N g^{-1/2} g_{ij} [(eE_T^i + \mu B_T^i)V^j + \frac{1}{2}(e^2 + \mu^2)V^iV^j] - \epsilon_{ijk} N^i V^j (eB_T^k - \mu E_T^k) \} \,. \tag{25}$$

For the mixed terms, it is clear that the field transformation is just compensated by the parameter rotation (24), while the VV term is invariant under the latter. To put it more formally, the extended duality invariance we have spelled out is one that links *different* systems, with different parameters:

$$I_M^{e,\mu}[\mathbf{E}_T, \mathbf{A}_T] = I_M^{e',\mu'}[\mathbf{E}'_T, \mathbf{A}'_T] \,, \tag{26}$$

where the primes denote the rotated values. As a special case, for the black holes without Maxwell excitations, we find equality of equally electrically and magnetic charge actions,

$$I_M^{e,0}[\mathbf{0}, \mathbf{0}] = I_M^{0,e}[\mathbf{0}, \mathbf{0}] \tag{27}$$

as also obtained, by explicit calculation of these actions, in [3]. This equality is thus not a special artifact, but reflects a general invariance property of the action appropriate to the variational principle considered here, in which the electric and magnetic fluxes are kept fixed.

It is a pleasure to acknowledge my collaborators, M. Henneaux and C. Teitelboim, as well as support from the National Science Foundation, under grant #PHY-9315811.

---

5) To discuss the surface terms that arise in the variation of the action, one must supplement the asymptotic behavior of the fields at infinity specified earlier by conditions at the horizon. These are especially obvious in the Euclidean continuation, where time becomes an angular variable with the horizon sitting at the origin of the corresponding polar coordinate system. Regularity then requires that $V \equiv A_0$ and the time derivatives $\dot{A}_i$, $\dot{E}^i$ all vanish at the horizon. We assume these conditions to be fulfilled throughout.

REFERENCES

1. S. Deser and C. Teitelboim, Phys. Rev. **D13**, 1592 (1976); S. Deser, J. Phys. **A15** 1053 (1982). For a recent review see D. I. Olive, *Exact Electromagnetic Duality*, hep-th/9508089.
2. S. Deser, M. Henneaux, and C. Teitelboim, "Electric-Magnetic Black Hole Duality," Phys. Rev. **D55** (Jan. 15, 1997), and to be published.
3. S. W. Hawking and S. F. Ross, Phys. Rev. **D52**, 5865 (1995).
4. R. Arnowitt, S. Deser, and C.W. Misner in *"Gravitation, An Introduction to Current Research"*, (L. Witten, ed.) (Wiley, NY, 1962).

Dirac Supersymmetry Applications

M. Moreno

Instituto de Física Universidad Nacional Autónoma de México, Ap. Postal 20-364, 01000 México, D. F., México.

Abstract. The basic ideas of Dirac supersymmetry are presented and three applications are developed. The systems to which this scheme is applied are: i. Dirac particle in an external potential, ii. superconductivity and iii. a nucleon in an arbitrary pseudoscalar or timelike pseudovector potential. It is shown that no bound states of a Dirac particle exist for these potentials when they vanish at infinity. The relevance of this result for the deuteron problem is discussed.

INTRODUCTION

The idea of Dirac supersymmety was initially developed as an extension of the so called Dirac oscillator. This system is a mathematically inspired generalization of the usual nonrelativistic harmonic oscillator [1–3]. In the free Dirac equation one makes the substitution $\vec{p} \rightarrow \vec{p} - im\omega\beta\vec{x}$, and one obtains a system in which the positive energy states have a spectrum similar to the nonrelativistic harmonic oscillator. It can be shown [4] that such a system can be obtained in a less arbitrary way in the following way: Consider a neutral, spin $\frac{1}{2}$, particle with purely anomalous magnetic moment; introduce it into a sphere of constant electric charge density and then make the radius of the sphere very large.

For our purposes the important thing about this construction is that it is a non trivial example of an surprisingly large class of systems. These have been named Dirac supersymmetric systems [5–7] because they include at least half of the time independent potentials to which a Dirac particle can be subjected. Additionally if the appropriate limit is taken these problems correspond to the usual *quantum mechanical* supersymmetry [8]. Furthermore this limit usually, but not necessarily, corresponds to the nonrelativistic approximation.

Interestingly enough the realm of applicability of these scheme is larger than Dirac equation. The fundamental mechanism that underlies the Dirac supersymmetry scheme is also responsible for the pairing that leads to, at least, BCS superconductivity [9] and to generalizations of BCS to multicomponent systems.

CP400, *First Latin American Symposium on High Energy Physics/ VII Mexican School of Particles and Fields,* edited by D'Olivo/Klein-Kreisler/Mendez

In this work I will review a few relevant applications of the Dirac supersymmetric scheme. In the next section a description of the basic general concepts of the Dirac supersymmetic scheme is presented. Then the application to Dirac particles in external fields is obtained as well as the superconductivity example. The final and application that we discuss is that of a fermion in a pion like potential; this is conceptually close to the deuteron problem.

I DIRAC SUPERSYMMETRY

By definition, a quantum mechanical supersymmetric (Susy) Hamiltonian H_S satisfies

$$[Q, H_S] = 0 = [Q^\dagger, H_S], \tag{1}$$

with Q and $Q^\dagger$ fermionic operators in the sense that $Q^2 = 0 = Q^{\dagger 2}$. Such a Hamiltonian can be of the form

$$H_S = \{Q, Q^\dagger\} \equiv QQ^\dagger + Q^\dagger Q, \tag{2}$$

where the $QQ^\dagger$ and $Q^\dagger Q$ commute with H_S and between themselves. If the ground state is nondegenerate and normalizable the supersymmetry will remain unbroken. The states $|n\rangle$, $|n_+\rangle \equiv Q^\dagger|n\rangle$ and $|n_-\rangle \equiv Q|n\rangle$ are degenerate, but either $|n_+\rangle$ or $|n_-\rangle$ are null.

A connection between the Dirac equation and supersymmetric quantum mechanics appears [5] if the Dirac Hamiltonian H can be written as

$$H = Q + Q^\dagger + \Lambda, \tag{3}$$

with Λ a hermitean operator, which together with Q and $Q^\dagger$ satisfies the anticommutation relations

$$\{Q, \Lambda\} = \{Q^\dagger, \Lambda\} = 0. \tag{4}$$

Then, the Hamiltonian (3) commutes with $QQ^\dagger$, and $Q^\dagger Q$, and we get for the squared Hamiltonian

$$H^2 = \{Q, Q^\dagger\} + \Lambda^2 = h^2 + \Lambda^2, \tag{5}$$

where h is required to be a positive even root of this operator. Defining $\hat{\Lambda} \equiv \frac{\Lambda}{\sqrt{\Lambda^2}}$, the Foldy-Wouthuysen (FW) Hamiltonian H_{FW} can be written

$$H_{FW} = \hat{\Lambda}\sqrt{\{Q, Q^\dagger\} + \Lambda^2}, \tag{6}$$

and for this form to hold Λ must possess non-null eigenvalues, otherwise $\hat{\Lambda}$ is not well defined.

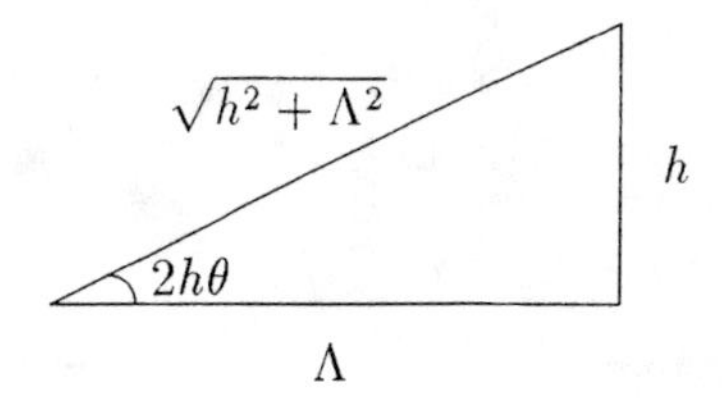

FIGURE 1. Triangle construction for the Foldy-Wouthuysen transformation.

From the hermiticity of Λ and conditions (4) the FW Hamiltonian (6) can be shown to anticommute with both Q and $Q^\dagger$; by (1) H_{FW} is thus a Susy Hamiltonian. The anticommutation condition (4) implies that, in order to construct observables which are constants of motion together with the Dirac-Susy Hamiltonian Eq. (3), one must construct bosonic operators from the fermionic ones.

A FW transformation is generated by [5–7]

$$iS \equiv \hat{\Lambda}(Q + Q^\dagger)\theta_{FW}, \tag{7}$$

where θ_{FW} is defined through

$$\tan(2h\theta_{FW}) = \frac{h}{\sqrt{\Lambda^2}}. \tag{8}$$

In our notation θ_{FW} has units of h^{-1} and from (8) one can write

$$\sqrt{\Lambda^2}\,\theta_{FW} = f(\frac{h^2}{\Lambda^2}), \tag{9}$$

showing that θ_{FW} exists with the required properties and that it obeys the usual right triangle construction of Fig. 1.

Often one can give a matrix representation for Q and $Q^\dagger$ in the form

$$Q = \begin{pmatrix} 0 & 0 \\ A & 0 \end{pmatrix}, \quad Q^\dagger = \begin{pmatrix} 0 & A^\dagger \\ 0 & 0 \end{pmatrix}. \tag{10}$$

Because $\hat{\Lambda}$ commutes with H_{FW} and its eigenvalues are ± 1, Eq.(6) implies that, in general, the energy spectra has two branches; between them there exists a gap determined by the lowest eigenvalues of Λ^2 and h^2. If $\hat{\Lambda}$ has negative eigenvalues and H^2 is not bounded from above one gets an unstable ground state for the Susy Hamiltonian. The usual remedy for this dilemma is to introduce a Dirac sea. The important feature of the Dirac Susy interactions is that the definition of this sea is independent of the size of the couplings, a property we refer to as a "stable Dirac sea".

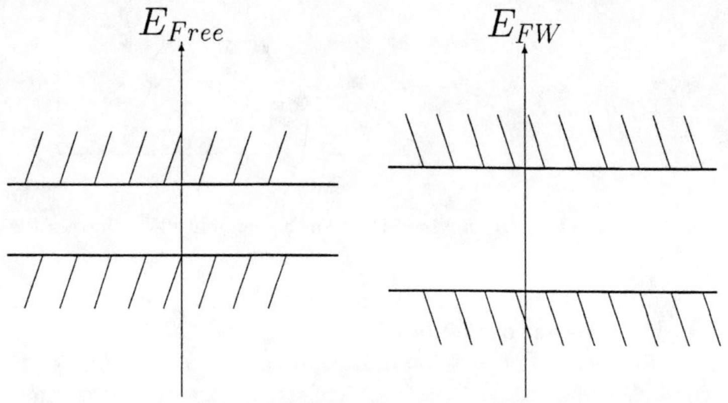

FIGURE 2. Schematic Dirac seas for the free case and Dirac Susy potentials.

Also important is that the positive and negative energy states do not mix, and the gap size is not decreased [5–7]. Fig. 2 displays the Dirac sea in both free and interaction (FW) cases. After second quantizing, one gets a stable Dirac sea.

II DIRAC SUPERSYMMETRIC POTENTIALS

Let us use the ideas presented in last section to study Dirac equation itself. To do this the first thing is to notice that the kinetic term in the Dirac Hamiltonian is of the form

$$H_0 = \vec{\alpha} \cdot \vec{p} + \beta m \tag{11}$$

it is well known that a single FWT can block diagonalize this to

$$H_{0FW} = \beta \sqrt{\vec{p} \cdot \vec{p} + m^2} \tag{12}$$

in terms of the development of last section this can be seen as a straightforward result of the standard representation of the Dirac matrices, indeed, using the standard representation

$$H_0 = \begin{pmatrix} m & \vec{\sigma} \cdot \vec{p} \\ \sigma \cdot \vec{p} & -m \end{pmatrix} \tag{13}$$

and from this it is simple to identify

$$\Lambda = \begin{pmatrix} m & 0 \\ 0 & -m \end{pmatrix} , \; Q = \begin{pmatrix} 0 & \vec{\sigma} \cdot \vec{p} \\ 0 & 0 \end{pmatrix} \tag{14}$$

From this structure one can immediately verify that a modification of the form

$$Q = \begin{pmatrix} 0 & \vec{\sigma} \cdot (\vec{p} + \vec{C}) + C_0 \\ 0 & 0 \end{pmatrix} \tag{15}$$

does not change the anticommutation properties of Q or Λ and that as a consequence of this a closed FWT can be performed also for the modified Hamiltonian. The four complex velocity dependent potentials $C_i(\vec{x}, \vec{p})$ can be easily related to the following interactions:

1. one pseudoscalar interaction: $Im(C_0) \leftrightarrow \gamma_5$,
2. one time-like pseudovector potential: $Re(C_0) \leftrightarrow \gamma_0 \gamma_5$,
3. three components of the spacelike vector potential: $Re(\vec{C}) \leftrightarrow \vec{\gamma}$
4. three components of the spacelike-spacelike tensor potential: $Im(\vec{C}) \leftrightarrow \sigma_{ij}$

The Dirac oscillator potential mentioned in the introduction corresponds to the particular case $\vec{C} = im\omega\vec{r}, C_0 = 0$. The study of Dirac supersymmetry in Dirac equation in dimension different form $3+1$ has been performed elsewhere [10].

III PAIRING PHENOMENA AND DYNAMICAL SYMMETRY BREAKING

Let us now look at a different physical phenomena that also features Dirac supersymmetry. In the pairing theory of superconductivity the instability in the Fermi sea is produced by an effective *attractive* electron-electron (or hole-hole) interaction. For the paradigmatical BCS theory the attractive part of the force arises from the electron-phonon interaction and for electrons or holes close to the Fermi surface. Cooper pairs with zero total spin and center of mass momenta play a vital role in the emergence of the superconducting phase out of the *"normal"* non paired phase.

The essential feature of the Bogoliubov-Valatin (BV) theory is to exhibit the Fermi sea instability through a canonical transformation which mixes electron and hole states of opposite momentum and spin quantum numbers ($\mathbf{k}\uparrow$ and $-\mathbf{k}\downarrow$). Basically one is showing that the true ground state of the system is *not* the one described Fermi surface. The physical basis of this transformation relies on the picture that a certain fraction of the electrons forming the Fermi sea escape it and become bound into Cooper pairs. The new quasi particles so-called bogolons, are related to the electron states $a_k^\dagger$, a_k by

$$A_k = \begin{pmatrix} a_k \\ a_{-k}^\dagger \end{pmatrix} = \begin{pmatrix} u_k & v_k \\ -v_k & u_k \end{pmatrix} \begin{pmatrix} \alpha_{\mathbf{k}} \\ \beta_{-\mathbf{k}} \end{pmatrix}, \tag{16}$$

in this formula we have extended the notation and the subindices k, $-k$ denote any pair of fermions that are coupled into Cooper pairs, in the usual BCS case: $k = \mathbf{k}\uparrow$ and $-k = -\mathbf{k}\downarrow$ of course. The transformation matrix must be unitary in order to preserve the anticommutation relations for the bogolon operators, α and β. One usually further restricts the transformation to be orthogonal. In what follows we will follow the notation of Fetter and Walecka Ref. [11]. The BV transformation can be shown to describe the instability of the electron Fermi sea [12]. Formally this manifests by a new vacuum state $|\mathbf{0}>$ that satisfies $\alpha_{\mathbf{k}}|\mathbf{0}>= \beta_{\mathbf{k}}|\mathbf{0}>= 0$. This can be achieved in a minimal way if $|\mathbf{0}>= \mathcal{N}\prod_i(1+g_i a_i^\dagger a_{-i}^\dagger)|\phi>$

The thermodynamic potential at zero temperature (or effective Hamiltonian) is

$$\begin{aligned}\hat{K} &= \hat{H} - \mu\hat{N} \\ &= \hat{T} - \hat{V}\end{aligned} \tag{17}$$

with

$$\begin{aligned}\hat{T} &= \sum_{\mathbf{k}\lambda} a_{\mathbf{k}\lambda}^\dagger a_{\mathbf{k}\lambda}(\epsilon_k^0 - \mu) \\ \hat{V} &= -\frac{1}{2}\sum_{\substack{\mathbf{k}_1+\mathbf{k}_2=\mathbf{k}_3+\mathbf{k}_4 \\ \lambda_1,\lambda_2,\lambda_3,\lambda_4}} <\mathbf{k}_1\lambda_1\mathbf{k}_2\lambda_2|V|\mathbf{k}_3\lambda_3\mathbf{k}_4\lambda_4> a_{\mathbf{k}_1\lambda_1}^\dagger a_{\mathbf{k}_2\lambda_2}^\dagger a_{\mathbf{k}_4\lambda_4} a_{\mathbf{k}_3\lambda_3}\end{aligned} \tag{18}$$

Let us now define

$$B_k := \begin{pmatrix} \check{\alpha}_{\mathbf{k}} \\ \check{\beta}_{-\mathbf{k}}^\dagger \end{pmatrix} = \begin{pmatrix} \check{u}_k & \check{v}_k \\ -\check{v}_k & \check{u}_k^* \end{pmatrix} A_k = UA_k, \tag{19}$$

where $\nu \times \nu$ matrices and ν vectors are denoted with a check above. The commutation relations for the operators $a_{k\lambda}$ can then be written in matrix form as

$$A_k A_{k'}^\dagger \pm A_{k'}^* A_k^T = 1\delta_{kk'}. \tag{20}$$

The canonicity condition requires that

$$B_k B_{k'}^\dagger \pm B_{k'}^* B_k^T = 1\delta_{kk'} \tag{21}$$

which is satisfied if U in (19) is an orthogonal matrix as can be easily verified by sandwiching Eq.(20) between U and $U^\dagger$.

When (19) to (21) are applied to the Hamiltonian (17) one obtains an operator of the form

$$\hat{K} = H_O + H_I + H_{II} \tag{22}$$

where the three terms in the rhs correspond to the zero- (c-number), one- and two-bogolon terms, respectively. The BV transformation is defined by the requirement that H_I, the one-bogolon term, be diagonal. One has

$$H_I = \sum_{\mathbf{k}} \begin{pmatrix} \breve{\alpha}_{\mathbf{k}} \\ \breve{\beta}^{\dagger}_{-\mathbf{k}} \end{pmatrix}^{\dagger} \begin{pmatrix} \breve{u}_k & \breve{v}_k \\ -\breve{v}_k & \breve{u}_k \end{pmatrix} \begin{pmatrix} \breve{\xi}_k & \breve{\Delta}_k \\ \breve{\Delta}_k & -\breve{\xi}_k \end{pmatrix} \begin{pmatrix} \breve{u}_k & -\breve{v}_k \\ \breve{v}_k & \breve{u}_k \end{pmatrix} \begin{pmatrix} \breve{\alpha}_{\mathbf{k}} \\ \breve{\beta}^{\dagger}_{-\mathbf{k}} \end{pmatrix}. \tag{23}$$

where the relevant structure for our purposes is

$$\begin{pmatrix} \breve{\xi}_k & \breve{\Delta}_k \\ \breve{\Delta}_k & -\breve{\xi}_k \end{pmatrix}. \tag{24}$$

If one identifies

$$Q = \begin{pmatrix} 0 & 0 \\ A & 0 \end{pmatrix} = \begin{pmatrix} 0 & 0 \\ \breve{\Delta}_k & 0 \end{pmatrix}, \tag{25}$$

$$\Lambda = \begin{pmatrix} \breve{\xi}_k & 0 \\ 0 & -\breve{\xi}_k \end{pmatrix}, \tag{26}$$

one can easily confirm our central result that the standard BV operator and its extrapolation to a ν-component system have in fact a Dirac supersymmetric structure. The BV transformation is completely equivalent in this case to the Foldy-Wouthuysen transformation, and its generator can be read form Eqs. (7) and (8). Quite generally the condition (4) implies that

$$[\breve{\Delta}, \breve{\xi}] = 0, \tag{27}$$

for example with $\breve{\xi}$ proportional to the ν times ν unit matrix

Making extensive use of the Dirac Susy structure of this operator one can show that a generalized version of the gap equation exists for this multicomponent case. One obtains

$$\Delta = \sum VW = \sum_k V \begin{pmatrix} \breve{\Delta}^{\dagger} & -\breve{\mathcal{E}} \\ \breve{\mathcal{E}} & \breve{\Delta} \end{pmatrix} \mathcal{T}\Big(\frac{\beta}{2}, \begin{pmatrix} \breve{E}_k & 0 \\ 0 & -\breve{E}_k \end{pmatrix}\Big) \tag{28}$$

where V and W are $2\nu \times 2\nu$ matrices and $\mathcal{T}(a,x) = \tanh(a\sqrt{x^2})/\sqrt{x^2}$ and $\breve{E}_k$ is the bogolon eigenvalue matrix. In order to obtain such equation it is necessary to demand $[\mathcal{E}, \Delta] = 0$, a generic interaction term thus will requires a kinetic term proportional to the unit matrix. Equivalently the mass term should be independent of the component index. With this condition taken into account one can easily show that for a multicomponent system with a BCS s-channel factorizable interaction a dramatic increase in the critical transition temperature can be achieved. Indeed for an interaction of the type

$$V_{k\lambda_1\lambda_2,l\lambda_3\lambda_4} = -V_0\theta(\Delta\epsilon - |\mu - \epsilon_k|)\theta(\Delta\epsilon - |\mu - \epsilon_l|)V_{\lambda_1\lambda_2}V_{\lambda_3\lambda_4}, \tag{29}$$

one gets an increase in the effective coupling $\lambda = \nu V_0 N(0) = \nu\lambda^{BCS}$ if all components interact among them with the BCS strength, *i.e.* $V_{\lambda_1\lambda_2} = \delta_{\lambda_1\lambda_2}$. This implies larger critical transition temperatures because in BCS theory this temperature scales as $e^{-\lambda}$.

IV THE PSEUDOSCALAR PION AND THE DEUTERON

According to nuclear physics textbooks [13] a deuteron is a $J^P = 1^+$ bound state of a proton and a neutron. The binding energy is very small $\Delta = m_d - m_p - m_n = -2.22 Mev/c^2$; has a quotient this is more evident $\Delta/(m_p + m_n) = -0.001185$. The magnetic moment moment is almost the magnetic moment of the antialigned proton and neutron magnetic moments, $\delta_\mu = \mu_d - \mu_p - \mu_n = 0.857 - 2.793 - (-1.913) = -0.022 nm$. The deuteron has a small quadrupolar moment $Q = +0.00286 b$.

The weakly bound nature of the deuteron is confirmed by the absence of other bound states. That the magnetic moment is simply the sum of the proton and neutron magnetic moments $\delta_\mu/(\mu_p + \mu_n) = 0.00475$ means that their spins are aligned in an $S = 1$ state with almost pure s-wave orbital angular momentum. The admixture of the d-wave can be estimated from the value of the quadrupolar moment to be of the order of 5%.

Almost any reasonable potential model in a nonrelativistic scheme gives results in agreement with the deuteron binding energy. From a physical perspective it seems more adequate to introduce a model potential based on the fact that the loosely bound nature of the deuteron indicates that this nucleus is large and that a one particle exchange (OPE) should give a good description. In any case the deuteron is the most promising nucleus in which a OPE approximation might work. That this is the case is known since long ago [13].

The standard treatment is to assume that the pion exchange dominates the long range nature of the potential. The pseudoscalar or pseudovector nature of the pion coupling require that in the nonrelativistic *and* perturbative approximation the potential seen by a nucleon be of the form

$$V(r) = \vec{\sigma} \cdot \nabla \phi, \tag{30}$$

where ϕ is the pi-meson cloud. This meson field is for a static pointlike source a solution to the Klein-Gordon equation

$$\nabla^2 \phi - m_\pi^2 \phi = \delta(\vec{r}) \tag{31}$$

Scattering data with large impact parameters fix the strength of the Yukawa potential to

$$\phi(r) = f \frac{e^{-m_\pi r}}{r} \quad , \quad f = -\frac{g^2}{4\pi} \approx -14. \tag{32}$$

Isospin can be easily taken into account if the change $\phi \to \vec{\tau} \cdot \vec{\phi}$ is done, this will not concern us here. The nucleon-nucleon potential derived from this can be shown to have correct spin-isospin behavior [13] in the sense that a tensor and a spin-spin terms are obtained for the two-body problem.

In this work it is sufficient to observe that the potential of a nucleon in a pion cloud is *not* given correctly by Eq.(30). This is physically clear from the singular behavior of the Yukawa potential when substituted into this equation which must be fixed with the introduction of a "realistic" hard core potential. To see this we simply check that in the notation of Eq.(15) a pseudoscalar interaction corresponds to $C_0 = i\phi$ and that this leads to a FWT Dirac equation of the form

$$E\Psi = \pm\sqrt{M^2 + p^2 \pm \vec{\sigma}\cdot\nabla\phi + \phi^2}\,\Psi, \tag{33}$$

where positive (negative) signs apply to (anti)particles. If in this equation we make a nonrelativistic expansion one verifies that the correct non-relativistic potential gets an apparently harmless addition ϕ^2. The usual argument to drop this term is that it is of higher order in the coupling and that a complete treatment of higher order terms will be necessary in order to assess its importance.

The exact FWT procedure that we have followed shows us differently: as long as the pion is treated in a static field and the quenched approximation is applied to the meson an fermion fields, *i.e.* as long as the potential approximation makes sense, the potential to be considered in a non-relativistic procedure should be

$$V_{FWT}(r) = \pm\vec{\sigma}\cdot\nabla\phi + \phi^2, \tag{34}$$

and a consistent expansion in powers of

$$\frac{(p^2 \pm \vec{\sigma}\cdot\nabla\phi + \phi^2)}{M^2} \tag{35}$$

can be performed. This is clearly unnecessary because it is just as simple to solve the auxiliary problem

$$\epsilon_\pm^2\Psi^\pm = (p^2 \pm \vec{\sigma}\cdot\nabla\phi + \phi^2)\Psi^\pm \tag{36}$$

and in a second step compute the exact quenched energies

$$E = \pm\sqrt{M^2 + \epsilon_\pm^2} \tag{37}$$

and the eigenstates will be the same for Eqs. (33) and (36).

The presence of the ϕ^2 term has the immediate consequence that it generates a welcomed repulsive short range core with a point-like potential of the form (32), this term is added *by hand* in so called realistic potentials. This additional terms are introduced both to cure the short distance singularities and the agreement with nucleon scattering. A schematic form of the effective potential is shown in Fig. 3. We note that the range of the repulsive potential ϕ^2 is much larger than the ρ on ω contributions. Thus the ϕ^2 contribution must

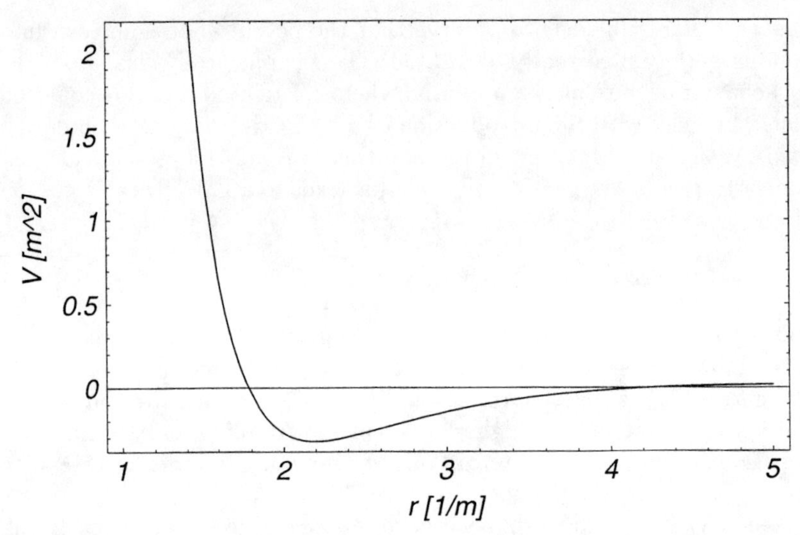

FIGURE 3. Generic form of the full Yukawa potential of a point source, m is the pion mass. The fermion spin direction is oriented to give a minimal energy in the angular direction.

dominate the repulsive core contribution. The long range part of this potential has been studied by D'Olivo *et al.* [14]

The problem that arises in this analysis stems from the fact that the potential that we are discussing is a supersymmetric term, $\{Q^\dagger, Q\} = p^2 \pm \vec{\sigma} \cdot \nabla\phi + \phi^2$. It is positive definite by construction and cannot lead to bound states if it falls, as it must, to zero at large distances. Thus a *generic pseudoscalar potential* cannot bind a nucleon. This has the puzzling consequence that the pseudoscalar nature of the Yukawa meson, at least in the quenched static approximation that has been used, cannot bind the deuteron.

One might wonder if a pseudovector coupling modifies this conclusions. The answer to this is negative at least for the timelike part of the pseudovector interaction which has also a supersymmetric nature. On the other hand the spacelike components of the pseudovector interaction that before the FWT have the form $\vec{\sigma} \cdot \nabla\phi'$ are good candidates to give a binding potential.

CONCLUSION

In this work we have presented several physical examples of Dirac supersymmetric hamiltonians. This structure generalizes the supersymmetric quantum mechanics framework. In a certain limit, often but not necessarily related to the non-relativistic approximation, the usual supersymmetric quantum mechanics is recovered. This formalism provides a mathematical link between a variety of physical phenomena as diverse as superconductivity and relativistic fermions in many external potentials. An explicit example of the later was developed in connection with the deuteron problem. It was shown that a pseudoscalar coupling of the pion to a fermion in an arbitrary pion field that vanishes at long distances cannot bind a fermion. Thus the simplest renormalizable coupling of the pion to nucleons cannot adequately describe the most important property of the deuteron: its bound nature.

ACKNOWLEDGMENT

Work partially supported by Consejo Nacional de Ciencia y Teconolgía, México D. F., México.

REFERENCES

1. D. Itô, K. Mori and E. Carriere, *Nuovo Cimento* **51A,** 1119 (1967).
2. P.A. Cook, *Lett. Nuovo Cimento,* **1,** 419 (1971) .
3. M. Moshinsky and A. Szczepaniak, J. Phys. **A5,** L817 (1989).
4. M. Moreno and A. Zentella, *J. Phys.* **A22,** L821 (1989).
5. M. Moreno, R. Martínez and A. Zentella, *Mod. Phys. Lett.* **A5,** 949 (1990).
6. R. Martínez, M. Moreno and A. Zentella, *Phys. Rev.* **D43,** 2036 (1991).
7. M. Moreno and R. M. Méndez-Moreno, *Generalized Supersymmetric Quantum Mechanics,* in *Proc. of the Workshop on High Energy Phenomenology,* Ed. by M. A. Pérez and R. Huerta, (World Scientific, Singapore,) 365 (1992). *ibid. Relativistic Equations in External Fields* in *in Proc. in honor of M. Moshinsky,* Ed. by A. Frank, T. H. Seligman and B. Wolf, (Springer-Verlag,) 185 (1992).
8. C. A. Blockley and G. Stedman, *Eur. J. Phys.* **6,** 218 (1985), and references therein.
9. M. Moreno *et al. Phys. Lett.* **A 208** 113 (1995).
10. M. Moreno and R. M. Méndez-Moreno, *Sace-time compactification and Dirac-supersymmetry,* to be published in *Proc. of the International Conference on the theory of electron* , Ed. by J. Keller (Cuautitlán, México,) (1996).
11. A. L. Fetter and J. D. Walecka, *Quantum Theory of Many-Particle Systems,* McGraw-Hill, New-York (1971).
12. J. R. Schrieffer, *Theory of Superconductivity,* W. A. Benjamin, Inc., New York, 1964.
13. S. DeBenedetti, *Nuclear Interactions,* John Wiley & Sons, New-York (1964). Mostly chapter VII and references therein
14. J.C.D'Olivo, L. Urrutia and F. Zertuche, *Phys. Rev.* **D32,** 2174 (1985).

Global Analysis of Duality Maps in Quantum Field Theory

A. Restuccia[1]

Universidad Simon Bolivar, Departamento de Fisica, Caracas, Venezuela. e-mail: arestu@usb.ve

Abstract. A global analysis of duality transformations is presented. Global constraints are introduced in order to have the correct structure of the configuration spaces. This global structure is completely determined from the quantum equivalence of dual actions. Applications to S-dual actions and to T duality of string theories and D-branes are briefly discussed. It is shown that a new topological term in the dual open string actions is required.

Duality transformations were introduced by Dirac and extended later on by Montonen and Olive. More recently were used by Seiberg and Witten [1] to relate the weak and strong coupling regime in the analysis of the low energy effective action of the N=2 SUSY SU(2) Yang-Mills. This approach to non-perturbative QFT was then introduced in string theory with spectacular success. It has been shown that the strong coupling regime of one string theory can be mapped to the weak coupling regime of another perturbatively different string theory, giving rise to a possible unification of all string theories in the context of an hypothetical M-theory. In this lecture we analyse the duality transformations from a global point of view. This approach requires the introduction of a more general geometrical structure than the associated to line bundles over a general euclidean base manifold. We describe [2] the general structure of higher order line bundles, and define over them dual maps between theories described locally by p-forms. Duality maps for theories with p-forms have been discussed in [3] and appear naturally in the description of D-brane theories. However all these interesting analysis were local arguments, the global aspects of the configuration space of these local p-forms was never described. We give a close definition of them. The interesting result related to this global structure is that duality between theories of local p-forms and D-p-forms not only imply the quantization of couplings, the known generalized Dirac quantization condition, but also determine completely from a global

[1] Plenary talk at SILAFAE96, Merida,Mexico.

CP400, *First Latin American Symposium on High Energy Physics/ VII Mexican School of Particles and Fields,* edited by D'Olivo/Klein-Kreisler/Mendez

point of view the confuguration space of these local p-forms. These spaces being defined in terms of local p-forms with transitions defined on higher order bundles. In the first part of the lecture we explain the global approach for the Maxwell theory formulated over a general base manifold. We then give the general results concerning the higher order bundles and discuss some applications to D-brane theories. To do so we first consider the duality analysis for open bosonic strings. We prove that the dual open string action requires a new topological term in order to obtain the correct dual boundary conditions.

The action of Maxwell theory over a 4-dim base manifold X, compact euclidean and orientable, is

$$I(F(A)) = \frac{1}{8\pi}\int_X d^4x\sqrt{g}[\frac{4\pi}{e^2}F^{mn}F_{mn} + i\frac{\theta}{4\pi}\frac{1}{2}\epsilon_{mnpq}F^{mn}F^{pq}] \tag{1}$$

where F is the curvature of a 1-form connection A on a U(1) bundle. This action may be rewritten in terms of the complex coupling $\tau = \frac{\theta}{2\pi} + \frac{4i\pi}{e^2}$

$$I_\tau(A) = \frac{i}{8\pi}\int_X d^4x\sqrt{g}[\bar{\tau}F^+_{mn}F^{+mn} - \tau F^-_{mn}F^{-mn}]\ , \tag{2}$$

In order to construct the dual map we introduce an equivalent problem to (1) or (2). We consider the action

$$I(\Omega) = \frac{i}{4\pi}[\bar{\tau}(\Omega^+,\Omega^+) - \tau(\Omega^-,\Omega^-)] \tag{3}$$

where Ω is a global 2-form satisfying the constraints

$$d\Omega = 0 \tag{4}$$

$$\oint_{\Sigma_2^I}\Omega = 2\pi n^I \tag{5}$$

The motivation to introduce the global constraint (5) is that by Weil's theorem constraints (4) and (5) ensure the existence of a unique complex line bundle and a connection on it -not necessarily unique- whose curvature is Ω. (3), (4) and (5) is then an equivalent formulation to (1). It is relevant to determine how many connections giving Ω can be constructed for a given line bundle. In this case it is given by $H^1(X.R)/H^1(X.Z)$. The cohomology classes take into account all canonical gauge equivalent connections, while $H^1(X.Z)$ counts for the equivalence under "large" gauge transformations.

Having determine the exact correspondence between connections over line bundles and global 2-forms constrained by (4) and (5), we have to introduce now the correct Lagrange multiplier. It must have also a precise global structure in order to account for the global constraint (5). It can be shown [2] that it may be expressed in terms of a 1-form connection V over the dual line bundle, provided summation over all dual bundles and all gauge inequivalent connections over every line bundle is performed in the functional integral.

The resulting functional integral in terms of A must also be an integral on all line bundles and all gauge inequivalent connections on every line bundle. The important point to emphasize here is that the configuration space for the 1-form connections A and V is uniquely determined from the duality equivalence. In this sense the requirement to theory (1) of having a dual formulations determines completely the global structure of its configurations space. The quantization of magnetic charge is then only one consequence of this global structure. The resulting action after the introduction of the Lagrange multiplier is given by

$$\mathcal{I}(\Omega, V) = I(\Omega) + \frac{i}{2\pi}\int_X W(V)\wedge\Omega \tag{6}$$

¿From (6) one may integrate on V and regain (3), (4) and (5) and after solving the constraints (4) and (5) one obtains (1). We can also integrate on ω and obtain the dual action in terms of V. The partition function of both quantum equivalent formulations is obtained in the standart way, with the known [3] result

$$\mathcal{Z}(\tau) = \mathcal{N}\tau^{-\frac{1}{2}B_2^-}\bar{\tau}^{-\frac{1}{2}B_2^+}\mathcal{Z}(-\frac{1}{\tau}) \tag{7}$$

where B_k^+ and B_k^- are the dimensions of the spaces of selfdual and antiselfdual k forms.

The generalization of the above construction may be analysed [2] by considering a globally defined p-form over X satisfying

$$\begin{aligned} dL_p &= 0 \\ \oint_{\Sigma_p^I} L_p &= 2\pi n^I. \end{aligned} \tag{8}$$

Let us consider $p = 3$. We take an open covering of $X : \{U_i, i \in I\}$. Without loosing generality we may consider every open set and its intersections to be contractible to a point by a retraction. On U_i we have

$$L_3 = dB_j \tag{9}$$

and on $U_i \cap U_j \neq \phi$

$$\begin{aligned} dB_i &= dB_j \\ B_i &= B_j + d\eta_{ij} \end{aligned} \tag{10}$$

where B_i is a 2-form with transitions given by (10), η_{ij} being a local 1-form defined on $U_i \cap U_j \neq \phi$. On $U_i \cap U_j \cap U_k$ we obtain

$$L_1 \equiv \eta_{ij} + \eta_{jk} + \eta_{ki}$$
$$dL_1 = 0 \tag{11}$$

¿From (8) we have

$$\int_{\Sigma_1} L_1 = 2\pi n, \tag{12}$$

where $\sum_1$ is a close curve on $U_i \cap U_j \cap U_k$. From (11) and (12) we obtain a 1-form L_1 defined over $U_i \cap U_j \cap U_k$ satisfying

$$dL_1 = 0 \tag{13}$$
$$\int_{\Sigma_1} L_1 = 2\pi n \tag{14}$$

which yields an uniform map from

$$U_i \cap U_j \cap U_k \to U(1) \tag{15}$$

The interesting property not present in the previous discussion is that the 1-cochain is now defined as

$$g : (i,j) \to g_{ij}(P, \mathcal{C}) \equiv \exp i \int_{\mathcal{C}} \eta_{ij} \tag{16}$$

where $\mathcal{C}$ is an open curve with end points O (a reference point) and P. g associates to(i,j) a map $g_{ij}(P,\mathcal{C})$ from the path space over $U_i \cap U_j$ to the structure group U(1).

Notice that the 1-form η_{ij} cannot be integrated out to obtain a transition function as in the case of a line bundle. However, we have

$$\delta g_{ijk} = g_{ij} g_{jk} g_{ki} = \exp i \int_O^P L_1 \tag{17}$$

which is precisely the uniform map M previously defined in (15). (16) explicitly shows that the geometrical structure we are dealing with is not that of an usual $U(1)$bundle since the cocycle condition on the intersection of three open sets of the covering is not satisfied. Starting from transitions functions g_{ij} defined on the space of paths over $U_i \cap U_j$, and acting with the coboundary operator δ we obtain the 2-cochain (17) which is properly defined in the sense of *Čech*. We may go further and consider in the intersection of four open sets the action of the coboundary operator δ on 2-cochains.We obtain on $U_i \cap U_j \cap U_k \cap U_l$ a 3-cocycle condition in the sense of *čech*:

$$\delta g_{ijkl} = g_{ijk} g_{ijl}^{-1} g_{ikl} g_{jkl}^{-1} = \mathbb{1} \tag{18}$$

The construction leads then to local p-forms with non-trivial transitions defined by the higher order bundle. Having extended the geometrical structure of line bundles we may then formulate over them duality maps generalizing the electromagnetic duality.

The action for the local p-form A_p defined over open sets of a covering of X and with transitions defined over a higher order bundle is

$$S(A_p) = \frac{1}{2}\int_X F_{p+1} \wedge *F_{p+1} \tag{19}$$

where F_{p+1} is the curvature of A_p.

Let us now consider its dual formulation. We introduce now the globally defined $p+1-form\ L_{p+1}$ satisfaying

$$dL_{p+1} = 0 \tag{20}$$

$$\oint_{\Sigma_p^I} L_{p+1} = \frac{2\pi n^I}{g_p}. \tag{21}$$

with action

$$S = \frac{1}{2}\int_X L_{p+1} \wedge *L_{p+1} \tag{22}$$

(20) and (21) ensure the existence of a bunble of order p+1 and a local p-form on it whose curvature is L_{p+1}.

The off-shell Lagrange problem of the above constrained system may be given by the action

$$S(L_{p+1}, V_{d-p-2}) = S(L_{p+1}) + i\int_X L_{p+1} \wedge W_{d-p-1}(V) \tag{23}$$

where V_{d-p-2} is a local $d-p-2$ form with transitions over a higher order bundle satisfying a $d-p-1$ cocycle condition and with coupling g_{d-p-2}.

Funtional integration on L_{p+1} yields

$$*L_{p+1} = -iW_{d-p-1} \tag{24}$$

where W_{d-p-1} is the curvature of V_{d-p-2}, and the dual action

$$S(V_{d-p-2}) = \frac{1}{2}\int_X W_{d-p-1}(V) \wedge *W_{d-p-1} \tag{25}$$

The quantum equivalence between the two dual actions follows once we integrate on all bundles of order p generalysing the electromagnetic duality previously shown. The quantization of charges arises directly from the global constraints needed for having a globally well defined higher order bundle. The configuration space of the local p-forms A_p and its dual are globally determined, they are defined over higher order bundles with cocycle condition of order $p+1$ which are classified by the integer numbers n^I associated to a basis of integer homology $\sum_I$ on X. For a given bundle of order p+1 the different local antisymmetric fields up to gauge transformations are given by $H^p(X, U(1))/H^p(X, Z)$.

These local antisymmetric fields with non trivial transitions appear naturally in the description of D-branes. For example it has been [4] conjectured that the d=11 5-brane action is given by

$$S = -\frac{1}{2}\int_X d^6\xi\sqrt{-\gamma}[\gamma^{ij}\partial_i x^M \partial_j x^N \eta_{MN} + \frac{1}{2}\gamma^{il}\gamma^{jm}\gamma^{kn}F_{ijk}F_{lmn} - 4] \tag{26}$$

where $F = dA$ is the self dual 3 form field strength of a local 2-form potential A which has to be defined over a bundle of order 3 if non trivial topological effects are expected. It would be interesting to determine completely from a geometrical point of view the moduli space of the self dual potentials over this higher order bundle. This problem is under study.

It is interesting to notice that dealing with D-brane theories, there are two different duality transformations involved. One is obtain by following the approach we have described previously with respect to the local two form A in (26). Because of the selfduality condition the curvature F_3 may be identified to W_3. The other duality arises by following the same approach but with a different interpretation for the global constraint, it is now related to the compactification condition on some of the coordinates on the target space. To show it in some detail we explain the duality transformation on the world-sheet of the string theory, and finally comment on the D=11 supermembrane, D=10 IIA Dirichlet supermembrane duality transformation which involve a compactification of χ^{11}, say, on S^1 and nontrivial line bundles over the world-volume from the other side. That is, both kind of global constraints appears in the duality map.

We discuss now the duality maps between first quantized string theories emphasizing the global constraint in the construction

The string action is

$$S(\chi) = \frac{1}{2\alpha'}\int_{\sum} d^2\xi\sqrt{g}g^{ij}\partial_i\chi^\mu\partial_j\chi_\mu \tag{27}$$

where g^{ij} is the world sheet metric and ξ^i , i=1,2 are the local coordinates of the Riemann surface $\sum$ of a fixed topology. We analyse first the closed string theory with one coordinate χ compactified over S^1. Associated to that coordinate we introduce a contrained 1-form $L = L_i d\xi^i$ satisfying

$$dL = 0 \tag{28}$$

$$\oint_{C^I} L = 2\pi n^I R \tag{29}$$

where C^I denotes a basis of the integer homology of dimension 1 over the worldsheet. Contraint (28) implies L is a closed 1-form, while (29) ensures the compactification over S^1, R is the compactification radius. The solution to (28) and (29) is the string map $\chi(\xi^1,\xi^2)$. We introduce Lagrange multipliers associated to constraints (28) and (29) and obtain the quantum equivalent action

$$S(L,V) = \frac{1}{2\alpha'}\int_\Sigma L\wedge^* L + \frac{i}{\alpha'}\int_\Sigma L\wedge W(V) \tag{30}$$

where $V(\xi^1,\xi^2)$ is the dual map to $\chi(\xi^1,\xi^2)$:

$$W(V) = dV \tag{31}$$

$$\oint_{C^J} = 2\pi m^J R' \tag{32}$$

(32) is uniquely determined to obtain quantum equivalence between $S(L,V)$ and $S(\chi)$.

Following the same arguments as in the S duality approach we obtain, in order to recover (29) from (30), after summation on all n in the functional integral,

$$R' = \frac{\alpha'}{R} \tag{33}$$

That is the dual radius arises directly from the off-shell construction of the dual action. ¿From (30) we obtain the standart on-shell duality relation

$$^*L + iW = 0 \tag{34}$$

From (30) after functional integration on L we obtain

$$S(V) = \frac{1}{2\alpha'}\int_\Sigma W(V)\wedge^* W(V). \tag{35}$$

The duality between $L = d\chi$ and $W = dV$ is resumed in the global constraints (29), (32) and (34). Notice that (32) is uniquely determined from the off-shell construction while (29) implies the compactification of χ on S^1 of radius R. The quantum equivalence between (27) and (35) has been shown for any compact Rienmann surface Σ hence the T-duality is valid order by order in the perturbative expansion of closed string theories. We now discuss the duality of open string theories. The standart open string boundary condition arises from (27) by considering the stationary points of $S(\chi)$. Its variation yields a boundary term

$$\left(\delta\chi\partial_i\chi.n^i\right)\Big|_{\partial\Sigma} \tag{36}$$

It can be anhilated by assuming

$$\partial_i \chi^\mu . n^i = 0 \tag{37}$$

This boundary condition together with the usual string field equation gives an stationary point of (27) with respect to the space of variations $\delta\chi$ which are arbitrary even on the boundary. If instead we consider the space of maps χ restricted by a boundary condition and look for a stationary point of (27) restricted to that space, then

$$\chi^\mu|_{\partial\Sigma} = cte \tag{38}$$

would be also a solution, since then $\delta\chi = 0$. In this case one can have even a mixture of Dirichlet and Newmann conditions on the boundary as an acceptable solution. We will discuss the construction of the dual string action on the first case and show that a topological action term has to be added to (35) in order to have a dual action whose stationary points yields the dual boundary condition to (37). Notice that from the duality relation (34) one obtains

$$n.L = 0 \rightarrow t.W = 0 \tag{39}$$

where t is tangent to the boundary. However from (35) if we consider arbitrary variations on the boundary we get

$$n.W = 0 \tag{40}$$

We thus must modify (35) and consequently (30). We consider

$$\tilde{S}(L, V, Y) = S(L, V) + \frac{i}{\alpha'} \int_\Sigma F(Y) \wedge W(V) \tag{41}$$

where $F = dY$ and Y is a map onto S^1. The new term in the action is a pure topological one. It does not modify the field equations, only contributes to the boundary terms. All the local dependence of $Y(\sigma)$ can be gauged away, only the boundary contribution remains. The boundary terms in the variation of (41) are

$$(\delta V(L + F)|_{\partial\Sigma} = 0 \tag{42}$$

$$(\delta Y W)|_{\partial\Sigma} = 0 \tag{43}$$

which imply V=cte over any connected part of the boundary, and

$$(L + F(Y)).t|_{\partial\Sigma} \tag{44}$$

(44) does not add any restriction to L. It only determines $F(Y)$ on $\partial\Sigma$. After integration on L we obtain

$$\tilde{S}(V, Y) = \frac{1}{2\alpha'} \int_\Sigma W(V) \wedge^* W(V) + \frac{i}{\alpha'} \int_\Sigma F(Y) \wedge W(V) \tag{45}$$

We will consider now that V and χ are maps onto S^1 with compactification radius R' and R respectively. This implies that V=C on the $\sigma = 0$ boundary and $V = C + 2\pi n R'$ in the $\sigma = \pi$ boundary. We will show quantum equivalence between (27) with boundary condition $d\chi.n = 0$ and (45) with boundary condition $dV.t = 0$. Starting from (41) integration on V yields (27) and we are left with the boundary terms

$$\frac{i}{\alpha'} C \int_{\partial\Sigma} [L + F(Y)] + \frac{i}{\alpha'} 2\pi n R' \int_{\sigma=\pi} [L + F(Y)] \tag{46}$$

integration on C and sunmation on n yield

$$\delta\left(\int_{\partial\Sigma} [L + F(Y)]\right) . \sum_m \delta\left(\frac{R'}{\alpha'} \int_{\sigma=\pi} [L + F(Y)] + 2\pi m\right) \tag{47}$$

They imply that

$$\int_{\sigma=\pi} L = [-Y(t_f) + Y(t_i)]_{\sigma=\pi} - 2\pi m R \tag{48}$$

which is the condition that χ is a map from the world sheet to S^1 with radius R. The construction yields

$$R' = \frac{\alpha'}{R}, \tag{49}$$

The global restriction is implemented here through the boundary conditions. We have shown that the dual action to the open string theory requires an additional topological term in the action in order to obtain the correct boundary condition.

The construction of global duality maps required then the implementation of a global constraint which in the case of S-duality ensures the existence of local p-forms with nontrivial transitions on a higher order bundle. In the case of T duality the global constraint is related to the compactification of one or several of the target coordinates. In the duality equivalence of the d=11 supermembrane and the d=10 IIA Dirichlet supermembrane the global constraint for the d=11 supermembrane is the compactification condition while the global constraint for the Dirichlet supermembrane ensures that the local 1-form A is a connection on a nontrivial line bundle over the world-volume. In the construction of duality maps between p-forms and d-p-2 forms the difficult but crucial step in the construction is the converse theorem that ensures that given a globallydefined p+1 form L_{p+1} there exists a bundle of a order p and a local antisymmetric field with non-trivial transitions whose curvature is L_{p+1}. In the case of p=2 there is a very elegant construction of the higher order bundle in terms of Dixmier-Douady Sheaves of groupoids [5].

The main result of the global analysis we have considered is that the existence of a quantum equivalent dual theory completely determines the configuration space of the potentials A_p and of its local dual forms V_{d-p-2}. The global constraint we have introduced are just the correct ones to describe the global structure of the configuration spaces. The geometrical description of these spaces allow an explicit formulation of the D-brane theories in terms of the potentials A_p, a necessary step for the quantization of the these theories.

REFERENCES

1. N. Seiberg and E. Witten, Nucl. Phys. **B426** (1994) 19,**B431** (1994) 484.
2. M. Caicedo, I. Martin and A. Restuccia, hep-th/9701010.
3. E. Witten, hep-th/9505186; E. Verlinde, Nucl.Phys. **B455** (1995) 211 ; Y. Lozano, Phys. Lett. **B364** (1995) 19; J.L.F. Barbon, Nucl. Phys. **B452** (1995) 313; A. Kehagias, hep-th/9508159.
4. P.K. Townsend, hep-th/9512062.
5. Jean-Luc Brylinski, *Loop Spaces, Characteristic Classes and Geometric Quantization*, Birkhäuser, 1993.

Rigged Hilbert Space, Duality, and Cosmology

Mario A. Castagnino

Instituto de Atronomía y Física del Espacio,
Casilla de Correos 67, Sucursal 28, 1428 Buenos Aires, Argentina

I INTRODUCTION

This is a conceptual essay, practically with no equations. The corresponding mathematical development can be found in the references.

We will try to answer the fundamental question: *Why the universe evolution, and those of most of its subsystems, are time-asymmetric if the laws of physics are time-symmetric?* (of course, we exclude from this analysis the law of weak interactions , because usually these interactions are considered too weak to explain the time-asymmetry of the universe [1]).

The usual answer is *coarse-graining:* Quantum-microscopic universe would be time-symmetric while classical-macroscopic universe would become time-asymmetric, due to an average process [2], [3]. Nevertheless every day quantum mechanics is, in fact, time-asymmetric, because even if the evolution equation of the theory (Schroedinger equation) is time-symmetric, the theory incorporates time-asymmetric features too, like *causality, and therefore its experimental outcome: dispersion relations* (and also the collapse or reduction of the wave function that, nevertheless, must be considered as a consequence of the irreversibility, introduced either by coarse-graining or by causality). Thus, as this causal quantum mechanics is already time-asymmetric, the average process *"in order to obtain irreversibility"* seems, at least as a non aesthetic addition, or even more, a misleading concept.

So the aim of this paper is to prove that, from causality and the *global nature of time-asymmetry* we can derive this asymmetry at the fundamental quantum level, defining a growing entropy and the outcome of equilibrium, explaining the decaying processes, etc.

II DUALITY

In a theory with time-symmetric evolution equations, like quantum mechanics, it is obviously impossible to obtain time irreversibility making just

CP400, *First Latin American Symposium on High Energy Physics/ VII Mexican School of Particles and Fields,* edited by D'Olivo/Klein-Kreisler/Mendez

rigorous mathematical manipulations. The maximum we can obtain, at least in some cases, is a couple of *dual structures* such that one is obtained from the other by a time inversion. This *duality property* seems the main feature that a time-symmetric evolution must have in order to incorporate time-asymmetry in a canonical way. E. g., in our formalism, if $\mathcal{L}$ is the Liouville space of density matrices, the dual structure is a couple of two dense subspaces of $\mathcal{L}$: (Φ_-, Φ_+), such that:

$$K : \Phi_- \to \Phi_+ \neq \Phi_- \tag{1}$$

where K is the Wigner time-inversion operator [4]. As we will see Φ_- can be considered as the set of real physical states (such that the entropy will never decrease in these states) while Φ_+ can be considered as the set of time-inverted physical states (such that the entropy never grows in these states) [5], [6], [7]. Many other authors have introduced different dual structures to obtain time asymmetry, as Balescu's projectors [5], [8], Courbage singular measures [9], Lax and Phillips semigroups [10], doublets [11], etc.

Once the duality is defined we can choose one of the two structures and the symmetry breaking so produced creates time-asymmetry. As this choice seems arbitrary we could say that we have introduced time-asymmetry by hand. It is not so if we consider that the arrow of time is a global structure, as we will prove in section V.

III CAUSALITY

We will see how causality motivates the choice of the dual structure of eq. (1). Let us remember the following theorems:

1. Tichmarsh theorem [12], [13].

Let $f(\omega)$ be an square integrable function on the real axis ($f(\omega) \in \mathcal{H}$ the Hilbert space of the theory) . Then the four following statements are equivalent:

1.- The Fourier transform $F(t)$ of $f(\omega)$ vanishes for $t < 0$.

2.- $f(\omega) \in H_+^2$, the Hardy class functions from above [14].

3.-

$$\mathrm{Re} f(\omega) = \frac{P}{\pi} \int_{-\infty}^{+\infty} \frac{\mathrm{Im} f(\omega')}{\omega' - \omega} d\omega' \tag{2}$$

4.-

$$\mathrm{Im} f(\omega) = -\frac{P}{\pi} \int_{-\infty}^{+\infty} \frac{\mathrm{Re} f(\omega')}{\omega' - \omega} d\omega' \tag{3}$$

Eqs. (2) and (3) are known as *unsubstracted dispersion relations* and $f(\omega)$ is called a *casual transform.*

2. Theorem [13].

If $f(\omega)$ is a bounded function $|f(\omega)| < const.$ the following statements are equivalent:

1.- The Fourier transform $F(t)$ of $f(\omega)$ vanishes for $t < 0$.

2.- $f(\omega)$ is the limit of an analytic function $f(z) = f(\omega + i\gamma)$ when $\gamma \to 0$, regular in the half-plane $\gamma > 0$ and $|f(z)| \leq const.$ there.

3 and 4 will correspond to the *substracted dispersion relations* version of eqs. (2) and (3).

$f(\omega)$ is called a *casual factor.*

3. Theorem [13].

If $g_1(\omega)$ is a causal transform and $f(\omega)$ is a causal factor then:

$$g_2(\omega) = f(\omega)g_1(\omega) \tag{4}$$

is a causal transform.

So let $\varphi_1(t)$ be a physical input and $\varphi_2(t')$ be a physical output , so $t > t'$ (e.g.: $\varphi_1(t)$ and $\varphi_2(t)$ are the same wave function at times t and t' in our case) related by the general integral transform:

$$\varphi_2(t) = \frac{1}{\sqrt{2\pi}} \int_{-\infty}^{+\infty} F(t - t')\varphi_1(t')dt' \tag{5}$$

Causality in a heuristic way can be stated as: *no input can occur before the output.* Then:

$$F(t) = 0, \text{ for } t < 0 \tag{6}$$

and as $\varphi_1(t)$, $\varphi_2(t) \in \mathcal{H}$, it can be proved that $F(t)$ is a causal factor. But in this heuristic definition we have used the word *before* which implies the knowledge of the arrow of time not yet defined. So in a more rigorous way we can say that causality can be defined saying that the output at time t depends either on the values of the data in the period $(-\infty, t)$ or in the period (t, ∞) *but not in both periods unless these data would be artificially fine tuned.* In the second case (namely for (t, ∞)) $F(t)$ would be a casual factor described by the theorem 2 and eq. (6). In the first case (namely for $(-\infty, t)$) it will be a casual factor corresponding to the time inverted case, e. g. in (2) we would have $\gamma < 0$ instead of $\gamma > 0$, in eq. (6) it would be $t > 0$, instead of $t < 0$, etc.

Now, always considering the second case, if we make the Fourier transform of eq. (5), we obtain:

$$\widehat{\varphi_2(\omega)} = f(\omega)\widehat{\varphi_1(\omega)} \tag{7}$$

where these factors are the Fourier transform of those of eq. (5) and ω is the energy. Now H_+^2 is dense in $\mathcal{H}$, (the Hilbert space realized by the square integrable functions on the energy spectrum, i. e.: the positive real axis [14]) so we can approximate $\widehat{\varphi_1(\omega)}$ as accurately as we want by a $g_1(\omega) \in H^2_+$. So using the theorem 3 $\widehat{\varphi_2(\omega)}$ can be approximated in a similar way by a function $g_2(\omega) \in H^2_+$. We cannot measure a wave function in an infinite set of points, so physically $\widehat{\varphi_1(\omega)}$ is equivalent to $g_1(\omega)$ and $\widehat{\varphi_2(\omega)}$ is equivalent to $g_2(\omega)$. Thus eq. (7) is equivalent to eq. (4) and we can conclude that causality (in the second case or in an heuristic way) is equivalent to choose the wave functions of the physical states in the space H^2_+. In the more rigorous way (and considering now both cases) we can say that causality is equivalent to choose the physical wave functions *either in* H^2_- *or in* H^2_+ *but never in both subspaces.* As one of this subspaces can be obtained from the other by a time inversion we have obtained a dual structure based on causality.

If instead of the integral evolution equation (5) the evolution equation would be a differential equation, we can repeat the reasoning and the result would be the same. If fact, every differential equation can be represented by an integral equation using a convenient Green kernel [10], [13].

Therefore, for pure states, we can say that a real physical state $|\psi> \in \phi_-$ iif $<\psi|\omega> \in H^2_+$ where $|\omega>$ is the energy eigenvector corresponding to the eigenvalues ω [14].

In the case of mixed states we must consider an eigenbasis $\rho(\omega, \omega') = |\omega><\omega'|$, that, if we use Riesz quantum numbers $\sigma = \frac{1}{2}(\omega + \omega')$, $\nu = \omega - \omega'$, it can be also symbolized by $\beta(\sigma, \nu) = \rho(\omega, \omega') = |\omega><\omega'|$. These states are the eigenvectors of the Liouville operator corresponding to the eigenvalue ν, which takes the role of ω, the eigenvalue of the hamiltonian in the pure state case. Then the physical states would be $\rho \in \Phi_-$ iif $Tr[\rho, \beta(\sigma, \nu)] \in H^2_+(\nu)$ [7]. It is easy to prove that Φ_- and its time inverted states Φ_+ satisfy eq. (1) thereby defining a duality.

IV RIGGED HILBERT SPACES

From the above definitions two rigged Hilbert spaces can be introduced:

$$\Phi_- \subset \mathcal{L} \subset \Phi_-^\times, \; \Phi_+ \subset \mathcal{L} \subset \Phi_+^\times \tag{8}$$

Using these spaces, their bases, and the corresponding spectral decomposition, a convenient conditional entropy [15] can be defined, such that it never decreases in the states of Φ_-, and eventually it goes to zero when the state reach the equilibrium at $t \to \infty$, and it never grows for the states of Φ_+, and eventually this entropy goes to zero when $t \to -\infty$ [7]. Then Φ_- can be considered as the space of physically admissible states.

In this way we have obtained a duality but the arrow of time is not yet defined. In fact, Φ_- and Φ_+ are only conventionally different, because with a sign change in the definition of the entropy Φ_+ could be considered as the space of physical states. This is not surprising since it is the essential property of the dual structures. They must only be conventionally different, in such a way that anyone of them can be chosen as the space of physical states. Anyhow, when the choice is made the time-symmetry is broken and we have an irreversible quantum mechanics. Which one must be chosen? And if we choose one or the other, are we putting time-asymmetry by hand? The solution of these problems is that the real nature of the arrow of time is global, so we are forced to consider cosmological models.

V COSMOLOGY

The arrow of time cannot be a local concept. If not we could conceive two isolated laboratories with different arrows of time, and this fact has never being observed [16]. So, in order to study this problem we must adopt a cosmological model and the simplest of all is the Branch System of Reichenbach [17], [18]. In this model it is considered that every irreversible process begins in an unstable state originated, not in a very unlikely fluctuation, but in an unstable state created by the energy coming from other irreversible processes. E. g., the famous Gibbs ink drop in the glass of water, was originated in an ink factory, where unstable coal was burned in an oven to extract energy. Coal was originated in geological ages using the energy of the light coming from the sun, where unstable H was burned, and the energy necessary to create H comes from the unstable initial state of the universe, the origin and source of energy of the whole branch system. We can represent this branch system, at the classical level by fig. 1.

To go to the quantum level let us consider a usual scattering process (fig. 2) and let us cut this process at time $t = 0$ [14], into a creation of unstable states process (fig. 3) and a decaying of unstable states process (fig. 4). This last process corresponds to states in the space Φ_- that naturally decay into an equilibrium state at $t \to \infty$ with a growing of entropy. On the contrary the created states correspond to space Φ_+ . Really this space is not realized in the real physical world as such, because before $t = 0$ the system is not just the scattering system, but a more complete one, that includes the acceleration apparatus and the source of energy (like the one of the dotted box "B" of fig. 5). Thus really the evolution of the universe can be symbolized as a sequence of states in local Φ_- spaces, as shown if fig 5, all of them coordinated because their energy comes from the unique unstable initial state, namely the cut box in the far left of fig. 5. The whole process can be described using just a global space Φ_-^G, (as in the simple cosmological models of ref. [19]). Fig. 5 can be considered the quantum image of Reichenbach branch system. It shows that:

1.- How and why the local arrows of time are coordinated in a global one.

2.-As entropy grows in fig 4, it also grows in every isolated laboratory of the universe endowed with a source of energy, like the dotted box "A" of fig. 5. The second law of thermodynamic thus follows.

3.- As fig. 5 corresponds to Φ_-^G its specular image corresponds to the time inverted space Φ_+^G . But physics choosing Φ_-^G is *identical* to physics choosing Φ_+^G, because, as there is nothing exterior to the universe, nobody can tell the difference. Then choosing either Φ_-^G or Φ_+^G we would obtain the same time-asymmetric physics with a growing entropy when we go from the initial unstable state in, what we will call, the "past", to the equilibrium final state in, what we will call, the "future". A realistic model of the universe is thus obtained.

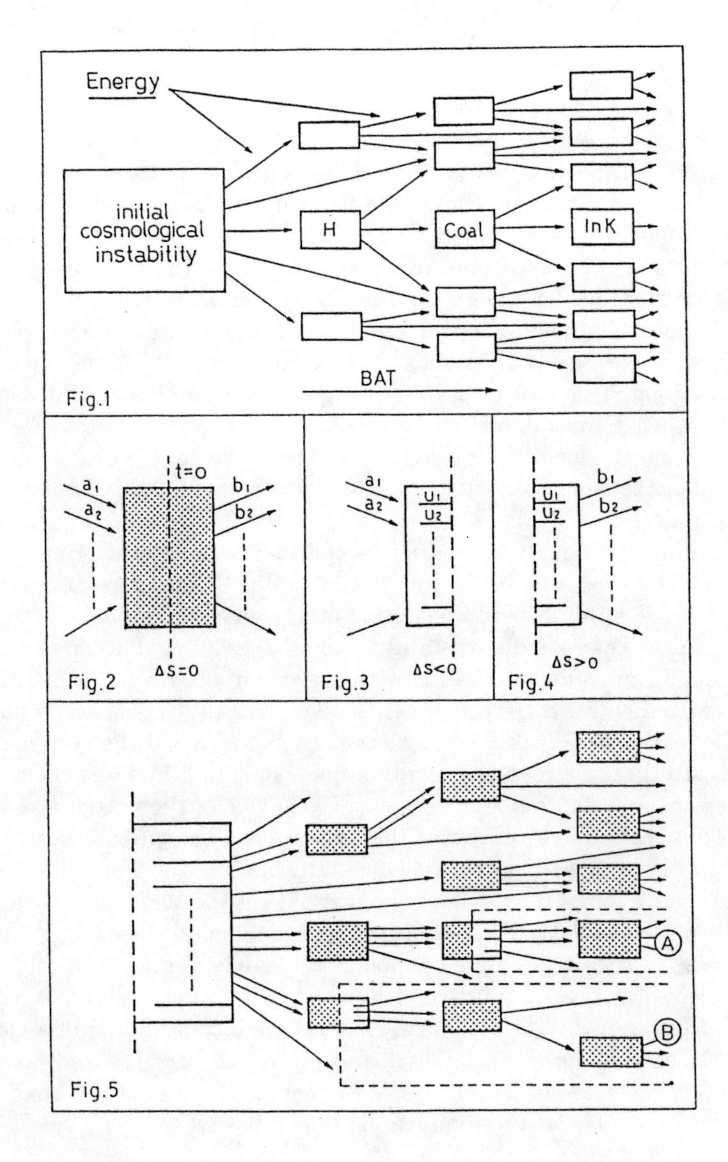

VI CONCLUSION

Based on first principles: causality and the global nature of the arrow of time we have explained the time-asymmetry of the universe, without the use of non-essential features as coarse graining, lost of information, etc.

REFERENCES

1. Sach R. G., *The physics of time reversal,* Univ. Chicago Press, Chicago, 1987.
2. Zwanzig R. W., Chem Phys., **33,** 1338, 1960.
3. Zurek W. H., Physic Today, **44,** (10), 36, 1991.
4. Messiah A., *Quantum mechanics,* North-Holland, Amsterdam, 1962.
5. Castagnino M., Gaioli F., Gunzig E., Found. Cos. Phys., **16,** 221, 1996.
6. Castagnino M., Gunzig E., *A landscape on time asymmetry,* Int. Jour. Theo Phys, in press, 1996.
7. Castagnino M., Laura R., *A minimal irreversible quantum mechanics: the axiomatic formalism,* submitted to Phys. Rev. A, 1996.
8. Balescu R., *Equilibrium and non equilibrium statistical mechanics,* J. Willey & Sons, New York, 1963.
9. Courbage M., Physica A, **122,** 459, 1983.
10. Lax P. D., Phillips R. S., *Scattering theory,* Acad. Press, New York, 1967.
11. Castagnino M., Domenech G., Levinas M. L., Umerez N., Jour. Math. Phys., **37,** 2107, 1996.
12. Tichmarsh E. C., *Theory of Fourier integrals,* Clarendon Press, Oxford, 1948.
13. Roman P., *Advanced quantum mechanics,* Addison-Wesley, Reading, 1965.
14. Bohm A., *Quantum mechanics: foundations and applications,* Springer Verlag, Berlin, 1986.
15. Mackey M. C., Rev. Mod. Phys., **61,** 981, 1989.
16. Castagnino M.,*The global nature of the arrow of time and the Bohm-Reichenbach diagram,* Proceeding of G21, Goslar, 1996.
17. Reichenbach H., *The direction of time,* Univ. of California Press, 1956.
18. Davies P., *Stirring up trouble,* in Physical origin of time-asymmetry, Halliwell J. J. et al. ed., Cambridge Univ. Press, Cambridge, 1994.
19. Castagnino M., Gunzig E., Nardone P., Prigogine I., Tasaki S., *Quantum cosmology and large Poincaré systems,* in Quantum, Chaos, and Cosmology, Namiki M. ed., AIP book division, in press, 1996.
    Castagnino M., Gunzig E., Lombardo F., Gen. Rel. and Grav.,**27,** 257, 1995.
    Castagnino M., Lombardo F., Gen. Rel. and Grav., **28,** 263, 1996.
    Castagnino M., Gaioli F., Sforza D., Gen. Rel. and Grav. **28,** 1129, 1966.

Heterotic-Type II Duality

Anamaría Font

Departamento de Física, Facultad de Ciencias
Universidad Central de Venezuela
A.P. 20513, Caracas 1020-A
Venezuela

Abstract. We discuss how transitions in the space of heterotic $K3 \times T^2$ compactifications are mapped by duality into transitions in the space of Type II compactifications on Calabi-Yau manifolds.

INTRODUCTION

Recent results indicate that seemingly different string theories are actually related by strong-weak coupling duality transformations. In this paper we explore four-dimensional $N=2$ string vacua in connection with the conjectured duality [1,2] between compactifications of the $E_8 \times E_8$ heterotic string on $K3 \times T^2$ and the type IIA string compactified on Calabi-Yau (CY) manifolds that are $K3$ fibrations [3,4]. Evidence in favor of this duality is based for instance on the identification of dual pairs with same spectrum [1,5] and same gauge symmetries [6].

The moduli spaces of CY manifolds form a web in which continuous transitions between different manifolds occur due to the shrinking to zero of certain homology two cycles and three cycles as the parameters of the manifold are varied. Indeed, recent studies [7] indicate that all CY manifolds are connected by processes of this type. These studies exploit the relation, pointed out by Batyrev [8], between CY manifolds and reflexive polyhedra, together with the observation that connectivity of manifolds corresponds to nesting of polyhedra. Duality requires that a web structure exist also on the heterotic side. Indeed, the space of heterotic vacua also forms a web in which different models are connected along branches parametrized by vacuum expectation values of scalars in vector and hypermultiplets. Thus, in virtue of duality there should be a dictionary that translates between the language of reflexive polyhedra and that of heterotic dynamics. A first step towards finding such dictionary was taken in ref. [5] where it was noticed that un-Higgsing of $SU(r)$ groups in certain heterotic models matched into a chain of known $K3$ fibrations.

CP400, *First Latin American Symposium on High Energy Physics/ VII Mexican School of Particles and Fields,* edited by D'Olivo/Klein-Kreisler/Mendez

Motivated by the observations of [5], we point out that corresponding to a Higgsing chain of heterotic models, there is a chain of CY manifolds with a simple structure revealed by their description in terms of reflexive polyhedra [9]. These manifolds are $K3$ fibrations. Indeed the four-dimensional polyhedron contains the polyhedron of the $K3$ in a simple way and it is this nesting of polyhedra that motivates much of our analysis. The heterotic chains of [5] end on models with maximal symmetry breaking that can be associated to type II compactifications on certain $K3$ fibrations. Starting at these 'irreducible models', we then study the type II description of symmetry restoration of different group factors. We also analyze type II processes that correspond to non-perturbative heterotic transitions in which the number of tensor multiplets jumps by one and an instanton shrinks to zero.

HETEROTIC WEB

The starting point is a $D=6$, $N=1$ heterotic $E_8 \times E_8$ compactification on $K3$ with $SU(2)$ bundles with instanton numbers (d_1, d_2) such that $d_1+d_2 = 24$. When both $d_i \geq 4$, the resulting group is the commutant of the instantons which is $E_7 \times E_7$ with massless hypermultiplets transforming as **56**s and/or singlets under each E_7. Using the index theorem we find the hypermultiplet spectrum

$$\frac{1}{2}(d_1 - 4)(\mathbf{56}, \mathbf{1}) + \frac{1}{2}(d_2 - 4)(\mathbf{1}, \mathbf{56}) + 62(\mathbf{1}, \mathbf{1}) \ . \tag{1}$$

When $d_1 = 24$, $d_2 = 0$, the gauge group is $E_7 \times E_8$ and the massless hypermultiplets include

$$10(\mathbf{56}, \mathbf{1}) + 65(\mathbf{1}, \mathbf{1}) \ . \tag{2}$$

Without loss of generality we can take $12 \leq d_1 \leq 20$, unless $d_1 = 24$. We also find it convenient to define $k \equiv (d_1 - 12)/2$. Since the **56** of E_7 is a pseudoreal representation, the d_i can be odd and k can be half-integer.

An initial heterotic model can be deformed by vevs of hypermultiplets thereby breaking the gauge group that in general is of the form $G = G_1 \times G_2$, with G_1 and G_2 coming from the first and second E_8's. Since $d_1 \geq 12$, the number of $(\mathbf{56}, \mathbf{1})$'s is such that the first E_7 can be completely broken. In particular, it can be broken through the chain

$$E_7 \rightarrow E_6 \rightarrow SO(10) \rightarrow SU(5) \rightarrow SU(4) \rightarrow SU(3) \rightarrow SU(2) \rightarrow SU(1) \ , \tag{3}$$

where $SU(1)$ denotes the trivial group consisting of the identity only. On the other hand, the group arising from the second E_8 can only be broken to some terminal group $G_2^{(0)}$ that depends on k. For example, when $k = 5/2$ the breaking can proceed to $G_2^{(0)}(5/2) = F_4$. The cases $k = 9/2, 5, 11/2$, that would require $d_2 < 4$, can be considered if we modify the $N = 1,\ D = 6$ construction so as to include effects seen in the compactification of M-theory [10] and F-theory [11]. More precisely we consider vacua with n_T tensor multiplets so that the condition $d_1 + d_2 = 24$ is replaced by

$$d_1 + d_2 + n_T - 1 = 24 \ . \tag{4}$$

We immediately see, for instance, that $n_T = 3$ permits $d_1 = 22,\ d_2 = 0$, so that $k = 5$. Moreover, the initial gauge group $E_7 \times E_8$ can be completely broken to a matter-free E_8. Results for all values of k are summarized in Table 1.

TABLE 1. Terminal groups and the numbers of tensor multiplets for the different values of k. The entry E_7^- corresponding to $k = \frac{7}{2}$ denotes E_7 with a half-multiplet of **56**.

k	0	$\frac{1}{2}$	1	$\frac{3}{2}$	2	$\frac{5}{2}$	3	$\frac{7}{2}$	4	$\frac{9}{2}$	5	$\frac{11}{2}$	6
$G_2^{(0)}$	SU_1	SU_1	SU_1	SU_3	SO_8	F_4	E_6	E_7^-	E_7	E_8	E_8	E_8	E_8
n_T	1	1	1	1	1	1	1	1	1	4	3	2	1

The hypermultiplet content at each stage of breaking can be derived using group theory. For instance, if the first E_7 is broken to $SU(2)$, the number of doublets turns out to be $n_2 = 12k + 16$. It is also useful to consider the restrictions on the spectrum imposed by cancellation of anomalies. In particular, there exists a condition

$$n_H - n_V = 273 - 29n_T \ , \tag{5}$$

where n_H and n_V denote the number of hyper and vector multiplets. Now, if the original group G is broken to $SU(2) \times G_2^{(0)}(k)$, using (5) and the number of doublets previously determined, we find the number of singlets

$$n_1 = 244 + \dim G_2^{(0)}(k) - 29n_T(k) - 24k \ , \tag{6}$$

assuming that $G_2^{(0)}(k)$ is free of charged matter. It is straightforward to repeat this sort of analysis for other breaking patterns.

Up to now we have focused on six-dimensional models. Upon further compactification on T^2, the $N=1$, $D=6$ hyper and vector multiplets of G give rise to $N=2$, $D=4$ hyper and vector multiplets also of G, in numbers n_H and n_V that still must fulfill (5). Each tensor multiplet produces an extra $U(1)$ vector multiplet and for generic 2-torus shape, there also appear two extra $U(1)$ vector multiplets. Another new feature is the existence of a Coulomb branch parametrized by expectation values of the adjoint scalars in the $N=2$ vector multiplets. At a generic point, the gauge group is $U(1)^{\mathrm{rank}\, G + n_T + 2}$, excluding the graviphoton, and the massless hypermultiplets include those n_{sing}^G fields originally neutral under G. In general, considering different breaking patterns $G_1 \to H$, while G_2 remains broken at the terminal $G_2^{(0)}(k)$, gives

$$n_{sing}^G = 273 + \dim G_2^{(0)}(k) + \dim H - 29 n_T(k) - a_H - b_H k \ . \tag{7}$$

The coefficients a_H and b_H encode the number of H-charged fields that disappear in the Coulomb phase. Our basic strategy is to start with $H = E_7$ and Higgs along the sequence (3). In this way, for each k, we generate a chain of heterotic models ending in an 'irreducible model' with maximal symmetry breaking.

TYPE II WEB

Compactification of the type IIA string on a CY manifold produces an $N = 2$, $D = 4$ theory with group $U(1)^{h_{11}}$, excluding the graviphoton, and $(h_{12} + 1)$ neutral hypermultiplets [12]. Here (h_{11}, h_{12}) are the Hodge numbers that characterize the manifold. Hence, such a type IIA model is potentially dual to a given $N=2$, $D=4$ heterotic model with group $U(1)^{\mathrm{rank}\,G + n_T + 2}$ (in the Coulomb phase) and n_{sing}^G neutral hypermultiplets provided that $h_{11} = \mathrm{rank}\,G + n_T + 2$ and $h_{12} = n_{sing}^G - 1$. In particular if the heterotic model has group $H \times G_2^{(0)}(k)$ before transition to the Coulomb phase, it is necessary that

$$\begin{aligned} h_{11} &= \mathrm{rank}\, G_2^{(0)}(k) + \mathrm{rank}\, H + n_T(k) + 2 \\ h_{21} &= 272 + \dim G_2^{(0)}(k) + \dim H - 29 n_T(k) - a_H - b_H k \ . \end{aligned} \tag{8}$$

Duality also requires that the CY manifold be a $K3$ fibration [3,4]. This property guarantees that one of the type IIA vector multiplets can be identified with the heterotic dilaton vector multiplet.

For $k = 6, 4, 3, 2, 1$ and maximal symmetry breaking, *i.e.* $H = SU(1)$, eqs. (8) yield Hodge numbers that match those of known $K3$ fibrations given by hypersurfaces $\mathcal{M}_0(k) = \mathbb{P}_4^{(1,1,2k,4k+4,6k+6)}[12k+12]$ [1,5,11]. These CY manifolds then correspond to irreducible heterotic models. Remarkably enough, sequentially un-Higgsing $SU(r)$ factors ($r = 2, \cdots, 3$) leads to Hodge numbers that also match into those of $K3$ fibrations realized as hypersurfaces in $\mathbb{P}_4$ [5]. Hence, there is also a chain of type IIA models corresponding to heterotic Higgsing along the last steps of the sequence (3). For $k = 0$ the manifold $\mathcal{M}_0(k)$ makes no sense since the weight of the third coordinate would be zero, whereas for $k = 1/2$ we would have $\mathbb{P}_4^{(1,1,1,6,9)}[18]$ which is not a $K3$-fibration. The cases $k = 0$ and $k = 1/2$ are however covered by the construction of [11] which realizes the lowest member of each chain as an elliptic fibration X_{2k} over the Hirzebruch surface $\mathbb{F}_{2k}$. For other half-integer values of k, $\mathcal{M}_0(k)$ is well defined and coincides with X_{2k}.

We now set out to analyze the X_{2k} manifolds following Batyrev's toric approach. (For a concise summary of Batyrev's construction in a form accesible to physicists see, for example, [13].) To a CY manifold (of any dimension) defined as a hypersurface in a weighted projective space one can associate its Newton polyhedron, which we denote by Δ. The Newton polyhedron is often (perhaps always) reflexive and when it is we may define the dual or polar polyhedron which we denote by ∇. In particular, we denote the dual polyhedron of X_{2k} by ${}^4\nabla^{k,SU(1)}$ since it corresponds to $H = SU(1)$. The ${}^4\nabla^{k,SU(1)}$ have the same properties for all k. Apart from the two points $(-1,0,2,3)$ and $(1,2k,2,3)$, the points lie in the plane $x_1 = 0$ forming the polyhedron, ${}^3\nabla$, of the $K3$. The polyhedron of the $K3$ is itself divided into a top and a bottom by the polyhedron ${}^2\nabla$ which is the triangle shown in Figure 1. This ${}^2\nabla$ is the dual polyhedron of the torus $\mathbb{P}_2{}^{(1,2,3)}[6]$, a sign of the elliptic fibration structure.

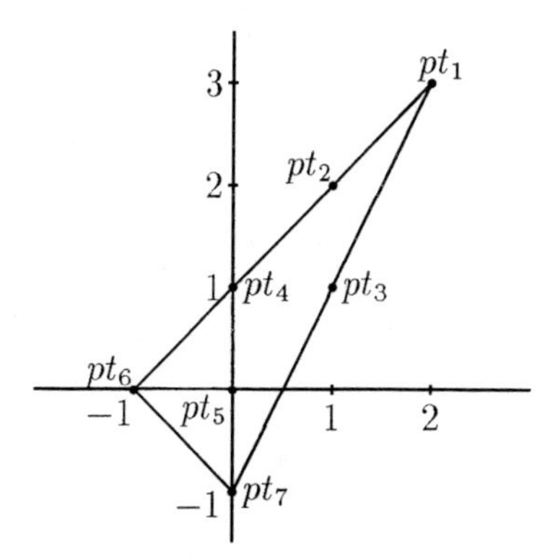

FIGURE 1. The polyhedron ${}^2\nabla$.

Our next task is to determine ${}^4\nabla^{k,H}$ for the groups H in the chain (3). We have found that the process of un-Higgsing in the first group factor correponds to modifying the bottom of the $K3$ sub-polyhedron. In fact, we can write

$$ {}^4\nabla^{k,H} = {}^3\nabla^{k,H} \cup \{(-1,0,2,3),\ (1,2k,2,3)\} \,, \tag{9} $$

and moreover,

$$ {}^3\nabla^{k,H} = \nabla^H_{\text{bot}} \cup \nabla^k_{\text{top}} \,, \tag{10} $$

where ∇^k_{top} depends only on k while ∇^H_{bot} depends only on the group H that is perturbatively restored in the heterotic side.

We can describe the tops and bottoms of ${}^3\nabla^{k,H}$ quite simply. If we denote by $\mathcal{T}^k$ the tetrahedron with base ${}^2\nabla$ and top vertex $(0,k,2,3)$, the top polyhedron may be specified as follows:

$$\nabla^0_{\rm top} = \nabla^{\frac{1}{2}}_{\rm top} = \nabla^1_{\rm top} = \mathcal{T}^1 \quad , \quad \nabla^{\frac{3}{2}}_{\rm top} = \mathcal{T}^1 \cup (0,1,1,2)$$
$$\nabla^2_{\rm top} = \mathcal{T}^2 \quad , \quad \nabla^{\frac{5}{2}}_{\rm top} = \nabla^3_{\rm top} = \mathcal{T}^3$$
$$\nabla^{\frac{7}{2}}_{\rm top} = \nabla^4_{\rm top} = \mathcal{T}^4 \cup \{(0,1,0,0),\ (0,2,0,1),\ (0,3,1,2)\}$$
$$\nabla^{\frac{9}{2}}_{\rm top} = \nabla^5_{\rm top} = \nabla^{\frac{11}{2}}_{\rm top} = \nabla^6_{\rm top} = \mathcal{T}^6 \ . \tag{11}$$

We can also easily describe $\nabla^H_{\rm bot}$ for $H = SU(r)$, $r = 1,2,3$. Consider first the polyhedron ${}^2\nabla$ shown in Figure 1 and let pt'_r be the point of the lattice that is directly below the pt_r and more, generally, let $pt_r^{(j)}$ be the point that is j levels below. We find that

$$\nabla^{SU(r)}_{\rm bot} = {}^2\nabla \cup \bigcup_{j=1}^{r} pt'_j \ . \tag{12}$$

Moreover, we have verified that including the point pt'_4 leads to a reflexive polyhedron that gives the expected Hodge numbers for an enhanced $SU(4)$. Further adding pt'_5 corresponds to an enhanced $SU(5)$. Hence, eq. (12) is actually valid for $r = 1, \cdots, 5$. Considering the possibility of adding combinations of the points $pt_i^{(j)}$ in all possible ways consistent with reflexivity leads to the polyhedra for other enhanced groups. In particular we find

$$\nabla^{SO(10)}_{\rm inf} = {}^2\nabla \cup \{pt''_1, pt''_2, pt'_3, pt'_4, pt'_5\}$$
$$\nabla^{E_6}_{\rm inf} = {}^2\nabla \cup \{pt'''_1, pt''_2, pt''_3, pt'_4, pt'_5\}$$
$$\nabla^{E_7}_{\rm inf} = {}^2\nabla \cup \{pt''''_1, pt'''_2, pt''_3, pt''_4, pt'_5\} \ . \tag{13}$$

The Hodge numbers computed from the polyhedra [9] coincide with those obtained from eqs. (8). All groups found by sequential Higgsing do appear and, as a matter of consistency, we see the inclusions

$$E_7 \supset E_6 \supset SO(10) \supset SU(5) \supset SU(4) \supset SU(3) \supset SU(2) \supset SU(1) \ , \tag{14}$$

as inclusions of the respective polyhedra.

Notice then that symmetry restoration in a terminal heterotic model corresponds to adding points in the ${}^3\nabla^k$ piece of ${}^4\nabla^k$. In this process the $K3$ fibers are modified, in agreement with the fact that the non-Abelian structure of the group that is perturbatively visible is related in turn to the structure of the $K3$ fibers [6]. The arguments of [4,6] also suggest to look for non-perturbative dynamics by including the effect of degenerate fibers, *i.e.* by modifying the $\mathbb{P}_1$ part of the fibration. Some of this can be done torically and amounts to adding points outside the plane of the $K3$. In fact, we have noticed that adding points $(1, 2k-j, 2, 3)$, $j = 1, \cdots, 2k+2$, to ${}^4\nabla^{k,SU(1)}$ is always consistent with reflexivity. Moreover, the Hodge numbers in this new sequence of polyhedra have a rather interesting pattern characterized by

$$\Delta h_{12} = -29 \quad ; \quad \Delta h_{11} = 1 \ . \tag{15}$$

Hence, as implied by eq. (5), in the new transitions $n_T \to n_T + 1$. Since, eq. (4) requires $(d_1 + d_2) \to (d_1 + d_2 - 1)$, this corresponds to shrinking of an instanton [10]. The result (15) is consistent with $d_1 \to d_1 - 1$ while d_2 is kept fixed. The first E_7 can be completely broken for $d_1 \geq 10$ so that to arrive at $SU(1)$ the original $d_1 = 12 + 2k$ can only be decreased in one unit $2 + 2k$ times as we have observed.

Our results can be extended in several directions. For instance, among the reflexive polyhedra constructed by modifying the bottoms of the terminal polyhedron ${}^4\nabla^{k,SU(1)}$ we find more examples of non-perturbative processes. In particular, we find [9] a description of the space with $(h_{11}, h_{12}) = (43, 43)$ associated to the heterotic compactification with $(d_1, d_2) = (0, 0)$ and $n_T = 25$. Also, adding points to the tops of ${}^3\nabla^k$ we can follow symmetry restoration in the gauge factor G_2 arising from the second E_8. Although we have emphasized the sequence (3), it is also possible to determine the polyhedra associated to other groups, including non-simply laced groups. In fact, the identification of the polyhedra has been verified analyzing the structure of singularities of the CY manifolds [14].

REFERENCES

1. S. Kachru and C. Vafa, Nucl. Phys. B450 (1995) 69.
2. S. Ferrara, J. Harvey, A. Strominger and C. Vafa, Phys. Lett. B361 (1995) 59.
3. A. Klemm, W. Lerche and P. Mayr, Phys. Lett. B357 (1995) 313.
4. P. S. Aspinwall and J. Louis, Phys. Lett. B369 (1996) 233.
5. G. Aldazabal, A. Font, L. E. Ibáñez and F. Quevedo, Nucl. Phys. B461 (1996) 85.
6. P. S. Aspinwall, Phys. Lett. B371 (1996) 231.
7. A. C. Avram, P. Candelas, D. Jančić and M. Mandelberg, Nucl. Phys. B465 (1996) 458; T.-M. Chiang, B. R. Greene, M. Gross and Y. Kanter, hep-th/9511204.
8. V. Batyrev, Duke Math. Journ. 69 (1993) 349.
9. P. Candelas and A. Font, hep-th/9603170.
10. N. Seiberg and E. Witten, Nucl. Phys. B471 (1996) 121.
11. D.R. Morrison and C. Vafa, Nucl. Phys. B473 (1996) 74, Nucl. Phys. B476 (1996) 437.
12. N. Seiberg, Nucl. Phys. B303 (1988) 286; M. Bodner, A. C. Cadavid and S. Ferrara, Class. Quantum Grav. 8 (1991) 789.
13. P. Candelas, X. de la Ossa and S. Katz, Nucl. Phys. B450 (1995) 267.
14. M. Bershadsky, K. Intriligator, S. Kachru, D.R. Morrison, V. Sadov and C. Vafa Nucl. Phys. B481 (1996) 215.

Bosonization in More than Two Spacetime Dimensions

C. D. Fosco[1]

Centro Atómico Bariloche
8400 Bariloche
Argentina

Abstract. A low-momentum path-integral bosonization method for more than two spacetime dimensions is extended in two different directions: We first consider the $2+1$ dimensional case. By keeping the full momentum-dependence of the one-loop vacuum polarization tensor, we can discuss both the massive and massless fermion cases on an equal footing. Results for massless fermions coincide with the ones of a different, (operatorial) approach to the problem of bosonization for a massless fermionic field. Secondly, the bosonization of a massless fermionic field coupled to both vector and axial-vector external sources in $3+1$ dimensions is developed. The resulting bosonic theory contains two antisymmetric tensor fields with non-local Kalb-Ramond-like terms plus interactions. Exact bosonization rules that take the axial anomaly into account are derived.

INTRODUCTION

Any physical system is suitable of many different mathematical descriptions. Different choices of variables are equivalent in the sense that they describe the same system. An extreme manifestation of this appears in some models which can be described in terms of either fermionic or bosonic variables. The equivalence between these two formulations is made explicit by the bosonization rules, that map fermionic into bosonic variables. There has been some progress in the program of extending the bosonization procedure to theories in more than two dimensions [1]- [10]. We present here the results obtained by applying the method of refs. [1], [2], [3], [6], [7], [8] (as presented in [8]), to two different problems: Abelian case in $2+1$ dimensions [9] [2], and Abelian case in $3+1$ dimensions with vector and axial-vector sources [11] [3].

---

1) Member of Conicet, Argentina.
2) In collaboration with D.G. Barci and L.D. Oxman.
3) In collaboration with F.A. Schaposnik.

CP400, *First Latin American Symposium on High Energy Physics/ VII Mexican School of Particles and Fields*, edited by D'Olivo/Klein-Kreisler/Mendez

MASSIVE AND MASSLESS FERMIONS IN $2+1$ DIMENSIONS

In Ref. [12], by applying canonical quantization methods, a bosonic non-local and gauge-invariant action for an Abelian vector field was constructed, the approximate bosonization rules (in Euclidean spacetime) being

$$\bar{\psi}\,\partial\!\!\!/\psi \leftrightarrow \frac{1}{4}\,F_{\mu\nu}(-\partial^2)^{-1/2}F_{\mu\nu} + \frac{i}{2}\,\theta\,\epsilon_{\mu\nu\lambda}A_\mu\partial_\nu A_\lambda + nqt$$

$$\bar{\psi}\gamma_\mu\psi \leftrightarrow \beta\,\epsilon_{\mu\nu\lambda}\partial_\nu A_\lambda - \beta\,\theta\,(-\partial^2)^{-1/2}\partial_\nu F_{\mu\nu} + nqt \qquad (1)$$

where ψ is a two-component Dirac spinor, A_μ is a $U(1)$ gauge field, and nqt means non-quadratic terms in A_μ. The parameter θ is regularization-dependent. This ambiguity, which manifests itself in the bosonization rules, already exists in the fermionic description [14].

In Ref. [6], functional methods were applied to derive low-momentum bosonization formulae for the free massive Thirring model, and in [7], the Abelian and non-Abelian cases in any dimension $d \geq 2$ were considered. Both the free massive Dirac field and the Thirring model (in 3 dimensions) are mapped to Chern-Simons theories by using the approximate bosonization rules

$$\bar{\psi}(\partial\!\!\!/ + m)\psi \leftrightarrow \pm\frac{i}{2}\epsilon_{\mu\nu\lambda}A_\mu\partial_\nu A_\lambda\,,\ \bar{\psi}\gamma_\mu\psi \leftrightarrow \pm i\sqrt{\frac{1}{4\pi}}\,\epsilon_{\mu\nu\lambda}\partial_\nu A_\lambda\,, \qquad (2)$$

obtained to leading order in $\frac{1}{m}$.

The question presents itself about how to improve the $\frac{1}{m}$ expansion in order to include cases where the derivative expansion is no longer valid, as it is indeed the case for massless fermions. We overcome this kind of limitation by including the full momentum dependence in the one-loop quadratic part of the effective action. Whence the results will also be valid for the massless case, without spoiling the proper low-momentum features. This introduces a non-locality in the bosonized action, a property shared with the approach of [12].

We will show that, by keeping the full momentum dependence of the vacuum polarization tensor in the approach of [8], one can reproduce [12] if the mass of the Dirac field is set equal to zero. The result of [8] will survive in the low-momentum (or $m \to \infty$) limit.

We start by constructing a bosonized version of the generating functional of current correlation functions in the case of a free fermionic field in three dimensions, reviewing the procedure followed in [8]. This method builds upon the functional representation of the fermionic generating functional

$$Z(s) = \int [d\psi][d\bar{\psi}]\ \exp\left[-\int d^3x\ \bar{\psi}(\partial\!\!\!/ + i\ s\!\!\!/ + m)\,\psi\right] \qquad (3)$$

by performing the change of variables

$$\psi(x) \to e^{i\alpha(x)}\,\psi(x) \quad , \quad \bar{\psi}(x) \to e^{-i\alpha(x)}\,\bar{\psi}(x) \tag{4}$$

to obtain

$$Z(s) = \int [d\psi][d\bar{\psi}]\, \exp\left[-\int d^3x\, \bar{\psi}(\not\partial + i(\not s + \not\partial\alpha) + m)\,\psi\right] . \tag{5}$$

Defining $b_\mu = \partial_\mu \alpha$ ($\Rightarrow F_{\mu\nu}(b) = \partial_\mu b_\nu - \partial_\nu b_\mu = 0$), as $Z(s)$ does not depend on b_μ, the pure-gauge field b_μ can be integrated with an arbitrary (non-singular) weight functional $f(b)$, yielding (up to a normalization factor)

$$\begin{aligned} Z(s) = \int [db][d\psi][d\bar{\psi}] f(b) \delta(F_{\mu\nu}(b))\, \exp - \int d^3x \bar{\psi}(\not\partial + i(\not s + \not b) + m)\psi \\ = \int [db][d\psi][d\bar{\psi}] f(b-s) \delta(F_{\mu\nu}(b-s)) \\ \times\, \exp - \int d^3x \bar{\psi}(\not\partial + i \not b + m)\psi , \end{aligned} \tag{6}$$

where the last equation follows from the first one by shifting $b \to b - s$. Introducing a Lagrange multiplier A_μ to exponentiate the δ-functional, integrating over the fermion fields and setting the weight functional equal to one, it yields

$$Z(s) = \int [dA]\, [db]\, \exp\left[-T(b) - i \int d^3x\, A_\mu(\epsilon_{\mu\nu\lambda}\partial_\nu b_\lambda - \epsilon_{\mu\nu\lambda}\partial_\nu s_\lambda)\right] \tag{7}$$

where $T(b)$ denotes the fermionic effective action in the presence of an external vector field

$$T(b) = -\log\det(\not\partial + i \not b + m) . \tag{8}$$

We now make the approximation of retaining up to quadratic terms in b_μ in (8). The quadratic part of $T(b)$ may be split as

$$\begin{aligned} T(b) &= T_{PC}(b) + T_{PV}(b) \\ T_{PC}(b) &= \int d^3x\, \frac{1}{4}\, F_{\mu\nu}(b)\, F(-\partial^2)\, F_{\mu\nu}(b) \\ T_{PV}(b) &= \int d^3x\, \frac{i}{2}\, b_\mu\, G(-\partial^2)\, \epsilon_{\mu\nu\lambda}\partial_\nu b_\lambda \end{aligned} \tag{9}$$

where T_{PC} and T_{PV} come from the parity-conserving and parity-violating pieces of the vacuum-polarization tensor, respectively [13]. The function F in (9) is regularization-independent, and a standard one-loop calculation yields

$$\tilde{F}(k^2) = \frac{|\, m\,|}{4\pi k^2}\left[1 - \frac{1 - \frac{k^2}{4m^2}}{(\frac{k^2}{4m^2})^{\frac{1}{2}}}\, \arcsin(1 + \frac{4m^2}{k^2})^{-\frac{1}{2}}\right] , \tag{10}$$

where here and in what follows we shall always denote momentum-space representation by putting a tilde over the corresponding coordinate-space representation quantity. The function $\tilde{G}$ in (9) is regularization *dependent*, and can be written as

$$\tilde{G}(k^2) \;=\; \frac{q}{4\pi} + \frac{m}{2\pi \mid k \mid} \arcsin(1 + \frac{4m^2}{k^2})^{-\frac{1}{2}} \,, \tag{11}$$

where q can assume any integer value [14,15], and may be thought of as the effective number of Pauli-Villars regulators, namely, the number of regulators with positive mass minus the number of negative mass ones. Adding a gauge-fixing term $\frac{\lambda}{2}(\partial\cdot b)^2$, the b - dependent part of the path integral (in momentum-space) reads:

$$I = \int [db] e^{-\frac{1}{2}\int \frac{d^3k}{(2\pi)^3}(\tilde{b}^\dagger \tilde{M}\,\tilde{b} + i\tilde{b}^\dagger |k|\,(\tilde{P}_+ - \tilde{P}_-)\,\tilde{A})} \,. \tag{12}$$

We introduced an obvious matrix notation, where the fields are represented by column vectors, the matrix $\tilde{M}$ is given by

$$\tilde{M}(k) \;=\; (\tilde{F}k^2 + i\,\tilde{G} \mid k \mid)\,\tilde{P}_+ + (\tilde{F}k^2 - i\,\tilde{G} \mid k \mid)\,\tilde{P}_- + \lambda k^2\,\tilde{L} \,, \tag{13}$$

and we introduced a complete set of hermitian orthogonal projectors

$$(\tilde{P}_\pm)_{\mu\nu} \;=\; \frac{1}{2}\left(\delta_{\mu\nu} - \frac{k_\mu k_\nu}{k^2} \pm i\epsilon_{\mu\lambda\nu}\frac{k_\lambda}{\mid k \mid}\right) \;,\; \tilde{L}_{\mu\nu} \;=\; \frac{k_\mu k_\nu}{k^2} \,, \tag{14}$$

which verify $\tilde{P}_\pm^2 = \tilde{P}_\pm$, $\tilde{L}^2 = \tilde{L}$; $\tilde{P}_\pm\tilde{L} = 0$, $\tilde{P}_+\tilde{P}_- = 0$; and $\tilde{P}_+ + \tilde{P}_- + \tilde{L} = 1$.

The bosonization formulae are obtained by integrating out the $\tilde{b}$-field

$$I \;=\; \exp\left[-\frac{1}{2}\int \frac{d^3k}{(2\pi)^3}\,\tilde{A}^\dagger\,(\tilde{P}_+ - \tilde{P}_-)\,k^2\,\tilde{M}^{-1}(\tilde{P}_+ - \tilde{P}_-)\,\tilde{A}\right] \tag{15}$$

The inverse of $\tilde{M}$, needed in (15) is computed from (13),

$$\tilde{M}^{-1}(k) \;=\; (\tilde{F}k^2 + i\,\tilde{G} \mid k \mid)^{-1}\,\tilde{P}_+ + (\tilde{F}k^2 - i\,\tilde{G} \mid k \mid)^{-1}\,\tilde{P}_- + (\lambda k^2)^{-1}\,\tilde{L} \,, \tag{16}$$

and by further use of the projectors' properties, we can write

$$Z(s) = \int [d\tilde{A}] \exp \int \frac{d^3k}{(2\pi)^3}[\frac{1}{2}k^2\tilde{A}^\dagger$$

$$(\frac{1}{\tilde{F}k^2 + i\tilde{G}|k|}\,\tilde{P}_+ + \frac{1}{\tilde{F}k^2 - i\tilde{G}|k|}\,\tilde{P}_-)\tilde{A} - i\,\tilde{s}^\dagger\,|k|\,(\tilde{P}_+ - \tilde{P}_-)\,\tilde{A}] \,. \tag{17}$$

There is still freedom to write the partition function (17) in different ways, namely, we can always redefine the field $\tilde{A}_\mu$ by performing a non-singular transformation on it. This will, of course, change both the quadratic and linear parts of the action, thus affecting both the bosonized action and the mapping between fermionic currents and bosonic fields, but in such a way that the current correlation functions are not modified, since we are just changing a dummy variable. It is however, necessary to do this in order to show explicitly the connection with the approach of [12]. A general redefinition of $\tilde{A}_\mu$ may be written as $\tilde{A} \rightarrow (\tilde{u}_+ P_+ + \tilde{u}_- P_- + \tilde{u}_L L)\tilde{A}$, where the $\tilde{u}$'s are functions of the momentum. Note that the effect of $\tilde{u}_L$ disappears as a consequence of gauge-invariance.

$$Z(s) = \int [d\tilde{A}] \exp - \int \frac{d^3k}{(2\pi)^3} (\frac{1}{2} k^2 \tilde{A}^\dagger [\frac{|\tilde{u}_+|^2}{\tilde{F}k^2 + i\tilde{G}|k|} P_+ + \frac{|\tilde{u}_-|^2}{\tilde{F}k^2 - i\tilde{G}|k|} P_-] \tilde{A} - i\tilde{s}^\dagger |k| (\tilde{u}_+ P_+ - \tilde{u}_- P_-) \tilde{A}) . \tag{18}$$

In what follows we shall restrict ourselves to the constant-$\tilde{u}_\pm$ case. Expression (18) can be put in coordinate space representation as follows:

$$Z(s) = \int [dA] \exp - \int d^3x [\frac{1}{4} F_{\mu\nu} C_1 F_{\mu\nu} - \frac{i}{2} A_\mu C_2 \epsilon_{\mu\nu\lambda} \partial_\nu A_\lambda + i (\frac{u_+ - u_-}{2}) s_\mu \frac{1}{\sqrt{-\partial^2}} \partial_\nu F_{\nu\mu} - i (\frac{u_+ + u_-}{2}) s_\mu \epsilon_{\mu\nu\lambda} \partial_\nu A_\lambda] \tag{19}$$

where

$$C_1 = \frac{1}{2} \frac{|u_+|^2 (F - iG) + |u_-|^2 (F + iG)}{-\partial^2 F^2 + G^2}$$
$$C_2 = \frac{i}{2} \frac{|u_+|^2 (F - iG) - |u_-|^2 (F + iG)}{-\partial^2 F^2 + G^2} \tag{20}$$

Let us discuss now the explicit form adopted by (20) for the cases $m \rightarrow \infty$ and $m \rightarrow 0$. When $m \rightarrow \infty$, C_1 tends to a constant which multiplies the Maxwell term. This is neglected to leading order in a derivative expansion, since there is also a Chern-Simons term, multiplied by the constant factor C_2:

$$C_2 \rightarrow 4\pi \, |u|^2 \times (q + \frac{m}{|m|}) . \tag{21}$$

C_2 is regularization-dependent, and its ambiguity is reflected here by the undefined constant q. To compare with [8], we partially fix q by the condition $q + \mathrm{sgn}(m) = \pm 1$, and chosing $u_+ = u_- = u = \frac{1}{2\pi}$, we see that the bosonized action (denoted S_{bos}), in the partition function (19) reduces to

$$S_{bos} = \int d^3x \left(\pm \frac{i}{2} A_\mu \epsilon_{\mu\nu\lambda} \partial_\nu A_\lambda - \frac{i}{\sqrt{4\pi}} s_\mu \epsilon_{\mu\nu\lambda} \partial_\nu A_\lambda \right) , \tag{22}$$

which agrees with the result of [8].

Now we discuss the limit $m \to 0$. In this case we have for F and G the behaviours $F(k^2) \to \frac{e^2}{16}|k|^{-1}$, $G(k^2) \to \dfrac{q}{4\pi}$ which imply for C_1 and C_2 $C_1 \to \frac{16|u|^2}{|k|}$, $C_2 \to \frac{4\pi|u|^2}{q}$. By taking then $\tilde{u}_+ = \tilde{u}_- = \dfrac{1}{4}e^{i\alpha}$, the bosonized action in coordinate space assumes the form

$$
\begin{aligned}
S_{bos} = \int d^3x \, (\frac{1}{4} F_{\mu\nu} \frac{1}{\sqrt{-\partial^2}} F_{\mu\nu} - \frac{i}{2}\frac{\pi}{4q} \epsilon_{\mu\nu\lambda} A_\mu \partial_\nu A_\lambda \\
- \frac{\sin\alpha}{4} s_\mu \frac{\partial_\nu F_{\nu\mu}}{\sqrt{-\partial^2}} - i\frac{\cos\alpha}{4} \epsilon_{\mu\nu\lambda}\partial_\nu A_\lambda) \,, \qquad (23)
\end{aligned}
$$

thus with the identifications $\theta = \frac{\pi}{4q}$, $\alpha = \arctan\frac{\pi}{4q}$, $\beta = \frac{\cos\alpha}{4}$, the bosonized action becomes identical with the one of Equation (1), which is the Euclidean version of the one of Ref. [12].

VECTOR AND AXIAL VECTOR SOURCES IN 3 + 1 DIMENSIONS

In this section we are concerned with the bosonization rules for a fermionic field in the presence of vector and axial vector sources in 3+1 dimensions. This will allow us to find the bosonization rules for the vector and axial fermionic currents, in agreement with the axial anomaly [17] in the presence of an external vector gauge field.

We start with the generating functional for a massless Dirac field in 3 + 1 (Euclidean) dimensions, coupled to Abelian vector (s_μ) and axial-vector (t_μ) external sources [4]

$$
\begin{aligned}
\mathcal{Z}(s_\mu, t_\mu) &= \int \mathcal{D}\bar{\psi}\mathcal{D}\psi \, \exp\left[-S(\bar{\psi}, \psi; s_\mu, t_\mu)\right] \\
S(\bar{\psi}, \psi; s_\mu, t_\mu) &= -i \int d^4x \, \bar{\psi}\,(i\not{\partial} - \not{s} - \gamma_5 \not{t})\,\psi \qquad (24)
\end{aligned}
$$

$$
\gamma_\mu^\dagger = \gamma_\mu \ , \ \gamma_5^\dagger = \gamma_5 \ , \ \{\gamma_\mu \, , \gamma_\nu\} = 2\,\delta_{\mu\nu} \, . \qquad (25)
$$

We perform in (24) the change of variables

$$
\psi(x) = e^{i\theta(x) - i\gamma_5\alpha(x)}\psi'(x) \, , \ \bar{\psi}(x) = \bar{\psi}'(x) e^{-i\theta(x) - i\gamma_5\alpha(x)} \, . \qquad (26)
$$

In terms of the new variables, (24) reads

[4] Note that the vector and axial vector currents are independent fermionic bilineals in 3 + 1 dimensions.

$$\mathcal{Z}(s_\mu, t_\mu) = \int \mathcal{D}\bar{\psi}\mathcal{D}\psi\, J(\alpha; s_\mu, t_\mu) \exp\left[-S(\bar{\psi}, \psi\, ;\, s_\mu + \partial_\mu \theta, t_\mu + \partial_\mu \alpha)\right] \quad (27)$$

where the primed fermionic fields have been renamed as unprimed, and J is the anomalous Jacobian corresponding to this change of variables. This Jacobian is evaluated by using the standard Fujikawa's recipe.

The natural choice of regularization in this case is the consistent one, since it assures the conservation of the vector current [19]. To obtain the Jacobian for this finite transformation, we use the techniques described in ref. [20]. The anomalous Jacobian for the transformation (26) becomes

$$J(\alpha; s_\mu, t_\mu) = \exp\left[\frac{1}{4\pi^2}\int d^4x\, \alpha(x)\, \epsilon_{\mu\nu\rho\sigma}(\partial_\mu s_\nu \partial_\rho s_\sigma + \frac{1}{3}\partial_\mu t_\nu \partial_\rho t_\sigma)\right]. \quad (28)$$

The next step follows from realizing that $\mathcal{Z}$ does not depend on either θ or α, which can be integrated out without other effect than the introduction of an irrelevant constant factor in $\mathcal{Z}$. This is equivalent to integrating over two flat Abelian vector fields $\theta_\mu = \partial_\mu \theta$, $\alpha_\mu = \partial_\mu \alpha$.

It is convenient to define

$$J(\alpha_\mu; s_\mu, t_\mu) = \exp\left[-\frac{1}{4\pi^2}\int d^4x\, \alpha_\mu(x)\, \epsilon_{\mu\nu\rho\sigma}(s_\nu \partial_\rho s_\sigma + \frac{1}{3} t_\nu \partial_\rho t_\sigma)\right]. \quad (29)$$

Thus (27) becomes

$$\mathcal{Z} = \int \mathcal{D}\theta_\mu\, \mathcal{D}\alpha_\mu\, \mathcal{D}\bar{\psi}\, \mathcal{D}\psi\; \delta[f_{\mu\nu}(\theta)]\, \delta[f_{\mu\nu}(\alpha)]\; J(\alpha_\mu; s_\mu, t_\mu)$$

$$\exp\left[-S(\bar{\psi}, \psi; s_\mu + \theta_\mu, t_\mu + \alpha_\mu)\right]. \quad (30)$$

Formally integrating out the fermionic fields and making the shift of variables $\theta_\mu \rightarrow \theta_\mu - s_\mu$, $\alpha_\mu \rightarrow \alpha_\mu - t_\mu$, (30) leads to

$$\mathcal{Z}(s_\mu, t_\mu) = \int \mathcal{D}\theta_\mu\, \mathcal{D}\alpha_\mu\, \delta[f_{\mu\nu}(\theta - s)]\delta[f_{\mu\nu}(\alpha - t)]\, J(\alpha_\mu - t_\mu; s_\mu, t_\mu)$$

$$\times \det(\partial\!\!\!/ + i\, \theta\!\!\!/ + i\gamma_5\, \alpha\!\!\!/). \quad (31)$$

Exponentiating the functional delta functions in (31) by using two antisymmetric tensor fields $A_{\mu\nu}$ and $B_{\mu\nu}$ as Lagrange multipliers yields

$$\mathcal{Z}(s_\mu, t_\mu) = \int \mathcal{D}A_{\mu\nu}\, \mathcal{D}B_{\mu\nu}\, \mathcal{D}\theta_\mu\, \mathcal{D}\alpha_\mu\;\; J(\alpha_\mu - t_\mu; s_\mu, t_\mu)$$

$$\exp\left(i\int d^4x[\epsilon_{\mu\nu\rho\sigma} A_{\mu\nu}(\partial_\rho \theta_\sigma - \partial_\rho s_\sigma) + \epsilon_{\mu\nu\rho\sigma} B_{\mu\nu}(\partial_\rho \alpha_\sigma - \partial_\rho t_\sigma)]\right)$$

$$\times\;\; \det(\partial\!\!\!/ + i\, \theta\!\!\!/ + i\gamma_5\, \alpha\!\!\!/). \quad (32)$$

The bosonized form of $\mathcal{Z}$ can then be obtained by integrating out θ_μ and α_μ in (32). This produces a generating functional with the tensor fields $A_{\mu\nu}$ and $B_{\mu\nu}$ as dynamical variables. This step requires the evaluation of the fermionic determinant, which in four dimensions is necessarily non-exact. Before embarking on such calculation, we derive the rules that map the vector and axial-vector currents into functions of the bosonic fields $A_{\mu\nu}$ and $B_{\mu\nu}$. This correspondence requires no approximation and may well be called 'exact'. These rules follow from elementary functional differentiation

$$j_\mu = \langle \bar{\psi}\gamma_\mu\psi\rangle = -i\frac{\delta}{\delta s_\mu}\log \mathcal{Z}|_{s_\mu=0} = -\epsilon_{\mu\nu\rho\sigma}\partial_\nu A_{\rho\sigma} \tag{33}$$

$$j_{5\mu} = \langle \bar{\psi}\gamma_5\gamma_\mu\psi\rangle = -i\frac{\delta}{\delta t_\mu}\log \mathcal{Z}|_{t_\mu=0} = -\epsilon_{\mu\nu\rho\sigma}\partial_\nu B_{\rho\sigma} - \frac{i}{4\pi^2}\,\epsilon_{\mu\nu\rho\sigma}s_\nu\partial_\rho s_\sigma\,. \tag{34}$$

¿From the antisymmetry of the tensors $A_{\mu\nu}$ and $B_{\mu\nu}$, we are entitled to derive the equations for the divergencies of the currents:

$$\partial_\mu j_\mu = 0 \quad \partial_\mu j^5_\mu = -\frac{i}{8\pi^2}\,\tilde{F}_{\mu\nu}(s)F_{\mu\nu}(s)\,. \tag{35}$$

with $\tilde{F}_{\mu\nu} = (1/2)\epsilon_{\mu\nu\alpha\beta}F_{\alpha\beta}$. We then see that the bosonization rule (34) correctly reproduces the axial anomaly.

As stated above, although the bosonization recipe (33)-(34) for associating the fermionic currents with expressions written in terms of bosonic fields is exact, the bosonic action governing the boson field dynamics cannot be evaluated in an exact form in $d > 2$ dimensions. Different approximations for computing the fermionic determinant would yield alternative effective bosonic actions valid in different regimes.

The fermionic determinant in (32) is evaluated to second order in the fields θ_μ and α_μ. The use of this quadratic approximation can be motivated by the same kind of arguments (see in particular the 'quasi-theorem') used in ref. [21].

As usual, it is convenient to work in terms of W, the generating functional of connected Green's functions of the fermionic current

$$\det(\not\partial + i\not\theta + i\gamma_5\not\alpha) = \exp\left[W(\theta_\mu,\alpha_\mu)\right]\,. \tag{36}$$

The renormalized functional W becomes becomes:

$$W(\theta_\mu,\alpha_\mu) = -\frac{1}{2}\int d^4x d^4y\,\Big[\theta_\mu(x)\delta^{\perp}_{\mu\nu}F(x-y)\theta_\nu(y)$$

$$+\alpha_\mu(x)\delta^{\perp}_{\mu\nu}G(x-y)\alpha_\nu(y) + m^2\,\alpha_\mu(x)\delta^{||}_{\mu\nu}\delta(x-y)\alpha_\nu(y)\Big] \tag{37}$$

where

$$F(x-y) = \int\frac{d^4k}{(2\pi)^4}\,e^{ik\cdot(x-y)}\tilde{F}(k)$$

$$\tilde{F}(k) = \frac{k^2}{2\pi^2} \int_0^1 dx\, x(1-x) \log\left[1 + x(1-x)\frac{k^2}{\mu^2}\right] . \tag{38}$$

$$G(x-y) = F(x-y) + m^2\delta(x-y) . \tag{39}$$

Inserting (37) into (32), we see that the functional integral is Gaussian with respect to θ_μ and α_μ:

$$\mathcal{Z}(s_\mu, t_\mu) = \int \mathcal{D}A_{\mu\nu}\, \mathcal{D}B_{\mu\nu}\, \mathcal{D}\theta_\mu\, \mathcal{D}\alpha_\mu$$

$$\exp\left[-\frac{1}{4\pi^2}\int d^4x\,(\alpha_\mu - t_\mu)\,\epsilon_{\mu\nu\rho\sigma}(s_\nu\partial_\rho s_\sigma + \frac{1}{3}t_\nu\partial_\rho t_\sigma)\right]$$

$$\exp\left\{i\int d^4x[\epsilon_{\mu\nu\rho\sigma}A_{\mu\nu}(\partial_\rho\theta_\sigma - \partial_\rho s_\sigma) + \epsilon_{\mu\nu\rho\sigma}B_{\mu\nu}(\partial_\rho\alpha_\sigma - \partial_\rho t_\sigma)]\right\}$$

$$\exp\left\{-\frac{1}{2}\int d^4x d^4y\,[\theta_\mu(x)\delta^\perp_{\mu\nu}F(x-y)\theta_\nu(y)\right.$$

$$\left. + \alpha_\mu(x)\delta^\perp_{\mu\nu}G(x-y)\alpha_\nu(y) + m^2\,\alpha_\mu(x)\delta^{||}_{\mu\nu}\delta(x-y)\alpha_\nu(y)]\right\} . \tag{40}$$

Performing the Gaussian integration over θ_μ and α_μ in (40) we get

$$\mathcal{Z}(s_\mu, t_\mu) = \exp[\mathcal{C}(s_\mu, t_\mu)] \int \mathcal{D}A_{\mu\nu}\, \mathcal{D}B_{\mu\nu} \times$$
$$\exp\left\{-i\int d^4x[s_\mu\epsilon_{\mu\nu\rho\sigma}\partial_\nu A_{\rho\sigma} + t_\mu(\epsilon_{\mu\nu\rho\sigma}\partial_\nu B_{\rho\sigma} + \frac{i}{4\pi^2}\epsilon_{\mu\nu\rho\sigma}s_\nu\partial_\rho s_\sigma)]\right\} \times$$
$$\exp\left\{-\frac{1}{3}\int d^4x d^4y[A_{\mu\nu\rho}(x)F^{-1}(x-y)A_{\mu\nu\rho}(y) + \right.$$
$$\left. B_{\mu\nu\rho}(x)G^{-1}(x-y)B_{\mu\nu\rho}(y)]\right\} \times \exp\left\{-\frac{i}{4\pi^2}\int d^4x d^4y \partial_\mu B_{\nu\rho}(x) \times\right.$$
$$\left. G^{-1}(x-y)\delta_{\mu\nu\rho,\alpha\beta\gamma}(s_\alpha\partial_\beta s_\gamma + \frac{1}{3}t_\alpha\partial_\beta t_\gamma)\right\} \tag{41}$$

where

$$A_{\mu\nu\rho} = \partial_\mu A_{\nu\rho} + \partial_\nu A_{\rho\mu} + \partial_\rho A_{\mu\nu} \ , \quad B_{\mu\nu\rho} = \partial_\mu B_{\nu\rho} + \partial_\nu B_{\rho\mu} + \partial_\rho B_{\mu\nu}$$

$$\delta_{\mu\nu\rho,\alpha\beta\gamma} = \det\begin{pmatrix} \delta_{\mu\alpha} & \delta_{\mu\beta} & \delta_{\mu\gamma} \\ \delta_{\nu\alpha} & \delta_{\nu\beta} & \delta_{\nu\gamma} \\ \delta_{\rho\alpha} & \delta_{\rho\beta} & \delta_{\rho\gamma} \end{pmatrix} \tag{42}$$

and

$$\mathcal{C}(s_\mu, t_\mu) = \frac{1}{2(2\pi)^4}\int d^4x d^4y \left\{[s_\mu(x)\partial_\nu s_\lambda(x) + \frac{1}{3}t_\mu(x)\partial_\nu t_\lambda(x)]\right.$$

$$\delta_{\mu\nu\rho,\alpha\beta\gamma} G^{-1}(x-y)[s_\alpha(y)\partial_\beta s_\gamma(y) + \frac{1}{3}t_\alpha(y)\partial_\beta t_\gamma(y)]$$

$$+\frac{1}{2(2\pi)^4}\int d^4x d^4y\, \mathcal{G}(x)\partial^{-2}G^{-1}(x-y)\mathcal{G}(y)$$

$$+\frac{1}{2m^2(2\pi)^4}\int d^4x d^4y\, \mathcal{G}(x)\partial^{-2}(x-y)\mathcal{G}(y)\Big\} \tag{43}$$

where $\mathcal{G} = \epsilon_{\mu\nu\rho\lambda}(\partial_\mu s_\nu \partial_\rho s_\lambda + \frac{1}{3}\partial_\mu t_\nu \partial_\rho t_\lambda)$.

SUMMARY AND CONCLUSIONS

We have obtained a bosonization recipe for the free Dirac field valid over the whole range of distances. It enables us to treat the massive and massless cases in an equal footing, leading to the bosonization formulae for a massless Dirac field (Eq. (1)), obtained by following the canonical method.

We have obtained bosonization rules for massless Dirac fermions in four dimensions with both vector and axial-vector sources. This has allowed us to find the bosonization rules for both fermionic currents, eqs.(33)-(34), in terms of Kalb-Ramond bosonic fields. While the bosonization rule for the vector current can be written in a natural and compact form, reminiscent of the well-known two-dimensional bosonization rule,

$$\bar{\psi}\gamma_\mu\psi \to -\epsilon_{\mu\nu\rho\sigma}\partial_\nu A_{\rho\sigma}\,, \tag{44}$$

the result for the axial current is more involved and includes the vector source

$$\bar{\psi}\gamma_5\gamma_\mu\psi \to -\epsilon_{\mu\nu\rho\sigma}\partial_\nu B_{\rho\sigma} - \frac{i}{4\pi^2}\,\epsilon_{\mu\nu\rho\sigma}s_\nu\partial_\rho s_\sigma\,. \tag{45}$$

This is a consequence of the anomalous behaviour of the fermionic measure under axial gauge transformations and in this way the bosonic form of the axial current correctly yields its anomalous divergence.

As stressed above, recipes (44) and (45) can be considered exact apart from the fact that if one is to work in the bosonic version one has to use an approximate expression for the bosonic action. The one we proposed is based in a quadratic approximation and leads to the bosonic generating functional presented in eqs.(41)-(43).

REFERENCES

1. C.P. Burgess and F. Quevedo, Nucl. Phys. **B421** (1994) 373.
2. C.P. Burgess an, C.A. Lütken and F. Quevedo, Phys. Lett. **B326** (1994) 18.
3. C.P. Burgess and F. Quevedo, Phys. Lett. **B329** (1994) 457.

4. A. Kovner and B. Rosenstein, Phys. Lett. **B342** (1985) 381.
A. Kovner and P. Kurzepa, Phys. Lett. **B328** (1994) 506.
A. Kovner and P. Kurzepa, Int. J. Mod. Phys. **A9** (1994) 129.
5. J.L. Cortés, E. Rivas and L. Velázquez, Phys. Rev. **D53** (1996) 5952.
6. E. Fradkin and F.A. Schaposnik, Phys. Lett. **B338** (1994) 253.
7. N. Bralić, E. Fradkin, M.V. Manías and F.A. Schaposnik, Nucl.Phys. **446** (1995) 144.
8. F.A. Schaposnik, Phys. Lett. **B356** (1995) 39.
9. D.G. Barci, C.D. Fosco and L.D. Oxman, Phys. Lett. **B375** (1996), 267.
10. J.C. Le Guillou, C.Núñez and F.A. Schaposnik, Ann. Phys. (N.Y.) **251** (1996) 426-441.
11. C.D. Fosco, F.A. Schaposnik, 'Bosonization of Vector and Axial-Vector Currents in (3+1)-dimensions', e-Print Archive: hep-th/9608057 (to appear in Phys. Lett. B).
12. E. C. Marino, Phys. Lett. **B 263** (1991) 63.
13. R. Jackiw and S. Templeton, Phys. Rev. **D 23** (1981) 2291; S. Deser, R. Jackiw and S. Templeton, Phys. Rev. Lett. **48** (1982) 975;
14. J. Fröhlich and T. Kerler, Nucl. Phys. **B 354** (1991) 369;
15. 'The ζ-function answer to parity violation in three dimensional gauge theories', R. E. Gamboa Saravi, G. L. Rossini and F. A. Schaposnik, hep-th/9411238.
16. I. J. R. Aitchison, C. D. Fosco and J. A. Zuk, Phys. Rev. **D 48** (1993) 5895.
17. J.S. Bell and R. Jackiw, Nuov. Cim. A**60** (1969) 47; S. L. Adler, Phys. Rev. **177** (1969), 2426.
18. K. Fujikawa, Phys. Rev. Lett. **42** (1979) 1195;
Phys. Rev. D**21** (1980) 2848; erratum-*ibid.* D**22** (1980) 1499;
Phys. Rev. D**29** (1984) 285.
19. K.Fujikawa, *Recent Developements in the Path-Integral Approach to Anomalies*, published in Vancouver Theory Workshop 1986, p.209.
20. R.E. Gamboa-Saraví, M.A. Muschietti, F.A. Schaposnik and J.E. Solomin, Ann. of Phys. **157** (1984) 360.
21. J. Frohlich, R. Gotschmann and P.A. Marchetti, J.Phys. **A 28** (1995) 1169.

The Origin of Matter in the Universe: A Brief Review

Marcelo Gleiser

Department of Physics and Astronomy, Dartmouth College[1]
Hanover, NH 03755

Abstract. In this talk I briefly review the main ideas and challenges involved in the computation of the observed baryonic excess in the Universe.

I THE SAKHAROV CONDITIONS AND GUT BARYOGENESIS

Given that the observational evidence is for a Universe with a primordial baryon asymmetry [1,2], we have two choices; either this asymmetry is the result of an initial condition, or it was attained through dynamical processes that took place in the early Universe. In 1967, just a couple of years after the discovery of the microwave background radiation, Sakharov wrote a ground-breaking work in which he appealed to the drastic environment of the early stages of the hot big-bang model to spell out the 3 conditions for dynamically generating the baryon asymmetry of the Universe [3]. Here they are, with some modifications:
i) Baryon number violating interactions: Clearly, if we are to generate any excess baryons, our model must have interactions which violate baryon number. However, the same interactions also produce antibaryons at the same rate. We need a second condition;
ii) C and CP violating interactions: Combined violation of charge conjugation (C) and charge conjugation combined with parity (CP) can provide a bias to enhance the production of baryons over antibaryons. However, in thermal equilibrium $n_{\rm b} = n_{\bar{\rm b}}$, and any asymmetry would be wiped out. We need a third condition;
iii) Departure from thermal equilibrium: Nonequilibrium conditions guarantee that the phase-space density of baryons and antibaryons will not be the same.

---

1) NSF Presidential Faculty Fellow.

CP400, *First Latin American Symposium on High Energy Physics/ VII Mexican School of Particles and Fields*,
edited by D'Olivo/Klein-Kreisler/Mendez

Hence, provided there is no entropy production later on, the net ratio n_B/s will remain constant.

Given the above conditions, we have to search for the particle physics models that both satisfy them and are capable of generating the correct asymmetry. The first models that attempted to compute the baryon asymmetry dynamically were Grand Unified Theory (GUT) models [4]. A typical mechanism of GUT baryogenesis is known as the "out-of-equilibrium decay scenario"; one insures that the heavy X bosons have a long enough lifetime so that their inverse decays go out of equilibrium as they are still abundant. Baryon number is produced by the free decay of the heavy Xs, as the inverse rate is shut off.

Interesting as they are, GUT models of baryogenesis have serious obstacles to overcome. Here I mention only the obstacle related to electroweak scale phenomena. The vacuum manifold of the electroweak model exhibits a very rich structure, with degenerate minima separated by energy barriers (in field configuration space). Different minima have different baryon (and lepton) number, with the net difference between two minima being given by the number of families. Thus, for the standard model, each jump between two adjacent minima leads to the creation of 3 baryons and 3 leptons, with net $B-L$ conservation and $B+L$ violation. At $T=0$, tunneling between adjacent minima is mediated by instantons, and, as shown by 't Hooft [5], the tunneling rate is suppressed by the weak coupling constant ($\Gamma \sim e^{-4\pi/\alpha_W} \sim 10^{-170}$). That is why the proton is stable. However, as pointed out by Kuzmin, Rubakov, and Shaposhnikov, at finite temperatures ($T \sim 100$ GeV), one could hop over the barrier, tremendously enhancing the rate of baryon number violation [6]. The height of the barrier is given by the action of an unstable static solution of the field equations known as the sphaleron [7].

Being a thermal process, the rate of baryon number violation is controlled by the energy of the sphaleron configuration, $\Gamma \sim \exp[-\beta E_S]$, with $E_S \simeq M_W/\alpha_W$, where M_W is the W-boson mass. Note that $M_W/\alpha_W = \langle\phi\rangle/g$, where $\langle\phi\rangle$ is the vacuum expectation value of the Higgs field. For temperatures above the critical temperature for electroweak symmetry restoration, it has been shown that sphaleron processes are not exponentially suppressed, with the rate being roughly $\Gamma \sim (\alpha_W T)^4$ [8]. Even though this opens the possibility of generating the baryonic asymmetry at the electroweak scale, it is bad news for GUT baryogenesis. Unless the original GUT model was $B-L$ conserving, any net baryon number generated then would be brought to zero by the efficient anomalous electroweak processes. There are several alternative models for baryogenesis invoking more or less exotic physics. The interested reader is directed to the review by Olive, listed in Ref. 1. I now move on to discuss the promises and challenges of electroweak baryogenesis.

II ELECTROWEAK BARYOGENESIS

As pointed out above, temperature effects can lead to efficient baryon number violation at the electroweak scale. Can the other two Sakharov conditions be satisfied in the early Universe so that the observed baryon number could be generated during the electroweak phase transition? The short answer is that in principle yes, but probably not in the context of the minimal standard model. Let us first see why it is possible to satisfy all conditions for baryogenesis in the context of the standard model.

Departure from thermal equilibrium is obtained by invoking a first order phase transition. After summing over matter and gauge fields, one obtains a temperature corrected effective potential for the magnitude of the Higgs field, ϕ. The potential describes two phases, the symmetric phase with $\langle\phi\rangle = 0$ and massless gauge and matter fields, and the broken-symmetric phase with $\langle\phi\rangle = \phi_+(T)$, with massive gauge and matter fields. The loop contributions from the gauge fields generate a cubic term in the effective potential, which creates a barrier separating the two phases. This result depends on a perturbative evaluation of the effective potential, which presents problems for large Higgs masses as I will discuss later. At 1-loop, the potential can be written as [9]

$$V_{\rm EW}(\phi, T) = D\left(T^2 - T_2^2\right)\phi^2 - ET\phi^3 + \frac{1}{4}\lambda_T\phi^4, \tag{1}$$

where the constants D and E are given by $D = \left[6(M_W/\sigma)^2 + 3(M_Z/\sigma)^2 + 6(M_T/\sigma)^2\right]/24 \sim 0.17$, and $E = \left[6(M_W/\sigma)^3 + 3(M_Z/\sigma)^3\right]/12\pi \sim 0.01$, where I used, $M_W = 80.6$ GeV, $M_Z = 91.2$ GeV, $M_T = 174$ GeV [10], and $\sigma = 246$ GeV. The (lengthy) expression for λ_T, the temperature corrected Higgs self-coupling, can be found in Ref. 9. At the critical temperature, $T_C = T_2/\sqrt{1 - E^2/\lambda_T D}$, the minima have the same free energy, $V_{\rm EW}(\phi_+, T_C) = V_{\rm EW}(0, T_C)$. As $E \to 0$, $T_C \to T_2$ and the transition is second order. Since E and D are fixed, the strength of the transition is controlled by the value of the Higgs mass, or λ.

Assuming that the above potential (or something close to it) correctly describes the two phases, as the Universe cools belows T_C the symmetric phase becomes metastable and will decay by nucleation of bubbles of the broken-symmetric phase which will grow and percolate completing the transition. Departure from equilibrium will occur in the expanding bubble walls. This scenario relies on the assumption that the transition is strong enough so that the usual homogeneous nucleation mechanism correctly describes the approach to equilibrium. As I will discuss later, this may not be the case for "weak" transitions. For now, we forget this problem and move on to briefly examine how to generate the baryonic asymmetry with expanding bubbles.

The last condition for generating baryon number is C and CP violation. It is known that C and CP violation are present in the standard model. However, the CP violation from the Kobayashi-Maskawa (KM) phase is too small to generate the required baryon asymmetry. Even though the debate is still going on, efficient baryogenesis within the standard model is a remote possibility.

For many, this is enough motivation to go beyond the standard model in search of extensions which have an enhanced CP violation built in. Several models have been proposed so far, although the simplest invoke either more generations of massive fermions, or multiple massive Higgs doublets with additional CP violation in this sector of the theory. Instead of looking into all models in detail, I will just briefly describe the essential ingredients common to most models.

The transition is assumed to proceed by bubble nucleation. Outside the bubbles the Universe is in the symmetric phase, and baryon number violation is occurring at the rate $\Gamma \sim (\alpha_{\rm W} T)^4$. Inside the bubble the Universe is in the broken symmetric phase and the rate of baryon number violation is $\Gamma \sim \exp[-\beta E_{\rm S}]$. Since we want any net excess baryon number to be preserved in the broken phase, we must shut off the sphaleron rate inside the bubble. This imposes a constraint on the strength of the phase transition, as $E_{\rm S} \simeq \langle\phi(T)\rangle/g$; that is, we must have a large "jump" in the vacuum expectation value of ϕ during the transition, $\langle\phi(T)\rangle/T \geq 1$, as shown by Shaposhnikov [11].

Inside the bubble wall the fields are far from equilibrium and there is CP violation, and thus a net asymmetry can be induced by the moving wall. In practice, computations are complicated by several factors, such as the dependence on the net asymmetry on the bubble velocity and on its thickness [12]. Different charge transport mechanisms based on leptons as opposed to quarks have been proposed, which enhance the net baryonic asymmetry produced [13]. However, the basic picture is that as matter traverses the moving wall an asymmetry is produced. And since baryon number violation is suppressed inside the bubble, a net asymmetry survives in the broken phase. Even though no compelling model exists at present, and several open questions related to the complicated nonequilibrium dynamics remain, it is fair to say that the correct baryon asymmetry may have been generated during the electroweak phase transition, possibly in some extension of the standard model. However, I would like to stress that this conclusion has two crucial assumptions built in it; that we know how to compute the effective potential reliably, and that the transition is strong enough to proceed by bubble nucleation. In the next Section I briefly discuss some of the issues involved and how they may be concealing interesting new physics.

III CHALLENGES TO ELECTROWEAK BARYOGENESIS

A The Effective Potential

A crucial ingredient in the computation of the net baryon number generated during the electroweak phase transition is the effective potential. In order to trust our predictions, we must be able to compute it reliably. However, it is well known that perturbation theory is bound to fail due to severe infrared problems. It is easy to see why this happens. At finite temperatures, the loop expansion parameter involving gauge fields is $g^2T/M_{\mathbf{gauge}}$. Since $M_{\mathbf{gauge}} = g\langle\phi\rangle$, in the neighborhood of $\langle\phi\rangle = 0$ the expansion diverges. This behavior can be improved by summing over ring, or daisy, diagrams [14].

Another problem that appears in the evaluation of the effective potential is due to loop corrections involving the Higgs boson. For second order phase transitions, the vanishing of the effective potential's curvature at the critical temperature leads to the existence of critical phenomena characterized by diverging correlation lengths. Even though there is no infrared-stable fixed point for first order transitions, for large Higgs masses the transition is weak enough to induce large fluctuations about equilibrium; the mean-field estimate for the correlation length $\xi(T) = M^{-1}(T)$ is certainly innacurate. This behavior has led some authors [15,16] to invoke ε-expansion methods to deal with the infrared divergences. Another alternative is to go to the computer and study the equilibrium properties of the standard model on the lattice [17]. Recent results are encouraging inasmuch as they seem to be consistent with perturbative results in the broken phase for fairly small Higgs masses. Furthermore, they indicate how the transition becomes weaker for large values of the Higgs mass, $M_{\mathrm{H}} \geq 60$ GeV.

B Weak vs. Strong First Order Transitions

In order to avoid the erasure of the produced net baryon number inside the broken-symmetric phase, the sphaleron rate must be suppressed within the bubble. As mentioned earlier, this amounts to imposing a large enough "jump" on the vacuum expectation value of ϕ during the transition. In other words, the transition cannot be too weakly first order. But what does it mean, really, to be "weakly" or "strongly" first order?

This is a very important point which must not be overlooked (although it often is!); the vacuum decay formalism used for the computation of nucleation rates relies on a semi-classical expansion of the effective action. That is, we assume we start at a *homogeneous* phase of false vacuum, and evaluate the rate by summing over small amplitude fluctuations about the metastable state [18]. This approximation must break down for weak enough transitions, when we expect large fluctuations to be present within the metastable phase. An explicit example of this breakdown was recently discussed, where the extra free energy available due to the presence of large-amplitude fluctuations was incorporated into the computation of the decay rate [19].

In Ref. 15, it was suggested that weak transitions may evolve by a different mechanism, characterized by substantial mixing of the two phases as the critical temperature is approached from above (*i.e.* as the Universe cools to T_C). They estimated the fraction of the total volume occupied by the broken-symmetric phase by assuming that the dominant fluctuations about equilibrium are subcritical bubbles of roughly a correlation volume which interpolate between the two phases. Their approach was later refined by the authors of Ref. [21] who found, within their approximations, that the 1-loop electroweak potential shows considerable mixing for $M_H \geq 55$ GeV. Clearly, the presence of large-amplitude, nonperturbative thermal fluctuations compromises the validity of the effective potential, since it does not incorporate such corrections.

In order to understand the shortcomings of the mean-field approximation in this context, numerical simulations in 2d [23] and 3d [24] were performed, which focused on the amount of "phase mixing" promoted by thermal fluctuations.

The results show that the problem boils down to how well localized the system is about the symmetric phase as it approaches the critical temperature. If the system is well localized about the symmetric phase, it will become metastable as the temperature drops below T_C and the transition can be called "strong". In this case, the mean-field approximation is reliable. Otherwise, large-amplitude fluctuations away from the symmetric phase rapidly grow, causing substantial mixing between the two phases. This will be a "weak" transition, which will not evolve by bubble nucleation. Defining $\langle\phi\rangle_V$ as the volume averaged field and $\phi_{\rm inf}$ as the inflection point nearest to the $\phi = 0$ minimum, the criterion for a strong transition can be written as [23]

$$\langle\phi\rangle_V < \phi_{\rm inf} \quad . \tag{2}$$

Recently, an analytical model, based on the subcritical bubbles method, was shown to qualitatively and *quantitatively* describe the results obtained by the 3d simulation [25]. The fact that subcritical bubbles successfully model the effects of thermal fluctuations promoting phase mixing and the breakdown of the mean-field approximation with subsequent symmetry restoration, supports previous estimates which showed that the assumption of homogeneous nucleation is incompatible with standard model baryogenesis for $M_H \leq 55$ GeV [21,22]. It is straightforward to adapt these computations to extensions of the standard model. Thus, the requirement that the transition proceeds by bubble nucleation can be used, together with the subcritical bubbles method, to constrain the parameters of the potential.

ACKNOWLEDGMENTS

I am grateful to my collaborators Rocky Kolb, Andrew Heckler, Graciela Gelmini, Mark Alford, Julian Borrill, and Rudnei Ramos for the many long discussions on bubbles and phase transitions. I am also grateful to Juan Carlos D'Olivo and the local organizing committee for their warm hospitality during this Conference. This work was partially supported by the National Science Foundation through a Presidential Faculty Fellows Award no. PHY-9453431 and by a NASA grant no. NAGW-4270.

REFERENCES

1. For reviews see, L. Yaffe, hep-ph/9512265, A. G. Cohen, D. B. Kaplan, and A. E. Nelson, Annu. Rev. Nucl. Part. Sci. **43**, 27 (1993); A. Dolgov, Phys. Rep. **222**, 311 (1992); K. A. Olive, in "Matter under extreme conditions", eds. H. Latal and W. Schweiger (Springer-Verlag, Berlin, 1994).
2. G. Steigman, Ann. Rev. Astron. Astrophys. **14**, 339 (1976).
3. A. D. Sakharov, JETP Lett. **5**, 24 (1967).
4. For a review see E. W. Kolb and M. S. Turner, Ann. Rev. Nucl. Part. Sci. **33**, 645 (1983); *ibid.* The Early Universe, (Addison-Wesley, Redwood, CA, 1990).
5. G. t'Hooft, Phys. Rev. Lett. **37**, 8 (1976); Phys. Rev. D**14**, 3432 (1976).
6. V. A. Kuzmin, V. A. Rubakov, and M. E. Shaposhnikov, Phys. Lett. **B155**, 36 (1985).
7. N. S. Manton, Phys. Rev. D**28**, 2019 (1983); F. R. Klinkhammer and N. S. Manton, Phys. Rev. D**30**, 2212 (1984).
8. P. Arnold and L. McLerran, Phys. Rev. D**36**, 581 (1987); *ibid.* D**37**, 1020 (1988); J. Ambjorn and A. Krasnitz, Phys. Lett. **B362**, 97 (1995).
9. G. W. Anderson and L. J. Hall, Phys. Rev. D **45**, 2685 (1992); M. Dine, P. Huet, and R. Singleton, Nucl. Phys. **B375**, 625 (1992).
10. F. Abe et al. (CDF Collaboration), Phys. Rev. Lett. **73**, 225 (1994).
11. C. Jarlskog, Z. Physik C**29**, 491 (1985); Phys. Rev. Lett. **55**, 1039 (1985); M. E. Shaposhnikov, Nucl. Phys. B**287**, 757 (1987); B**299**, 797 (1988).
12. B. Liu, L. McLerran, and N. Turok, Phys. Rev. **D46**, 2668 (1992); M. Dine and S. Thomas, Phys. Lett. **B328**, 73 (1994); G.D. Moore and T. Prokopec, Phys. Rev. Lett. **75**, 777 (1995).
13. A. G. Cohen, D. B. Kaplan, and A. E. Nelson, Phys. Lett. B**245**, 561 (1990); Nucl. Phys. B**349**, 727 (1991); B**373**, 453 (1992); Phys. Lett. **B336**, 41 (1994); M. Joyce, T. Prokopec, and N. Turok, Phys. Rev. Lett. **75**, 1695 (1995).
14. P. Arnold and O. Espinosa, Phys. Rev. **D47**, 3546 (1993); M. Dine, P. Huet, R.G. Leigh, A. Linde, and D. Linde, Phys. Rev. **D46**, 550 (1992); C.G. Boyd, D.E. Brahm, and S. Hsu, Phys. Rev. **D48**, 4963 (1993); M. Quiros, J.R. Spinosa, and F. Zwirner, Phys. Lett. **B314**, 206 (1993); W. Buchmüller, T. Helbig, and D. Walliser, Nucl. Phys. **B407**, 387 (1993); M. Carrington, Phys. Rev. **D45**, 2933 (1992).

15. M. Gleiser and E.W. Kolb, Phys. Rev. **D48**, 1560 (1993).
16. P. Arnold and L. Yaffe, Phys. Rev. **D49**, 3003 (1994).
17. K. Farakos, K. Kajantie, K. Rummukainen, and M. Shaposhnikov, hep-lat/9510020; Nucl. Phys. **B407**, 356 (1993); *ibid.* **B425**, 67 (1994); *ibid.***B442**, 317 (1995); B. Bunk, E.-M. Ilgenfritz, J. Kripfganz, and A. Schiller, Phys. Lett. **B284**, 372 (1992); Nucl. Phys. **B403**, 453 (1993); Z. Fodor et al. Nucl. Phys. **B439**, 147 (1995).
18. J. S. Langer, Ann. Phys. (NY) **41**, 108 (1967); *ibid.* **54**, 258 (1969); M. B. Voloshin, I. Yu. Kobzarev, and L. B. Okun', Yad. Fiz. **20**, 1229 (1974) [Sov. J. Nucl. Phys. **20**, 644 (1975); S. Coleman, Phys. Rev. **D15**, 2929 (1977); C. Callan and S. Coleman, Phys. Rev. **D16**, 1762 (1977); A. D. Linde, Nucl. Phys. **B216**, 421 (1983); [Erratum: **B223**, 544 (1983)]; M. Gleiser, G. Marques, and R. Ramos, Phys. Rev. **D48**, 1571 (1993); D. Brahm and C. Lee, Phys. Rev. **D49**, 4094 (1994).
19. M. Gleiser and A. Heckler, Phys. Rev. Lett. **76**, 180 (1996).
20. M. Gleiser and E. W. Kolb, Phys. Rev. Lett. **69**, 1304 (1992); M. Gleiser, E. W. Kolb, and R. Watkins, Nucl. Phys. **B364**, 411 (1991); N. Tetradis, Z. Phys. **C57**, 331 (1993).
21. G. Gelmini and M. Gleiser, Nucl. Phys. B**419**, 129 (1994).
22. M. Gleiser and R. Ramos, Phys. Lett. B**300**, 271 (1993).
23. M. Gleiser, Phys. Rev. Lett. **73**, 3495 (1994).
24. J. Borrill and M. Gleiser, Phys. Rev. **D51**, 4111 (1995).
25. M. Gleiser, A. Heckler, and E.W. Kolb, cond-mat/9512032, submitted to Physical Review Letters.

Geon Statistics and UIR's of the Mapping Class Group

Rafael D. Sorkin*† and Sumati Surya†

*Instituto de Ciencias Nucleares, UNAM, A. Postal 70-543, D.F. 04510, Mexico[1]
†Department of Physics, Syracuse University, Syracuse, NY 13244-1130, U.S.A.[2]

Abstract. Quantum Gravity admits topological excitations of microscopic scale which can manifest themselves as particles — topological geons. Non-trivial spatial topology also brings into the theory free parameters analogous to the θ-angle of QCD. We show that these parameters can be interpreted in terms of geon properties. We also find that, for certain values of the parameters, the geons exhibit new patterns of particle identity together with new types of statistics. Geon indistinguishability in such a case is expressed by a *proper subgroup* of the permutation group and geon statistics by a (possibly projective) representation of the subgroup.

I INTRODUCTION

This talk attempts to answer two questions concerning the effect of topology in generally covariant theories: "how many free parameters are there in quantum gravity?" and "do topological geons really act like particles?" By way of comparison, consider the standard model, which contains both continuous parameters (like the masses of the quarks, the strong coupling constant, and $\theta_{\rm QCD}$) and discrete parameters (like the handedness of the neutrinos, taken relative to $\theta_{\rm QCD}$, say). In that flat space theory, all of the many parameters could be interpreted in terms of the properties of the particles the theory describes, though in some cases the interpretation would be rather indirect. The question here is what additional parameters arise from *spatial topology*, and can they be interpreted in terms of the properties of the topological particles (geons) that quantum gravity describes? To further focus the question for this talk, we will concentrate on the question of particle *statistics*.

To our two questions we will encounter the following answers. There exist in fact very many additional parameters, both continuous and discrete. And

1) email: sorkin@nuclecu.unam.mx
2) email: ssurya@suhep.syr.edu

CP400, *First Latin American Symposium on High Energy Physics/ VII Mexican School of Particles and Fields*, edited by D'Olivo/Klein-Kreisler/Mendez

the topological excitations show themselves consistently as particle-like, in the sense that all of the new parameters can be interpreted as telling us about one of the following:

- internal geon "qualities" or "quantum numbers"
- geon collision parameters
- statistics of identical geons

This positive outcome may be taken to bolster the interpretation of the topological excitations as particles (geons).

In addition we will see that geons can manifest new types of indistinguishability and statistics, beyond the (very familiar) bosonic and fermionic types and the (somewhat less familiar) para-statistical types. These novel statistics include:

- A statistics based on cyclic subgroups of the permutation group $\mathcal{S}_N$
- Possibly a new statistics based on *projective* representations of $\mathcal{S}_N$ or its subgroups.

However, all of these results assume that the topology is "frozen", in the sense that they presuppose a spacetime of the product form ${}^3M \times \mathbb{R}$. A key question then is ¿which quantum sectors will survive when topology changing processes are incorporated into the theory?. (Here, we mean by sector a specific set of values of the parameters. Each such sector is "superselected" in the approximation that topology change is ignored, but it can be expected to communicate with other sectors in a more general setting.)

II PRIMES AND GEONS

Before we begin the formal analysis, let us recall the definition of a geon, and the mathematical decomposition theorem on which it is based. (For more about this theorem, and about the subject of this talk in general, see the more complete exposition in [1], as well as the references therein.) According to this theorem, an arbitrary 3-manifold (without boundary) which is asymptotically $\mathbb{R}^3$ can be expressed as

$$ {}^3M = \mathbb{R}^3 \# P_1 \# P_2 \# \cdots \# P_N, \tag{1} $$

where the P_i are *primes* — manifolds that cannot be built up by "sticking together" smaller manifolds. In fact we will assume further that the primes are all "irreducible", that is, we will exclude the orientable and nonorientable handles (or "worm-holes") from consideration, as they do not seem to be particle-like in the same sense as other primes. By a *geon* then, we mean a "quantized prime" regarded as a particle, or in other words, that object in the quantum theory to which the irreducible prime submanifold corresponds.

To list all possible geons is unfortunately impossible, because there exist an infinite number of primes, not all of which are known. Nevertheless, it is not difficult to visualize an arbitrary geon in a general sort of way: it is the result of excising a polyhedron from $\mathbb{R}^3$, and then performing appropriate identifications on the boundary faces created by the excision. One knows that every prime can be made in this manner.

At this point, we must recall also that the presence of non-Euclidean spatial topology implies the existence of distinct quantum sectors of any theory that includes gravity. Moreover, if we assume a fixed spatial topology, then these sectors do not communicate. Specifically, if we assume first that

$$^4M = \mathbb{R} \times M$$

(M being the topology of the spatial slices), and if we assume further that the meaningful assertions of the theory are all diffeomorphism invariant ("general covariance"), then we get *a distinct quantum sector for each UIR of the MCG of M*, where 'UIR' stands for 'unitary irreducible representation' and 'MCG' stands for 'mapping class group'. In fact, the mapping class group of a manifold M (also called "homeotopy group" or "group of large diffeos") is the analog for gravity of the group of large gauge transformations in a gauge theory. Its formal definition is

$$G = \pi_0(\mathrm{Diff}^\infty(M)) := \mathrm{Diff}^\infty(M)/\mathrm{Diff}_0^\infty(M),$$

where $\mathrm{Diff}^\infty(M)$ is the group of diffeomorphisms of M that are trivial at infinity and $\mathrm{Diff}_0^\infty(M)$ is its connected subgroup. (An asymptotically flat approximation should be good for all but cosmological considerations.) In order to understand the different quantum sectors, we thus have to understand the structure of the homeotopy group G and then to use this information to analyze and interpret the different possible UIR's of G.

III THE STRUCTURE OF THE HOMEOTOPY GROUP

In this task a major help is that fact that G is generated[3] by only three types of diffeomorphism, each with a clear physical meaning. The three categories

[3] This assertion is fully established in the mathematical literature for the orientable case, and appears to be true in the nonorientable case as well. More generally various statements we will make about the structure of G are in some cases known only under the assumption that the Poincaré conjecture is true, or that "homotopy implies isotopy" for the primes in question, or that the primes in question are "sufficiently large". All our statements are in any case known to be true for large families of primes, and are plausibly true in general. If they fail for certain primes, then the analysis given here will remain true as long as those primes are absent from the decomposition (1).

of generators are the *exchanges*, the *internal diffeomorphisms* and the *slides.* Like every diffeomorphism, each of these generators can be viewed as the result of a certain *process* (a "development" [2]), with the nature of the process being suggested by the name of the category. Thus, an exchange is the result of a process in which two identical primes continuously change places and a slide is the result of a process in which one prime travels around a loop threading through one or more other primes, while an internal diffeomorphism is a diffeomorphism whose support is restricted to a single prime. For example, to visualize an exchange of two handles in 2D, imagine the manifold as a rubber sheet, and imagine taking hold of the handles and dragging them around until they have changed places.

We spoke just now of "generators", but actually the slides and internals are already complete subgroups of G, while the exchanges generate a subgroup of G isomorphic to the group of all permutations of the identical (diffeomorphic) primes among themselves. The basic group theoretical fact we will need for our analysis is then that

$$G = (slides) \ltimes (internals) \ltimes (perms) \tag{2}$$

where the symbol $\ltimes$ denotes semidirect product (with the normal subgroup on the left). What this says more concretely is that every element of G is uniquely a product of three diffeomorphism-classes, one from each subgroup, and that each subgroup is invariant under conjugation by elements of the subgroups standing to its right in (2).

IV THE UIR'S OF A SEMIDIRECT PRODUCT GROUP

That fact that G is a semidirect product lets us analyze its UIR's in terms of representations of its factor groups and their subgroups. Indeed, there exists a very general analysis of the UIR's of a semidirect product, which is explained in detail in [1]. In essence it says the following. Let

$$G = N \ltimes K$$

be a semidirect product with N being the normal subgroup. A finite dimensional UIR of G is then determined by the following data

- Γ = a UIR of N
- T = a PUIR of $K_0 \subseteq K$,

where K_0 is the subgroup of K that remains "unbroken by Γ". (Often one calls K_0 "the little group".)

Perhaps the meaning of "unbroken" here can be illustrated most easily with the classic example of irreducible representations of the Poincaré group.

There G is the Poincaré group itself, $N \simeq \mathbb{R}^4$ is its translation subgroup, and the quotient $K \simeq G/N$ is the Lorentz group. A choice of UIR Γ of $\mathbb{R}^4$ is then nothing but a choice of four-momentum P^μ, and (assuming that P^μ is timelike) the subgroup $K_0 \subseteq K$ left unbroken by this choice is $SO(3)$, the group of spatial rotations in the center of mass frame of P^μ. The structure theorem for UIR's of semidirect products therefore tells us that we get a UIR of the Poincaré group (in this case an infinite dimensional UIR) by choosing a 4-momentum and a UIR T of $K_0 = SO(3)$. In particular language, we get a UIR by choosing two parameters, a mass[4] and a spin. In particular the "internal" properties of the particle (it's spin) are determined by the representation T of K_0.

For geons the formal situation is analogous, and the key interpretive point for us will be that the *statistics* of the geons will be determined by a representation of the "unbroken subgroup" of the group of permutations of identical primes.

Finally a comment on the "P" which occurs above in the phrase "PUIR of K_0." It stands for "projective", and reflects the fact that, even if one is seeking only ordinary representations of G, one may have to consider projective UIR's of K_0, i.e. representations up to a phase. Whether or not this occurs depends on the case in question; and when it does occur, the particular equivalence class of projective multipliers σ for K_0 which one must use is determined by the properties of the UIR Γ and how K_0 acts on it. (In the case of the Poincaré group, it does not occur, which is why we have to consider just integer spins, unless we want a spinorial representation of the overall group G itself.)

With these preparations complete, we can proceed to analyze the UIR's of the MCG of our spatial manifold M, and thereby the quantum sectors of gravity on $\mathbb{R} \times M$. One may distinguish two situations, according to whether the slide subgroup is represented trivially or not.

V THE SECTORS WITH TRIVIAL SLIDES

In the simpler case of UIR's G which annihilate the slides, we may give in effect a complete classification. In this case, the mathematical problem is reduced to finding the UIR's of the quotient group, $G/(slides)$, which by (2) is just the semidirect product

$$(internals) \ltimes (perms). \tag{3}$$

Let us find the finite dimensional UIR's of this group (sometimes called the "particle group"), when only a single type of prime P is present in the decom-

[4] Only the mass counts, according to the theorem, because UIR's Γ of N which belong to the same K-orbit (in this case P^μ's lying on the same mass shell) yield equivalent representations of G.

position (1). In this situation the permutation subgroup is just

$$K = (perms) = \mathcal{S}_N,$$

the full permutation group on N elements (N being the number of copies of P which are present), and the group of internal diffeos is the direct product of N copies of the corresponding group $G^{(1)}$ for a single prime P:

$$N = (internals) = G^{(1)} \times G^{(1)} \times \cdots \times G^{(1)}.$$

¿From this last equation it follows immediately that the most general UIR of the normal subgroup N is itself a product, namely the tensor product

$$\Gamma = \Gamma_a \otimes \Gamma_a \cdots \Gamma_b \otimes \Gamma_b \cdots \Gamma_c. \tag{4}$$

Notice here that although all the underlying primes are all identical, there is no reason for all the factors in (4) to be so. Rather, we can choose an independent UIR of $G^{(1)}$ for each prime summand, and in the above formula, the subscripts $a, b, \ldots c$ label the different equivalence classes among them. Physically a given Γ_a specifies a certain "internal structure" for the corresponding geon, and is therefore a "species parameter" or "quantum number", analogous in many ways to the spin of a rigid nucleus or molecule in its body-centered frame.

With respect to the choice (4), it is intuitively clear (and true as well) that the unbroken subgroup $K_0 \subseteq (perms)$ reduces to a product of permutation groups,

$$K_0 = \mathcal{S}_{N_a} \times \mathcal{S}_{N_b} \times \cdots \times \mathcal{S}_{N_c}, \tag{5}$$

where N_a is the number of occurrences of the representation Γ_a, N_b of Γ_b, etc. The statistics is then given by a UIR T of K_0, that is to say by an independent UIR T_a, $T_b, \cdots$ for each of the subgroups $\mathcal{S}_{N_a}$, $\mathcal{S}_{N_b}, \cdots$. Each of these T's in turn, can be specified by a choice of a Young tableau, and determines whether the corresponding geons will manifest Bose statistics, Fermi statistics or some particular parastatistics. Since there is no restriction on the choice of T, there is no restriction on which combinations of these possible statistics can occur.

Notice that these conclusions (deriving from the structure of K_0, as given in (5)) are entirely consistent with our interpretation of different choices of internal UIR in (4) as yielding physically distinct geons — geons of different "species". In this sense, we can say that there occurs a *quantum breaking of indistinguishability* conditioned by the choice of representation of (*internals*). This phenomenon, together with the possibility of assigning an arbitrary statistics to each such resulting species, exhausts the possibilities inherent in UIR's of the group (3). Thus, all possible sectors with trivial slides are accounted for by specifying

- a *species* for each geon (i.e. a UIR of $G^{(1)}$)
- a *statistics* for each resulting set of identical geons

VI SOME SECTORS WITH NONTRIVIAL SLIDES

When the slide subgroup is represented nontrivially, we are unable to give a full classification of the possible UIR's of G, due primarily to the difficulty of analyzing the UIR's of $(slides)$, but also due in part to the relative complexity of the manner in which $(internals)$ acts on these UIR's. Instead, let us consider a special case which avoids most of these complications, by choosing a prime that lacks internal diffeomorphisms, and then limiting ourselves mainly to abelian UIR's of the slides.

The prime in question is $\mathbb{R}P^3$, which can be visualized as a region of M produced by excising a solid ball and then identifying antipodal pairs of points on the resulting S^2 boundary. Since the internal group is trivial for this prime, we can concentrate on the effects of the slides. For each pair of $\mathbb{R}P^3$'s, one can slide one through the other, with the square of this slide being trivial (since $\pi_1(\mathbb{R}P^3) = \mathbf{Z}_2$), making a total of $N(N-1)$ independent order 2 generators. The complete group $(slides)$ is then generated by products of these elementary slides, subject (when $N > 2$) to certain geometrically evident commutation relations, like the fact that slides involving disjoint subsets of the primes commute with each other. For abelian representations of $(slides)$, all of the commutation relations will of course be satisfied trivially.

Since for $\mathbb{R}P^3$, $(internals)$ is trivial, the MCG reduces to

$$G = (slides) \ltimes (perms)$$

Hence, according to the general scheme outlined earlier, we get a UIR of G by choosing first a UIR Γ of $N = (slides)$, and then a PUIR T (with the correct projective multiplier σ) of the resulting unbroken subgroup $K_0 \subseteq (perms)$. As before, we may interpret K_0 as describing the surviving indistinguishability of the geons, and T as describing the statistics within each set of identical geons. Here we will just quote the results of this analysis, referring the listener to [1] for more details.

VII A SINGLE PAIR OF $\mathbb{R}P^3$'S

This case is simple enough that we can classify all the UIR's of G, without limiting ourselves to abelian representations Γ. The quantum sectors comprise a 1-parameter continuous family together with a handful of discrete cases. Aside from the long-known violation of the spin-statistics correlation which occurs in one of the sectors, the intriguing new result is that there exist other sectors where the Bose-Fermi distinction becomes ambiguous in a certain sense. In these sectors the permutation group remains unbroken ($K_0 = \mathbf{Z}_2$), but there is no natural way to say which of its two representations describes a pair of bosons and which a pair of fermions! This happens because the UIR's

Γ of (*slides*) and T of (*perms*) mix in such a way that it apparently becomes meaningless to identify either UIR of T as the trivial one.

VIII A TRIO OF $\mathbb{R}P^3$'S

Now we revert to the special case where Γ is abelian, meaning in effect that it merely associates a sign with each ordered pair of primes. We can represent the various such Γ pictorially by drawing three dots to represent the three primes and an arrow to represent each ordered pair that receives a minus sign (meaning the a slide of the first prime through the second produces a phase-factor of -1). Each distinct diagram of this type then gives rise to a different class of UIR's of (*slides*), and therefore furnishes a different building block for constructing UIR's of G.

Perhaps the most interesting abelian UIR of (*slides*) comes from the cyclic graph in which the three dots and arrows form a circle. Clearly this pattern leaves $\mathbf{Z}_3 \subseteq (perms)$ as the unbroken subgroup K_0, so we acquire three distinct UIR's of G, corresponding to the three possible UIR's of $\mathbf{Z}_3$. What is remarkable here is first of all the pattern of geon identity, which is expressed not by a permutation group $\mathcal{S}_n$ at all, but by the cyclic group $\mathbf{Z}_3$. With this new type of group comes a new type of statistics, in which a cyclic permutation of the geons produces the complex phase q or $\bar{q}$, $q = 1^{1/3}$ being a cube-root of unity.[5]

Although this pattern of identity is unusual for the simple type of particle that physics usually deals with, it has obvious precedents in the social world of human beings. There it might happen, for example, that three people could stand in a triangle so that each was the teacher (in some different subject) of the one to his/her right. The pattern would then be preserved by a cyclic permutation, but not if two of the people changed places while the third stayed put.

IX MORE THAN THREE $\mathbb{R}P^3$'S

An interesting possibility in this case is that of "projective statistics", meaning a type of statistics expressed by a properly projective representation of the permutation group or one of its subgroups.[6] We do not have an example yet, but there seems to be no good reason why one shouldn't exist. We would need at least four geons because $\mathcal{S}_n$ possesses properly projective representations

---

[5]) The same UIR of $\mathbf{Z}_3$ also occurs in connection with parastatistics, where it represents only a proper subspace of the full state-space. Indeed, any UIR of any finite group can be realized in connection with parastatistics, since any finite group is a subgroup of some permutation group.

[6]) Unlike the "cyclic statistics" we just met, a projective statistics would not be realizable in connection with parastatistics.

only for $n \geq 4$. We would also need a non-abelian UIR Γ of $(slides)$, because in the contrary case, the projective multiplier σ will always be trivial. Thus, the simplest example one might try to construct, would employ a two dimensional representation Γ, so chosen that the unbroken subgroup $K_0 \subseteq \mathcal{S}_4$ would come out as the subgroup $\mathcal{S}_4^{even}$ of even permutations, and σ would come out as the projective multiplier for the spin=1/2 representation of the symmetry group of a regular tetrahedron (to which $\mathcal{S}_4^{even}$ is isomorphic).

From the above examples it emerges clearly, we believe, that the possibility of non-Euclidean spacetime topology introduces a once unexpected richness into quantum gravity. In particular it brings with it

- topological particles (geons)
- half integer spin in pure gravity
- distinct quantum sectors (both continuous and discrete families of them)
- quantum multiplicity

We have not really discussed the second of these, and we did not mention the third at all before now, but we list them here to help illustrate the fertility of topology in the quantum context.

We have seen, moreover, that the mathematical structure theorems for representations of semidirect product groups provide a remarkably natural physical description of the different sectors which can occur. Indeed these theorems read almost as if they had been expressly designed to describe the representations in the language of quantum particles and their properties! In this language identical geon statistics is expressed by a (possibly projective) unitary irreducible representation of the unbroken subgroup of the group of permutations of identical primes. Two noteworthy features of the resulting interpretation are that

- All the familiar types of statistics can occur, together with some new, unexpected types.
- New patterns of particle identity occur, in which not all permutations of the identical particle leave the physics invariant.

It is interesting that the most novel of these features are associated with the slide diffeomorphisms, which correspond physically to processes in which one geon "slides thru" another. For this reason it seems possible that analogous condensed matter effects could occur with objects like vortices, which also can slide through one another in an obvious manner [3].)

Beyond the existence of sectors associated with new types of particle statistics and indistinguishability, it seems that continuous parameters will also occur (we saw an example in the case of two $\mathbb{R}P^3$ geons), so that a great multiplicity of sectors can be expected to exist, even with a given spatial topology. This answers our initial question about new topological parameters, and it seems that the answer is that there are many —probably too many in fact, since they can be chosen so that the spin-statistics correlation fails, indicating[7] that quantum gravity is more akin to a phenomenological theory than a fundamental one. One can expect that unfreezing the topology will remove some of these unwanted sectors, but we would conjecture that a deeper, discrete theory will be needed to restore a physically reasonable degree of uniqueness to quantum gravity.

X ACKNOWLEDGEMENTS

This research was partly supported by NSF grant PHY-9600620.

REFERENCES

1. R.D. Sorkin and S. Surya, "An Analysis of the Representations of the Mapping Class Group of a Multi-geon Three-manifold" ⟨e-print archive: gr-qc/9605050⟩.
2. R.D. Sorkin, "Introduction to Topological Geons", in P.G. Bergmann and V. de Sabbata (eds.), *Topological Properties and Global Structure of Space-Time*, pp. 249-270 (Plenum, 1986).
3. A.P. Balachandran, private communication.

[7] If geons propagating in a flat ambient metric admit an approximate description in terms of an effective flat-space quantum field theory, then they must satisfy the standard spin-statistics correlation.

The Structures Underlying Soliton Solutions in Integrable Hierarchies

Luiz A. Ferreira[1]

Instituto de Física Teórica - IFT/UNESP
Rua Pamplona 145
01405-900, São Paulo - SP, Brazil

Abstract. We point out that a common feature of integrable hierarchies presenting soliton solutions is the existence of some special "vacuum solutions" such that the Lax operators evaluated on them, lie in some abelian subalgebra of the associated Kac-Moody algebra. The soliton solutions are constructed out of those "vacuum solitons" by the dressing transformation procedure.

This talk is concerned with the structures responsible for the appearance of soliton solutions for a large class of non linear differential equations. In spite of the great variety of types of equations presenting soliton solutions, some basic features seem to be common to all of them. Practically all such theories have a representation in terms of a zero curvature condition [1], and the corresponding Lax operators lie in some infinite dimensional Lie algebra, in general a Kac-Moody algebra $\hat{\mathcal{G}}$. We argue that one of the basic ingredients for the appearance of soliton solutions in such theories is the existence of "vacuum solutions" corresponding to Lax operators lying in some abelian (up to central term) subalgebra of $\hat{\mathcal{G}}$. Using the dressing transformation procedure [2] we construct the solutions in the orbit of those vacuum solutions, and conjecture that the soliton solutions correspond to some special points in those orbits. The talk is based on results obtained in collaboration with J.L. Miramontes and J. Sanchez Guillén and reported in ref. [3].

We consider non-linear integrable hierarchies of equations which can be formulated in terms of a system of first order differential

$$\mathcal{L}_N \Psi = 0 \,, \qquad \mathcal{L}_N \equiv \frac{\partial}{\partial t_N} - A_N \tag{1}$$

where the variables t_N are the various "times" of the hierarchies, and their

[1]) Partially supported by a CNPq research grant

CP400, *First Latin American Symposium on High Energy Physics/ VII Mexican School of Particles and Fields*, edited by D'Olivo/Klein-Kreisler/Mendez

number may be finite or infinite. The equations of the hierarchies are then equivalent to the integrability or zero-curvature conditions of (1)

$$[\mathcal{L}_N \, , \, \mathcal{L}_M] = 0 \, . \tag{2}$$

Therefore, the Lax operators are "flat connections"

$$A_N = \frac{\partial \Psi}{\partial t_N} \Psi^{-1} \, . \tag{3}$$

The type of integrable hierarchy considered here is based on a Kac-Moody algebra $\hat{\mathcal{G}}$, furnished with an integral gradation

$$\hat{\mathcal{G}} = \bigoplus_{i \in \mathbb{Z}} \hat{\mathcal{G}}_i \qquad \text{and} \qquad [\hat{\mathcal{G}}_i, \hat{\mathcal{G}}_j] \subseteq \hat{\mathcal{G}}_{i+j} \, . \tag{4}$$

The connections are of the form

$$A_N = \sum_{i=N_-}^{N_+} A_{N,i} \, , \quad \text{where} \quad A_{N,i} \in \hat{\mathcal{G}}_i \tag{5}$$

where N_- and N_+ are non-positive and non-negative integers, respectively.

We assume that the hierarchy possesses at least one vacuum solution such that the connections evaluated on such solution has the form

$$A_N^{(\mathrm{vac})} \; = \; \sum_{i=N_-}^{N_+} c_N^i b_i + f_N(t) \, C \; \equiv \; \varepsilon_N + f_N(t) \, C \, . \tag{6}$$

where C is the central element of $\hat{\mathcal{G}}$ and $b_i \in \hat{\mathcal{G}}_i$ are generators of a subalgebra of $\hat{\mathcal{G}}$ satisfying

$$[b_j, b_k] = j \; \beta_j \; C \; \delta_{j+k,0} \tag{7}$$

with β_j being some complex numbers ($\beta_{-j} = \beta_j$), c_N^i are constants, and $f_N(t)$ are functions ot the times t_N. [2] These vacuum potentials correspond to the solution of the associated linear problem (1)-(3), given by the group element

$$\Psi^{(\mathrm{vac})} = \exp\left(\sum_N \varepsilon_N t_N \; + \; \gamma(t) \, C \right) \tag{8}$$

where the numeric function $\gamma(t)$ is a solution of the equations

$$\frac{\partial \gamma(t)}{\partial t_N} = f_N(t) + \frac{1}{2} \sum_{M,i} i \; \beta_i \; c_N^i \; c_M^{-i} \, t_M \, . \tag{9}$$

---

2) As a consequence of (2), those functions have to satisfy $\frac{\partial f_N(t)}{\partial t_M} - \frac{\partial f_M(t)}{\partial t_N} = \sum_i i \beta_i c_M^i c_N^{-i}$

We now consider the dressing transformations which map known solutions of the hierarchy into new solutions [2]. Denote by $\hat{G}_-$, $\hat{G}_+$, and $\hat{G}_0$ the subgroups of the Kac-Moody group $\hat{G}$ formed by exponentiating the subalgebras $\hat{\mathcal{G}}_{<0} \equiv \bigoplus_{i<0} \hat{\mathcal{G}}_i$, $\hat{\mathcal{G}}_{>0} \equiv \bigoplus_{i>0} \hat{\mathcal{G}}_i$, and $\hat{\mathcal{G}}_0$, respectively. According to Wilson [4], the dressing transformations can be described in the following way. Consider a solution Ψ of the linear problem (1), and let $\rho = \rho_- \, \rho_0 \, \rho_+$ be a constant element in the "big cell" of $\hat{G}$, *i.e.*, in the subset $\hat{G}_- \, \hat{G}_0 \, \hat{G}_+$ of $\hat{G}$, such that

$$\Psi \, \rho \, \Psi^{-1} = (\Psi \, \rho \, \Psi^{-1})_{<0} \, (\Psi \, \rho \, \Psi^{-1})_0 \, (\Psi \, \rho \, \Psi^{-1})_{>0} \, . \tag{10}$$

Notice that these conditions are equivalent to say that both ρ and $\Psi \, \rho \, \Psi^{-1}$ admit a generalized Gauss decomposition with respect to the gradation (4). Define

$$\begin{aligned} \Psi^\rho &= \Theta_-^{(0)} [(\Psi \, \rho \, \Psi^{-1})_{<0}]^{-1} \, \Psi \rho \equiv \Theta_- \, \Psi \, \rho \\ &= \Theta_+^{(0)} \, (\Psi \, \rho \, \Psi^{-1})_{>0} \, \Psi \; \equiv \Theta_+ \, \Psi \end{aligned} \tag{11}$$

where

$$\Theta_-^{(0)^{-1}} \, \Theta_+^{(0)} = \left(\Psi \rho \Psi^{-1} \right)_0 . \tag{12}$$

Then, Ψ^ρ is another solution of the linear problem. Indeed, one can check that by exploring the gradation of $\hat{\mathcal{G}}$ and the fact that the dressing transformation is written in two ways (11), that the transformed connection lie in the same subspace of $\hat{\mathcal{G}}$ as the original one, i.e.

$$A_N^\rho = \frac{\partial \Psi^\rho}{\partial t_N} \, (\Psi^\rho)^{-1} \in \bigoplus_{i=N_-}^{N_+} \hat{\mathcal{G}}_i \, , \tag{13}$$

We now consider the orbit of the vacuum solution (8) under the group of dressing transformations. For each constant group element ρ one gets a new solution, out of the vacuum solution, by the transformation $\Psi^{(\mathrm{vac})} \mapsto \Psi^\rho = \Theta_- \, \Psi^{(\mathrm{vac})} \rho = \Theta_+ \, \Psi^{(\mathrm{vac})}$. The the vacuum connection $A_N^{(\mathrm{vac})}$ becomes

$$\begin{aligned} A_N^\rho - f_N(t) \, c &= \Theta_- \, \varepsilon_N \, {\Theta_-}^{-1} \; + \partial_N \Theta_- \, {\Theta_-}^{-1} \; \in \bigoplus_{i \leq N_+} \hat{\mathcal{G}}_i \\ &= \Theta_+ \, \varepsilon_N \, {\Theta_+}^{-1} \; + \partial_N \Theta_+ \, {\Theta_+}^{-1} \; \in \bigoplus_{i \geq N_-} \hat{\mathcal{G}}_i, \end{aligned} \tag{14}$$

The components $A_{N,i}$ in (5), of the connection are functionals of the fields of the hierarchy. Then one can consider (14) as a local change of variables, and use it to relate the parameters of the group elements Θ_+ and Θ_-, to the fields of the hierarchy. In fact, one can choose a suitable set of parameters to write the fields in terms of them. The value of that particular set of parameters evaluated on the solution can be obtained by considering matrix elements of the form

$$\langle \mu \mid \Theta_-{}^{-1}\, \Theta_+ \mid \mu' \rangle = \langle \mu \mid e^{\sum_N \varepsilon_N t_N}\, \rho\, e^{-\sum_N \varepsilon_N t_N} \mid \mu' \rangle \,, \tag{15}$$

where $\mid \mu\rangle$ and $\mid \mu'\rangle$ are vectors in a given representation of $\hat{\mathcal{G}}$. The appropriate set of vectors is specified by the condition that all the required components of Θ_- and Θ_+, used to parametrize the fields, can be expressed in terms of the resulting matrix elements. It turns out [3] that the required matrix elements, considered as functions of the group element ρ, constitute the generalization of the Hirota's tau-functions for these hierarchies [5]. Moreover, Eq. (15) is the analogue of the, so called, solitonic specialization of the Leznov-Saveliev solution proposed in [6–9] for the affine (abelian and non-abelian) Toda theories.

Consider now the common eigenvectors of the adjoint action of the ε_N's that specify the vacuum solution (6). Then, the important class of multi-soliton solutions is conjectured to correspond to group elements ρ which are the product of exponentials of eigenvectors

$$\rho = e^{F_1}\, e^{F_2} \ldots e^{F_n} \,, \qquad [\varepsilon_N \,,\, F_k] = \omega_N^{(k)}\, F_k \,, \quad k = 1, 2, \ldots n \,. \tag{16}$$

In this case, the dependence of the solution upon the times t_N can be made quite explicit

$$\langle \mu \mid \Theta_-^{-1}\, \Theta_+ \mid \mu' \rangle = \langle \mu \mid \prod_{k=1}^{n} \exp(e^{\sum_N \omega_N^{(k)} t_N} F_k) \mid \mu' \rangle \,. \tag{17}$$

We emphasize that not all solutions of the type (17) are soliton solutions, but we conjecture that the soliton and multi-soliton solutions are among them. The conjecture that multi-soliton solutions are associated with group elements of the form (16) naturally follows from the well known properties of the multi-soliton solutions of affine Toda equations and of hierarchies of the KdV type, and, in the sine-Gordon theory, it has been explicitly checked in ref. [10]. Actually, in all these cases, the multi-soliton solutions are obtained in terms of representations of the "vertex operator" type where the corresponding eigenvectors are nilpotent. Then, for each eigenvector F_k there exists a positive integer number m_k such that $(F_k)^m \neq 0$ only if $m \leq m_k$. This remarkable property simplifies the form of (17) because it implies that $e^{F_k} = 1 + F_k + \cdots + (F_k)^{m_k}/m_k!$, which provides a group-theoretical justification of Hirota's method [5].

An interesting feature of the dressing transformations method is the possibility of relating the solutions of different integrable equations. Consider two different integrable hierarchies whose vacuum solutions are compatible, in the sense that the corresponding vacuum Lax operators commute. Then, one can consider the original integrable equations as the restriction of a larger hierarchy of equations. Consequently, the solutions obtained through the group of dressing transformations can also be understood in terms of the solutions of the larger hierarchy, which implies certain relations among them. (see section 4 of [8] for more details).

The developments described here lead to a quite general definition of tau functions for such hierachies, in terms of integrable highest weight representations of the associated Kac-Moody algebra [3].

ACKNOWLEDGEMENTS

The author is very grateful to the organizers of I SILAFAE and VII EMPC, and to CLAF (Centro Latino Americano de Física) for the financial support.

REFERENCES

1. V.E. Zakharov, A.B. Shabat, *Functional Analysis and its Application* **13** (1979) 166.
2. E. Date, M. Jimbo, M. Kashiwara, T. Miwa, *Proc. Japan. Acad.* **57** (1981) 3806; *Physica* 4 (1982) 343; *Publ. RIMS Kyoto University* **18**, (1982) 1077; M. Semenov-Tian-Shansky, *Functional Analysis and its Application* **17** (1983) 259; *Publ. RIMS Kyoto Univ.* **21** (1985) 1237.
3. L.A. Ferreira, J.L. Miramontes and J. Sánchez Guillén, *Journal of Mathematical Physics* **38** # 2, (1997) , hep-th/9606066.
4. G. Wilson, *Phil. Trans. R. Soc. Lond.* **A 315** (1985) 383; *Habillage et fonctions τ C. R. Acad. Sc. Paris* **299 (I)** (1984) 587; *The τ-Functions of the gAKNS Equations*, in "Verdier memorial conference on integrable systems" (O. Babelon, P. Cartier, and Y. Kosmann-Schwarzbach, eds.), Birkhauser (1993) 131-145.
5. R. Hirota, *Direct methods in soliton theory*, in "Soliton" (R.K. Bullough and P.S. Caudrey, eds.), (1980) 157.; J. Phys. Soc. Japan **33** (1972) 1459.
6. D. Olive, N. Turok and J.W.R. Underwood, *Nucl. Phys.* **B401** (1993) 663-697.; *Nucl. Phys.* **B409** (1993) 509-546, hep-th/9305160.
7. D. Olive, M.V. Saveliev and J.W.R. Underwood, *Phys. Lett.* **B311** (1993) 117-122, hep-th/9212123.
8. L.A. Ferreira, J.L. Miramontes and J. Sánchez Guillén, *Nucl. Phys.* **449** (1995) 631-679, hep-th/9412127.
9. L.A. Ferreira, J-L. Gervais, J. Sánchez Guillén and M.V. Saveliev, *Affine Toda Systems Coupled to Matter Fields, Nucl. Phys.* **B470** (1996) 236, hep-th/9512105.
10. O. Babelon and D. Bernard, *Int. J. Mod. Phys.* **A8** (1993) 507-543, hep-th/9206002.

The Nambu-Goto model, perturbation theory at finite temperature and the ϕ^4 model

G. Germán

Laboratorio Cuernavaca, Instituto de Física, UNAM
Apdo. Postal 48-3,
62251 Cuernavaca, Mor., Mexico

Abstract. We suggest that the ϕ^4 model is only a polynomial approximation to a more fundamental theory. We illustrate our conjecture by using the Nambu-Goto string model as an example. Here, we compare a two-loop calculation with an exact derivation (in the large-d limit) of the free energy. We find why is that the perturbative approach fails to reproduce important features of the exact result and suggest a possible analogous behavior in the ϕ^4 model.

The purpose of this note is to discuss the validity of the ϕ^4 model at high temperature. We suggest that the ϕ^4 model is only a polynomial approximation to a more fundamental theory. Although no proof of this is given we use the Nambu-Goto string model to exemplify how this can happen. For this we compare a two-loop calculation of the free energy or static quark-antiquark potential at finite temperature with an exact calculation in the large-d limit, where d is the number of dimensions of the space were the string evolves. We are able to understand why is that the perturbative calculation fails to reproduce important features of the exact calculation and suggest that perhaps something similar occurs with the ϕ^4 model. The Nambu-Goto model in Euclidean space is given by the following action

$$A = M^2 \int d^2\xi \sqrt{g}, \tag{1}$$

where g is the determinant of the metric

$$g_{ij} = \partial_i x^\mu(\xi_i) \partial_j x^\mu(\xi_i), \qquad i = 0, 1, \tag{2}$$

and $x^\mu(\xi_i)$ are the string coordinates. We choose a gauge called *physical gauge*

$$x^\mu(\xi_i) = (t, r, u^a), \tag{3}$$

CP400, *First Latin American Symposium on High Energy Physics/ VII Mexican School of Particles and Fields,* edited by D'Olivo/Klein-Kreisler/Mendez

where $u^a = u^a(t,r)$, $a = 2,3,...(d-2)$ are the $(d-2)$ transverse oscillations of the string. For the two-loop calculation we are interested in, the Nambu-Goto model Eq. (1) can be written as

$$A = M^2 \int dt dr [1 + \frac{1}{2}\vec{u}_i^2 + \frac{1}{8}\vec{u}_i^4 - \frac{1}{4}(\vec{u}_i \cdot \vec{u}_j)^2]. \tag{4}$$

We then evaluate the quadratic part of the action by using the functional integral

$$Z = \int D[u^a] e^{-A}, \tag{5}$$

and the quartic order terms by Wick's theorem for a system with thermodynamic boundary conditions. These are required by a string with fixed ends at finite temperature; the momenta and frequencies are

$$k_n = \frac{n\pi}{R} \qquad \omega = 2m\pi T \qquad n = 1,2,..., \qquad m = 0, \pm 1, \pm 2, ..., \tag{6}$$

where R is the (extrinsic) length of the string and T the temperature. The static potential is then given by

$$V(R,T) = V_0 + V_1 + V_2, \tag{7}$$

where V_0 is the classical part, V_1 and V_2 are the one and two-loop contributions, respectively (see [1] for details). The important point is that for temperatures T bigger than the deconfinement temperature T_{dec} there is a region of R values, which we denote $\mathcal{R}$, where $V_2 > V_1$. This result clearly invalidates the perturbative calculation. However outside the region $\mathcal{R}$, $V_2 < V_1$ even for $T > T_{dec}$, thus apparently giving reliable results everywhere else even for high temperatures. In the exact calculation [2] no expansion of Eq. (1) is required and we find that for $T > T_{dec}$ the potential exists although only for strings up to certain length i.e., up to certain value R_{max}. We understand this phenomena as a possible indication that the string decays, with the potential becoming imaginary. The interesting observation is that the values of R_{max} for which the potential stops in the exact calculation lie, in the perturbative approach, in the region where $V_2 > V_1$ for a given temperature $T > T_{dec}$ with V_2 becoming again less than V_1 for R bigger than R_{max}. Thus one could say that the perturbative approach is well behaved except in a small region of R-values. Having the exact result, however, we know that this is not the case; it is an artefact of the perturbative calculation. The point is that the perturbative approach is jumping the singularity of the exact model where the string decays. The reason why this occurs turns out to be very simple i.e., to do the perturbative calculation we had to expand, from the very beginning, the model Eq. (1) obtaining Eq. (4) thus missing the posibility of encountering the singularity of the exact model. There is, however, some trace of this singularity (smoothened by the expansion in the perturbative approach) and this is the region $\mathcal{R}$ where V_2 becomes bigger than V_1. In conclusion had we only had the perturbative result (as is the case for most models) we would probably believe that high temperature calculations make sense everywhere except in the small region $\mathcal{R}$ where $V_2 > V_1$. In the ϕ^4 model it is usually said that the perturbative approach at finite temperature is fine everywhere except in a small region of ϕ values around the critical temperature where the effective potential diverges. Here the problem is dealt with by including an infinite set of daisy and superdaisy diagrams. On the other hand it has been argued recently that the ϕ^4 model has no particle interpretation for $m^2 > 0$ (since there is no ϕ^2 term in the effective potential) [3]. For large temperatures the ϕ particle probably disappears [4]. This looks very similar to the problem just discussed in the Nambu-Goto although, of course, these are completely different models. Thus it is very tempting to suggest that the problems in the ϕ^4 model at temperatures $T \geq T_c$ are only an indication of the possibility that we are working with an approximate expanded model of a more fundamental theory. The ϕ^4 model as is known at present would be playing the role in field theory of the polynomial expansion given by Eq. (4) of the exact model Eq. (1), and some of the previous discussion in the string model could find an equivalent expression in the ϕ^4. If this speculation turns out to be true, and it should be investigated, then critical temperatures as well as symmetry restoration at high temperatures and high temperature expansions would have to be reconsidered.

ACKNOWLEDGMENT

This work was partially supported by CONACyT through project 1199-E9203.

REFERENCES

1. G. Germán, Perturbative Approach at Finite Temperature and the ϕ^4 model. Preprint, 1996.
2. A. Antillón and G. Germán, Phys. Rev. D47,4567(1993).
3. H.A. Al-Kuwari, Phys. Lett. B375,217(1996).
4. B.A. Campbell, J. Ellis and K.A. Olive, Phys. Lett. B235,325(1990).

Gauge-Yukawa Unification in $SU(5)$ Models

J. Kubo[a], M. Mondragón[b] and G. Zoupanos[c]

[a] *Faculty of Nat. Sciences, Kanazawa University, 920-11 Kanazawa, Japan*
[b] *Instituto de Física, UNAM, Apdo. Postal 20-364, México 01000 D.F.*
[c] *Physics Dept., Nat. Tech. Univ., GR-157 80 Zografou, Athens, Greece*

Abstract. We achieve Gauge-Yukawa Unification (GYU) by applying the principles of reduction of couplings and finiteness to two supersymmetric $SU(5)$ models. As a result we obtain, among others, predictions for the top quark mass which have passed successfully the tests of progressively more accurate measurements.

INTRODUCTION

Although the Standard Model (SM) has been very successful in describing the elementary particles interactions, it has a large number of parameters whose values are determined only experimentally. The usual way to reduce the number of free parameters, and thus render the theory more predictive, is to invoke a symmetry. Grand Unified Theories (GUTs) are good examples of this approach. Still, enlarging the symmetry does not always lead to more predictions at low energies.

A natural extension of the GUT idea is to relate the gauge and Yukawa sectors of a theory, that is, to have Gauge-Yukawa Unification (GYU). A natural way of achieving this is by introducing supersymmetry (SUSY), in particular $N = 2$ supersymmetry. However, $N = 2$ SUSY theories have serious phenomenological problems. The same applies to superstrings and composite models, which could in principle lead to relations among the gauge and Yukawa couplings.

Recently [1–3] we have considered the GYU which is based on the principles of reduction of couplings [2–4] and in addition finiteness [1,5–7]. These principles formulated in perturbation theory, are not explicit symmetry principles, although they might imply symmetries. The former principle is based on the existence of renormalization group invariant (RGI) relations among couplings which preserve perturbative renormalizability. The latter one is based on the fact that it is possible to find RGI relations among couplings that keep

CP400, *First Latin American Symposium on High Energy Physics/ VII Mexican School of Particles and Fields,*
edited by D'Olivo/Klein-Kreisler/Mendez

finiteness in perturbation theory, even to all orders [7]. Applying these principles, one can relate the gauge and Yukawa couplings, thereby improving the predictive power of a model.

GAUGE-YUKAWA UNIFICATION

Let us now sketch briefly the tools of this GYU scheme.

A RGI relation among couplings can be expressed in an implicit form $\Phi(g_1, \cdots, g_N) = 0$, which has to satisfy the partial differential equation

$$\mu \, d\Phi/d\mu = \sum_{i=1}^{N} \beta_i \, \partial\Phi/\partial g_i = 0, \tag{1}$$

where β_i is the β-function of g_i. There exist $(N-1)$ independent Φ's, and finding the complete set of these solutions is equivalent to solve the so-called reduction equations [4],

$$\beta_g \frac{dg_i}{dg} = \beta_i \ , \ i = 1, \cdots, N \ , \tag{2}$$

where g is the primary coupling and β_g its β-function, and i does not include g. Using all the $(N-1)$ Φ's to impose RGI relations, one can in principle express all the couplings in terms of a single coupling g. The complete reduction, which formally preserves perturbative renormalizability, can be achieved by demanding a power series solution

$$g_i = \sum_{n=0} \kappa_i^{(n)} \, g^{2n+1} \ , \tag{3}$$

whose uniqueness can be investigated at the one-loop level [4]. The completely reduced theory contains only one independent coupling with the corresponding β-function. In supersymmetric Yang-Mills theories with a simple gauge group it is possible to have the vanishing of the β-function to all orders in perturbation theory, if all the one-loop anomalous dimensions of the matter fields in the completely and uniquely reduced theory vanish identically [7], thus leading to a finite theory.

This possibility of coupling unification is attractive, but it can be too restrictive and hence unrealistic. To overcome this problem, one may use fewer Φ's as RGI constraints. This is the idea of partial reduction, [2–4] and the power series solution (3) becomes in this case

$$g_i = \sum_{n=0} \kappa_i^{(n)}(g_a/g) \, g^{2n+1}, \quad i = 1, \cdots, N' \ , \ a = N'+1, \cdots, N \ . \tag{4}$$

The coefficient functions $\kappa_i^{(n)}$ are required to be unique power series in g_a/g so that the g_a's can be regarded as perturbations to the completely reduced

TABLE 1. The predictions of the FUT $SU(5)$

m_{SUSY} [GeV]	$\alpha_3(M_Z)$	$\tan\beta$	M_{GUT} [GeV]	m_b [GeV]	m_t [GeV]
200	0.123	53.7	2.25×10^{16}	5.2	184.0
500	0.118	54.2	1.45×10^{16}	5.1	184.4

system in which the g_a's identically vanish. In the following, we would like to consider two very interesting models which are also representative of the two mentioned possibilities [1,2].

THE $SU(5)$ FINITE UNIFIED THEORY

This is a $N = 1$ supersymmetry Yang-Mills theory based on $SU(5)$ [6] which contains one **24**, four pairs of $(\mathbf{5} + \overline{\mathbf{5}})$-Higgses and three $(\overline{\mathbf{5}} + \mathbf{10})$'s for three fermion generations. A complete reduction has been done of the dimensionless parameters of the theory in favour of the gauge coupling g, and the unique power series solution [1] corresponds to the Yukawa matrices without intergenerational mixing. In the one-loop approximation it yields

$$g_t^2 = g_c^2 = g_u^2 = 8/5 g^2 \ , \tag{5}$$
$$g_b^2 = g_s^2 = g_d^2 = 6/5 g^2 \ , \tag{6}$$
$$g_\tau^2 = g_\mu^2 = g_e^2 = 6/5 g^2 \ , \tag{7}$$

where g_i's stand for the Yukawa couplings. At first sight, this GYU seems to lead to unacceptable predictions for the fermion masses. However, this is not the case, since each generation has its own pair of $(\overline{\mathbf{5}} + \mathbf{5})$-Higgses so that one may assume [6,1] that after the diagonalization of the Higgs fields the effective theory is exactly MSSM, where the pair of the Higgs supermultiplets mainly stems from the $(\mathbf{5} + \overline{\mathbf{5}})$ which couples to the third fermion generation. The Yukawa couplings of the first two generations can be regarded as free parameters. The predictions of m_t and m_b for various m_{SUSY} are given in Table 1.

THE MINIMAL SUPERSYMMETRIC $SU(5)$ MODEL

The field content of this model is minimal. We start with six Yukawa and two Higgs couplings, and neglect the CKM mixing. We then require GYU to occur among the Yukawa couplings of the third generation and the gauge coupling, and the theory to be completely asymptotically free. In the one-loop approximation, the GYU yields $g_{t,b}^2 = \sum_{m,n=1}^{\infty} \kappa_{t,b}^{(m,n)} \, h^m \, f^n \, g^2$ (h and f are

TABLE 2. The predictions of the minimal SUSY $SU(5)$

$m_{\rm SUSY}$ [GeV]	g_t^2/g^2	g_b^2/g^2	$\alpha_3(M_Z)$	$\tan\beta$	$M_{\rm GUT}$ [GeV]	m_b [GeV]	m_t [GeV]
300	0.97	0.57	0.120	47.7	1.8×10^{16}	5.4	179.7
500	0.97	0.57	0.118	47.7	1.39×10^{16}	5.3	178.9

related to the Higgs couplings). Where h is allowed to vary from 0 to 15/7, while f may vary from 0 to a maximum which depends on h and vanishes at $h = 15/7$. As a result, we obtain [2]

$$\begin{aligned} 0.97\, g^2 \lesssim\ & g_t^2 \lesssim 1.37\, g^2 \ , \\ 0.57\, g^2 \lesssim\ & g_b^2 = g_\tau^2 \lesssim 0.97\, g^2 \ . \end{aligned} \tag{8}$$

We found [8] that consistency with proton decay requires g_t^2, g_b^2 to be very close to the left hand side values in the inequalities. In Table 2 we give the predictions for representative values of m_{SUSY}.

To perform the above analyses we have used the RG technique and regarded the GYU relations as boundary conditions holding at the unification scale $M_{\rm GUT}$. We have assumed that it is possible to arrange the SUSY mass parameters along with the soft breaking terms in such a way that the desired symmetry breaking pattern occurs, all the superpartners are unobservable at present energies, there is a unique threshold $m_{\rm SUSY}$ for all the superpartners, and there is no contradiction with proton decay and other experimental constraints.

Using the updated experimental data on the SM parameters, we have reexamined the m_t prediction of the two GYU $SU(5)$ models described above [8], and obtained

$$\text{FUT}: \quad m_t = (183 + \delta_{m_t}^{MSSM} \pm 5)\ GeV \tag{9}$$

$$\text{Min. SUSY } SU(5): \quad m_t = (181 + \delta_{m_t}^{MSSM} \pm 3)\ GeV \tag{10}$$

where $\delta_{m_t}^{MSSM}$ stands for the MSSM threshold corrections. To obtain an idea about the magnitude of the correction we considered the case that all superpartners have the same mass m_{SUSY} and $m_{SUSY} \gg \mu_H$, where μ_H describes the mixing of the two Higgs doublets in the superpotential, and found [8] $\delta_{m_t}^{MSSM} \sim -1\%$.

CONCLUSIONS

We have achieved GYU by applying the principles of reduction of couplings and finiteness. In this way, we have shown that it is possible to construct some supersymmetric GUTs with GYU in the third generation that predict the bottom and top quark masses in accordance with recent experimental data, thus explaining the top-bottom hierarchy in these models.

The GYU scenario is the most predictive scheme as far as the mass of the top quark is concerned. We have analyzed [8] the infrared quasi-fixed-point behaviour of the m_t prediction in some detail. In particular we have seen that the *infrared value* for large $\tan\beta$ depends on $\tan\beta$ and its lowest value is $\sim$ 188 GeV. Comparing this with the experimental value $m_t = (176.8 \pm 6.5)$ GeV we may conclude that the present data on m_t cannot be explained from the infrared quasi-fixed-point behaviour alone.

To exclude or verify different GYU models, the experimental as well as theoretical uncertainties have to be further reduced. One of the largest theoretical uncertainties in FUT results from the not-yet-calculated threshold effects of the superheavy particles. Bringing these threshold effects under control will reduce the uncertainty of the m_t prediction. We have taken $\delta^{\rm MSSM} m_t$ as unknown because we do not have sufficient information on the superpartner spectra. Recently, however, we have demonstrated [9] how to extend the principle of reduction of couplings in a way as to include the dimensionfull parameteres. As a result, it is in principle possible to predict the superpartner spectra as well as the rest of the massive parameters of a theory.

REFERENCES

1. D. Kapetanakis, M. Mondragón and G. Zoupanos, *Zeit. f. Phys.* **C60** 181 (1993); M. Mondragón and G. Zoupanos, *Nucl. Phys.* **37C** 98 (1995).
2. J. Kubo, M. Mondragón and G. Zoupanos, *Nucl. Phys.* **B424** 291 (1994).
3. J. Kubo, M. Mondragón, N.D. Tracas and G. Zoupanos, *Phys. Lett.* **B342** 155 (1991).
4. W. Zimmermann, *Commun. Math. Phys.* **97** 211 (1985); R. Oehme and W. Zimmermann, *Commun. Math. Phys.* **97** 569 (1985); J. Kubo, K. Sibold and W. Zimmermann, *Nucl. Phys.* **B259** 331 (1985).
5. A.J. Parkes and P.C. West, *Phys. Lett.* **B138** 99 (1984); *Nucl. Phys.* **B256** 340 (1985); D.R.T. Jones and A.J. Parkes, *Phys. Lett.* **B160** 267 (1985); D.R.T. Jones and L. Mezinescu, *Phys. Lett.* **B136** 242 (1984); *Phys. Lett.* **138** 293 (1984); A.J. Parkes, *Phys. Lett.* **B156** 73 (1985); I. Jack and D.R.T. Jones, *Phys. Lett.* **B333** 372 (1994).
6. S. Hamidi and J.H. Schwarz, *Phys. Lett.* **B147** 301 (1984); D.R.T. Jones and S. Raby, *Phys. Lett.* **B143** 137 (1984); J.E. Björkman, D.R.T. Jones and S. Raby, *Nucl. Phys.* **B259** 503 (1985); J. León et al, *Phys. Lett.* **B156** 66 (1985).
7. D.I. Kazakov, *Mod. Phys. Lett.* **A2**663 (1987); *Phys. Lett.* **B179** 352 (1986); C. Lucchesi, O. Piguet and K. Sibold, *Helv. Phys. Acta.* **61** 321 (1988).
8. J. Kubo, M. Mondragón, M. Olechowski and G. Zoupanos, *Nucl. Phys.* **B479** 25 (1996).
9. J. Kubo, M. Mondragón and G. Zoupanos, *Phys. Lett.* **B389** (1996) 252.

Bound States in Quantum Field Theory[1]

Juan Carlos López[2], Axel Weber[3], C. R. Stephens[4], Peter O. Hess[5]

Instituto de Ciencias Nucleares, UNAM
Circuito Exterior C.U., A. Postal 70–543, 04510 México, D.F., Mexico

Abstract. We present a new technique for the calculation of bound states in relativistic quantum field theories using renormalization group methods. As a simple example, we show our results for the mass of the lowest–lying bound state in a scalar model and compare with the Bethe–Salpeter approach.

While there have been a lot of interesting and promising developments in many areas of quantum field theory during the last 25 years, very little progress has been made on the fundamental issue of bound state calculations. The main difficulty is, of course, that this is an intrinsically non–perturbative problem. Essentially the only known tool for the calculation of bound states for a given theory without additional inputs or approximations is the Bethe–Salpeter (BS) equation [1,2].

Here we propose to use renormalization group methods for bound state calculations, and we will illustrate the approach with the help of a simple scalar model given by the Lagrangian (in four–dimensional Minkowski space)

$$L = \sum_{a=1}^{2} \left(\partial_\mu \phi_a^\dagger \partial^\mu \phi_a - m^2 \phi_a^\dagger \phi_a \right) + \frac{1}{2} \left(\partial_\mu \varphi \, \partial^\mu \varphi - \mu^2 \varphi^2 \right) - g \left(\phi_1^\dagger \phi_1 + \phi_2^\dagger \phi_2 \right) \varphi \, . \quad (1)$$

We thus have two complex fields ϕ_1, ϕ_2 of equal mass, interacting via a real field φ.

1) This work was partially supported by Conacyt grant 3298P–E9608.
2) e–mail: vieyra@nuclecu.unam.mx
3) Supported by fellowships of the DAAD and the Mexican Government; e–mail: axel@nuclecu.unam.mx
4) e–mail: stephens@nuclecu.unam.mx
5) e–mail: hess@nuclecu.unam.mx

CP400, *First Latin American Symposium on High Energy Physics/ VII Mexican School of Particles and Fields,* edited by D'Olivo/Klein-Kreisler/Mendez

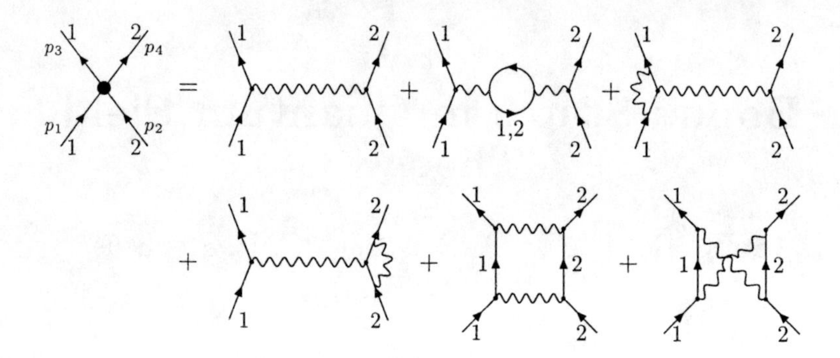

FIGURE 1. The tree and one–loop diagrams contributing to the four–point function under consideration.

We are interested in bound states of the two "charged" particles corresponding to ϕ_1 and ϕ_2, respectively, so we consider the dressed four–point function in fig. 1, giving the amplitude for the scattering of these particles. A possible bound state is associated with a pole of this four–point function of the form

$$\frac{\bar{\chi}(p_3, p_4)\chi(p_1, p_2)}{s - M^2} \tag{2}$$

(without external propagators), where $s = (p_1 + p_2)^2$. The position of the pole determines the mass M of the bound state ($M < 2m$), and χ and $\bar{\chi}$ are the Bethe–Salpeter wave functions.

The BS equation in principle allows to calculate M, χ and $\bar{\chi}$. However, there are several serious problems related to the BS approach [2], on which we cannot elaborate here. As an alternative to the BS equation, we propose the use of "environmentally friendly" renormalization [3], which generalizes usual renormalization group techniques.

We consider the (truncated) tree and one–loop diagrams in fig. 1, which contribute to the four–point function in question. These diagrams are precisely the (1,2)–irreducible contributions to the respective orders, where (1,2)–irreducible stands for connected and one–particle irreducible with respect to ϕ_1 and ϕ_2, but not φ.

For technical reasons we Wick–rotate to Euclidean space–time. The parameter s is then given by $s = -(p_1 + p_2)^2$ in terms of the Euclidean momenta. There are (off shell) six independent Lorentz invariants, say, the Mandelstam variables s, $t = -(p_1 - p_3)^2$, $u = -(p_1 - p_4)^2$, and three of the p_i^2. For the moment we are only interested in the s–dependence of the four–point function, so we are free to choose convenient values for the other invariants. Here we take $t = u = 0$, which implies $p_1 = p_2 = p_3 = p_4$, so there is only one kinematical variable left.

For this choice of momenta the diagrams of fig. 1 give

$$\Gamma^{(4)}(s, t{=}u{=}0) = \frac{g^2}{\mu^2} + g^4 f(s) \,. \tag{3}$$

We have combined the first two diagrams to replace the bare mass of the "neutral" particle by the renormalized mass μ. The analytic function f is obtained by an explicit calculation of the four remaining one–loop diagrams. In order to compare with the BS equation in the ladder approximation, we omit in the following the contributions of the third and fourth diagram on the r.h.s. of fig. 1, which correspond to the vertex corrections.

We define the four–point coupling constant λ at a renormalization scale κ by

$$\lambda(\kappa) = \Gamma^{(4)}(s, t{=}u{=}0)\Big|_{s=-\kappa^2} \,. \tag{4}$$

The β–function for λ reads

$$\beta(\kappa) = \kappa \left(\frac{\partial}{\partial\kappa}\lambda\right)_{g,\mu} = \mu^4 \lambda^2 \, \kappa \frac{\partial}{\partial\kappa} f(-\kappa^2) \,. \tag{5}$$

Integration of (5) with respect to κ yields

$$\lambda(\kappa) = \frac{1}{\dfrac{1}{\lambda_\infty} - \mu^4 f(-\kappa^2)} \,, \tag{6}$$

where $\lambda_\infty = \lambda(\kappa_i = \infty)$, and we have used the fact that $f(-\kappa^2)$ vanishes in the limit $\kappa \to \infty$.

Returning to the definition (4) of λ, we see that a pole of $\lambda(\kappa)$ for a certain value of κ corresponds to a pole of the four–point function at $s = -\kappa^2$, which gives us the mass of the bound state. Putting $m^2 = \mu^2 = 1$ for simplicity, we find the pole of λ from (6) at

$$f(s) = \frac{1}{\lambda_\infty} \qquad (m^2 = \mu^2 = 1) \,. \tag{7}$$

For a given value of λ_∞, this equation determines the bound state mass M through $s = M^2$. The function f is plotted in fig. 2 (to the left). At $s = 4$, it has a cusp. We read off from the plot that there is no (stable) bound state for small couplings λ_∞. At the threshold value indicated in fig. 2, the bound state has a mass of $M = 2 = 2m$, i.e. there is no binding energy. For larger values of the coupling constant, the bound state mass becomes smaller, until at $\lambda_\infty/(4\pi)^2 = 3$ we have $M = 0$, i.e. the binding energy is equal to $2m$. For even larger values of λ_∞, the bound state formally becomes tachyonic, which means that we have vacuum condensation.

In fig. 2 to the right we compare our (analytic) results for λ_∞ as a function of the corresponding bound state mass (curve labeled RG in the plot) with the

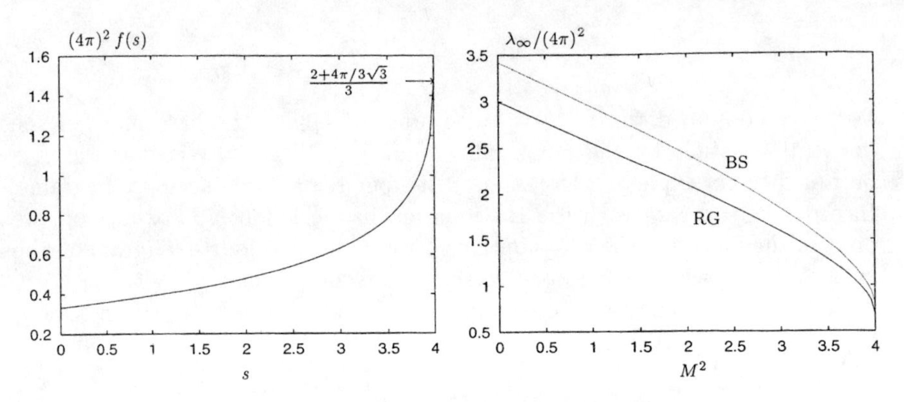

FIGURE 2. The function f and our results for the bound state mass.

numerical results from the BS equation in the ladder approximation [4] (curve BS). We find qualitative agreement, which assures us that our approach yields sensible results. If we include the vertex corrections in f, we find a similar behaviour with smaller values of λ_∞ for the corresponding bound state masses due to the running of the fundamental coupling.

In conclusion, we have presented a new and very promising method for bound state calculations, which is free from the problems encountered in the BS framework. Of course, the formalism has still to be developed further to describe, e.g., excited states and the Bethe–Salpeter wave functions.

ACKNOWLEDGEMENTS

A.W. would like to thank the ICN and the CIC, UNAM, for financial support during the VII–EMPC and the I–SILAFAE in Merida, where this work was presented.

REFERENCES

1. Salpeter E. E., and Bethe H. A., *Phys. Rev.* **84**, 1232 (1951);
   Gell–Mann M., and Low F., *Phys. Rev.* **84**, 350 (1951).
2. Nakanishi N., *Prog. Phys. Suppl.* **43**, 1 (1969);
   Itzykson C., and Zuber J.–B., *Quantum Field Theory*, New York: McGraw–Hill, 1980.
3. O'Connor D., and Stephens C. R., *Nucl. Phys.* **B360**, 297 (1991);
   Stephens C. R., "Why Two Renormalization Groups are Better than One", UNAM preprint ICN–UNAM–96–11, hep–th/9611062.
4. Zur Linden E., and Mitter H., *Nuovo Cim.* **61B**, 389 (1969);
   Kaufmann W. B., *Phys. Rev.* **187**, 2051 (1969).

Non Abelian Duality

I. Martín

Departamento de Física, Universidad Simón Bolívar, Apartado 89000, Caracas, Venezuela. e-mail: isbeliam@usb.ve

Abstract. We study an extension of the procedure to construct duality transformations among abelian gauge theories to the non abelian case.

Duality maps between abelian gauge theories given by $U(1)$ connections on line bundles over a manifold X can be shown to exist by using a quantum equivalent formulation of the original theory in terms of closed 2-forms. This is expressed as a functional on the space of abelian 2-forms which must be constrained by non-local restrictions, namely, the requeriment of being closed and with integral periods, ensuring the existence of a 1-1 correspondence between the space of constrained 2-forms and the line bundles over X [1].This procedure has been successfully applied even to more general U(1) bundles [2] based on an extension of Weil's theorem to complex p-forms [3].Once the equivalence between the formulation in terms of the configuration space of abelian connections and that of the space of closed 2-forms is achieved, the latter is used to construct at the quantum level the dual gauge theory, by introducing dual Hodge-$\star$ forms through Lagrange multipliers proving the existence of non trivial relations between the partition functions of the abelian theory and its dual.

The purpose of this talk is to inquire on the possibility of extending the above procedure to the non abelian case. In the first place, we will begin by asking what conditions should be imposed on matrix-valued 2-forms over a manifold X so that we could produce something similar to Weil's theorem for non-abelian 2-forms, so that we could achieve an equivalence between the formulation of the theory on the configuration space of connections and the formulation on the space of 2-forms. This actually is a formidable problem still not solved but only to the level of conjectures [8]. In any case, we could try to see where failures lie.

The Bianchi identity for a matrix-valued 2-form Ω

$$\mathcal{D}\,\Omega = 0$$

CP400, *First Latin American Symposium on High Energy Physics/ VII Mexican School of Particles and Fields,*
edited by D'Olivo/Klein-Kreisler/Mendez

is the first condition that comes to mind when looking for restrictions to implement, since curvatures for connections on fiber bundles satisfy it. But this, in general, does not assure even that Ω may be expressed locally in terms of any 1-form connection A as

$$\Omega = dA + A \wedge A$$

something equivalent to a Poincaré's lemma for "covariantly closed" forms does not hold. Moreover, even when we could express Ω in terms of A as above, on open sets U_i of a covering of the manifold X, compatibility of the curvature-like 2-forms $\Omega(A_i)$ and $\Omega(A_j)$ on the intersection of two open sets U_i and U_j should imply that A_i and A_j are related by a well defined gauge transformation on the intersection of open sets. Simple calculations show that this is not the case. A_i and A_j could be related by some other more general transformations that no doubt include the mentioned gauge transformations i.e.

$$\Omega(A_i) = g^{-1}\Omega(A_j)\, g \overset{not}{\Longrightarrow} A_i = g^{-1}A_j\, g + g^{-1}dg$$

Obviously, we need more restrictive conditions to arrive to the necessary compatibility glueing for constructing globally well defined non abelian vector bundles.

It is well known [4], that a formulation of non abelian gauge theories has a rather simple expression on the space of loops as a trivial flat gauge theory. The main ingredient in this formulation is the use of the holonomy associated to each class of non abelian Lie algebra valued connections on a vector bundle. The use of holonomies is quite adequate since its non local character as a geometrical object carries a lot more information about the bundle than curvatures or connections. So, we should go to the loop space formulation and see whether it is possible to write some conditions that could characterize the non abelian bundles and look for a procedure to build the duality maps. In what follows, we suceed in proving half the task, for a more detailed discussion see [5].

For our purpose, instead of using the space of closed curves [4,6], we will consider a space of open curves $\mathcal{C}$ with fixed endpoints O, P over a compact manifold X. This will allow the construction of smoothly behaving mathematical objects like functionals, variations of functionals, 1-form connection functionals and so on, on open neighborhoods of the space of curves. Particularly, we avoid regularization problems in the definition of the gauge "potential" on loop space.

First, any functional over this space will be denoted $\tilde{\Phi}(\mathcal{C}_{\mathrm{O,P}})$ and a variation or increment of this functional due to a deformation on the curve leaving the endpoints fixed is defined as

$$\tilde{\Delta}\tilde{\Phi}(\mathcal{C}_{\mathrm{O,P}}) \equiv \tilde{\Phi}(\mathcal{C}_{\mathrm{O,P}} + \delta\mathcal{C}_{\mathrm{O,P}}) - \tilde{\Phi}(\mathcal{C}_{\mathrm{O,P}})$$

Deformations on the curves are smooth vector fields on open neighborhoods of X where the curve $\mathcal{C}_{\mathrm{O,P}}$ lies, tending to zero on the endpoints of the curve. We could relax this definition allowing non zero deformations on one of the endpoints but then we would need to impose a non linear condition to get the compatibility requirement on the patching of the vector bundle [5]. Our version of holonomy is H_A, the path ordered exponential of a 1-form connection A over X integrated over the open curve $\mathcal{C}$ i.e

$$H_A(\mathrm{O},\mathrm{P},\mathcal{C}) \equiv exp: \int_{\mathrm{O}}^{\mathrm{P}} A:$$

it becomes the ordinary holonomy when O and P are identified. $\tilde{\mathcal{A}}(\mathcal{C}_{\mathrm{O,P}})$ denotes the 1-form connection functional acting on deformations S. It is obtained from H_A

$$\tilde{\mathcal{A}}(\mathcal{C}_{\mathrm{O,P}}) = -\tilde{\Delta} H_A \cdot H_A^{-1}$$

and may be expressed in terms of $F_{p(t')}$, the ordinary pointwise defined curvature 2-form associated to the connection A, as

$$\tilde{\mathcal{A}}(\mathcal{C}_{\mathrm{O,P}})[S] = \int_{\mathrm{O}}^{\mathrm{P}} H_A(\mathrm{O},p(t'),\mathcal{C}) F_{p(t')}[T,S] H_A(\mathrm{O},p(t'),\mathcal{C}^{-1}) dt'$$

where T is a vector field tangent to the curve $\mathcal{C}$, t' is a parameter along the curve and $p(t')$ is an ordinary point on the curve.

$\tilde{\mathcal{A}}(\mathcal{C})$ is defined for classes of equivalence of ordinary connections under gauge transformations, i.e it is gauge invariant up to elements of the structure group on the endpoints of the curve. This is a rather nice feature of working in loop spaces.

We could continue and define also the curvature functional $\tilde{\mathcal{F}}(\mathcal{C},A)$ for the connection functional $\tilde{\mathcal{A}}$ in the usual manner

$$\tilde{\mathcal{F}}(\mathcal{C},A) = \tilde{\Delta}\tilde{\mathcal{A}}(\mathcal{C}) + \tilde{\mathcal{A}}(\mathcal{C}) \wedge \tilde{\mathcal{A}}(\mathcal{C})$$

for this free formulation of non abelian gauge theories, calculations show that

$$\tilde{\mathcal{F}}(\mathcal{C},A) = 0$$

and it is a gauge invariant statement.

The "covariant" derivative $\tilde{\mathcal{D}}$ may be also introduced as

$$\tilde{\mathcal{D}} \cdot \equiv \tilde{\Delta} \cdot + \tilde{\mathcal{A}} \wedge \cdot$$

Now on the space of curves where non zero deformations are allowed on one endpoint [5], we may impose a condition on general 2-forms functionals $\tilde{\Omega}$ that can lead only to curvature functionals, this is

$$\tilde{\mathcal{D}}\tilde{\Omega} = 0$$

the only possibility for solving this equation is

$$\tilde{\Omega} = \tilde{\mathcal{F}}$$

On intersecting neiborhoods U_i and U_j the compatibility condition means, in general, that

$$\tilde{\mathcal{F}}(\mathcal{C}, A_i) = \tilde{\mathcal{F}}(\mathcal{C}, A_j)$$

must hold on the space of curves of $U_i \bigcap U_j$. Calculations show that this can only be achieved if and only if

$$A_i = g^{-1} A_j \, g + g^{-1} d \, g.$$

In the case of the space of curves with two fixed endpoints, we need only to require compatibility of $\tilde{\mathcal{A}}(\mathcal{C}_{\,\mathrm{O,P}})$ on the intersecting neigborhoods

$$\tilde{\mathcal{A}}(\mathcal{C}_{\,\mathrm{O,P}})_i \;\; = \;\; \tilde{\mathcal{A}}(\mathcal{C}_{\,\mathrm{O,P}})_j$$

and get, in the same manner, that this is only possible if $A_i = g^{-1} A_j \, g + g^{-1} d \, g$.

So we have suceeded in the first step towards the construction of dual non abelian fields suggesting that the natural space for building up the dual maps are loop spaces or open curve spaces. Now, it rests to find a global condition equivalent to that of integral periods of the curvature 2-form for abelian gauge theories, that actually labels the different line bundles, i.e an equivalent Dirac quantization condition. We know that for particular $SU(2)$ bundles, there may be a splitting into the direct sum of two line bundles, for those bundles the usual Dirac quantization may suffice. In the space of loops then the restriction to be imposed would be that the ordinary curvature 2-form appearing in $\tilde{\mathcal{A}}(\mathcal{C}_{\,\mathrm{O,P}})$ would belong to the set of "diagonalizable" 2-form curvatures through a condition involving the intersection form and the non abelian topological charge associated to the second Chern class. This suggests that perhaps the global condition needed for non abelian gauge theories, at least for the case of $SU(N)$, involves the "quantization" of the topological charge associated to the second Chern class. Once we find the exact condition to be imposed on the loop space, the dual map for constructing dual gauge theories should be no problem since the operation that generalizes the Hodge-$\star$ operation for loop spaces has already been defined [7]. The partition function is also easily implemented in our formulation. A characterization of matrix-valued 2-forms for being curvatures of non abelian bundles has also recently been conjectured using partial differential equations on a loop space [8].

ACKNOWLEDGEMENTS

I wish to thank the organizers of the I-SILAFAE for inviting me to give this talk in such a splendid site of Yucatán.

REFERENCES

1. E. Verlinde, Nucl. Phys. **B455**, 211 (1995); Y. Lozano, Phys. Lett. **B364**, 19 (1995); J. L. F. Barbón, Nucl. Phys. **B452**, 313 (1995); A. Kehagias, hep-th/9508159.
2. M. Caicedo, I. Martin, and A. Restuccia, hep-th/9701010.
3. J. L. Brylinski, Prog. in Math. Vol.107, *Loop Spaces, Characteristic Classes and Geometric Quantization*, Birkhäuser, Boston, 1993, where the extension is proven for curvature 3-forms associated to sheaf of grupoids. Also, it has recently been proven for the case of 4-forms by Brylinski and McLaughlin (private communication).
4. A.M.Polyakov, Nucl. Phys. **B164**,171 (1980).
5. I. Martin, L. Recht, and A.Restuccia, in preparation.
6. J. W. Barrett, Int. J. Theo. Phys. **30**,1171 (1991).
7. C. Hong-Mo, J. Faridani, and T. Sheung Tsun, Phys. Rev. **D52**, 6134 (1995).
8. J. L. Brylinski, Gelfand Seminars, Vol.2 (private communication).

Reality Conditions as Second Class Constraints

Hugo A. Morales–Técotl *, Luis F. Urrutia † and J. David Vergara†

* Departamento de Física
Universidad Autónoma Metropolitana–Iztapalapa
Apartado Postal 55-534, 09340 México D.F., México
† Instituto de Ciencias Nucleares
Universidad Nacional Autónoma de México
Apartado Postal 70-543, 04510 México D.F., México

Abstract. We show that the reality conditions to be imposed on Ashtekar variables to recover real gravity can be implemented as second class constraints *à la* Dirac. Thus, counting gravitational degrees of freedom follows accordingly. Some constraints of the real theory turn out to be non-polynomial, regardless of the form, polynomial or non-polynomial, taken for the reality conditions.

The aim of the present work is to reanalyze classical Ashtekar complex gravity plus the reality conditions [1], in the light of the canonical Dirac approach to constrained systems. In particular, we want to elucidate the status of the reality conditions viewed as Dirac constraints. A naive analysis of the standard reality conditions: q^{ab} real and $\dot{q}^{ab}$ real, as second class constraints, is misleading because the count of the number of degrees of freedom in the configuration space turns out to be -2 per spatial point, which is obviously wrong! In this work we start with reality conditions, also imposed by hand, but considering them as primary constraints in the Ashtekar self-dual action. Then, we systematically apply the Dirac procedure, showing that these reality conditions end up by being second-class constraints and recovering the standard Hamiltonian formulation of general relativity in its Palatini canonical form [1]. In this way, we show that it is possible to extract a satisfactory canonical description of general relativity starting from the complex theory. In fact, we prove that the original complexified phase-space, both in the driebein and in the connection, together with the self-dual Ashtekar action plus the reality conditions, lead to the standard phase space of real Einstein gravity, via the Dirac method.

CP400, *First Latin American Symposium on High Energy Physics/ VII Mexican School of Particles and Fields*,
edited by D'Olivo/Klein-Kreisler/Mendez

Let us proceed now to analyze the canonical form of the self–dual action [2], [3],

$$S = \int dt d^3x \left\{ -i\tilde{e}^{ai}\dot{A}_{ai} - N\mathcal{S} - N^a\mathcal{V}_a - N^i\mathcal{G}_i \right\}, \tag{1}$$

where

$$\mathcal{S} := \epsilon_{ijk}\tilde{e}^{ai}\tilde{e}^{bj}F_{ab}{}^k, \quad \mathcal{V}_a := \tilde{e}^b{}_j F_{ab}{}^j, \quad \mathcal{G}_i := \mathcal{D}_a\tilde{e}^a{}_i\,, \tag{2}$$

are the constraints of the theory and N, N^a, N^i are Lagrange multipliers. The *complex* canonical variables are: i) $\tilde{e}^{ai} := ee^{ai}$, with e^{ai} being the triad ($e^{ai}e^b{}_i := q^{ab}$, q^{ab} is the spatial three–metric), and $a, b, \ldots = 1, 2, 3$ are spatial indices, whereas $i, j, \ldots = 1, 2, 3$ are so(3) internal indices. Also $e := \det e_{bj}$ with e_{bj} being the inverse of e^{ai}. ii) A_{ai} is the three–dimensional projection of the selfdual connection [1] with associated covariant derivative $\mathcal{D}_a\lambda_i = \partial_a\lambda_i + \epsilon_{ijk}A_a{}^j\lambda^k$ and $F_{ab}{}^i := \partial_a A_b{}^i - \partial_b A_a{}^i + \epsilon^i{}_{jk}A_a{}^j A_b{}^k$ is the corresponding curvature. The above complex pair of canonically conjugated variables satisfies

$$\{\tilde{e}^{ai}(x), A_{bj}(y)\} = i\delta_b{}^a\delta_j{}^i\delta^3(x,y)\,. \tag{3}$$

We shall proceed with the canonical analysis of the action (1), together with the necessary reality conditions, by splitting the complex canonical variables $A_{ai}, \tilde{e}^{bj}$ into real and imaginary parts according to

$$\tilde{e}^{ai} = E^{ai} + i\mathcal{E}^{ai}, \quad A_{bj} = \gamma_{bj} - iK_{bj}\,. \tag{4}$$

Most importantly, we will consider all of them (36 real variables) as configuration variables to begin with. The reality conditions we start with are

$$\psi^{ai} := \mathcal{E}^{ai} = 0, \quad \chi_{ai} := \gamma_{ai} - f_{ai}(E) = 0, \tag{5}$$

which are subsequently taken as primary constraints ψ^{ai}, χ_{bj}, supplementing the action (1). Based upon the compatibility condition between a real torsion-free connection $\overset{0}{\gamma}_{ai}$ and a real densitized triad E^{bj}

$$D_a E^{bj} = \partial_a E^{bj} + \Gamma_{ac}{}^b E^{cj} + \epsilon^j{}_{kl}\overset{0}{\gamma}_a{}^k E^{bl} - \Gamma_{da}{}^d E^{bj} = 0, \tag{6}$$

where $\Gamma_{ab}{}^c$ is the Levi-Civita connection defined by the metric $h^{ab} := E^{ai}E^b{}_i$, we choose the form of f_{ai} as

$$f_{ai} = \frac{1}{2}[E_{ai}E_c{}^j\epsilon_{jrs} - 2E_{aj}E_c{}^j\epsilon_{irs}]E^{dr}\partial_d E^{cs}. \tag{7}$$

In this way we guarantee that γ_{ai} is the real torsion–free connection compatible with E^{bj}, i.e. $\gamma_{ai} = \overset{0}{\gamma}_{ai}$. Since E_{ai} is the inverse of E^{bj}, (7) is non-polynomial in E^{bj} and so it is χ_{ai}.

The definition of the canonical momenta, denoted generically by Π, in the action (1) leads to the following constraints

$$\phi_{\mathcal{E}ai} = \Pi_{\mathcal{E}ai}, \quad \phi_{\gamma}{}^{ai} = \Pi_{\gamma}{}^{ai} + iE^{ai}, \quad \phi_{K}{}^{ai} = \Pi_{K}{}^{ai} + E^{ai}, \quad \phi_{Eai} = \Pi_{Eai}, \tag{8}$$

where we note that $\Pi_{\gamma}{}^{ai}$ is purely imaginary. In this way, the full set of primary constraints is formed by (2), (5) and (8), written in terms of the real variables (4). The next step is to analyze the time evolution of the primary constraints, together with the classification of the resulting set of constraints. It turns out to be appropriate to redefine

$$\phi_{Eai} \rightarrow \phi'_{Eai} = \phi_{Eai} + \alpha_{aibj}\phi_{\gamma}{}^{bj} + \beta_{ai}{}^{bj}\chi_{bj} + \eta_{aibj}\phi_{K}{}^{bj}, \tag{9}$$

with

$$\alpha_{aibj}(x,y) = \frac{\delta f_{bj}(y)}{\delta E^{ai}(x)}, \qquad \beta_{ai}{}^{bj}(x,y) = i\delta_a{}^b\delta_i{}^j\delta^{(3)}(x,y),$$

$$\eta_{aibj}(x,y) - \eta_{bjai}(y,x) = i\left(\frac{\delta f_{ai}(x)}{\delta E^{bj}(y)} - \frac{\delta f_{bj}(y)}{\delta E^{ai}(x)}\right), \tag{10}$$

where $\frac{\delta f(y)}{\delta E(x)}$ denotes the corresponding functional derivative. Our notation, introduced in Eq.(9), is that the summation over repeated indices involves also an integral over the three dimensional space. In this way, for example $\phi_A = \alpha_{AB}\chi^B$ stands for $\phi_A(\mathbf{x},t) = \int d^3\mathbf{y}\ \alpha_{AB}(\mathbf{x},\mathbf{y})\chi^B(\mathbf{y},t)$.

The resulting Poisson brackets matrix for the subset of constraints (5), (8) and (9) is

	$\phi_{\mathcal{E}}$	ψ	ϕ_γ	χ	ϕ_K	ϕ'_E
$\phi_{\mathcal{E}}$	0	$-\delta_a{}^b\delta_i{}^j\delta^{(3)}$	0	0	0	0
ψ	$\delta_b{}^a\delta_j{}^i\delta^{(3)}$	0	0	0	0	0
ϕ_γ	0	0	0	$-\delta_b{}^a\delta_j{}^i\delta^{(3)}$	0	0
χ	0	0	$\delta_a{}^b\delta_i{}^j\delta^{(3)}$	0	0	0
ϕ_K	0	0	0	0	0	$\delta_a{}^b\delta_j{}^i\delta^{(3)}$
ϕ'_E	0	0	0	0	$-\delta_b{}^a\delta_i{}^j\delta^{(3)}$	0

, (11)

where $\delta^{(3)} = \delta^{(3)}(x,y)$. From the above matrix we conclude that this subset, which includes the reality conditions, is second class. Since the original constraints (2) generate the gauge symmetries of the system, we would like to keep them as first–class constraints. To see whether this is tenable, we start by searching a redefinition of them so that they have zero Poisson brackets with the second–class set. Notice that the constraints (2) only depend on the configuration variables and not on the momenta. For any of them, say $\mathcal{R}$, we construct

$$\mathcal{R}' = \mathcal{R} + \{\phi_{\mathcal{E}bj}, \mathcal{R}\}\psi^{bj} + \{\phi_\gamma^{bj}, \mathcal{R}\}\chi_{bj} + \{\phi'_{Ebj}, \mathcal{R}\}\phi_K{}^{bj} - \{\phi_K{}^{bj}, \mathcal{R}\}\phi'_{Ebj}, \tag{12}$$

which we can verify to fulfill the above requirement. Summarizing, our system is conveniently described by the following set of primary constraints: $\phi_{\mathcal{E}ai}$, $\phi_\gamma{}^{ai}$, $\phi_K{}^{ai}$, ϕ'_{Eai}, ψ^{ai}, χ_{ai}, $\mathcal{S}'$, $\mathcal{V}'_a$, $\mathcal{G}'_i$. Next, we impose the conservation in time of this whole set, using the total hamiltonian density

$$\mathcal{H}_{\text{Total}}(\mathbf{x}) \approx \mu_{\mathcal{E}}{}^{ai}\phi_{\mathcal{E}ai} + \mu_{\gamma ai}\phi_\gamma{}^{ai} + \mu_{Kai}\phi_K{}^{ai} + \mu_E{}^{ai}\phi'_{Eai} + \lambda_{ai}\psi^{ai} + \omega^{ai}\chi_{ai}$$
$$+N\mathcal{S}' + N^a\mathcal{V}'_a + N^i\mathcal{G}'_i, \tag{13}$$

The properties of the constraints $\mathcal{S}', \mathcal{V}'_a, \mathcal{G}'_i$, guarantee that their Poisson brackets with $H_{\text{Total}} = \int d^3\mathbf{x}\ \mathcal{H}_{\text{Total}}(\mathbf{x})$ is weakly zero. By virtue of the Poisson brackets matrix (11) together with the definition (12), the calculation of the corresponding time evolution fixes each of the Lagrange multipliers $\mu_{\mathcal{E}}{}^{ai}$, $\mu_{\gamma ai}$, μ_{Kai}, $\mu_E{}^{ai}$, λ_{ai}, ω^{ai} equal to zero. We conclude that there are no secondary constraints and that the set $\mathcal{S}', \mathcal{V}'_a, \mathcal{G}'_i$ is first class while the set $\phi_{\mathcal{E}ai}$, $\phi_\gamma{}^{ai}$, $\phi_K{}^{ai}$, ϕ'_{Eai}, ψ^{ai}, χ_{ai}, which includes the reality conditions, is second class. To make sure that ultimately we are dealing with real gravity we still have to get the correct number of degrees of freedom, using the Dirac prescription. Also we must recover the specific form of the constraints characterizing real gravity. To do so we start by examining the complex character of each constraint. By simple inspection of (5) and (8) we realize that each of the second class constraints is either purely imaginary or purely real. To proceed with the count we impose them strongly. This leaves us with $2 \times 9 \times 4 - 6 \times 9 = 18$ as the dimension of partially reduced real phase space. The first class constraints remain first class in the resulting Dirac brackets. As shown below, this first class set contains seven constraints each of which is either purely real or purely imaginary. In this way the final count yields $\frac{1}{2}[18 - 2 \times 7] = 2$ real configuration degrees of freedom per spatial point.

An extended version of the present work has been reported elsewhere [4]. HAMT, LFU and JDV were partially supported by grant CONACyT-3141P and LFU and JDV by grant DGAPA-UNAM-IN100694.

REFERENCES

1. Ashtekar, A.,*Phys. Rev. Letts.* **57** 2244 (1986); *Phys. Rev.* **D36** 1587 (1987); *Lectures on Non-Perturbative Canonical Gravity (Notes prepared in collaboration with R S Tate)*, Singapore: World Scientific, (1991) and references therein.
2. Samuel, J., *Pramana J. Phys.* **L429**, 28 (1987).
3. Jacobson, T. and Smolin, L., *Phys. Lett.* **B196**, 39 (1987); *Class. Quant. Grav.* **5**, 583 (1988).
4. Morales-Técotl, H. A., Urrutia, L. F. and Vergara, J. D., *Class. Quant. Grav.* **13**, 2933 (1996).

Axial Couplings in the World Line Formalism

Lukas Nellen

Instituto de Física, UNAM, Apdo. Postal 20–364, México 01000 DF, MEXICO

Abstract. We present the first-quantized version of the coupling of a Dirac particle to scalar, pseudo-scalar, vector, and axial background fields in the language of one-dimensional field theory. This way, all the standard-model couplings can be presented in a world-line formulation. Such a formulation can be used to write down a generating functional for all one-loop N-point functions as a Feynman path integral over a one-dimensional world line. The evaluation of its expansion in N-point functions produces expressions more efficient than those from ordinary field theory.

As everybody knows, the complexity of calculations in quantum field theory tends to grow rapidly with increased order of loops or external legs. For certain classes of Feynman diagrams, first quantized methods can help to simplify the calculation. For a long time, first quantized methods for point particles have been used for the calculation of one-loop determinants and propagators [1–4]. The development of the Bern-Kosower formalism [5–7] provided a set of rules for the efficient calculation of tree and one-loop diagrams. This set of rules was derived taking the field theory (infinite tension) limit of string theory. Strassler showed [8] that their one-loop rules can be derived directly form the first quantized description of a relativistic particle, *i.e.*, from the treatment of a one-dimensional QFT on the world-line of a relativistic particle. This way, the calculation of one-loop effective action can be performed significantly more efficiently than in the usual heat-kernel approach [9–11]. Also, generalizations to higher-loop calculations have been found [12–14]. Initially, the work done in the the world-line formalism was either for a scalar particle or for a Dirac fermion with gauge couplings [8,9,15]. Recently, the extensions have been presented which generalise the world-line formalism to spin-one [16] particles, and to Dirac particles with scalar, pseudo-scalar, and axial couplings [17–20].

In the following we want to discuss the treatment of a Dirac particle with arbitrary coupling. The central object we have to study is the one-loop effective action of the Dirac fermion in an arbitrary background. In ordinary

CP400, *First Latin American Symposium on High Energy Physics/ VII Mexican School of Particles and Fields*, edited by D'Olivo/Klein-Kreisler/Mendez

quantum field theory, we can write it as the logarithm of the determinant of the Dirac operator $\mathcal{O}$, where

$$\mathcal{O} = (\partial_\mu + igV_\mu + ig_5\gamma_5 A_\mu)\gamma^\mu - im - i\lambda\phi + \gamma_5\lambda'\phi' \tag{1}$$

with the convention that $\{\gamma_\mu, \gamma_\nu\} = -2\delta_{\mu\nu}$ and $\gamma_\mu^\dagger = -\gamma_\mu$.

To connect the field theory expression to the world-line formulation requires some manipulations of the expression for the one-loop effective action. This becomes clear as soon as one looks at the propagator of the Dirac particle. In the usual field theory formulation, the Dirac propagator is $i/(\not{p} + m)$. The world-line expression, however, is second order in the momentum, *i.e.*, the propagator is of Klein-Gordon type $1/(p^2 + m^2)$, where the Dirac equation enters a constraint on the allowed states [21,22]. The same can also be done in an ordinary field theory formulation, writing down a second-order formulation of the Dirac particle [23,7,24].

For the one-loop effective action, this can be achieved by rewriting

$$\Gamma = \mathrm{Tr}\log\mathcal{O} = \frac{1}{2}\left\{\mathrm{Tr}(\log\mathcal{O} + \log\mathcal{O}^\dagger) + \mathrm{Tr}(\log\mathcal{O} - \log\mathcal{O}^\dagger)\right\}. \tag{2}$$

This rewriting automatically splits the one-loop effective action into its real and imaginary parts The real part can be rewritten as $\mathrm{Tr}\log\mathcal{O}\mathcal{O}^\dagger$. Extracting the free part from

$$\begin{aligned}\mathcal{O}\mathcal{O}^\dagger = &-D_\mu D^\mu + g\sigma^{\mu\nu}V_{\mu\nu} + g_5\gamma_5\sigma^{\mu\nu}A_{\mu\nu} + i\lambda\gamma^\mu\partial_\mu\phi - \gamma_5\lambda'\gamma^\mu\partial_\mu\phi' \\ &+ 2(im + i\lambda\phi - \gamma_5\lambda'\phi')ig_5\gamma_5 A_\mu\gamma^\mu + m^2 + 2m\lambda\phi + \lambda^2\phi^2 + \lambda'^2\phi'^2\end{aligned} \tag{3}$$

we see that the resulting propagator is of Klein-Gordon type. We define $D_\mu = \partial_\mu + igV_\mu + ig_5\gamma_5 A_\mu$, and we use $V_{\mu\nu}$ and $A_{\mu\nu}$ to denote the field-strength tensors.

In the absence of γ_5-couplings the relation $\mathcal{O}^\dagger = -\gamma_5\mathcal{O}\gamma_5$. This eliminates the second term in (2), which implies that in this case the one-loop effective action is real.

To bring the second term in (2) into second-order form is less straightforward. We first have to differentiate wrt. one of the fields to get

$$\frac{\delta}{\delta U}\mathrm{Tr}(\log\mathcal{O} - \log\mathcal{O}^\dagger) = \mathrm{Tr}\left(\left\{\frac{\delta\mathcal{O}}{\delta U}\mathcal{O}^\dagger - \mathcal{O}\frac{\delta\mathcal{O}^\dagger}{\delta U}\right\}\frac{1}{\mathcal{O}\mathcal{O}^\dagger}\right), \tag{4}$$

where $U = A_\mu$ or ϕ' (the cases $U = V_\mu$ or ϕ can be ignored since each diagram in the imaginary part of Γ contains at least one γ_5-coupling). The second term is of the form of an insertion times the full second order propagator.

The second order expression for the one-loop effective action can be written as a path integral over (second order) space-time fermions. This way, one can derive a set of second order Feynman rules. The vertices of the resulting

Feynman rules bear a very close relation to the vertex operators in the world-line formalism.

The world-line description of a Dirac particle is given by the supersymmetric generalisation of the proper length of the world-line [21,22], where the embedding coordinate x_μ is complemented by a fermionic super-partner ψ_μ. These fields suffice to represent a massless Dirac particle coupled to a vector field. To include a mass-term [25] or generalised couplings [17–20], however, one has to extend the set of world-line fields to include extra fermions $\psi_{5,6}$ which are accompanied by auxiliary bosonic fields $x_{5,6}$. Using this field content, one can write down the world-line action for a particle with general coupling as

$$\begin{aligned} S = \int \mathrm{d}\tau \Big(& \frac{\dot{x}^2}{4} + \frac{x_5^2}{4} + \frac{x_6^2}{4} + \frac{1}{2}\psi\dot{\psi} + \frac{1}{2}\psi_5\dot{\psi}_5 + \frac{1}{2}\psi_6\dot{\psi}_6 \\ & + 2i\lambda(x_5\Phi - 2\psi_5\psi\cdot\partial\Phi) + 2i\lambda'(x_6\Phi' - 2\psi_6\psi\cdot\partial\Phi') \\ & + g_5 2\big[\psi_5\psi_6(\dot{x}_\mu A_\mu + 2\psi_\mu\psi_\nu\partial_\nu A_\mu) + (\psi_5 x_6 - \psi_6 x_5)\psi_\mu A_\mu\big] \\ & - ig(\dot{x}_\mu V_\mu + 2\psi_\mu\psi_\nu\partial_\nu V_\mu)\Big). \end{aligned} \tag{5}$$

The one-dimensional path-integral $-n_F/2\int \mathcal{D}x_i\mathcal{D}\psi_i \exp(-S)$ generates the real part of the one-loop effective action. For the imaginary part, we relate to the once differentiated form also used in the presentation of the second order formalism. Its path integral representation is

$$\Gamma'_U = \frac{\delta}{\delta U} i\,\Im m\,\Gamma = -\frac{n_F}{2}\int_0^\infty \frac{\mathrm{dT}}{T}\int \mathcal{D}X\mathcal{D}\bar{X}\mathcal{D}X'(-1)^F\Omega_U e^{-S} \tag{6}$$

where we have $U = A_\mu$ or ϕ'. Corresponding to the insertions in (4), we have insertions into the world-line path integral

$$\begin{aligned} \Omega_A &= -g_5\int_0^T \mathrm{d}\tau\left(\psi_\mu\psi_\nu\dot{\mathrm{x}}_\nu - \frac{\mathrm{i}}{2}\mathrm{p}_\mu\right)\psi_5\psi_6, \\ \Omega_{\phi'} &= -i\lambda'\int_0^T \mathrm{d}\tau\left(\frac{1}{2}\psi_\nu\dot{\mathrm{x}}_\nu + \frac{1}{2}\psi_5\mathrm{x}_5\right)\psi_6. \end{aligned} \tag{7}$$

The equivalent of γ_5 in the Dirac algebra is achieved by the inclusion of the operator $(-1)^F$ in (6). This operator changes the boundary conditions of the world-line fermions from odd to even. On the circle, fermions with even boundary conditions have zero-modes which have to be treated separately from the rest of the path integral. Integrating the zero-modes $\psi_\mu^{(0)}$ of the ψ_μ

$$\int \prod_{i=1}^{d}\left(\mathrm{d}\psi_{\mu_i}\psi_{\mu_i}^{(0)}\right) = \epsilon_{\mu_1,\ldots,\mu_d} \tag{8}$$

generates the ϵ-tensor in d dimensions which is part of all diagrams generated by the imaginary part of the effective action. The rest of the evaluation of the path integral is done as in the case of even boundary conditions using Wick

contractions. Calculations done this way yield the same results as calculations done using the second order formalism. This way, and comparing world-line vertex operators to the vertices in the second order formalism, one can establish the link between the world-line formulation and ordinary field theory.
Acknowledgments: I would like to thank M. Mondragón, M.G. Schmidt and C. Schubert for their collaboration on this paper.

REFERENCES

1. R. P. Feynman, Phys. Rev. **80** (1950) 440.
2. R. P. Feynman, Phys. Rev. **84** (1951) 108.
3. M. B. Halpern, A. Jevicki, and P. Senjanović, Phys. Rev. **D16** (1977) 2476.
4. A. M. Polyakov, *Gauge Fields and Strings*, Harwood Academic Publishers, 1987.
5. Z. Bern and D. A. Kosower, Phys. Rev. Lett. **66** (1991) 1669.
6. Z. Bern and D. A. Kosower, Nucl. Phys. **B379** (1992) 451.
7. Z. Bern and D. C. Dunbar, Nucl. Phys. **B379** (1992) 562.
8. M. J. Strassler, Nucl. Phys. **B385** (1992) 145.
9. M. G. Schmidt and C. Schubert, Phys. Lett. **B318** (1993) 438.
10. D. Fliegner, P. Haberl, M. G. Schmidt, and C. Schubert, Preprint HD-THEP-94-25 (hep-th/9411177), Heidelberg, 1994.
11. D. Fliegner, P. Haberl, M. G. Schmidt, and C. Schubert, Preprint HD-THEP-96/54 (hep-th/9702092), Heidelberg, 1996.
12. M. G. Schmidt and C. Schubert, Phys. Lett. **B331** (1994) 69.
13. M. G. Schmidt and C. Schubert, Phys. Rev. **D53** (1996) 2150.
14. P. D. Vecchia, L. Magnea, A. Lerda, R. Marotta, and R. Russo, Phys. Lett. **B388** (1996) 65.
15. D. Fliegner, M. G. Schmidt, and C. Schubert, Z. Phys. **C64** (1994) 111.
16. M. Reuter, M. G. Schmidt, and C. Schubert, Preprint IASSNS-HEP-96/90 (hep-th/9610191), Princeton, 1996.
17. M. Mondragón, L. Nellen, M. G. Schmidt, and C. Schubert, Phys. Lett. **B351** (1995) 200, (hep-th/9502125).
18. M. Mondragón, L. Nellen, M. G. Schmidt, and C. Schubert, Phys. Lett. **B366** (1996) 212, (hep-th/9510036).
19. E. D'Hoker and D. G. Gagné, Nucl. Phys. **B467** (1996) 272, (hep-th/9508131).
20. E. D'Hoker and D. G. Gagné, Nucl. Phys. **B467** (1996) 297, (hep-th/9512080).
21. L. Brink, S. Deser, B. Zumino, P. DiVecchia, and P. Howe, Phys. Lett. **64B** (1976) 435.
22. L. Brink, P. DiVecchia, and P. Howe, Nucl. Phys. **B118** (1977) 76.
23. L. C. Hostler, J. Math. Phys. **26** (1985) 1348.
24. A. G. Morgan, Phys. Lett. **B351** (1995) 249.
25. J. C. Henty, P. S. Howe, and P. K. Townsend, Class. Quant. Grav. **5** (1988) 807.

Gauge and Gravitational Interactions with Local Scale Invariance

Christopher Pilot* and Subhash Rajpoot†[1]

*Department of Physics, Maine Maritime Academy, Castine, ME 04420
†Department of Physics and Astronomy, California State University, Long Beach, CA 90840

Abstract. Local scale invariance as a fundamental symmetry of Nature is proposed. This necessitates the existence of a new vector boson associated with local scale invariance. The new boson is referred to as the Weylon. The local gauge action is taken to be that of the Standard Model with $SU(3) \otimes SU(2) \otimes U(1)$ gauge symmetry. For gravitational interactions, two actions are considered, one linear in curvature and the other quadratic in torsion. In both types of theories breaking of the local gauge and scale invariance is achieved through the usual doublet of scalars and leads to massive vector bosons with the remarkable result that all Higgs degrees of freedom are entirely eliminated. In both models, the Weylon acquires a mass of around $10^{19} GeV$.

I INTRODUCTION

This talk is dedicated to the memory of Professor Abdus Salam. We learnt of his passing away in the course of writing up this talk. The material presented here is a reflection of the diverse activities Professor Salam's school was engaged in, both in phenomenology and field theory. He will be remembered for his many contributions and deep insights.

The realm of elementary particle interactions is governed by strong, weak, electromagnetic and gravitational interactions. In practical calculations gravitational interactions are usually ignored. This fact stems from the apparent feebleness of the gravitational force. On the scale where the strong fine structure constant α_s is of order unity, the analogue of the gravitational fine structure constant is of order 10^{-37} which adequately justifies the neglect of the gravitational force. However gravity may contribute indirectly towards a more complete theory of particle interactions. In spite of the successes of the Standard Model [1] in describing phenomena probed thus far there remains

[1] Talk presented by S. Rajpoot

CP400, *First Latin American Symposium on High Energy Physics/ VII Mexican School of Particles and Fields,* edited by D'Olivo/Klein-Kreisler/Mendez

the unattractive feature of the leftover Higgs meson. As we shall see, marrying gravity with the Electronuclear force [2] eliminates the last left-over scalar degree of freedom conventionally known as the Higgs boson of the standard model. Assuming the mass of the top is given by $m_t = 174 \pm 10^{+14}_{-12} GeV$ the phenomenological lower bound of the mass of the Higgs is around 200 GeV. The upper bound is around 1 TeV from unitarity arguments. The Higgs meson may well be discovered at the upcoming LHC or a similar accelarator. However there exists the bleak possibility of no Higgs discovery at the LHC or for that matter at one of its future upgrades. In that case the Standard Model will need to undergo major conceptual alterations. In this talk we entertain the possibility that there is no real leftover elementary scalar in the standard model. We achieve this by ressurecting the original work of Weyl [3] on local scale invariance. Weyl's idea was to unify electromagnetism with gravity through local scale invariance. We also follow in the same vein, but instead unify strong and weak interactions with gravity through local scale invariance.

Weyl's geometry is a generalization of Reimannian Geometry. In Weyl's geometry a vector transported parallelly around a closed loop not only changes its phase but also its magnitude. Formally, under scale transformations the four-dimensional elemental length ds and ds' are related as

$$ds' = \Lambda ds \tag{1}$$

Local quantities, Y and Y', in the two coordinate systems are related by

$$Y = \Lambda^n Y \tag{2}$$

This implies the following tranformations [4] for the various fields of interest,

$$\begin{array}{lcll}
g^{\mu\nu} & \longrightarrow & \Lambda^{-2} g^{\mu\nu} & n=-2 \\
g_{\mu\nu} & \longrightarrow & \Lambda^{2} g_{\mu\nu} & n=2 \\
\sqrt{-g} & \longrightarrow & \Lambda^{4}\sqrt{-g} & n=4 \\
\text{Vierbein} & \longrightarrow & \Lambda\text{Vierbein} & n=1 \\
(\text{Scalar field}) & \longrightarrow & \Lambda^{-1}(\text{Scalar field}) & n=-1 \\
W^{\pm}, W^{o}, B^{o} & \longrightarrow & W^{\pm}, W^{o}, B^{o} & n=0 \\
(\text{Fermion field})_\mu & \longrightarrow & \Lambda^{-3/2}(\text{Fermion field})_\mu & n=-3/2 \\
(\text{Weyl field})_\mu & \longrightarrow & (\text{Weyl field})_\mu - f^{-1}\partial_\mu(\ln\Lambda) & \\
D_\mu(\text{field}) & \longrightarrow & D_\mu^{gauge}(\text{field}) + nf(\text{Weyl field})_\mu & \\
\tilde{\Gamma}_{\mu\nu}{}^{\rho} & \longrightarrow & \Gamma_{\mu\nu}{}^{\rho} + f(U_\mu g^\rho_\nu + U_\nu g^\rho_\mu - U^\rho g_{\mu\nu}) & \\
D_{\mu\nu}{}^{\alpha} & \longrightarrow & \partial_\mu g^\alpha_\nu - \tilde{\Gamma}_{\mu\nu}{}^{\rho} g^\alpha_\rho + f(\text{Weyl field}_\mu) g^\alpha_\nu &
\end{array} \tag{3}$$

In the following, we use the above tranformations to construct two models of particle interactions.

II GRAVITY LINEAR IN CURVATURES

The Lagrangian for the model with gravity linear in curvatures is taken to be

$$L = \int d^4x\sqrt{-g}[L_1 + L_0 + L_{1/2} + L_{Yukawa} + L_{gravity}] \quad (4)$$

where

$$L_1 = -\frac{1}{4}g^{\mu\rho}g^{\nu\sigma}\left[G_{\mu\nu}\cdot G_{\rho\sigma} + W_{\mu\nu}\cdot W_{\rho\sigma} + B_{\mu\nu}B_{\rho\sigma} + U_{\mu\nu}U_{\rho\sigma}\right] \quad (5)$$

$$L_0 = g^{\mu\nu}(D_\mu\phi)(D_\nu\phi)^\dagger - \frac{1}{4}\lambda(\phi^+\phi)^2 \quad (6)$$

$$L_{1/2} = \overline{\psi}i\gamma^c E_c^\mu\left[D_\mu\psi - \frac{1}{2}\sigma_{ab}E^{b\nu}D_{\mu\nu}{}^{\alpha}E_\alpha^a\psi\right] \quad (7)$$

$$L_{Yukawa} = -y_{ij}\tilde{\psi}_L\phi\psi_A^i + h.c. \quad (8)$$

$$L_{gravity} = \frac{1}{2}\beta\phi^\dagger\phi R - \rho\left(R^{\mu\nu}R_{\mu\nu} - \frac{1}{3}R^2\right) \quad (9)$$

where β and ρ are dimensionless coupling constants and the dot in L_1 represents the approperiate summation over the SU(2) and SU(3) gauge bosons. Note that the quadratic term $\mu^2\phi^\dagger\phi$ is absent due to scale invariance. The curvatures are all cast in terms of $\tilde{\Gamma}$'s and ψ is generic field representing quarks and leptons. Neutrinos are assumed to be massless. Breaking of scale invariance can be achieved by either choosing vacuum expectation value for the doublet, i.e.

$$<\phi>= \frac{1}{\sqrt{2}}\begin{pmatrix} 0 \\ v \end{pmatrix} \quad (10)$$

or inserting a soft scale breaking term $\mu^2\phi^\dagger\phi$ into the lagrangian (Eq.4) which then develops the desired vacuum expectation value. The former procedure is conceptually on par with the conventional mechanism of spontaneous symmetry breaking in which one chooses an appropriate signature for the quadratic term $\mu^2\phi^\dagger\phi$. Gauge symmetries are also simultaneously broken. The four scalar degrees of freedom are absorbed by the four vector bosons to become massive with the remarkable result that there are no left over Higgs degrees of freedom. This can readily be demonstrated by making the ansatz $g_{\mu\nu} \sim \eta_{\mu\nu} + h_{\mu\nu}$ where $h_{\mu\nu}$ is the graviton field and $\eta_{\mu\nu}$ is the Minkowski metric, $\eta_{\mu\nu} = \text{diag}(+1,-1,-1,-1)$. In the present model, the masses of the conventional fermions and vector bosons are given by the same expressions as in the standard model,i.e,

$$
\begin{aligned}
M_W &= \frac{g}{2} v \\
M_w &= M_Z \cos\theta_W \\
\tan\theta_W &= \frac{g'}{g} \\
M_{ij}^{quark} &= Y_{ij}^{quark} \frac{v}{\sqrt{2}} \\
M_{ij}^{lepton} &= Y_{ij}^{lepton} \frac{v}{\sqrt{2}}
\end{aligned} \tag{11}
$$

where $v = 246 GeV$. Newton gravitational constant is given by

$$
G_N = \frac{1}{4\pi\beta v^2} \tag{12}
$$

Since $v = 246 GeV$, one finds $\beta = 10^{32}$. The mass of the Weylon is given by

$$
M_U = \left(f^2 + 3\beta f^2\right)^{1/2} v \sim 10^{19} f GeV \tag{13}
$$

If $\frac{f^2}{4\pi} \sim \alpha_{em}$, then $M_U \sim M_{Planck}$. Thus it may be a while before the Weylon U_μ makes its appearence. It can readily be deduced that the Weylon U_μ of the present model does not interact either with the fermions or the vector bosons $(W_\mu^\pm, Z_\mu^o)$. It is to be noted that the model under consideration derives its name from the fact that it is only the term linear in R that is relevant to the analysis presented.

III GRAVITY QUADRATIC IN TORSION

As our second example we consider gravity as a gauge theory of the Poincare group. The gravitational interactions are generated from the translation symmetry part of the Poincare group [5]. The action for gravity is determined to be

$$
L_{gravity} = \int d^4x \sqrt{-g} \beta \phi^\dagger \phi \left[-\frac{1}{4} \tilde{T}_{\mu\nu\lambda} \tilde{T}^{\mu\nu\lambda} - \frac{1}{2} \tilde{T}_{\mu\nu\lambda} \tilde{T}^{\mu\lambda\nu} + \tilde{T}_{\mu\nu}{}^{\nu} \tilde{T}^{\mu\lambda}{}_{\lambda} \right] \tag{14}
$$

where, $\tilde{\Gamma}^\lambda_{\mu\nu}$ is given by

$$
\tilde{\Gamma}^\lambda_{\mu\nu} = \left(\left(\partial_\mu + f U_\mu \right) E^\alpha_\nu \right) E^\lambda_\alpha - \left(\left(\partial_\nu + f U_\nu \right) E^\alpha_\mu \right) E^\lambda_\alpha \tag{15}
$$

and β is a dimensionless coupling constant. The fermions interact with the graviton via the spin connections of the Lorentz group. It can readily be established that the resulting Lagrangian is scale and gauge invariant. The numerical coefficients of the various terms in (Eq.14) are chosen so as to reproduce the conventional gravitational interactions as given by the Einstein-Hilbert action. With the ansatz $g_{\mu\nu} \sim \eta_{\mu\nu}$, and $<\phi>= \frac{1}{\sqrt{2}} \begin{pmatrix} 0 \\ v \end{pmatrix}$, Newton's gravitational constant is again given by (Eq.12) and the mass of the Weylon by (Eq.13). All scalar degrees of freedom are again eliminated entirely. Renormalization is a problem for both models considered. In the latter model, due to the quadratic nature of gravity, improvement in overall renormalization is expected. One may regard the present effort as initial strides towards a more complete theory.

REFERENCES

1. S. Glashow, Nucl. Phys. **22**, 579 (1961); S. Weinberg Phys. Rev. Lett. **19**, 1264 (1967); A. Salam, in *Elementary Particle Theory: Relativistic Groups and Analycity*, edited by N Svartholm, Nobel Symposium No. 8 (Wiley, New York, 1968), p. 369.
2. S. Rajpoot, A. Salam, ICTP internal report (1980, unpublished)
3. H. Weyl, Z. Phys. **56**, 330 (1929).
4. P. A. M. Dirac, Proc. Roy. Soc. London A **333**, 403 (1973); R. Utiyama, Prog. Theor. Phys. **50** 2080 (1973), and **53**, 565 (1975).
5. C.Pilot and S. Rajpoot, CSU-Long Beach Preprint (1997), submitted for publication to Physical Review D.

The Quantum Group of Seiberg-Witten Theory

M. Ruiz-Altaba

Instituto de Física, UNAM, A. P. 20-364, 01000 México, D.F.

Abstract. A brief overview of some ideas concerning the realization of a quantum group symmetry of an $N = 2$ supersymmetric Yang-Mills theory in the Coulomb phase, through the identification of the Seiberg-Witten variables $a(u)$ and $a_D(u)$ as solutions to the Picard-Fucks equation solving the Riemann-Hilbert problem on the moduli space. We also comment on differences and relationships with other approaches, namely the original one, those based on integrable spin chains, and those from string-string duality and the projection of six-dimensional null strings.

A Bocha, de recuerdo.

First of all, a quote by Juan José Giambiagi: "Soy pesimista respecto al futuro de A.L. Toda la politica científica está en manos de economistas neoliberales. No creo en el neoliberalismo. Con este modelo económico, los científicos no hacemos falta. Las multinacionales se encargarán de todo. El problema es que tampoco creo en la física de partículas y de campos (en especial las supercuerdas, etc.) durante los próximos 50 años. Puedo, naturalmente, estar equivocado."

The discovery of Γ_2 duality in $N = 2$ supersymmetric gauge theories by Seiberg and Witten (SW) [1] has had enormous significance for a wide variety of problems in theoretical physics. It fits in beautifully with the heterotic-type II duality of string theory [2], where the auxiliary torus of the SW construction is geometrized to become a piece of a fundamental Calabi-Yau manifold [3,4]. The Riemann sphere of the modulus of the theory remains the set of labels for different effective theories, and indeed the only role of the vacuum expectation value u is to characterize the holomorphic one-form on the torus, whose cycles give the masses of the BPS states. It is intriguing, neverthelesss, that this Riemann sphere can be viewed as a conformal plane with punctures (the singularities where monopoles and dyons become massless and the $SU(2)$

CP400, *First Latin American Symposium on High Energy Physics/ VII Mexican School of Particles and Fields,* edited by D'Olivo/Klein-Kreisler/Mendez

symmetry is restored), in such a way that the monodromies yield uniquely, via the Riemann-Hilbert problem, a second order differential equation customary in conformal field theory. As pointed out by Bilal [5], this differential equation should be viewed, very naturally, as the conformal Ward identity or null vector decoupling condition. Although it is perhaps puzzling at first, and somewhat contradictory with the stringy picture of these effective field theories, to consider the moduli space as the dynamical arena for a two-dimensional field theory, the issue seems worth being looked at. After all, the relevant physical information from the purely four-dimensional point of view is the set of monodromies around the singularities (including the perturbative infinity), which generate the duality group Γ_M.

The purpose of this talk is to sketch the interpretation of the lowering operators of the associated quantum group as null strings on the moduli space.

We shall think of a_D and a not as cycles over Σ_u, but simply as solutions Π to the Picard-Fuchs equations on Σ_u. Then, we shall interpret the differential equation as a null vector decoupling condition, that is, a Ward identity for some conformal field theory. The differential equation can be determined uniquely, via the solution to the Riemann-Hilbert problem, from the monodromies M_i, defined from $\begin{pmatrix} a_d \\ a \end{pmatrix} (e^{2\pi i}(u - e_i)) = m_i \begin{pmatrix} a_d \\ a \end{pmatrix} (u)$ Then, we can use the contour picture of quantum groups [6] to identify the quantum symmetry.

For example, in the case of pure $N = 4$ supersymmetric Yang-Mills theory, the β function vanishes and the duality group is the full $SL(2, Z)$ [7–9] whereby $a_D = \tau\sqrt{2u}$ and $a = \sqrt{2u}$, with $\tau = \theta/2\pi + i4\pi/g^2$ the constant coupling. The Picard-Fuchs equation is thus $\left(-\frac{\mathrm{d}}{\mathrm{d}z} + \frac{1}{2z}\right)\Pi = 0$, which can be interpreted as the decoupling at level one of the identity ($c = 1$ free boson). This theory is too perfect for our purposes.

To proceed, let us briefly recall the contour representation of quantum groups. On the complex plane $\{z\}$ we evaluate a correlation function of conformal fields by computing the OPE of their vertex operators. Nevertheless, because the quantum symmetry of conformal field theories is hidden or confined, we must ensure that the correlation function is a quantum group singlet. Each vertex operator of a primary field is the highest weight of a quantum group irrep, and the lowering operators are the screening charges $Q_i = \int dz J_i(z)$, of vanishing conformal weight. Thus, the physical correlator of the conformal primary fields of highest weights $\lambda_1, \lambda_2, \cdots, \lambda_n$ is of the form $G(\lambda_1, z_1; \cdots : \lambda_n, z_n) = \left\langle \prod_{i+1}^n V_{\lambda_i}(z_i) \prod_j Q_j^{n_j} \right\rangle$, where the various possible independent choices of integration contours reflect the multiplicity of internal (factorizing) channels. In this representation, the chiral vertex operators can be constructed by inserting two vertex operators at finite z's and a certain number of screening charges between them. Once a complete collection of chiral vertex operators is at hand, any correlation function can be constructed using the sewing procedure.

Acting with a screening charge (that is, with a quantum group lowering operator) on two vertex operators is not quite the same as the sum of acting on each of them separately. The decomposition goes through because the contour can be deformed and split in two pieces, one around each vertex operator. Naturally, since the integrand of Q_i is an operator $J_i(z)$ with non-trivial braiding with respect to the vertex operators V_{λ_i}, the extra braiding factor makes the coproduct non-commutative.

The well-known variables $a(u)$ and $a_D(u)$, which SW so ingeniously constructed as cycles of a particular one-form over an auxiliary torus, are of primordial physical relevance because they parametrize the spectrum of BPS states: $M^{\mathrm{BPS}}_{n_e,n_m}(u) = |n_e a(u) + n_m a_D(u)|$. They can be written as $a(u) = \frac{\sqrt{2}}{\pi}\int_{-1}^{1} dt\sqrt{\frac{t-u}{t^2-1}}$ and $a_D(u) = \frac{\sqrt{2}}{\pi}\int_{1}^{u} dt\sqrt{\frac{t-u}{t^2-1}}$ and thus they are simply independent solutions to the differential equation $\left(\frac{\mathrm{d}^2}{\mathrm{d}u^2} + \frac{1}{4}\frac{1}{u^2-1}\right)\Pi = 0$. Since this differential equation is of second order, it could be the conformal Ward identity derived from the decoupling of a level-two Virasoro null vector, where Π is interpreted as the correlator of primary fields at the singularities (1, -1 and ∞) and at the location (u) in moduli space of our physical theory. This correlator satisfies the Picard-Fucks equation iff the vertex operator at u has a null descendant at level two. Also, in order to be a quantum group singlet, it needs the insertion of one screening charge $\int dx J(x)$.

What conformal field theory has the above equation as one of its Ward identities? If we construct it out of one single Feigin-Fuks field $X(z)$, so that all the vertex operators are of the form $: \exp i\alpha X :$, then the central charge turns out to be 28 and all the α's pure imaginary, so it is some sort of weird Liouville-like theory. It is more rewarding to allow for a richer conformal field theory, for instance a Wess-Zumino-Witten model. Full details of this proposal will be presented elsewhere.

Let us turn now to strings. String duality tells us, among other results and conjectures, that the heterotic string compactified over $K3 \times T2$ is dual to a type IIX string compactified over a Calabi-Yau manifold C_X, provided all the non-perturbative contributions (in the infrared) are taken into account (and, of course, C_A and C_B are mirrors of each other). A beautiful result is that if the Calabi-Yau manifold C_B is a $K3$-fibration, then in the limit of infinite Planck mass it becomes the SW auxiliary manifold Σ_u. Accordingly, the volume 3-form of the Calabi-Yau, wrapped over an appropriate 2-cycle which vanishes in the limit, becomes the SW 1-form λ. The 1-cycle over which λ gets integrated is either a circle (coming from $S^2 \times S^1$) , giving rise to a gauge vector in the $D = N = 2$ SYM language, or a path between two singularities (from D^3), resulting in a hypermultiplet. The mirror dictionary goes like this: IIB near an ALE singularity is like IIA with 5-branes. The 3-branes over vanishing 2-cycles correspond to 2-branes ending on a 5-brane. The left-over 1-brane mirrors to the boundary of the 2-brane on the 5-brane.

Now, the important step is to interpret the 5-brane as spacetime $\times\Sigma_u$. Then, the SW interpretation is complete: BPS states correspond to the 2-branes ending on Σ_u.

Simple 1+1 dimensional integrable models carry the same mathematical structure as the SW solution to $N = 2$ supersymmetric Yang-Mills. Pure $SU(2)$ is Toda [10], with matter in the adjoint it is the Calogero-Moser system [11], with fundamentals it is another integrable spin chain [12]. The spectral variety for the one-soliton periodic solution to relativistic KdV is precisely the SW torus, and the Jacobi differential determining the dynamics of the soliton is the SW one-form whose integrals yield a and a_D and thus, the BPS spectrum.

Of course, the main pending problem is to identify more reasonable conformal field theories whose Ward identities reproduce the Picard-Fuchs equations of the SW theory. Once this is settled, it will be interesting to pull back to the stringy level the quantum symmetry in the vacua of moduli. One might conjecture that the null strings of the six-dimensional five-brane will be created by acting with the quantum group lowering operator on the punctured Riemann sphere. ¿From a mathematical perspective, this pull-back ought to yield a hint towards the generalization of the quantum group structure to dimensions higher than two. The asymmetric coproduct follows naturally, in the generalization of the contour representation, from wrapping higher branes on bigger cycles.

This work is supported in part by CONACYT. Conversations with J.L. Lucio, M. Moreno, A. Pérez-Martínez and C. Stephens are gratefully acknowledged.

REFERENCES

1. N. Seiberg, E. Witten, Nucl. Phys. B426 (1994) 19; ibid. B430 (1994) 485; ibid. B431 (1994) 484.
2. S. Kachru, C. Vafa, Nucl. Phys. B450 (1995) 69.
3. C. Gomez, R. Hernandez, E. Lopez, hep-th/9512017.
4. A. Klemm, W. Lerche, P. Mayr, C. Vafa, N. Warner, Nucl. Phys. B477 (1996) 746.
5. A. Bilal, hep-th/9601007.
6. C. Gómez, G. Sierra, M. Ruiz-Altaba, *Quantum Groups in Two-Dimensional Physics*, Cambridge University Press (1996) Cambridge.
7. C. Montonen D.I. Olive, Phys. Lett. 72B (1977) 117.
8. A. Sen, Phys. Lett. 329B (1994) 217.
9. C. Hull, P. Townsend, Nucl. Phys. B438 (1995) 109.
10. A. Gorsky, I. Krichever, A. Marshakov, A. Mironov, A. Morozov, Phys. Lett. 355B (1995) 466.
11. R. Donagi, E. Witten, Nucl. Phys. B460 (1996) 299.
12. A. Gorsky, A. Marshakov, A. Mironov, A. Morozov, hep-th-9603140.

Environmentally Friendly Renormalization in Finite Temperature QCD

C. R. Stephens

ICN, UNAM, Circuito Exterior, A. Postal 70-543, México D.F. 04510[1]

Abstract. A brief review is given of some of the applications of environmentally friendly renormalization to finite temperature field theory. In particular I show the benefits of using more than one renormalization group in the context of a gauge invariant coupling constant in the magnetic sector of QCD. It is shown that at high temperatures the magnetic sector shows confining properties characteristic of a three dimensional theory in accord with recent lattice results.

Phase transitions in field theory, both relativistic and non-relativistic are of enormous theoretical and practical interest, with applications ranging from the formation of structure in the early universe to the exploitation of high temperature superconductors. One of the most challenging areas is that of finite temperature behaviour in models that contain non-abelian gauge fields, e.g., QCD where the transition between hadronic matter and a quark-gluon plasma is of such interest.

One of the key difficulties in describing a system as a function of temperature is that its characteristic effective degrees of freedom may change qualitatively as a function of "scale". For instance, in QCD there exists a critical temperature above which quarks are effectively "free" whereas below it they are confined, forming mesons and baryons. There is also a qualitative change in the effective degrees of freedom in simpler theories, such as $\lambda\phi^4$. In the latter if one uses a zero temperature renormalization of λ one finds that near the critical point the loop expansion breaks down due to infrared divergences. Intuitively one can understand this from the point of view that, at least in the imaginary time formalism, the effective degrees of freedom are now three dimensional and not four. Environmentally friendly renormalization (EFR) [1] offers a quite general methodology for tackling such problems, the epithet "environmentally" being associated with the fact that very often the effective

1) This work was partially supported by Conacyt grant 3298P-E9608

CP400, *First Latin American Symposium on High Energy Physics/ VII Mexican School of Particles and Fields,* edited by D'Olivo/Klein-Kreisler/Mendez

degrees of freedom change due to the effect of some "environmental" parameter, such as temperature. Using EFR it is possible to describe for $\lambda\phi^4$ the complete crossover between $T = 0$ and $T = \infty$ in a controllable, systematic way irrespective of whether or not there is an intermediate critical point [2]. Here I will show how to implement an (apparently) EFR in the context of QCD.

As Lorentz invariance is broken at finite temperature there is an asymmetry between the electric and magnetic sectors. Here I will consider only the magnetic sector utilizing the normalization condition that the static (i.e. zero energy), spatial three-gluon vertex equals the tree-level vertex in the symmetric momentum configuration

$$\Gamma^{abc}_{ijk}(p^0_i = 0, \vec{p}_i, g_{\kappa,\tau}, T = \tau)\Big|_{\substack{\text{symm.}\\ \kappa}} = g_{\kappa,\tau} f^{abc}\left[g_{ij}(p_1 - p_2)_k + \text{cycl.}\right]. \tag{1}$$

Notice that this condition depends on two parameters, the momentum scale κ, and the temperature scale τ. Thus one can perform a RG analysis with respect to both parameters, i.e. we can run more than one environmental parameter at the same time. The advantages of this are discussed in [3].

For the calculation we used the Landau gauge Background Field Feynman rules resulting from the Vilkovisky-de Witt effective action in order to get rid of ambiguities arising from gauge dependence (more details can be found in [4]). Due to the corresponding Ward Identities the calculation is simplified in that one only needs to calculate the transverse gluon self energy Π^{Tr} in the static limit. In terms of the coupling $\alpha_{\kappa,\tau} := g^2_{\kappa,\tau}/4\pi^2$ the β functions are then

$$\kappa\frac{d\alpha_{\kappa,\tau}}{d\kappa} = \alpha_{\kappa,\tau}\,|\vec{p}|\,\frac{d\Pi^{\text{Tr}}}{d\,|\vec{p}\,|}\Bigg|_{\substack{|\vec{p}\,|=\kappa\\ T=\tau}}, \qquad \tau\frac{d\alpha_{\kappa,\tau}}{d\tau} = \alpha_{\kappa,\tau}\,T\frac{d\Pi^{\text{Tr}}}{dT}\Bigg|_{\substack{|\vec{p}\,|=\kappa\\ T=\tau}}. \tag{2}$$

The τ RG is needed to draw conclusions about the temperature dependence of the coupling. This can not be done using the κ-scheme alone without assuming something about the temperature dependence of the initial value of the coupling used in solving the differential equation.

The result is

$$\kappa\frac{d\alpha_{\kappa,\tau}}{d\kappa} = \beta_{vac} + \beta_{th}, \qquad \tau\frac{d\alpha_{\kappa,\tau}}{d\tau} = -\beta_{th}, \tag{3}$$

where the vacuum contribution is, as usual,

$$\beta_{vac} = \alpha^2_{\kappa,\tau}\left(-\tfrac{11}{6}N_c + \tfrac{1}{3}N_f\right), \tag{4}$$

and the thermal contribution is given by

$$\beta_{th} = \alpha^2_{\kappa,\tau}\Big[\left(\tfrac{21}{16}F^1_0 + \tfrac{3}{4}F^1_2 - \tfrac{3}{2}G^1_0 - \tfrac{25}{8}G^1_1 - G^1_2\right)N_c + \\ \left(\tfrac{1}{4}F^{-1}_0 + \tfrac{3}{4}F^{-1}_2 - \tfrac{3}{2}G^{-1}_0 - G^{-1}_1\right)N_f\Big]. \tag{5}$$

in terms of the IR and UV convergent integrals

$$F_n^\eta = \int_0^\infty dx \frac{x^n}{e^{\kappa x/2\tau} - \eta} \left[\log \left| \frac{x+1}{x-1} \right| - 2 \sum_{k=0}^{\frac{n}{2}-1} \frac{x^{2k+1}}{2k+1} \right] \tag{6}$$

and

$$G_n^\eta = \int_0^\infty dx \, \frac{1}{e^{\kappa x/2\tau} - \eta} \, \mathrm{P} \, \frac{x}{(x^2-1)^n}, \tag{7}$$

Because the two beta functions (3) are not exactly each other's opposite the RG improved coupling is not just a function of the ratio κ/τ. There is another dimensionful scale (such as Λ_{QCD}) that comes from an initial condition for these differential equations. The solution of the set of coupled differential equations can be written in the form

$$\alpha_{\kappa,\tau} = \frac{1}{\left(\frac{11}{6}N_c - \frac{1}{3}N_f\right) \ln \frac{\kappa}{\Lambda_{QCD}} - f\left(\frac{\kappa}{\tau}\right)} \tag{8}$$

where

$$f = \left(\tfrac{21}{16}F_0^1 + \tfrac{1}{4}F_2^1 + \tfrac{7}{8}G_1^1\right) N_c + \left(\tfrac{1}{4}F_0^{-1} + \tfrac{1}{4}F_2^{-1}\right) N_f. \tag{9}$$

and satisfies $\beta_{th} = \alpha_{\kappa,\tau}^2 \kappa df/d\kappa$ with the initial condition $\lim_{\tau \downarrow 0} f = 0$ so that we can identify Λ_{QCD} with the usual zero-temperature QCD scale.

The high-temperature behaviour (i.e. for $\tau \gg \kappa$) is determined by

$$f \longrightarrow N_c \tfrac{21\pi^2}{16} \tfrac{\tau}{\kappa} + \left(\tfrac{11}{6}N_c - \tfrac{1}{3}N_f\right) \ln \tfrac{\kappa}{\tau} + O(1). \tag{10}$$

and shows that in spite of the original belief [5] that high-temperature QCD would be asymptotically free as in the high-momentum situation in fact it is strongly coupled in the infrared where in contradistinction to the zero temperature case the strong coupling region is that associated with a three dimensional theory. This confirms in a purely continuum QCD setting the lattice based calculations and simulations of [6]. Here I have neglected the possibility of a non-perturbatively generated magnetic mass, however unless this mass increases quickly enough with temperature in order to act as a sufficient IR cutoff, one cannot get around this problem without actually solving confinement. I believe this to be an important consideration when considering phase transitions which involve non-abelian gauge fields.

If one allows the momentum-scale to change with temperature the high-temperature limit can be taken in many ways. In the region $\tau \gg \kappa$ a contour of constant coupling is given by $\tau \sim \kappa \ln \frac{\kappa}{\Lambda_{QCD}}$. This characterizes exactly along which paths in the (τ, κ)-plane the coupling increases or decreases. For example at a fixed ratio τ/κ (no matter what this ratio is) one eventually finds a coupling that decreases like $1/\ln\kappa$. This is a natural contour to consider for a weak-coupling regime [7] where one could treat the quark-gluon plasma as a perfect gas, as then the thermal average of the momentum of massless quanta at temperature T is proportional to the temperature. However at low momenta the assumption of weak coupling breaks down. Furthermore, instead of considering quantities at the average momentum it is more appropriate to use thermal averages of the quantities themselves as a weighted integral over all momenta [8]. But once again one runs into problems at low-momentum due to the strong-coupling.

The key problem is that the magnetic sector still shows remnants of confinement even above the critical temperature. Intuitively, the correct effective degrees of freedom in this sector are "bound" states, thus the above RG is not sufficiently environmentally friendly because it does not take this into account. Preliminary work on an environmentally friendly RG that does take the crossover between bound and unbound effective degrees of freedom can be found in this volume [9].

REFERENCES

1. Denjoe O'Connor and C.R. Stephens, *Nuc. Phys.* **B360** (1991) 237; *Int. Jou. Mod. Phys.* **A9** (1994) 2805.
2. Denjoe O'Connor, C.R. Stephens and F. Freire, *Mod. Phys. Lett* **A25** (1993) 1779; M.A. van Eijck and C.R. Stephens, *Proceedings of the Banff/CAP Workshop on Thermal Field Theory* (World Scientific, Singapore, 1994); M.A. van Eijck, D.J. O'Connor and C.R. Stephens, *Int. Jou. Mod. Phys.* **A 10**, (1995) 3343.
3. C.R. Stephens, *"Why Two Renormalization Groups Are Better than One"*, preprint ICN-UNAM-96-11, hep-th/9611062 (to be published in *RG '96*, Dubna (1996).
4. M.A. van Eijck, C.R. Stephens and C.W. van Weert, *Mod. Phys. Lett.* **A9** (1994) 309.
5. J.C. Collins and M.J. Perry, *Phys. Rev. Lett.* **34** (1975) 1353.
6. C. Borgs, *Nucl. Phys.* **B261** (1985) 455; E. Manousakis and J. Polonyi, *Phys. Rev. Lett.* **58** (1987) 847; F. Karsch, E. Laermann and M. Lütgemeier, *Phys. Lett.* **B346** (1995) 94.
7. K. Enquist and K. Kajantie, *Mod. Phys. Lett.* **A2** (1987) 479.
8. K. Enquist and K. Kainulainen, *Z. Phys.* **C53** (1992) 87.
9. J.C. López, A. Weber, C.R. Stephens and P.O. Hess, *"Bound States in Quantum Field Theory"*, this volume.

Gauge Fixing and Gribov Problems in a Solvable Model[1]

V. M. Villanueva[2†], Jan Govaerts[3*] and J. L. Lucio[4†]

†*Instituto de Física de la Universidad de Guanajuato, P.O. Box E-143, 37150, León, Gto., México*
**Institut de Physique Nucléaire, Université catholique de Louvain, B-1348 Louvain-la-Neuve, Belgium*

Abstract. The consistency of gauge fixing in a quantum mechanical model with Gribov ambiguities recently proposed and studied by Friedberg, Lee, Pang and Ren (FLPR) [1] is analyzed.

INTRODUCTION

In a recent paper, Friedberg, Lee, Pang and Ren (FLPR) [1] analyzed a quantum mechanical model involving gauge equivalent copies and Gribov ambiguities [2]. According to FLRP a correct description of the system is obtained by including all gauge equivalent copies. We reanalyze the model using the Teichmüller space approach emphasized by one of the authors [3]. We show that the prescription suggested by FLPR is not enough to guarantee an admissible gauge fixing. The model is analyzed at the classical level, including its gauge fixing, and its canonical quantization is worked out.

I CLASSICAL ANALYSIS

The FLPR model [1] is defined by the lagrangian

$$L = \frac{1}{2}[(\dot{X} + g\xi Y)^2 + (\dot{Y} - g\xi X)^2 + (\dot{Z} - \xi)^2] - U(X^2 + Y^2)\,, \tag{1}$$

where (X, Y, Z, ξ) denote four cartesian coordinates. This lagrangian is invariant under the following local gauge transformations,

1) Work supported by CONACyT under contract 4918-E
2) victor@ifug2.ugto.mx;
3) govaerts@fynu.ucl.ac.be;
4) lucio@ifug.ugto.mx.

CP400, *First Latin American Symposium on High Energy Physics/ VII Mexican School of Particles and Fields*, edited by D'Olivo/Klein-Kreisler/Mendez

$$X' = X\cos\alpha - Y\sin\alpha\,, \quad Z' = Z + g^{-1}\alpha\,,$$
$$Y' = X\sin\alpha + Y\cos\alpha\,, \quad \xi' = \xi + g^{-1}\dot{\alpha}\,. \tag{2}$$

Working in cylindrical coordinates and following Dirac's canonical formulation [4] for constrained systems, the fundamental first order action [3] is found to be given by,

$$S = \int_{t_i}^{t_f} dt\Big[\dot{\rho}P_\rho + \dot{\varphi}P_\varphi + \dot{Z}P_Z - \frac{1}{2}P_\rho^2 - \frac{1}{2}\frac{P_\varphi^2}{\rho^2} - \frac{1}{2}P_Z^2 - U(\rho) - \xi\phi\Big]\,, \tag{3}$$

where ϕ denotes the first class constraint,

$$\phi = P_Z + gP_\varphi \approx 0\,. \tag{4}$$

Note that the degree of freedom ξ now plays rather the role of a Lagrange multiplier. A physical, *i.e.* gauge invariant, choice of boundary conditions is obtained by fixing the values of $X(t), Y(t)$ and $Z(t)$, or equivalently by fixing the values of $\rho(t)$, $\varphi(t)$ and $Z(t)$, at the end points in time, namely $\rho(t_{i,f}) = \rho_{i,f}$, $\varphi(t_{i,f}) = \varphi_{i,f}$, $Z(t_{i,f}) = Z_{i,f}$, and requiring that the gauge parameter $\alpha(t)$ vanishes at these end points, $\alpha(t = t_i) = \alpha(t = t_f) = 0$. With this choice of boundary conditions, the classical solutions to the equations of motion are given by $p_\rho(t) = \dot{\rho}(t)$, $p_\varphi(t) = L$, $p_Z(t) = -gL$, as well as,

$$\begin{aligned}
t &= t_i + \int_{\rho(t_i)}^{\rho(t)} d\rho \frac{\pm 1}{\sqrt{2(E - U(\rho) - \frac{L^2}{2\rho^2})}}\,, \\
\varphi(t) &= \varphi_i + L\int_{t_i}^{t} dt' \frac{1}{\rho^2(t')} + g\int_{t_i}^{t} dt'\xi(t')\,, \\
Z(t) &= Z_i - gL(t - t_i) + \int_{t_i}^{t} dt'\xi(t')\,,
\end{aligned} \tag{5}$$

with the additional condition that $\gamma \equiv \int_{t_i}^{t_f} dt\,\xi(t) = (Z_f - Z_i) + gL(t_f - t_i)$. In these expressions, L and E stand for the constants of motion to be identified with the angular momentum and energy of the system, respectively.

Following Ref. [3], let us introduce the parameter defined by $\gamma = \int_{t_i}^{t_f} dt\xi(t)$. This quantity is invariant under local gauge tranformations which satisfy the above boundary conditions. Futhermore, given these boundary conditions, any pair of Lagrange multipliers related by a gauge transformation, ($\xi_2 = \xi_1 + \frac{1}{g}\frac{d\alpha(t)}{dt}$) lead to identical parameters γ_1 and γ_2. Moreover, once the Lagrange multiplier $\xi(t)$ is specified, no further gauge freedom is operative.

Consequently [3], γ parametrizes the space of gauge orbits in the space of Lagrange multipliers. In other words, γ is the Teichmüller parameter of Teichmüller space [3] which in the present case is identified with the space of real numbers . Furthermore, gauge fixing in the space of Lagrange multipliers entails the specification of a section $\xi(t;\gamma)$ which to a value of γ associates a Lagrange multiplier $\xi(t;\gamma)$. An admissible gauge fixing in Teichmüller space is such that the section $\xi(t;\gamma)$ defines one and only one Lagrange multiplier for each of all the possible real values of γ. Moreover, in the case of a closed algebra, it may be shown [3] that an admissible gauge fixing $\xi(t;\gamma)$ in Teichmüller space implies an admissible gauge fixing of the complete system itself. A non admissible gauge fixing is either not complete (*i.e.* a given gauge orbit is selected more than once), or not global (*i.e.* not all gauge orbits are selected), or both.

The previous paragraph briefly indicated that a gauge fixing of the system is admissible if the associated set of Teichmüller parameter values γ covers once and only once every point of Teichmüller space, which in the present case is identified with the real line. With this understanding in hand, let us now turn to examples of gauge fixings.

Gauge Fixing

1) If $\xi(t) = \xi_0$ constant, then $\gamma = \xi_0 \Delta t$. This implies that the Teichmüller parameter γ takes only one single value. Hence, such a condition cannot define an admissible gauge fixing. Indeed, given the chosen boundary conditions, not all possible gauge orbits of the system are accounted for. In other words, there are certain physically acceptable configurations of the system which cannot be reached, since one must satisfy the constraint $\gamma = \Delta Z + gL\Delta t$. Even though this gauge fixing is complete (no gauge freedom left), it is not global.

2) The condition $\Omega = Z = 0$. In the conventional approach à la Faddeev [1,3], such a condition is said to be admissible, which is a misnomer since even though it is complete, it is not global. Indeed, the conditions $\{\Omega, \phi\} = 1$ and $\dot{\Omega} = P_Z + \xi = 0$ determine the Lagrange multiplier $\xi(t) = -P_Z$ uniquely, leaving no further gauge freedom. This gauge fixing implies a specific time dependence of $\xi(t)$ and also a specific value for the Teichmüller parameter. Clearly, this gauge fixing cannot be admissible, since only those configurations for which $Z(t)$ vanishes at all times are included, which is only a subset of all possible configurations, hence also of all possible gauge orbits of the system.

3) Gauge fixing conditions of the type $\dot{\xi} = F(\xi)$ were first proposed in Ref. [5], and investigated in Ref. [3]. In particular the choices $F(\xi) = \beta$ and $F(\xi) = \beta\xi$, β being a constant, are examples [3] of admissible gauge fixing functions leading to one-to-one relations between ξ and the Teichmüller parameter γ.

The gauge fixing conditions in examples 1 and 2 were analyzed by FLPR [1]. Notice that they select different sets of gauge orbits, *i.e.* they are not gauge equivalent. The physics that each of these conditions describe is different, and even though they lead to gauge invariant descriptions, they do not provide the correct description of the physical system, since none of them includes all possible gauge orbits and accounts for all physically consistent choices of boundary conditions.

II QUANTIZATION AND PHYSICAL STATES

In order to avoid ordering ambiguities, canonical quantization of the system is worked out in cartesian coordinates. The Hamiltonian and the constraint are given by,

$$\hat{H} = \frac{1}{2}[\hat{P}_X^2 + \hat{P}_Y^2 + \hat{P}_Z^2] + U(\hat{X}^2 + \hat{Y}^2), \quad \hat{\phi} = \hat{P}_Z + g(\hat{X}\hat{P}_Y - \hat{Y}\hat{P}_X) = 0. \quad (6)$$

For simplicity, we consider the harmonic oscillator potential $U(X^2 + Y^2) = \frac{\omega^2}{2}(X^2 + Y^2)$. By analogy with the usual harmonic oscillator, one introduces the operators

$$a_i = \sqrt{\omega/2}[\hat{X}_i + i\hat{P}_{X_i}/\omega], \quad A_\pm = 1/\sqrt{2}(a_1 \mp ia_2) \quad (7)$$

where $X_1 = X$ and $X_2 = Y$. The $A_\pm$ operators are such that $[A_\pm, A_\pm^\dagger] = 1$ with all other commutators vanishing. In terms of the $A_\pm$ operators (6) can be written as

$$\hat{H} = \omega(A_+^\dagger A_+ + A_-^\dagger A_- + 1) + \frac{1}{2}\hat{P}_Z^2, \quad \hat{\phi} = \hat{P}_Z + g(A_+^\dagger A_+ - A_-^\dagger A_-). \quad (8)$$

Solving for the ground state, it is then possible to obtain the complete basis of gauge invariant physical states annihilated by the constraint, with the Z dependence included,

$$<u,\bar{u},Z|k,n-k> = \frac{1}{\sqrt{2\pi}} \exp\Big(igZ(n-2k)\Big)(\tfrac{\omega}{\pi})^{\frac{1}{2}} \frac{1}{\sqrt{k!(n-k)!}} \exp\big(-\tfrac{1}{2}\omega u\bar{u}\big)\big(-1\big)^k \sum_{p=0}^{n-k} \binom{n-k}{p} \times$$
$$\times (-1)^p k(k-1)(k-2)...(k-(n-k-p-1))\big(\sqrt{\omega}u\big)^{k-(n-k-p)}\big(\sqrt{\omega}\bar{u}\big)^p, \quad (9)$$

where $u = X + iY$ and $\bar{u} = X - iY$.

REFERENCES

1. Friedberg R., Lee T. D., Pang Y. and Ren H. C., *Ann. Phys.* **246**, 381 (1996).
2. Gribov V. N., *Nucl. Phys.* **B139**, 1 (1978).
3. Govaerts J., *Hamiltonian Quantisation and Constrained Dynamics* (Leuven University Press, Leuven, 1991).
4. Dirac P. A. M., *Lectures on Quantum Mechanics* (Belfer Graduate School of Science, Yeshiva University, New York, 1964).
5. Teitelboim C., *Phys. Rev.* **D25**, 3159 (1982).

Graded External Symmetries and r-associativity

Luis Alberto Wills-Toro

Departamento de Física, Universidad de Antioquia, AA. 1226 Medellín, Colombia, e-mail: law@fisica.udea.edu.co

Abstract. We discuss the construction of parameter algebras compatible with the underlying graded structure of Special Relativity. The existence of associative and nonassociative $\mathbb{Z}_2\times(\mathbb{Z}_{4\Lambda}\times\mathbb{Z}_{4\Lambda})$-graded parameter algebras is thoroughly analyzed. The corresponding enlarged Lie algebraic structures are presented.

in Memoriam **J. J. Giambiagi**

EXTERNAL SYMMETRY GROUPS AND GENERATOR ALGEBRAS

The symmetries play a central role in contemporary physics. The knowledge of the most fundamental structure of the microworld seems to be closely related to the determination of (exact, approximated, broken, global, local) symmetries. Every symmetry transformation is mathematically described by the concept of group of transformations. We recognize two classes: the discrete and the continuous groups of transformations. In the latter class the number of elements is not countable. It turns out, that every continuous group of transformations is the product of a discrete and a Lie group, being this Lie group the connected component which contains the identity transformation. Every element g_α of the Lie group can be realized in terms of the exponential of a linear combination of linearly independent generators G_j ; $j = 1, \ldots, N$,

$$g_\alpha = \exp\{\mathrm{i}\alpha^j G_j\} \tag{1}$$

where α^j; $j = 1, \ldots, N$ are commutative numeric parameters. The generators G_j ; $j = 1, \ldots, N$ build a Lie algebra $L = \mathrm{gen}\{G_j\}$ with a Lie product "$[\cdot,\cdot]$" which fulfils for every $X, Y, Z \in L$: *Closure* $[X, Y] \in L$, *Linearity* $[X + Y, Z] = [X, Z] + [Y, Z]$, *Antisymmetry* $[X, Y] = -[Y, X]$, and *Jacobi Associativity* $[X, [Y, Z]] = [[X, Y], Z] + [Y, [X, Z]]$.

CP400, *First Latin American Symposium on High Energy Physics/ VII Mexican School of Particles and Fields*, edited by D'Olivo/Klein-Kreisler/Mendez

TABLE 1. $\mathbb{Z}_2$-Group

$(+)^{\mathbb{Z}_2}$	B	F
B	B	F
F	F	B

TABLE 2. $q^{\mathbb{Z}_2}$-factors

$q^{\mathbb{Z}_2}$	B	F
B	1	1
F	1	-1

Supersymmetry offered the very smart extension of the concept of symmetry transformations in which *Grassmann or fermionic* parameters $\theta^\beta, \bar{\theta}_{\dot{\beta}}, \ldots$ are introduced besides the *commutative or bosonic* parameters α^j $, j = 1, \ldots, N$:

$$g_{\alpha,\theta,\bar{\theta}} = \exp\{\mathrm{i}\alpha^j G_j + \mathrm{i}\theta^\beta Q_\beta + \mathrm{i}\bar{\theta}_{\dot{\beta}}\bar{Q}^{\dot{\beta}}\}. \tag{2}$$

The novel fermionic generators $Q_\beta, \bar{Q}^{\dot{\beta}}, \ldots$ transform as Weyl spinors which are allowed spin-$\frac{1}{2}$ representations of the Poincaré algebra with fermionic statistic according to spin-statistics theorem. In fact, the no-go theorems of S. Coleman & J. Mandula [1] and of R. Haag, J. T. Łopuszański & M. F. Sohnius [2] provide constraints on the external symmetries formulated in terms of Lie- and $\mathbb{Z}_2$-graded Lie- algebras respectively.

The Grassmann parameters are characterized by its nilpotency $\theta_\beta\theta_\beta = 0$, and by the anticommutativity $\theta_\beta\theta_\rho = -\theta_\rho\theta_\beta$. The bosonic "B" and the fermionic "F" behaviour build a $\mathbb{Z}_2$-graded algebra. The product of parameters fulfills

$$\kappa_{\tilde{a}}\pi_{\tilde{e}} = q^{\mathbb{Z}_2}_{\tilde{a},\tilde{e}}\ \pi_{\tilde{e}}\kappa_{\tilde{a}} = \sigma_{\tilde{a}+\tilde{e}}, \tag{3}$$

where the tilded indices indicate the bosonic or fermionic character of the parameter , and $\tilde{a}, \tilde{e} \in \{B, F\}$. The parameters α^j $, j = 1, \ldots, N$ are bosonic and the parameters , and $\theta^\beta, \bar{\theta}_{\dot{\beta}}, \ldots$ are fermionic. The product of parameters and the corresponding q-factors are described by the tables 1 and 2. This implies in particular:

$$\alpha^j\alpha^k = \alpha^k\alpha^j, \quad \alpha^j\theta^\beta = \theta^\beta\alpha^j, \quad \theta^\beta\theta^\rho = -\theta^\rho\theta^\beta. \tag{4}$$

TOWARDS Q-COMMUTATIVE SYMMETRY PARAMETERS

The study extensions of the Lie algebraic structure associated with the external symmetries of Special Relativity can involve gradings beyond the $\mathbb{Z}_2$-

gradings of supersymmetry [3] [4]. The assumptions of the mentioned no-go theorems are then relaxed to include generators which are *not necessarily* acted by commutators under Poincaré transformations. The graded structures involve generalized commutation properties (q-commutativity) for the parameters of the corresponding symmetry group.

The grading groups $\mathbb{Z}_2\times(\mathbb{Z}_{4\Lambda}\times\mathbb{Z}_{4\Lambda})$; $\Lambda \in \mathbb{R}$ are consistent with the underlying grading structure of the Poincaré algebra [3]. The elements $\tilde{a}, \tilde{e} \in \mathbb{Z}_2\times(\mathbb{Z}_{4\Lambda}\times\mathbb{Z}_{4\Lambda})$ and its corresponding applications "+" and "q" can be written as follows:

$$\tilde{a} = (a_0, (a_1, a_2)), \quad \tilde{e} = (e_0, (e_1, e_2)), \tag{5}$$

$$a_0, e_0 \in \mathbb{Z}_2 = \{0, 1\}, \tag{6}$$

$$(a_1, a_2), (e_1, e_2) \in \mathbb{Z}_{4\Lambda}\times\mathbb{Z}_{4\Lambda} = \{(n, m) \ : \ n, m \in \{0, 1, \cdots, 4\Lambda - 1\}\}, \tag{7}$$

$$\tilde{a} + \tilde{e} = ((a_0 + e_0)\mathrm{mod}\ 2, ((a_1 + e_1)\mathrm{mod}\ 4\Lambda, (a_2 + e_2)\mathrm{mod}\ 4\Lambda)), \tag{8}$$

$$q^{\mathbb{Z}_2}_{a_0,e_0} = \exp\{i\pi a_0 e_0\}, \tag{9}$$

$$q^{\mathbb{Z}_{4\Lambda}\times\mathbb{Z}_{4\Lambda}}_{(a_1,a_2),(e_1,e_2)} = \exp\{i\pi(a_1 e_2 - e_1 a_2)/2\Lambda\}, \tag{10}$$

$$q_{\tilde{a},\tilde{e}} = q^{\mathbb{Z}_2}_{a_0,e_0}\ q^{\mathbb{Z}_{4\Lambda}\times\mathbb{Z}_{4\Lambda}}_{(a_1,a_2),(e_1,e_2)}. \tag{11}$$

The first part provides the standard $\mathbb{Z}_2$-grading of tables 1 and 2, with B$\leftrightarrow$ 0, F$\leftrightarrow$ 1. The $\mathbb{Z}_{4\Lambda}\times\mathbb{Z}_{4\Lambda}$-grading factor provides non-symmetric phase contributions to the function q. The $\mathbb{Z}_2\times(\mathbb{Z}_{4\Lambda}\times\mathbb{Z}_{4\Lambda})$-graded parameters fulfill *q-commutativity*:

$$\kappa_{\tilde{a}}\pi_{\tilde{e}} = q_{\tilde{a},\tilde{e}}\pi_{\tilde{e}}\kappa_{\tilde{a}} = \sigma_{\tilde{a}+\tilde{e}}. \tag{12}$$

TOWARDS R-ASSOCIATIVE SYMMETRY PARAMETERS

The introduction of the group gradings beyond supersymmetry offers a further and surprising possibility: the introduction of symmetry parameters with *generalized associative properties* (r-associativity) which is also compatible with the underlying Lie algebraic Poincaré subalgebra [5]. We call the group parameters *r-associative* if the change of parenthesis produces a numeric factor:

$$\kappa_{\tilde{a}}(\pi_{\tilde{e}}\gamma_{\tilde{c}}) = r_{\tilde{a},\tilde{e},\tilde{c}}(\kappa_{\tilde{a}}\pi_{\tilde{e}})\gamma_{\tilde{c}}. \tag{13}$$

Besides the trivial case in which associativity holds ($r_{\tilde{a},\tilde{e},\tilde{c}} = 1$ for all $\tilde{a}, \tilde{e}, \tilde{c}$), there is a model for the r-factors of the $\mathbb{Z}_2\times(\mathbb{Z}_{4\Lambda}\times\mathbb{Z}_{4\Lambda})$-graded group parameters which is as well consistent with the underlying grading structure of the Poincaré subalgebra:

$$r_{\tilde a,\tilde e,\tilde c} = r_{(a_0,(a_1,a_2)),(e_0,(e_1,e_2)),(c_0,(c_1,c_2))} =$$
$$= \exp\{i\pi(a_1c_2 - c_1a_2)\{(a_1+c_1)e_2 - e_1(a_2+c_2)\}/2\Lambda\}. \tag{14}$$

A concrete basis for $\mathbb{Z}_2\times(\mathbb{Z}_{4\Lambda}\times\mathbb{Z}_{4\Lambda})$-graded parameters includes *self-bosonic* and *self-fermionic* elements

$$\epsilon_{\tilde a}\ ;\ \tilde a \in \{0\}\times(\mathbb{Z}_{4\Lambda}\times\mathbb{Z}_{4\Lambda}) \quad \text{and} \quad \epsilon_{\tilde e}\ ;\ \tilde e \in \{1\}\times(\mathbb{Z}_{4\Lambda}\times\mathbb{Z}_{4\Lambda}). \tag{15}$$

The product among these basis elements is characterized by *multiplication constants* $\chi^{\tilde c+\tilde u}_{\tilde c,\tilde u} \in \mathbb{C}$:

$$\epsilon_{\tilde c}\epsilon_{\tilde u} = \chi^{\tilde c+\tilde u}_{\tilde c,\tilde u}\,\epsilon_{\tilde c+\tilde u}. \tag{16}$$

The consistency of the q-commutativity and r-associativity properties for arbitrary monomial of parameters leads to the following conditions on the multiplication constants:

$$\chi^{\tilde c+\tilde u}_{\tilde c,\tilde u} = q_{\tilde c,\tilde u}\chi^{\tilde c+\tilde u}_{\tilde u,\tilde c}, \quad \chi^{\tilde a+\tilde c+\tilde u}_{\tilde a,\tilde c+\tilde u}\chi^{\tilde c+\tilde u}_{\tilde c,\tilde u} = r_{\tilde a,\tilde c,\tilde u}\chi^{\tilde a+\tilde c}_{\tilde a,\tilde c}\chi^{\tilde a+\tilde c+\tilde u}_{\tilde a+\tilde c,\tilde u}. \tag{17}$$

A suitable choice for the multiplication constants in order to satisfy (14) and (17) is given by:

$$\chi^{\tilde a+\tilde e}_{\tilde a,\tilde e} = \exp\{i\pi(a_0e_0/2 + (a_1e_2 - e_1a_2)/4\Lambda)\}\cdot$$
$$\cdot\exp\{i\pi(a_1e_2 - e_1a_2)^2/4\Lambda)\}\rho^{\tilde a+\tilde e}_{\tilde a,\tilde e}, \tag{18}$$

where the *branching constants* $\rho^{\tilde c+\tilde u}_{\tilde c,\tilde u}$ should satisfy:

$$\rho^{\tilde c+\tilde u}_{\tilde c,\tilde u} = \rho^{\tilde c+\tilde u}_{\tilde u,\tilde c}, \quad \rho^{\tilde a+\tilde c+\tilde u}_{\tilde a,\tilde c+\tilde u}\rho^{\tilde c+\tilde u}_{\tilde c,\tilde u} = \rho^{\tilde a+\tilde c}_{\tilde a,\tilde c}\rho^{\tilde a+\tilde c+\tilde u}_{\tilde a+\tilde c,\tilde u}. \tag{19}$$

(In the case in which associativity holds ($r_{\tilde a,\tilde e,\tilde c} = 1$ for all $\tilde a, \tilde e, \tilde c$), the second exponential in right-hand side of equation (18) drops.) The self-fermionic parameters are nilpotent,

$$\epsilon_{\tilde u}\epsilon_{\tilde u} = q_{\tilde u,\tilde u}\epsilon_{\tilde u}\epsilon_{\tilde u} = -\epsilon_{\tilde u}\epsilon_{\tilde u} = 0\ ;\ \tilde u \in \{1\}\times(\mathbb{Z}_{4\Lambda}\times\mathbb{Z}_{4\Lambda}). \tag{20}$$

And since either $q_{\tilde c,\tilde c} = 1$ or $q_{\tilde c,\tilde c} = -1$, then the branching constants fulfill

$$\rho^{\tilde c+\tilde u}_{\tilde c,\tilde u} = \tfrac{1}{2}(1 + q_{\tilde c,\tilde c}\delta_{\tilde c,\tilde u})\rho^{\tilde c+\tilde u}_{\tilde c,\tilde u}. \tag{21}$$

TOWARDS GENERALIZED LIE ALGEBRAIC STRUCTURES

The parameters $\epsilon_{\tilde u}$; $\tilde u \in \mathbb{Z}_2\times(\mathbb{Z}_{4\Lambda}\times\mathbb{Z}_{4\Lambda})$ and its products are the starting point for the search of $\mathbb{Z}_2\times(\mathbb{Z}_{4\Lambda}\times\mathbb{Z}_{4\Lambda})$-graded extensions of the Poincaré group [5] [6] with elements $\exp\{i\pi^{j}_{-\tilde{u}}G^{j}_{\tilde{u}}\}$. The corresponding algebraic structure $\mathbb{L}$ is called $(\mathbb{Z}_2\times(\mathbb{Z}_{4\Lambda}\times\mathbb{Z}_{4\Lambda}); q, r)$*-graded Lie algebra over* $\mathbb{C}$ which has an internal product "$[\![\cdot,\cdot]\!]$" that fulfills for every $X_{\tilde{a}}, Y_{\tilde{e}}, Z_{\tilde{c}}, X'_{\tilde{a}} \in \mathbb{L}$:

$$\textbf{Closure and group grading: } \text{exists } U_{\tilde{a}+\tilde{e}} \in \mathbb{L} : [\![X_{\tilde{a}}, Y_{\tilde{e}}]\!] = U_{\tilde{a}+\tilde{e}}. \tag{22}$$

$$\textbf{Linearity: } [\![X_{\tilde{a}} + X'_{\tilde{a}}, Y_{\tilde{e}}]\!] = [\![X_{\tilde{a}}, Y_{\tilde{e}}]\!] + [\![X'_{\tilde{a}}, Y_{\tilde{e}}]\!]. \tag{23}$$

$$\textbf{q-antisymmetry: } [\![X_{\tilde{a}}, Y_{\tilde{e}}]\!] = -q_{\tilde{a},\tilde{e}}[\![Y_{\tilde{e}}, X_{\tilde{a}}]\!]. \tag{24}$$

$$\begin{aligned}\textbf{(q,r)-Jacobi associativity: } [\![X_{\tilde{a}}, [\![Y_{\tilde{e}}, Z_{\tilde{c}}]\!]]\!] &= r_{\tilde{a},\tilde{e},\tilde{c}}[\![[\![X_{\tilde{a}}, Y_{\tilde{e}}]\!], Z_{\tilde{c}}]\!] \\ &\quad + r_{\tilde{a},\tilde{c},\tilde{e}}q_{\tilde{a},\tilde{e}}[\![Y_{\tilde{e}}, [\![X_{\tilde{a}}, Z_{\tilde{c}}]\!]]\!].\end{aligned} \tag{25}$$

Observe that addition of elements of $\mathbb{L}$ in condition (23) is only defined among elements carrying *identical* group indices. The Lorentz covariance forbids the addition of a fermionic with a bosonic object. The exclusion of the addition among objects with different statistic behaviour seems to be very well motivated in physical applications.

A particular model (analog to the commutators for Lie algebras) for the product "$[\![\cdot,\cdot]\!]$" is given by the *q-commutator* made out of a r-associative product "$\clubsuit$":

$$[\![X_{\tilde{a}}, Y_{\tilde{e}}]\!] = X_{\tilde{a}}\clubsuit Y_{\tilde{e}} - q_{\tilde{a},\tilde{e}}Y_{\tilde{e}}\clubsuit X_{\tilde{a}}. \tag{26}$$

ACKNOWLEDGMENT

The financial support of *COLCIENCIAS-Colombia*, and the organizers of the *VII-EMPC* and the *I-SILAFAE* is acknowledged.

REFERENCES

1. S. Coleman and J. Mandula, Phys. Rev. **159**, 1251 (1967).
2. R. Haag, J.T. Łopuszański and M.F. Sohnius, Nucl. Phys. **B88**, 257 (1975).
3. L. A. Wills Toro, *(I, q)-graded Lie Algebraic Extensions of the Poincaré Algebra, Constraints on I and q*, Journal of Mathematical Physics **36** (4), 2085–2112 (1995).
4. L.A. Wills Toro, *Grading beyond supersymmetry*, Workshops Puebla Mexico 1995, AIP Conference Proceedings **359**, J.C. D'Olivo, A. Fernandez, M.A. Pérez eds, AIP-Press, 508-512 (1996).
5. L.A. Wills-Toro, *(I;q,r)-graded pseudoalgebras over a field*, Submitted for publication (1997).
6. L.A. Wills-Toro and J.I. Zuluaga, *Extended superspace for gradings beyond supersymmetry*, in preparation (1997).

LIST OF PARTICIPANTS

ACOSTA, Milenis	Cinvestav-UM, México
AGUAYO, Aaron	Cinvestav-UM, México
ALFARO, Jorge	Universidad Católica de Chile, Chile
ANJOS, Joao dos	CBPF, Brazil
ARANDA-SÁNCHEZ, Jorge	Cinvestav, México
ASTORGA, Francisco	IF-UMSNH, México
AVILA, Manuel	FC-UAEM, México
AYALA, Alejandro	Univ. of Illinois at Urbana-Champaign, USA
BEDIAGA, Ignacio	CBPF, Brazil
BESPROSVANY, Jaime	IF-UNAM, México
BIANCO, Stefano	Laboratori Nazionali di Frascati, INFN, Italy
BIJKER, Roelof	ICN-UNAM, México
BINÉTRUY, Pierre	LPTHE, Université Paris-Sud, France
BRANDENBERGER, Robert	Brown University, USA
BROCKWAY, Jack	Wake Forest University, USA
BUTLER, Joel N.	Fermilab, USA
CABO, Alejandro	IF-UG/ICIMAF, Cuba
CANTORAL, Eduardo	FCFM-BUAP, México
CARRILES, Ramón	UAM, México
CARRILLO, Iván	FC-UNAM, México
CARRILLO, Salvador	Cinvestav, México
CASTAGNINO, Mario	Universidad de Buenos Aires, Argentina
CASTILLA, Heriberto	Cinvestav, México
CÁZAREZ, Federico Jesús	FC-UNAM, México
CIFUENTES, Edgar	Universidad de San Carlos, Guatemala
COTTI, Umberto	Cinvestav, México
CUAUTLE, Eleazar	FCFM-BUAP, México
CUMALAT, John P.	University of Colorado, USA
D'OLIVO, Juan Carlos	ICN-UNAM, México
DE LA MACORRA, Axel	IF-UNAM, México
DECOSS GÓMEZ, Maritza	Cinvestav-UM, México
DEL OOSO ACEVEDO, José	FC-UNAM, México
DESER, Stanley	Brandeis University, USA
DÍAZ-CRUZ, Lorenzo	IF-BUAP, México
DOVA, María Teresa	Univ. Nacional de la Plata, Argentina
ESCALERA SANTOS, Gerardo	Cinvestav-UM, México
FEARNLEY, Tom	Niels Bohr Institute/CERN, Denmark
FERNÁNDEZ, Arturo	FCFM-BUAP, México
FERREIRA, Luiz Agostinho	IFT/UNESP, Brazil
FLORES REYES, Angel	FC-UNAM, México
FLORES, Rubén	Cinvestav, México
FONT, Anamaría	Univ. Central de Venezuela, Venezuela
FOSCO, César	Centro Atómico Bariloche, Argentina

GARCÍA, Augusto — Cinvestav, México
GARCÍA, José Luis — Cinvestav, México
GARCÍA ZENTENO, Antonio — ICN-UNAM, México
GAITÁN, Ricardo — Cinvestav, México
GERMÁN, Gabriel — IF-UNAN, México
GIUBELLINO, Paolo — INFN, Torino, Italy
GLEISER, Marcelo — Dartmouth College, USA
GOBEL B. DE MELLO, Carla — CBPF, Brazil
GÓMEZ, Sandra — Universidad de Antioquia, Colombia
GONZÁLEZ, Hernando — Cinvestav, México
GUPTA, Virendra — Cinvestav-UM, México
GUTIÉRREZ, Alejandro — IF-BUAP, México
HERNÁNDEZ MONTOYA, Raúl — Cinvestav, México
HERNÁNDEZ, Albino — Cinvestav, México
HERNÁNDEZ, Eva — UAM, México
HERNÁNDEZ, Javier M. — Cinvestav, México
HERRERA, Gerardo — Cinvestav, México
HERRERO, María José — Universidad Autónoma de Madrid, Spain
HOJVAT, Carlos — Fermilab, USA
HUERTA, Dora — Cinvestav-UM, México
HUERTA, Rodrigo — Cinvestav-UM, México
JARAMILLO-ARANGO, Daniel — Cinvestav, México
JUÁREZ, Rebeca — ESFM-IPN, México
KANE, Gordon L. — University of Michigan, USA
KIELANOWSKI, Piotr — Cinvestav, México
KLEIN, Martin — IF-UNAM, México
KONIGSBERG, Jacobo — University of Florida, USA
LANGACKER, Paul — University of Pennsylvania, USA
LARIOS, Francisco — Michigan State Univ./Cinvestav, México
LÓPEZ, Angel — Universidad de Puerto Rico, Puerto Rico
LÓPEZ FALCÓN, Dennys A. — Cinvestav-UM, México
LÓPEZ VIEYRA, Juan Carlos — ICN-UNAM, México
LÓPEZ, Rebeca — BUAP, México
LUCIO, José Luis — IF-UG, México
MA, Ernest — University of California, USA
MAGAÑA, Leonel — Cinvestav, México
MALBOUISSON, Adolfo — CBPF, Brazil
MARTÍN, Isbelia — Universidad Simón Bolívar, Venezuela
MARTÍNEZ, Jesús — ESFM-IPN, México
MARTÍNEZ LEDEZMA, Juan — FC-UNAM, México
MARTÍNEZ, Roberto — Universidad Nacional de Bogotá, Colombia
MÉNDEZ, Héctor — Cinvestav, México
MÉNDEZ-GALAIN, Ramón — Universidad de la República, Uruguay
MONDRAGÓN, Myriam — IF-UNAM, México
MONTANO, Luis — Cinvestav, México
MORENO, Gerardo — IF-UG, México

MORENO, Matías — IF-UNAM, México
MOTA PORRAS, Gregorio — FC-UNAM, México
MURGUIA, Gabriela — FC-UNAM, México
MUÑOZ ÑUNGO, José H. — Cinvestav, México
MUÑOZ SALAZAR, Laura — FCFM-BUAP, México
MUSTRE, José — Cinvestav-UM, México
NAHMAD, Yuri — Cinvestav-UM, México
NAPSUCIALE, Mauro — IF-UG, México
NELLEN, Lukas — IF-UNAM, México
NIEVES, José F. — Universidad de Puerto Rico, Puerto Rico
OGURI, Vitor — Univ. do Estado do Rio de Janeiro, Brazil
OROZCO, Luis A. — State University of New York, USA
PEÑA ARELLANO, Fabián — UAM, México
PÉREZ, Aurora — ICIMAF, Cuba
PÉREZ, Bolivia — FCFM-BUAP, México
PÉREZ, Gabriel — Cinvestav-UM, México
PÉREZ, Hugo — ICIMAF, Cuba
PÉREZ, Miguel Angel — Cinvestav, México
PÉREZ-LORENZANA, Abdel — Cinvestav, México
PICCINELLI, Gabriella — IA-UNAM, México
QUINTERO, Lilian — Cinvestav-UM, México
QUIROZ, Norma — Cinvestav, México
RAJPOOT, Subhash — California State University, USA
RAMIREZ, Carlos A. — Univ. Industrial de Santander, Colombia
RESTUCCIA, Alvaro — Universidad Simón Bolivar, Venezuela
REYES-SANTOS, Marco Antonio — IF-UG, México
RIQUER RAMÍREZ, Verónica — IF-UNAM, México
RITTO, Pavel — Cinvestav-UM, México
RODRÍGUEZ, Ezequiel — IF-UNAM, México
RUIZ, Jesús — Cinvestav-UM, México
RUIZ, Salvador — ICN-UNAM, México
RUIZ-ALTABA, Martí — IF-UNAM, México
SÁ BORGES, José de — Univ. do Estado do Rio de Janeiro, Brazil
SALAZAR IBARGUEN, Humberto — FCFM-BUAP, México
SAMPAYO, Oscar Alfredo — Universidad de Mar del Plata, Argentina
SÁNCHEZ, Sully — Cinvestav, México
SÁNCHEZ COLÓN, Gabriel — Cinvestav-UM, México
SÁNCHEZ-HERNÁNDEZ, Alberto — Cinvestav, México
SASSOT, Rodolfo — Universidad de Buenos Aires, Argentina
SASTRE, Francisco — Cinvestav-UM, México
SCHAT, Carlos Luis — CNEA, Buenos Aires, Argentina
SCHMIDT, Iván — Universidad Federico Santa María, Chile
SHEAF, Marleigh — Cinvestav, México
SIMÃO, Fernando — CBPF, Brazil
SOCOLOVSKY, Miguel — ICN-UNAM, México
SORKIN, Rafael — ICN-UNAM, México

STEPHENS, Christopher	ICN-UNAM, México
STICKLAND, David	CERN, Switzerland
STUART, Robin G.	Randall Laboratory of Physics, USA
SWAIN, John	Northeastern University, USA
SÁNCHEZ TOLEDO, Genaro	Cinvestav, México
TORRES, Manuel	IF-UNAM, México
TÚTUTI, Eduardo	IF-UNAM, México
URRUTIA, Luis F.	ICN-UNAM, México
VALENCIA, Mario	Cinvestav-UM, México
VÁZQUEZ, Fabiola	Cinvestav, México
VÁZQUEZ-BELLO, José Luis	Univ. Autónoma de Yucatán, México
VEGA, Belia	FCFM-BUAP, México
VERGARA, José David	ICN-UNAM, México
VILLANUEVA, Víctor	IF-UG, México
VINIEGRA, Fermín	FC-UNAM, México
WEBER MULLER, Axel	ICN-UNAM, México
WELDON, H. Arthur	West Virginia University, USA
WILLS-TORO, Luis Alberto	Universidad de Antioquia, Colombia
WISE, Mark B.	California Inst. of Tech., USA
ZEPEDA, Arnulfo	Cinvestav, México

Author Index

AIP Conference Proceedings

	Title	L.C. Number	ISBN
No. 276	Very High Energy Cosmic-Ray Interactions: VIIth International Symposium (Ann Arbor, MI 1992)	93-71342	1-56396-038-9
No. 277	The World at Risk: Natural Hazards and Climate Change (Cambridge, MA 1992)	93-71333	1-56396-066-4
No. 278	Back to the Galaxy (College Park, MD 1992)	93-71543	1-56396-227-6
No. 279	Advanced Accelerator Concepts (Port Jefferson, NY 1992)	93-71773	1-56396-191-1
No. 280	Compton Gamma-Ray Observatory (St. Louis, MO 1992)	93-71830	1-56396-104-0
No. 281	Accelerator Instrumentation Fourth Annual Workshop (Berkeley, CA 1992)	93-072110	1-56396-190-3
No. 282	Quantum 1/f Noise & Other Low Frequency Fluctuations in Electronic Devices (St. Louis, MO 1992)	93-072366	1-56396-252-7
No. 283	Earth and Space Science Information Systems (Pasadena, CA 1992)	93-072360	1-56396-094-X
No. 284	US-Japan Workshop on Ion Temperature Gradient-Driven Turbulent Transport (Austin, TX 1993)	93-72460	1-56396-221-7
No. 285	Noise in Physical Systems and 1/f Fluctuations (St. Louis, MO 1993)	93-72575	1-56396-270-5
No. 286	Ordering Disorder: Prospect and Retrospect in Condensed Matter Physics: Proceedings of the Indo-U.S. Workshop (Hyderabad, India 1993)	93-072549	1-56396-255-1
No. 287	Production and Neutralization of Negative Ions and Beams: Sixth International Symposium (Upton, NY 1992)	93-72821	1-56396-103-2
No. 288	Laser Ablation: Mechanismas and Applications-II: Second International Conference (Knoxville, TN 1993)	93-73040	1-56396-226-8
No. 289	Radio Frequency Power in Plasmas: Tenth Topical Conference (Boston, MA 1993)	93-72964	1-56396-264-0
No. 290	Laser Spectroscopy: XIth International Conference (Hot Springs, VA 1993)	93-73050	1-56396-262-4

	Title	L.C. Number	ISBN
No. 291	Prairie View Summer Science Academy (Prairie View, TX 1992)	93-73081	1-56396-133-4
No. 292	Stability of Particle Motion in Storage Rings (Upton, NY 1992)	93-73534	1-56396-225-X
No. 293	Polarized Ion Sources and Polarized Gas Targets (Madison, WI 1993)	93-74102	1-56396-220-9
No. 294	High-Energy Solar Phenomena: A New Era of Spacecraft Measurements (Waterville Valley, NH 1993)	93-74147	1-56396-291-8
No. 295	The Physics of Electronic and Atomic Collisions: XVIII International Conference (Aarhus, Denmark, 1993)	93-74103	1-56396-290-X
No. 296	The Chaos Paradigm: Developments an Applications in Engineering and Science (Mystic, CT 1993)	93-74146	1-56396-254-3
No. 297	Computational Accelerator Physics (Los Alamos, NM 1993)	93-74205	1-56396-222-5
No. 298	Ultrafast Reaction Dynamics and Solvent Effects (Royaumont, France 1993)	93-074354	1-56396-280-2
No. 299	Dense Z-Pinches: Third International Conference (London, 1993)	93-074569	1-56396-297-7
No. 300	Discovery of Weak Neutral Currents: The Weak Interaction Before and After (Santa Monica, CA 1993)	94-70515	1-56396-306-X
No. 301	Eleventh Symposium Space Nuclear Power and Propulsion (3 Vols.) (Albuquerque, NM 1994)	92-75162	1-56396-305-1 (set) 156396-301-9 (pbk. set)
No. 302	Lepton and Photon Interactions/ XVI International Symposium (Ithaca, NY 1993)	94-70079	1-56396-106-7
No. 303	Slow Positron Beam Techniques for Solids and Surfaces Fifth International Workshop (Jackson Hole, WY 1992)	94-71036	1-56396-267-5
No. 304	The Second Compton Symposium (College Park, MD 1993)	94-70742	1-56396-261-6
No. 305	Stress-Induced Phenomena in Metallization Second International Workshop (Austin, TX 1993)	94-70650	1-56396-251-9
No. 306	12th NREL Photovoltaic Program Review (Denver, CO 1993)	94-70748	1-56396-315-9

	Title	L.C. Number	ISBN
No. 307	Gamma-Ray Bursts Second Workshop (Huntsville, AL 1993)	94-71317	1-56396-336-1
No. 308	The Evolution of X-Ray Binaries (College Park, MD 1993)	94-76853	1-56396-329-9
No. 309	High-Pressure Science and Technology—1993 (Colorado Springs, CO 1993)	93-72821	1-56396-219-5 (set)
No. 310	Analysis of Interplanetary Dust (Houston, TX 1993)	94-71292	1-56396-341-8
No. 311	Physics of High Energy Particles in Toroidal Systems (Irvine, CA 1993)	94-72098	1-56396-364-7
No. 312	Molecules and Grains in Space (Mont Sainte-Odile, France 1993)	94-72615	1-56396-355-8
No. 313	The Soft X-Ray Cosmos ROSAT Science Symposium (College Park, MD 1993)	94-72499	1-56396-327-2
No. 314	Advances in Plasma Physics Thomas H. Stix Symposium (Princeton, NJ 1992)	94-72721	1-56396-372-8
No. 315	Orbit Correction and Analysis in Circular Accelerators (Upton, NY 1993)	94-72257	1-56396-373-6
No. 316	Thirteenth International Conference on Thermoelectrics (Kansas City, Missouri 1994)	95-75634	1-56396-444-9
No. 317	Fifth Mexican School of Particles and Fields (Guanajuato, Mexico 1992)	94-72720	1-56396-378-7
No. 318	Laser Interaction and Related Plasma Phenomena 11th International Workshop (Monterey, CA 1993)	94-78097	1-56396-324-8
No. 319	Beam Instrumentation Workshop (Santa Fe, NM 1993)	94-78279	1-56396-389-2
No. 320	Basic Space Science (Lagos, Nigeria 1993)	94-79350	1-56396-328-0
No. 321	The First NREL Conference on Thermophotovoltaic Generation of Electricity (Copper Mountain, CO 1994)	94-72792	1-56396-353-1
No. 322	Atomic Processes in Plasmas Ninth APS Topical Conference (San Antonio, TX)	94-72923	1-56396-411-2
No. 323	Atomic Physics 14 Fourteenth International Conference on Atomic Physics (Boulder, CO 1994)	94-73219	1-56396-348-5

	Title	L.C. Number	ISBN
No. 324	Twelfth Symposium on Space Nuclear Power and Propulsion (Albuquerque, NM 1995)	94-73603	1-56396-427-9
No. 325	Conference on NASA Centers for Commercial Development of Space (Albuquerque, NM 1995)	94-73604	1-56396-431-7
No. 326	Accelerator Physics at the Superconducting Super Collider (Dallas, TX 1992-1993)	94-73609	1-56396-354-X
No. 327	Nuclei in the Cosmos III Third International Symposium on Nuclear Astrophysics (Assergi, Italy 1994)	95-75492	1-56396-436-8
No. 328	Spectral Line Shapes, Volume 8 12th ICSLS (Toronto, Canada 1994)	94-74309	1-56396-326-4
No. 329	Resonance Ionization Spectroscopy 1994 Seventh International Symposium (Bernkastel-Kues, Germany 1994)	95-75077	1-56396-437-6
No. 330	E.C.C.C. 1 Computational Chemistry F.E.C.S. Conference (Nancy, France 1994)	95-75843	1-56396-457-0
No. 331	Non-Neutral Plasma Physics II (Berkeley, CA 1994)	95-79630	1-56396-441-4
No. 332	X-Ray Lasers 1994 Fourth International Colloquium (Williamsburg, VA 1994)	95-76067	1-56396-375-2
No. 333	Beam Instrumentation Workshop (Vancouver, B. C., Canada 1994)	95-79635	1-56396-352-3
No. 334	Few-Body Problems in Physics (Williamsburg, VA 1994)	95-76481	1-56396-325-6
No. 335	Advanced Accelerator Concepts (Fontana, WI 1994)	95-78225	1-56396-476-7 (set) 1-56396-474-0 (Book) 1-56396-475-9 (CD-Rom)
No. 336	Dark Matter (College Park, MD 1994)	95-76538	1-56396-438-4
No. 337	Pulsed RF Sources for Linear Colliders (Montauk, NY 1994)	95-76814	1-56396-408-2
No. 338	Intersections Between Particle and Nuclear Physics 5th Conference (St. Petersburg, FL 1994)	95-77076	1-56396-335-3

	Title	L.C. Number	ISBN
No. 339	Polarization Phenomena in Nuclear Physics Eighth International Symposium (Bloomington, IN 1994)	95-77216	1-56396-482-1
No. 340	Strangeness in Hadronic Matter (Tucson, AZ 1995)	95-77477	1-56396-489-9
No. 341	Volatiles in the Earth and Solar System (Pasadena, CA 1994)	95-77911	1-56396-409-0
No. 342	CAM -94 Physics Meeting (Cacun, Mexico 1994)	95-77851	1-56396-491-0
No. 343	High Energy Spin Physics Eleventh International Symposium (Bloomington, IN 1994)	95-78431	1-56396-374-4
No. 344	Nonlinear Dynamics in Particle Accelerators: Theory and Experiments (Arcidosso, Italy 1994)	95-78135	1-56396-446-5
No. 345	International Conference on Plasma Physics ICPP 1994 (Foz do Iguaçu, Brazil 1994)	95-78438	1-56396-496-1
No. 346	International Conference on Accelerator-Driven Transmutation Technologies and Applications (Las Vegas, NV 1994)	95-78691	1-56396-505-4
No. 347	Atomic Collisions: A Symposium in Honor of Christopher Bottcher (1945-1993) (Oak Ridge, TN 1994)	95-78689	1-56396-322-1
No. 348	Unveiling the Cosmic Infrared Background (College Park, MD, 1995)	95-83477	1-56396-508-9
No. 349	Workshop on the Tau/Charm Factory (Argonne, IL, 1995)	95-81467	1-56396-523-2
No. 350	International Symposium on Vector Boson Self-Interactions (Los Angeles, CA 1995)	95-79865	1-56396-520-8
No. 351	The Physics of Beams Andrew Sessler Symposium (Los Angeles, CA 1993)	95-80479	1-56396-376-0
No. 352	Physics Potential and Development of $\mu^+ \mu^-$ Colliders: Second Workshop (Sausalito, CA 1994)	95-81413	1-56396-506-2
No. 353	13th NREL Photovoltaic Program Review (Lakewood, CO 1995)	95-80662	1-56396-510-0
No. 354	Organic Coatings (Paris, France, 1995)	96-83019	1-56396-535-6
No. 355	Eleventh Topical Conference on Radio Frequency Power in Plasmas (Palm Springs, CA 1995)	95-80867	1-56396-536-4

	Title	L.C. Number	ISBN
No. 356	The Future of Accelerator Physics (Austin, TX 1994)	96-83292	1-56396-541-0
No. 357	10th Topical Workshop on Proton-Antiproton Collider Physics (Batavia, IL 1995)	95-83078	1-56396-543-7
No. 358	The Second NREL Conference on Thermophotovoltaic Generation of Electricity	95-83335	1-56396-509-7
No. 359	Workshops and Particles and Fields and Phenomenology of Fundamental Interactions (Puebla, Mexico 1995)	96-85996	1-56396-548-8
No. 360	The Physics of Electronic and Atomic Collisions XIX International Conference (Whistler, Canada, 1995)	95-83671	1-56396-440-6
No. 361	Space Technology and Applications International Forum (Albuquerque, NM 1996)	95-83440	1-56396-568-2
No. 362	Two-Center Effects in Ion-Atom Collisions (Lincoln, NE 1994)	96-83379	1-56396-342-6
No. 363	Phenomena in Ionized Gases XXII ICPIG (Hoboken, NJ, 1995)	96-83294	1-56396-550-X
No. 364	Fast Elementary Processes in Chemical and Biological Systems (Villeneuve d'Ascq, France, 1995)	96-83624	1-56396-564-X
No. 365	Latin-American School of Physics XXX ELAF Group Theory and Its Applications (México City, México, 1995)	96-83489	1-56396-567-4
No. 366	High Velocity Neutron Stars and Gamma-Ray Bursts (La Jolla, CA 1995)	96-84067	1-56396-593-3
No. 367	Micro Bunches Workshop (Upton, NY, 1995)	96-83482	1-56396-555-0
No. 368	Acoustic Particle Velocity Sensors: Design, Performance and Applications (Mystic, CT, 1995)	96-83548	1-56396-549-6
No. 369	Laser Interaction and Related Plasma Phenomena (Osaka, Japan 1995)	96-85009	1-56396-445-7
No. 370	Shock Compression of Condensed Matter-1995 (Seattle, WA 1995)	96-84595	1-56396-566-6
No. 371	Sixth Quantum 1/f Noise and Other Low Frequency Fluctuations in Electronic Devices Symposium (St. Louis, MO, 1994)	96-84200	1-56396-410-4

	Title	L.C. Number	ISBN
No. 372	Beam Dynamics and Technology Issues for + - Colliders 9th Advanced ICFA Beam Dynamics Workshop (Montauk, NY, 1995)	96-84189	1-56396-554-2
No. 373	Stress-Induced Phenomena in Metallization (Palo Alto, CA 1995)	96-84949	1-56396-439-2
No. 374	High Energy Solar Physics (Greenbelt, MD 1995)	96-84513	1-56396-542-9
No. 375	Chaotic, Fractal, and Nonlinear Signal Processing (Mystic, CT 1995)	96-85356	1-56396-443-0
No. 376	Chaos and the Changing Nature of Science and Medicine: An Introduction (Mobile, AL 1995)	96-85220	1-56396-442-2
No. 377	Space Charge Dominated Beams and Applications of High Brightness Beams (Bloomington, IN 1995)	96-85165	1-56396-625-7
No. 378	Surfaces, Vacuum, and Their Applications (Cancun, Mexico 1994)	96-85594	1-56396-418-X
No. 379	Physical Origin of Homochirality in Life (Santa Monica, CA 1995)	96-86631	1-56396-507-0
No. 380	Production and Neutralization of Negative Ions and Beams / Production and Application of Light Negative Ions (Upton, NY 1995)	96-86435	1-56396-565-8
No. 381	Atomic Processes in Plasmas (San Francisco, CA 1996)	96-86304	1-56396-552-6
No. 382	Solar Wind Eight (Dana Point, CA 1995)	96-86447	1-56396-551-8
No. 383	Workshop on the Earth's Trapped Particle Environment (Taos, NM 1994)	96-86619	1-56396-540-2
No. 384	Gamma-Ray Bursts (Huntsville, AL 1995)	96-79458	1-56396-685-9
No. 385	Robotic Exploration Close to the Sun: Scientific Basis (Marlboro, MA 1996)	96-79560	1-56396-618-2
No. 386	Spectral Line Shapes, Volume 9 13th ICSLS (Firenze, Italy 1996)		1-56396-656-5
No. 387	Space Technology and Applications International Forum (Albuquerque, NM 1997)	96-80254	1-56396-679-4 (Case set) 1-56396-691-3 (Paper set)

	Title	L.C. Number	ISBN
No. 388	Resonance Ionization Spectroscopy 1996 Eighth International Symposium (State College, PA 1996)	96-80324	1-56396-611-5
No. 389	X-Ray and Inner-Shell Processes 17th International Conference (Hamburg, Germany 1996)	96-80388	1-56396-563-1
No. 390	Beam Instrumentation Proceedings of the Seventh Workshop (Argonne, IL 1996)	97-70568	1-56396-612-3
No. 391	Computational Accelerator Physics (Williamsburg, VA 1996)	97-70181	1-56396-671-9
No. 392	Applications of Accelerators in Research and Industry: Proceedings of the Fourteenth International Conference (Denton, TX 1996)	97-71846	1-56396-652-2
No. 393	Star Formation Near and Far Seventh Astrophysics Conference (College Park, MD 1996)	97-71978	1-56396-678-6
No. 394	NREL/SNL Photovoltaics Program Review Proceedings of the 14th Conference— A Joint Meeting (Lakewood, CO 1996)	97-72645	1-56396-687-5
No. 395	Nonlinear and Collective Phenomena in Beam Physics (Arcidosso, Italy 1996)	97-72970	1-56396-668-9
No. 396	New Modes of Particle Acceleration— Techniques and Sources (Santa Barbara, CA 1996)	97-72977	1-56396-728-6
No. 397	Future High Energy Colliders (Santa Barbara, CA 1997)	97-73333	1-56396-729-4
No. 398	Advanced Accelerator Colliders Seventh Workshop (Lake Tahoe, CA 1996)	97-72788	1-56396-697-2 (set) 1-56396-727-8 (cloth) 1-56396-726-X (CD-Rom)
No. 399	The Changing Role of Physics Departments (College Park, MD 1996)		1-56396-698-0
No. 400	High Energy Physics First Latin Symposium (Yucatan, Mexico 1996)	97-073971	1-56396-686-7